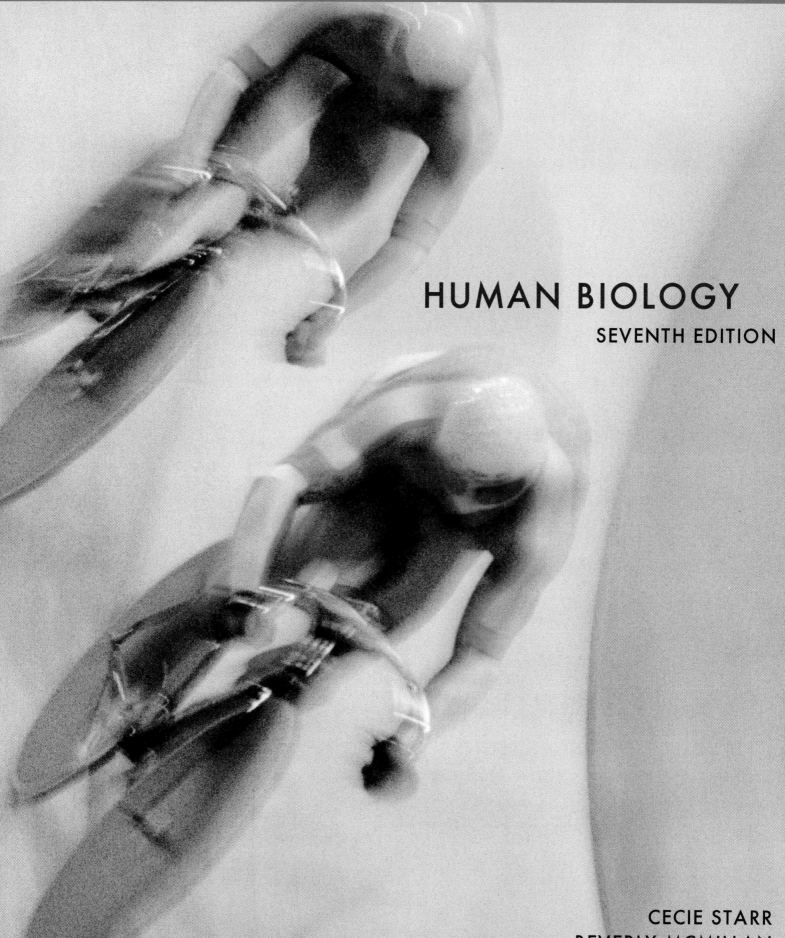

HUMAN BIOLOGY

SEVENTH EDITION

CECIE STARR

BEVERLY MCMILLAN

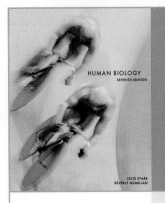

Here's your *Issues and Resources Integrator* for Starr and McMillan's *Human Biology*, Seventh Edition

At a glance—text applications, news articles, videos, and websites, all organized around relevant issues

With its strong issues orientation, Starr and McMillan's **Human Biology, Seventh Edition** encourages students to investigate—and vote on—current topics. By making it easy for you to find, integrate, and use these topics from the text and in supporting resources, this *Issues and Resources Integrator* saves you time preparing lectures and assignments. You'll find a chapter-by-chapter guide to the book's most interesting issues and applications. You'll also find references to these real-world examples from the news, magazines, and the web—all available to work hand in hand with the book's content.

The one-page grid for each text chapter includes these resources for each section:

- **Selected applications** from the text with page references, organized by chapter section.
- **Titles of brief, high-interest video segments**—presenting coverage of current issues and historic raw footage in digital formats— that bring the chapter-related issues and applications to life. (See page 30 of this guide for a list of segments.)
- **Applied readings** that are accessible through InfoTrac® College Edition with InfoMarks® (the vast online library). See pages 31 and 32 of this guide for a listing of the hundreds of biology-related publications available to you and your students through InfoTrac College Edition.
- **URLs for websites** with information related to issues and applications in the text.
- **A listing of the *How Would You Vote?* question** (inspired by the *Impacts, Issues* essay on a human biology-related issue that opens the chapter). Listed below the question are InfoTrac College Edition articles and websites where students will find information that helps them to research the question and develop their position.

For your convenience . . .
An electronic copy of this guide is available to both instructors and students on the Book Companion Website: www.thomsonedu.com/biology/starr/humanbio.

CHAPTER 1: LEARNING ABOUT HUMAN BIOLOGY

Section	Applications	Videos	InfoTrac® Readings	Weblink Examples
1.0 *Impacts, Issues:* **What Kind of World Do We Live In?**	"Bird flu," p. 1 / Coral reefs, p. 1 / Earthquakes, p. 1 / Global warming, p. 1 / Infectious diseases, p. 1 / Pollution, p. 1	• Environmental Science, 2005, *Bird Flu* (3:19)	• The Great Plague. Rene Skelton. *National Geographic World*, Mar. 1999. • Avian Flu: Why All the Squawk? Linda S. Nield. *Consultant*, Feb. 1, 2006.	• U.S. EPA Global Warming Site: http://yosemite.epa.gov/oar/globalwarming.nsf/content/index.html
1.1 **The Characteristics of Life**	Deoxyribonucleic acid (DNA), p. 2 / Energy, p. 2 / Environment, p. 2		• The Meaning of Life. Lin Chao. *BioScience*, Mar. 2000. • The Mystery of Life. Carl Sagan. *UNESCO Courier*, Sept. 1984.	• DNA: http://www.pbs.org/wnet/dna/
1.2 **Our Place in the Natural World**	Animals, p. 3 / Bacteria, p. 3 / Bones, p. 3 / Brain, p. 3 / Evolution, p. 3 / Fungi, p. 3 / Human evolution, p. 3	• Biology, 2004, *Earliest Homo Sapiens* (1:42)	• The Human Genus. Bernard Wood, Mark Collard. *Science*, April 2, 1999. • A Tale Told by DNA. Nell Boyce. *U.S. News & World Report*, Dec. 16, 2002.	• The Tree of Life Web Project: http://phylogeny.arizona.edu/tree/phylogeny.html
1.3 **Life's Organization**	Animals, p. 4 / Bacteria, pp. 4–5 / Biodiversity, p. 5 / Ecosystems, pp. 4–5 / Fungi, pp. 4–5 / Global warming, p. 5 / Solar energy, p. 4	• Environmental Science, 2003, *Biodiversity* (2:29)	• Lost at Sea: Coral Reefs, Considered the 'Rainforests' of the Marine World, Now Have Even More in Common with Those Fragile Ecosystems. Scott Kirkwood. *National Parks*, Spring 2006.	• Ecological Society of America: http://www.esa.org/
1.4 **Science Is a Way of Learning about the Natural World**			• Water Saver: Could a Population Boom Cause a Decline in One Bay's Water Quality? An Environmentally Concerned Teen Finds Out. Mona Chiang. *Science World*, Sept. 19, 2005. • The Real Method of Scientific Discovery … Often Involves a Creative, Imaginative Leap. Burton S. Guttman. *Skeptical Inquirer*, Jan.–Feb. 2004.	• Discovery, Chance, and the Scientific Method: http://www.accessexcellence.org/AE/AEC/CC/chance.html • The Scientific Method: http://physics.ucr.edu/~wudka/Physics7/Notes_www/node5.html
1.5 **Science in Action: Cancer, Broccoli, and Mighty Mice**	Anticancer strategy, p. 8 / Cancer, experiments involving, p. 8 / Carcinogens, p. 8 / Genetic mutations, p. 8 / Health, p. 8 / Mutations, p. 8 / Ovarian tumors, p. 8 / Tumors, p. 8 / Uterine tumors, p. 8		• My Life as a Guinea Pig: Clinical Trials Are Inherently Risky, But One Saved Me. Jamie Reno. *Newsweek*, Aug. 6, 2001. • Anti-cancer Veggies. *Natural Life*, May–June 2006.	• Cancer Research Institute: Cervical Cancer Vaccine Approved: http://www.cancerresearch.org/news/news.html#HPVVaccineApproved
1.6 **Science in Perspective**	Animals, scientific research and, p. 9 / Antibiotics, p. 9 / Genetic engineering, p. 9 / Internet, p. 9 / Morality, p. 9 / Ozone issues, p. 9 / Pollution, p. 9	• Environmental Science, 2004, *Smog Pollution* (1:40)	• Scientists at War. Leif J. Robinson. *Odyssey*, Feb. 2002. • Above All, Do No Harm. Stephen Jay Gould. *Natural History*, Oct. 1998. • Bioethics and the Stem Cell Research Debate. Robyn S. Shapiro. *Social Education*, May–June 2006.	• Bioethics.net: http://ajobonline.com/ • NOVA—The Stem Cell Debate: http://www.pbs.org/wgbh/nova/miracle/stemcells.html
1.7 **Critical Thinking in Science and Life**			• Question Authority: Kids Need to Be Skeptical of the Curriculum. It's the Only Way to Develop a Balanced View of the World. Glenn DeVoogd. *School Library Journal*, April 2006.	• Smart People Believe Weird Things: http://www.sciam.com/article.cfm?articleID=0002F4E6-8CF7-1D49-90FB809EC5880000
1.8 **Are Herbal Supplements Safe?**	AMA, p. 11 / Cholesterol, p. 11 / Cold, common, p. 11 / Depression, p. 11 / Echinacea, p. 11 / Ephedra, p. 11 / Heart attacks, p. 11 / Herbal supplements, p. 11 / Stress, p. 11 / Stroke, p. 11 / Weight-loss, p. 11	• Biology, 2003, *Ephedra Dangers* (2:00)	• Any Science Behind the Hype of 'Natural' Dietary Supplements? (ce series). Teri Capriotti. *Dermatology Nursing*, Oct. 2005. • Ephedra/Ephedrine Dangers (clinical clips) (brief article). David Nicklin. *Patient Care for the Nurse Practitioner*, June 2003.	• U.S. FDA: Dietary Supplements: http://www.cfsan.fda.gov/~dms/supplmnt.html • Medicinal Herb Research Using the Internet: http://lib.colostate.edu/research/medherbs/
How Would You Vote?	Middle Eastern seas support spectacular coral reef ecosystems. Should the United States provide funding to help preserve the reefs? • WWW: Humans Are Not the Only Victims of War: http://www.uga.edu/srel/ecoview1-27-03.html • WWW: Regional Perspectives: Seas of the Middle East: http://www.ngdc.noaa.gov/paleo/outreach/coral/sor/sor_mideast.html • InfoTrac: Should the Arabian (Persian) Gulf Become a Marine Sanctuary? *Oceanus*, Fall 1993.			

CHAPTER 2: MOLECULES OF LIFE

Section	Applications	Videos	InfoTrac® Readings	Weblink Examples
2.0 *Impacts, Issues:* It's Elemental	Arsenic, p. 15 / Birth defects, p. 15 / Bones, damage in, p. 15 / Calcium, p. 15 / Fluoride, p. 15 / Fluorine, p. 15 / Iron, p. 15 / Oxygen, p. 15		•Probing Dietary Copper's Healthy Limits. Judith R. Turnlund. *Agricultural Research*, Jan. 2006.	•USGS National Analysis of Trace Elements: http://water.usgs.gov/nawqa/trace/
2.1 Atoms, the Starting Point	Environment, p. 16		•Making It Visual: Creating a Model of the Atom. Rose M. Pringle. *Science Activities*, Winter 2004.	•A Look Inside the Atom: http://www.aip.org/history/electron/jjhome.htm
2.2 Medical Uses for Radioisotopes	Neurological disorders, p. 17 / Positron-emission tomography (PET), p. 17 / Radiation therapy, p. 17		•A Career in Nuclear Medicine. John Frank. *Student BMJ*, Jan. 2004.	•Let's Play PET: http://www.crump.ucla.edu/software/lpp/shocked/lppshocked.html
2.3 What Is a Chemical Bond?			•Do You Need to Believe in Orbitals to Use Them? Zack Jenkins. *Philosophy of Science*, Dec. 2003.	•Overview of Chemical Bonding: http://www.launc.tased.edu.au/online/sciences/physsci/pschem/metals/Bond.htm
2.4 Important Bonds in Biological Molecules	Deoxyribonucleic acid (DNA), p. 21		•Chemists Look to Follow Biology Lead. Joe Alper. *Science*, Mar. 29, 2002.	•ChemSketch (Freeware): http://www.acdlabs.com/products/chem_dsn_lab/chemsketch/
2.5 Antioxidants	Alpha carotene, p. 22 / Antioxidants, p. 22 / Carotenoids, p. 22 / Cigarette smoke, p. 22 / DNA, p. 22 / Fats, p. 22 / Fiber, p. 22 / Ultraviolet radiation, p. 22 / Vitamin C, p. 22 / Vitamin E, p. 22	•Human Biology, 2006, *The Wine of Life* (3:37)	•Successful Aging and the Role of Nutrients. *Prepared Foods*, June 2006.	•Antioxidants and Free Radicals: http://www.rice.edu/~jenky/sports/antiox.html
2.6 Life Depends on Water	Blood, pp. 22–23		•The Rarest Element. Sidney Perkowitz. *The Sciences*, Jan. 1999.	•The Hydrophobic Effect: http://www.princeton.edu/~lehmann/BadChemistry.html#hydrophobic
2.7 Acids, Bases, and Buffers: Body Fluids in Flux	Acidosis, p. 25 / Alkalosis, p. 25 / Calcium, p. 25 / Fossil fuels, p. 25 / Nitrogen fertilizers, p. 25 / pH scale, p. 24 / Tetany, p. 25		•Acid Rain Causing Decline of Sugar Maples. *UPI NewsTrack*, June 2, 2006.	•Acids and Bases—pH Chemistry: http://chemistry.about.com/od/acidsbases/
2.8 Molecules of Life	Estrogen, p. 26 / Lipids, p. 26 / Organic compounds, pp. 26–27 / Oxygen, p. 26 / Sex hormones, pp. 26–27 / Testosterone, p. 26		•Tropical Forest CO_2 Emissions Tied to Nutrient Increases. *Ascribe Higher Education News Service*, June 19, 2006.	•The Chemistry of Carbon: http://www.nyu.edu/pages/mathmol/modules/carbon/carbon1.html
2.9 Carbohydrates: Plentiful and Varied	Cellulose, p. 29 / Deoxyribonucleic acid (DNA), p. 28 / Digestive enzymes, p. 29 / Fiber, p. 29 / High fructose corn syrup, p. 28 / "Junk" foods, p. 28 / Vitamin C, p. 28		•Too Many Carbs May Spoil Eyesight. A. Taylor, J. Mayer. *Agricultural Research*, Feb. 2006.	•Biochemistry of Carbohydrates: http://web.indstate.edu/thcme/mwking/carbohydrates.html
2.10 Lipids: Fats and Their Chemical Kin	Cholesterol, p. 31 / Development, p. 31 / Fats, pp. 30–31 / Heart disease, pp. 30–31 / Liver, p. 31 / Steroid hormones, p. 31 / Vitamin D, p. 31	•Biology, 2004, *Fat Facts* (2:29)	•'Fat Gene' Is Discovered by Rutgers Researchers. *Record (Hackensack, NJ)*, Mar. 21, 2006.	•Biochemistry of Lipids: http://web.indstate.edu/thcme/mwking/lipids.html
2.11 Proteins: Biological Molecules with Many Roles	Hormones, p. 32		•Can Jellyfish Help Fight Alzheimer's? Michael Arndt. *Business Week*, June 5, 2006.	•The Chemistry of Amino Acids: http://www.biology.arizona.edu/biochemistry/problem_sets/aa/aa.html
2.12 A Protein's Function Depends on Its Shape	Albumin, p. 35 / Cholesterol, p. 35 / Hormones, pp. 34–35 / Insulin, pp. 34–35 / Iron, p. 35 / Keratin, p. 35 / pH scale, p. 35		•Study Lends Support to Mad Cow Theory. Sandra Blakeslee. *The New York Times*, July 30, 2004.	•Protein Structure and Diversity: http://www.rothamsted.ac.uk/notebook/courses/guide/prot.htm
2.13 Nucleotides and Nucleic Acids	Hormones, p. 36 / Organic compounds, p. 36		•DNA or RNA? Versatile Player Takes a Leading Role. *The New York Times*, June 20, 2006.	•Molecular Expressions— The Nucleotide Gallery: http://micro.magnet.fsu.edu/micro/gallery/nucleotides/nucleotides.html
2.14 Food Production and a Chemical Arms Race	Asthma, p. 37 / Learning disabilities, p. 37 / Pathogens, p. 37 / Pesticides, p. 37 / Rashes, p. 37 / Toxins, p. 37 / U.S. Environmental Protection Agency, p. 37	•Biology, 2004, *Mold Exposure* (2:43)	•Agricultural Health Study Links Pesticides, Prostate Cancer in Iowa Farmers. *Waterloo Courier*, Mar. 2, 2004.	•U.S. EPA Pesticide Programs: http://www.epa.gov/pesticides/ •The Killer in Your Yard: http://magazine.audubon.org/backyard/backyard0005.html
How Would You Vote?	Many communities add fluoride to drinking water supplies. Do you want it in yours? •InfoTrac: What about Fluoride? *E*, Sept.–Oct. 2003. •InfoTrac: BDA Rejects Green Party Fluoride Claims. *M2 Presswire*, Aug. 19, 2003. •InfoTrac: Ingested Fluoride Needless; Children Dangerously Overdosed, Studies Show. *PR Newswire*, May 15, 2003.			

CHAPTER 3: CELLS AND HOW THEY WORK

Section	Applications	Videos	InfoTrac® Readings	Weblink Examples
3.0 *Impacts, Issues:* **When Mitochondria Spin Their Wheels**	Defective cell organelle, p. 41 / Drugs, p. 41 / Luft's syndrome, p. 41 / Mitochondrial disorders, p. 41 / Pharmaceutical companies, p. 41		• Damaged Mitochondria in Parkinson's. *GP*, April 14, 2006.	• Friedreich's Ataxia: http://www.mda.org.au/specific/mdafa.html
3.1 **What Is a Cell?**	Lipids, in cells, p. 43 / Metabolism, p. 42		• Celle Fantastyk. Jennifer Ackerman. *Natural History*, May 2001.	• What Is a Cell?: http://library.thinkquest.org/5420/cellwhat.html
3.2 **The Parts of a Eukaryotic Cell**	Toxins, p. 44		• Lysosomes: The Cell's Recycling Centres. *Biological Sciences Review*, Nov. 2004.	• Cells Alive: http://www.cellsalive.com
3.3 **The Plasma Membrane: A Double Layer of Lipids**	Cholesterol, p. 46 / Lipids, p. 46		• Scratching the Cell Surface: Total Internal Reflection Fluorescence Microscopy Exposes Events at the Cell Membrane. *The Scientist*, Nov. 22, 2004.	• The Plasma Membrane: http://www.beyondbooks.com/lif71/4b.asp
3.4 **How Do We See Cells?**			• Confocal Microscopy Used to Detect and Quantitate HIV-1 Particles. *Medical Devices & Surgical Technology Week*, Oct. 17, 2004.	• Electron Microscopy: http://www.unl.edu/CMRAcfem/em.htm
3.5 **The Nucleus**			• Scientists 'Operate' on a Cell's Nucleus. *Current Health 2, a Weekly Reader publication*, April–May 2005.	• The Cell Nucleus: http://www.cellnucleus.com/
3.6 **The Endomembrane System**	Alcohol, p. 51 / Digestion, p. 51 / Drugs, p. 50 / Metabolism, p. 50 / Toxins, p. 51		• Scientists Give Golgi Apparatus Its Own Identity. *Cancer Weekly*, Dec. 5, 2000.	• The Golgi Complex: http://www.cytochemistry.net/Cell-biology/golgi.htm
3.7 **Mitochondria: The Cell's Energy Factories**	Liver, p. 52		• Power Failure: What Happens When Muscle Cells Run Out of Fuel. Kathleen Fackelmann. *Science News*, Sept. 27, 1997.	• Mitochondria: http://www.scripps.edu/mem/biochem/ayagi/mito.html
3.8 **The Cell's Skeleton**	Human sperm, p. 53		• Slime and the Cytoskeleton: How the Defensive Ooze of a Hagfish Sheds Light on Cellular Structure. Adam Summers. *Natural History*, Oct. 2004.	• The Cytoskeleton of Cells: http://sun.menloschool.org/~birchler/cell
3.9 **How Diffusion and Osmosis Move Substances across Membranes**			• Life Puts Random Motions to Work. Ronald F. Fox. *World and I*, Oct. 2001.	• Diffusion and Osmosis: http://hyperphysics.phy-astr.gsu.edu/hbase/kinetic/diffus.html#c3
3.10 **Other Ways Substances Cross Cell Membranes**	Antibiotics, p. 57 / Calcium, p. 56 / Hurricane Katrina, p. 57 / Public sanitation, p. 57 / Rehydration, p. 57 / *Vibrio cholerae*, p. 57		• Molecular Analysis Shows Complexity behind Cargo Delivery System of Mammalian Cells. *Genomics & Genetics Weekly*, Mar. 16, 2001.	• CDC: Cholera: http://www.cdc.gov/ncidod/dbmd/diseaseinfo/cholera_g.htm
3.11 **Metabolism: Doing Cellular Work**	Blood clotting, p. 59 / Food proteins, p. 59 / Hormones, p. 59 / Superoxide dismutase, p. 59 / Vitamins, p. 59		• Cooperation and Competition in the Evolution of ATP-Producing Pathways. *Science*, April 20, 2001.	• Cellular Metabolism: http://www.mhhe.com/biosci/ap/holehaap/student/olc2/chapterindex04.htm
3.12 **How Cells Make ATP**	Lipids, ATP production and, p. 60		• Saliva: A Window to Health Status. Jan Suszkiw. *Agricultural Research*, May 2006.	• ATP and Biological Energy: http://www.emc.maricopa.edu/faculty/farabee/biobk/BioBookATP.html
3.13 **Summary of Cellular Respiration**			• The Advantages of Togetherness. *Science*, April 20, 2001.	• Cellular Respiration: http://users.rcn.com/jkimball.ma.ultranet/BiologyPages/C/CellularRespiration.html
3.14 **Alternative Energy Sources in the Body**	Adipose tissues, p. 63 / Bones, p. 63 / Fats, p. 63 / Insulin, p. 63 / Lactic acid, p. 63 / Liver, p. 63 / Muscles p. 63 / Urea, p. 63		• A Prospective Study of Psychological Predictors of Body Fat Gain among Children at High Risk for Adult Obesity. *Pediatrics*, April 2006.	• Lactic Acidosis: http://www.emedicine.com/EMERG/topic291.htm
How Would You Vote?	Should pharmaceutical companies receive financial incentives (such as tax breaks) to search for cures for diseases that affect only a small number of people? • InfoTrac: Drug Development for Neglected Diseases: A Deficient Market and a Public-Health Policy Failure. *The Lancet*, June 22, 2002. • InfoTrac: Collaborators against Cancer: Pharmaceutical Companies Work with Their Rivals to Produce Cancer Drugs. *The Scientist*, June 30, 2003. • InfoTrac: NORD: Proposed CMS Rule Jeopardizes Care for Seniors with Rare Diseases; Congressional Action Needed to Remedy Problem. *US Newswire*, Aug. 26, 2003.			

CHAPTER 4: TISSUES, ORGANS, AND ORGAN SYSTEMS

Section	Applications	Videos	InfoTrac® Readings	Weblink Examples
4.0 *Impacts, Issues*: Stem Cells	Alzheimer's disease, p. 67 / Disorders, p. 67 / Embryos, p. 67 / Ethics, p. 67 / Heart attacks, p. 67 / Paralysis, p. 67 / Reeve, Christopher, p. 67 / Spinal cord injuries, p. 67 / Stem cells, p. 67	•Biology, 2005, *New Nerves* (2:37)	•Stem Cells Offer Promise—Not More—for Heart Disease. *Harvard Heart Letter*, July 2006.	•Stem Cell Research Foundation: http://www.stemcell researchfoundation.org/ •The Stem Cell Debate: http://www.time.com/time/2001/stemcells/
4.1 **Epithelium: The Body's Covering and Linings**	Digestive enzymes, p. 68 / Lungs, pp. 68, 69 / Oil, p. 68 / Sinuses, p. 68 / Sweat glands, p. 68 / Thyroid, p. 68		•New Treatments Target Causes of Acne. Diana Gorgos. *Dermatology Nursing*, Feb. 2006.	•Simulation and Reconstruction of 3-D Epithelial Structure: http://members.tripod.com/~Gensav/index.htm
4.2 **Connective Tissue: Binding, Support, and Other Roles**	Elastic cartilage, p. 71 / Embryos, pp. 70, 71 / Fats, p. 71 / Kidneys, pp. 70, 71 / Liver, p. 70 / Lungs, p. 70 / Metabolism, p. 71 / Spleen, p. 70 / Stress, p. 71		•Catch and Release: Sea Cucumbers Might Put a Torn Achilles Tendon Back Together Again. Adam Summers. *Natural History*, Nov. 2003.	•Connective Tissue Disorders: http://www.ctds.info/
4.3 **Muscle Tissue: Movement**	Dystrophin, p. 72 / Genetic disorders, p. 72 / Huard, Johnny, p. 72 / Muscular dystrophy, p. 72 / Stem cells, p. 72	•Genetics, 2005, *Producing Human "Replacement Cells"* (2:22)	•The Muscular Dystrophies: From Genes to Therapies. *Physical Therapy*, Dec. 2005. •Shortcut to Big Heart. Susan Milius. *Science News*, Mar. 5, 2005.	•Visual Histology: Muscle: http://www.visualhistology.com/Visual_Histology_Atlas/VHA_Chpt7_Muscle.html
4.4 **Nervous Tissue: Communication**	Biotechnology, p. 73 / Bone marrow, p. 73 / Defective stem cells, p. 73 / Memorial Sloan-Kettering Cancer Center, p. 73 / Sickle cell anemia, p. 73	•Biology, 2005, *New Nerves* (2:37)	•Astrocytes May Play Role in Alzheimer's. *UPI NewsTrack*, Jan. 6, 2006.	•Neurons and Neurotransmitters: http://www.utexas.edu/research/asrec/neuron.html
4.5 **Cell Junctions: Holding Tissues Together**	Peptic ulcers, p. 74		•Estrogen Accelerates Occludin Modulation at Tight Junctions. *Medical Devices & Surgical Technology Week*, Dec. 26, 2004.	•Gap Junctions: http://www.vivo.colostate.edu/hbooks/cmb/cells/pmemb/junctions_g.html
4.6 **Tissue Membranes: Thin, Sheetlike Covers**			•Keeping a Watch on Vision: Use of Antioxidants and Nutrients. Ross Pelton. *American Druggist*, May 1999.	•Membranes: http://training.seer.cancer.gov/module_anatomy/unit2_3_membranes.html
4.7 **Organs and Organ Systems**	Aging hearts, p. 76 / Artificial organs, p. 76 / Cultured muscle cells, p. 76 / Dehydration, p. 77 / Diseased hearts, p. 76 / Epithelial cells, p. 76 / Microbes, p. 77 / Tissue damage, p. 77	•Anatomy & Physiology, 2004, *Immortality Industry* (2:22)	•Homegrown Defender: Urinary Infections Face Natural Guard. *Science News*, June 10, 2006.	•Basic Anatomy: Tissues and Organs: http://web.jjay.cuny.edu/~acarpi/NSC/14-anatomy.htm
4.8 **The Integument—Example of an Organ System**	Artificial skin, p. 78 / Bacteria, p. 79 / Cholecalciferol, p. 78 / Cuts, p. 78 / Organs, growing, p. 78 / Solar radiation, p. 78 / Squamous cell carcinoma, p. 79 / Virus particles, p. 79 / Vitamin C, p. 79	•Biology, 2003, *Skin Sun Damage* (1:23)	•Fetal Skin Cells Found to Be a Promising Treatment for Burns. *The New York Times*, Aug. 18, 2005.	•Anatomy of the Skin: http://www.telemedicine.org/science.htm •Dermatology Cinema: http://www.skinema.com/
4.9 **Homeostasis: The Body in Balance**	Brain, homeostasis and, p. 80 / Fetuses, p. 81 / Hormones, p. 81 / Labor, p. 81 / Oxytocin, p. 81 / Uterus, p. 81		•Reversal of Type 1 Diabetes by Engineering a Glucose Sensor in Skeletal Muscle. *Diabetes*, June 2006.	•Homeostasis: http://www3.fhs.usyd.edu.au/bio/homeostasis/Introduction.htm
4.10 **How Homeostatic Feedback Maintains the Body's Core Temperature**	Blood, p. 83 / Blood pressure, p. 83 / Cold stress, pp. 82–83 / Coma, p. 83 / Death, p. 83 / Frostbite, p. 83 / Heat exhaustion, p. 83 / Heat stroke, p. 83 / Hyperthermia, p. 83 / Hypothermia, p. 83 / Mental confusion, p. 83 / Shivering, pp. 82, 83 / Water, p. 83		•Totally Cool: Trainers Are Turning to a New Device to Help Depleted Muscles Recover and Improve Performance. *Sports Illustrated*, Nov. 14, 2005.	•Effect of Excessive Environmental Heat on Core Temperature in Critically Ill Patients: http://bja.oxfordjournals.org/cgi/content/full/94/1/39 •Temperature Homeostasis: http://www.biologymad.com/master.html?http://www.biologymad.com/Homeostasis/Homeostasis.htm
How Would You Vote?	Should researchers be allowed to start embryonic stem cell lines from human embryos that are not used for in vitro fertilization? •InfoTrac: Human Embryonic Stem-Cell Research: Science and Ethics. *American Scientist*, July–Aug. 1999. •InfoTrac: Holy Grail or Pandora's Box?: Evaluating Human Embryonic Stem Cell Research. *World and I*, Nov. 1999. •InfoTrac: The Politics of Stem Cells: Why Do Some Scientists and Politicians Insist on Exploiting Embryos? *Christianity Today*, Nov. 2004.			

CHAPTER 5: THE SKELETAL SYSTEM

Section	Applications	Videos	InfoTrac® Readings	Weblink Examples
5.0 *Impacts, Issues:* **Creaky Joints**	Arthritis, p. 87 / Cartilage lining, p. 87 / Diseases, p. 87 / Disorders, p. 87 / Herbal remedies, p. 87 / Joints, p. 87 / Obesity, p. 87 / Osteoarthritis, p. 87 / Pain relievers, p. 87 / Steroid drugs, p. 87	• Anatomy & Physiology, 2004, *Juvenile Arthritis* (2:02)	• New Treatments Give Arthritis Sufferers Relief. *Daily Oklahoman*, April 13, 2006.	• Scientific Research Validates Use of Many Popular Herbs: http://bastyrcenter.org/content/view/655/
5.1 **Bone – Mineralized Connective Tissue**	Bone loss, p. 89 / Growth hormones (GH), p. 88 / Mechanical stress, pp. 88, 89 / Menopause, p. 89 / Osteoporosis, p. 89 / Sedentary lifestyle, p. 89 / Smoking, p. 89		• Our Paleolithic Ancestors Had Stronger Bones. *GP*, June 16, 2006.	• Normal Bone Anatomy: http://cal.vet.upenn.edu/saortho/chapter_01/01mast.htm
5.2 **The Skeleton: The Body's Bony Framework**			• Skull Science. *New York State Conservationist*, April 2006. • Bones Hold Key to Blood Renewal. *Ascribe Higher Education News Service*, June 20, 2006.	• Bone Tissues and Bone Types Tutorial: http://www.zoology.ubc.ca/~lacombe/biomania/tutorial/bonets/bnts01.htm
5.3 **The Axial Skeleton**	Aging, p. 93 / Painkilling drugs, p. 93 / Slipped disk, p. 93 / Surgery, p. 93 / Weight, excess, p. 93	• Human Biology, 2006, *Painful Painkillers* (1:48)	• Clearing Up the Mysteries Surrounding 'Slipped Disks.' *Executive Health's Good Health Report*, Sept. 1998.	• The Axial Skeleton: http://bioweb.uwlax.edu/APLab/Table_of_Contents/Lab_03/lab_03.html
5.4 **The Appendicular Skeleton**	Anterior cruciate ligament (ACL) injury, p. 95 / Bone marrow, p. 95 / Broken bones, p. 94 / Carpal tunnel syndrome, p. 94 / Childbearing, p. 95 / Jumping, p. 95 / Ligaments, p. 95 / Running, p. 95 / Stem cells, p. 95 / Stress, p. 95 / Tufts University, p. 95		• Winning Workers' Comp Claims for Carpal Tunnel Syndrome; Understanding the Symptoms and Causes … Thomas M. Domer. *Trial*, June 2002. • All about Your ACL. *Ski*, Nov. 1, 2005.	• The Appendicular Skeleton: http://www.botany.uwc.ac.za/SCI_ED/grade10/manphys/appendicular.htm
5.5 **Joints – Connections between Bones**	Childbirth, p. 96		• Glucosamine and Chondroitin: Update 2006. *Alternative Medicine Alert*, July 1, 2006.	• Skeleton: Joint Tutorial: http://www.zoology.ubc.ca/~biomania/tutorial/bonejt/outline.htm
5.6 **Disorders of the Skeleton**	Amputation, p. 99 / Antibiotics, p. 99 / Anti-inflammatory drugs, p. 98 / Arthritis, p. 98 / Cancer, bone, p. 99 / Child abuse, p. 99 / Collagen, p. 99 / Collision sports, p. 99 / Degenerative joint conditions, p. 98 / Diseases, p. 98 / Dislocations, pp. 98–99 / Fractures, pp. 98, 99 / Growth, stunted, p. 99 / Infections, p. 99 / Inflammation, p. 98 / Injuries, pp. 98, 99 / Osteogenesis imperfecta (OI), p. 99 / Osteosarcoma, p. 99 / Prosthesis, p. 99 / Rheumatoid arthritis, p. 98 / Skeletal disorders, pp. 98–99 / Smoking, p. 99 / Sprains, pp. 98–99 / Strains, pp. 98–99 / Stress, p. 99 / Surgery, pp. 98, 99 / Tendonitis, p. 98 / Tennis elbow, p. 98	• Anatomy & Physiology, 2004, *Fighting Sarcoma* (2:43)	• New Knees, Hips Are Hot Items: As Americans Age, Demand for Artificial Joints Expected to Soar. *Chicago Tribune*, Mar. 24, 2006. • Key to Rare 'Skeleton' Disease Discovered. *UPI NewsTrack*, April 24, 2006.	• American Academy of Orthopaedic Surgeons: Fractures: http://orthoinfo.aaos.org/brochure/thr_report.cfm?Thread_ID=9&top_category=About%20Orthopaedics • American Academy of Orthopaedic Surgeons: Arthritis: http://orthoinfo.aaos.org/brochure/thr_report.cfm?Thread_ID=2&top_category=Arthritis
How Would You Vote?	Should claims about "medicinal" exotic plant extracts have to be backed up by independent scientific testing? • WWW: Herbal Supplements Are Regulated: http://www.herbalgram.org/default.asp?c=092000press • InfoTrac: Registrar: MRCGP Exam Update – Are Herbal Medicines Safe? *GP*, Mar. 31, 2006. • InfoTrac: Britain Warns against Lead Danger in Indian Herbal Medicines. *AsiaPulse News*, Aug. 18, 2005.			

CHAPTER 6: THE MUSCULAR SYSTEM

Section	Applications	Videos	InfoTrac® Readings	Weblink Examples
6.0 *Impacts, Issues*: **Pumping Up Muscles**	Androstenedione (andro), p. 103 / Athletes, p. 103 / Creatine, p. 103 / Drugs, performance-enhancing, p. 103 / Food and Drug Administration, p. 103 / Food supplements, p. 103 / Liver damage, p. 103 / McGwire, Mark, p. 103 / Sex hormones, p. 103 / Steroids, p. 103 / Tetrahydrogestrinone (THG), p. 103		• Supplement Off Limits. *Airman*, Mar. 2005. • FDA Is Cracking Down on Manufacturers of 'Andro.' *Clinical Psychiatry News*, June 2004.	• Choices in Sports: http://www.drugfreesport.com/choices/ • Drugs in Sport News: http://www.drugsinsport.net/
6.1 **The Body's Three Kinds of Muscle**			• Mouse Study: New Muscle-Building Agent Beats All Previous Ones. *Ascribe Higher Education News Service*, Dec. 8, 2005.	• A Comparison between Three Types of Muscle: http://www.erin.utoronto.ca/~w3lange/BIO210/tutorials/tutorial05.htm
6.2 **The Structure and Function of Skeletal Muscles**	Athletes, p. 107 / Distance swimmer, p. 107 / Sprinter, p. 107		• Running Generates Fiber Type-Specific Angiogenesis in Skeletal Muscles. *Obesity, Fitness & Wellness Week*, Dec. 25, 2004.	• The Muscular System: http://www.faqs.org/health/Body-by-Design-V1/The-Muscular-System.html
6.3 **How Muscles Contract**	Rigor mortis, p. 109		• Power at the Tip of the Tongue. *Science*, April 9, 2004.	• Actin-Myosin Crossbridge 3D Animation: http://www.sci.sdsu.edu/movies/actin_myosin_gif.html
6.4 **How the Nervous System Controls Muscle Contraction**	Acetylcholine (ACh), p. 111 / Botox, p. 111 / *Clostridium botulinum*, p. 111 / Droopy eyelids, p. 111 / Facial wrinkles, p. 111 / Muscle contraction, p. 111 / Muscle-relaxing effect, p. 111 / Side effects, p. 111		• Therapeutic Uses of Botulinum Toxin. Anthony H. Wheeler. *American Family Physician*, Feb. 1, 1997.	• The Nervous System and Muscle Contraction: http://www.emc.maricopa.edu/faculty/farabee/biobk/BioBookNERV.html
6.5 **How Muscle Cells Get Energy**	Exercise, p. 112		• Acidity Helps Prevent Muscle Fatigue. *UPI NewsTrack*, Aug. 19, 2004.	• NISMAT Exercise Physiology Corner: Muscle Physiology Primer: http://www.nismat.org/physcor/muscle.html
6.6 **Properties of Whole Muscles**	Exercise, p. 113 / Health, muscle, p. 113 / Muscle fatigue, p. 113 / Tetanus, pp. 112–113		• Blindsided by Tetanus. Claire Panosian Dunavan. *Discover*, Jan. 2000.	• BBC Science & Nature: Muscle Tone: http://www.bbc.co.uk/science/humanbody/body/factfiles/tone/adductor_animation.shtml
6.7 **Muscle Disorders**	Adenosine triphosphate (ATP), exercise and, p. 115 / Aerobic exercise, p. 115 / Aging, p. 115 / Anti-inflammatory drugs, p. 114 / Anxiety, p. 114 / Atrophy, p. 115 / Blood circulation, p. 115 / Creatine, p. 115 / Diseases, p. 114 / Duchenne muscular dystrophy (DMD), pp. 114, 115 / Exercise, p. 115 / Fitness experts, p. 115 / Glycolysis, p. 115 / Ibuprofen, p. 114 / Injuries, p. 114 / Involuntary twitches, p. 114 / Jogging, p. 115 / Mitochondria, p. 115 / Muscle contraction, pp. 114–115 / Muscle disorders, pp. 114–115 / Muscular dystrophies, pp. 114, 115 / Mutant genes, p. 114 / Myofibrils, p. 115 / Myotonic muscular dystrophy, p. 114 / Nerve damage, p. 115 / Oxygen, p. 115 / Potassium deficiency, p. 114 / Resistance training, p. 115 / Scar tissue, p. 114 / Sedentary lifestyle, p. 115 / Strains, p. 114 / Strength training, p. 115 / Swimming, p. 115 / Walking, p. 115 / Weight lifting, p. 115	• Genetics, 2003, *Muscular Dystrophy* (2:25)	• The Muscular Dystrophies: From Genes to Therapies. *Physical Therapy*, Dec. 2005. • Effect of Exercise on Muscle Function Decline with Aging. *The Western Journal of Medicine*, May 1991.	• Muscular Dystrophy Information Page: http://www.ninds.nih.gov/disorders/md/md.htm • Mitochondrial Myopathy: An Energy Crisis in the Cells: http://www.mdausa.org/publications/quest/q64mito.html
How Would You Vote?	Dietary supplements are largely unregulated. Should they be subject to more stringent testing for effectiveness and safety? • WWW: The Power of Creatine: It's Real but Subtle: http://bioweb.usc.edu/courses/2002-fall/documents/bisc438-supp_creatine.pdf • WWW: The Need for Regulation of Dietary Supplements – Lessons from Ephedra: http://jama.ama-assn.org/cgi/content/full/289.12.1568v1 • WWW: Regulation of Dietary Supplements Is Misrepresented by the New England Journal of Medicine: http://www.herbs.org/current/nejmresponse.html • InfoTrac: New York Task Force Calls for State Regulation of Supplements. *Food Chemical News*, Dec. 5, 2005.			

CHAPTER 7: DIGESTION AND NUTRITION

Section	Applications	Videos	InfoTrac® Readings	Weblink Examples
7.0 *Impacts, Issues:* **Hormones and Hunger**	Adipose tissues, p. 119 / Appetite, p. 119 / Calories, p. 119 / Cancer, p. 119 / Diabetes, p. 119 / Ghrelin, p. 119 / Heart disease, p. 119 / Leptin, p. 119 / Nutrition, p. 119 / Obesity, p. 119 / Stomach, p. 119	• Anatomy & Physiology, 2003, *Obesity and Infertility* (2:16)	• A Curb on Calories: Long-Term Studies Begin to Show Positives of Restrictive Eating. *Milwaukee Journal Sentinel,* April 17, 2006.	• Why We're So Fat – And the French Are Not: http://www.findarticles.com/p/articles/mi_m1175/is_6_33/ai_66278317/pg_3
7.1 **The Digestive System: An Overview**	Nutrition, p. 120		• Researchers: Natural Birth Could Be Good for Baby's Digestion. *Xinhua News Agency,* April 30, 2006.	• Digestive System Tutorial: http://www.innerbody.com/image/digeov.html
7.2 **Chewing and Swallowing: Food Processing Begins**			• The Science of Gum Chewing Studied. *Confectioner,* June 2006.	• TMJ: http://www.entnet.org/healthinfo/topics/tmj.cfm
7.3 **The Stomach: Food Storage, Digestion, and More**	Alcohol, p. 124 / Antibiotics, p. 124 / Gastrin, p. 124 / Obesity, p. 124 / Osteoporosis, p. 124 / Smoking, p. 124 / Ulcers, p. 124 / Vitamin B$_{12}$, p. 124		• Stomachs out of Africa. Brian G. Spratt. *Science,* Mar. 7, 2003.	• *H. Pylori* and Peptic Ulcer: http://digestive.niddk.nih.gov/ddiseases/pubs/hpylori/
7.4 **The Small Intestine: A Huge Surface Area for Digestion and Absorption**			• New Genes in Inflammatory Bowel Disease: Lessons for Complex Diseases? *The Lancet,* April 15, 2006.	• Anatomy of the Small Intestine: http://www.vivo.colostate.edu/hbooks/pathphys/digestion/smallgut/anatomy.html
7.5 **Accessory Organs: The Pancreas, Gallbladder, and Liver**	Alcohol, p. 127 / Cirrhosis, p. 127 / Gallstones, pp. 126, 127 / Hepatitis, p. 127 / Inflammation, p. 127 / Tissue regeneration, p. 127		• Liver Health and Diet Linked in Film 'Super Size Me.' *Internet Wire,* May 7, 2004.	• Cirrhosis of the Liver: http://digestive.niddk.nih.gov/ddiseases/pubs/cirrhosis/
7.6 **Digestion and Absorption in the Small Intestine**			• Lowering the Glycemic Index of Food with Carbohydrate Enzyme Inhibitors. *Original Internist,* Dec. 2004.	• Absorption in the Small Intestine: http://www.vivo.colostate.edu/hbooks/pathphys/digestion/smallgut/absorb.html
7.7 **The Large Intestine**	Appendicitis, p. 130 / Coliform bacteria, p. 130 / *Escherichia coli (E. coli),* p. 130 / Fecal contamination, p. 130 / Peritonitis, p. 130 / Vitamin K, p. 130		• New Study Shows Colon Cancer Related to Body Measures. *Xinhua News Agency,* July 5, 2006.	• Colonoscopy, Sigmoidoscopy, and Polyps: http://www.hemorrhoid.net/colonoscopy.php
7.8 **Managing Digestion and the Processing of Nutrients**			• Coffee Prevents Liver Cirrhosis in Alcoholics. *GP,* June 16, 2006.	• Pregastric Digestion: http://biology.about.com/library/organs/blpathodigest.htm
7.9 **Digestive System Disorders**	Cancer, pp. 132, 133, / Diseases, pp. 132, 133 *Escherichia coli (E. coli),* p. 132 / Gastritis, p. 132 / *H. pylori,* p. 132 / Lactose intolerance, p. 133 / *Salmonella,* p. 132	• Biology, 2005, *The Problem with Pork* (2:42)	• Clinical: Ulcerative Colitis and Crohn's Disease. *GP,* May 26, 2006.	• Lactose Intolerance: http://digestive.niddk.nih.gov/ddiseases/pubs/lactoseintolerance/
7.10 **The Body's Nutritional Requirements**	Calories, p. 135 / Cancer, pp. 134, 135 / Cholesterol, p. 134 / Fats, pp. 134, 135 / Fiber, p. 134 / Glycemic index, p. 134 / Vitamins, p. 134 / Weight loss, p. 135	• Biology, 2004, *Fat Facts* (2:29)	• Pharmacy Update: Fishing for Fat Facts. *Chemist & Druggist,* April 29, 2006.	• BBC: Fats and Sugars: http://www.bbc.co.uk/health/healthy_living/nutrition/basics_fatsugar.shtml
7.11 **Vitamins and Minerals**	Bones, pp. 136, 137 / Depression, p. 136 / Fats, p. 136 / FDA, pp. 136, 137 / Growth, pp. 136, 137 / Metabolism, pp. 136, 137 / Paralysis, p. 137 / Pellagra, p. 136 / Scurvy, p. 136 / Supplements, p. 137 / Vitamins, pp. 136–137		• Vitamins and Minerals Lower Risk of Macular Degeneration. *Patient Care,* Mar. 2006.	• Dietary Supplement Fact Sheet: http://riley.nal.usda.gov/nal_display/index.php?info_center=4&tax_level=3&tax_subject=274&topic_id=1323&level3_id=5147
7.12 **Calories Count: Food Energy and Body Weight**	Basal metabolic rate (BMR), pp. 138, 139 / Body mass index (BMI), p. 138 / Calories, pp. 138–139 / Exercise, p. 139 / Fats, pp. 138, 139 / Genes, p. 139 / Heart disease, p. 138 / Metabolic activities, p. 139 / Sedentary lifestyle, p. 138 / Smoking, p. 138 / Weight control, p. 139	• Genetics, 2003, *Fat Hormone* (2:33)	• Weighing Your Options. *Journal-World,* July 5, 2006. • Western Lifestyle, Lack of Sleep, Low Vitamin D Linked to Obesity. *PR Newswire,* June 27, 2006.	• Obesity Threatens Life Expectancy: http://www.usatoday.com/news/health/2005-03-16-obesity-lifespan_x.htm
How Would You Vote?	Obesity may soon replace smoking as the main cause of preventable deaths in the United States. Fast food is contributing to the problem. Should fast-food items be required to carry health warnings? • InfoTrac: McDonald's Suit Dismissed, but Judge Leaves Door Open for Plaintiffs to Re-File. *Food Chemical News,* Jan. 27, 2003. • InfoTrac: Think Tanks Wrap-Up. *United Press International,* April 24, 2003. • WWW: Food Fight: http://www.law.harvard.edu/alumni/bulletin/2002/fall/bf_05.html			

CHAPTER 8: BLOOD

Section	Applications	Videos	InfoTrac® Readings	Weblink Examples
8.0 *Impacts,* *Issues:* **Chemical** **Questions**	Cigarette smoke, p. 143 / Homeostasis, p. 143 / Lead, p. 143 / Phthalates, p. 143 / Pollutants, p. 143 / Reproductive system abnormalities, p. 143 / Synthetic chemicals, p. 143 / Urine, p. 143		• Elevated Blood Lead Levels in Refugee Children – New Hampshire, 2003–2004. *Morbidity and Mortality Weekly Report*, Jan. 21, 2005.	• Environmental Scorecard – In Your Community: http://www.scorecard.org/
8.1 **Blood:** **Plasma, Blood** **Cells, and** **Platelets**	Blood clotting, p. 145 / Edema, p. 144 / Infections, p. 145 / Inflammation, p. 144 / Parasitic worms, p. 144 / Therapeutic drugs, p. 144	• Anatomy & Physiology 2004, *Immortality Industry* (2:22)	• FDA to Hear Artificial Blood Test Proposal. *UPI NewsTrack*, July 6, 2006.	• Blood Tutorial: http://anthro.palomar.edu/ blood/default.htm
8.2 **How Blood** **Transports** **Oxygen**			• Hemoglobin Levels Are Testy Issue. *New York Daily News*, Feb. 12, 2006.	• Transport of Oxygen in the Blood: http://www.chemsoc.org/networks/ LearnNet/cfb/transport.htm
8.3 **Hormonal** **Control of Red** **Blood Cell** **Production**	Blood doping, p. 147 / Muscles, red blood cells and, p. 147		• Blood-Doping Crackdown Will Remove Corruption from Cycling. Ailene Voisin column, *Sacramento Bee*, July 3, 2006.	• Formation of Blood Cells: http://www.merck.com/mmhe/ sec14/ch169/ch169c.html
8.4 **Blood Types –** **Genetically** **Different Red** **Blood Cells**	ABO blood typing, p. 148 / Agglutination, pp. 148–149 / Blood transfusions, pp. 148, 149 / Clumping, pp. 148–149 / Donation, blood, pp. 148, 149		• Babies Accept Hearts with All Blood Types. *UPI NewsTrack*, April 5, 2005.	• Blood Types Tutorial: http://www.biology.arizona.edu/ Human_Bio/problem_sets/blood_ types/Intro.html
8.5 **Rh Blood** **Typing**	Antibodies, p. 150 / Anti-Rh gamma globulin (RhoGam), p. 150 / Childbirth, p. 150 / Cross-matching, p. 150 / Hemolytic disease of the newborn, p. 150		• Rh Disease: It's Still a Threat. *Contemporary OB/GYN*, May 2004.	• Rh Blood Types: http://anthro.palomar.edu/ blood/Rh_system.htm
8.6 **New Frontiers** **of Blood** **Typing**	AIDS, p. 151 / Autologous transfusion, p. 151 / Blood substitutes, p. 151 / Hepatitis viruses, p. 151 / HIV, p. 151 / Side effects, of Oxygent™, p. 151		• Autologous Blood Transfusion: Clinical: Expert Opinion – A Safer Way of Transfusion. *GP*, Feb. 24, 2006.	• Blood Typing in Paternity Tests: http://www.paternity-answers.com/ history-paternity-test.html
8.7 **Hemostasis** **and Blood** **Clotting**	Aggregation, p. 153 / Blood clotting, pp. 152–153 / Collagen, p. 153 / Embolism, p. 153 / Genetic disorders, p. 153 / Hemophilia, p. 153 / Infections, p. 153 / Intrinsic clotting mechanism, p. 152 / Neutrophils, p. 153 / Paralysis, p. 153 / Physical therapy, p. 153 / *Serotonin,* p. 152 / Speech therapy, p. 153 / Strokes, p. 153 / Thrombosis, p. 153		• Factor V Leiden as a Common Genetic Risk Factor for Venous Thromboembolism. *Journal of Nursing Scholarship*, Spring 2006.	• Blood Coagulation: http://web.indstate.edu/thcme/ mwking/blood-coagulation.html • Hypercoagulation: http://familydoctor.org/244.xml#1
8.8 **Blood** **Disorders**	AIDS, p. 155 / Anemias, pp. 154, 155 / Antibiotics, p. 155 / Bacteria, p. 155 / Blood clotting, p. 155 / Bone marrow, pp. 154–155 / Carbon monoxide poisoning, p. 154 / Chills, p. 154 / Epstein-Barr virus (EBV), p. 154 / Fever, pp. 154, 155 / Folic acid, p. 154 / HIV, p. 155 / Internal bleeding, p. 155 / Iron, p. 154 / Kidney disease, p. 155 / / Leukemias, pp. 154–155 / Lymphocytes, pp. 154, 155 / Malaria, p. 154 / Metabolic poisons, p. 155 / Mononucleosis, pp. 154–155 / Nutrient deficiencies, p. 154 / Radiation, p. 154 / Red blood cell disorders, p. 154 / Septicemia, p. 155 / Sickle-cell anemia, p. 154 / *Staphylococcus aureus* (staph A), p. 155 / Stem cells, pp. 154, 155 / Thalassemia, p. 154 / Toxemia, p. 155 / Viruses, p. 154 / Vitamin B$_{12}$, p. 154 / Weight loss, p. 155 / White blood cells, pp. 154–155	• Biology, 2003, *Global AIDS* (2:18)	• When Mono Strikes. Nina M. Riccio. *Current Health 2*, Mar. 2000. • Anemia: That Run-Down Feeling. Shiela Globus. *Current Health 2*, Mar. 1999.	• Leukemia & Lymphoma Society: http://www.leukemia.org/hm_lls • Blood Diseases and Resources Information: http://www.nhlbi.nih.gov/health/ public/blood/index.htm
How Would **You Vote?**				

How Would You Vote? Government regulation of substances such as lead seems to be effective: In recent years the levels of several pollutants in the general population have fallen. Should other suspect industrial chemicals be regulated?
• WWW: CDC's Second National Report on Human Exposure to Environmental Chemicals: http://www.cdc.gov/exposurereport/
• WWW: Environmental Working Group: Body Burden: http://www.ewg.org/reports/bodyburden/index.php
• WWW: Healthy Milk, Healthy Baby – Chemical Pollution and Mother's Milk: http://www.nrdc.org/breastmilk/govt.asp

CHAPTER 9: CIRCULATION—THE HEART AND BLOOD VESSELS

Section	Applications	Videos	InfoTrac® Readings	Weblink Examples
9.0 *Impacts, Issues*: The Breath of Life	Breathing, p. 157 / Cardiac arrest, p. 157 / CPR, p. 157 / Defibrillator, p. 157 / Kardia, p. 157 / Pregnancy, p. 157 / Ventricular fibrillation (VF), p. 157	• Anatomy & Physiology, 2004, *New Heart Treatments* (1:59)	• For CPR, Machines Come in Second to Humans. *The New York Times*, June 27, 2006.	• CPR and First Aid Simulator: http://www.cprsim.com/
9.1 The Cardiovascular System – Moving Blood through the Body	Cholesterol, coronary artery disease and, p. 159 / Coronary artery disease, p. 159 / Gene mutation, coronary artery disease and, p. 159 / Heart attacks, p. 159 / MEF2A, coronary heart disease and, p. 159		• 34-Year-Old Woman's Heart Recovers while Supported by ABIOMED's AB5000™ Circulatory Support System. *PR Newswire*, Feb. 9, 2005.	• The Heart and the Circulatory System: http://www.accessexcellence.org/AE/AEC/CC/heart_background.html
9.2 The Heart: A Double Pump			• Heart Surgery Data May Go Public. *Boston Globe*, July 4, 2006.	• Cardiovascular System: http://webschoolsolutions.com/patts/systems/heart.htm
9.3 The Two Circuits of Blood Flow	Bloodstream, p. 163 / Blood vessels, p. 163 / Cholesterol, HDL, p. 163, LDL, p. 163 / Heart attacks, p. 163 / Lipitor, p. 163 / Statins, p. 163 / Strokes, brain, p. 163 / Triglycerides, p. 163 / Zocor, p. 163	• Human Biology, 2006, *Heart Healthy* (1:43)	• Artery Cleanup: Blocked Blood Flow Causes Pain. *Herald & Review*, June 14, 2006.	• The Pulmonary and Systemic Circuits: http://biology.about.com/library/organs/blcircsystem5.htm
9.4 How Cardiac Muscle Contracts	Pacemakers, p. 164		• Tethering of Protein Kinase A May Be Essential for Cardiac Muscle Contraction. *Heart Disease Weekly*, Dec. 12, 2004.	• How Cardiac Muscle Contracts: http://library.thinkquest.org/C003758/Function/How%20Cardiac%20Muscle%20Contracts.htm
9.5 Blood Pressure	Atherosclerosis, p. 165 / Heart attacks, p. 165 / Hypertension, p. 165 / Hypotension, p. 165 / Obesity, p. 165 / Smoking, p. 165 / Stress, p. 165 / Strokes, p. 165		• High Blood Pressure? Don't Blame Your Job. *The New York Times*, May 16, 2006.	• Treatment of High Blood Pressure: http://www.nhlbi.nih.gov/hbp/treat/treat.htm
9.6 Structure and Functions of Blood Vessels	Obesity, p. 167 / Pregnancy, p. 167 / Varicose vein, p. 167		• Cellular Spaces Become Blood Vessels. *UPI NewsTrack*, June 21, 2006.	• Blood Vessels: http://www.botany.uwc.ac.za/SCIED/grade10/manphys/vessel.htm
9.7 Capillaries: Where Blood Exchanges Substances with Tissues	Blood pressure, p. 169 / Blood vessels, p. 169 / Heat, body, p. 169 / High blood pressure, p. 169 / Impotence, p. 169 / Kidneys, p. 169 / Nicotine, p. 169 / Smooth muscles, p. 169 / Strokes, p. 169	• Human Biology, 2006, *Heart Healthy* (1:43)	• Nicotine in Pregnancy Linked to Infant Death. *GP*, April 14, 2006.	• Capillary Anatomy Animation: http://www.innerbody.com/anim/blood.html
9.8 Cardiovascular Disorders	Aging, pp.170, 171 / Alcohol, p. 171 / Aneurysm, p. 171 / Aorta, pp. 170, 171 / Arrhythmias, p. 171 / Arteriosclerosis, p. 170 / Atherosclerosis, pp. 170, 171 / Balloon angioplasty, p. 171 / Blood clotting, p. 171 / Bradycardia, p. 171 / Cardiac rate, p. 171 / Cardiovascular disease, p. 170 / Cholesterol, pp. 170, 171 / Circulation, pp. 170–171 / Coronary bypass, p. 170 / C-reactive protein, p. 170 / Diet, proper, p. 170 / Drugs, p. 171 / Exercise, pp. 170, 171 / Fatigue, p. 171 / Heart damage, p. 171 / Homocysteine, p. 170 / Hypertension, pp. 170, 171 / Infections, p. 170 / Inflamed tissues, p. 170 / Laser angioplasty, p. 171 / Lipoproteins, p. 170 / Low blood pressure, p. 171 / Lumen, p. 170 / Obesity, p. 170 / Plaques, p. 170 / Shock, p. 171 / Shunt, p. 170 / Statins, p. 171 / Stent, p. 171 / Stress, p. 171 / Tachycardia, p. 171 / Thyroid hormone, p. 171 / Viruses, p. 170	• Anatomy & Physiology, 2003, *Super Heart Scaffold* (2:10) • Genetics, 2005, *Gene Therapy on Ailing Hearts* (3:11)	• The Arsenal. Avery Comarow. *U.S. News & World Report*, Dec. 1, 2003. • Chronic Heart Failure and Depression. *Australian Nursing Journal*, May 2006.	• American Heart Association – Arrhythmias: http://www.americanheart.org/presenter.jhtml?identifier=4469 • American Heart Association – Congestive Heart Failure: http://www.americanheart.org/presenter.jhtml?identifier=4585
How Would You Vote?	Some advocates think that CPR training should be a required mini-course in high schools. People who learn CPR also must be periodically recertified. Would you favor mandatory CPR training in high schools? • InfoTrac: A Breath-Saving Rescue: A Teenage Daughter Saves Her Dad's Life with the Help of Her Studies. *Current Health 2*, a Weekly Reader publication, Oct. 2002. • InfoTrac: Revived by Teammates' CPR, Player Feels 'Lucky,' Eager to Recover. Knight Ridder/Tribune News Service, Jan. 9, 2003. • InfoTrac: American Red Cross, American Heart Association Emphasize CPR Training; CPR with Rescue Breathing, Chest Compressions Benefits Victims. *US Newswire*, Mar. 12, 2004.			

Chapter 10: Immunity

Section	Applications	Videos	InfoTrac® Readings	Weblink Examples
10.0 *Impacts, Issues*: The Face of AIDS	AIDS, p. 175 / Cancer, rare, p. 175 / HIV, p. 175 / Tissue damage, p. 175 / Vaccines, AIDS, p. 175	• Biology, 2003, *Global AIDS* (2:18)	• People Who Pass on AIDS Virus May Be Sued. *The New York Times*, July 4, 2006.	• World Health Organization: http://www.who.int/
10.1 Overview of Body Defenses	Bacteria, p. 176 / Cancer cells, p. 176 / Diseases, p. 176 / Fever, p. 176 / Fungi, p. 176 / Infections, pp. 176, 177 / Inflammation, pp. 176, 177 / Parasitic worms, pp. 176, 177 / Pathogens, p. 176 / Protozoa, p. 176 / Toxins, p. 176 / Viruses, p. 176		• General Preventative Maintenance of the Immune System. *Townsend Letter for Doctors and Patients*, Feb.–Mar. 2002.	• Understanding the Immune System: http://www.thewellproject.org/ Treatment_and_Trials/First_Things_First/Understanding_the_Immune_System.jsp
10.2 The Lymphatic System			• The Lymphatic System. *Townsend Letter for Doctors and Patients*, July 2005.	• The Lymphatic System: http://www.innerbody.com/image/lympov.html
10.3 Surface Barriers	Athlete's foot, p. 180 / Dehydration, p. 180 / *Lactobacillus*, p. 180 / *P. acnes*, p. 180 / Plaque, dental, p. 180 / *S. mutans*, p. 180 / Urine, pH of, p. 180	• Biology, 2003, *Salon Infections* (2:05)	• Vulvovaginal Candidiasis. *Townsend Letter for Doctors and Patients*, Nov. 2005.	• The Bacterial Flora of Humans: http://www.textbookofbacteriology.net/normalflora.html
10.4 Innate Immunity	Bacteria, pp. 180, 181 / Edema, p. 181 / Fevers, pp. 180, 181 / Pain, p. 181 / Pathogens, pp. 180, 181 / Phagocytosis, pp. 180, 181 / Temperature, core, p. 181		• Defective Acute Inflammation in Crohn's Disease. *The Lancet*, Feb. 25, 2006.	• Fever Overview: http://www.mayoclinic.com/health/fever/DS00077
10.5 Overview of Adaptive Defenses	Common cold, p. 182 / Flu virus, p. 182 / Infections, p. 182 / Inflammation, p. 182 / Pathogens, p. 182 / Stem cells, immunity and, p. 182		• New Model Can Aid in Understanding Immune System Diseases. *Obesity, Fitness & Wellness Week*, Sept. 11, 2004.	• The Adaptive Defense System: http://www.cancerresearch.org/immunology/imdefenseadapt.html
10.6 Antibody-Mediated Immunity: Defending against Threats outside Cells	Allergic reactions, p. 185 / Asthma, p. 185 / Bacterial invasion, p. 184 / Cholera, p. 185 / Gonorrhea, p. 185 / Hay fever, p. 185 / Hives, p. 185 / Influenza, p. 185 / Parasitic worms, p. 185 / Pathogens, pp. 184, 185 / Placenta, antigen, p. 185 / *Salmonella*, p. 185 / Viruses, p. 185	• Human Biology, 2006, *Germs in Pakistan* (3:13)	• Computer Models Aid Understanding of Antibody-Dependent Enhancement in Spread of Dengue Fever. *Ascribe Higher Education News Service*, Oct. 14, 2005.	• How Lymphocytes Produce Antibody: http://www.cellsalive.com/antibody.htm
10.7 Cell-Mediated Responses – Defending against Threats inside Cells	Antibiotics, p. 187 / Antibodies, p. 186 / B cells, p. 187 / Cancer cells, p. 186 / Drugs, p. 187 / Infections, p. 187 / Lymphocytes, p. 187 / Stress markers, p. 187 / T cell-mediated immune response, p. 186 / Tissue grafts, p. 187 / Transplants, p. 187 / Viruses, p. 186 / Xenotransplantation, p. 187	• Human Biology, 2006, *A Saving Graft* (2:23)	• Induction of Human T Cell-Mediated Immune Responses after Primary and Secondary Smallpox Vaccination. *Journal of Infectious Diseases*, Oct. 1, 2004.	• The Cytotoxic T-Lymphocyte: http://www.cellsalive.com/ctl.htm • What Is the Role of T Cells in Cell-Mediated Immunity?: http://w3.dwm.ks.edu.tw/bio/activelearner/43/ch43intro.html
10.8 Immunological Memory	Memory, immunological, p. 188 / Pathogens, immunological memory and, p. 188 / T cells, immunological memory and, p. 188 / Viruses, immunological memory and, p. 188	• Anatomy & Physiology, 2003, *Custom Cancer Vaccine* (3:10)	• HIV-1-Infected Children Have Impaired Immunological Memory. *Biotech Week*, Sept. 3, 2003.	• La Jolla Institute for Allergy & Immunology: Research: http://www.liai.org/research/faculty/1873/downloads/SemImm_04.pdf
10.9 Applications of Immunology	AIDS, p. 189 / Cancer, p. 189 / Genetic engineering, p. 189 / Hepatitis, pp. 188, 189 / Herceptin, p. 189 / HIV, p. 189 / Measles, pp. 188, 189 / Polio, pp. 188, 189 / Tetanus, pp. 188, 189 / Vaccines, pp. 188–189	• Biology, 2005, *Polio Scare* (1:35) • Genetics, 2003, *Transgenic Tobacco* (2:00)	• Scientists Derive Crucial Immune Cells from Stem Cells: Report. *Xinhua News Agency*, July 5, 2006.	• Immunotherapy Cancer Treatment: http://www.cancersupportivecare.com/immunotherapy.html
10.10 Disorders of the Immune System	AIDS, p. 191 / Anaphylactic shock, pp. 190–191 / Cancer, p. 191 / Diabetes, p. 191 / Effector cells, p. 190 / Gene therapy, p. 191 / Helper cells, p. 191 / HIV, p. 191 / Insulin, p. 191 / Lymphocytes, p. 191 / Macrophages, p. 191 / Severe combined immune deficiency (SCID), p. 191 / Viral infections, p. 191	• Anatomy & Physiology, 2004, *Peanut Allergies* (1:33)	• Growth of Peanut Allergies Forces Changes at Schools. *Monitor*, June 16, 2006. • The Dirty Truth about Allergies. *Australasian Business Intelligence*, June 27, 2006.	• Arthritis Foundation: http://www.arthritis.org/ • American Autoimmune Related Diseases Association: http://www.aarda.org/
How Would You Vote?	Drugs that extend the life of AIDS patients are unaffordable for people in most developing countries. Should the federal government offer incentives to companies to discount the drugs for developing countries? What about AIDS patients at home? Who should pay for their drugs? • InfoTrac: Drug Companies and AIDS in Africa. *America*, Nov. 25, 2002. • InfoTrac: The Pill Machine. *Newsweek International*, Nov. 19, 2001. • InfoTrac: A Global Market Isn't as Easy as It Looks. *Business Week*, Sept. 3, 2001.			

CHAPTER 11: THE RESPIRATORY SYSTEM

Section	Applications	Videos	InfoTrac® Readings	Weblink Examples
11.0 *Impacts, Issues*: Down in Smoke	Asthma, p. 195 / Blood pressure, p. 195 / Bronchitis, p. 195 / Cancer, p. 195 / Cigarette smoke, p. 195 / Common cold, p. 195 / HDL cholesterol, p. 195 / Microbes, p. 195 / White blood cells p. 195	• Human Biology, 2006, *Heart Healthy* (1:43)	• Patterns of Global Tobacco Use in Young People and Implications for Future Chronic Disease Burden in Adults. *The Lancet*, Mar. 4, 2006.	• Smoking's Impact on the Lungs: http://whyquit.com/joel/Joel_02_17_smoke_in_lung.html
11.1 The Respiratory System – Built for Gas Exchange	Allergies, p. 196 / Choking, p. 197 / Common cold, p. 196 / Nasal spray, p. 196		• Decorate a Nursery with Baby's Lungs, Health in Mind. *South Florida Sun-Sentinel*, April 27, 2006.	• The Voice Doctor: http://www.voicedoctor.net/
11.2 Respiration = Gas Exchange			• Cell Respiration Process Is Identified. *UPI NewsTrack*, April 6, 2006.	• BBC: Respiration and Gas Exchange: http://www.bbc.co.uk/schools/ks3bitesize/science/biology/respiration2_intro.shtml
11.3 The "Rules" of Gas Exchange	Decompression sickness, p. 199 / Hyperventilation, p. 199 / Nitrogen narcosis, p. 199		• Breathing on the Edge. Jessica Gorman. *Science News*, Mar. 31, 2001.	• Brain Anoxia or Hypoxia: http://healthlink.mcw.edu/article/921384224.html
11.4 Breathing – Air In, Air Out	Choking, p. 201 / Collapsed lung, p. 201 / Heimlich maneuver, p. 201 / Illnesses, p. 201 / Pneumothorax, p. 201		• Wood Smoke and Women's Health. *Environment*, May 2006.	• Lungs Animations and Interactives: http://www.smm.org/heart/lungs/top.html
11.5 How Gases Are Exchanged and Transported	Infant respiratory distress syndrome, p. 202 / Premature babies, p. 202		• Gas Exchange in the Lungs. *Biological Sciences Review*, Sept. 2003.	• Blood Gases Manual: http://www.madsci.com/manu/indexgas.htm
11.6 Homeostasis Depends on Controls over Breathing	Alzheimer's, apnea and, p. 205 / Snoring, apnea and, p. 205		• Sleeping without Breath: One Woman's Frightening Story. *Vibrant Life*, Mar.–April 2004.	• PET Images Reveal Brain's Response to Hunger for Air: http://biad02.uthscsa.edu/projects/mario-air/
11.7 Disorders of the Respiratory System	Aerosol inhalers, p. 207 / Air pollution, p. 207 / Allergens, p. 207 / American Cancer Society, p. 206 / Antibiotics, p. 207 / Antifungal drugs, p. 207 / Asthma, p. 207 / Bacteria, pp. 206–207 / Bronchitis, p. 207 / Cancer, p. 206 / Carcinogens, p. 206 / Cigarette smoke, pp. 206–207 / Collagen fibers, p. 206 / Common cold, p. 207 / Coronary heart disease, p. 206 / Coughing, pp. 206–207 / Edema, p. 206 / Emphysema, pp. 206–207 / Exercise, p. 207 / Fever, p. 207 / Flu symptoms, p. 207 / Fungi, pp. 206–207 / Gases, p. 207 / Genetic damage, p. 206 / Heart disease, p. 206 / Histoplasmosis, p. 207 / Immune system, pp. 206–207 / Infections, pp. 206–207 / Inflammation, pp. 206–207 / Influenza, p. 206 / Irritants, p. 207 / Mucus, pp. 206–207 / Oxygen, p. 206 / Pathogens, pp. 206–207 / Phagocytes, p. 206 / Pneumonia, p. 206 / Pregnancy, p. 206 / Scar tissue, p. 207 / Secondhand smoke, p. 206 / Smooth muscles, p. 207 / Steroids, asthma and, p. 207 / Stillbirth, p. 206 / Stress, p. 207 / Toxic fumes, p. 206 / Tuberculosis (TB), p. 207 / Viruses, p. 206 / Vitamin C, p. 206	• Biology, 2004, *Mold Exposure* (2:43) • Environmental Science, 2005, *Clean Air Act* (1:34)	• Nicotine Metabolism May Spawn Carcinogen. J. Travis. *Science News*, Oct. 28, 2000. • Cannabis Lifts Emphysema Risk. *Australasian Business Intelligence*, Mar. 27, 2006.	• Oncology Channel – Lung Cancer: http://oncologychannel.com/lungcancer/ • CDC – Tobacco: Information and Prevention Source: http://www.cdc.gov/tobacco
How Would You Vote?	Tobacco is both a worldwide threat to health and a profitable product for American companies. As tobacco use by its citizens declines, should the United States encourage international efforts to reduce tobacco use? • WWW: Towards Health with Justice: Litigation and Public Inquiries as Tools for Tobacco Control: http://www.who.int/tobacco/media/en/final_jordan_report.pdf • WWW: Tobacco and the Rights of the Child: http://whqlibdoc.who.int/hq/2001/WHO_NMH_TFI_01.3_Rev.1.pdf • WWW: Global Tobacco Control: http://www.apha.org/wfpha/tob.htm			

CHAPTER 12: THE URINARY SYSTEM

Section	Applications	Videos	InfoTrac® Readings	Weblink Examples
12.0 *Impacts, Issues*: Truth in a Test Tube	Athletes, urine tests for, p. 211 / Bacterial infection, p. 211 / Blood, illegal drugs and, p. 211 / Cancer, p. 211 / Cocaine, p. 211 / Drug testing, p. 211 / Ecstasy, p. 211 / Marijuana, p. 211 / Menopause, p. 211 / Metabolic problems, p. 211 / National Collegiate Athletic Association (NCAA), p. 211 / Pregnancy, tests for, p. 211 / Steroids, p. 211 / Urinalysis, p. 211		• NCAA Steps Up Steroid Tests: Organization Is Making a Far-Reaching Effort to Ensure That College Athletes Stay Clean. *Kansas City Star,* June 22, 2006.	• Early Prostate Cancer Detected in Molecular-Based Urine Test: http://www.sciencedaily.com/releases/2000/04/000406090853.htm
12.1 The Challenge: Shifts in Extracellular Fluid	Joints, gout in, p. 213 / Research, water, p. 213		• Gouty Arthritis: A Primer on Late-Onset Gout. *Geriatrics,* July 2005.	• Electrolytes: http://www.medicinenet.com/electrolytes/article.htm
12.2 The Urinary System – Built for Filtering and Waste Disposal	Vitamin D, p. 214		• Renal Anatomy and Overview of Nephron Function. *Nephrology Nursing Journal,* April 2003.	• "How Your Kidneys Work": http://health.howstuffworks.com/kidney.htm
12.3 How Urine Forms: Filtration, Reabsorption, and Secretion	Drug testing, p. 217 / Vitamins, p. 216		• Renal Hemodynamics: An Overview. *Nephrology Nursing Journal,* Aug. 2003. • Glomerular Filtration: An Overview. *Nephrology Nursing Journal,* June 2003.	• Urine Formation: http://distance.stcc.edu/AandP/AP/AP2pages/Units24to26/urinary/urinform.htm
12.4 How Kidneys Help Manage Fluid Balance and Blood Pressure	Blood, thirst and, p. 219 / Brain, thirst and, p. 219 / Salivary glands, thirst and, p. 219 / Salt(s), thirst and, p. 219 / Thirst center, p. 219		• Urinary Concentration and Dilution. *Nephrology Nursing Journal,* May–June 2004.	• Fluid and Electrolyte Balance: http://mcb.berkeley.edu/courses/mcb135e/kidneyfluid.html
12.5 Removing Excess Acids and Other Substances in Urine	Albumin, p. 220 / Cancer, urinary system and, p. 220 / Cirrhosis, p. 220 / Diabetes, p. 220 / Diseases, kidney, p. 220, urinalysis and, p. 220 / Hepatitis, p. 220 / Hypertension, p. 220 / Infections, bladder, p. 220, urinary tract, p. 220 / Injuries, p. 220 / Pus, p. 220		• Urine Albumin Considered Independent Marker for Kidney, Cardiovascular Disease. *Heart Disease Weekly,* July 4, 2004.	• Urine pH: http://www.rnceus.com/ua/uaph.html • Acidosis: http://www.merck.com/mmhe/sec12/ch159/ch159b.html
12.6 Kidney Disorders	Arteries, p. 221 / Bladder, p. 221 / Blood, p. 221 / Bloodstream, p. 221 / Chlamydia, p. 221 / Cysts, p. 221 / Diabetes, p. 221 / Dialysis, p. 221 / Diseases, kidney, p. 221, polycystic kidney, p. 221 / Disorders, inherited, p. 221, kidney, p. 221 / Extracellular fluid (ECF), p. 221 / Fatigue, p. 221 / Glomerulonephritis, p. 221 / Glucose, p. 221 / Hemodialysis, p. 221 / Hypertension, p. 221 / Immune response, faulty, p. 221 / Infections, bacterial, p. 221, bladder, p. 221, urinary tract, p. 221 / Inflammation, p. 221 / Kidney stones, p. 221 / Kidney transplants, p. 221 / Lithotripsy, p. 221 / Memory loss, p. 221 / Nausea, p. 221 / Nephritis, p. 221 / Proteins, p. 221 / Pyelonephritis, p. 221 / Renal pelvis, p. 221 / Salt(s), p. 221 / Solutes, p. 221 / Tissues, p. 221 / Urinary system, p. 221 / Urination, p. 221 / Urine, disorders in, p. 221, kidney disorders and, p. 221 / Veins, p. 221		• Filter Fault Is Kidney Failure. *The Economic Times,* April 24, 2006. • Battle Well Fought: *Miami Herald,* July 13, 2006.	• National Kidney and Urologic Diseases Information Clearinghouse: http://kidney.niddk.nih.gov/ • A Patient's Guide to Kidney Transplant Surgery: http://www.kidneytransplant.org/patientguide/index.html
How Would You Vote?	Many companies use urine testing to screen prospective employees for drug and alcohol use. Some people say this is an invasion of privacy. Do you think employers should be allowed to require a person to undergo urine testing before being hired? • InfoTrac: Reasons to Consider: Drug Testing. Industrial Safety & Hygiene News, April 2002. • InfoTrac: Urine – Or You're Out: Drug Testing Is Invasive, Insulting, and Generally Irrelevant to Job Performance. Why Do So Many Companies Insist on It? Reason, Nov. 2002. • InfoTrac: Street Smarts: Just Say Yes. Inc., Nov. 1, 2004.			

CHAPTER 13: THE NERVOUS SYSTEM

Section	Applications	Videos	InfoTrac® Readings	Weblink Examples
13.0 *Impacts, Issues:* In Pursuit of Ecstasy	Amphetamines, p. 225 / Depression, p. 225 / Ecstasy, p. 225 / MDMA, p. 225 / Memory, loss of, p. 225 / Overdose, drug, p. 225 / Panic, p. 225 / Seizures, p. 225 / Serotonin, p. 225		• Man Still Alive after 40,000 Ecstasy Hits. *UPI NewsTrack*, April 4, 2006.	• NIDA – Ecstasy: http://www.nida.nih.gov/ Infofacts/ecstasy.html
13.1 Neurons – The Communication Specialists	Active transport, p. 227 / Cocaine, p. 227 / "Drug memory" effect, p. 227 / Learning, scientific studies of, p. 227 / Memory, scientific studies of, p. 227 / Stimulus/stimuli, neurons and, p. 226		• Hopkins Scientists Use Embryonic Stem Cells, New Cues to Awaken Latent Motor Nerve Repair. *Ascribe Higher Education News Service*, June 25, 2006.	• Basic Neural Processes Tutorials: http://psych.hanover.edu/ Krantz/neurotut.html • The Biological Neuron: http://vv.carleton.ca/~neil/ neural/neuron-a.html
13.2 Action Potentials = Nerve Impulses	Active transport, p. 229 / Adenosine triphosphate (ATP), p. 229 / Ion pumps, pp. 228–229 / Sodium-potassium pumps, p. 229 / Stimulation, action potentials and, pp. 228, 229		• The Where and When of Intention. *Science*, Feb. 20, 2004.	• Propagation of an Action Potential: http://www.accessexcellence.org/ RC/VL/GG/action_Potent.html
13.3 Chemical Synapses: Communication Junctions	Disorders, nervous system, p. 230 / Drugs, antidepressant, p. 231 / Inhibitory postsynaptic potentials (IPSPs), p. 231 / Learning, p. 231 / Memory, p. 231 / Pain, p. 231 / Prozac, p. 231 / Serotonin, pp. 230, 231 / Sexual behavior, p. 231	• Biology, 2003, *Conquering Depression* (3:10)	• Neurotransmitter Orexin Linked to Drug-Seeking … *Alcoholism & Drug Abuse Weekly*, Sept. 5, 2005.	• Biochemistry of Neurotransmitters: http://web.indstate.edu/thcme/ mwking/nerves.html
13.4 Information Pathways	Contraction(s), pp. 232, 233 / Ions, p. 232 / Reflex, stretch, p. 232 / Reflex arcs, pp. 232, 233 / Reflex pathways, p. 232 / Sensory receptors, p. 233 / Stimulation, pp. 232, 233		• Critical for Coating: Protein Directs Nerve-Sheath Construction. *Science News*, Sept. 10, 2005.	• The Stretch Reflex: http://psychlab1.hanover.edu/ Classes/Neuro/powerpoint/ StretchReflex_files/frame.htm
13.5 Overview of the Nervous System	Brain, damage to, p. 235 / Brain stem, p. 235 / Consciousness, p. 235 / Pregnancy, mercury and, p. 235 / Wastes, industrial, p. 235		• Report Shows CDC Ignored Autism-Mercury Data. *Pain & Central Nervous System Week*, Dec. 20, 2004.	• Overview of the Nervous System: http://quest.nasa.gov/neuron/ background/nervsys.html
13.6 Major Expressways: Peripheral Nerves and the Spinal Cord	Adrenal gland, p. 236 / Brain stem, p. 236 / Constriction, eyes and, p. 236 / Dilation, eyes and, p. 236 / Meninges, p. 237 / Reflex, p. 237 / Spinal reflexes, p. 237 / Stimulation, nerves and, p. 237 / Stimulus/stimuli, spinal cord and, p. 237	• Biology, 2005, *New Nerves* (2:37)	• Knitting Nerves Back Together. *Technology Review*, May–June 2006.	• Nerve Structures of the Spine: http://www.spineuniverse.com/ displayarticle.php/article1275. html
13.7 The Brain – Command Central	Barbiturates, p. 239 / Brain stem, p. 238 / Caffeine, p. 239 / Cerebrospinal fluid (CSF), pp. 238, 239 / Heroin, p. 239 / Meningitis, p. 239 / Pia mater, p. 238 / Sexual behavior, p. 239		• Meningitis Survivor Pushes Vaccination. *News-Sentinel*, April 11, 2006.	• The Whole Brain Atlas: http://www.med.harvard.edu/ AANLIB/home.html
13.8 A Closer Look at the Cerebrum	Association areas, pp. 240–241 / Conscious behavior, p. 240 / EEG, p. 240 / Memory, p. 241 / Positron-emission tomography (PET), pp. 240, 241		• Fear Factor Function in Brain. *USA Today*, June 2005.	• The Cerebrum: http://www.ship.edu/~ cgboeree/genpsycerebrum.html
13.9 Memory and Consciousness	Amnesia, p. 242 / Coma, p. 243 / Consciousness, pp. 242, 243 / EEG, p. 243 / Memory, p. 242 / Memory loss, p. 242 / Positron-emission tomography (PET), p. 243 / Serotonin, p. 243 / Sleep, p. 243	• Biology, 2004, *Brain Fitness* (2:16)	• Do I Know You? *Time*, July 17, 2006.	• Exploratorium – The Memory Exhibition: http://www.exploratorium.edu/ memory/
13.10 Disorders of the Nervous System	Aging, Alzheimer's and, p. 244 / Amyloid protein, p. 244 / Autoimmune disease, p. 244 / Bacteria, p. 244 / Concussion, p. 244 / Confusion, p. 244 / Encephalitis, p. 244 / Epilepsy, p. 244 / Herpes, p. 244 / HIV, p. 244 / Memory loss, p. 244 / Meningitis, p. 244 / Migraines, p. 244 / Multiple sclerosis (MS), p. 244 / Nausea, p. 244 / Painkillers, p. 244 / Paralysis, p. 244 / Parkinson's disease (PD), p. 244 / Seizures, p. 244 / Thalamus, p. 244 / Trauma, birth, p. 244 / Viruses, p. 244 / Vision, blurred, p. 244 / Vomiting, p. 244	• Anatomy & Physiology, 2003, *Brain-washing Toxins Away* (2:45) • Genetics, 2003, *Alzheimer's Mutation* (2:05)	• Knocked Cold: A Concussion Discussion. *Harvard Men's Health Watch*, April 2002. • Sleep Disorder Early Sign of Brain Disease. *GP*, June 23, 2006.	• Alzheimer's Brain Tour: http://www.med.harvard.edu/ AANLIB/cases/case3/mr1/ 040.html • What Happens in Parkinson's Disease: http://www.parkinsonsinfo.com/ about_parkinsons/what happens.html
13.11 The Brain on Drugs	Amphetamines, p. 245 / Barbiturates, p. 245 / Cocaine, p. 245 / Depressants, p. 245 / Dopamine, p. 245 / Hallucinogen, p. 245 / Heroin, p. 245 / Marijuana, p. 245 / Morphine, p. 245 / Nervous system, drugs and, p. 245 / Painkillers, p. 245 / Psychoactive drugs, p. 245		• Drug Abuse May Increase Risk of Later Schizophrenia. *The Brown University Digest of Addiction Theory and Application*, Nov. 2003.	• Drugs and the Brain Tutorial: http://www.csusm.edu/DandB/ • Crack and Cocaine: http://www.nida.nih.gov/ Infofacts/cocaine.html
How Would You Vote?	Would you support legislation that forces nonviolent drug offenders to enter drug rehab programs as an alternative to jail? • InfoTrac: Rehab, Not Jail. *State Government News*, Sept. 2001. • InfoTrac: When Smart Women Make Foolish Choices. Knight Ridder/Tribune News Service, April 24, 2001. • WWW: A Familiar Debate: Jail or Treatment?: http://news.minnesota.publicradio.org/features/2004/06/14_steilm_methpenalties/			

CHAPTER 14: SENSORY SYSTEMS

Section	Applications	Videos	InfoTrac® Readings	Weblink Examples
14.0 *Impacts, Issues:* Private Eyes	Iris scanning technology, p. 249 / Pain, p. 249 / Security, p. 249 / Sensations, p. 249 / Sensory systems, p. 249 / Smooth muscles, iris scanning technology and, p. 249		• Iris-Scanning Devices in N.J. Schools. *UPI NewsTrack*, Feb. 23, 2006.	• Iris Recognition Technology: http://www.biometricgroup.com/reports/public/reports_iris-scan.html
14.1 Sensory Receptors and Pathways	Cones, p. 250 / Perceptions, p. 250 / Reflex, p. 250 / Sensory adaptation, p. 251 / Sensory receptors, pp. 250–251 / Sensory systems, sensory receptors and, pp. 250–251 / Sinuses, p. 250 / "Special senses," p. 251 / Stimulation, p. 250 / Temperature, p. 250		• Inability to Smell the Lilacs May Predict AD. *Geriatrics*, Feb. 2005.	• Seeing, Hearing and Smelling the World: http://www.hhmi.org/senses/ • The Fetal Senses: http://www.birthpsychology.com/lifebefore/fetalsense.html
14.2 Somatic Sensations	Amputation, p. 253 / Childbirth, p. 253 / Heart attacks, p. 253 / Hypnosis, p. 253 / Injuries, p. 252 / Itching, p. 252 / Morphine, p. 253 / Opiates, p. 253 / Pain, pp. 252, 253 / Perceptions, p. 253 / Sensations, pp. 252–253 / Sensory receptors, pp. 252–253 / Spasms, p. 252 / Stimulation, p. 252		• Are There Gender Differences in Pain Perception? *Journal of Neuroscience Nursing*, June 2006.	• Somatosensation and the Perception of Pain: http://www.fleshandbones.com/readingroom/pdf/716.pdf • Referred Pain: http://www.merck.com/mmhe/sec06/ch078/ch078a.html
14.3 Taste and Smell: Chemical Senses	Organs, vomeronasal, p. 255 / Pheromones, p. 255 / "Sexual nose," p. 255 / Stored memory, p. 254 / Tongue, p. 254 / Tonsils, p. 254 / *Umami*, pp. 254, 255	• Human Biology, 2006, *Tongue Tied* (3:55)	• The Vomeronasal (Jacobson's) Organ. *Ear, Nose and Throat Journal*, July 2005.	• Physiology of Taste: http://www.sff.net/people/mberry/taste.htm
14.4 A Tasty Morsel of Sensory Science	Artificial sweeteners, p. 255 / Morphine, p. 255 / Quinine, p. 255 / Receptors, taste, p. 255 / Salts, p. 255 / Sucrose, p. 255 / Toxic chemicals, p. 255 / *Umami*, p. 255		• Salty Taste Preference Linked to Birth Weight. *Nutrition Today*, Jan.–Feb. 2006.	• Taste – A Brief Tutorial: http://www.cf.ac.uk/biosi/staff/jacob/teaching/sensory/taste.html
14.5 Hearing: Detecting Sound Waves	"Anvil," in ear, p. 256 / Eustachian tube, p. 257 / Pitch, pp. 256, 257 / Sensory systems, pp. 256–257 / Stimulation, pp. 256, 257		• Noise-Induced Hearing Loss in Young People. *Pediatrics*, Jan. 2006.	• The Physiology of Hearing: http://www.avatar.com.au/courses/PPofM/hearing/hearing1.html
14.6 Balance: Sensing the Body's Natural Position	Brain stem, p. 258 / Deceleration, p. 258 / Motion sickness, p. 259 / Overstimulation, p. 259 / Reflex centers, p. 258 / Sensory systems, pp. 258–259 / Vomiting, p. 259		• Vestibulo-Ocular Physiology Underlying Vestibular Hypofunction. *Physical Therapy*, April 2004.	• Vestibular Disorders Association: http://www.vestibular.org/
14.7 Disorders of the Ear	Aspirin, p. 259 / Deafness, p. 259 / Ear infections, p. 259 / Hearing loss, p. 259 / Noise, p. 259 / Pus, p. 259 / Respiratory infection, p. 259 / Tinnitus, p. 259		• Ear Protection: Combo Vaccine Prevents Some Infections. *Pediatrics*, June 2006.	• Types and Causes of Deafness: http://library.thinkquest.org/15390/causes.htm
14.8 Vision: An Overview	Blind spots, p.260 / Choroid, p. 260 / Farsightedness, p. 261 / Iris, p. 260 / Nearsightedness, p. 261 / Retinal stimulation, p. 261 / Visual accommodation, p. 261	• Biology, 2005, *To See Again* (3:05)	• Ocular Anatomy: Front-to-Back Tour of the Basics. *Contemporary Pediatrics*, Aug. 2002.	• How Vision Works: http://health.howstuffworks.com/eye.htm
14.9 From Visual Signals to "Sight"	Choroid, p. 262 / Cones, pp. 262–263 / Pigments, p. 262 / Rhodopsin, p. 262 / Stimulation, p. 262 / Vitamin A, p. 262		• Biological Clock Linked to Network of Non-Rod, Non-Cone Cells. *Ophthalmology Times*, April 15, 2002.	• Ophthalmologists and Physicists Team Up to Design 'Bionic Eye': http://www.stanford.edu/dept/news/pr/2005/pr-retina-033005.html
14.10 Disorders of the Eye	Aging, pp. 264, 265 / Astigmatism, p. 264 / Bacteria, p. 264 / Blindness, pp. 264, 265 / Cancer, pp. 264–265 / Cataracts, p. 265 / Chlamydia, p. 265 / Choroid, p. 265 / Chronic glaucoma, p. 265 / Color blindness, p. 264 / Conductive keratoplasty (CK), p. 265 / Conjunctivitis, pp. 264, 265 / Diabetes, p. 265 / Farsightedness, p. 264 / Herpes, p. 265 / Lasik, p. 265 / Macular degeneration, p. 265 / Malignant melanoma, p. 265 / Nearsightedness, pp. 264, 265 / Peripheral vision, loss of, p. 265 / Retinoblastoma, p. 265 / Scar tissue, p. 265 / Sores, p. 265 / Trachoma, p. 265 / Viruses, pp. 264, 265 / Vision, blurred, p. 265	• Biology, 2005, *To See Again* (3:05) • Anatomy & Physiology, 2004, *CK Eye Surgery* (1:39)	• Traffic Signal Color Recognition Is a Problem for Both Protan and Deutan Color-Vision Deficients. *Human Factors*, Fall 2003. • Global Increase in Glaucoma Foreseen: Early Detection Could Curb Rise in Global Glaucoma Cases. *The Futurist*, July–Aug. 2006.	• Anatomy of the Eye and Macular Degeneration: http://www.macula.org/anatomy/ • Anatomy, Physiology and Pathology of the Human Eye: http://www.tedmontgomery.com/the_eye/
How Would You Vote?	Do you favor laws that allow employers and others to collect the information required for iris scanning – and to be protected from liability if the scans are misused? • WWW: "The Eyes Have It: Body Scans at the ATM": http://www.aclu.org/privacy/consumer/15393prs19990621.html • InfoTrac: Behold the Body Biometric; Before Long, We May Use Fingerprints, Iris Scans, and Voice Recognition to Log Onto Computers, Buy Groceries — Even When Picking Up Kids from School. *Business Week Online*, May 1, 2006. • InfoTrac: Biometrics: Payments at Your Fingertips; Fingerprints and Iris Scans Will Replace Keys and Credit Cards If Outfits Like Pay by Touch Succeed in Their Biologic Mission. *Business Week Online*, Mar. 28, 2006.			

CHAPTER 15: THE ENDOCRINE SYSTEM

Section	Applications	Videos	InfoTrac® Readings	Weblink Examples
15.0 ***Impacts, Issues*:** **Hormones in the Balance**	Arsenic, p. 269 / Atrazine, p. 269 / Blood sugar levels, p. 269 / Cancer, p. 269, / Endocrine disrupters, p. 269 / Environment, PCBs and, p. 269 / Herbicides, p. 269 / Poisons, p. 269 / Sperm, low count in, p. 269		•Environment/Public Health: New Report Points Up Growing Evidence of Endocrine Disrupters. *European Report*, May 4, 2006.	•EPA — Endocrine Disruptors Research Initiative: http://www.epa.gov/endocrine/ •Our Stolen Future: http://www.ourstolenfuture.org/
15.1 **The Endocrine System: Hormones**	Agent Orange, p. 270 / Diabetes, p. 270 / Insulin, pp. 270, 271 / Parathyroid hormone (PTH), p. 271 / Somatotropin (STH), p. 271 / Thymosins, p. 271 / Thyroxine (T$_4$), p. 271	•Genetics, 2005, *Gene Therapy for Diabetes* (2:16)	•Review of the Endocrine System. Deirdre G. Bauer. *MedSurg Nursing*, Oct. 2005.	•Endocrine System Tutorial: http://users.stlcc.edu/dwallner/AdultChildII/Endocrine/
15.2 **Types of Hormones and Their Signals**	Growth hormone (GH), p. 272 / Insulin, pp. 272, 273 / Prolactin (PRL), p. 272 / Steroid hormones, pp. 272, 273 / Thyroid hormone (TH), p. 272 / Vitamin D, p. 272		•Can PYY Cure Obesity? (hormone Peptide YY3-36). *U.S. News & World Report*, Sept. 15, 2003.	•Steroid Hormones: http://web.indstate.edu/thcme/mwking/steroid-hormones.html
15.3 **The Hypothalamus and Pituitary Gland: Major Controllers**	Adrenal cortex, p. 274 / Blood, loss of, p. 275 / Blood pressure, p. 275 / Growth hormone (GH), pp. 274, 275 / Pregnancy, p. 275 / Prolactin (PRL), pp. 274, 275 / Sexual behavior, p. 274 / Somatotropin (STH), pp. 274, 275 / Stimulation, p. 275 / Thyrotropin (TSH), pp. 274, 275 / Vasopressin, p. 275		•The Scent of Trust (oxytocin, hormone secreted by the posterior pituitary gland). *Prevention*, Oct. 2005.	•The Pituitary Society: http://www.pituitarysociety.org/public
15.4 **Factors That Influence Hormone Effects**	Acromegaly, p. 276 / Dwarfism, p. 276 / Gigantism, p. 276 / Hormone replacement therapy (HRT), p. 276 / Thymosins, p. 277 / Thyrotropin-releasing hormones (TRH), p. 276 / Triiodothyronine (T$_3$), p. 277 / Vitamin D3, p. 277	•Anatomy & Physiology, 2003, *Hormone Replacement Therapy* (2:49)	•Acromegaly (diagnosis and treatment). Dr. Ana Pokrajac-Simeunovic and Dr. Peter Trainer. *Chemist & Druggist*, Nov. 19, 2005.	•Growth Hormone Deficiency: http://www.schneiderchildrenshospital.org/peds_html_fixed/peds/diabetes/ghd.htm
15.5 **The Thymus, Thyroid, and Parathyroid Glands**	Goiter, pp. 278, 279 / Hyperparathyroidism, p. 279 / Hyperthyroidism, p. 278 / Kidney stones, p. 279 / Overweight, p. 278 / Parathyroid hormone (PTH), p. 279 / Rickets, p. 279 / Thyroid hormone (TH), p. 278 / Vitamin deficiency, p. 279		•Primary Hyperparathyroidism (The Effective Physician). William E. Golden; Robert H. Hopkins. *Internal Medicine News*, Dec. 1, 2005.	•The American Thyroid Association: http://www.thyroid.org/
15.6 **Adrenal Glands and Stress Responses**	Adrenal gland, pp. 280–281 / Allergic reactions, p. 280 / Cardiovascular disease, p. 281 / Cortisone, p. 280 / Epinephrine, pp. 280, 281 / Exercise, stress and, p. 281 / Hypoglycemia, p. 280 / Insomnia, p. 281 / Steroid hormones, p. 280 / Stress, pp. 280–281	•Anatomy & Physiology, 2004, *Peanut Allergies* (1:33)	•Prenatal Exposure to Stress and Stress Hormones Influences Child Development. Elysia Poggi Davis; Curt A. Sandman. *Infants & Young Children*, July–Sept. 2006.	•Glucocorticoids: http://arbl.cvmbs.colostate.edu/hbooks/pathphys/endocrine/adrenal/gluco.html
15.7 **The Pancreas: Regulating Blood Sugar**	Blood sugar levels, p. 282 / Growth hormone (GH), p. 282 / Insulin, p. 282 / Secretion, pancreas and, p. 282 / Somatostatin, p. 282 / Stimulation, endocrine system and, p. 282	•Genetics, 2005, *Gene Therapy for Diabetes* (2:16)	•Perfecting a 'Pancreas': Scientists Fine-Tune a Device to Be Used by Diabetics. Jamie Talan. *Newsday*, May 15, 2006.	•Journal of the Pancreas: http://www.joplink.net/
15.8 **Disorders of Glucose Homeostasis**	Atherosclerosis, p. 283 / Autoimmune response, p. 283 / Blindness, p. 283 / Blood sugar levels, p. 283 / Cardiovascular disease, p. 283 / Cholesterol, p. 283 / Dehydration, p. 283 / Diabetes, p. 283 / Diet, proper, p. 283 / Exercise, p. 283 / Fats, p. 283 / Gangrene, p. 283 / Gum disease, p. 283 / Heart attacks, p. 283 / Kidney disease, p. 283 / Low blood pressure, p. 283 / Obesity, p. 283 / Pancreas, p. 283 / pH, p. 283 / Strokes, p. 283 / Ulcers, p. 283 / Urinalysis, p. 283 / Vomiting, p. 283		•Pancreas and Islet Transplantation in Type 1 Diabetes. *Diabetes Care*, April 2006. •Preventing Diabetes. Kathy Doheny. *Natural Health*, April 2004.	•American Diabetes Association: http://www.diabetes.org/ •CDC Health Topic: Diabetes: http://www.cdc.gov/doc.do/id/0900f3ec802723eb
15.9 **Some Final Examples of Integration and Control**	Cerebrospinal fluid (CSF), p. 284 / Cramping, p. 284 / Epidermal growth factor (EGF), p. 285 / Epinephrine, p. 284 / Heart attacks, p. 285 / Labor, p. 284 / Malignant tumors, p. 285 / Nerve growth factor (NGF), p. 285 / Obesity, childhood, p. 285 / Pesticides, p. 285 / Research, p. 284, cancer, p. 285, endocrine disrupter, p. 285, melatonin, p. 284, pheromone, p. 285 / Seasonal affective disorder (SAD), p. 284 / Sleep, p. 284 / Spinal cord, damage to, p. 284		•Drug of Darkness: Can a Pineal Hormone Head Off Everything from Breast Cancer to Aging? *Science News*, May 13, 1995. •The Chemistry of Love. Sanjay Gupta. *Time*, Feb. 18, 2002.	•Are Pretty Products Causing Early Puberty?: http://www.center4research.org/puberty.html •HHMI – Biological Clockworks: http://www.hhmi.org/biointeractive/museum/exhibit00/index.html
How Would You Vote?	Some pesticides may disrupt hormone function in humans and other animals. Should they remain in use while researchers study their safety? •InfoTrac: Endocrine Disruption Study on Atrazine Disputed. *Pesticide & Toxic Chemical News*, Jan. 13, 2003. •WWW: Popular Pesticide Faulted for Frogs' Sexual Abnormalities: http://www.mindfully.org/Pesticide/2003/Atrazine-Frog-Sexual19jun03.htm •WWW: Chemical Affecting Frogs' Sexuality; Males Are Acquiring Female Attributes after Exposure to a Common Weedkiller, Study Says: http://www.waterconserve.info/articles/reader.asp?linkid=1711			

CHAPTER 16: REPRODUCTIVE SYSTEMS

Section	Applications	Videos	InfoTrac® Readings	Weblink Examples
16.0 *Impacts, Issues:* **Sperm with a Nose for Home?**	Artificial insemination, p. 289 / Sexually transmitted diseases (STDs), p. 289		• Identification of a Testicular Odorant Receptor Mediating Human Sperm Chemotaxis. Marc Spehr et al. *Science,* Mar. 28, 2003.	• Sniff and Swim: How Sperm Find Eggs: http://www.livescience.com/humanbiology/060519_sperm_attraction.html
16.1 **The Male Reproductive System**	Cancer, breast, p. 291, prostate, p. 291, testicular, p. 291 / Contraction(s), p. 291 / Secretion, male reproductive system and, pp. 290, 291 / Sexual arousal, pp. 290, 291 / Temperature, scrotum, p. 290	• Anatomy & Physiology, 2003, *Sperm Count and Age* (1:27)	• Alcohol and the Male Reproductive System. Mary Ann Emanuele, Nicholas Emanuele. *Alcohol Research & Health,* Winter 2001.	• Does Smoking During Pregnancy Affect Sons' Sperm Counts?: http://www.protectingourhealth.org/newscience/infertility/2003/2003-0519storgaardetal.htm
16.2 **How Sperm Form**	Interstitial cells, p. 292 / Leydig cells, pp. 292, 293 / Secretion, p. 292 / Sexual traits, p. 292 / Spermatogenesis, p. 293 / Spermatogonia, pp. 292, 293 / Spermatozoa, p. 292		• Key Gene for Sperm, Egg Formation Identified. *Genomics & Genetics Weekly,* Dec. 26, 2003.	• Spermiogenesis: Sperm Cell Differentiation: http://www.mcb.arizona.edu/wardlab/spermio.html
16.3 **The Female Reproductive System**	Menopause, p. 295 / Menstrual cycle, pp. 294–295 / Placenta, p. 295 / Pregnancy, pp. 294, 295 / Scar tissue, p. 295 / Secretion, pp. 294, 295 / Sexual traits, p. 294		• Introduction. Joan C. Chrisler, Ingrid Johnston-Robledo. *Sex Roles: A Journal of Research,* Jan. 2002.	• Female Reproductive System: http://training.seer.cancer.gov/module_anatomy/unit12_3_repdt_female.html
16.4 **The Ovarian Cycle: Oocytes Develop**	Bacteria, p. 297 / Menopause, p. 296 / Menstrual cycle, pp. 296, 297 / Pregnancy, pp. 296, 297 / Research, ovaries, p. 296 / Stimulation, ovarian cycle and, p. 297		• Intercellular Communication in the Mammalian Ovary. Martin M. Matzuk et al. *Science,* June 21, 2002.	• Oogenesis: Mammalian Ovary: http://www.uoguelph.ca/zoology/devobio/210labs/ovary4.html
16.5 **Visual Summary of the Menstrual and Ovarian Cycles**	Menstrual cycle, p. 298 / Menstruation, p. 298 / Ovarian cycle, p. 298 / Pregnancy, menstrual cycle and, p. 298		• Menstruation on Hold? More Women Favor Option (using hormones for menstruation regulation). *Contraceptive Technology Update,* May 2003.	• Menstrual Cycle: http://www.merck.com/mmhe/sec22/ch241/ch241e.html
16.6 **Sexual Intercourse**	Copulation, p. 299 / Orgasm, p. 299 / Pelvic thrusts, p. 299 / Pregnancy, p. 299 / Secretion, p. 299 / Sensory receptors, p. 299 / Stimulation, mechanical, p. 299 / Vasodilation, p. 299		• The Physiology of Sexual Arousal in the Human Female. Roy J. Levin. *Archives of Sexual Behavior,* Oct. 2002.	• Sex: The Science of Sexual Arousal: http://www.apa.org/monitor/apr03/arousal.html
16.7 **Controlling Fertility**	Abortion, p. 301 / Contraceptive methods, p. 300 / "Fertility awareness," p. 300 / Morning-after pills, p. 301 / Research, birth control, p. 301 / Side effects, p. 301 / Smoking, oral contraceptives and, p. 301 / Tubal ligation, p. 300	• Biology, 2005, *Bonus for a Baby* (2:26)	• Rate, Extent, and Modifiers of Spermatogenic Recovery after Hormonal Male Contraception. Peter Y. Liu et al. *The Lancet,* April 29, 2006.	• Planned Parenthood: http://www.plannedparenthood.org/pp2/portal/
16.8 **Options for Coping with Infertility**	Artificial insemination, pp. 302–303 / Fertility drugs, p. 302 / Gamete intrafallopian transfer (GIFT), p. 303 / In vitro fertilization (IVF), pp. 302–303 / Multiple births, p. 303 / Sperm bank, p. 302 / Stem cells, embryonic, p. 302 / Surrogate mother, p. 303 / Zygote intrafallopian transfer (ZIFT), p. 303	• Anatomy & Physiology, 2004, *IVF Anniversary* (2:24)	• What Affects Sperm Quality? *Contemporary OB/GYN,* Mar. 2006. • Nationwide Success Rates at Fertility Clinics Are Increasing. Bill Zepf. *American Family Physician,* Feb. 1, 2005.	• Infertility Resources for Consumers: http://www.ihr.com/infertility/ • Fertility Treatment Options: http://www.babycentre.co.uk/preconception/fertilityproblems/treatmentoptions/
16.9 **A Trio of Common Sexually Transmitted Diseases**	Antibiotics, pp. 304, 305 / Chlamydia, p. 304 / Gonorrhea, pp. 304–305 / Hysterectomy, p. 304 / Lesions, p. 305 / Miscarriages, p. 305 / Penicillin, p. 305 / "Shanker," pp. 305, 305 / Syphilis, pp. 304, 305 / Vaginal discharge, pp. 304, 305		• Recent STD Trend Data. Dore Hollander. *Perspectives on Sexual and Reproductive Health,* Mar. 2006.	• STIs: Common Symptoms & Tips on Prevention: http://familydoctor.org/165.xml • CDC – Sexually Transmitted Diseases: http://www.cdc.gov/std/
16.10 **The Human Immunodeficiency Virus and AIDS**	AIDS pp. 306–307 / Helper cells, pp. 306, 307 / HIV, pp. 306–307 / Kaposi's sarcoma, p. 306 / Killer cells, p. 306 / Macrophages, pp. 306, 307 / Retroviruses, p. 306 / Synthetic virus, p. 307	• Biology, 2003, *Global AIDS* (2:18)	• Romanian Parents Keep HIV a Secret from Infected Children. Carmiola Ionescu. *The Lancet,* May 13, 2006.	• CDC – HIV and Its Transmission: http://www.cdc.gov/hiv/pubs/facts/transmission.htm
16.11 **A Rogue's Gallery of Other STDs**	Candidiasis, p. 309 / Cervical cancer, p. 308 / Drugs, abuse of, p. 308, antiparasitic, p. 309, antiviral, p. 308 / Hepatitis, p. 308 / Herpes, p. 308 / Human papillomavirus (HPV), p. 308 / Pap smear, p. 308 / Sores, p. 308 / Trichomoniasis, p. 309 / Vaginal discharge, p. 308 / Warts, p. 308 / Yeast infections, p. 309		• HPV Prevalence Peaks in 14- to 19-Year-Olds. Damian McNamara. *Family Practice News,* June 15, 2006.	• CDC – Viral Hepatitis: http://www.cdc.gov/ncidod/diseases/hepatitis/index.htm • CDC – Pubic Lice Infestation Fact Sheet: http://www.cdc.gov/ncidod/dpd/parasites/lice/factsht_pubic_lice.htm
16.12 **Eight Steps to Safer Sex**	Alcohol, p. 309 / Celibacy, p. 309 / Condom, p. 309 / Drugs, p. 309 / Gonorrhea, p. 309 / Herpes, p. 309 / Sexual intercourse, p. 309 / Sores, p. 309 / Spermicide, p. 309		• What They Say Isn't What They Do. *MMR,* May 9, 2005.	• Safer Sex Guidelines – Information for Students: http://www.stdservices.on.net/std/prevention/safersex_guidelines.htm
How Would You Vote?	In recent years the rate of teenage pregnancies in the United States has fallen slightly. One contributing factor has been greater access to contraception. Some people feel strongly that teens should not use contraception without parental consent. What's your opinion? • InfoTrac: "Should Minors Have Over-the-Counter Access to Plan B Emergency Contraception?" *Internal Medicine News,* Mar. 15, 2004. • WWW: Parental Consent and Notice for Contraceptives Threatens Teen Health and Constitutional Rights: http://www.reproductiverights.org/pub_fac_parentalconsent.html • WWW: "Whose Children? Bill Would Return Parental Authority for Medical Treatment": http://www.mfc.org/contents/article.asp?id=1119			

CHAPTER 17: DEVELOPMENT AND AGING

Section	Applications	Videos	InfoTrac® Readings	Weblink Examples
17.0 *Impacts, Issues:* **Fertility Factors and Mind-Boggling Births**	Caesarian section delivery, p. 313 / Drugs, fertility, p. 313 / Heart failure (HF), p. 313 / Lung failure, p. 313 / Miscarriages, p. 313 / Multiple births, p. 313 / Newborns, p. 313 / Premature delivery, p. 313	•Anatomy & Physiology, 2004, *C-Section Choice* (1:24)	•Fertility Drugs Cited as Preterm Births Rise. *San Jose Mercury News*, July 14, 2006.	•Multiple Gestation: http://www.asrm.org/Patients/topics/mutiple.html
17.1 **The Six Stages of Early Development: An Overview**	Cells, embryonic, p. 314 / Conjoined twins, p. 315 / Twins, p. 315	•Anatomy & Physiology, 2003, *Conjoined Twins* (1:39)	•Distinctly Human: The When, Where & How of Life's Beginnings. John Collins Harvey. *Commonweal*, Feb. 8, 2002.	•The Multi-Dimensional Human Embryo: http://embryo.soad.umich.edu/
17.2 **The Beginnings of You – Fertilization to Implantation**	Ectopic/tubal pregnancy, p. 317 / Home pregnancy tests, p. 317 / Human chorionic gonadotropin (HCG), p. 317 / Trophoblast, pp. 316, 317 / "Twin studies," p. 317		•Calcium at Fertilization and in Early Development. Michael Whitaker. *Physiological Reviews*, Jan. 2006.	•Attachment and Implantation: http://www.vivo.colostate.edu/hbooks/pathphys/reprod/placenta/implant.html
17.3 **How the Early Embryo Takes Shape**	Development, embryo, pp. 318–319 / Menstrual periods, p. 318 / Organ systems, prenatal development and, pp. 318, 319 / Prenatal development, pp. 318, 319		•Genetic Control of Branching Morphogenesis. Ross J. Metzger, Mark A. Krasnow. *Science*, June 4, 1999.	•Embryo Images: http://www.med.unc.edu/embryo_images/
17.4 **Vital Membranes Outside the Embryo**	Allantois, p. 320 / Amniotic fluid, pp. 320, 321 / Nutrients, pp. 320–321 / Placenta, membranes and, pp. 320–321 / Prenatal development, pp. 320, 321 / Trophoblast, p. 320		•What to Tell Patients about Banking Cord Blood Stem Cells. Kenneth J. Moise Jr. *Contemporary OB/GYN*, April 15, 2006.	•Development of the Fetal Membranes and Placenta: http://www.bartleby.com/107/12.html
17.5 **The First Eight Weeks – Human Features Emerge**	Abortion, p. 323 / Genetic disorders, p. 323 / Miscarriages, p. 323 / Pigments, retinal, p. 322 / Prenatal development, pp. 322, 323 / Spontaneous abortion, p. 323 / Ultrasound, p. 323		•Legs, Eyes, or Wings – Selectors and Signals Make the Difference. Markus Affolter; Richard Mann. *Science*, May 11, 2001.	•3D First Trimester Ultrasound Scan Photos: http://www.layyous.com/ultasound/early_pregnancy_ultrasound_photos2.htm
17.6 **Development of the Fetus**	Abrasion, p. 324 / Arterial duct, p. 325 / Metabolic wastes, p. 324 / Placenta, pp. 324, 325 / Prenatal development, pp. 324, 325 / Respiratory distress syndrome, p. 324		•Neonatal Respiratory Distress Syndrome. Sven Montan, Sabaratnam Arulkumaran. *The Lancet*, June 10, 2006.	•The Second Trimester of Pregnancy: http://www.gynob.com/2ndtrime.htm
17.7 **Birth and Beyond**	Cloning, p. 327 / Labor, pp. 326–327 / Lactation, p. 327 / Premature babies, p. 327 / Prolactin (PRL), p. 327 / Vitamin A, p. 327	•Genetics, 2005, *Cloning Humans*, (2:04)	•Nightshifts Linked to Premature Birth. *Work & Family Newsbrief*, Jan. 2006.	•Labor Resources: http://www.childbirth.org/articles/labor.html
17.8 **Potential Disorders of Early Development**	Drugs, illegal, p. 328, prescription, p. 329 / Fetal alcohol syndrome (FAS), p. 329 / Neural tube defects, p. 328 / Retardation, p. 329 / Spina bifida, p. 328 / Teratogens, pp. 328, 329	•Human Biology, 2006, *Mermaid Baby* (3:07)	•Prevention and Diagnosis of Fetal Alcohol Syndrome. Anne D. Walling. *American Family Physician*, May 15, 2006.	•Spina Bifida Information: http://www.ninds.nih.gov/disorders/spina_bifida/spina_bifida.htm
17.9 **Prenatal Diagnosis: Detecting Birth Defects**	Amniocentesis, p. 330 / Chorionic villus sampling (CVS), p. 330 / Hemophilia, p. 330 / Preimplantation diagnosis, p. 330 / Prenatal diagnosis, p. 330 / Ultrasound, p. 330	•Genetics, 2005, *Pre-implantation Genetics* (2:24)	•Prenatal Genetics. Carolyn Slack. *Journal of Perinatal & Neonatal Nursing*, Jan.–Mar. 2006.	•Prenatal Diagnostic Testing: http://www.merck.com/mmhe/sec22/ch256/ch256c.html
17.10 **From Birth to Adulthood**	Deterioration, p. 331 / Growth spurts, p. 331 / Puberty, p. 331 / Senescence, p. 331 / Sensory systems, child, p. 331 / Sexual maturity, p. 331 / Sexual traits, p. 331 / Wrinkles, p. 331		•The Age of Senescence: Judy Campisi's Work on Cancer May Reveal the Secrets of (Not) Getting Older. Karen Hopkin. *The Scientist*, Feb. 2006.	•Why Puberty Now Begins at Seven: http://news.bbc.co.uk/2/hi/health/4530743.stm
17.11 **Time's Toll: Everybody Ages**	Arthritis, p. 332 / Cancer cells, p. 332 / Cells, embryonic, p. 332 / Hair, graying, p. 332 / Life span, p. 332 / Research, aging, p. 332 / Werner's syndrome, p. 332	•Anatomy & Physiology, 2003, *Healthy Aging* (2:33)	•What Can Progeroid Syndromes Tell Us about Human Aging? David Kipling et al. *Science*, Sept. 3, 2004.	•Low Calorie Diet Affects Aging-Related Factors: http://www.nih.gov/news/pr/apr2006/nia-04.htm
17.12 **Aging Skin, Muscle, Bones and Reproductive Systems**	Atrophy, p. 333 / Cancer, p. 333 / Connective tissues, p. 333 / Exercise, p. 333 / Fertility, p. 333 / Heart disease, p. 333 / Joints, p. 333 / Menopause, p. 333 / Osteoporosis, p. 333 / Sex hormones, p. 333 / Therapeutic drugs, p. 333	•Anatomy & Physiology, 2003, *Sperm Count and Age* (1:27)	•Human Sexual Development. John DeLamater, William N. Friedrich. *The Journal of Sex Research*, Feb. 2002.	•The North American Menopause Society: http://www.menopause.org/
17.13 **Age-Related Changes in Some Other Body Systems**	Alzheimer's disease, p. 334 / Arthritis, p. 334 / Basal metabolic rate (BMR), p. 335 / Cardiovascular disease, p. 335 / Dementia, p. 334 / Immune response, p. 335 / Memory loss, p. 334 / Nutrition, p. 335 / Oxygen, p. 335 / Reflex, p. 335 / Respiratory system, p. 335 / Smoking, p. 335	•Biology, 2004, *Is Aging Treatable?* (2:25) •Biology, 2004, *Brain Fitness* (2:16)	•Glossaries in Public Health: Ider People. A. Bowling, S. Ebrahim. *Journal of Epidemiology & Community Health*, April 2001. •A Key to Preventing Alzheimer's Found. *UPI NewsTrack*, Mar. 10, 2005.	•PBS – Secret Life of the Brain: http://www.pbs.org/wnet/brain/ •Healthy Aging News: http://www.sciencedaily.com/news/health_medicine/healthy_aging/
How Would You Vote?	Should we restrict the use of fertility drugs to conditions that could limit the number of embryos that form? •InfoTrac: The Limits of Science. *Time*, April 15, 2002. •InfoTrac: In Vitro Fertilization Technology Much Improved after First Birth 25 Years Ago. *Knight Ridder/Tribune Business News*, July 25, 2003. •WWW: Multiple Births Not Necessarily a Consequence of Conception Drugs: http://www.findarticles.com/p/articles/mi_m1272/is_2657_128/ai_59473796			

CHAPTER 18: LIFE AT RISK: INFECTIOUS DISEASE

Section	Applications	Videos	InfoTrac® Readings	Weblink Examples
18.0 *Impacts, Issues:* **Virus, Virus Everywhere**	Avian (bird) flu, p. 339 / Brain, encephalitis and, p. 339, infectious diseases and, p. 339 / Fevers, p. 339 / Global impacts, p. 339 / H5N1 virus, p. 339 / Paralysis, p. 339 / Sexually transmitted diseases (STDs), p. 339 / West Nile virus, p. 339	• Anatomy & Physiology, 2003, *West Nile Virus* (3:23) • Environmental Science, 2005, *Bird Flu* (3:19)	• Avian Flu, West Nile Virus, and Lyme Disease. Robert Charles Moellering Jr. et al. *Patient Care for the Nurse Practitioner*, April 2006.	• CDC – West Nile Virus: http://www.cdc.gov/ncidod/dvbid/westnile/ • CDC – Avian Influenza: http://www.cdc.gov/flu/avian/
18.1 Some General Principles of Infectious Disease	Antibiotics, p. 341 / Avian (bird) flu, p. 341 / Botulism, p. 340 / Cold, common, pp. 340, 341 / Dengue fever, p. 341 / Ebola virus, p. 341 / Emerging diseases, p. 341 / HIV, p. 341 / Measles, p. 341 / Neurotoxin, p. 340 / Parasites, pp. 340, 341 / Protozoa, infectious, p. 340 / Rabies, p. 341 / SARS virus, p. 341 / Septic shock, p. 340 / West Nile virus, p. 341	• Human Biology, 2006, *Germs in Pakistan* (2:13) • Human Biology, 2006, *Mask of Technology* (2:49)	• Antibiotics Alter the Normal Bacterial Flora in Humans. *Biotech Week*, April 7, 2004. • Botulism. Jeremy Sobel. *Clinical Infectious Diseases*, Oct. 15, 2005.	• Centers for Disease Control and Prevention: http://www.cdc.gov/ • CDC – Emerging Infectious Diseases: http://www.cdc.gov/ncidod/eid/
18.2 Viruses and Infectious Proteins	Bacteriophages, p. 342 / Bovine spongiform encephalitis (BSE), p. 343 / Cold sores, p. 342 / Flu virus, p. 342 / Hepatitis, p. 342 / Herpes, p. 342 / HIV, pp. 342, 343 / "Mad cow disease," p. 343 / Variant Creutzfeldt-Jakob disease (vCJD), p. 343	• Anatomy & Physiology, 2003, *Mad Cow Victim* (3:25)	• Cannibals to Cows: The Path of a Deadly Disease *Newsweek*, Mar. 12, 2001. • Cold Sore Virus Can Evade Immune System. *UPI NewsTrack*, July 17, 2006.	• Big Picture Book of Viruses: http://www.tulane.edu/%7Edmsander/Big_Virology/BVHomePage.html • Virus Replication: http://www.tulane.edu/~dmsander/WWW/224/Replication.html
18.3 Bacteria – The Unseen Multitudes	Antibiotic-resistant microbes, p. 345 / Dysentery, p. 345 / *Escherichia coli (E. coli)*, p. 344 / Infections, ear, p. 345 / Penicillin, p. 344 / Pneumonia, p. 345 / *Salmonella*, p. 344 / *Staphylococcus aureus* (staph A), p. 345 / Traits, bacterial, p. 344 / Tuberculosis (TB), p. 345 / Vitamin K, p. 344	• Biology, 2003, *Salon Infections* (2:05) • Human Biology, 2006, *Fingerprinting E. coli* (3:04)	• Overdoing Antibiotics. *Harvard Health Letter*, Nov. 2002. • Chronic Middle Ear Infections Linked to Resistant Biofilm Bacteria. *PR Newswire*, July 11, 2006.	• Introduction to the Bacteria: http://www.ucmp.berkeley.edu/bacteria/bacteria.html • Bacteria – Life History and Ecology: http://www.ucmp.berkeley.edu/bacteria/bacterialh.html
18.4 Infectious Fungi, Protozoa and Worms	African sleeping sickness, p. 346 / Amoebic dysentery, p. 346 / *Ascaris* roundworms, p. 346 / Cryptosporidiosis, p. 346 / *Entamoeba histolytica*, p. 346 / Fungi, infectious, p. 346 / *Giardia*, p. 346 / Hookworms, p. 346 / Ringworm, p. 346 / Sanitation, p. 346 / Tapeworms, p. 346		• Dark Fungi Emerging as Cause of Lethal Infections. Nancy Walsh. *Family Practice News*, May 15, 2006.	• African Trypanosomiasis: http://www.sgpp.org/african_sleeping_sickness.shtml • What is Amoebic Dysentary?: http://www.netdoctor.co.uk/travel/diseases/amoebic_dysentery.htm
18.5 Malaria: Efforts to Conquer a Killer	Anemia, p. 347 / Cells, infectious disease principles and, p. 347 / Chills, p. 347 / Enlarged liver, p. 347 / Fevers, p. 347 / Immune response, p. 347 / Infectious diseases, malaria as, p. 347 / Parasites, p. 347 / Pathogens, malaria and, p. 347 / *Plasmodium*, p. 347 / Red blood cells (RBCs), p. 347 / Salivary glands, p. 347 / Shaking, p. 347 / Sickle-cell anemia, p. 347 / Sweat, p. 347 / Vaccinations, p. 347		• Fatal Inaction: There Is a Silver Bullet for Africa's Malaria Epidemic. Why the Bush Administration Won't Pull the Trigger. Joshua Kurlantzick. *Washington Monthly*, July–Aug. 2006. • Push for New Tactics as War on Malaria Falters. Celia W. Dugger. *The New York Times*, June 28, 2006.	• CDC – Malaria: http://www.cdc.gov/malaria/ • Malaria: http://www-micro.msb.le.ac.uk/224/Malaria.html • WHO – Malaria: http://www.who.int/topics/malaria/en/
18.6 Patterns of Infectious Diseases	AIDS, global cases of, p. 348 / Antibiotic resistance, p. 349 / Bubonic plague, p. 348 / Cholera, p. 348 / Coughing, p. 348 / Disease vector, p. 348 / Diseases, as endemic, p. 348, preventing, p. 349 / Epidemic, p. 348 / Hepatitis, p. 348 / Infectious diseases, p. 348, global health threats of, p. 348, protection from, p. 349 / Pandemic, p. 348 / Pathogens, p. 348 / Sneezing, p. 348 / Sores, p. 348 / Tuberculosis (TB), p. 348 / Vaccinations, p. 349 / Virulence, p. 349 / Viruses, pp. 348, 349 / Water, pp. 348, 349 / Whooping cough, p. 348 / Worms, infectious, p. 348	• Anatomy & Physiology, 2004, *Whooping Cough Immunization* (1:52) • Biology, 2003, *Global AIDS* (2:18)	• The Science of Clean Water. Elettra Ronchi. *OECD Observer*, Mar. 2003. • Pandemic Dilemma: Who Gets the Shot? *Chicago Tribune*, June 27, 2006.	• Pathogenicity vs. Virulence: http://www.tulane.edu/~wiser/protozoology/notes/Path.html • Methods of Disease Transmission: http://microbiology.mtsinai.on.ca/faq/transmission.shtml
How Would You Vote?	Killing mosquitoes is the best defense against West Nile virus. Many local agencies now spray pesticides wherever mosquitoes are likely to breed. Some people object to spraying, fearing harmful effects on health or wildlife. Would you support a spraying program in your area? • InfoTrac: Suit over West Nile Spraying Goes Forward against City. *New York Law Journal*, June 13, 2005. • InfoTrac: Study Claims Risks of West Nile Spraying Exaggerated. *Pesticide & Toxic Chemical News*, May 3, 2004. • WWW: Overkill: Why Pesticide Spraying for West Nile Virus May Cause More Harm Than Good: http://www.meepi.org/wnv/overkillma.htm			

CHAPTER 19: CELL REPRODUCTION

Section	Applications	Videos	InfoTrac® Readings	Weblink Examples
19.0 *Impacts, Issues*: Henrietta's Immortal Cells	Cancer, research for, p. 353 / Diseases, research for, p. 353	• Genetics, 2005, *Producing Human "Replacement Cells"* (2:22)	• Disease Model Uses 3-D Cancer Cells Grown in Space. *Cancer Weekly*, Aug. 21, 2001. • Stem Cells That Kill. Alice Park. *Time*, April 24, 2006.	• Cell Culture: http://www-micro.msb.le.ac.uk/ Video/culture.html
19.1 Dividing Cells Bridge Generations	Sexual reproduction, p. 354 / Spermatogonia, pp. 354, 355		• Learning about Sex from an Elegant Worm: Three Major Papers Advance Understanding of Meiosis in *C. Elegans*. *Ascribe Higher Education News Service*, Dec. 15, 2005. • Cell Growth, Division Genes Are Identified. *UPI NewsTrack*, Feb. 22, 2006.	• Talking Glossary of Genetic Terms: http://www.genome.gov/ glossary.cfm • Mitosis/Meiosis Tutorial: http://biog-101-104.bio.cornell .edu/biog101_104/tutorials/ cell_division.html
19.2 A Closer Look at Chromosomes		• Genetics, 2003, *Creating Chromosomes* (2:53)	• Major Advance Made on DNA Structure. *Ascribe Higher Education News Service*, May 2, 2005.	• Images – Supercoiling of DNA: http://web.siumed.edu/~ bbartholomew/images/ chapter29/F29-17.jpg
19.3 The Cell Cycle	Bone marrow, cancer of, p. 357		• Dutch Team Finds Aspects of Cell Cycle Checkpoints. Sharon Kingman. *BioWorld International*, Dec. 14, 2005.	• The Cell Cycle – An Interactive Animation: http://www.cellsalive.com/ cell_cycle.htm
19.4 The Four Stages of Mitosis			• Mitosis through the Microscope: Advances in Seeing inside Live Dividing Cells. *Science*, April 4, 2003.	• Stages of Mitosis and Confocal Microscope Images: http://www.itg.uiuc.edu/technology /atlas/structures/mitosis/
19.5 How the Cytoplasm Divides	Cancer, cervical, p. 360 / Cancer cells, p. 360 / Human papillomavirus (HPV), p. 360 / Viruses, HeLa cells and, p. 360		• Measuring Protein Concentrations in Live Cells: Yale Group Demonstrates New Method Using Cytokinesis. Sarah Rothman. *The Scientist*, Dec. 5, 2005.	• Observing Mitosis with Fluorescence Microscopy – Cytokinesis: http://micro.magnet.fsu.edu/ cells/fluorescencemitosis/ cytokinessmall.html
19.6 Concerns and Controversies over Irradiation	Anemia, p. 361 / Epithelial cells, p. 361 / Food-borne illness, p. 361 / Fungi, p. 361 / Ionizing radiation, p. 361 / Irradiation, p. 361 / Magnetic resonance imaging (MRI), p. 361 / Microbes, radiation-resistant, p. 361 / Positron-emission tomography (PET), p. 361 / Radiation therapy, p. 361	• Anatomy & Physiology, 2003, *Breast Cancer Treatment* (2:28)	• Food Irradiation May Start to Take Off: The Process, Which Can Kill Most Bacteria in Ground Beef and Poultry, Is Still Not a Widespread Food Treatment. *The Wichita Eagle*, Jan. 8, 2006.	• DOE – Office of Human Radiation Experiments: http://tis.eh.doe.gov/ohre/ • The Health Effects of Exposure to Indoor Radon: http://www.nsc.org/EHC/ radon/public.htm
19.7 Meiosis – The Beginnings of Eggs and Sperm	Puberty, p. 363 / Sexual reproduction, p. 362 / Spermatogenesis, pp. 362, 363 / Spermatogonia, p. 363		• Neonatal Outcome of Preimplantation Genetic Diagnosis by Polar Body Removal. Charles M. Strom et al. *Pediatrics*, Oct. 2000.	• Meiosis Tutorial: http://www.biology.arizona. edu/cell_bio/tutorials/ meiosis/main.html
19.8 A Visual Tour of the Stages of Meiosis			• Association between Maternal Age and Meiotic Recombination for Trisomy 21. Neil E. Lamb et al. *American Journal of Human Genetics*, Jan. 2005.	• Meiosis Tutorial: http://www.biology.arizona. edu/cell_bio/tutorials/ meiosis/page3.html
19.9 The Second Stage of Meiosis – New Combinations of Parents' Traits	Recombination, genetic, p. 367		• Sex, Not Genotype, Determines Recombination Levels in Mice. Audrey Lynn et al. *American Journal of Human Genetics*, Oct. 2005.	• Meiosis and Genetic Recombination: http://web.mit.edu/esgbio/ www/mg/meiosis.html
19.10 Meiosis and Mitosis Compared	Cancer, research for, p. 368 / *Chlamydomonas*, p. 369 / Colchicine, p. 368 / *Giardia intestinalis*, p. 369 / Poisons, p. 368		• Absence of Age Effect on Meiotic Recombination between Human X and Y Chromosomes. Qinghua Shi et al. *American Journal of Human Genetics*, Aug. 2002.	• PBS – How Cells Divide: Mitosis vs. Meiosis: http://www.pbs.org/wgbh/ nova/miracle/divide.html
How Would You Vote?	It is illegal to sell your organs, but you can sell your cells, including eggs, sperm and blood cells. Descendants of HeLa cells are sold all over the world by cell culture firms. Should the family of Henrietta Lacks share in the profits? • InfoTrac: Companies Covet Genes; Ethics and Profits Compete in the Patenting of Human Genetic Materials. *Alternatives Journal*, Summer 1997. • WWW: Life Itself: Exploring the Realm of the Living Cell: http://www.washingtonpost.com/wp-srv/style/longterm/books/chap1/lifeitself.htm • WWW: Henrietta's Dance: http://www.jhu.edu/~jhumag/0400web/01.html			

CHAPTER 20: OBSERVING PATTERNS IN INHERITED TRAITS

Section	Applications	Videos	InfoTrac® Readings	Weblink Examples
20.0 *Impacts, Issues*: Designer Genes?	Cloning, p. 373 / Ethics, p. 373 / Eugenic engineering, p. 373 / "Genetic technology," p. 373 / Inherited traits, p. 373 / Morality, p. 373 / Research, gene, p. 373	•Genetics, 2005, *Genetics in Sports* (2:48)	•Designer Genes: Will DNA Technology Let Parents Design Their Kids? Ingrid Wickelgren. *Current Science*, Dec. 3, 2004.	•Beyond Steroids: Designer Genes For Unscrupulous Athletes: http://www.washingtonpost.com/wp-dyn/articles/A42270-2005Mar16.html
20.1 Basic Concepts of Heredity	Chromosomes, heredity and, p. 374 / Genes, heredity and, p. 374 / Inherited traits, heredity concepts and, p. 374 / Research, heredity, p. 374		•Darwin Would Have Loved It. Michael J. Novacek. *Time*, April 17, 2006.	•Investigations Bearing on Heredity and Variation: http://www.genetics.org/
20.2 One Chromosome, One Copy of a Gene	Chromosomes, heredity and, p. 375 / Ethics, p. 375 / Gametes, heredity and, p. 375 / Heredity, p. 375 / Inherited traits, genes and, p. 375 / Segregation, p. 375		•Human Chromosome 3 Is Sequenced. *UPI NewsTrack*, April 27, 2006.	•The Monohybrid Cross Problem Set: http://www.biology.arizona.edu/mendelian_genetics/problem_sets/monohybrid_cross/monohybrid_cross.html
20.3 Figuring Genetic Probabilities	Cystic fibrosis (CF), p. 376 / Genetic disorders, p. 376 / Inherited traits, genetic probabilities and, pp. 376–377 / Ratio, pp. 376, 377 / Segregation, p. 377 / Sickle-cell anemia, p. 376	•Genetics, 2003, *Cystic Fibrosis* (2:16)	•Advances in Preconception Genetic Counseling. Marta C. Wille. *Journal of Perinatal & Neonatal Nursing*, Jan.–Mar. 2004.	•Punnett Squares: Just Another Probability Game?: http://knowledgene.com/public/view.php3?db=fun_stuff&uid=50
20.4 How Genes for Different Traits Are Sorted into Gametes	Inherited traits, independent assortment and, pp. 378–379 / Probability, p. 379 / Ratio, p. 379 / Spindles, inherited traits and, p. 378		•Germline Susceptibility to Colorectal Cancer Due to Base-Excision Repair Gene Defects. Susan M. Farrington et al. *American Journal of Human Genetics*, July 2005.	•Independent Assortment Animation: http://www.sumanasinc.com/webcontent/anisamples/majorsbiology/independent_assortment.html
20.5 Single Genes, Varying Effects	ABO blood typing, p. 381 / Amino acids, p. 381 / Antibiotics, p. 380 / Birth, sickle-cell anemia at, p. 380 / Bone marrow, p. 381 / Cartilage-hair hypoplasia (CHH), p. 380 / Gastrointestinal (GI) tract, p. 381 / Genetic disease, p. 380 / Heart failure (HF), p. 381 / Hemoglobin (Hb), pp. 380, 381 / Immune system, p. 380 / Inherited traits, genes effects on, pp. 380–381 / Kidney failure, p. 381 / Lymphomas, p. 380 / Malaria, p. 380 / Paralysis, p. 381 / Pleiotropy, p. 380 / Proteins, p. 381 / Red blood cells (RBCs), p. 380, sickled, p. 381 / Research, gene, p. 380 / Rheumatism, p. 381 / Sickle-cell anemia, pp. 380, 381 / Stem cells, p. 380 / Traits, sickle-cell, p. 380 / Transfusions, p. 380		•Pleiotropy and the Genomic Location of Sexually Selected Genes. Mark J. Fitzpatrick. *The American Naturalist*, June 2004. •Bone Area and Bone Mineral Content Deficits in Children with Sickle Cell Disease. Anne M. Buison et al. *Pediatrics*, Oct. 2005.	•Incomplete Dominance & Codominance, Pleiotropy and Multialleles: http://www.synapses.co.uk/genetics/tsg14.html •Cartilage-Hair Hypoplasia: http://www.emedicine.com/ped/topic329.htm
20.6 Other Gene Impacts and Interactions	Behavior problems, p. 383 / Blood pressure, drugs for, p. 383 / Campodactyly, p. 382 / Cystic fibrosis (CF), p. 382 / Inherited traits, genes effects on, pp. 382–383 / Mental illness, p. 383 / Penetrance, p. 382 / Pigments, eye, p. 383 / Probability, p. 382 / Sexual orientation, p. 383 / Traits, polygenic, pp. 382, 383 / Twins, p. 383	•Genetics, 2003, *Cystic Fibrosis* (2:16)	•Disease Versus Disease: How One Disease May Ameliorate Another. E. Richard Stiehm. *Pediatrics*, Jan. 2006. •Mitochondrial Disease. Anthony H.V. Schapira. *The Lancet*, July 1, 2006.	•Searching for Genes That Explain Our Personalities: http://www.apa.org/monitor/sep02/genes.html •Genes and Behavior: http://www.dnafiles.org/resources/res06.html
20.7 Searching for Custom Cures	Asthma, p. 383 / Drugs, asthma, p. 383, blood pressure, p. 383, therapeutic, p. 383 / Pharmacogenetics, p. 383 / Research, drug, p. 383 / Side effects, p. 383 / Therapeutic drugs, p. 383		•Scientific, Ethical Questions Temper Pharmacogenetics. Karen Young Kreeger. *The Scientist*, June 11, 2001. •A Target for Iressa: The Fall and Rise (And Fall) of a Pharmacogenetics Poster Child. David Secko. *The Scientist*, April 2006.	•Pharmacogenetics: http://www.merck.com/mrkshared/mmanual/section22/chapter301/301a.jsp •Genes and Drugs: http://genesanddrugs.dnadirect.com/patients/what_is_pgx/genes_and_drugs.jsp
How Would You Vote?	Would you favor legislation that limits or prohibits engineering genes except for health reasons? •InfoTrac: The Science and Politics of Genetically Modified Humans: Will New Genetic Technologies Be Carefully Controlled for Their Benefits – Or Will They Inadvertently Destroy Civil Society? Richard Hayes. *World Watch*, July–Aug. 2002. •InfoTrac: Who Gets the Good Genes? Robert Wright. *Time*, Jan. 11, 1999. •InfoTrac: To Build a Baby. Fred Guterl. *Newsweek International*, June 30, 2003.			

CHAPTER 21: CHROMOSOMES AND HUMAN GENETICS

Section	Applications	Videos	InfoTrac® Readings	Weblink Examples
21.0 *Impacts,* *Issues:* **Menacing** **Mucus**	Abortion, p. 387 / Antibiotics, p. 387 / Cystic fibrosis (CF), p. 387 / Genetic testing, p. 387 / Inherited diseases, p. 387 / Lung failure, p. 387 / Pathogens, p. 387 / Physiotherapy, p. 387 / Secretion, p. 387	• Genetics, 2003, *Cystic Fibrosis* (2:16)	• Constant Battle: Kerri Marks Is One of the Survivors, Living Under a Disease's Invisible Clock. Jennifer Becknell. *Herald*, July 10, 2006.	• Cystic Fibrosis Foundation: http://www.cff.org/ • Gene Therapy and CF: http://www.cff.org/about cf/gene_therapy_and_cf/
21.1 **Genes and** **Chromosomes**	Heredity, p. 388 / Human genetics, p. 388, genes/chromosomes and, p. 388 / Segregation, p. 388	• Genetics, 2003, *Creating Chromosomes* (2:53)	• Cord-Blood Storage Is Big Business, But Is It Worth It? Blythe Bernhard. *Orange County Register (Santa Ana, CA)*, June 2, 2006.	• Genetic Behaviors Go Beyond X and Y Chromosomes: http://www.rocklintoday.com/news/templates/health_news.asp?articleid=3632&zoneid=7
21.2 **Picturing** **Chromosomes** **with Karyotypes**	Colchicine, p. 389 / Human genetics, karyotypes with, p. 389		• Microarray Analysis of Cell-Free Fetal DNA in Amniotic Fluid: A Prenatal Molecular Karyotype. Paige B. Larrabee et al. *American Journal of Human Genetics*, Sept. 2004.	• Karyotyping Activity: http://www.biology.arizona.edu/human_bio/activities/karyotyping/karyotyping.html
21.3 **How Sex Is** **Determined**	Anhidrotic ectodermal dysplasia, p. 391 / Blood clotting, determining sex and, p. 390 / Research, Y chromosome, p. 391 / Sexual traits, p. 390 / Traits, nonsexual, p. 390		• The Genetic Legacy of the Mongols. Tatiana Zerjal et al. *American Journal of Human Genetics*, Mar. 2003.	• Gender Selection Technology: http://www.havingbabies.com/gender-selection.html
21.4 **Human Genetic** **Analysis**	Amish microencephaly, p. 392 / Bacteria, p. 393 / Brain defects, p. 392 / Cystic fibrosis (CF), p. 393 / Huntington's disease (HD), pp. 392, 393 / Pregnancy, predicting disorders in, p. 393	• Genetics, 2003, *Treating Hungtington's Disease* (2:18)	• LCT Announces Results for Treating Huntington's Disease. *AsiaPulse News*, Aug. 2, 2005.	• Amish Lethal Microcephaly: http://www.genetests.org/query?dz=amish-mcph • Genetics Education Center: http://www.kumc.edu/gec/
21.5 **Inheritance of** **Genes on** **Autosomes**	Achondroplasia, p. 395 / Autosomal recessive condition, p. 394 / Cystic fibrosis (CF), pp. 394, 395 / Familial hypercholesterolemia, p. 395 / Marfan syndrome, p. 395 / Pleiotropy, p. 395 / Tay-Sachs disease, p. 394 / Tyrosine, p. 394	• Genetics, 2003, *Engineering Perfect People* (5:07)	• Geographic Distribution of Disease Mutations in the Ashkenazi Jewish Population Supports Genetic Drift over Selection. Neil Risch et al. *American Journal of Human Genetics*, April 2003.	• Phenylketonuria Home Page: http://www.pkunews.org/ • Marfan Syndrome: http://www.healthline.com/adamcontent/marfan-syndrome
21.6 **Inheritance of** **Genes on the X** **Chromosome**	Androgen insensitivity, p. 397 / Disorders, X chromosome and, p. 396 / Duchenne muscular dystrophy (DMD), p. 397 / Faulty enamel trait, p. 397 / Hemophilia A, pp. 396, 397 / Hemophilia B, p. 396 / Testicular feminizing syndrome, p. 397	• Genetics, 2003, *Muscular Dystrophy* (2:25)	• Significant Improvement in Spirometry after Stem Cell Transplantation in One Duchenne Muscular Dystrophy Patient. Zhiping Li et al. *Chest*, Oct. 2005.	• Hemophilia: "The Royal Disease": http://www.sciencecases.org/hemo/hemo.asp • Androgen Insensitivity Syndrome: http://www.emedicine.com/PED/topic2222.htm
21.7 **Sex-Influenced** **Inheritance**	Baldness, p. 398 / Human genetics, sex-influenced inheritance and, p. 398		• Genetic Variation in the Human Androgen Receptor Gene Is the Major Determinant of Common Early-Onset Androgenetic Alopecia. Axel M. Hillmer et al. *American Journal of Human Genetics*, July 2005.	• Sex-Influenced Inheritance: http://www.il-st-acad-sci.org/geneticsbook/gene0010.html • Male Pattern Baldness: http://nj.essortment.com/malepatternbal_rcad.htm
21.8 **Changes in a** **Chromosome** **or Its Genes**	Blood cell cancer, p. 399 / Cri-du-chat, p. 398 / Cystic fibrosis (CF), p. 399 / Human genetics, chromosome number changes and, pp. 398, 399, gene changes and, pp. 398–399 / Irradiation, p. 398 / Leukemias, p. 399 / Stem cells, p. 399 / Viruses, p. 398	• Genetics, 2003, *Cystic Fibrosis* (2:16)	• Brazilian Researchers Connect Gene Mutation to Blood Conditions. *Xinhua News Agency*, June 20, 2006.	• 5P-Society – Cri du Chat Syndrome: http://www.fivepminus.org/ • Translocation: http://www.bbc.co.uk/health/conditions/chromosomal translocation1.shtml
21.9 **Changes in** **Chromosome** **Number**	Aging, p. 401 / Aneuploidy, p. 400 / Down syndrome, pp. 400, 401 / Fertility, p. 400 / Human genetics, chromosome number changes and, pp. 400, 401 / Klinefelter syndrome, p. 401 / Mental retardation, pp. 400, 401 / Miscarriages, pp. 400, 401 / Pregnancy, p. 401 / Probability, p. 401 / Puberty, p. 401 / Sexual traits, p. 401 / Spontaneous abortion, p. 400 / Sterility, p. 401 / Trisomy 21, p. 400 / Turner syndrome, pp. 400–401 / X chromosomes, pp. 400, 401	• Genetics, 2003, *Genetic Sex* (2:30)	• A Very Special Wedding. Claudia Wallis. *Time*, July 24, 2006. • Common Age Misconception about Down Syndrome. *PR Newswire*, May 2, 2006.	• Down Syndrome – Medical Essays and Information: http://www.ds-health.com/ • The Turner Center: http://www.aaa.dk/TURNER/ENGELSK/INDEX.HTM
How Would **You Vote?**	Do we as a society want to encourage women to give birth only to offspring who will not develop serious gene-based medical problems? • InfoTrac: Ohio Court Limits Relief for Fetal Testing Error; Compensation Denied for Costs of Raising Disabled Child. Judy Greenwald. *Business Insurance*, Mar. 13, 2006. • InfoTrac: In New Tests for Fetal Defects, Agonizing Choices for Parents. Amy Harmon. *The New York Times*, June 20, 2004. • InfoTrac: Offer All Pregnant Women Fetal Genetic Testing. Sherry Boschert. *OB GYN News*, Dec. 15, 1999.			

CHAPTER 22: DNA, GENES, AND BIOTECHNOLOGY

Section	Applications	Videos	InfoTrac® Readings	Weblink Examples
22.0 *Impacts, Issues:* **Ricin and Your Ribosomes**	Antidotes, p. 405 / Biotechnology, p. 405 / Cell/plasma membranes, p. 405 / DNA, p. 405 / Enzymes, p. 405 / Plants, castor, p. 405 / Proteins, p. 405 / Ricin, p. 405		• UT Scientists Develop Vaccine for Ricin. *UPI NewsTrack*, Jan. 30, 2006.	• CDC – Facts about Ricin: http://www.bt.cdc.gov/agent/ricin/faq/index.asp
22.1 **DNA: A Double Helix**	Gel electrophoresis, p. 407 / Heredity, p. 407 / Proteins, DNA and, pp. 406, 407 / Research, DNA structure, pp. 406, 407 / Thymine, pp. 406, 407		• Double-Teaming the Double Helix. Ellen F. Licking. *U.S. News & World Report*, Aug. 17, 1998.	• Introduction to DNA Structure: http://www.blc.arizona.edu/Molecular_Graphics/DNA_Structure/DNA_Tutorial.HTML
22.2 **Passing on Genetic Instructions**	Fragile X syndrome, p. 409 / Huntington's disease (HD), p. 409 / Sickle-cell anemia, p. 409 / Ultraviolet light, p. 408 / Xeroderma pigmentosum, pp. 408, 409	• Genetics, 2003, *Treating Huntington's Disease* (2:18)	• Arsenic Inhibits DNA Repair. *National Driller*, July 2006.	• Xeroderma Pigmentosum Society: http://www.xps.org/xp.htm#XP
22.3 **DNA into RNA – The First Step in Making Proteins**	Introns, p. 411 / Steroid hormones, p. 411		• DNA or RNA? Versatile Player Takes a Leading Role in Molecular Research. Nicholas Wade. *The New York Times*, June 20, 2006.	• Animations – DNA Structure, Transcription, Translation: http://www.johnkyrk.com/DNAanatomy.html
22.4 **Reading the Genetic Code**	Anticodon, pp. 412, 413 / Enzymes, ricin and, p. 413 / Genes, genetic code and, p. 412 / Mutations, p. 413 / Research, p. 413 / Ricin, p. 413 / Tryptophan (trp), p. 412 / Wobble effect, p. 413		• Scientists Discover a Genetic Code for Organizing DNA within the Nucleus. *Ascribe Higher Education News Service*, July 19, 2006.	• The Genetic Code: http://psyche.uthct.edu/shaun/SBlack/geneticd.htm
22.5 **Translating the Genetic Code into Protein**	Anticodon, pp. 414, 415 / Termination, p. 414		• New Insights into Protein Synthesis and Hepatitis C Infections. *Ascribe Higher Education News Service*, Dec. 2, 2005.	• Protein Synthesis: http://www.emc.maricopa.edu/faculty/farabee/biobk/BioBookPROTSYn.html
22.6 **Tools for "Engineering" Genes**	Antibodies, p. 417 / Chromosomes, p. 416 / Cloning, p. 416 / Genetic engineering, p. 416 / Mutations, p. 416 / Recombinant DNA technology, p. 416 / Ribosomes, p. 417 / Ricin, p. 417 / Traits, engineering, p. 416	• Anatomy & Physiology, 2004, *New Heart Treatments* (1:59)	• 2021: You'll Grow a New Heart. Elizabeth Svoboda. *Popular Science*, June 2006.	• Recombinant DNA: http://web.mit.edu/esgbio/www/rdna/rdnadir.html • Making PCR: http://sunsite.Berkeley.EDU/pcr/
22.7 **"Sequencing" DNA**	Nucleotides, engineering/exploring genes and, p. 418 / Probe, p. 418 / Recombinant DNA technology, p. 418 / Research, p. 418		• Visible DNA Sequencing. *Advanced Imaging*, Jan. 2006.	• The Wellcome Trust Sanger Institute: http://www.sanger.ac.uk/
22.8 **Mapping the Human Genome**	Alzheimer's disease (AD), p. 419 / Amyotrophic lateral sclerosis (ALS), p. 419 / Ethics, p. 419 / Genes, cancer and, p. 419 / Leukocyte adhesion deficiency, p. 419 / Single nucleotide polymorphism (SNP), pp. 418–419	• Genetics, 2003, *Alzheimer's Mutation* (2:05) • Genetics, 2003, *p53 Gene Therapy* (2:08)	• Daiichi Pure Chemicals, Toshiba and Toshiba Hokuto Electronics agree to develop DNA chip diagnostics business. *M2 Presswire*, July 11, 2006.	• DOE – Human Genome Project: http://www.doegenomes.org/ • DNA Sequencing Animation: http://www.biostudio.com/case_freeman_dna_sequencing.html
22.9 **Some Applications of Biotechnology**	Cancer, engineering/exploring genes and, pp. 420–421, / Cloning, p. 420 / Cystic fibrosis (CF), p. 420 / Gel electrophoresis, p. 421 / Gene therapy, pp. 420–421 / Leukemias, pp. 420, 421 / Malignant melanoma, p. 420 / Restriction fragment length polymorphisms (RFLPs), p. 421 / Stem cells, p. 420	• Genetics, 2003, *Cystic Fibrosis* (2:16) • Genetics, 2005, *Cloning Humans* (2:04)	• Keeping Secrets. Dana Hawkins. *U.S. News & World Report*, Dec. 2, 2002. • Gene Therapy Trials: Parents in UK Say Include CF Kids. *Medical Ethics Advisor*, July 1, 2006.	• The Dolan DNA Learning Center: http://www.dnalc.org/ • The [Criminal] Innocence Project: http://innocenceproject.org
22.10 **Issues for a Biotechnological Society**	Bacteria, p. 422 / Bioremediation, p. 422 / Biotechnology, p. 422 / DNA, p. 422 / Environment, p. 422 / Genetically modified (GM) food, p. 422 / Green Revolution, p. 422 / Herbicides, p. 422 / Pesticides, p. 422 / Plants, genetic modification of, p. 422 / Research, p. 422 / Viruses, p. 422	• Environmental Science, 2003, *Organic Farming* (1:37) • Genetics, 2005, *Improving Rice in China* (2:22)	• Unapproved GM Corn in U.S. Food Chain. *Oils & Fats International*, May 2005. • Concerns over GM Crops in Food Aid. *Oils & Fats International*, Sept. 2002.	• U.S. Geological Survey Bioremediation: Nature's Way to a Cleaner Environment: http://water.usgs.gov/wid/html/bioremed.html
22.11 **Engineering Bacteria, Animals and Plants**	Bacteria, p. 423 / Blood-borne diseases, p. 423 / Cystic fibrosis (CF), p. 423 / Genes, p. 423 / Hemophilia A, p. 423 / Herbicides, p. 423 / Human growth hormone (GH), p. 423 / Plants, genetic modification of, p. 423 / Recombinant DNA technology, p. 423 / Strokes, p. 423 / Tissue plasminogen activator (tPA), p. 423 / Traits, p. 423	• Biology, 2003, *Tastier Tomatoes* (2:53) • Genetics, 2003, *Transgenic Tobacco* (2:00)	• Nexia Receives a $3.94 Million Award from the US Army to Develop Protexia™ to Prevent Toxic Effects from Nerve Agents. *PR Newswire*, April 4, 2003.	• Animal and Plant Transformation: The Application of Transgenic Organisms in Agriculture: http://www.ag.uiuc.edu/~vista/html_pubs/irspsm91/transfor.html
How Would You Vote?	There is evidence that in recent years terrorists have explored developing ricin-based weapons. Scientists are working to develop a vaccine to protect against it. Given the threat of biochemical warfare, would you be willing to be vaccinated – or does the threat seem too remote? • InfoTrac: Lethal Toxin from a Bean: Could the Simple Toxin Ricin Be Used as a Terrorist Weapon? Ted Allbeury. *Chemistry and Industry*, Jan. 20, 2003. • InfoTrac: The Growing Threat of Biological Weapons. *American Scientist*, Jan. 2001. • InfoTrac: Researchers Create Human Ricin Vaccine. *The New York Times*, Jan. 31, 2006.			

CHAPTER 23: GENES AND DISEASE: CANCER

Section	Applications	Videos	InfoTrac® Readings	Weblink Examples
23.0 *Impacts, Issues:* **Between You and Eternity**	American Cancer Society, p. 427 / Cancer, p. 427, breast, p. 427, colon, p. 427, ovarian, p. 427, research for, p. 427, skin, p. 427 / Genes, cancer and, p. 427 / Mastectomy, p. 427 / Prosthesis, p. 427	• Anatomy & Physiology, 2003, *Preventing Cancer Surgically* (2:28)	• Monster Tumors Show Scientific Potential in War against Cancer. Elizabeth Svoboda. *The New York Times*, June 6, 2006.	• Breast Cancer Susceptibility Genes: Overstating the Risk?: http://www.genomenews network.org/articles/09_02/ cancer_breast.shtml
23.1 **Cancer: Cell Controls Go Awry**	Angiogenin, p. 429 / Cancer, cell controls and, pp. 428–429, in connective tissues, p. 428, Cardiovascular system, p. 429 / Cell division, pp. 428–429 / Cysts, p. 428 / Dysplasia, p. 428 / Human chorionic gonadotropin (HCG), p. 429 / Hyperplasia, p. 428 / Tumors, pp. 428, 429 / White blood cells (WBCs), p. 429	• Anatomy & Physiology, 2004, *Fibroid Tumors* (2:28)	• Gene Signature Helps Predict Melanoma Outcome. Diana Mahoney. *Internal Medicine News*, Mar. 1, 2006.	• Human Chorionic Gonadotropin Decreases Proliferation and Invasion of Breast Cancer MCF-7 Cells: http://www.ihop-net. org/UniPub/iHOP/pm/ 10277364.html?pmid=15044447
23.2 **The Genetic Triggers for Cancer**	Aging, p. 431 / Cancer, liver, p. 431, lymphocytes and, p. 431 / Carcinogens, pp. 430, 431 / Cells, normal *vs.* cancer, pp. 430–431 / Gene therapy, p. 430 / Heredity, pp. 430, 431 / Immune system, pp. 430, 431 / Mutations, pp. 430, 431 / Natural killer (NK) cells, p. 430 / Oncogenes, pp. 430, 431 / Retinoblastoma, p. 430 / Tumor suppressor gene, pp. 430, 431		• Study IDs Gene Triggers to Liver Cancer. Jamie Talan. *Newsday (Melville, NY)*, June 28, 2006. • Scientists Link Gene to Metastatic Cancer. *UPI NewsTrack*, July 3, 2006.	• The p53 Home Page: http://p53.free.fr/ • Retinoblastoma in Children: http://www.cancerbackup.org. uk/Cancertype/Childrens cancers/Typesofchildrens cancers/Retinoblastoma
23.3 **Assessing the Cancer Risk from Environmental Chemicals**	American Cancer Society, p. 432 / Cancer, chemicals and, p. 432, diagnosing, p. 432, nasal epithelium, p. 432 / Carcinogens, p. 432 / DDT, p. 432 / Fungicides, p. 432 / Herbicides, p. 432 / Insecticides, p. 432 / Pesticides, p. 432 / Radiation, UV, p. 432		• Science That Sticks: Chemist's Work behind U.S. Decision to Ban PFOA. *Canadian Chemical News*, Mar. 2006.	• EPA – Carcinogens: http://www.epa.gov/ebt pages/pollcarcinogens.html • Your Cancer Risk: http://www.yourdiseaserisk. harvard.edu/
23.4 **Diagnosing Cancer**	Biopsy, p. 433 / Breast self-examination, p. 433 / Cancer, screening tests for, p. 433, warning signs for, p. 433 / Computerized tomography (CT), p. 433 / Magnetic resonance imaging (MRI), p. 433 Sigmidoscopy, p. 433 / Testicle self-examination, p. 433 / Tumor markers, p. 433 / Ultrasound, p. 433		• Early Cancer Diagnosis: Present and Future. Andre Baron et al. *Patient Care for the Nurse Practitioner*, Sept. 2005.	• Cancer Symptoms: http://www.cancer symptoms.org/ • Diagnosing Cancer: http://www.mayoclinic.com/ health/cancer-diagnosis/ CA00028
23.5 **Some Major Types of Cancer**	Adenocarcinoma, p. 434 / Cancer, brain, p. 434, breast, p. 434, colon, p. 434, kidney, p. 434, lung, p. 434, ovarian, p. 434, pancreatic, p. 434, prostate, p. 434, uterine, p. 434 / Carcinomas, p. 434 / Leukemias, p. 434 / Lymphomas, p. 434 / Malignant melanoma, p. 434 / Multiple myeloma, p. 434 / Non-Hodgkin lymphoma, p. 434 / Sarcomas, p. 434	• Anatomy & Physiology, 2004, *Fighting Sarcoma* (2:28)	• Sarcomas Most Common Cancer Type in Teenagers. *GP*, April 7, 2006. • Mutations Point the Way to New Leukemia Drugs. *Ascribe Higher Education News Service*, July 17, 2006.	• Oncolink: http://www.oncolink.com • Types of Cancer: http://www.acor.org/types.html
23.6 **Treating and Preventing Cancer**	Adjuvant therapy, p. 435 / Antibodies, monoclonal, p. 435 / Anticancer drugs, p. 435 / Antioxidants, p. 435 / Cancer, chemicals and, p. 435, weight and, p. 435 / Chemotherapy, p. 435 / Cytotoxic drugs, p. 435 / Interferon therapy, p. 435 / Radiation, p. 435 / Smoking, p. 435 / Stem cells, p. 435 / Surgery, p. 435 / Vitamin E, p. 435	• Anatomy & Physiology, 2004, *Prostate Cancer Prevention* (2:28) • Genetics, 2003, *p53 Gene Therapy* (2:28)	• HPV DNA Testing and HPV Vaccines Described by Experts as Combination That Offers Best Hope of Preventing Cervical Cancer. *PR Newswire*, April 27, 2006.	• Clinical Trials: http://www.cancer.gov/ clinicaltrials • Cancer Research Institute: http://www.cancerresearch.org/
23.7 **Cancers of the Breast and Reproductive System**	Biopsy, p. 437 / Blood tests, p. 437 / Breast self-examination, p. 436 / Cancer, breast, p. 436, cervical, p. 437, endometrial, p. 437, ovarian, p. 437, prostate, p. 437, testicular, p. 437, uterine, p. 437, / Genital warts, p. 437 / Lumpectomy, p. 436 / Mastectomy, p. 436 / Menopause, pp. 436, 437 / Obesity, p. 436 / Pap smear, p. 437 / Taxol, p. 437 / Tumor suppressor gene, p. 436	• Anatomy & Physiology, 2003, *Breast Cancer Treatment* (2:28)	• Tylenol Lowers Ovarian Cancer Risk. *UPI NewsTrack*, July 10, 2006.	• Breast Cancer: http://www.breastcancer.org/ • A Testicular Cancer Resource Center: http://tcrc.acor.org/
23.8 **A Survey of Other Common Cancers**	American Cancer Society, p. 438 / Cancer, bone, p. 438, colorectal, p. 438, lung, p. 438, lymphatic, pp. 438–439, oral, p. 438, pancreatic, p. 438, skin, p. 439 / Carcinomas, pp. 438, 439 / Chemotherapy, pp. 438, 439 / Hodgkin's disease, p. 438 / Leukemias, p. 439 / Lymphomas, p. 438 / Malignant melanoma, p. 439 / Non-Hodgkin lymphoma, p. 438 / Secondhand smoke, p. 438 / Wilms tumor, p. 438	• Biology, 2003, *Skin Sun Damage* (1:23) • Anatomy & Physiology, 2003, *Custom Cancer Vaccine* (3:10)	• American Academy of Dermatology Warns of Skin Cancer Risks This Summer; More Than 1 Million New Cases of Skin Cancer to Be Diagnosed This Year in U.S. *Internet Wire*, July 7, 2006.	• Pancreatic Cancer: http://www.mayoclinic. com/health/pancreatic cancer/DS00357 • All about Colon and Rectum Cancer: http://www.cancer.org/ docroot/cri/cri_2x.asp? sitearea=lrn&dt=10
How Would You Vote?	Some young women with an elevated genetic risk of developing breast cancer have chosen to have their breasts removed even before cancer develops. Should health insurers help pay for the cost of this surgery? • InfoTrac: Breast Removal: The Latest in Cancer Prevention. *HealthFacts*, Feb. 1999. • InfoTrac: Is a 0.2% Chance of Death from Breast Cancer Worth a Bilateral Prophylactic Mastectomy? *Contemporary OB/GYN*, June 2005. • WWW: Insurance May Not Cover Preventive Surgery for High-Risk Women: http://www.cancer.org/docroot/NWS/content/NWS_1_1x_Insurance_May_Not_Cover_Preventive_Surgery_for_High_Risk_Women.asp			

CHAPTER 24: PRINCIPLES OF EVOLUTION

Section	Applications	Videos	InfoTrac® Readings	Weblink Examples
24.0 *Impacts, Issues:* **Measuring Time**	Evolution, p. 443 / Human evolution, p. 443 / Iridium, p. 443	• Biology, 2005, *Asteroid Menace* (1:25)	• Reporting Asteroid Impact Risks. David L. Chandler. *Sky & Telescope*, Jan. 2003.	• Asteroid and Comet Impact Hazards: http://impact.arc.nasa.gov/
24.1 **A Little Evolutionary History**	Evolution, basics of, p. 444, defined, p. 444 / Natural selection, p. 444 / Organisms, evolution of, p. 444 / Population, evolution and, p. 444 / Species, p. 444 / Traits, evolution and, p. 444	• Environmental Science, 2004, *Creation vs. Evolution* (2:29)	• The Voyage of the Beagle (excerpt from Charles Darwin's journal). Charles Darwin. *International Wildlife*, Nov.–Dec. 1996.	• Darwin and Natural Selection: http://anthro.palomar.edu/evolve/evolve_2.htm
24.2 **A Key Evolutionary Idea: Individuals Vary**	Evolution, individuals varying and, p. 445 / Genes, p. 445 / Macroevolution, p. 445 / Microevolution, p. 445 / Mutations, p. 445 / Natural selection, p. 445 / Phenotype, p. 445 / Traits, behavioral, p. 445, evolution and, p. 445, morphological, p. 445, physiological, p. 445		• Genetic Variation of the Y Chromosome in Chibcha-Speaking Amerindians of Costa Rica and Panama. Edward A. Ruiz-Narvaez et al. *Human Biology*, Feb. 2005.	• Macroevolution in the 21st Century: http://www.nhm.ac.uk/hosted_sites/paleonet/paleo21/mevolution.html
24.3 **Microevolution: How New Species Arise**	Chance, evolutionary, p. 446 / Gametes, p. 446 / Genes, pp. 446, 447 / Hemophilia, p. 446 / Inheritance, p. 446 / Mutations, pp. 446, 447 / Natural selection, pp. 446, 447 / Speciation, p. 447 / Species, defined, p. 447, microevolution and, p. 447 / Traits, p. 446		• On Misconceptions of Evolutionary Archaeology: Confusing Macroevolution and Microevolution. R. Lee Lyman; Michael J. O'Brien. *Current Anthropology*, June 2001.	• Allopatric Speciation: http://www.evotutor.org/Speciation/Sp1A.html • Observed Instances of Speciation: http://www.talkorigins.org/faqs/faq-speciation.html
24.4 **Looking at Fossils and Biogeography**	Diversity, pp. 448, 449 / Evolution, evidence for, p. 448 / Iridium, p. 449 / Natural selection, p. 448 / Plate tectonics, p. 449 / Population, evolution and, p. 448 / Radiometric dating, p. 449 / Sedimentary rock, pp. 448, 449 / Species, biogeography and, p. 449, fossils and, p. 448		• Insular Carnivore Biogeography: Island Area and Mammalian Optimal Body Size. Shai Meiri et al. *The American Naturalist*, April 2005.	• Radiocarbon Dating: http://www.c14dating.com/index.html • Continental Drift Animation: http://library.thinkquest.org/12232/data/ContinentalDrift.html
24.5 **Comparing the Form and Development of Body Parts**	Development, pp. 450–451 / Diversity, p. 451 / Environment, p. 450 / Evolution, body form/development and, pp. 450–451 / Fingers, p. 450 / Genes, p. 451 / Hearts, p. 451 / Limbs, pp. 450, 451 / Mutations, pp. 450–451 / Organisms, evolution of, pp. 450, 451 / Veins, p. 451	• Biology, 2004, *Lizard Diversity* (1:33)	• Out on a Limb: Parallels in Vertebrate and Invertebrate Limb Patterning and the Origin of Appendages. Clifford J. Tabin et al. *American Zoologist*, June 1999.	• UCMP – Phylogeny Wing: http://www.ucmp.berkeley.edu/clad/clad4.html • Vestigial Structures: http://www.txtwriter.com/backgrounders/Evolution/EVpage12.html
24.6 **Comparing Biochemistry**	DNA, p. 452 / Genes, evolution and, p. 452 / Hemoglobin (Hb), evolution and, p. 452 / Mutations, evolution and, p. 452 / Proteins, p. 452 / Ribonucleic acid (RNA), p. 452 / Species, p. 452 / Traits, evolution and, p. 452		• The Etruscans: A Population-Genetic Study. Cristiano Vernesi et al. *American Journal of Human Genetics*, April 2004.	• Gene Hunters Sifting through Evolution's Shadows: http://sciencereview.berkeley.edu/articles.php?issue=5&article=briefs_1
24.7 **How Species Come and Go**	Evolution, evidence for, pp. 452–453 / Mass extinction, p. 452 / Pollution, p. 453 / Population, *Neandertals*, p. 453 / Radiation, adaptive, pp. 452–453 / Species, pp. 452–453, endangered, p. 453		• Speciation's Defining Moment: Genetics and Genomics Enliven an Old Controversy on Reinforcement. Nick Atkinson. *The Scientist*, April 11, 2005.	• Mass Extinction Underway: http://www.well.com/~davidu/extinction.html
24.8 **Endangered Species**	Endangered species, p. 453 / Endemic, p. 453 / Evolution, evidence for, p. 453 / Mass extinction, p. 453 / Wildlife trade, p. 453	• Biology, 2004, *Gorilla Poaching* (1:59)	• Illegal Fishing : More Cooperation in Fight against Illegal Caviar Trade. *European Report*, July 5, 2006.	• Endangered Species Resource Center: http://www.endangeredspecie.com/
24.9 **Evolution from a Human Perspective**	Natural selection, p. 454 / Power grip, p. 454 / Precision grip, p. 454 / Social behaviors, of humans, p. 455 / Species, human evolution and, p. 454 / Traits, p. 454, human evolution and, p. 454 / Variations, human evolution and, p. 455	• Biology, 2004, *Earliest* Homo sapiens (1:42)	• New Fossils Add Link to the Chain of the Evolution of Humans. John Noble Wilford. *The New York Times*, April 13, 2006.	• Becoming Human: http://www.becominghuman.org/ • Evolution of Modern Humans: http://anthro.palomar.edu/homo2/
24.10 **Emergence of Early Humans**	African emergence model, p. 457 / Evolution, biological, p. 457, cultural, p. 457 / Research, human evolution, p. 457 / *Sahelanthropus tchadensis*, p. 456 / Speciation, p. 457, human evolution and, pp. 456, 457	• Biology, 2003, *Chad Fossil Find* (2:20)	• The First Exodus. Stephen Opperheimer. *Geographical*, July 2002.	• Human Ancestors Hall: http://www.mnh.si.edu/anthro/humanorigins/ha/a_tree.html
24.11 **Earth's History and the Origin of Life**	Cells, pp. 458–459 / DNA, p. 459 / Enzymes, p. 459 / Evolution, chemical, p. 459, Earth history and, pp. 458–459 / Hydrogen, p. 458 / Nucleotides, p. 459 / Organic compounds, pp. 458, 459 / Research, origin of life, pp. 458–459 / RNA, p. 459 / Water, pp. 458, 459		• How Life Began: Microbes at the Extremes May Tell Us. *Time*, July 29, 2002.	• Chemical Evolution: http://www.ibiblio.org/astrobiology/index.php?page=origin03
How Would You Vote?	A large asteroid impact could obliterate civilization and much of Earth's biodiversity. Should we spend millions, even billions, of dollars to search for and track asteroids? • InfoTrac: Target Earth: Chances of an Asteroid or Large Meteoroid Impacting Earth Are Low, But the Results Could Be Disastrous. *Astronomy*, Feb. 2002. • InfoTrac: Asteroid Impact: Mountains or Molehills: Sizing Up the Impact Hazard. *Mercury*, Nov.–Dec. 2002. • InfoTrac: Killer Impact: NASA Has Detailed Plans for Saving the Planet in Case an Asteroid Is Discovered Coming Our Way, Right? *Astronomy*, Dec. 2004.			

CHAPTER 25: ECOLOGY AND HUMAN CONCERNS

Section	Applications	Videos	InfoTrac® Readings	Weblink Examples
25.0 *Impacts, Issues:* **The Human Touch**	Ecology, Easter Island and, p. 463 / Ecosystems, p. 463 / Environmental concerns, p. 463 / Food, p. 463 / Human biology, p. 463 / Population, p. 463	• Environmental Science, 2005, *Easter Island* (1:41)	• Overpopulation and the Fate of Easter Island. *World and I*, Sept. 2001.	• Association of Ecosystem Research Centers: http://www.ecosystem research.org/
25.1 **Some Basic Principles of Ecology**	Diversity, p. 464 / Ecological nutrients, p. 464 / Ecosystems, examples of, pp. 464, 465 / Pioneer species, p. 464 / Specialist species, p. 464 / Temperature, ecology and, p. 464	• Biology, 2005, *Frogs Galore* (2:18)	• Prescription for a Burn. Margaret A. Hill. *Odyssey*, Sept. 2005.	• Ecology.com – An Ecological Source of Information: http://www.ecology.com/
25.2 **Feeding Levels and Food Webs**	Cycling, p. 466 / Ecological nutrients, p. 466 / Energy, in ecosystems, p. 466 / Plants, food webs and, pp. 466, 467 / Sun, energy from, p. 466		• Study Shows Parasites Form the Thread of Food Webs. *Ascribe Higher Education News Service*, July 12, 2006.	• From the Top of the World to the Bottom of the Food Web: http://www.bigelow.org/foodweb/
25.3 **Energy Flow through Ecosystems**	Ecology, food webs and, p. 468 / Ecosystems, energy in, p. 468 / Photosynthesis, p. 468 / Plants, food webs and, p. 468 / Temperature, ecology and, p. 468		• Taking the Measure of Life in the Ice, Kathryn S. Brown. *Science*, April 18, 1997.	• Earth Observatory – Primary Productivity: http://earthobservatory. nasa.gov/Observatory/ Datasets/psn.modis.html
25.4 **Chemical Cycles – An Overview**	Ecological nutrients, p. 469, water cycle and, p. 469 / Global water cycle, p. 469 / Hydrogen, p. 469 / Nitrogen, p. 469 / Oxygen, p. 469 / Phosphorus, p. 469 / Sedimentary cycles, p. 469 / Water, p. 469		• Ice Core Records of Atmospheric N_2O Covering the Last 106,000 Years. Todd Sowers et al. *Science*, Aug. 15, 2003.	• Breakthrough Made in Dating of the Geological Record: http://www.agu.org/sci soc/eos96336.html
25.5 **The Water Cycle**	Cycling, p. 470 / Ecological nutrients, water cycle and, p. 470 / Global water cycle, p. 470 / Water, in ecosystems, p. 470		• NASA Satellites Find Balance in S. America's Water Cycle. *Ascribe Higher Education News Service*, July 5, 2006.	• Global Energy and Water Cycle Experiment: http://www.gewex.org/
25.6 **Cycling Chemicals from the Earth's Crust**	Cycling, p. 471 / Ecological nutrients, eutrophication and, p. 471 / Minerals, p. 471 / Phosphorus, p. 471 / Potassium, p. 471 / Sedimentary cycles, p. 471		• Eutrophication: An Ecological Vision. Fareed A. Khan; Abid Ali Ansari. *The Botanical Review*, Dec. 1, 2005.	• Phosphorous Cycle: http://www.enviroliteracy. org/article.php/480.php
25.7 **The Carbon Cycle**	Carbon cycle, p. 473 / Ecology, food webs and, p. 472, fossil fuels and, p. 473 / Global warming, p. 473 / Greenhouse effect, p. 473 / Photosynthesis, carbon cycle and, pp. 472–473	• Environmental Science, 2003, *Global Environment Outlook* (2:37)	• Ancient Lessons for Our Future Climate. Daniel P. Schrag; Richard B. Alley. *Science*, Oct. 29, 2004.	• Earth Observatory – The Carbon Cycle: http://earthobservatory.nasa. gov/Library/CarbonCycle/ carbon_cycle.html
25.8 **Global Warming**	Chlorofluorocarbons (CFCs), pp. 474, 475 / Greenhouse effect, pp. 474–475 / Ozone, p. 474 / Water, pp. 474–475	• Environmental Science, 2004, *Kyoto Protocol* (2:14)	• Thawing Permafrost Might Release Carbon. *UPI NewsTrack*, June 15, 2006.	• Cow Gas – Methane: http://www.csiro.au/promos/ ozadvances/Series4Cow.html
25.9 **The Nitrogen Cycle**	Plants, nitrogen cycle and, p. 476		• Fungi Important in Arctic Nitrogen Cycle. *UPI NewsTrack*, May 9, 2006.	• EPA – Nitrogen: http://www.epa.gov/maia/ html/nitrogen.html
25.10 **Biological Magnification**	Biological magnification, p. 477 / DDT, p. 477 / Malaria, p. 477 / Mosquitoes, malaria from, p. 477 / Pesticides, p. 477 / Plasmodium, p. 477 / Rickettsia, p. 477 / Typhus, p. 477	• Environmental Science, 2003, *Biomagnification* (2:11)	• Study Links DDT, Skill Problems in Babies. *Monterey County Herald*, July 6, 2006.	• Silent Spring Institute: http://www.silentspring.org/
25.11 **Human Population Growth**	Age structure, p. 479 / AIDS, p. 478 / Fertility rate, pp. 478, 479 / HIV, p. 478 / Infant mortality, pp. 478, 479 / Population density, p. 479 / Population growth, pp. 478–479 / Total fertility rate (TFR), pp. 478, 479	• Biology, 2003, *Global AIDS* (2:18)	• The Numbers Game: Myths, Truths and Half-Truths about Human Population Growth and the Environment. Jim Motavalli. *E*, Jan.–Feb. 2004.	• Population Reference Bureau: http://www.prb.org/ • World Population Information: http://www.census.gov/ipc/ www/world.html
25.12 **Nature's Controls on Population Growth**	Bubonic plague, p. 480 / Carrying capacity, p. 480 / Fungi, p. 480 / Parasites, p. 480 / Pneumonic plague, p. 480 / Pollution, p. 480 / Population density, p. 480	• Human Biology, 2006, *People Explosion* (2:45)	• Malthus Revisited. John Attarian. *World and I*, Nov. 2002.	• A Population Crash Resources Page: http://dieoff.org/
25.13 **Assaults on Our Air**	Acid rain, p. 481 / Fossil fuels, p. 481 / Industrial smog, p. 481 / Ozone thinning, p. 481 / Photochemical smog, p. 481	• Environmental Science, 2005, *Clean Air Act* (1:34)	• Air Pollution and Very Low Birth Weight Infants. Rogers & Dunlop. *Pediatrics*, July 2006.	• EPA – Local Air Quality Index: http://www.epa.gov/airnow/
25.14 **Water, Wastes and Other Problems**	Biosphere, p. 483 / Deforestation, p. 483 / Desertification, p. 483 / Green Revolution, p. 482 / Pollution, NIMBY approach and, p. 482 / Rain forest, p. 483 / Water pollution, p. 482	• Environmental Science, 2005, *Desertification in China* (3:42)	• Rocky Mountain Dry. Ed Quillen. *Planning*, Jan. 2003.	• Environmental Scorecard for Your Community: http://www.scorecard.org/
25.15 **Concerns about Energy**	Energy, alternative sources for, p. 484, / Nuclear power, p. 484 / Radioactive wastes, p. 484 / Solar energy, p. 484	• Environmental Science, 2005, *Nuclear Energy* (1:50)	• Renewable Energy Industry. Michael Eckhart. *Power Engineering*, Jan. 2006.	• Education for Sustainability: http://www.secondnature.org/
25.16 **Loss of Biodiversity**	Anticancer drugs, p. 485 / Coral reefs, p. 485 / Deforestation, extinction and, p. 485 / Plants, drugs from, p. 485, / Population growth, wildlife extinction from, p. 485	• Environmental Science, 2004, *Global Extinction* (1:42)	• Biodiversity Loss Expected to Increase. *UPI NewsTrack*, May 20, 2005.	• Convention on Biological Diversity: http://www.biodiv.org/default.shtml
How Would You Vote?	Goods can be manufactured in ways that protect the environment, but often cost more than comparable goods produced without regard for their ecological impact. Are you willing to pay extra for "green" products? • InfoTrac: It Doesn't Pay to Go Green When Consumers Are Seeing Red. Lorne Manly. *ADWEEK Eastern Edition*, Mar. 23, 1992. • WWW: Consumers Willing to Pay More for rBGH-Free Milk:			

Animations and Interactions—Available on ThomsonNOW™

Students have access to more than 300 high-quality animations and interactions through **ThomsonNOW™ for Starr and McMillan's** *Human Biology,* **Seventh Edition.** As students work through the text, an "Animated!" notation next to a figure number directs them to these animated resources. These and other resources, including video segments, are available to instructors on **PowerLecture for** *Human Biology,* **Seventh Edition: A One Stop Microsoft® PowerPoint® Tool.**

Chapter 1: Learning About Human Biology
Life's diversity
Life's levels of organization
Energy flow and materials
  cycling in the biosphere
Bacteriophage experiments:
  an example of the scientific method
Building blocks of life
Sampling error

Chapter 2: Molecules of Life
Subatomic particles
Atomic number, mass number
Positron-emission tomography (PET)
Electron arrangements in atoms
The shell model of the distribution
  of electrons in atoms
Shell models of common elements
Electron distribution
Chemical bookkeeping
Bonds in biological molecules
Structure of water
Dissolution of sodium chloride
The pH scale
Functional groups
Condensation and hydrolysis
Cellulose and starch
Fatty acid saturation
Triglyceride formation
Phospholipid structure
Cholesterol
Structure of an amino acid
Peptide bond formation
Secondary and tertiary structure
Globin and hemoglobin structure
Structure of ATP
Nucleotide subunits of DNA

Chapter 3: Cells and How They Work
Overview of cells
Lipid bilayer organization
Parts of a eukaryotic cell
Structure of the plasma membrane
How a light microscope works
How an electron microscope works
Nuclear envelope
The nucleus and
  endomembrane system
Structure of a mitochondrion
Cytoskeletal components
Flagella structure
Selective permeability
Diffusion
Solute concentration and osmosis
Tonicity and diffusion of water
Passive transport across
  a cell membrane
Active transport across
  a cell membrane
The ATP–ADP cycle
Induced fit model
Overview of glycolysis
Third stage of aerobic respiration
Overview of aerobic respiration

Chapter 4: Tissues, Organs, and Organ Systems
Structure of epithelium
Types of simple epithelium
Examples of soft connective tissues
Examples of specialized
  connective tissues
Muscle tissues
Cell junctions
Major body cavities
Directional terms and planes
  of symmetry
Organ systems of the human body
Structure of human skin
Negative feedback at the organ level
Homeostatic control of temperature
Human thermoregulation
Heat denaturation of enzymes

Chapter 5: The Skeletal System
Structure of the human thigh bone
How a long bone forms
The human skeletal system
Axial skeleton
Appendicular skeleton

Chapter 6: The Muscular System
Major skeletal muscles
Two opposing muscle groups
  in human arms
Structure of skeletal muscle
Banding patterns and
  muscle contraction
Muscle contraction
Pathway from nerve signal to
  skeletal muscle contraction
Actin, troponin, and tropomyosin
  in a skeletal muscle cell
Energy sources for muscle contraction
Effects of stimulation on muscle

Chapter 7: Digestion and Nutrition
Human digestive system
Location and anatomy of human teeth
Swallowing
Peristalsis
Small intestine structure
Digestive enzymes
Digestion and absorption
  in the small intestine
Structure of the large intestine
Vitamins
Body mass calculator
Caloric requirements

Chapter 8: Blood
Cellular components of blood
Globin and hemoglobin structure
Transfusions and blood types
Genetics of the ABO blood types
Rh blood type and
  complications during pregnancy
Hemostasis

Chapter 9: Circulation—The Heart and Blood Vessels
Major human blood vessels
Anatomy of the heart
Cardiac cycle
Blood circulation
Cardiac conduction
Blood pressure
Structure of blood vessels
Skeletal muscles and fluid
  pressure in a vein
Capillary forces
Examples of ECGs

Chapter 10: Immunity
White blood cells
Lymphoid organs
Human lymphatic system
Formation of membrane
  attack complexes
Inflammatory response
Molecular cues
Immune responses
Antibody structure
Clonal selection
Generating antibody diversity
Antibody-mediated immune response
Cell-mediated immune response
Immunological memory

Chapter 11: The Respiratory System
Human respiratory system
Larynx function
Respiratory cycle
Lung volume
Heimlich maneuver
Structure of an alveolus
Partial pressure gradients

Chapter 12: The Urinary System
Water and solute balance
Human urinary system
Kidney structure
Structure of the nephron
Urine formation
Reabsorption and secretion
Kidney dialysis

Chapter 13: The Nervous System
Neuron structure and function
Ion concentrations
Acton potential propagation
Threshold level
Chemical synapse
Nerve structure
Stretch reflex
Direction of information flow
Divisions of the human nervous system
Autonomic nerves
Spinal cord structure
The human brain
Receiving and integrating areas
Primary motor cortex
Structures involved in memory

Animations and Interactions—Available on ThomsonNOW™ (continued)

Chapter 14: Sensory Systems
Mechanoreceptors
Somatosensory cortex
Sensory receptors
Referred pain
Taste receptors
Olfactory pathway
Wavelike properties of sound
Ear structure and function
The vestibular apparatus
  and equilibrium
Dynamic equilibrium
Eye structure
Visual accommodation
Organization of cells in the retina
Pathway to the visual cortex
Focusing problems

Chapter 15: The Endocrine System
Major human endocrine glands
Posterior pituitary function
Anterior pituitary function
Parathyroid hormone action
Control of cortisol secretion
Hormones and glucose metabolism

Chapter 16: Reproductive Systems
Male reproductive system
Route sperm travel
Spermatogenesis
Hormonal control of sperm production
Female reproductive system
Ovarian function
Hormones and the menstrual cycle
Menstrual cycle summary
HIV replication cycle

Chapter 17: Development and Aging
Stages of development
Fertilization
Cleavage and implantation
Weeks 3 to 4 of development
Neural tube formation
Formation of human fingers
First two weeks of development
Structure of the placenta
Fetal development
Birth
Anatomy of the breast
Sensitivity to teratogens
Amniocentesis
Proportional changes
  during development

Chapter 18: Life at Risk: Infectious Diseases
Prokaryotic body plan
Prokaryotic fission
Life cycle of the beef tapeworm
Life cycle of Plasmodium

Chapter 19: Cell Reproduction
Human chromosomes
Duplicating chromosome
Chromosome structural organization
Spindle apparatus
The cell cycle
Mitosis step-by-step
Cytoplasmic division
Two meiotic divisions
Sperm formation

Egg formation
Meiosis step-by-step
Crossing over
Random alignment
Comparing mitosis and meiosis

Chapter 20: Observable Patterns in Inherited Traits
Crossing garden pea plants
Chromosome segregation
Monohybrid cross
Three to one ratio
Genotypes variation calculator
Testcross
Independent assortment
Dihybrid cross
Overview of sickle cell anemia
Symptoms of sickle cell anemia
Height bar graph

Chapter 21: Chromosomes and Human Genetics
Crossover review
Karyotype preparation
Human sex determination
Effects of the SRY gene
Pedigree diagrams
Autosomal recessive inheritance
Autosomal dominant inheritance
Marfan syndrome
X-linked inheritance
Duplication
Deletion
Translocation
Nondisjunction

Chapter 22: DNA, Genes, and Biotechnology
Subunits of DNA
DNA close up
DNA replication
Base-pair substitution
Deletion
Overview of transcription
  and translation
Comparing DNA and RNA subunits
Gene transcription
Pre-mRNA transcript processing
Genetic code
Structure of a tRNA
Structure of a ribosome
Translation
Restriction enzymes
Base-pairing of DNA fragments
Formation of recombinant DNA
Polymerase chain reaction (PCR)
Automated DNA sequencing
Use of a radioactive probe
DNA fingerprinting
Transferring genes into plants
Protein synthesis summary
How Dolly was create

Chapter 23: Genes and Disease: Cancer
Cancer and metastasis
Characteristics of cancer

Chapter 24: Principles of Evolution
The Galapagos Islands
Finches of the Galapagos

Directional selection
Simulation of genetic drift
Reproductive isolating mechanisms
Models of speciation
Radioisotope decay
Radiometric dating
Plate tectonics
Geologic time scale
Morphological divergence
Comparative embryology
Mutation and proportional changes
Comparative pelvic anatomy
Genetic distance between
  human groups
Primate skeletons
Primate evolutionary tree
Fossils of australopiths
Homo skulls
Miller's reaction chamber experiment
Milestones in the history of life

Chapter 25: Ecology and Human Concerns
Levels of organization
Major biomes
Terrestrial biomes
Two types of ecological succession
Energy flow
The role of organisms in an ecosystem
Trophic levels in a simple food chain
Prairie trophic levels
Prairie food web
Rainforest food web
Categories of food webs
Hydrologic cycle
Phosphorus cycle
Carbon cycle
Greenhouse effect
Greenhouse gases
Carbon dioxide and temperature
Nitrogen cycle
Pesticide examples
Exponential growth
Current and projected
  population sizes by region
Age structure diagrams
U.S. age structure
Demographic transition model
Thermal inversion and smog
Formation of photochemical smog
Acid deposition
Effect of air pollution in forests
How CFCs destroy ozone
Stream pollution
Threats to aquifers
Effects of deforestation
Energy use
Humans affect biodiversity
Biodiversity hot spots
Habitat loss and fragmentation
Resource depletion and degradation

THOMSON BROOKS/COLE VIDEOS FOR THE LIFE SCIENCES
Featuring Human Biology in the Headlines from ABC News

Looking for an engaging way to launch your lectures? Thomson Brooks/Cole's series of videos for the Life Sciences—available digitally—includes more than 120 brief clips. These short, high-interest segments include coverage of current events as well as historic footage. In the coming months, Brooks/Cole will be continually adding to its portfolio of videos with new segments from ABC News.

 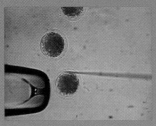  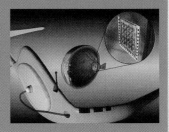

A Saving Graft*	2:23
AIDS Conference in Brazil*	2:35
Air Pollution	2:07
Air Pollution in China*	3:46
Aneurysm Choices	3:42
Arctic National Wildlife Refuge*	3:13
Asteroid Menace*	1:25
Bachelor Pad at the Zoo*	3:03
Biodiversity	2:29
Biomagnification	2:11
Bird Flu*	3:49
Bonus for a Baby*	2:26
Brain Fitness	2:16
Brainwashing Toxins Away	2:45
Breast Cancer Treatment	2:28
California Energy Crisis	2:01
Chad Fossil Find	2:20
CK Eye Surgery	1:39
Cloned Pooch*	3:13
Conjoined Twins	1:39
Conquering Depression	3:10
Creation vs. Evolution	2:31
C-Section Choice	1:24
Custom Cancer Vaccine	1:49
Deadly Animals	2:33
Deforestation	2:18
Deforestation	2:35
Desertification in China*	3:35
Desertification	2:45
Dinosaur Discovery*	2:29
Dolphin Sixth Sense	1:04
Drought	1:40
Earliest Homo Sapiens	1:42
Easter Island*	3:28
El Niño	3:47

Ephedra Dangers	3:08
Fat Facts	2:29
Fat Stem Cells	2:03
Fibroid Tumors	1:51
Fighting Sarcoma	2:43
Fish Farming	1:37
Fish On Prozac	1:57
Fishery Management	2:07
Frogs Galore*	2:35
Germs in Pakistan*	3:13
Global AIDS	2:18
Global Environment Outlook	2:37
Global Extinction	1:42
Global Warming	1:57
Global Warming	3:08
Gorilla Poaching	1:59
Healthy Aging	2:33
Healthy Forests Initiative	2:27
Heart Healthy*	1:43
Hormone Replacement Therapy	2:13
How Pfiesteria Kills	2:30
Hsing Hsing Dies*	1:16
Immigration	2:13
Immortality Industry	2:22
Insect Robots	2:46
Is Aging Treatable?	2:30
IVF Anniversary	2:24
Juvenile Arthritis	2:02
Kyoto Protocol	2:14
Lake Pollution	1:44
Land Management	1:45
Lizard Diversity	1:33
Lorenzo's Oil	3:33
Mad Cow Victim	3:52
Melting Glaciers	1:16

Mermaid Baby*	3:07
Mold Exposure	2:43
Monkeypox	2:25
New Hands*	2:50
New Heart Treatments	1:59
New Nerves*	2:37
New Parts for Old Hearts	2:31
Nuclear Energy*	2:00
Nuclear Fallout	2:01
Obesity and Infertility	2:16
Organic Farming	1:37
Overfishing	2:39
Painful Painkillers*	1:48
Peanut Allergies	1:33
Penguin Rescue*	2:16
People Explosion*	2:45
Pfiesteria Outbreak	1:46
Polio Scare*	1:35
Pork Danger*	2:42
Preventing Cancer Surgically	4:01
Prostate Cancer Prevention	1:33
Psoriasis Drug	1:31
Recycling	2:17
Religion and Health	2:09
Salon Infections	2:05
Seahorse Pregnancy	2:36
Sexy Singers	2:02
Skin Sun Damage	1:23
Slithery Stowaways	2:52
Smog Pollution	1:40
Solar Energy	2:38
Sperm Count and Age	1:27
State of the Air 2002	2:06
State of the Beach	1:55
Stressed Salamanders	3:23
Super Heart Scaffold	2:10

Taller and Taller*	2:40
Tastier Tomatoes	3:03
Teen Obesity Surgery	2:00
The Mask of Technology*	2:49
The Throw Away Society	2:26
The Wine of Life*	3:37
The Working Poor	3:30
To See Again*	3:05
Tongue Tied*	3:55
Turning To Bone	2:21
Vietnam Virus*	3:51
West Nile Virus	2:26
Western Drought	2:11
Whaling	1:28
Whooping Cough Immunization	1:52
Winged Reptiles	1:25

ORDERING INFORMATION

Human Biology in the Headlines
2006 DVD: 0-534-40481-2

Biology in the Headlines
2005 DVD: 0-534-40589-4
2006 DVD: 0-495-01602-0

CNN Anatomy & Physiology
2003: 0-534-39812-X
2004: 0-534-39842-1

CNN Biology
2003: 0-534-39185-0
2004: 0-534-39896-0

CNN Environmental Science
2003: 0-534-39811-1
2004: 0-534-99736-8

* An asterisk denotes segments that are available on DVD only. All other clips are available in VHS and digital CD formats while supplies last.

Due to contractual reasons certain ancillaries are available only in higher education or U.S. domestic markets. Minimum purchases may apply to receive the ancillaries at no charge. For more information, please contact your local Thomson sales representative.

A Sampling of Publications Available Through InfoTrac® College Edition

Selected General Interest Publications
(Consumer, Leisure, News, Sports)
Advocate
American Fitness
Atlantic Monthly
Audubon
Backpacker
Bicycling
Camping Magazine
Consumer Reports
Cross Country Skier
Digital Cinema
Discover
DVD News
Ebony
Entertainment Weekly
Esquire
Field & Stream (West ed.)
Harper's Magazine
Leisure Week
Life
Mountain Bike
Nation
National Geographic
New York Times Upfront
New Yorker
Newsweek
Outdoor Life
Popular Mechanics
Prevention
Runner's World
Saturday Evening Post
Science
Science News
Skiing
Sports Illustrated
Sports Illustrated for Women
Tennis
Time
Travel Weekly
U.S. News & World Report
USA Today (Magazine)

Biological Sciences
American Journal of Human Genetics
American Midland Naturalist
American Naturalist
Annual Review of Genetics
Annual Review of Microbiology
Biological Bulletin
Biological Sciences
Biological Sciences Review
BioScience
Botanical Review
Ecological Monographs
Ecology
Evolution
Human Biology
Indian Journal of Cancer
Integrative Physiological
   and Behavioral Science
Life Science Weekly
Monkeyshines on Health
   and Science: Biology

Physiological Reviews
Quarterly Review of Biology
Scientist

Biotechnology Industry
Applied Genetics News
Biomedical Products
Biomolecular Diagnostics News
Bioscience Technology
Biotech Business
BIOTECH Patent News
Biotech Week
Bioworld Financial Watch
BioWorld International
BioWorld Week
Cell Therapy News
Food Traceability Report
Industrial Bioprocessing
Japanese Biotechnology
Life Sciences & Biotechnology
   Update
Nano/Bio Convergence News
Proteomics Weekly
Worldwide Biotech

Environmental Issues
Air/Water Pollution Report
American Forests
Archives of Environmental Health
Buzzworm
Buzzworm's Earth Journal
Clean Air Today
Clean Water Report
Conservation Matters
Conservationist
Earth Island Journal
Ecological Applications
Ecologist
Endangered Species Update
Energy Conservation News
Environment
Environment and Behavior
Environment Bulletin
Environmental Action Magazine
Environmental Health Perspectives
Environmental Law
Environmental Technology
Environments
EPA Journal
Global Environmental
Global Warming Today
Ground Water Monitor
Journal of Environment
   & Development
Journal of Soil and Water
   Conservation
Journal of the Air & Waste
   Management Association
Management of World Wastes
Mobile Emissions Today
U.S. Environments
National Parks
Natural Life
New York State Conservationist
Oil Spill Intelligence Report

OnEarth
Sierra
Whole Earth
Wilderness
World Watch
Your Public Lands

Environmental Services
Air Pollution Consultant
Asbestos & Lead Abatement Report
Environment Business News Briefing
Europe Environment
Fire
Greener Management International
Hazardous Waste Consultant
Hazardous Waste News
Hazardous Waste Superfund Week
Heating, Piping, Air Conditioning
Heating/Piping/Air Conditioning
   Engineering
Municipal & Industrial Water
   & Pollution Control
Noise Regulation Report
Nuclear Waste News
Ozone Depletion Network
Ozone Depletion Today
Plants Sites & Parks
Pollution Engineering
Recycling Today
Safety Director's Report
Solid Waste Report
Sulphur
Superfund Week
SWT—Solid Waste Technologies
Waste Age
Waste News
Waste Treatment Technology News
Water and Waste Water International
Water Technology News
Water World

Health
AIDS Alert
AIDS Treatment News
AIDS Vaccine Week
AIDS Weekly
AIDS Weekly Plus
Alcohol Research & Health
Alcoholism & Addiction Magazine
American Journal of Health Studies
Anti-Infectives Week
Biomechanics
Bioterrorism Week
Bio-Terrorism.Info
Bulletin of the World
   Health Organization
Cancer Gene Therapy Week
Cancer Vaccine Week
Cancer Weekly
Cardiovascular Week
Clinical Infectious Diseases
Clinical Oncology Alert
Clinician Reviews
Emerging Infectious Diseases
Family Health

A Sampling of Publications Available Through InfoTrac® College Edition (continued)

FDA Consumer
Gene Therapy Weekly
Genomics & Genetics Weekly
Harvard Health Letter
Health & Medicine Week
Health News
Health Science
Heart Disease Weekly
Hospital Infection Control
Immunotherapy Weekly
Infectious Disease Alert
Journal of Alcohol & Drug Education
Journal of Clinical Pathology
Journal of Community Health
Journal of Environmental Health
Journal of Epidemiology
   & Community Health
Journal of Family Practice
Journal of Infectious Diseases
Journal of Medical Genetics
Journal of Musculoskeletal Medicine
Journal of Neurology,
   Neurosurgery and Psychiatry
Journal of Studies on Alcohol
Journal of the American
   Dietetic Association
Journal of Toxicology
Lancet
Medical World News
Medicine
Medicine & Health

Molecular Pathology
Morbidity and Mortality
Weekly Report
Muscle & Fitness
National Women's Health Report
Perspectives in Biology
   and Medicine
Radiologic Technology
Stem Cell Week
Tufts University Health
   & Nutrition Letter
Vaccine Weekly
Virus Weekly
Weekly Epidemiological Record
World Disease Weekly
World Disease Weekly Plus
World Health
World Health Report

Science & Technology
American Scientist
AScribe Science News Service
Bulletin (Southern California
   Academy of Sciences)
Chemistry and Industry
Current Science, a Weekly
   Reader publication
Discover
Issues in Science and Technology
Journal of the Alabama
   Academy of Science

Journal of the Colorado-
   Wyoming Academy of Science
Journal of the Idaho
   Academy of Science
Journal of the Mississippi
   Academy of Sciences
Journal of the Tennessee
   Academy of Science
Micro Engineering
   & Nanotechnology News
MIT's Technology Review
Natural History
Odyssey
Omni
Pacific Science
Perspectives on Science
PharmaGenomics
R & D
Science
Science News
Science Progress
Science Weekly
Science World
Science, Technology,
   & Human Values
Scienceland
Sciences
Scientist
Technology Review

THOMSON
BROOKS/COLE

ISBN 0-495-12556-3

90000

9 780495 125563

Should lecture prep be easier?

PowerLecture—all lecture resources on one easy-to-use DVD

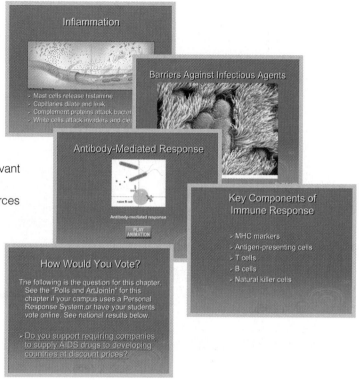

CREATE CUSTOMIZED LECTURES WITH THIS INTEGRATED RESOURCE

The new **PowerLecture for *Human Biology*, Seventh Edition: A One Stop Microsoft® PowerPoint® Tool** is a more feature-rich version of our Multimedia Manager. Each chapter's Microsoft PowerPoint lecture now consolidates all relevant resources—illustrations, all chapter photographs, animations, key concepts, links to the web, and other chapter-related resources listed below. No more hassling with media files, multiple discs, and limited formats. Everything is on one DVD; a simple click on a lecture slide accesses the resource. If you don't have a DVD drive, we can provide you with CDs.

NEW! MICROSOFT POWERPOINT LECTURE SLIDES FOR EACH CHAPTER INCLUDE:

- All diagrams and photos from the chapter, plus bonus photos
- Links to video clips from ABC News
- Links to the animations and interactions from **ThomsonNOW™ for Starr and McMillan's *Human Biology*, Seventh Edition** (formerly BiologyNow™)
- Links to all websites listed in the *Issues and Resources Integrator* (available with the Instructor's Edition of this text)
- *The How Would You Vote?* question plus a link to the online polling site
- Slides of book-specific questions that also appear on **JoinIn™ on TurningPoint®** (a CD-ROM with software and content for student classroom response systems)
- Bulleted points on key content
- Links to InfoTrac® College Edition articles listed in the *Issues and Resources Integrator*

EDITING FEATURES AND OTHER RESOURCES:

- Use slide sorter views to select, copy, and paste your chosen slides onto your chapter lecture file.
- Click a slide's green button to launch a related animation without leaving Microsoft PowerPoint.
- Edit text on the ready-made lecture slides to fit your needs.
- Present art in segments, or select and enlarge portions of a figure.
- Edit labels, remove labels, or present one label at a time.
- Use text figures in JPG format with labels and leaders (lines that point to key parts of a figure).
- JPG figures with leaders only are available for in-class quizzing or labeling by the instructor.
- Additional JPG figures without labels or leaders and 1,000 bonus photos are included.
- Chapter menus include relevant Microsoft® Word files of the **Instructor's Resource Manual and Test Bank.**

Due to contractual reasons, certain ancillaries are available only in higher education or U.S. domestic markets. Minimum purchases may apply to receive the ancillaries at no charge. For more information, please contact your local Thomson sales representative.

Would you like biology to matter more to your students?

An issues-oriented approach engages students' interest.

IMPACTS, ISSUES—HUMAN BIOLOGY'S RELEVANCE TO EVERYDAY LIFE

Each chapter opens with *Impacts, Issues*—a brief, interesting essay about a biology-related issue or research finding that relates to the chapter topic. Topics include the obesity epidemic, unregulated food supplements, hormone disrupters in food and water supplies, and the effects of Ecstasy on the nervous system. The chapter discussions show students how understanding basic concepts of human biology can help them think critically and make informed decisions.

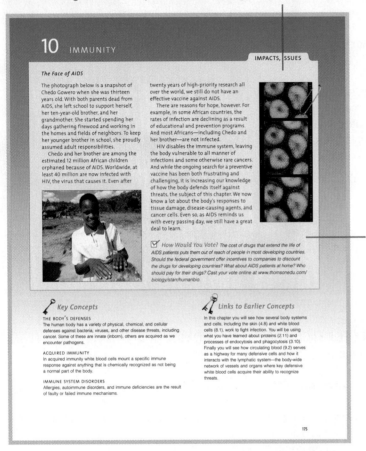

HOW WOULD YOU VOTE?

Each chapter's introductory essay is followed by a related *How Would You Vote?* question. The question invites students to explore the issue in more depth before forming an opinion about it.

MINI-BOXES TO KEEP THE ISSUE "TOP OF MIND"

The authors return to the chapter-opening issue with a brief marginal segment on a topic related to health, research, or environmental concerns.

How Would You Vote? questions get students personally involved.

WAKE-UP CALLS FOR TODAY'S HOT BUTTON ISSUES

There is nothing abstract about the chapter *How Would You Vote?* question. Should fast foods carry warning labels? Should teens have access to contraception without parental consent? How far should government go in regulating emissions of industrial chemicals? Students decide for themselves.

> ☑ *How Would You Vote?* *The cost of drugs that extend the life of AIDS patients puts them out of reach of people in most developing countries. Should the federal government offer incentives to companies to discount the drugs for developing countries? What about AIDS patients at home? Who should pay for their drugs? Cast your vote online at www.thomsonedu.com/ biology/starr/humanbio.*

STUDENTS BROADEN THEIR UNDERSTANDING . . .

Students access the *How Would You Vote?* exercise as an assignment via **ThomsonNOW**™ or at the Book Companion Website at www.thomsonedu.com/biology/starr/humanbio. The exercise directs them to articles from the InfoTrac® College Edition university library, which includes such publications as *The New York Times, Discover,* and *Science News.* Students review relevant articles, weigh information on both sides of the issue, and develop their position. (To access **ThomsonNOW** and InfoTrac College Edition, students first register at www.thomsonedu.com. Contact your representative to package an access card with each new text.)

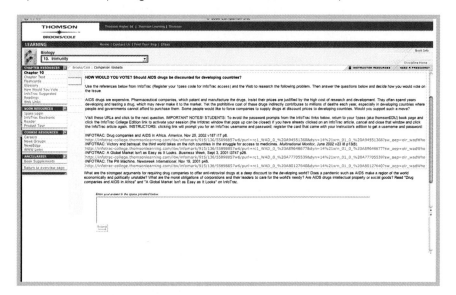

. . . COME TO TENTATIVE CONCLUSIONS . . .

After reflecting on the issue, students answer a few questions, including the reasoning for their conclusion. They can submit their answers to you via **ThomsonNOW,** WebCT,® or Blackboard,® or by email if they accessed the feature via the website.

. . . AND CAST THEIR VOTE.

As part of the exercise, students log their vote for the class tally. They also visit the worldwide survey site, where they can view a tally of how their peers voted on the issue.

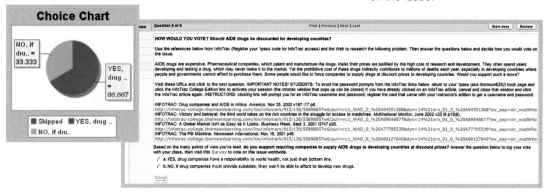

Due to contractual reasons, certain ancillaries are available only in higher education or U.S. domestic markets. Minimum purchases may apply to receive the ancillaries at no charge. For more information, please contact your local Thomson sales representative.

Are students aware of connections between concepts in different chapters?

A new feature identifies connections to important material presented earlier. Cross-references appear within chapter sections.

A SIMPLE LIST OF CONCEPTS RELATED TO THE CHAPTER CONTENT
To master any new concept, students need to build on their understanding of related concepts. At the beginning of each chapter, *Links to Earlier Concepts* identify related material that was covered earlier in the book. These lists also help instructors convey how all aspects of biology are interconnected.

Links to Earlier Concepts

In this chapter you will see how several body systems and cells, including the skin (4.8) and white blood cells (8.1), work to fight infection. You will be using what you have learned about proteins (2.11) and processes of endocytosis and phagocytosis (3.10). Finally you will see how circulating blood (9.2) serves as a highway for many defensive cells and how it interacts with the lymphatic system—the body-wide network of vessels and organs where key defensive white blood cells acquire their ability to recognize threats.

LINKS TO
SECTIONS
2.6, 2.11, 3.9,
AND 7.5

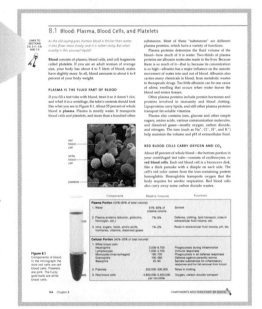

ON-PAGE CROSS-REFERENCES TO PREVIOUS DISCUSSIONS
Key icons with cross-references to earlier sections help students find relevant bits of information quickly—precisely when they need it. Section 8.1 focuses on blood as a mix of watery plasma, blood cells, and platelets. The linking icons in the margin remind students that they learned about water's key biological roles in Section 2.6, and about various proteins, including albumin, in Section 2.11. The link to Section 3.9 refers to the discussion of how substances may move between the blood and tissues by diffusion and osmosis. Later sections, including cardiovascular loops and gas transport in the respiratory system, link students back to this section.

Step-by-step art, also animated on ThomsonNOW,™ introduces and reinforces important concepts.

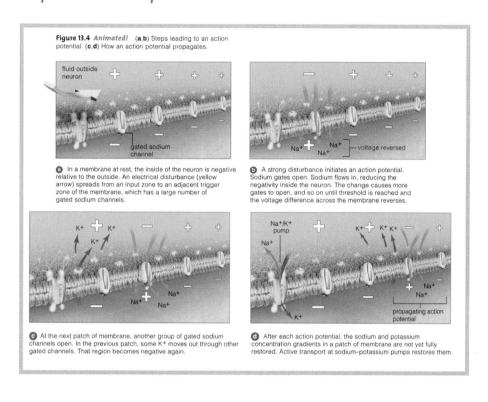

Figure 13.4 *Animated!* **(a,b)** Steps leading to an action potential. **(c,d)** How an action potential propagates.

ⓐ In a membrane at rest, the inside of the neuron is negative relative to the outside. An electrical disturbance (yellow arrow) spreads from an input zone to an adjacent trigger zone of the membrane, which has a large number of gated sodium channels.

ⓑ A strong disturbance initiates an action potential. Sodium gates open. Sodium flows in, reducing the negativity inside the neuron. The change causes more gates to open, and so on until threshold is reached and the voltage difference across the membrane reverses.

ⓒ At the next patch of membrane, another group of gated sodium channels open. In the previous patch, some K+ moves out through other gated channels. That region becomes negative again.

ⓓ After each action potential, the sodium and potassium concentration gradients in a patch of membrane are not yet fully restored. Active transport at sodium–potassium pumps restores them.

ANNOTATED ART TO HELP STUDENTS "GET" CONCEPTS

Visual learners use diagrams with step-by step summaries within the diagram as a preview of text topics. The art program—including vivid illustrations, photographs, and micrographs—highlights every major body system and function in human biology. New diagrams illustrate such topics as the shell model, organ systems, and controls over glucose homeostasis.

ANIMATIONS OF TEXT ART

Through **ThomsonNOW,** students access narrated animations of hundreds of the illustrations in the book. An *Animated!* notation next to a figure number directs them to the animated or interactive version.

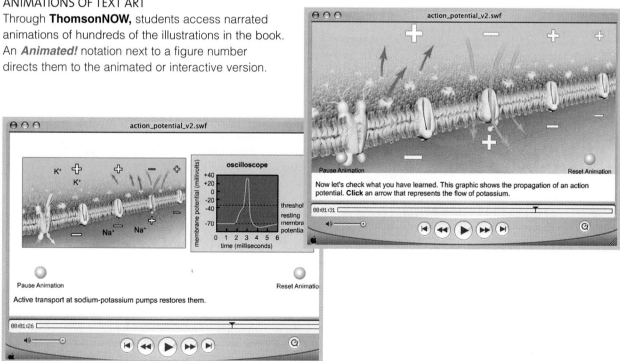

Would you like to monitor student progress quickly and easily?

ThomsonNOW™ and JoinIn™ on TurningPoint help you assess student progress and make learning more interactive.

Just what you need to do now!

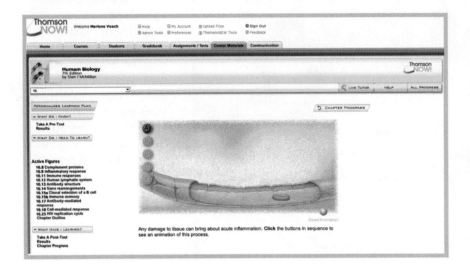

PERSONALIZED STUDY PLANS—AUTOMATIC GRADING

ThomsonNOW for Starr and McMillan's *Human Biology*, Seventh Edition saves you time through automatic grading, and provides your students with an efficient way to study. After reading a chapter, students can go to **ThomsonNOW** and take the assignable pre-test. A Personalized Study plan evaluates their answers and directs them to animations and text sections they still need to review. A post-test measures their mastery. And with **ThomsonNOW**, students can choose how they want to read the textbook—via an integrated eBook or by reading the print version.

EXTRA HELP TO GET STUDENTS UP TO SPEED

A *How Do I Prepare?* feature in **ThomsonNOW** presents tutorials that help students brush up on the skills they need to succeed in biology. Topics include Animal Systems and Structure, Atoms and Cells, Ecology, Energy, Evolution, Genetics and Inheritance, and Study Skills. Within these topic areas, students can also get help with:

- Math (ratios; fractions; simple unit conversions; fractions, decimals, percents; scientific notation; logarithms; area and volume; probability; binomial expressions; logistic growth; exponential growth)
- Chemistry (atomic particles; atomic number and mass number; isotopes; atomic shell model; atomic orbitals; notating atomic characteristics; ionic bonding; covalent bonding; hydrogen bonding; balancing chemical equations; acids, bases, and pH)
- Graphing (Cartesian graphing basics, interpreting and creating graphs)
- Latin and Greek Word Roots
- Study Skills (study habits, memory, getting the most out of lectures, getting ready for labs, reading effectively, taking tests, critical thinking and writing)

To include access to **ThomsonNOW** with each new text, contact your Thomson Brooks/Cole representative or visit us online at www.thomsonedu.com. For a demonstration of **ThomsonNOW for Starr and McMillan's *Human Biology,*** visit www.thomsonedu.com/biology/starr/hb7demo.

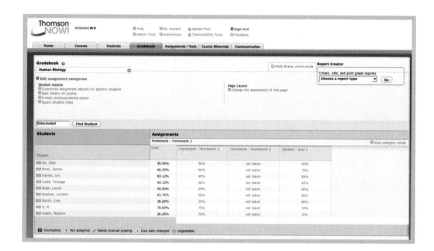

A BUILT-IN GRADEBOOK FOR INSTRUCTORS

The **ThomsonNOW** instructor gradebook makes it easy for you to track student grades and progress. You can create your own exams using questions from the Test Bank. In addition, you have the option to assign prebuilt homework, including electronic versions of the text's end-of-chapter questions and hands-on exercises from Bres and Weisshaar's *Cooperative Learning.* **ThomsonNOW** also can be easily integrated with your WebCT® or Blackboard® gradebook. Or, if you prefer a hands-off approach, students can benefit from the intelligent study system without any instructor setup or involvement.

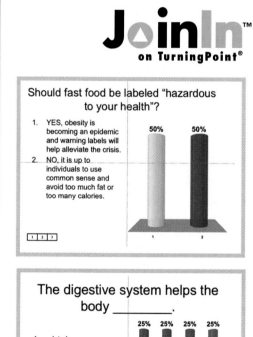

ASSESS STUDENT PROGRESS IMMEDIATELY IN CLASS

Here's an easy way to motivate students to come to class—and to focus their attention. Thomson Brooks/Cole is pleased to offer you book-specific **JoinIn™ on TurningPoint®** content for classroom response systems tailored to **Human Biology, Seventh Edition. JoinIn** allows you to transform your classroom and assess the progress of individual students with instant in-class quizzes and polls.

Our exclusive agreement to offer **TurningPoint®** software lets you pose book-specific questions and display students' answers seamlessly within the Microsoft PowerPoint slides of your own lecture, in conjunction with the "clicker" hardware of your choice. We provide the software and questions for each chapter of the text on ready-to-use Microsoft PowerPoint slides.

Due to contractual reasons, certain ancillaries are available only in higher education or U.S. domestic markets. Minimum purchases may apply to receive the ancillaries at no charge. For more information, please contact your local Thomson sales representative.

Would you like a book that clearly conveys core concepts of human biology in 488 pages?

In response to instructor feedback, this text's presentation is more focused than ever.

In revising this edition, the authors took a fresh look at every concept and topic, honing discussions to improve their clarity. Some topics are condensed and simplified, some reorganized or reframed, and still others are expanded. In addition, concept discussions are strengthened with new step-by-step art. Here are a few topics that received particular attention.

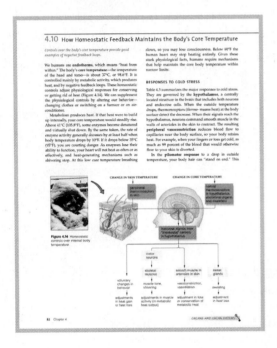

■ *Coverage of diseases, disorders, and related health issues* is expanded in all body systems chapters, with:
- More examples of disorders of the skeletal, muscular, digestive, and nervous systems
- Two full pages on blood disorders (e.g., bacterial toxins, carbon monoxide, leukemias, and anemias)
- Updated treatment of the causes, symptoms, and prevention of diabetes (Chapter 15)
- New material on general principles that govern infection (Chapter 18)

■ *A simplified treatment of enzymes and substrates* in the discussion of cell metabolism in section 3.11 includes a single, simple diagram showing how the molecules interact at each step.

■ *An expanded discussion of homeostasis* in Chapter 4 goes beyond the usual "thermostat" analogy to include regulation of core body temperature.

■ *A reorganized chapter on the muscular system* leads students through this challenging material from a "macro" to "micro" perspective.

■ *A new integrated section on accessory organs* appears in the digestion and nutrition chapter.

■ *In Section 9.7, elimination of mathematical calculations of pressure changes* that govern fluid movement into and out of capillaries provides students with a simple, cleanly illustrated overview of capillary functions.

■ *An overhauled chapter on immunity* presents basic concepts in the easy-to-understand framework of innate and adaptive defenses, and includes a simplified discussion and illustration of the complement system that emphasizes major points without confusing detail.

■ *A reorganized presentation of the endocrine system* in Chapter 15 includes a more focused discussion of adrenal hormones and describes pituitary gland malfunctions as examples of factors that influence hormone effects.

■ *Sexually transmitted diseases are now covered in Chapter 16,* where they fit as a logical extension of the discussion of the reproductive system.

HUMAN BIOLOGY

SEVENTH EDITION

CECIE STARR

Belmont, California

BEVERLY McMILLAN

Gloucester, Virginia

THOMSON

BROOKS/COLE

Australia • Brazil • Canada • Mexico • Singapore • Spain
United Kingdom • United States

THOMSON

BROOKS/COLE

Human Biology, Seventh Edition
Cecie Starr/Beverly McMillan

BIOLOGY PUBLISHER: Jack Carey

DEVELOPMENT EDITOR: Suzannah Alexander

ASSISTANT EDITOR: Kristina Razmara

TECHNOLOGY PROJECT MANAGER: Fiona Chong

MARKETING MANAGER: Kara Kindstrom

MARKETING ASSISTANT: Catie Ronquillo

MARKETING COMMUNICATIONS MANAGER: Jessica Perry

PROJECT MANAGER, EDITORIAL PRODUCTION: Andy Marinkovich

CREATIVE DIRECTOR: Rob Hugel

PRINT BUYER: Judy Inouye

PERMISSIONS EDITOR: Sarah d'Stair

PRODUCTION SERVICE: Lachina Publishing Services, Inc.

TEXT DESIGNER: Gary Head, Gary Head Design

PHOTO RESEARCHER: Dena Digilio-Betz

COPY EDITOR: Lachina Publishing Services, Inc., Amy Mayfield

ILLUSTRATOR: Lachina Publishing Services, Inc., Matt Fornadel

COVER DESIGNER: Irene Morris/Morris Design

COVER IMAGE: © David Madison/Getty Images

COVER PRINTER: Courier-Kendallville

COMPOSITOR: Lachina Publishing Services, Inc.

PRINTER: Courier-Kendallville

For more information about our products, contact us at:
Thomson Learning Academic Resource Center
1-800-423-0563
For permission to use material from this text or product, submit a request online at **http://www .thomsonrights.com**.
Any additional questions about permissions can be submitted by e-mail to **thomsonrights@thomson.com**.

Thomson Higher Education
10 Davis Drive
Belmont, CA 94002-3098
USA

Library of Congress Control Number: 2006906897

ISBN: 0-495-01596-2

CONTENTS IN BRIEF

DETAILED CONTENTS

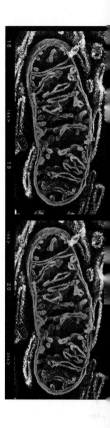

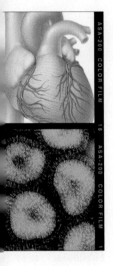

17 Development and Aging

18 Life at Risk: Infectious Disease

24 Principles of Evolution

25 Ecology and Human Concerns

Preface

Over the years, six editions of this textbook have been tested by many thousands of students and scrutinized by hundreds of instructors. As we embarked on this seventh edition of *Human Biology*, we were determined both to keep faith with that unstinting appraisal and to shape a textbook that continues to be grounded in the real world of learning and teaching. Setting the stage for this work was the stark fact that today students are not only enormously pressed for time, but bombarded with a steady stream of half-the-story news snippets about familiar and emerging diseases, medical "breakthroughs," unfiltered blogs on health issues, and a host of other information related to the structure and operation of that natural wonder, the human body. As a result, the environment in which human biology is taught and must be learned is more dynamic and challenging than ever before.

We realized at the outset that our approach to meeting the challenge had to be equally dynamic. First, we knew it was vital to do even more to frame an easy-to-follow presentation of basic concepts of human anatomy and physiology in the context of current issues and their impacts on human life and concerns. Critical thinking skills, and genuine opportunities to use them, had to be part of the mix because students are or will become participants in major decisions in their communities and the larger society. Another crucial element was to keep pace with the changing understanding of the genetic basis of body functions, offering nontechnical explanations while avoiding unnecessary details that can make learning a chore instead of an awakening. Finally, although *Human Biology* has always drawn on Thomson Brooks/Cole's pioneering electronic learning tools for students, we knew

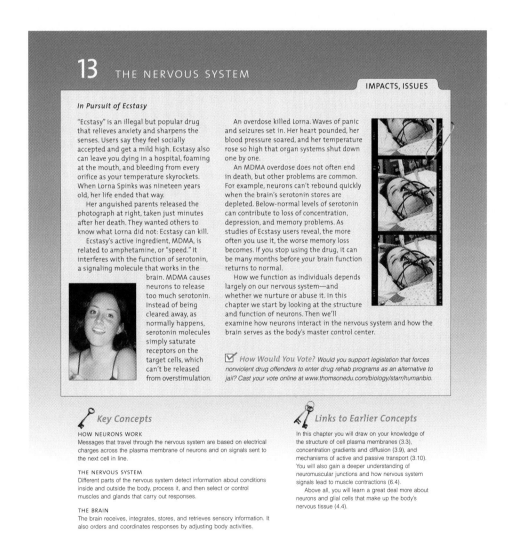

13   THE NERVOUS SYSTEM

IMPACTS, ISSUES

In Pursuit of Ecstasy

"Ecstasy" is an illegal but popular drug that relieves anxiety and sharpens the senses. Users say they feel socially accepted and get a mild high. Ecstasy also can leave you dying in a hospital, foaming at the mouth, and bleeding from every orifice as your temperature skyrockets. When Lorna Spinks was nineteen years old, her life ended that way.

Her anguished parents released the photograph at right, taken just minutes after her death. They wanted others to know what Lorna did not: Ecstasy can kill.

Ecstasy's active ingredient, MDMA, is related to amphetamine, or "speed." It interferes with the function of serotonin, a signaling molecule that works in the brain. MDMA causes neurons to release too much serotonin. Instead of being cleared away, as normally happens, serotonin molecules simply saturate receptors on the target cells, which can't be released from overstimulation.

An overdose killed Lorna. Waves of panic and seizures set in. Her heart pounded, her blood pressure soared, and her temperature rose so high that organ systems shut down one by one.

An MDMA overdose does not often end in death, but other problems are common. For example, neurons can't rebound quickly when the brain's serotonin stores are depleted. Below-normal levels of serotonin can contribute to loss of concentration, depression, and memory problems. As studies of Ecstasy users reveal, the more often you use it, the worse memory loss becomes. If you stop using the drug, it can be many months before your brain function returns to normal.

How we function as individuals depends largely on our nervous system—and whether we nurture or abuse it. In this chapter we start by looking at the structure and function of neurons. Then we'll examine how neurons interact in the nervous system and how the brain serves as the body's master control center.

☑ *How Would You Vote?* Would you support legislation that forces nonviolent drug offenders to enter drug rehab programs as an alternative to jail? Cast your vote online at www.thomsonedu.com/biology/starr/humanbio.

🔑 *Key Concepts*

HOW NEURONS WORK
Messages that travel through the nervous system are based on electrical charges across the plasma membrane of neurons and on signals sent to the next cell in line.

THE NERVOUS SYSTEM
Different parts of the nervous system detect information about conditions inside and outside the body, process it, and then select or control muscles and glands that carry out responses.

THE BRAIN
The brain receives, integrates, stores, and retrieves sensory information. It also orders and coordinates responses by adjusting body activities.

🔑 *Links to Earlier Concepts*

In this chapter you will draw on your knowledge of the structure of cell plasma membranes (3.3), concentration gradients and diffusion (3.9), and mechanisms of active and passive transport (3.10). You will also gain a deeper understanding of neuromuscular junctions and how nervous system signals lead to muscle contractions (6.4).

Above all, you will learn a great deal more about neurons and glial cells that make up the body's nervous tissue (4.4).

225

we could do even more to meet our student users on their turf, engage their minds, and facilitate their educational journey.

With these core ideas to guide us, the next step was to enlist the help of thoughtful, dedicated instructor-reviewers who teach and interact with a broad spectrum of students in a variety of educational settings. They read and dissected every chapter and their help was invaluable. This collaboration produced the book you hold in your hands—a lively, relevant, beautifully illustrated, and user-friendly textbook that we believe is a major step forward in helping students understand the basic principles and key concepts of human biology.

MAKING IT CLEAR AND RELEVANT

Current Issues Most students who take this course will not become biologists, but biological research will influence their lives in many ways. Each chapter starts with a current issue that relates to its content. For instance, the chapter on digestion and nutrition opens with the role of hormones in determining appetite, hunger, and personal eating patterns, discussing these topics in the context of the global obesity epidemic. The cardiovascular system chapter opens with the dramatic true tale of how a young woman's wildly erratic heart rhythm was restored by an automated external defibrillator—a device that is increasingly saving lives in shopping malls, airports, arenas, and other public places. Our innovative chapter on infectious disease opens with the hypothesis that Alexander the Great died after becoming infected with West Nile virus, a pathogen that now reaches around the globe. Each introduction concludes with *How Would You Vote?*, which gives students an opportunity to weigh in on research or an application

IMPACTS, ISSUES

Several large clinical trials are under way to test the hypothesis that creatinine can slow the deterioration of muscle function in people with muscular dystrophy. In one trial, children with Duchenne muscular dystrophy are being given creatinine or a placebo for several months. They will be tested periodically for muscle strength. The results will be compared to determine whether creatinine increased muscle strength in the experimental groups.

related to the story. Since we introduced this feature in the previous edition of our book, thousands of students throughout the country have voted *on-line* and accessed campus-wide, statewide, and nationwide tallies. By doing so, they may sense how individual actions can make a difference. With a nod to students raised on music videos, there are close to 100 CNN video clips, available to instructors on either VHS or CD-ROMs while supplies last. Thomson Brooks/Cole has also expanded its portfolio of videos with numerous current segments from ABC News. These two- to three-minute video clips are available on DVD. We don't raise chapter opener issues and then forget them. One or more *Impacts/Issues* mini-boxes are presented in the chapter's page margins, next to the relevant text. These features sample topics related to health, research, and environmental issues. A sample is shown to the left.

HELPING STUDENTS TRACK CONNECTIONS

Key Concepts The first page of each chapter also lists the *Key Concepts* students will be studying, as well as *Links to Earlier Concepts* presented in previous chapters. On appropriate text pages, a key icon in the margin shows *Links to Sections* in previous chapters where students can review topics they're unsure about or refresh their memory of a related discussion. The links also serve as an ongoing reminder of the big picture—that everything they are learning is part of a larger whole.

Highlighted Key Points Students learn best when they can stay focused on a text section's "story" without worrying about highlighting something they might be tested on. To make that process easier, we have summarized key points and placed them in blue boxes at the end of each section. All chapters end with a *section-by-section summary*.

Animated Art In this edition, dozens of *animated diagrams* help students who are more comfortable with visual learning than with reading. These diagrams are easily noticed by the *Animated!* icon placed at the beginning of the figure caption. For a preview of where their reading will take them, students can first walk step-by-step through the art (as shown on the next page) and listen to a narrated animation of it. These diagrams also are available on ThomsonNOW for students and on the PowerLecture DVD for instructors.

Explore on Your Own With this edition we have expanded our popular *Explore on Your Own* exercises, which now appear at the end of every chapter. These real-life applications of human biology are simple but informative activities such as learning to take a pulse, measuring your "Blood IQ," and calculating personal water use over the course of a day or week. They can be used by students independently or assigned for extra credit.

Explore on Your Own

As described in Section 9.6, a pulse is the pressure wave created during each cardiac cycle as the body's elastic arteries expand and then recoil. Common pulse points—places where an artery lies close to the body surface—include the inside of the wrist, where the radial artery travels, and the carotid artery at the front of the neck. Monitoring your pulse is an easy way to observe how a change in your posture or activity affects your heart rate.

To take your pulse, simply press your fingers on a pulse point and count the number of "beats" during one minute. For this exercise, take your first measurement after you've been lying down for a few minutes. If you are a healthy adult, it's likely that your resting pulse will be between 65 and 70 beats per minute. Now sit up, and take your pulse again. Did the change in posture correlate with a change in your pulse? Now run in place for 30 seconds and take your pulse rate once again. In a short paragraph, describe what changes in your heart's activity led to the pulse differences.

MAKING IT BRIEF, WITH CLEAR EXPLANATIONS

To keep the book length to less than 500 pages as in previous editions, we were selective about which

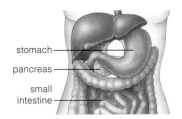

stomach

pancreas

small
intestine

Figure 15.10 *Animated!* How cells that secrete insulin and glucagon respond to a change in the level of glucose in blood. These two hormones work antagonistically to maintain the glucose level in its normal range.

(**a**) *After* a meal, glucose enters blood faster than cells can take it up and its level in blood increases. In the pancreas, the increase (**b**) stops alpha cells from secreting glucagon and (**c**) stimulates beta cells to secrete insulin. In response to insulin, (**d**) adipose and muscle cells take up and store glucose, and cells in the liver synthesize more glycogen. As a result, insulin *lowers* the blood level of glucose (**e**).

(**f**) *Between* meals, the glucose level in blood falls. The decrease (**g**) stimulates alpha cells to secrete glucagon and (**h**) slows the insulin secretion by beta cells. (**i**) In the liver, glucagon causes cells to convert glycogen back to glucose, which enters the blood. As a result, glucagon *raises* the blood level of glucose (**j**).

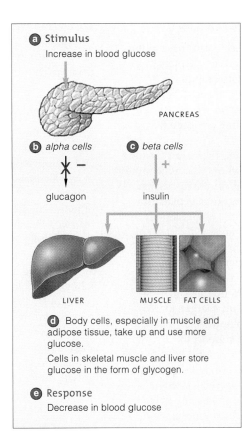

a Stimulus
Increase in blood glucose

PANCREAS

b *alpha cells*

glucagon

c *beta cells*

insulin

LIVER   MUSCLE   FAT CELLS

d Body cells, especially in muscle and adipose tissue, take up and use more glucose.

Cells in skeletal muscle and liver store glucose in the form of glycogen.

e Response
Decrease in blood glucose

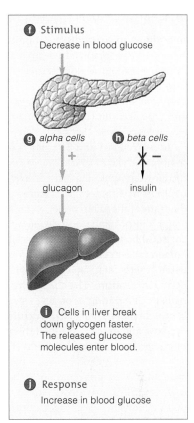

f Stimulus
Decrease in blood glucose

g *alpha cells*

glucagon

h *beta cells*

insulin

i Cells in liver break down glycogen faster. The released glucose molecules enter blood.

j Response
Increase in blood glucose

topics to include. An arena where we did *not* skimp was clear explanation. If something is worth reading about, why reduce it to a factoid? Isolated facts just invite mind-numbing memorization, a study habit that won't help students learn to think critically about the world and their place in it.

For instance, we know that non-science majors taking a human biology course do not want to memorize each step of the Krebs cycle. They do want to know about what researchers are discovering about the biological basis of weight control, and many female students want to know what will be going on inside their body if they become pregnant. Good explanations can help students make their own informed decisions on many biology-related issues, including stem cell research, STDs, fertility drugs, prenatal diagnoses of genetic disorders, and genetically engineered food plants.

Our choices for which topics to condense, expand, or delete were shaped by a decade of feedback from the front lines—those who teach human biology on college campuses large and small.

OFFERING EASY-TO-USE MEDIA TOOLS

Each copy of *Human Biology* can come with an access code card that students register at www.thomsonedu.com to access all the electronic study aids for this text, including ThomsonNOW. For each text chapter, ThomsonNOW provides *animated figures* and *diagnostic pretest questions.* After reading a chapter, students can go to ThomsonNOW and take the assignable pretest. A *personalized study plan* evaluates their answers and directs them to animations and text sections they still need to review. To get more help, students can also access *vMentor,* a free tutoring service. Moreover, *How Do I Prepare* presents tutorials that help students brush up on the skills they need to succeed in human biology. *Post-test questions* can be used as a self-assessment tool or submitted to instructors electronically.

Our readers have free access to an exclusive online database—*InfoTrac® College Edition,* a full-text library of 5000+ periodicals. A pre-selected online reader, along with an annotated list of relevant websites, is also available on the student website, making research on the pros and cons of an issue a snap. After students research the *How Would You Vote?* question, they can cast their votes at the website and also see at once how others have voted. They can e-mail information about their research into the issue and their votes to instructors. The student website at www .thomsonedu.com/biology/starr/humanbio also contains *interactive flash cards* with audio pronunciation guides that define all the book's boldface terms and an online Issues and Resources Integrator that correlates chapter sections with applications, videos, InfoTrac articles, and websites to help instructors bring together all of our resources related to specific applications.

OTHER REVISION HIGHLIGHTS

In addition to strengthening our issue-oriented focus in this edition, we reorganized within and between chapters to improve the logical flow of ideas, updated topics throughout, and added more of the real-life applications that grab student interest. In every chapter we also revised and enhanced illustrations. The following are just a few examples of these changes.

Improved Art and Sharper Focus We find it impossible to hone the clarity and focus of text without doing the same for the accompanying art. For example, for the discussion of atoms in Chapter 2 we created a new, more explanatory diagram of the shell model, expanded the discussion of water's importance, and developed animated diagrams of chemical bonding and the pH scale. The discussion of homeostasis in Chapter 4 has been expanded beyond the usual "thermostat" analogy to include regulation of core body temperature. The chapter also offers a new, more realistic illustration of organ systems and a larger, more complete diagram of skin structure. Chapter 8, on blood, has new and expanded illustrations to support the reorganized text discussion of blood typing and agglutination. As another example, we rethought and then reorganized the presentation of the endocrine system in Chapter 15. We use growth hormone and pituitary gland malfunctions to discuss factors that influence hormone effects. Endocrine functions of the thyroid, parathyroid, and thymus now are integrated in a single section. A more focused discussion of adrenal hormones includes a new, animated illustration of feedback control mechanisms. We also created a clearer, more interpretive, animated illustration of controls over glucose homeostasis.

More Coverage of Diseases and Disorders We expanded coverage of diseases, disorders, and related health issues in all body system chapters. For instance, Chapter 5 now has two full pages on skeletal disorders, including genetic diseases, infections, and cancer. The discussion of blood disorders in Chapter 8 has been expanded to include bacterial toxins, carbon monoxide, leukemias, anemias, and other ailments. Chapter 7 has a new two-page section on GI tract disorders. Other highlights include an updated, more comprehensive discussion of cardiovascular system disorders in Chapter 9, an expanded section on nervous system injuries and disorders in Chapter 13, and a two-page section dealing with causes, symptoms, and prevention of diabetes in Chapter 15. We moved the topic of sexually transmitted diseases to Chapter 16, where it fits as a logical extension of the discussion of reproductive systems. Chapter 18 now focuses on other types of infectious disease and begins with a new section on general principles that govern the infection process. Its discussions of avian influenza, SARS, West Nile virus, and other emerging pathogens all have been updated.

Looking at the biology of the muscular system in Chapter 6, we reorganized the discussion to lead students through this challenging material from a "macro" to "micro" perspective, providing the big picture of how skeletal muscles operate before taking on the details of contraction. A critical look at Chapter 7 led us to create an all-new, integrated section on accessory organs with enhanced, more explanatory illustrations of the liver, pancreas, and gallbladder. Chapter 10, on immunity, has been completely overhauled. Introduced by a vignette on HIV and AIDS, it now includes the text's two-page spread on the lymphatic system and presents basic concepts in the easy-to-understand framework of innate and adaptive defenses. New art includes clearer diagrams of cell-mediated and antibody-mediated immunity. Improved art in other "system" chapters includes diagrams of kidney and nephron structure in Chapter 12 and a new overview illustration of the brain in Chapter 13.

In Chapter 19, reorganization improved the flow of discussions of chromosome structure and of the cell cycle, and enhanced art makes cytokinesis easier than ever to follow. Chapter 22 brings front and center not only the basics of DNA structure and function, but also what researchers are learning as they explore the sequence of the human genome. Chapter 24 presents recent discoveries that shed light on human evolution. In Chapter 25, narrated animations and updated sections on human population growth and global warming provide food for thought about our collective future.

ACKNOWLEDGMENTS

Our dedicated advisors, listed below, continue to shape and re-shape our thinking about the best way to reach students in human biology courses. Through focus group participation, reviews, and class testing, the instructors listed at right helped transform our desire for a maximally effective book into reality. The rethinking and honing of each chapter, the fresh look at the book's organization and emphases, the immense effort to create animations and interactions to help students learn, the custom videos—all are responses to their insights from the classroom.

Starting with Susan Badger, Thomson Brooks/Cole proved why it's one of the world's foremost educational publishers. We doubt that authors get finer support anywhere. Jack Carey, Sean Wakely, Michelle Julet, and Kathie Head, thank you yet again. Kristina Razmara and Suzannah Alexander provided editorial support far beyond the call of duty. Keli Amann and Fiona Chong created great Web components and managed the multimedia production. Both Andy Marinkovich and Mandy Hetrick at Lachina Publishing Services contributed greatly to the complex, sometimes harrowing process of shepherding this edition to completion. Gary Head created functional designs and great graphics for the book. The list could go on and on. Ultimately, though, no listing can convey how lucky we are to work with an amazing mix of talented people, all devoted to creating something extraordinary.

Cecie Starr
Beverly McMillan
July 2006

Reviewers

ALCOCK, JOHN   *Arizona State University*
ALFORD, DONALD K.   *Metropolitan State College of Denver*
ALLISON, VENITA F.   *Southern Methodist University*
ANDERSON, D. ANDY   *Utah State University*
ANDERSON, DAVID R.   *Pennsylvania State University, Fayette*
ARMSTRONG, PETER   *University of California, Davis*
AULEB, LEIGH   *San Francisco State University*
BAATH, KIRAT   *University of Southern Indiana*
BAKKEN, AIMÉE H.   *University of Washington, Seattle*
BARNUM, SUSAN R.   *Miami University*
BAUER, SALLY M.   *Hudson Valley Community College*
BEDECARRAX, EDMUND E.   *City College of San Francisco*
BENNETT, JACK   *Northern Illinois University*
BENNETT, SHELIA K.   *University of Maine, Augusta*
BOHR, DAVID F.   *University of Michigan, Ann Arbor*
BOOTH, CHARLES E.   *Eastern Connecticut State University*
BRAMMER, J. D.   *North Dakota State University*
BROADWATER, SHARON T.   *College of William and Mary*
BROWN, MELVIN K.   *Erie Community College, City Campus*
BRUSLIND, LINDA D.   *Oregon State University*
BUCHANAN, ALFRED B.   *Santa Monica College*
BURKS, DOUGLAS J.   *Wilmington College*
CHRISTENSEN, A. KENT   *University of Michigan*
CHRISTENSEN, ANN   *Pima Community College*
CHOW, VICTOR   *City College of San Francisco*
CONNELL, JOE   *Leeward Community College*
COX, GEORGE W.   *San Diego State University*
COYNE, JERRY   *University of Chicago*
CROSBY, RICHARD M.   *Treasure Valley Community College*
DANKO, LISA   *Mercyhurst College*
DELCOMYN, FRED   *University of Illinois, Urbana-Champaign*
DENNER, MELVIN   *University of Southern Indiana*
DENNISTON, KATHERINE J.   *Towson State University*
DENTON, TOM E.   *Auburn University, Montgomery*
DLUZEN, DEAN   *Northeastern Ohio University—College of Medicine*
EDLIN, GORDON J.   *University of Hawaii*
EMSLEY, MICHAEL   *George Mason University*
ERICKSON, GINA   *Highline Community College*
FAIRBANKS, DANIEL J.   *Brigham Young University*
FALK, RICHARD H.   *University of California, Davis*
FREY, JOHN E.   *Mankato State University*
FROEHLICH, JEFFERY   *University of New Mexico*
FULFORD, DAVID E.   *Edinboro University of Pennsylvania*
GARNER, JAMES G.   *Long Island University, C.W. Post Campus*
GENUTH, SAUL M.   *Case Western Reserve University*
GIANFERRARI, EDMUND A.   *Keene State College of the University of New Hampshire*
GOODMAN, H. MAURICE   *University of Massachusetts— Medical School*
GORDON, SHELDON R.   *Oakland University*
GREGG, KENNETH W.   *Winthrop University*
GREW, JOHN C.   *New Jersey City University*
HAHN, MARTIN   *William Paterson College*
HALL, N. GAIL   *Trinity College*

HARLEY, JOHN P.   *Eastern Kentucky University*
HASSAN, ASLAM S.   *University of Illinois*
HENRY, MICHAEL   *Santa Rosa Junior College*
HERTZ, PAUL E.   *Barnard College*
HILLE, MERRILL B.   *University of Washington*
HOEGERMAN, STANTON F.   *College of William and Mary*
HOHAM, RONALD W.   *Colgate University*
HOSICK, HOWARD L.   *Washington State University*
HUCCABY, PERRY   *Elizabethtown Community College*
HUNT, MADELYN D.   *Lamar University*
HUPP, EUGENE W.   *Texas Woman's University*
JOHNS, MITRICK A.   *Northern Illinois University*
JOHNSON, LEONARD R.   *University of Tennessee—College of Medicine*
JOHNSON, TED   *St. Olaf College*
JOHNSON, VINCENT A.   *St. Cloud State University*
JONES, CAROLYN K.   *Vincennes University*
JOSEPH, JANN   *Grand Valley State University*
KAREIVA, PETER   *University of Washington*
KAYE, GORDON I.   *Albany Medical College*
KEIPER, RONALD   *Valencia Community College*
KENYON, DEAN H.   *San Francisco State University*
KEYES, JACK   *Linfield College, Portland Campus*
KLEIN, KEITH K.   *Mankato State University*
KENNEDY, KENNETH A. R.   *Cornell University*
KIMBALL, JOHN W.
KROHNE, DAVID T.   *Wabash College*
KRUMHARDT, BARBARA   *Des Moines Area Community College, Urban Campus*
KUPCHELLA, CHARLES E.   *Southeast Missouri State University*
KUTCHAI, HOWARD   *University of Virginia*
LAMBERT, DALE   *Tarrant County College*
LAMMERT, JOHN M.   *Gustavus Adolphus College*
LAPEN, ROBERT   *Central Washington University*
LASSITER, WILLIAM E.   *University of North Carolina, Chapel Hill—School of Medicine*
LEVY, MATTHEW N.   *Mt. Sinai Hospital*
LITLLE, ROBERT C.   *Medical College of Georgia*
LUCAS, CRAN   *Louisiana State University, Shreveport*
MACHUNIS-MASUOKA, ELIZABETH A.   *University of Virginia*
MANIA-FARRELL, BARBARA   *Purdue University, Calumet*
MANN, ALAN   *University of Pennsylvania*
MANN, NANCY J.   *Cuesta College*
MARCUS, PHILIP I.   *University of Connecticut*
MARTIN, JOSEPH V.   *Rutgers University, Camden*
MATHIS, JAMES N.   *West Georgia College*
MATTHEWS, PATRICIA   *Grand Valley State University*
MAYS, CHARLES   *DePauw University*
MCCUE, JOHN F.   *St. Cloud State University*
MCMAHON, KAREN A.   *University of Tulsa*
MCNABB, F.M. ANNE   *Virginia Polytechnic Institute and State University*
MILLER, G. TYLER
MITCHELL, JOHN L. A.   *Northern Illinois University*
MITCHELL, ROBERT B.   *Pennsylvania State University*
MOHRMAN, DAVID E.   *University of Minnesota, Duluth*
MOISES, HYLAN   *University of Michigan*
MORK, DAVID   *St. Cloud University*

MORTON, DAVID *Frostburg State University*
MOTE, MICHAEL I. *Temple University*
MOWBRAY, ROD *University of Wisconsin, LaCrosse*
MURPHY, RICHARD A. *University of Virginia—Health Sciences Center*
MYKLES, DONALD L. *Colorado State University*
NORRIS, DAVID O. *University of Colorado, Boulder*
PARSON, WILLIAM *University of Washington*
PETERS, LEWIS *Northern Michigan University*
PIPERBERG, JOEL B. *Millersville University*
PLETH, HAROLD K. *Fullerton Community College*
PORTEOUS-GAFFARD, SHIRLEY *Fresno City College*
POZZI-GALLUZI, G. *Dutchess Community College*
QUADAGNO, DAVID *Florida State University*
REINER, MAREN *University of Richmond*
REZNICK, DAVID *University of California, Riverside*
ROBERTS, JANE C. *Creighton University*
RUBENSTEIN, ELAINE *Skidmore College*
SCALA, ANDREW M. *Dutchess Community College*
SAPP OLSON, SALLY
SEALY, LOIS *Valencia Community College*
SHEPHERD, GORDON M. *Yale University—School of Medicine*
SHERMAN, JOHN W. *Erie Community College—North*
SHERWOOD, LAURALEE *West Virginia University*
SHIPPEE, RICHARD H. *Vincennes University*
SLOBODA, ROGER D. *Dartmouth College*
SMIGEL, BARBARA W. *Community College of Southern Nevada*
SMITH, ROBERT L. *West Virginia University*
SOROCHIN, RON *State University of New York—College of Technology at Alfred*

SPORER, RUTH *Rutgers University-Camden*
STEELE, CRAIG W. *Edinboro University of Pennsylvania*
STEMPLE, JR., FRED E. *Tidewater Community College*
STEUBING, PATRICIA M. *University of Nevada, Las Vegas*
STEWART, GREGORY J. *State University of West Georgia*
STONE, ANALEE G. *Tunxis Community College*
SULLIVAN, ROVERT J. *Marist College*
SUN, ERIC *Macon State College*
TAUCK, DAVID *Santa Clara University*
THIEMAN, WILLIAM *Ventura College*
THOMPSON, ED W. *Winona State University*
TIZARD, IAN *Texas A&M University*
TROTTER, WILLIAM *Des Moines Area Community College—Sciences Center*
TUTTLE, JEREMY B. *University of Virginia—Health Sciences Center*
VALENTINE, JAMES W. *University of California, Berkeley*
VAN DE GRAAFF, KENT M. *Weber State University*
VAN DYKE, PETE *Walla Walla Community College*
VARKEY, ALEXANDER *Liberty University*
WALSH, BRUCE *University of Arizona*
WARNER, MARGARET R. *Purdue University*
WEISBRODT, DAN R. *William Paterson University*
WEISS, MARK L. *Wayne State University*
WHIPP, BRIAN J. *St. George's Hospital Medical School*
WHITTEMORE, SUSAN *Keene State College*
WILLIAMS, ROBERTA B. *University of Nevada, Las Vegas*
WISE, MARY *Northern Virginia Community College*
WOLFE, STEPHEN L. *University of California, Davis (Emeritus)*
YONENAKA, SHANNA *San Francisco State University*

What Kind of World Do We Live In?

Glance at a newspaper or click on your Web browser and you may wonder what kind of world you're living in. Headlines mingle news about various wars or political wrangles with tips for managing your love life or choosing food supplements. On any given day you'll read about how infectious diseases such as "bird flu" pose global threats, or about the devastation caused by a natural catastrophe such as an earthquake. Often there are stories about human activities that have major impacts on nature. We hear more and more about global warming, polar ice caps and glaciers melting, and other regions experiencing record storms, droughts, and heat waves. During the Persian Gulf War, pollution from intentional fires set in oil fields seriously damaged some of the Earth's most spectacular coral reefs,

and we still don't know today if they will ever recover.

But while coping with an environmental disaster or predicting the course of a flu epidemic are challenging, we humans have an ace in the hole. We learned a long time ago that it is possible to study nature, including ourselves, in a systematic way that may help us understand the natural world and our place in it. We can observe carefully, come up with ideas, and find ways to test them. With time we can learn a great deal about factors that affect our health, environmental concerns, and a host of other issues. That's what this book is for—to help give you a fuller understanding of how your body works and where we humans fit in the world around us.

Each chapter in this book will give you a chance to express your opinion on an issue that is challenging us today. When you cast your vote on this book's website you will be able to see how others feel about current concerns related to the environment, health, and ethical dilemmas.

 How Would You Vote? Middle Eastern seas support spectacular coral reef ecosystems. Should the United States provide funding to help preserve the reefs? Cast your vote online at www.thomsonedu.com/ biology/starr/humanbio.

Key Concepts

THE NATURE OF LIFE
All living things share some basic features, including the genetic material DNA and the ability to take in and use energy and materials. A cell is the smallest unit that can be alive.

LIFE'S DIVERSITY
We humans are one of many millions of kinds of living things. Nature is organized from simple to complex, starting with nonliving atoms. Whole organisms are part of this continuum. The entire living world, the biosphere, is the most encompassing level of life's organization.

STUDYING LIFE
Biology is a way of thinking critically about the natural world. Biologists make systematic observations, hypotheses, and predictions. They often test their predictions by way of experiments in nature and in the laboratory. Critical thinking provides all of us with a means to evaluate information objectively.

Links to Earlier Concepts

This book follows nature's levels of organization, from atoms to the biosphere. This first chapter provides a broad view of where we humans fit in the world of life. Later chapters will introduce you to the chemical foundations of life and how our body cells are built and operate. This background paves the way for a survey of how the body's tissues, organs, and organ systems function. You will also learn about genes, the principles that guide inheritance, and basic concepts of evolution and ecology.

You will discover that each chapter in this book builds upon previous ones. Keychain icons mark cross-references to earlier sections where you can review related topics.

1.1 The Characteristics of Life

Picture a group of hikers, carefully making their way up slippery stairs next to a waterfall. Without thinking about it, you know that the hikers are alive and the stairs are not. But could you explain why?

Living and nonliving things share some common features. For instance, both are made up of atoms, which are the smallest units of nature's fundamental substances. On the other hand, wherever we look in the living world we find that all living things share some characteristics that nonliving ones don't have. These basic features of life are:

1. **Living things take in energy and materials**. Like other animals, and many other kinds of organisms, we humans take in energy and materials by consuming food (Figure 1.1). Our bodies use the energy and materials to build and operate their parts in ways that keep us alive.

2. **Living things sense and respond to changes in their environment**. For example, a tulip's petals close up when night falls, and you might put on a sweater or turn up the heat on a chilly afternoon.

3. **Living things reproduce and grow**. Organisms can make more of their own kind, based on instructions in DNA, the genetic material. Only living things have DNA. Guided by the instructions in their DNA, most organisms develop through a series of life stages. For us humans, the basic life stages are infancy, childhood, adolescence, and adulthood.

4. **Living things consist of one or more cells**. A cell is an organized unit that can live and reproduce by itself, using energy, the required raw materials, and instructions in DNA. Cells are the smallest units that can be alive. The energy to power cell activities comes from another special chemical found only in living things, ATP.

5. **Living things maintain homeostasis**. The term **homeostasis** (hoe-me-oh-STAY-sis) means "staying the same." Homeostasis is a state in which the physical and chemical conditions of the environment *inside* the body

are being maintained within life-supporting limits. This internal environment consists of the fluid, including blood, that surrounds our cells. For now, the key point to keep in mind is that maintaining homeostasis is crucial for the survival of every living cell in the body. Later chapters will provide a more complete picture of how body systems help in this task.

> *Living organisms share characteristics that nonliving objects do not have.*
>
> *All living things take in and use energy and materials, and they sense and can respond to changes in their environment.*
>
> *Living things can reproduce and grow, based on instructions in DNA. The cell is the smallest unit that can be alive.*
>
> *Organisms maintain the internal state called homeostasis.*

Figure 1.1 A child taking in energy by eating food. His body will extract energy and raw materials from the food and use them for processes that are required to keep each of the child's cells, and the whole child, alive.

1.2 Our Place in the Natural World

Human beings arose as a distinct group of animals during an evolutionary journey that began billions of years ago.

HUMANS HAVE EVOLVED OVER TIME

In biology, **evolution** means change in the body plan and functioning of organisms through the generations. It is a process that began billions of years ago on the Earth and continues today. In the course of evolution, major groups of life forms have emerged.

Figure 1.2 provides a snapshot of how we fit into the natural world. Humans, apes, and some other closely related animals are primates (PRY-mates). Primates are mammals, and mammals make up one group of "animals with backbones," the vertebrates (VER-tuh-braytes). Of course, we share our planet with millions of other animal species, as well as with plants, fungi, countless bacteria, and other life forms. Biologists classify living things according to their characteristics, which in turn reflect their evolutionary heritage. Notice that Figure 1.2 shows three domains of life. Animals, plants, fungi, and microscopic organisms called protists are assigned to kingdoms in a domain called Eukarya. The other two domains are reserved for bacteria and some other single-celled life forms. Some biologists prefer different schemes. For

Figure 1.3 Bonobos (*left*) are our closest primate relatives. Like us, they walk upright and use tools. Only humans have a capacity for sophisticated language and technology.

example, for many years all living things were simply organized into five kingdoms—animals, plants, fungi, protists, and bacteria. The key point is that despite the basic features all life forms share, evolution has produced a living world of incredible diversity.

HUMANS ARE RELATED TO ALL OTHER ORGANISMS—AND HUMANS ALSO HAVE SOME DISTINCTIVE FEATURES

Because of evolution, humans are related to every other life form and we share characteristics with many of them. For instance, we and all other mammals have body hair, a feature that no other vertebrate has. We share the most characteristics with apes, our closest primate relatives (Figure 1.3). But humans also have some distinctive features that evolved as traits of our primate ancestors were modified. For example, we have great manual dexterity due to the arrangement of muscles and bones in our hands and the wiring of our nervous system to operate them. Even more astonishing is the human brain. Relative to overall body mass it is the largest brain of any animal, and it gives us the capacity for sophisticated language and analysis, for developing advanced technology, and for a remarkably wide variety of social behaviors.

Figure 1.2 *Animated!*
Classifying life. Humans are one of more than a million species in the Animal Kingdom, which is part of the domain Eukarya. Plants, fungi, and some other life forms make up other kingdoms in Eukarya. The domains Bacteria and Archaea contain vast numbers of single-celled organisms.

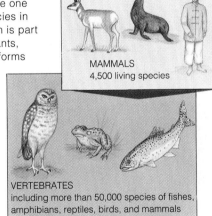

MAMMALS
4,500 living species

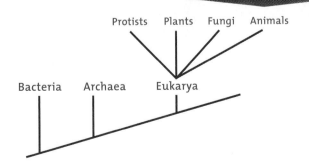

VERTEBRATES
including more than 50,000 species of fishes, amphibians, reptiles, birds, and mammals

Protists Plants Fungi Animals

Bacteria Archaea Eukarya

Like all life forms, human beings arose through evolution—change in the details of the body plan and functions of organisms through the generations.

Features that set humans apart from other complex animals include sophisticated verbal and analytical skills and exceptionally complex social behaviors.

1.3 Life's Organization

The world of life is organized from the simple to the complex. The illustration below gives you an overview of how these levels connect.

LIFE IS ORGANIZED ON MANY LEVELS

When you look closely at the living world, it doesn't take long to realize that nature is organized on many different levels (Figure 1.4). At the most basic level are atoms. Next come molecules, which are combinations of atoms. Atoms and molecules are the nonliving materials from which cells are built. In a multicellular organism such as a human, cells are organized into tissues—muscle, the epithelium of your skin, and so forth. Different kinds of tissues make up organs, and coordinated systems of organs make up whole complex organisms.

We can study the living world on any of its levels. Many courses in human biology focus on organ systems, and a good deal of this textbook explores their structure and how they function.

Nature's organization doesn't end with individuals. Each organism is part of a population, such as the Earth's whole human population. When we cast the net a little farther, populations of different organisms interact in communities, the populations of all species occupying the same area. Communities in turn interact in ecosystems. The most inclusive level of organization is the **biosphere**. This term refers to all parts of Earth's waters, crust, and atmosphere in which organisms live.

ORGANISMS ARE CONNECTED THROUGH THE FLOW OF ENERGY AND CYCLING OF MATERIALS

Organisms must take in energy and materials to keep their life processes going. Where do these essentials come from? Energy flows into the biosphere from the sun. This solar energy is captured by "self-feeding" life forms such as plants, which use a sunlight-powered process called photosynthesis to make fuel for building tissues, such as a corn kernel. Raw materials such as carbon that are needed to build the corn come from air, soil, and water. Thus self-feeding organisms are the living world's basic food producers. Animals, including humans, are the consumers: When we eat plant parts, or feed on animals that have done so, we take in materials and energy to fuel our body functions. You tap directly into stored energy when you eat corn on the cob, and you tap into it indirectly when you eat the meat of a chicken that fed on corn. Organisms such as bacteria and fungi obtain energy and materials when they decompose tissues, breaking down biological molecules to substances that

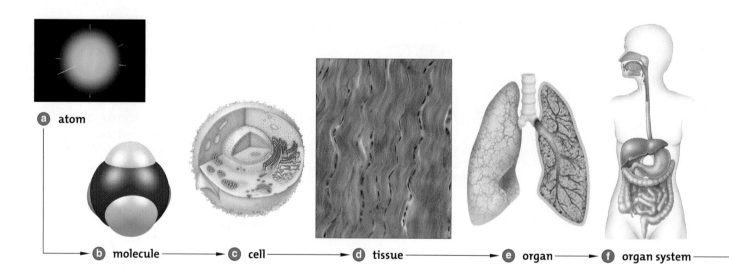

a atom → **b** molecule → **c** cell → **d** tissue → **e** organ → **f** organ system

Figure 1.4 *Animated!* An overview of the levels of organization in nature.

LIFE'S DIVERSITY

can be recycled back to producers. By way of this one-way flow of energy through organisms, and the cycling of materials among them, every part of the living world is linked to every other part. Figure 1.5 summarizes these relationships, which we'll return to in Chapter 25.

Because of the interconnections among organisms, it makes sense to think of ecosystems as webs of life. With this perspective, we can see that the effects of events in one part of the web will eventually ripple through the whole, and may even affect the entire biosphere. For example, we see evidence of large-scale impacts of human activities in phenomena such as global warming, the loss of biodiversity in many parts of the world, acid rain, and a host of other problems.

Nature is organized on several levels, starting with nonliving materials and culminating with the whole living world, the biosphere.

Life's organization is sustained by a flow of energy from the sun and the cycling of raw materials among organisms.

As a result of the interconnections between living things, ecosystems are webs of life. What happens in one part of the web ripples through the whole.

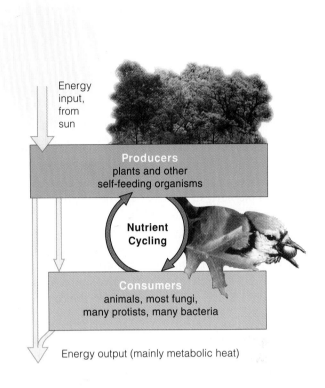

Figure 1.5 *Animated!* The flow of energy and the cycling of materials in the biosphere.

g multicellular organism — h population — i community — j ecosystem — k the biosphere

1.4 Science Is a Way of Learning about the Natural World

How do researchers go about discovering how anything in nature works? Said another way, what exactly do we mean when we talk about "science"?

SCIENCE IS AN APPROACH TO GATHERING KNOWLEDGE

The word "science" can conjure up images of white-coated researchers, test tubes, and sophisticated laboratories. In fact, however, the basic ingredients in "doing science" are simple curiosity and a healthy skepticism. Science blends these ingredients, and the result is a systematic way of gathering knowledge about the natural world. This information-gathering system is sometimes called the **scientific method**, but there is no rigid script for it (Figure 1.6). The following practices are common in scientific research. It's likely that some of them are commonsense things you yourself do.

1. **Observe some aspect of the natural world, then ask a question about it or identify a problem to explore**. For example, in the late 1990s, a fat substitute called Olestra® was approved for use in foods. Concocted from vegetable oil and sugar, Olestra is indigestible and has been touted as a dieter's dream. When potato chips made with Olestra came on the market, however, some consumers reported intestinal ailments, including gas, diarrhea, and cramps. Curious researchers at Johns Hopkins University began to wonder: Was Olestra causing the problems?

2. **Develop a hypothesis**. A hypothesis is a tentative explanation for an observation or how some natural process works. In the case of a scientific hypothesis, there must be some objective way of testing it, such as through experiments. The Johns Hopkins scientists hypothesized that Olestra can indeed cause cramps and they had an idea for an experiment to test this explanation.

3. **Make a prediction**. As a first step in testing their hypothesis, the scientists made a prediction: People who eat food containing Olestra are more likely to have intestinal side effects than people who do not. As in this example, a prediction states what you should observe about the question or problem if the hypothesis is valid.

4. **Test the prediction**. To see if their prediction was accurate, the researchers invited almost 1,100 people aged 13 to 38 to watch a movie in a Chicago theater. They were divided into two roughly equal groups and given unmarked bags of potato chips. One group received chips made with Olestra while the other group received regular chips. When the scientists later interviewed the participants, they found that about 15.8 percent of those

Figure 1.6 In the laboratory or in the field, scientists can pursue their research in different ways. One researcher may emphasize observing and describing phenomena. Another may focus on using experiments to test hypotheses.

in the Olestra group complained of intestinal problems—but so did 17.6 percent in the "regular" group. As a result, this experiment provided no evidence that eating Olestra-laced potato chips causes intestinal ills, at least after a one-time use. The experiment wasn't a "failure," however. A properly designed test is supposed to reveal flaws. If the findings don't support the prediction, then some factor that influenced the test may have been overlooked, or the hypothesis may simply have been wrong.

5. **Repeat the tests or develop new ones**—the more the better. Hypotheses that are supported by the results of repeated testing are more likely to be correct.

6. **Analyze and report the test results and conclusions**. Scientists typically publish their findings in scientific journals, with a detailed description of their methods so that other researchers can try the same test and see if they obtain the same result.

EXPERIMENTS ARE MAJOR SCIENTIFIC TOOLS

Experimenting is a time-honored way to test a prediction that flows from a hypothesis. An **experiment** is a test carried out under controlled conditions that a researcher can manipulate. Figure 1.7 shows the typical steps followed, using the Olestra study as an example.

To get meaningful test results, experimenters use safeguards. They begin by reviewing information that may bear on their project. The makers of Olestra had conducted tests on human subjects before their product was approved, and the Johns Hopkins study considered these reports. Then the researchers designed a **controlled experiment**, one that would test only a single prediction of a hypothesis at a time. In this case, it was the prediction that people who consume Olestra have a greater chance of developing intestinal side effects.

Almost any aspect of the natural world is the result of interacting variables. As the term suggests, a **variable** is a factor that can change with time or in different circumstances. Researchers design experiments to test one variable at a time. They also set up a **control group** to which one or more experimental groups can be compared. The control group was identical to the experimental one except for the variable being studied—chips containing Olestra. Identifying possible variables, and eliminating unwanted ones, is extremely important if an experiment is to produce reliable results. For instance, if some people in the Olestra study had had a prior history of unrelated intestinal difficulties, they could have skewed the study results. Likewise, if any of the participants included people who were already eating foods made with Olestra, it would have been impossible for the experimenters to determine if any reported side effects were due not to the single bag of chips but to long-term use.

Scientists usually can't observe all the individuals in a group they want to study. In studies of a food additive such as Olestra it certainly would be difficult to include every possible consumer. If the sample is too small, however, differences among research subjects might distort the findings. To avoid this problem, researchers use a sample group that is large enough to be representative of the whole. That is why the Olestra study included such a large number of participants.

SCIENCE IS AN ONGOING ENTERPRISE

Did the Johns Hopkins study give Olestra a green light as a trouble-free food additive? No. One reason is that "doing science" requires a researcher to draw logical conclusions about any findings. That is, the conclusion cannot be at odds with the evidence used to support it. Based on the results of their Olestra experiment, the Johns Hopkins scientists could not conclude that the promising "fake fat" did cause intestinal problems. On the other hand, their limited, one-time experiment also could not give Olestra a clean bill of health. In fact, in

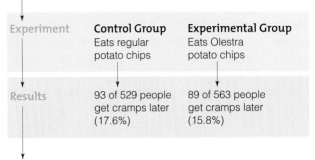

Hypothesis
Olestra® causes intestinal cramps.

↓

Prediction
People who eat potato chips made with Olestra will be more likely to get intestinal cramps than those who eat potato chips made without Olestra.

↓

Experiment	**Control Group** Eats regular potato chips	**Experimental Group** Eats Olestra potato chips
Results	93 of 529 people get cramps later (17.6%)	89 of 563 people get cramps later (15.8%)

↓

Conclusion
Percentages are about equal. People who eat potato chips made with Olestra are just as likely to get intestinal cramps as those who eat potato chips made without Olestra. These results do not support the hypothesis.

Figure 1.7 A typical sequence of steps followed in a scientific experiment.

the years since Olestra was first developed, the United States Food and Drug Administration (FDA) has received more than 20,000 consumer complaints alleging problems, and Olestra has been reformulated to reduce certain side effects. Today a variety of processed foods sold in the United States are made with Olestra, but the jury may still be out on its potential effects in some people, and some advocates say that more research is needed. For this reason regulators in Canada and Britain still have not approved it for use in human food.

Not all scientists perform experiments. In some cases careful observations, often made over a period of years, are the basis for predictions about natural events.

A scientific approach to studying nature is based on asking questions, formulating hypotheses, making predictions, devising tests, and then objectively reporting the results.

Controlled experiments are one way to test scientific ideas. This kind of experiment explores a single variable and uses a control group as a standard to which experimental results can be compared.

1.5 Science in Action: Cancer, Broccoli, and Mighty Mice

Scientific advances that benefit human health often come about when researchers ask creative questions and then perform experiments to get at the answers. In this section we look at a real-life example of the scientific method in action, step by step.

Scientists in the United States and Japan cooperated on the research, exploring this question: Does a substance found in plants of the cabbage family, such as broccoli, somehow boost the body's defenses against cancer? And if so, how? The teams worked with mice, which are easy to maintain in the laboratory. Mice also are mammals, so their cells function much like human cells, increasing the likelihood that experimental findings will be relevant to human beings.

We already know that some plants contain chemicals that can be useful against human cancers. The yew tree, for example, was the original source of the drug taxol, which combats ovarian and uterine tumors. Broccoli and its cabbage family relatives such as cauliflower have long been thought to contain substances that help protect against the changes in cells that can lead to cancer (Figure 1.8). One of those substances is a compound called sulforaphane, so that is where the research teams started.

The researchers' overall hypothesis was that a diet rich in sulforaphane would help protect against cancer. To test that idea carefully, however, would require a series of experiments. In their first study, following good scientific methodology, they established a control group of mice that would eat a normal diet while a test group would be fed food that contained large amounts of sulforaphane. If nothing about the test group changed with this diet, the experiments might have ended there. But while no changes were detected in the control group (the expected outcome), the test animals' cells began making abnormally large quantities of a protein that short-circuits the development of cancer. And when both groups of mice were then exposed to a *carcinogen*—a substance that can cause cancer—the test mice developed fewer cancers than the controls did.

Armed with evidence that sulforaphane boosts the body's natural cancer defenses, the teams asked another question: What would happen if mice that couldn't make the natural cancer-inhibiting protein were exposed to carcinogens? Their hypothesis was that such abnormal mice, which have a genetic mutation that prevents their cells from making the protein, would be especially prone to develop tumors. For their experimental tests the teams again used two groups of mice, a test group and a control. After thirty weeks, the stomachs of the mutant mice were full of tumors—a whopping fourteen,

Figure 1.8 Experiments with mice have supported a hypothesis that vegetables in the cabbage family contain substances that are protective against cancer when they work in concert with natural defenses in body cells.

on average. The normal mice also developed tumors, but fewer of them—an average of ten—and their tumors did not grow as large as those of the test group.

Now the researchers repeated the experiment, but altered a variable so they could test a new hypothesis: that sulforaphane provides protection against cancer even when body cells lack natural defenses. This time there were two test groups, normal and mutant mice that would both receive the treatments, and control groups of each type that would be exposed to the carcinogen but not the drug treatment. (The drug was one that is chemically similar to sulforaphane.) After another thirty-week trial, the results were thought-provoking. The normal mice had developed only half as many tumors as before, so the sulforaphane-like drug seemed to help them fend off cancer. But all the mutant mice developed just as many tumors as before. Based on these results, the scientists had to conclude that a diet rich in sulforaphane is not by itself an effective anticancer strategy. The body's cells must also be able to mount their own defenses. This finding also reinforces how important it is to try to avoid carcinogens that may be in our environment. This is a topic you will read more about in Chapter 23.

Every day, scientists develop hypotheses and make predictions that are testable, often through a process of experimentation.

1.6 Science in Perspective

Several decades ago our world began to be dominated by science, and biology has been a big part of that change. Day by day, knowing more about the nature and limits of science becomes crucial for "biology consumers."

Antibiotics. Insights into genetic disease, cancer, and global problems such as the thinning ozone layer and water pollution (Figure 1.9). Advances like these—not to mention technologies such as genetic engineering and the Internet—have changed our lives. So what is the overall role of science in human affairs? We know that the practice of science can yield powerful ideas, like the theory of evolution, that explain key aspects of life. At the same time, we also know that science is only one part of human experience.

A SCIENTIFIC THEORY EXPLAINS A LARGE NUMBER OF OBSERVATIONS

You've probably said, "I've got a theory about that!" Whatever "that" is, you are really saying that you have an untested idea about something. In science, a **theory** is exactly the opposite: It is an explanation of a broad range of related natural events and observations that is based on repeated, careful testing of hypotheses.

A hypothesis usually becomes accepted as a theory only after years of testing by many scientists. Then, if the hypothesis has not been disproved, scientists may feel confident about using it to explain more data or observations. The theory of evolution by natural selection—a topic we will look at in Chapter 24—is a prime example of a "theory" that is supported by tens of thousands of scientific observations.

Science demands critical thinking, so a theory can be modified, and even rejected, if results of new scientific tests call it into question. It's the same with other scientific ideas. Today, for instance, sophisticated technologies are giving us a new perspective on subjects such as how our immune system operates to defend the body against disease threats. Some "facts" in this textbook one day will likely be revised as we learn more about various processes. This willingness to reconsider ideas as new information comes to light is a major strength of science.

SCIENCE HAS LIMITS

Science requires an objective mind-set, and this means that scientists can only do certain kinds of studies. No experiment can explain the "meaning of life," for example, or why each of us dies at a certain moment. Those kinds of questions have subjective answers, shaped by our experiences and beliefs. Every culture and society has its own standards of morality and esthetics,

Figure 1.9 Science in our service: (**a**) Microbiologist Rita Colwell working to improve drinking water in Bangladesh. (**b**) Ozone thinning above Antarctica in 1996.

and there are hundreds or thousands of different sets of religious beliefs. All guide their members in deciding what is important and good and what is not. By contrast, the external world, rather than internal conviction, is the only testing ground for scientific views.

Because science does not involve value judgments, it sometimes has been or can be used in controversial pursuits. The discovery of atomic power in the early twentieth century, and its continuing use today, is one example; some people also are worried about issues such as the use of animals in scientific research and the consequences of genetic modification of food plants. The debate over the genetic modification of human beings grows stronger by the day. Meanwhile, whole ecosystems are being altered by technologies that allow millions of a forest's trees to be cut in a single year, and hundreds of millions of fishes to be taken from the sea. These are matters we can't leave to the scientific community alone to guide the wise use of scientific knowledge. That responsibility also belongs to us.

A scientific theory is a testable explanation about the cause or causes of a broad range of related natural phenomena. It remains open to tests, revision, and even rejection if new evidence comes to light.

Science has limits. Scientific beliefs can only be tested in nature, and responsibility for the wise use of scientific information must be shared by all.

1.7 Critical Thinking in Science and Life

Having a solid set of critical thinking skills will help you make the most of your study of human biology. As you'll now see, those skills can also help everyone make well-reasoned decisions about health issues, environmental controversies, and other concerns.

Have you ever tried a new or "improved" product and been disappointed when it didn't work as expected? Everyone learns, sometimes the hard way, how useful it can be to cast a skeptical eye on advertising claims or get an unbiased evaluation of, say, a used car you are considering buying. This sort of objective evaluation of information is **critical thinking**.

Scientists use critical thinking in their work, and when they review findings reported by others. One reason this approach is so important is that even when a scientist tries to be objective, there is always a chance that pride or bias will creep in. Critical thinking is a  smart practice in everyday life, too, because many decisions we face involve scientific information. Will an herbal food supplement really boost your immune system? Is it safe to eat irradiated food? The critical thinking skills listed in Figure 1.10 aren't complicated, and they can be helpful tools for assessing a wide range of issues.

CONSIDER THE SOURCE

An easy way to begin evaluating a piece of information is to look carefully at where it is coming from and how it is presented. Two simple strategies for sifting the factual wheat from the unreliable or biased chaff are the following:

LET CREDIBLE SCIENTIFIC EVIDENCE, NOT OPINIONS OR HEARSAY, DO THE CONVINCING For instance, if you are concerned about reports that heavy use of a cell phone might cause brain cancer, material published on the website of the American Cancer Society is much more likely to be reliable than something cousin Fred heard a friend say at work. The friend's information may be correct, but neither you nor your cousin can know for sure without investigating further.

QUESTION CREDENTIALS AND MOTIVES For example, if an advertisement is printed in the format of a news story, or a product is touted on TV by someone being paid to sing its praises, your critical thinking antennae should go up. Is the promoter merely trying to sell a product with the help of "scientific" window dressing? Can any facts presented be checked out? Responsible

A Critical Thinking Checklist
✓ **Do** gather information or evidence from reliable sources.
✓ **Don't** rely on hearsay.
✓ **Do** look for facts that can be checked independently and for signs of obvious bias (such as paid testimonials).
✓ **Don't** confuse *cause* with *correlation*.
✓ **Do** separate *facts* from *opinions*.

Figure 1.10 A critical thinking checklist.

scientists try to be cautious and accurate in discussing their findings and are willing to supply the evidence to back up their statements.

EVALUATE THE CONTENT

Even if information seems authoritative and unbiased, critical thinking is shaped by two other considerations. One is awareness of the difference between the *cause* of something and factors that are only *related* to an event or phenomenon. Suppose, for example, that a television program presents statistics correlating shark attacks on surfers in Florida with the cycles of tides during the summer months. Does this information, which in fact is gathered by scientists, mean that the cycles *cause* shark attacks? No, even though the two may indeed be *related* because sharks may come closer to shore under certain tide conditions. But pinning down the cause of shark attacks will require a great deal more understanding of shark behavior and other factors.

It also is important to keep in mind the difference between facts and opinions or speculation. A **fact** is a piece of information that can be verified, such as the price of a loaf of bread. An *opinion*—whether the bread tastes good—can't be verified because it involves a subjective judgment. Likewise, a marketer's prediction that many consumers will favor a new brand of bread is speculation, at least until there are statistics to back up the hype.

Thinking critically allows us to make decisions based on objective information. It is important to evaluate both the source and the content of information and to understand the difference between facts and opinions and between cause and correlation.

1.8 Are Herbal Supplements Safe?

Each year Americans spend more than $4 billion on herbal food supplements to ward off ailments ranging from the common cold and forgetfulness to depression, stress, and high cholesterol. Some of the biggest and most controversial sellers have been products containing ephedrine (Figure 1.11), derived from the plant ephedra. The FDA banned ephedra products in 2004 after 155 people died from strokes or heart attack while using the substance as a weight-loss "metabolism booster." Ephedra marketers sued, denying any link between their product and the deaths, and in 2005 the federal ban was overturned. More studies are under way.

Controversy has swirled around several other herbal supplements as well. In 2000 a study of echinacea's effectiveness in warding off colds found "no significant effect on either the occurrence of infection or the severity of illness." In 2002 research on St. John's wort, an herb touted for soothing depression, no biological effect was detected either.

With so much at stake, it's not surprising that the debate over herbal supplements is fierce. As the battle goes on, alert consumers are likely to be the ultimate winners. Recognizing that millions of people are buying and using herbal supplements, the National Institutes of Health has stepped up its support for scientifically rigorous testing of various herbal products. The American Medical Association recommends that anyone using herbal supplements check with a physician about possible harmful side effects. Consumers can also get reliable updates from the National Institutes of Health website at www.nih.gov.

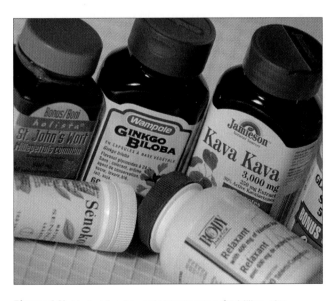

Figure 1.11 Herbal food supplements are a $4-billion-plus industry in the United States.

Summary

Section 1.1   Humans have the following characteristics found in all forms of life (Table 1.1):

a. They take in and use energy and materials from the environment.

b. They sense and respond to specific changes in the environment.

c. They can grow, develop, and reproduce, based on instructions contained in their DNA.

d. They are composed of cells. A cell is the smallest unit that can show the characteristics of life.

e. In homeostasis, internal physical and chemical conditions are maintained within limits that support the ongoing functioning of an organism's cells.

Section 1.2   Earth's life forms all have come about through a process of evolution. The evolutionary heritage of each type of organism is reflected in how biologists classify it. Moving from the most general category to the most specific, humans are classified as animals, vertebrates, mammals, and, finally, primates.

The defining features of humans include a large and well-developed brain that correlates with great manual dexterity and sophisticated skills for language and mental analysis. Humans also possess an advanced capacity for technological innovation and the ability to engage in a complex set of social behaviors.

Section 1.3   The living world is highly organized. Atoms, molecules, cells, tissues, organs, and organ systems make up whole, complex organisms. Each organism is a member of a population, populations live together in communities, and communities form ecosystems. The biosphere is the most inclusive level of biological organization. The organization of life is sustained by a continual flow of energy and cycling of raw materials.

 Explore levels of biological organization.

Table 1.1   Summary of Life's Characteristics
1. Living things take in and use energy and materials.
2. Living things sense and respond to changes in their surroundings.
3. Living things reproduce and grow based on information in DNA.
4. Living things consist of one or more cells.
5. Living things maintain the internal steady state called homeostasis.

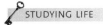 STUDYING LIFE

Section 1.4 Science is an approach to gathering knowledge. There are many versions of the scientific method. The following elements are important in all of them (Table 1.2):

a. Hypothesis: a possible explanation of a specific phenomenon in nature. A hypothesis is scientific only if it can be tested in ways that might disprove it.

b. Prediction: a claim about what an observer can expect to see in nature if a theory or hypothesis is correct.

c. Test: an effort to gather actual observations that support or contradict the predicted findings.

d. Conclusion: a statement about whether a hypothesis should be accepted, rejected, or modified, based on tests of the predictions it generated.

Predictions that flow from hypotheses can be tested by repeated observations or by experiments in nature or in the laboratory.

Experiments are tests that allow researchers to control and adjust the conditions under which they make observations.

Researchers design experiments to test one variable at a time. A variable is an aspect of an object or event that may differ over time and between subjects. Experimenters directly manipulate the variable they are studying in order to test their prediction.

A controlled experiment will have a control group that provides a baseline against which one or more experimental groups are compared. Ideally, it is the same as each experimental group except in the one variable being investigated.

A reputable scientist must draw logical conclusions about any findings. That is, conclusions cannot be at odds with the evidence used to support them.

Section 1.6 Systematic observations, hypotheses, predictions, and tests are the foundation of scientific theories. A theory is a thoroughly tested explanation of a broad range of related phenomena.

Table 1.2	Scientific Method Review
Hypothesis	Possible explanation of a natural event or observation
Prediction	Proposal or claim of what testing will show if a hypothesis is correct
Experimental test	Controlled procedure to gather observations that can be compared to prediction
Control group	Standard to compare test group against
Variable	Aspect of an object or event that may differ with time or between subjects
Conclusion	Statement that evaluates a hypothesis based on test results

Section 1.7 Critical thinking allows us to evaluate information objectively. Critical thinking skills include scrutinizing information sources for bias, seeking reliable opinions, and separating the causes of events from factors that may only be associated with them.

Review Questions

1. For this and all other chapters, make a list of the boldface terms in the text. Write a definition next to each, and then check it against the one in the text.

2. As a human, you are a living organism. List all the characteristics of life that you exhibit.

3. Why is the concept of homeostasis meaningful in the study of human biology?

4. What is meant by biological evolution?

5. Study Figure 1.4. Then, on your own, summarize what is meant by biological organization.

6. Explain what we mean by "the one-way flow of energy and the cycling of materials" through the biosphere.

7. Why does it make sense to think of ecosystems as webs of life?

8. Define and distinguish between:
   a. a hypothesis and a scientific theory
   b. an observational test and an experimental test
   c. an experimental group and a control group

Self-Quiz   *Answers in Appendix V*

1. Instructions in _____ govern how organisms are built and function.

2. A _____ is the smallest unit that can live and reproduce by itself using energy, raw materials, and DNA instructions.

3. _____ is a state in which an organism's internal environment is being maintained within a tolerable range.

4. Humans are _____ (animals with backbones); like other primates, they also are _____.

5. Starting with cells, nature is organized on at least _____ levels.

6. A scientific approach to explaining some aspect of the natural world includes all of the following except _____.

   a. a hypothesis   c. faith-based views
   b. testing   d. systematic observations

7. A controlled experiment should have all the features listed below except _____.
   a. a control group   c. a variable
   b. a test subject   d. several testable predictions

8. A related set of hypotheses that collectively explain some aspect of the natural world makes up a scientific _____.
   a. prediction   d. authority
   b. test   e. observation
   c. theory

9. The diagram below depicts the concept of _____.
   a. evolution
   b. reproduction
   c. levels of organization
   d. energy transfers in the living world

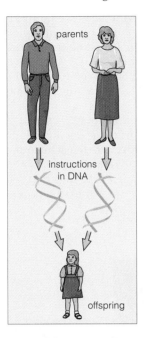

parents

instructions
in DNA

offspring

Critical Thinking

1. The following diagram shows ways that the same materials—here, a set of tiles—can be put together in different ways. How does this example relate to the role of DNA as the universal genetic material in organisms?

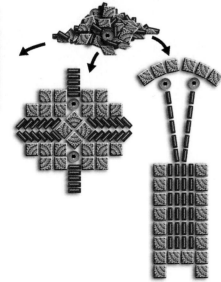

2. Court witnesses are asked "to tell the truth, the whole truth, and nothing but the truth." Research shows, however, that eyewitness accounts of crimes often are unreliable because even the most conscientious witnesses misremember details of what they observed. Can you think of other factors that might affect the "truth" a court witness presents?

3. Design a test (or series of tests) to support or refute this hypothesis: A diet that is high in salt is associated with hypertension (high blood pressure), but hypertension is more common in people with a family history of the condition.

4. In popular magazine articles on health-related topics, the authors often recommend a specific diet or dietary supplement. What kinds of evidence should the articles describe so you can decide whether you should accept their recommendations?

5. A scientific theory about some aspect of nature rests upon inductive reasoning. The assumption is that, because an outcome of some event has been observed to happen quite regularly, it will happen again. We can't know this for certain because there is no way to account for all possible variables that may affect the outcome. To illustrate this point, Garvin McCain and Erwin Segal offer a parable:

> Once there was a highly intelligent turkey. It lived in a pen attended by a kind, thoughtful master, and had nothing to do but reflect on the world's wonders and regularities. Morning always began with the sky getting light, followed by the clop, clop, clop of its master's friendly footsteps, then by the appearance of delicious food. Other things varied—sometimes the morning was warm and sometimes cold—but food always followed footsteps. The sequence of events was so predictable it became the basis of the turkey's theory about the goodness of the world. One morning, after more than 100 confirmations of the goodness theory, the turkey listened for the clop, clop, clop, heard it, and had its head chopped off.

Scientists understand that all thoroughly tested theories about nature have a high probability of being correct. They realize, however, that any theory is subject to being modified if contradictory evidence becomes available. The absence of absolute certainty has led some people to conclude that "facts are irrelevant—facts change." If that is so, should we just stop doing scientific research? Why or why not?

6. Some years ago Dr. Randolph Byrd and his colleagues started a study of 393 patients admitted to the San Francisco General Hospital Coronary Care Unit. In the experiment, born-again Christian volunteers were asked to pray daily for a patient's rapid recovery and for prevention of complications and death.

None of the patients knew if he or she was being prayed for. None of the volunteers or patients knew each other. Byrd categorized how each patient fared as "good," "intermediate," or "bad." He concluded that patients who had been prayed for fared a little better than those who had not. His was the first experiment that had documented statistically significant results that seemed to support the prediction that prayer might have beneficial effects for seriously ill patients.

His published results engendered a storm of criticism, mostly from scientists who cited bias in the experimental design. For instance, Byrd had categorized the patients after the experiment was over, instead of as they were undergoing treatment, so he already knew which ones had improved, stayed about the same, or gotten worse. Think about how experimenters' bias might play a role in how they interpret data. Why do you suppose the experiment generated a heated response from many in the scientific community? Can you think of at least one other variable that might have affected the outcome of each patient's illness?

Explore on Your Own

As you read in Section 1.4, having a sample of test subjects or observations that is too small can skew the results of experiments. This phenomenon is called *sampling error*. To demonstrate this for yourself, all you need is a partner, a blindfold, and a jar containing beans of different colors—jelly beans will do just fine (Figure 1.12). Have your partner stay outside the room while you combine 120 beans of one color with 280 beans of the other color in a bowl. This will give you a ratio of 30 to 70 percent. With the bowl hidden, blindfold your partner; then ask him or her to pick one bean from the mix. Hide the bowl again and instruct your friend to remove the blindfold and tell you what color beans are in the bowl, based on this limited sample. The logical answer is that all the beans are the color of the one selected.

Next repeat the trial, but this time ask your partner to select 50 beans from the bowl. Does this larger sample more closely approximate the actual ratio of beans in the bowl? You can do several more trials if you have time. Do your results support the idea that a larger sample size more closely reflects the actual color ratio of beans?

a Natalie, blindfolded, randomly plucks a jelly bean from a jar of 120 green and 280 black jelly beans, a ratio of 30 to 70 percent.

b The jar is hidden before she removes her blindfold. She observes a single green jelly bean in her hand and assumes the jar holds only green jelly beans.

c Still blindfolded, Natalie randomly picks 50 jelly beans from the jar and ends up with 10 green and 40 black ones.

d The larger sample leads her to assume one-fifth of the jar's jelly beans are green and four-fifths are black (a ratio of 20 to 80). Her larger sample more closely approximates the jar's green-to-black ratio. The more times Natalie repeats the sampling, the greater the chance she will come close to knowing the actual ratio.

Figure 1.12 An easy way to demonstrate sampling error.

2 MOLECULES OF LIFE

It's Elemental

Like everything else on Earth, your body consists of chemicals, some of them solids, others liquid, still others gases. Each of these chemicals consists of one or more elements. An **element** is a fundamental form of matter. No ordinary process can break it down to other substances. There are ninety-two natural elements on Earth, and researchers have created other, artificial ones.

Organisms consist mostly of just four elements: oxygen, carbon, hydrogen, and nitrogen. The human body also contains some calcium, phosphorus, potassium, sulfur, sodium, and chlorine, plus trace elements. A **trace element** is one that represents less than 0.01 percent of body weight. Trace elements are vital; for example, your

red blood cells can't carry oxygen without the trace element iron.

Atoms of elements can combine into molecules—the first step in biological organization. Molecules in turn can combine to form larger structures, as described shortly. The body's chemical makeup is finely tuned. For example, many trace elements found in our tissues—such as arsenic, selenium, and fluorine—are toxic in amounts larger than normal. Fluoride, which is added to many toothpastes and drinking water supplies, is a form of fluorine. Although it helps prevent tooth decay, too much of it can damage teeth and bones, cause birth defects, or even death.

As you will read often in this book, the body's ability to manage changes that disturb its chemistry is vital to maintaining the stable internal state of homeostasis.

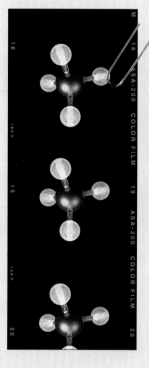

Human Body	
Oxygen	65%
Carbon	18
Hydrogen	10
Nitrogen	3
Calcium	2
Phosphorus	1.1
Potassium	0.35
Sulfur	0.25
Sodium	0.15
Chlorine	0.15
Magnesium	0.05
Iron	0.004

Earth's Crust	
Oxygen	46.6%
Silicon	27.7
Aluminum	8.1
Iron	5.0
Calcium	3.6
Sodium	2.8
Potassium	2.1
Magnesium	1.5

 How Would You Vote? Many communities add fluoride to drinking water supplies. Do you want it in yours? Cast your vote online at www.thomsonedu.com/biology/starr/ humanbio.

 ## Key Concepts

ATOMS BOND
Bonds between atoms form molecules. Both the number of an atom's electrons and how they are arranged determine whether it can bond with another atom.

WATER AND BODY FLUIDS
Life depends on key properties of water. Substances that are dissolved in the water of body fluids have important effects on body functions.

BIOLOGICAL MOLECULES
Cells build complex carbohydrates and lipids, proteins, and nucleic acids from simple subunits that contain the element carbon. All of these large molecules have a backbone of carbon atoms. Groups of atoms bonded to the backbone help determine a molecule's properties.

 ## Links to Earlier Concepts

Atoms are the nonliving raw materials from which living cells and organisms are built. The processes that harness atoms and assemble them into the many different parts of cells all are guided by the DNA we inherit from our parents—and that was inherited by all the generations that came before us. In addition to being constructed according to a DNA blueprint, each of our living cells is surrounded by watery fluid. That is why this chapter gives you some background about the properties of water, which are essential to the body's ability to maintain homeostasis in body fluids (1.1–1.3).

2.1 Atoms, the Starting Point

LINK TO
SECTION
1.3

A million atoms could fit on the period at the end of this sentence. The parts of atoms determine how the molecules of life are put together.

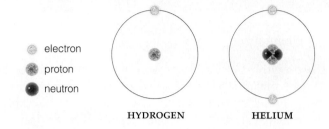

Figure 2.1 *Animated!* A simple model of atoms of hydrogen and helium. These sketches are simplified; at the scale used here, the nuclei of these atoms would be invisible specks.

ATOMS ARE COMPOSED OF SMALLER PARTICLES

An **atom** is the smallest unit that retains the properties of a given element. In spite of their tiny size, all atoms are composed of more than a hundred kinds of subatomic particles. The ones we are concerned with in this book are **protons**, **electrons**, and **neutrons**, illustrated in Figure 2.1.

All atoms have one or more protons, which carry a positive charge, marked by a plus sign (p^+). Except for hydrogen, atoms also have one or more neutrons, which have no charge. Neutrons and protons make up the atom's core, the atomic nucleus. Electrons move around the nucleus, occupying most of the atom's volume. They have a negative charge, which we write as e^-. An atom usually has an equal number of electrons and protons.

Each element is assigned its own "atomic number," which is the number of protons in its atoms. As Table 2.1 indicates, the atomic number for the hydrogen atom, which has one proton, is 1. For the carbon atom, which has six protons, the atomic number is 6.

Each element also has a "mass number"—the sum of the protons and neutrons in the nucleus of its atoms. For a carbon atom, with six protons and six neutrons, this number is 12.

Knowing atomic numbers and mass numbers allows a chemist to gauge whether and how substances will interact. This kind of information also helps us predict how substances will behave in cells, in the body, and in the environment.

ISOTOPES ARE VARYING FORMS OF ATOMS

All atoms of a given element have the same number of protons and electrons, but they may *not* have the same number of neutrons. When an atom of an element has more or fewer neutrons than the most common number, it is called an **isotope** (EYE-so-tope). For instance, while a "standard" carbon atom will have six protons and six neutrons, the isotope called carbon 14 has six protons and *eight* neutrons. These two forms of carbon atoms also can be written as ^{12}C and ^{14}C. The prefix *iso-* means

Table 2.1	Atomic Number and Mass Number of Elements Common in Living Things		
Element	Symbol	Atomic Number	Most Common Mass Number
Hydrogen	H	1	1
Carbon	C	6	12
Nitrogen	N	7	14
Oxygen	O	8	16
Sodium	Na	11	23
Magnesium	Mg	12	24
Phosphorus	P	15	31
Sulfur	S	16	32
Chlorine	Cl	17	35
Potassium	K	19	39
Calcium	Ca	20	40
Iron	Fe	26	56
Iodine	I	53	127

same, and all isotopes of an element interact with other atoms in the same way. Most elements have at least two isotopes. Cells can use any isotope of an element for their metabolic activities, because the isotopes behave the same as the standard form of the atom in chemical reactions.

Have you heard of radioactive isotopes? A French scientist discovered them in 1896, after he had set a chunk of rock on top of an unexposed photographic plate in a desk drawer. The rock contained some isotopes of uranium, which emit energy. This unexpected chemical behavior is what we today call radioactivity. A few days after the Frenchman's plate was exposed to the uranium emissions, he was astonished to see that a faint image of the rock appeared on it.

The nucleus of a **radioisotope** is unstable and stabilizes itself by emitting energy and particles (other than protons, electrons, and neutrons). This process, called radioactive decay, takes place spontaneously and it transforms a radioisotope into an atom of a different element at a known rate. For instance, over a predictable time span, carbon 14 becomes nitrogen 14. Scientists can use radioactive decay rates to determine the age of very old substances. This chapter's *Science Comes to Life* describes some ways radioisotopes are used in medicine.

An atom is the smallest unit of matter that has the properties of a particular element. Atoms have one or more positively charged protons, negatively charged electrons, and (except for hydrogen) neutrons.

Most elements have two or more isotopes—atoms that differ in the number of neutrons.

Figure 2.1 legend:
- electron
- proton
- neutron

HYDROGEN   **HELIUM**

ATOMS BOND

2.2 Medical Uses for Radioisotopes

To someone who is ill, radioisotopes can be of keen interest. For example, they can allow a physician to diagnose disease with exquisite precision—and without requiring the patient to undergo exploratory surgery. Radioisotopes also are part of the treatment of certain cancers. Regardless of the use, radioisotopes always are handled with great care. For safety's sake, clinicians use only radioisotopes with extremely short half-lives. **Half-life** is the time it takes for half of a quantity of a radioisotope to decay into a different, more stable isotope.

TRACKING TRACERS   Various devices can detect radioisotope emissions. Thus, radioisotopes can be employed in tracers. A **tracer** is a substance with a radioisotope attached to it, rather like a shipping label, that a physician can administer to a patient. The tracking device then follows the tracer's movement through a pathway or pinpoints its destination.

SAVING LIVES   In nuclear medicine, radioisotopes are used under carefully controlled conditions to diagnose and treat diseases. Serious illnesses involving the thyroid are a case in point. The thyroid is the only gland in the human body that takes up iodine. After a tiny amount of a radioactive isotope of iodine, iodine 123, is injected into a patient's bloodstream, the thyroid can be scanned, producing the kinds of images you see in Figure 2.2.

Treatments for some cancers rely on radiation from radioisotopes to destroy or impair the activity of cells that are not functioning properly. For instance, such

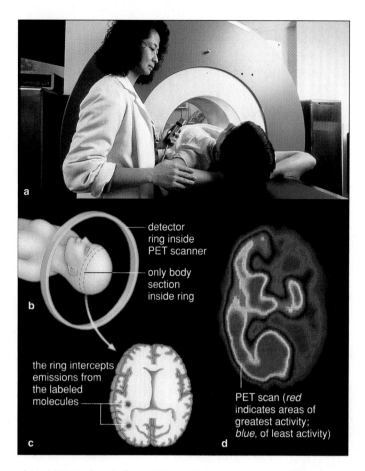

detector ring inside PET scanner

only body section inside ring

the ring intercepts emissions from the labeled molecules

PET scan (*red* indicates areas of greatest activity; *blue*, of least activity)

a

b

c   d

Figure 2.3 *Animated!*   (**a**) Patient being moved into a PET scanner. Inside (**b**), a ring of detectors intercepts the radioactive emissions from labeled molecules that were injected into the patient. Computers color-code the number of emissions from each location in the scanned body region. (**c**) Brain scan of a child who has a neurological disorder. (**d**) Different colors in a brain scan signify differences in metabolic activity. The right half of this brain shows little activity. By comparison, cells of the left half absorbed and used the labeled molecule at expected rates.

radiation therapy can be used to target a small, localized cancer—one that has not spread—and bombard it with radium 226 or cobalt 60.

Patients with irregular heartbeats use artificial pacemakers powered by energy emitted from plutonium 238. (This dangerous radioisotope is sealed in a case to block damaging emissions.) Used with positron-emission tomography (PET), radioisotopes provide information about abnormalities in the metabolic functions of specific tissues. The radioisotopes are attached to the sugar glucose or to some other biological molecule, then injected into a patient, who is moved into a PET scanner (Figure 2.3a,b). When cells in a target tissue absorb glucose, emissions from the radioisotope can be used to produce a vivid image of changes in metabolic activity. PET has been extremely useful for studying human brain activity (Figure 2.3d).

normal thyroid

enlarged

cancerous

Figure 2.2 Scans of the thyroid gland from three patients who have ingested radioactive iodine, which is taken up by the thyroid.

2.3 What Is a Chemical Bond?

LINK TO SECTION 2.1

Life requires chemical reactions, and in each one, atoms interact. This section explains what makes those essential interactions possible.

INTERACTING ATOMS: ELECTRONS RULE!

There are three ways atoms can interact: A given atom may share one or more of its electrons, it can accept extra ones, or it can donate electrons to another atom. *Which* of these events takes place depends on how many electrons an atom has and how they are arranged.

If you have ever played with magnets you know that like charges (++ or −−) repel each other and unlike charges (+−) attract. Electrons carry a negative charge, so they are attracted to the positive charge of protons. On the other hand, electrons repel each other. In an atom, electrons respond to these pushes and pulls by moving around the atomic nucleus in "shells." A shell is

not a flat, circular track around the nucleus; it has three dimensions, like the space inside a balloon, and the electron or electrons inside it travel in "orbitals." Figure 2.4 shows one way of visualizing these three-dimensional spaces.

You can think of an orbital as a room in an apartment building—the atom—that allows exactly two renters per room. Hence no more than two electrons can occupy an orbital. Because atoms of different elements differ in how many electrons they have, they also differ in how many of their "rooms" are filled.

Hydrogen is the simplest atom. It has one electron in a single shell (Figure 2.5a). In atoms of elements other than hydrogen, the first shell holds two electrons. Any additional electrons are in shells farther from the nucleus.

The shells around an atom's nucleus are equivalent to energy levels. The shell closest to the nucleus is the lowest energy level. Each shell farther out from the nucleus is at a progressively higher energy level. Because the atoms of different elements have different numbers of electrons, they also have different numbers of shells that electrons can occupy. A shell can have up to eight electrons, but not more. This means that larger atoms, which have more electrons than smaller ones do, also have more shells. The known elements, listed in Appendix II, include some that have many shells to hold all their electrons.

CHEMICAL BONDS JOIN ATOMS

A union between the electron structures of atoms is a **chemical bond**. We can think of bonds as the glue that joins atoms into molecules. How does this "glue" come about? An atom is most stable when its outer shell is filled. Atoms that have too few electrons to fill their outer shell tend to form chemical bonds with other atoms in order to do so. Atoms of oxygen, carbon, hydrogen, and nitrogen—the four most abundant elements in the human body—are like this. As shown in Figure 2.5, hydrogen and helium atoms have a single shell. It is full when it contains two electrons. Other kinds of atoms that have unfilled outer shells take part in chemical bonds that fill their outer shell with eight electrons. Check for electron vacancies in an atom's outer shell and you have a clue as to whether the atom will bond with others. When its outer shell has one or more vacant "slots," an atom may give up electrons, gain them, or share them.

In Figure 2.5 you can count the electron vacancies in the outer shell of each of the atoms pictured. Atoms like helium, which have no vacancies, are said to be *inert*. They usually don't take part in chemical reactions.

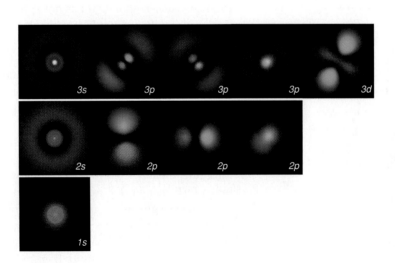

third energy level (second floor)

3s   *3p*   *3p*   *3p*   *3d*

second energy level (first floor)

2s   *2p*   *2p*   *2p*

first energy level (closest to the basement)

1s

...and so on if there are more electrons in the third level. There also may be more energy levels in large atoms.

Figure 2.4 *Animated!* One way to visualize the three-dimensional arrangements of an atom's electrons. A hydrogen atom, which has only one electron, has just a single orbital. The small letters s, p, and d indicate different paths that electrons travel in each energy level.

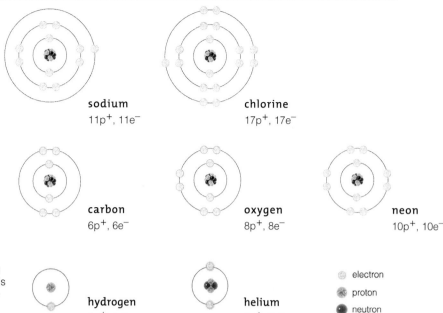

c The **third shell** shows the third energy level, a combined set of nine orbitals that have room for 18 electrons in total. Sodium has one electron in the third shell, and chlorine has seven. Both have vacancies, so both form chemical bonds.

sodium
$11p^+$, $11e^-$

chlorine
$17p^+$, $17e^-$

b The **second shell** shows the second energy level. It has four orbitals with room for up to eight electrons. Carbon has six electrons, two in the first shell and four in the second shell, so it has four vacancies. Oxygen has two. Both atoms form chemical bonds. Neon, with no vacancies, does not.

carbon
$6p^+$, $6e^-$

oxygen
$8p^+$, $8e^-$

neon
$10p^+$, $10e^-$

a The **first shell** shows the first energy level, a single orbital with room for two electrons. Hydrogen has one electron in this orbital. Hydrogen gives up its electron easily. A helium atom has two electrons in this orbital. Having no vacancies, helium does not usually form chemical bonds.

hydrogen
$1p^+$, $1e^-$

helium
$2p^+$, $2e^-$

electron
proton
neutron

Figure 2.5 *Animated!* Shell model. Using this model it is easy to see the vacancies in each atom's outer orbitals. Each circle represents all of the orbitals on one energy level. The larger the circle, the higher the energy level.

ATOMS CAN COMBINE INTO MOLECULES

When chemical bonding joins atoms, the new structure is a **molecule**. Many molecules contain atoms of only one element. Molecular nitrogen (N_2), with its two nitrogen atoms, is an example. Figure 2.6 explains how to read the notation used in representing chemical reactions that occur between atoms and molecules.

Many other kinds of molecules are **compounds**. That is, they consist of two or more elements in proportions that never vary. For example, water is a compound. Every water molecule has one oxygen atom bonded with two hydrogen atoms. No matter where water molecules are—in rain clouds or in a lake or in your bathtub—they *always* have twice as many hydrogen as oxygen atoms.

In a **mixture**, two or more kinds of molecules simply mingle. The proportions don't necessarily vary, although they may. For example, the sugar sucrose is a compound of carbon, hydrogen, and oxygen. If you swirl together molecules of sucrose and water, you'll get a mixture—sugar-sweetened water. If you increase the amount of sucrose in the mixture relative to water, you will still have a mixture—just an extremely sweet one, such as syrup.

We use symbols for elements when writing *formulas*, which identify the composition of compounds. For example, water has the formula H_2O. Symbols and formulas are used in *chemical equations*, which are representations of reactions among atoms and molecules.

In written chemical reactions, an arrow means "yields." Substances entering a reaction (reactants) are to the left of the arrow. Reaction products are to the right. For example, the reaction between hydrogen and oxygen that yields water is summarized this way:

$$2H_2 \ + \ O_2 \ \longrightarrow \ 2H_2O$$
4 hydrogens   2 oxygens   4 hydrogens, 2 oxygens

Note that there are as many atoms of each element to the right of the arrow as there are to the left. Although atoms are combined in different forms, none is consumed or destroyed in the process. The total mass of all products of any chemical reaction equals the total mass of all its reactants. All equations used to represent chemical reactions, including reactions in cells, must be balanced this way.

Figure 2.6 *Animated!* Chemical bookkeeping.

Electrons move in orbitals, volumes of space around an atom's nucleus. In a simple model, orbitals are arranged as a series of shells. The successive shells correspond to increasing levels of energy.

An orbital can contain only one or two electrons. Atoms with unfilled orbitals in their outermost shell tend to interact with other atoms; those with no vacancies do not.

In molecules of an element, all of the atoms are the same kind. In molecules of a compound, atoms of two or more elements are bonded together, in proportions that stay the same. In a mixture, the proportions can vary.

2.4 Important Bonds in Biological Molecules

The atoms in biological molecules are held together by a few kinds of bonds. Mainly, these are the bonds known as ionic, covalent, and hydrogen bonds, as summarized in Table 2.2.

AN IONIC BOND JOINS ATOMS THAT HAVE OPPOSITE CHARGES

Overall, an atom carries no charge because it has just as many electrons as protons. That balance can change if an atom has a vacancy—an unfilled orbital—in its outer shell. For example, a chlorine atom has one vacancy and therefore can gain one electron. A sodium atom, on the other hand, has a single electron in its outer shell, and that electron can be knocked out or pulled away. When an atom gains or loses an electron, the balance between its protons and its electrons shifts, so the atom becomes *ionized*; it has a positive or negative charge. An atom that has a charge is called an **ion**.

It's common for neighboring atoms to accept or donate electrons among one another. When one atom loses an electron and one gains, both become ionized. Depending on conditions inside the cell, the ions may separate, or they may stay together as a result of the mutual attraction of their opposite charges. An association of two ions that have opposing charges is called an **ionic bond**. Figure 2.7

shows how sodium ions (Na^+) and chloride ions (Cl^-) interact through ionic bonds, forming NaCl, or table salt.

ELECTRONS ARE SHARED IN A COVALENT BOND

In a **covalent bond**, atoms *share* two electrons. The bond forms when two atoms each have a lone electron in their outer shell and each atom's attractive force "pulls" on the other's unpaired electron. The tug is not strong enough to pull an electron away completely, so the two electrons occupy a shared orbital. Covalent bonds are stable and much stronger than ionic bonds.

In structural formulas, a single line between two atoms means they share a single covalent bond. Molecular hydrogen, a molecule that consists of two hydrogen atoms, has this kind of bond and can be written as H—H. In a *double* covalent bond, two atoms share two electron pairs, as in an oxygen molecule (O=O). In a *triple* covalent bond, two atoms share three pairs of electrons. A nitrogen molecule (N≡N) is this way. All three examples are gases. When you breathe, you inhale H_2, O_2, and N_2 molecules.

In a *nonpolar* covalent bond, the two atoms pull equally on electrons and so share them equally. The term "nonpolar" means there is no difference in charge at the two ends ("poles") of the bond. Molecular hydrogen is a

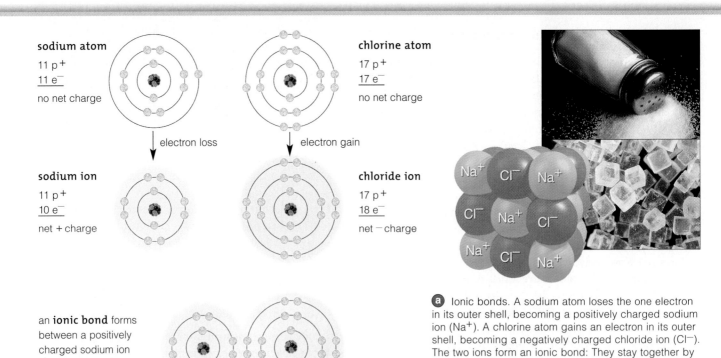

sodium atom
11 p$^+$
11 e$^-$
no net charge

→ electron loss

sodium ion
11 p$^+$
10 e$^-$
net + charge

chlorine atom
17 p$^+$
17 e$^-$
no net charge

→ electron gain

chloride ion
17 p$^+$
18 e$^-$
net − charge

an **ionic bond** forms between a positively charged sodium ion and a negatively charged chloride ion

a Ionic bonds. A sodium atom loses the one electron in its outer shell, becoming a positively charged sodium ion (Na^+). A chlorine atom gains an electron in its outer shell, becoming a negatively charged chloride ion (Cl^-). The two ions form an ionic bond: They stay together by the mutual attraction of opposite charges. A crystal of table salt, or NaCl, is a cube-shaped lattice of ionically bonded sodium and chloride ions.

Figure 2.7 *Animated!* Important bonds in biological molecules.

ATOMS BOND

simple example. Its two hydrogen atoms, each with one proton, attract the shared electrons equally.

In a *polar* covalent bond, two atoms do not share electrons equally. Why not? The atoms are of different elements, and one has more protons than the other. The one with the most protons pulls more, so its end of the bond ends up with a slight negative charge. We say it is "electronegative." The atom at the other end of the bond ends up with a slight positive charge. For instance, a water molecule (H—O—H) has two polar covalent bonds. The oxygen atom carries a slight negative charge, and each of the two hydrogen atoms has a slight positive charge.

A HYDROGEN BOND IS A WEAK BOND BETWEEN POLAR MOLECULES

A **hydrogen bond** is a weak attraction that has formed between a covalently bound hydrogen atom and an electronegative atom in a different molecule or in another part of the same molecule. The dotted lines in Figure 2.7c depict this link.

Hydrogen bonds are weak, so they form and break easily. Even so, they are essential in biological molecules. For example, the genetic material DNA is built of two parallel strands of chemical units, and the strands are

Table 2.2	Major Chemical Bonds in Biological Molecules
Bond	Characteristics
Ionic	Joined atoms have opposite charges.
Covalent	Strong; joined atoms share electrons. In a *polar* covalent bond one end is positive, the other negative.
Hydrogen	Weak; joins a hydrogen (H$^+$) atom in one polar molecule with an electronegative atom in another polar molecule.

held together by hydrogen bonds. In Section 2.6 you will read how hydrogen bonds between water molecules contribute to water's life-sustaining properties.

An ion forms when an atom gains or loses electrons, and so acquires a positive or negative charge. In an ionic bond, ions of opposite charge attract each other and stay together.

In a covalent bond, atoms share electrons. If the electrons are shared equally, the bond is nonpolar. If the sharing is not equal, the bond is polar—slightly positive at one end, slightly negative at the other.

In a hydrogen bond, a covalently bound hydrogen atom attracts a small, negatively charged atom in a different molecule or in another part of the same molecule.

molecular hydrogen (H—H)

Two hydrogen atoms, each with one proton, share two electrons in a single covalent bond that is nonpolar.

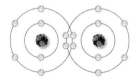

molecular oxygen (O=O)

Two oxygen atoms, each with eight protons, share four electrons in a double covalent bond, also nonpolar.

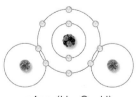

water (H—O—H)

Two hydrogen atoms each share an electron with oxygen. The resulting two covalent bonds form a water molecule. These bonds are polar. The oxygen exerts a greater pull on the shared electrons, so it bears a slight negative charge. Each of the hydrogens has a slight positive charge.

b Covalent bonds. Two atoms with unpaired electrons in their outer shell become more stable by sharing electrons. Two electrons are shared in each covalent bond. When the electrons are shared equally, the covalent bond is nonpolar. If one atom exerts more pull on the shared electrons, the covalent bond is polar.

hydrogen bond

water molecule   ammonia molecule

Two molecules interacting in one hydrogen (H) bond.

H bonds helping to hold part of two large molecules together.

Many H bonds hold DNA's two strands together. Individually, each H bond is weak, but collectively they stabilize DNA's large structure.

c Hydrogen bonds. Such bonds can form when a hydrogen atom is already covalently bonded in a molecule. The hydrogen's slight positive charge weakly attracts an atom with a slight negative charge that is already covalently bonded to something else. As shown above, this can happen between one of the hydrogen atoms of a water molecule and the nitrogen atom of an ammonia molecule.

ATOMS BOND

2.5 Antioxidants

The process in which an atom or molecule loses one or more electrons to another atom or molecule is called *oxidation*. It's what causes a match to burn and an iron nail to rust, and it is part of all kinds of important metabolic events in body cells. Unfortunately, the countless oxidations that go on in our cells also release highly unstable molecules called **free radicals**. Each one is a molecule (such as O_2^-) that includes an oxygen atom lacking a full complement of electrons in its outer shell. To fill the empty slot, a free radical can easily "steal" an electron from another, stable molecule. This theft disrupts both the structure and functioning of the affected molecule.

In large numbers, free radicals pose a serious threat to essential molecules, including a cell's DNA. Cigarette smoke and the ultraviolet radiation in sunlight produce additional free radicals in the body.

An **antioxidant** is a substance that can give up an electron to a free radical before the rogue does damage to DNA or some other vital cell component. The body makes some antioxidants, including the hormone melatonin (Chapter 15), that neutralize free radicals by giving up electrons to them. This home-grown chemical army isn't enough to balance the ongoing production of free radicals, however. This is why many nutritionists recommend adding antioxidants to the diet by eating lots of the foods that contain them, using supplements only in moderation.

Ascorbic acid—vitamin C—is an antioxidant, as is vitamin E. So are some carotenoids, such as alpha carotene, which are pigments in orange and leafy green vegetables, among other foods (Figure 2.8). Antioxidant-rich foods typically also are low in fat and high in fiber.

Figure 2.8 Antioxidants occur in many plant-derived foods, especially orange and green vegetables and fruits.

2.6 Life Depends on Water

Life on our planet probably began in water, and for all life forms it is indispensable. Our blood is more than 90 percent water, and water helps maintain the shape and internal structure of our cells. Many chemical reactions required for life processes require water, or occur only after other substances have dissolved in it. Three unusual properties of water suit it for its key roles in the body, starting with the fact that water is liquid at body temperature.

HYDROGEN BONDING MAKES WATER LIQUID

Any time water is warmer than about 32°F or cooler than about 212°F, it is a liquid. Therefore it is a liquid at body temperature; our watery blood flows and our cells have the fluid they need to maintain their structural integrity and to function properly. What keeps water liquid? You may recall that while a water molecule has no net charge, it does carry charges that are distributed unevenly. The water molecule's oxygen end is a bit negative and its hydrogen end is a bit positive (Figure 2.9*a*). This uneven charge distribution makes water molecules polar, able to attract other water molecules and form hydrogen bonds with them. Collectively, the bonds are so strong that they hold the water molecules close together (Figure 2.9*b* and 2.9*c*). This effect of hydrogen bonds is why water is always in a liquid state unless its temperature falls to freezing or rises to the boiling point.

Water attracts and hydrogen-bonds with other polar substances, such as sugars. Because polar molecules are

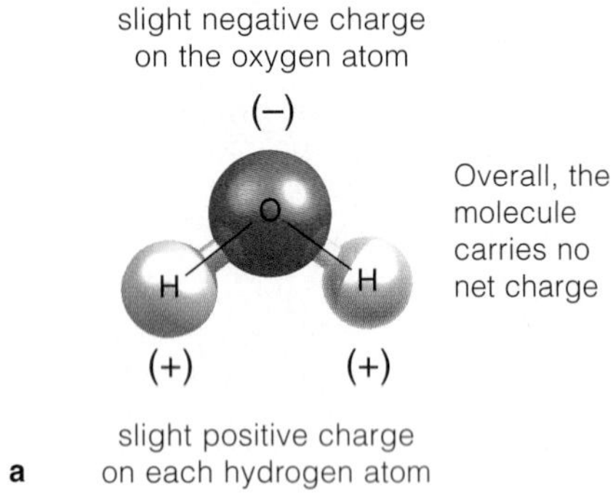

slight negative charge
on the oxygen atom

(−)

Overall, the molecule carries no net charge

(+)   (+)

a   slight positive charge
on each hydrogen atom

Figure 2.9 *Animated!* Water, a substance essential for life. (**a**) Polarity of a water molecule.

(**b**) Hydrogen bonds between molecules in liquid water (dashed lines).

(**c**) Water's cohesion. When a pebble strikes liquid water, the blow forces molecules away from the surface. The individual water molecules don't scatter every which way, however, because hydrogen bonds pull inward on those at the surface. As a result, the molecules tend to stay together in droplets.

attracted to water, they are said to be **hydrophilic**, or "water-loving." Water repels nonpolar substances, such as oils. Hence nonpolar molecules are **hydrophobic**, or "water-dreading." We will return to these concepts when we examine cell structure in Chapter 3.

WATER CAN ABSORB AND HOLD HEAT

Water's hydrogen bonds give it a high *heat capacity*—they enable water to absorb a great deal of heat energy before it warms significantly or evaporates. This is because it takes a large amount of heat to break the many hydrogen bonds that are present in water. Water's ability to absorb significant heat before becoming hot is the reason it was used to cool hot automobile engines in the days before alcohol-based coolants became available. In a similar way, water helps stabilize the temperature inside cells, which are mostly water. The chemical reactions in cells constantly produce heat, yet cells must stay fairly cool because their proteins can only function properly within narrow temperature limits.

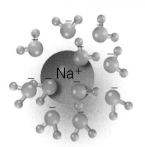

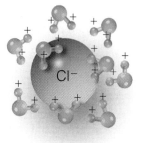

Figure 2.10 *Animated!* Clusters of water molecules around ions. The clusters are called "spheres of hydration."

When enough heat energy is present, hydrogen bonds between water molecules break apart and do not re-form. Then liquid water evaporates—molecules at its surface begin to escape into the air. When a large number of water molecules evaporate, heat energy is lost. This is why sweating helps cool you off on a hot, dry day. Your sweat is 99 percent water. When it evaporates from the millions of sweat glands in your skin, heat leaves with it.

WATER IS A BIOLOGICAL SOLVENT

Water also is a superb *solvent*, which means that ions and polar molecules easily dissolve in it. In chemical terms a dissolved substance is called a **solute**. When a substance dissolves, water molecules cluster around its individual molecules or ions and form "spheres of hydration." This is what happens to solutes in blood and other body fluids. Most chemical reactions in the body occur in water-based solutions.

Figure 2.10 shows what happens when you pour some table salt (NaCl) into a cup of water. After a while, the salt crystals separate into Na^+ and Cl^-. Each Na^+ attracts the negative end of some of the water molecules while each Cl^- attracts the positive end of others.

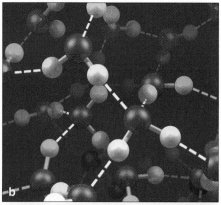

A water molecule is polar. This polarity allows water molecules to hydrogen-bond with one another and with other polar (hydrophilic) substances. Water molecules tend to repel nonpolar (hydrophobic) substances.

Among other effects, the hydrogen bonds in water help it stabilize temperature in body fluids and allow it to dissolve many substances.

2.7 Acids, Bases, and Buffers: Body Fluids in Flux

Ions dissolved in the fluids inside and outside cells influence cell structure and functioning. Especially important are hydrogen ions, which affect many body functions.

THE pH SCALE INDICATES THE CONCENTRATION OF HYDROGEN IONS

The water in the human body contains various ions. Some of the most important are **hydrogen ions**, which have far-reaching effects because they are chemically active and there are so many of them. At any instant, some water molecules are breaking apart into H^+ and **hydroxide ions** (OH^-). These ions are the basis for the **pH scale** (Figure 2.11), which measures the concentration (relative amount) of H^+ in water, blood, and other fluids. Pure water (not rainwater or tap water) always has just as many H^+ as OH^- ions. This state is neutrality, or pH 7, on the pH scale (Figure 2.11).

Starting at neutrality, each change by one unit of the pH scale corresponds to a tenfold increase or decrease in the concentration of H^+. One way to get a personal sense of range is to taste a bit of baking soda (pH 9), and then follow it with water (7), and then lemon juice (2.3).

ACIDS GIVE UP H^+ AND BASES ACCEPT H^+

You've probably heard of "acids" and "bases," but what are they, chemically? An **acid** is a substance that donates protons (H^+) to other solutes or to water molecules when it dissolves in water. A **base** accepts H^+ when it dissolves in water. When either an acid or a base dissolves, OH^- then forms in the solution as well. *Acidic* solutions, such as lemon juice and the gastric fluid in your stomach, release more H^+ than OH^-; their pH is below 7. *Basic* solutions, such as seawater, baking soda, and egg white, release more OH^- than H^+. Basic solutions are also called *alkaline* fluids; they have a pH above 7.

The fluid inside most human cells is about 7 on the pH scale. Body cells also are surrounded by fluids, and the pH values of most of those fluids are slightly higher, ranging between 7.3 and 7.5. The pH of the fluid portion of blood is in the same range.

To a chemist most acids are either weak or strong. Weak ones, such as carbonic acid (H_2CO_3), don't readily donate H^+. Depending on the pH, they just as easily accept H^+ after giving it up, so they alternate between acting as an acid and acting as a base. On the other hand, strong acids totally give up H^+ when they dissociate in water. Hydrochloric acid (HCl), nitric acid (HNO_3), and sulfuric acid (H_2SO_4) are examples.

High concentrations of strong acids or strong bases can be important in the body. For instance, when you eat, cells in your stomach are stimulated to secrete HCl, which separates into H^+ and Cl^- in water. The H^+ ions make your stomach fluid more acidic, and the increased acidity switches on enzymes that can digest (chemically break down) food particles. The acid also helps kill harmful bacteria. However, eating a meal that contains too much of certain kinds of foods can lead to "acid stomach." Antacids such as milk of magnesia are strong bases. In your stomach, milk of magnesia releases magnesium ions and OH^-, which combines with excess H^+ in your stomach fluid. This chemical reaction raises the fluid's pH, and your acid stomach goes away.

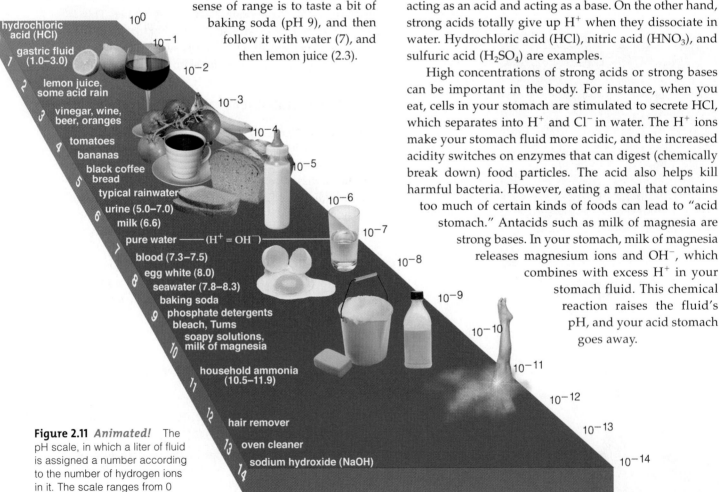

hydrochloric acid (HCl) 10^0
gastric fluid (1.0–3.0) 10^{-1}
lemon juice, some acid rain 10^{-2}
vinegar, wine, beer, oranges 10^{-3}
tomatoes bananas 10^{-4}
black coffee bread 10^{-5}
typical rainwater
urine (5.0–7.0) 10^{-6}
milk (6.6)
pure water ——— ($H^+ = OH^-$) 10^{-7}
blood (7.3–7.5)
egg white (8.0) 10^{-8}
seawater (7.8–8.3)
baking soda
phosphate detergents 10^{-9}
bleach, Tums
soapy solutions, milk of magnesia 10^{-10}
household ammonia (10.5–11.9) 10^{-11}
10^{-12}
hair remover 10^{-13}
oven cleaner
sodium hydroxide (NaOH) 10^{-14}

Figure 2.11 *Animated!* The pH scale, in which a liter of fluid is assigned a number according to the number of hydrogen ions in it. The scale ranges from 0 (most acidic) to 14 (most basic).

High concentrations of strong acids or bases also can have a harmful effect on the world around us. Read the labels on bottles of ammonia, drain cleaner, and other common household products and you'll learn that many of them can cause severe chemical burns. So can sulfuric acid in car batteries. Smoke from fossil fuels, exhaust from motor vehicles, and nitrogen fertilizers release strong acids, which alter the pH of rain (Figure 2.12). The resulting acid rain is an ongoing environmental problem considered in more detail in Chapter 25.

BUFFERS PROTECT AGAINST SHIFTS IN pH

Chemical reactions in cells are sensitive to even slight shifts in pH, because H^+ and OH^- can combine with many different molecules and change their functions. Normally, control mechanisms minimize undesirable shifts in pH, as they do when HCl enters the stomach in response to a meal. Many of the controls involve buffer systems.

A **buffer system** is a partnership between a weak acid and a base, which work together to counter slight shifts in pH. The system forms when the acid dissolves in water. Remember, when a strong base enters a fluid, the OH^- level rises. But a weak acid neutralizes part of the added OH^- by combining with it. By this interaction, the weak acid's partner forms. If a strong acid floods in later, the base will accept H^+ and become its partner in the system.

A key point to remember is that the action of a buffer system can't make new hydrogen ions or eliminate those that already are present. It can only bind or release them.

Several buffer systems operate in the blood and in tissue fluids—the internal environment of the body. For example, reactions in the lungs and kidneys help control the acid–base balance of this environment at levels suitable for life (Sections 9.5 and 10.8). For now, consider what happens when the blood level of H^+ falls and the blood is not as acidic as it should be. At times like this, carbonic acid that is dissolved in blood releases H^+ and becomes the partner base, bicarbonate:

$$H_2CO_3 \longrightarrow HCO_3^- + H^+$$
carbonic acid   bicarbonate

When the blood becomes more acidic, more H^+ is bound to the base, thus forming the partner acid:

$$HCO_3^- + H^+ \longrightarrow H_2CO_3$$
bicarbonate   carbonic acid

Figure 2.12 Sulfur dioxide emissions from a coal-burning power plant. Camera lens filters revealed the otherwise invisible emissions. Sulfur dioxide and other airborne pollutants dissolve in water vapor to form acidic solutions. They are a major component of acid rain.

Uncontrolled shifts in the pH of body fluids can be disastrous. If blood's pH (7.3–7.5) declines to even 7, a person will fall into a *coma*, a state of unconsciousness. An increase to 7.8 can lead to *tetany*, a potentially fatal condition in which the body's skeletal muscles contract uncontrollably. In *acidosis*, carbon dioxide builds up in the blood, too much carbonic acid forms, and blood pH plummets. The condition called *alkalosis* is an abnormal increase in blood pH. Left untreated, both acidosis and alkalosis can be lethal.

A SALT RELEASES OTHER KINDS OF IONS

Salts are compounds that release ions *other than* H^+ and OH^- in solutions. Salts and water often form when a strong acid and a strong base interact. Depending on a solution's pH value, salts can form and dissolve easily. Consider how sodium chloride forms, then dissolves:

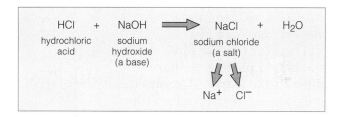

Many salts dissolve into ions that have key functions in cells. For example, the activity of nerve cells depends on ions of sodium, potassium, and calcium, and your muscles contract with the help of calcium ions.

Hydrogen ions (H^+) and other ions dissolved in the fluids inside and outside cells affect cell structure and function.

Acidic substances release H^+, and basic (alkaline) substances accept them. Certain acid–base interactions, such as in buffer systems, help maintain the pH value of a fluid—that is, its H^+ concentration.

A buffer system counters slight shifts in pH by releasing hydrogen ions when their concentration is too low or by combining with them when the concentration is too high.

Salts release ions other than H^+ and OH^-.

2.8 Molecules of Life

LINK TO SECTION 2.4

Carbohydrates, lipids, proteins, and nucleic acids—the four classes of biological molecules—all are built on atoms of the element carbon.

BIOLOGICAL MOLECULES CONTAIN CARBON

Each of the molecules of life is an **organic compound**: it contains the element carbon and at least one hydrogen atom. Chemists once thought organic substances were those obtained from animals and vegetables, as opposed to "inorganic" ones from minerals. Today, however, there is evidence that organic compounds were present on Earth before organisms were.

CARBON'S KEY FEATURE IS VERSATILE BONDING

Our bodies are mostly oxygen, hydrogen, and carbon. Their oxygen and hydrogen are mainly in the form of water. Remove that, and carbon makes up more than half of what's left.

Carbon's importance to life starts with its versatile bonding behavior. As the sketch below shows, *each carbon atom can share pairs of electrons with as many as four other atoms*. The covalent bonds are fairly stable, because the carbon atoms share pairs of electrons equally. Such bonds link carbon atoms together in chains. These form a backbone to which hydrogen, oxygen, and other elements can attach.

The angles of the covalent bonds help produce the three-dimensional shapes of organic compounds. A chain of carbon atoms, bonded covalently one after another, forms a backbone from which other atoms can project:

single covalent bond

carbon atom

atoms branching from backbone

carbon backbone

Often, the backbone coils back on itself in a ring, as in the following diagrams:

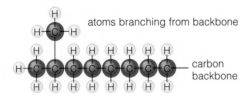

or

carbon rings

Functional group		Some locations
Hydroxyl —OH		Sugars, amino acids.
Carbonyl —CHO (aldehyde)	>CO (ketone)	Sugars, amino acids, nucleotides. An *aldehyde* if at end of a carbon backbone; a *ketone* if attached to an interior carbon of backbone
Carboxyl —COOH (non-ionized)	—COO⁻ (ionized)	Amino acids, fatty acids.
Amino —NH₂ (non-ionized)	—NH₃⁺ (ionized)	Amino acids and certain nucleotide bases.
Phosphate —O—P—O⁻	P icon	ATP, many proteins, phospholipids.

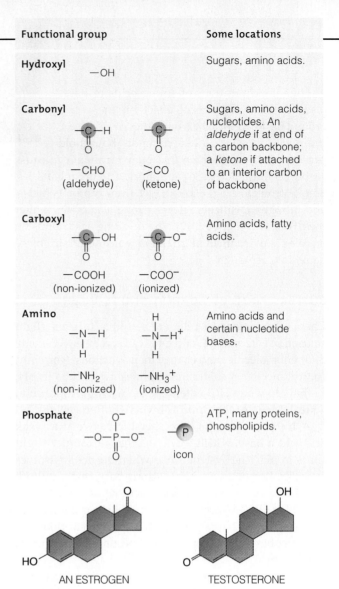

AN ESTROGEN   TESTOSTERONE

Figure 2.13 *Animated!* Examples of functional groups. The sex hormones estrogen and testosterone (bottom) differ chemically because their hydroxyl groups attach to different carbons.

FUNCTIONAL GROUPS AFFECT THE CHEMICAL BEHAVIOR OF ORGANIC COMPOUNDS

A carbon backbone with only hydrogen atoms attached to it is a hydrocarbon, which is a very stable structure. Besides hydrogen atoms, biological molecules also have **functional groups**, which are particular atoms or clusters of atoms that are covalently bonded to carbon and can influence the chemical behavior of organic compounds.

Figure 2.13 shows a few functional groups. Sugars and other organic compounds classified as alcohols have one or more hydroxyl groups (—OH). Water forms hydrogen bonds with hydroxyl groups, which is why sugars can dissolve in water. The backbone of a protein forms by reactions between amino groups and carboxyl groups. As you will see shortly, the backbone is the start

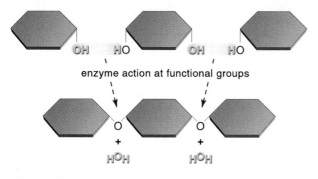

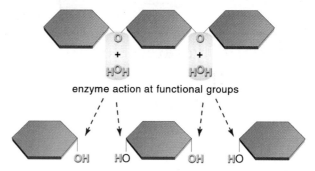

a Two condensation reactions. Enzymes remove an —OH group and an H atom from two molecules, which covalently bond as a larger molecule. Two water molecules form.

b Hydrolysis, a water-requiring cleavage reaction. Enzyme action splits a molecule into three parts, then attaches an H atom and an —OH group derived from a water molecule to each exposed site.

Figure 2.14 *Animated!* (*above*) Examples of metabolic reactions that build, rearrange, and break apart most biological molecules. (*right*) Hydrolysis reactions in the digestive system break starch molecules (polymers) in corn kernels into smaller chemical units (sugar monomers).

of bonding patterns that produce the protein's three-dimensional structure. Amino groups also can combine with hydrogen ions and act as buffers against decreases in pH. The functional groups of sex hormones shown in Figure 2.13 help account for differences between males and females.

CELLS HAVE CHEMICAL TOOLS TO ASSEMBLE AND BREAK APART BIOLOGICAL MOLECULES

How do cells put together the organic compounds they need for their structure and functioning? The details can fill whole books, but here you only need to focus on a few basic concepts. First, whatever happens in a cell requires energy. This rule holds regardless of whether a cell is building, rearranging, or even breaking apart large organic compounds needed for its operations. Chemical reactions in cells also require a class of proteins called **enzymes**, which make metabolic reactions take place faster than they would on their own. Different enzymes facilitate different kinds of reactions. Two reactions that go on constantly in cells are those called condensation and hydrolysis.

CONDENSATION REACTIONS As a cell builds or changes organic compounds, a common step is the **condensation reaction**. Often in this kind of reaction, enzymes remove a hydroxyl group from one molecule and an H atom from another, then speed the formation of a covalent bond between the two molecules (Figure 2.14*a*). The discarded hydrogen and oxygen atoms may combine to form a molecule of water (H_2O). Because this kind of reaction often forms water as a by-product, condensation is sometimes called *dehydration* ("un-watering") *synthesis*. Cells can use series of condensation reactions to assemble

polymers. *Poly-* means many, and a **polymer** is a large molecule built of three to millions of subunits. The individual subunits, called **monomers**, may be the same or different.

HYDROLYSIS REACTIONS **Hydrolysis** is like condensation in reverse (Figure 2.14*b*). In a first step, enzymes that act on particular functional groups split molecules into two or more parts; then they attach an —OH group and a hydrogen atom from a molecule of water to the exposed sites. With hydrolysis, cells can break apart large polymers into smaller units when these are required for building blocks or energy.

Carbohydrates, lipids, proteins, and nucleic acids are the main biological molecules. All are organic compounds.

Organic compounds have carbon backbones. The backbone allows bonding arrangements that help give organic compounds their three-dimensional shapes.

Functional groups are covalently bonded to the carbon backbone of organic compounds. The groups increase the structural and functional diversity of organic compounds.

Enzymes speed the chemical reactions cells use to build, rearrange, and break down organic compounds. These reactions include the combining or splitting of molecules, as in condensation and hydrolysis reactions.

2.9 Carbohydrates: Plentiful and Varied

Carbohydrates are the most abundant biological molecules. Cells use them in building cell parts, or package them for energy in forms that can be stored or transported elsewhere in the body.

Most **carbohydrates** consist of carbon, hydrogen, and oxygen atoms in a 1:2:1 ratio. Because of differences in their structure, chemists separate carbohydrates into three major classes, **monosaccharides**, **oligosaccharides**, and **polysaccharides**.

SIMPLE SUGARS—THE SIMPLEST CARBOHYDRATES

"Saccharide" comes from a Greek word meaning sugar. A *mono*saccharide, meaning "one monomer of sugar," is the simplest carbohydrate. It has at least two —OH groups joined to the carbon backbone plus an aldehyde or a ketone group. Monosaccharides usually taste sweet and dissolve easily in water. The most common ones have a backbone of five or six carbons; for example, there are five carbon atoms in deoxyribose, the sugar in DNA. Glucose, the main energy source for body cells, has six carbons, twelve hydrogens, and six oxygens. (Notice how it meets the 1:2:1 ratio noted above.) Glucose is a building block for larger carbohydrates (Figure 2.15*a*). It also is the parent molecule (precursor) for many compounds, such as vitamin C, which are derived from sugar monomers.

OLIGOSACCHARIDES ARE SHORT CHAINS OF SUGAR UNITS

Unlike the simple sugars, an *oligo*saccharide is a short chain of two or more sugar monomers that are united by dehydration synthesis. (*Oligo-* means a few.) The type known as *di*saccharides consists of just two sugar units. Lactose, sucrose, and maltose are examples. Lactose (a glucose and a galactose unit) is a milk sugar. Sucrose, the most plentiful sugar in nature, consists of one glucose and one fructose unit (Figure 2.15*c*). You consume sucrose when you eat fruit, among other plant foods. Table sugar is sucrose crystallized from sugar cane and sugar beets.

Proteins and other large molecules often have oligosaccharides attached as side chains to their carbon backbone. Some chains have key roles in activities of cell membranes, as you will read in Chapter 3. Others are important in the body's defenses against disease.

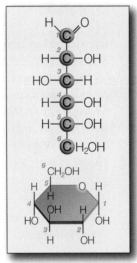

a Structure of glucose   **b** Structure of fructose

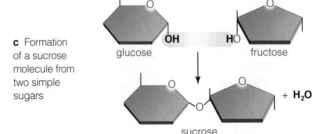

c Formation of a sucrose molecule from two simple sugars

glucose   fructose

sucrose   + H_2O

Figure 2.15 Straight-chain and ring forms of (**a**) glucose and (**b**) fructose. For reference purposes, the carbon atoms of simple sugars are commonly numbered in sequence, starting at the end closest to the molecule's aldehyde or ketone group. (**c**) Condensation of two monosaccharides (glucose and fructose) into a disaccharide (sucrose).

BIOLOGICAL MOLECULES

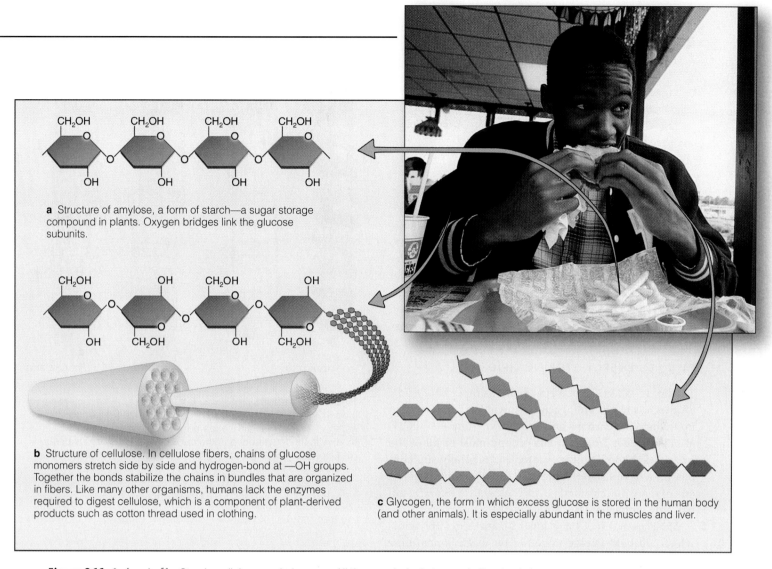

a Structure of amylose, a form of starch—a sugar storage compound in plants. Oxygen bridges link the glucose subunits.

b Structure of cellulose. In cellulose fibers, chains of glucose monomers stretch side by side and hydrogen-bond at —OH groups. Together the bonds stabilize the chains in bundles that are organized in fibers. Like many other organisms, humans lack the enzymes required to digest cellulose, which is a component of plant-derived products such as cotton thread used in clothing.

c Glycogen, the form in which excess glucose is stored in the human body (and other animals). It is especially abundant in the muscles and liver.

Figure 2.16 *Animated!*   Starch, cellulose, and glycogen. All three carbohydrates are built only of glucose monomers.

POLYSACCHARIDES ARE SUGAR CHAINS THAT STORE ENERGY

The "complex" carbohydrates, or polysaccharides, are straight or branched chains of sugar monomers. Often thousands have been joined by dehydration synthesis. A great deal of energy is stored in the many chemical bonds in polysaccharides. The energy is released to cells when these sugars are digested. Most of the carbohydrates humans eat are in the form of polysaccharides. The most common ones—glycogen, starch, and cellulose—consist only of glucose.

When you eat meat you are consuming glycogen (Figure 2.16c). Glycogen is one form in which animals store sugar, most notably in their muscles and the liver. When a person's blood-sugar levels fall, liver cells break down glycogen and release glucose to the blood. When you exercise, your muscle cells tap into their glycogen stores for quick access to energy.

Foods such as potatoes, rice, wheat, and corn are all rich in starch (Figure 2.16a), which is a storage form of glucose in plant cells. In starch the glucose subunits are arranged in a linear fashion. Plants also store glucose in the form of cellulose (Figure 2.16b). We humans lack digestive enzymes that can break down the cellulose in vegetables, whole grains, and other plant tissues. We do benefit from it, however, as undigested "fiber" that adds bulk and so helps move wastes through the lower part of the digestive tract.

Carbohydrates range from simple sugars (such as glucose) to molecules composed of many sugar units.

In order of their structural complexity, the three types of carbohydrates are monosaccharides, oligosaccharides, and polysaccharides.

The body uses carbohydrates in building cell parts and also packages them for energy in forms that can be stored or transported from one site to another.

2.10 Lipids: Fats and Their Chemical Kin

LINKS TO
SECTIONS
2.3, 2.4, AND 2.8

Cells use lipids to store energy, as structural materials (as in cell membranes), and as signaling molecules. Greasy or oily to the touch, major lipids in the body include fats, phospholipids, and sterols.

Oil and water don't mix. Why? Oils are a type of lipid, and a **lipid** is a nonpolar hydrocarbon. A lipid's large nonpolar region makes it hydrophobic, so it tends not to dissolve in water. Lipids easily dissolve in other nonpolar substances. For example, you can dissolve melted butter in olive oil. Here we focus on fats and phospholipids, both of which have chemical "tails" called fatty acids, and on sterols, which have a backbone of four carbon rings.

FATS ARE ENERGY-STORING LIPIDS

The lipids called **fats** have as many as three fatty acids, all attached to glycerol. Each **fatty acid** has a backbone of up to thirty-six carbons and a carboxyl group (—COOH) at one end. Hydrogen atoms occupy most or all of the remaining bonding sites. A fatty acid typically stretches out like a flexible tail (Figure 2.17). An *unsaturated* tail has one or more double bonds in its backbone. A *saturated* tail has only single bonds.

Most animal fats have lots of saturated fatty acids, which are held packed together by weak bonds. Like uncooked bacon fat, or lard, they are solid at room temperature. Most plant fats—"vegetable oils" such as canola, peanut oil, corn oil, and olive oil—stay liquid at room temperature. They stay pourable because the packing interactions in plant fats are not as stable due to rigid kinks in their fatty acid tails.

Butter, lard, plant oils, and other dietary fats consist mostly of **triglycerides**, so-called "neutral" fats that have three fatty acid tails attached to a glycerol backbone (Figure 2.18). Triglycerides are the most abundant lipids in the body and its richest source of energy. Compared to complex carbohydrates, they yield more than twice as much energy, gram for gram, when they are broken down. This is because triglycerides have more removable electrons than do carbohydrates—and energy is released when electrons are removed. In the body, cells of fat-storing tissues stockpile triglycerides as fat droplets.

The *trans fatty acids* are "hydrogenated," or partially saturated. They often are the main ingredient in solid margarines and occur in many packaged foods as well. Many people limit the amount of trans fatty acids in their diet because trans fatty acids have been implicated in the development of some types of heart disease.

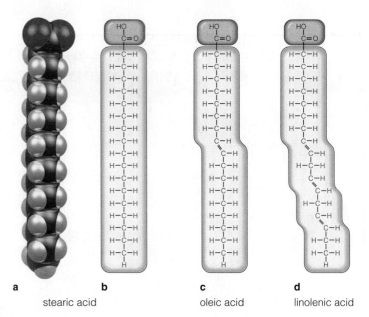

a stearic acid **b** **c** oleic acid **d** linolenic acid

Figure 2.17 *Animated!* Three fatty acids. (**a**,**b**) A space-filling model of stearic acid is shown next to its structural formula. Its carbon backbone is fully saturated with hydrogens. (**c**) Oleic acid, with its double bond in the carbon backbone, is unsaturated. (**d**) Linolenic acid, with three double bonds, is a "polyunsaturated" fatty acid.

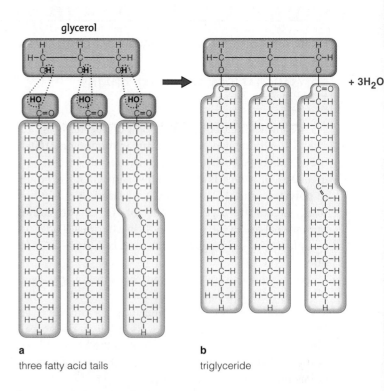

a three fatty acid tails **b** triglyceride

Figure 2.18 *Animated!* Condensation of (**a**) three fatty acids and a glycerol molecule into (**b**) a triglyceride.

BIOLOGICAL MOLECULES

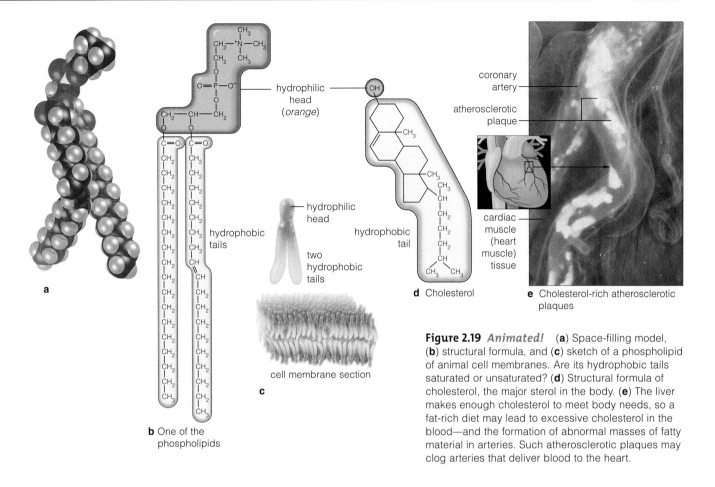

d Cholesterol

e Cholesterol-rich atherosclerotic plaques

coronary artery

atherosclerotic plaque

cardiac muscle (heart muscle) tissue

hydrophilic head (*orange*)

hydrophobic tails

hydrophilic head

two hydrophobic tails

cell membrane section

c

hydrophobic tail

b One of the phospholipids

a

Figure 2.19 *Animated!* (**a**) Space-filling model, (**b**) structural formula, and (**c**) sketch of a phospholipid of animal cell membranes. Are its hydrophobic tails saturated or unsaturated? (**d**) Structural formula of cholesterol, the major sterol in the body. (**e**) The liver makes enough cholesterol to meet body needs, so a fat-rich diet may lead to excessive cholesterol in the blood—and the formation of abnormal masses of fatty material in arteries. Such atherosclerotic plaques may clog arteries that deliver blood to the heart.

PHOSPHOLIPIDS ARE KEY BUILDING BLOCKS OF CELL MEMBRANES

A **phospholipid** has a glycerol backbone, two fatty acid tails, and a hydrophilic "head" with a phosphate group— a phosphorus atom bonded to four oxygen atoms—and another polar group (Figure 2.19*a*). Phospholipids are the main materials of cell membranes, which have two layers of lipids. The heads of one layer are dissolved in the cell's fluid interior, while the heads of the other layer are dissolved in the surroundings. Sandwiched between the two are all the fatty acid tails, which are hydrophobic.

STEROLS ARE BUILDING BLOCKS OF CHOLESTEROL AND STEROIDS

Sterols are among the lipids that have no fatty acid tails. Sterols differ in the number, position, and type of their functional groups, but they all have a rigid backbone of four fused-together carbon rings:

sterol backbone

Many people associate the sterol cholesterol (Figure 2.19*b* and 2.19*c*) with heart and artery disease. However, normal amounts of this sterol are crucial to the structure and proper functioning of cells. For instance, cholesterol is a vital component of membranes of every cell in your body. Important derivatives of cholesterol include vitamin D (essential for bone and tooth development), bile salts (which help with fat digestion in the small intestine), and steroid hormones such as estrogen and testosterone. Later chapters will look more closely at the ways steroid hormones influence reproduction, development, growth, and some other body functions.

Lipids are hydrophobic, greasy or oily compounds.
Important lipids in the human body include:

* *Triglycerides (neutral fats), which are major reservoirs of energy*

* *Phospholipids, the main components of cell membranes*

* *Sterols (such as cholesterol), which are components of membranes and precursors of steroid hormones and other vital molecules.*

2.11 Proteins: Biological Molecules with Many Roles

Of all the large biological molecules, proteins are the most diverse. Those called enzymes speed up chemical reactions. Structural proteins are building blocks of your bones, muscles, and other body elements. Transport proteins help move substances, while the regulatory proteins, including some hormones, help adjust cell activities. They help make possible activities such as waking, sleeping, and engaging in sex, to cite just a few. Still other proteins function as weapons against harmful bacteria and other invaders.

PROTEINS ARE BUILT FROM AMINO ACIDS

Amazingly, our body cells build thousands of different **proteins** from only twenty kinds of amino acids. An **amino acid** is a small organic compound that consists of an amino group, a carboxyl group (an acid), an atom of hydrogen, and one or more atoms called its R group. As you can see from the structural formula in Figure 2.20, these parts generally are covalently bonded to the same carbon atom. Figure 2.21 shows several amino acids that we will consider later in the book.

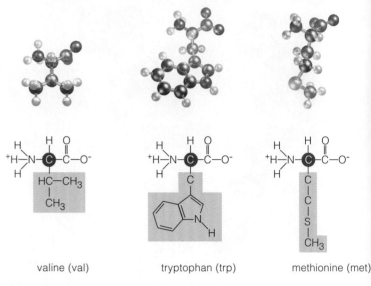

valine (val)   tryptophan (trp)   methionine (met)

Figure 2.21 Three amino acids shown as ball-and-stick models with structural formulas of their common ionized form. Green boxes indicate R groups, which are side chains that include functional groups. A side chain helps determine an amino acid's chemical and physical properties.

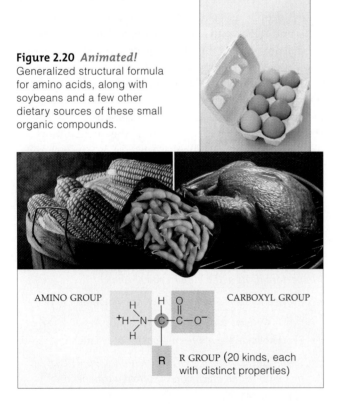

Figure 2.20 *Animated!* Generalized structural formula for amino acids, along with soybeans and a few other dietary sources of these small organic compounds.

AMINO GROUP   CARBOXYL GROUP

R GROUP (20 kinds, each with distinct properties)

THE SEQUENCE OF AMINO ACIDS IS A PROTEIN'S PRIMARY STRUCTURE

When a cell makes a protein, amino acids become linked, one after the other, by *peptide* bonds. As Figure 2.22 shows, this is the type of covalent bond that forms between one amino acid's amino group (NH_3^+) and the carboxyl group ($—COO^-$) of the next amino acid.

When peptide bonds join two amino acids together, we have a dipeptide. When they join three or more amino acids, we have a **polypeptide chain**. The backbone of each chain incorporates nitrogen atoms in this regular pattern: —N—C—C—N—C—C—.

For each kind of protein, different amino acids are added in a specific order, one at a time, from the twenty kinds available. As a later chapter describes more fully, DNA determines the order in which amino acids are "chosen" to be added to the growing chain. Once the chain is completed, the order of its linked amino acids is the unique sequence for that particular kind of protein (Figure 2.22*e*). Every other kind of protein in the body will have its own sequence of amino acids, linked one to the next like the links of a chain. This sequence is the *primary* structure of a protein. A large number of amino acids can be linked up this way. The primary structure of the largest known protein, which is a component of human muscle, is a string of some 27,000 amino acids!

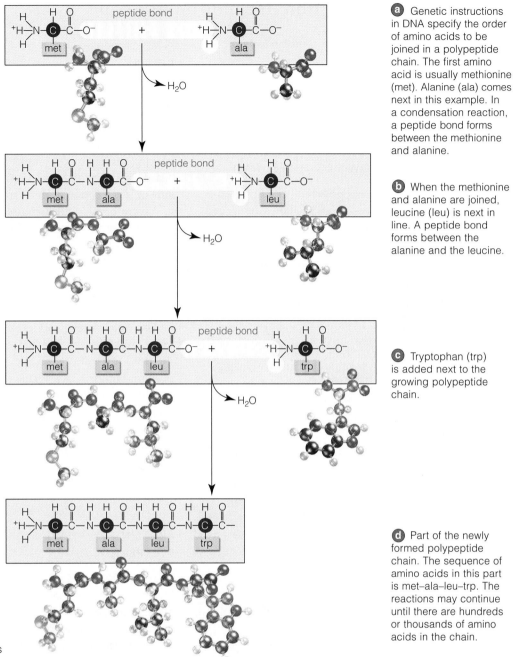

a Genetic instructions in DNA specify the order of amino acids to be joined in a polypeptide chain. The first amino acid is usually methionine (met). Alanine (ala) comes next in this example. In a condensation reaction, a peptide bond forms between the methionine and alanine.

b When the methionine and alanine are joined, leucine (leu) is next in line. A peptide bond forms between the alanine and the leucine.

c Tryptophan (trp) is added next to the growing polypeptide chain.

d Part of the newly formed polypeptide chain. The sequence of amino acids in this part is met–ala–leu–trp. The reactions may continue until there are hundreds or thousands of amino acids in the chain.

Figure 2.22 *Animated!* The formation of peptide bonds between amino acids as a protein is synthesized.

Different cells make thousands of different proteins. Many of the proteins are fibrous, with polypeptide chains organized as strands or sheets. They contribute to the shape and internal organization of cells. Other proteins are globular: They have one or more polypeptide chains folded into compact, rounded shapes. Most enzymes are globular proteins. So are the proteins that help cells, and cell parts, move.

As you will see next, a protein's final shape and chemical behavior arise from its primary structure. Said another way, information built into a protein's amino acid sequence determines its structure and function.

A protein consists of one or more polypeptide chains of amino acids. A sequence of amino acids makes up each kind of protein; this sequence is the protein's primary structure. It determines the protein's final structure, its chemical behavior, and its function.

2.12 A Protein's Function Depends on Its Shape

A key take-home message in this book is that in biology, structure equals function. Proteins are a prime example. How a protein functions (or malfunctions) depends on events that take place after a polypeptide has formed—that is, once the primary structure is in place—and the chain then bends, folds, and coils.

MANY PROTEINS FOLD TWO OR THREE TIMES

For the most part, a protein's primary structure guides its final shape in two ways. First, it allows hydrogen bonds to form between different amino acids in the chain. Second, it puts R groups in positions where they can interact. The interactions force the chain to bend and twist, as diagrammed in Figure 2.23.

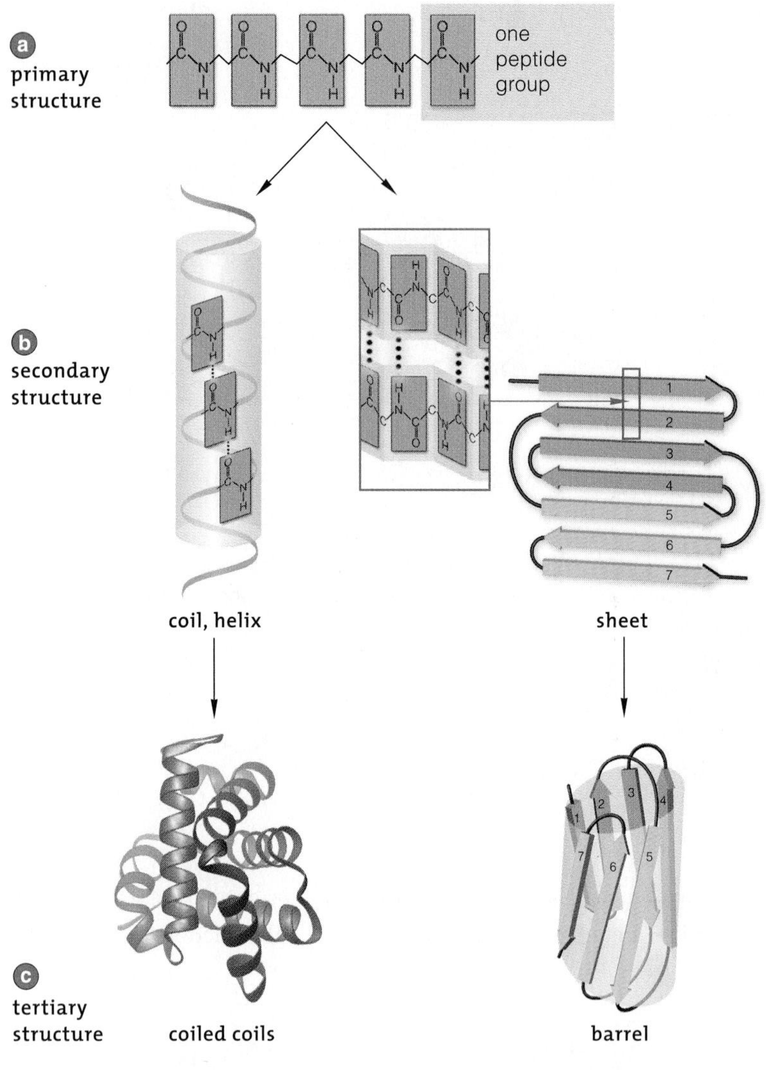

Hydrogen bonds form at regular, short intervals along a new polypeptide chain. They give rise to a coiled or extended pattern known as the *secondary structure* of a protein. Think of a polypeptide chain as a set of rigid playing cards joined by links that can swivel a bit. Each "card" is a peptide group (Figure 2.23a). Atoms on either side of it can rotate slightly around their covalent bonds and form bonds with neighboring atoms. For instance, in many chains a hydrogen bond forms after every third amino acid. The bonding pattern forces the peptide groups to coil helically, like a spiral staircase (Figure 2.23b). In other proteins, the hydrogen bonds hold two or more chains side by side, like sections of a folded sheet of paper.

The coils, sheets, and loops of a protein fold up even more, much like an overly twisted rubber band. This is the third level of organization, or *tertiary* structure, of a protein (Figure 2.23c). Tertiary structure is what makes a protein a molecule that can perform a particular function. For instance, some proteins fold into a hollow "barrel" that provides a channel through cell membranes.

Figure 2.24a shows how a polypeptide chain was folded into the tertiary structure of the protein globin. Functional groups along the chain caused the folding when hydrogen bonds formed between certain ones.

PROTEINS CAN HAVE MORE THAN ONE POLYPEPTIDE CHAIN

Imagine that bonds form between *four* molecules of globin and that an iron-containing functional group, a heme group, nestles near the center of each. The result is hemoglobin, an oxygen-transporting protein. At this moment, each of the millions of red blood cells in your body is transporting a billion molecules of oxygen, bound to 250 million molecules of hemoglobin.

With its four globin molecules, hemoglobin is a good example of a protein that is built of more than one polypeptide chain. The hormone insulin, which consists of two chains, is another. Proteins that are constructed this way are said to have *quaternary* structure (Figure 2.24b). Their polypeptide chains are joined by weak interactions (such as hydrogen bonds) and sometimes by covalent bonds between sulfur atoms of R groups. These bonds between two sulfur atoms are called disulfide bridges (di = two).

Disulfide bridges

a primary structure

b secondary structure

one peptide group

coil, helix

sheet

coiled coils

barrel

c tertiary structure

Figure 2.23 *Animated!* The first three levels of protein structure. (**a**) Primary structure is a chain of amino acids. (**b**) Hydrogen bonds (dotted lines) along a polypeptide chain form a coiled or sheetlike secondary structure. (**c**) Coils and sheets pack together to form the protein's third level of structure, which suits it for its particular function.

BIOLOGICAL MOLECULES

heme, an iron-containing, oxygen-transporting functional group

helically coiled, third structural level of one globin molecule

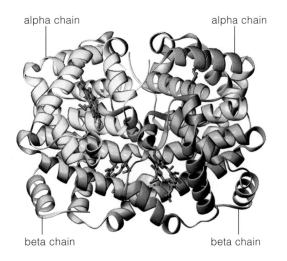

alpha chain

alpha chain

beta chain

beta chain

 a third level

b fourth level of protein structure

a

b

Figure 2.24 *Animated!* (**a**) A globin molecule. This coiled polypeptide chain attracts heme, a functional group containing iron. (**b**) Hemoglobin, an oxygen-transporting pigment in red blood cells. It has four globin chains and four heme groups. The two alpha chains and two beta chains have slightly different amino acid sequences. To make it easier to tell the four chains apart, here the coiled regions are color-coded. The heme groups are shown in red. Each one can bind a molecule of oxygen (O_2), and this is how red blood cells transport oxygen to cells.

Figure 2.25 The hair of actress Nicole Kidman (**a**) before and (**b**) after a permanent wave.

Hemoglobin and insulin are globular proteins; so are most enzymes. Many other proteins with quaternary structure are fibrous. Collagen, the most common protein in the body, is an example of this. (Your skin, bones, corneas, and other body parts depend on collagen's strength.) Keratin, a structural protein of hair, is another example. The chemicals used in a permanent wave break hydrogen bonds in disulfide bridges in the keratin chains in hair. After the hair is wrapped around curlers that hold polypeptide chains in new positions, a second chemical causes disulfide bridges to form between *different* sulfur-bearing amino acids. The rearranged bonding locks the hair in curls (Figure 2.25).

GLYCOPROTEINS HAVE SUGARS ATTACHED AND LIPOPROTEINS HAVE LIPIDS

Some proteins have other organic compounds attached to their polypeptide chains. For example, **lipoproteins** form when certain proteins circulating in blood combine with cholesterol, triglycerides, and phospholipids that were consumed in food. Most **glycoproteins** (from *glukus*, the Greek word for sweet) have oligosaccharides bonded to them. Most of the proteins found at the surface of cells are glycoproteins, as are many proteins in blood and those that cells secrete (such as protein hormones).

DISRUPTING A PROTEIN'S SHAPE DENATURES IT

When a protein or any other large molecule loses its normal three-dimensional shape, it is *denatured*. For example, hydrogen bonds are sensitive to increases or decreases in temperature and pH. If the temperature or pH exceeds a protein's tolerance, its hydrogen bonds break, polypeptide chains unwind or change shape, and the protein no longer functions. Cooking an egg destroys weak bonds that contribute to the three-dimensional shape of the egg white protein albumin. Some denatured proteins can resume their shapes when normal conditions are restored—but not albumin. There is no way to uncook a cooked egg white.

Proteins have a secondary structure, a coil or an extended sheet. It results from hydrogen bonds located at short, regular intervals along a polypeptide chain.

At the third level of protein structure, bonding at certain amino acids makes a coiled chain bend and loop. Interactions among R groups in the chain hold the loops in the proper positions.

We see a fourth level of structure in proteins that consist of more than one polypeptide chain. Hydrogen bonds and other interactions join the chains.

2.13 Nucleotides and Nucleic Acids

The fourth and final class of biological molecules consists of nucleotides and nucleic acids. Since nucleotides are the subunits (monomers) of which nucleic acids (polymers) are built, we begin with them.

NUCLEOTIDES ARE ENERGY CARRIERS AND HAVE OTHER ROLES

A **nucleotide** (NOO-klee-oh-tide) is a small organic compound with a sugar, at least one phosphate group, and a base. The sugar is ribose or deoxyribose; both have a five-carbon ring structure. The only difference is that ribose has an oxygen atom attached to carbon 2 in the ring and deoxyribose does not. The bases have a single or double carbon ring structure that incorporates nitrogen.

One nucleotide, **ATP** (for adenosine triphosphate), has three phosphate groups attached to its sugar, as you can see in Figure 2.26. In cells, ATP molecules link chemical reactions that *release* energy with other reactions that *require* energy. How? ATP can readily transfer a phosphate group to many other molecules in the cell, providing the acceptor molecules with the energy they need to enter into a reaction.

In later chapters, you'll read about other nucleotides. For instance, some are subunits of coenzymes. A **coenzyme** is a molecule that accepts hydrogen atoms and electrons that are being removed from other molecules and transfers them to sites where they can be used for other reactions. Some other nucleotides act as chemical messengers inside

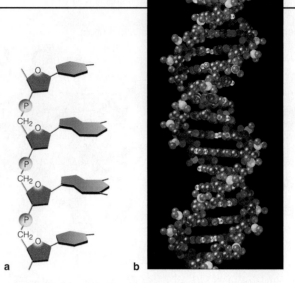

Figure 2.27 *Animated!* (**a**) Bonds between the bases in nucleotides. (**b**) Model of DNA, which has two strands of nucleotides joined by hydrogen bonds and twisted into a double helix. Here the nucleotide bases are blue.

and between cells. One of these is a nucleotide called cAMP (for cyclic adenosine monophosphate). It is a major player in events by which some hormones act.

NUCLEIC ACIDS INCLUDE DNA AND RNA

Nucleotides are building blocks for **nucleic acids**, which are single- or double-stranded molecules. In the backbones of the strands, each nucleotide's sugar is covalently bonded to a phosphate group of the neighboring nucleotides (Figure 2.27a). In this book you will read often about the nucleic acid **DNA** (deoxyribonucleic acid), which contains the sugar deoxyribose. DNA consists of two strands of nucleotides, twisted together in a double helix (Figure 2.27b). Hydrogen bonds between the nucleotide bases hold the strands together, and the sequence of bases encodes genetic information. In some parts of the molecule, the bases occur in a sequence that is unique to each species. Unlike DNA, **RNA** (short for ribonucleic acid) is usually a single strand of nucleotides. There are different kinds of RNA, but all have the sugar ribose. RNAs have crucial roles in processes that use genetic information to build proteins.

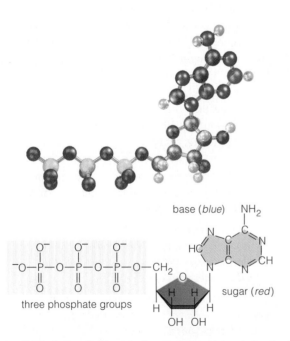

base (*blue*) NH₂

three phosphate groups

sugar (*red*)

OH OH

Figure 2.26 *Animated!* A ball-and-stick model and structural formula for ATP, the energy-carrying nucleotide in cells.

> *A nucleotide is an organic compound with a sugar, one or more phosphate groups, and a base. Nucleotides are building blocks for DNA and RNA. Some, including ATP, have key roles in energy transfers.*
>
> *DNA is a double-stranded nucleic acid. Its sequence of nucleotide bases carries genetic information. RNAs are single-stranded nucleic acids with roles in the processes by which DNA's genetic information is used to build proteins.*

2.14 Food Production and a Chemical Arms Race

The next time you shop for groceries, consider what it takes to provide you with your daily supply of organic compounds. For example, the lettuce for your salad most likely grew in fertilized cropland, and the grower may well have been concerned about invading weeds and attacks by insects. Each year, these food pirates and others ruin or eat nearly half of the food that people all over the world try to grow.

People—and plants—marshal various chemical defenses against the attackers (Figure 2.28). For instance, the tissues of many plants contain toxins that repel or kill harmful organisms. Humans encounter natural plant toxins in a wide range of foods—chili peppers, potatoes, figs, celery, and alfalfa sprouts, for instance. By and large, our bodies seem to be able to cope with those chemicals just fine—possibly because we have evolved our own biochemical ways of neutralizing them.

In 1945, the human race took a cue from the plant world as chemists began developing synthetic toxins that could improve our ability to protect crop yields, stored grains, ornamental plants, and even our pets. Since then researchers have developed a wide array of herbicides to kill weeds, insecticides to eradicate unwanted insects, and fungicides against harmful molds and other fungi.

Although extremely useful in some applications, pesticides are powerful chemicals and have become more so with the passing years. Some of them kill natural enemies of the targeted pest and others harm wildlife such as birds. Some, such as DDT, stay active for years. (DDT is banned in the United States, although not in many other countries.) When people are exposed to unsafe doses, either by accident or misuse, some pesticides can trigger rashes, hives, headaches, asthma, and joint pain. According to some authorities, young children who are exposed to pesticides applied to keep a lawn thick and green may be at risk of developing behavior problems, learning disabilities, and other problems. Although manufacturers dispute these claims, it is worth noting that according to the U.S. Environmental Protection Agency, homeowners in the United States use 10 times more pesticides on their lawns than farmers do in agricultural fields.

On the other hand, many studies show that used properly, modern pesticides increase food supplies and profits for farmers. They also save lives by killing disease-causing insects and other pathogens. And despite the natural worries of consumers, for now there is little evidence that the usual amounts of pesticides in or on food pose a significant health risk.

Figure 2.28 A low-flying crop duster with its rain of pesticides. Such pesticides may leave harmful residues in food.

Summary

Introduction An element is a fundamental substance that cannot be broken down to other substances by ordinary chemical means. The four main elements in the body are oxygen, carbon, hydrogen, and nitrogen. Small amounts of trace elements are also vital to many bodily functions.

Section 2.1 An atom is the smallest unit that has the properties of an element. Atoms are composed of protons, neutrons, and electrons. An element's atoms may vary in how many neutrons they contain. These various forms are isotopes. Whether an atom interacts with others depends on the number of electrons it has and how they are arranged in the atom.

 Learn how radioisotopes are used in a PET scan.

Section 2.3 We can think of electrons as moving in orbitals that are arranged in a series of shells around an atom's nucleus. An atom with one or more unfilled orbitals in its outer shell will tend to take part in chemical bonds.

A chemical bond is a union between the electron structures of atoms. Bonds join atoms into molecules. A chemical compound is a molecule that consists of atoms of two or more elements in proportions that do not vary. In a mixture, two or more kinds of molecules mingle, and their proportions *can* vary.

 Investigate electrons and the shell model.

Section 2.4 Atoms normally have no net charge. However, if an atom gains or loses one or more electrons it becomes an ion, which has a positive or negative charge.

In an ionic bond, positive and negative ions stay together by the mutual attraction of their opposite charges. In a covalent bond, atoms share one or more electrons. A hydrogen bond is a weak bond between polar molecules. At least one of the molecules contains a hydrogen atom that is part of a polar covalent bond.

 Compare the types of chemical bonds found in biological molecules.

Section 2.6 Water is crucial for the shape, internal organization, and chemical activities of cells. Hydrogen bonds between its molecules give water special properties, such as the ability to resist temperature changes and to dissolve other polar substances. A dissolved substance is a solute. Polar molecules are hydrophilic (attracted to water). Nonpolar substances, such as oils, are hydrophobic (repelled by water).

 Explore the structure and properties of water.

Section 2.7 The pH scale measures the concentration of hydrogen ions in a fluid. Acids release hydrogen ions (H^+), and bases release hydroxide ions (OH^-) that can combine with H^+. At pH 7, the H^+ and OH^- concentrations in a solution are equal; this is a neutral pH. A buffer system maintains pH values of blood, tissue fluids, and the fluid inside cells. A salt is a compound that releases ions other than H^+ and OH^-.

 Investigate the pH of common solutions.

Section 2.8 A carbon atom forms up to four covalent bonds with other atoms. Carbon atoms bonded together in linear or ring structures are the backbone of organic compounds. The chemical and physical properties of many of those compounds depend largely on functional groups of atoms attached to the carbon backbone.

Cells assemble and break apart most organic compounds by way of five kinds of reactions: transfers of functional groups, electron transfers, internal rearrangements, condensation reactions (dehydration synthesis), and cleavage reactions such as hydrolysis. Enzymes speed all these reactions. A polymer is a molecule built of three or more subunits; each subunit is called a monomer.

Cells have pools of dissolved sugars, fatty acids, amino acids, and nucleotides. These are small organic compounds with no more than about twenty carbon atoms. They are building blocks for the larger biological molecules—the carbohydrates, lipids, proteins, and nucleic acids (Table 2.3).

 Learn more about functional groups and watch animations that explain condensation, hydrolysis, and how a triglyceride forms.

Section 2.9 Cells use carbohydrates for energy or to build cell parts. Monosaccharides, or single sugar units, are the simplest ones. Chains of sugars linked by covalent bonds are oligosaccharides; common ones, such as glucose, are disaccharides built of two sugar units. Polysaccharides are longer chains that store energy in the bonds between the sugar units (Table 2.3).

Section 2.10 The body uses lipids for energy, to build cell parts, and as signaling molecules. The most important dietary fats are triglycerides. Phospholipids are building blocks of cell membranes; sterols also are constituents of membranes and various key molecules.

Table 2.3 Summary of the Main Carbon Compounds in the Human Body

Category	Main Subcategories	Examples	Functions
Carbohydrates *contain an aldehyde or a ketone group and one or more hydroxyl groups*	**Monosaccharides (simple sugars)** **Oligosaccharides**	Glucose Sucrose (a disaccharide)	Structural roles, energy source Form of sugar transported in plants
	Polysaccharides (complex carbohydrates)	Starch Cellulose	Energy storage Structural roles
Lipids *are largely hydrocarbon, generally do not dissolve in water but dissolve in nonpolar solvents*	**Lipids with fatty acids:** *Glycerides:* one, two, or three fatty acid tails attached to glycerol backbone *Phospholipids:* phosphate group, another polar group, and (often) two fatty acids attached to glycerol backbone	Fats (e.g., butter) Oils (e.g., corn oil) Phosphatidylcholine	Energy storage Key component of cell membranes
	Lipids with no fatty acids: *Sterols:* four carbon rings; the number, position, and type of functional groups vary	Cholesterol	Component of animal cell membranes, can be rearranged into other steroids (e.g., vitamin D, sex hormones)
Proteins *are polypeptides (up to several thousand amino acids, covalently linked)*	**Fibrous proteins:** Individual polypeptide chains, often linked into tough, water-insoluble molecules	Keratin Collagen	Structural element of hair, nails Structural element of bones and cartilage
	Globular proteins: One or more polypeptide chains folded and linked into globular shapes; many roles in cell activities	Enzymes Hemoglobin Insulin Antibodies	Increase in rates of reactions Oxygen transport Control of glucose metabolism Tissue defense
Nucleic Acids (and Nucleotides) *are chains of units (or individual units) that each consist of a five-carbon sugar, phosphate, and a nitrogen-containing base*	**Adenosine phosphates** **Nucleotide coenzymes**	ATP NAD^+, $NADP^+$	Energy carrier Transport of protons (H^+) and electrons from one reaction site to another
	Nucleic acids: Chains of thousands to millions of nucleotides	DNA, RNAs	Storage, transmission, translation of genetic information

Sections 2.11, 2.12 Proteins are built of amino acids and have many roles in the body; each one's function depends on its structure. Three or more amino acids (linked by peptide bonds) form a polypeptide chain. The linear sequence of the amino acids is a protein's primary structure. A protein's final shape comes about as the polypeptide chain bends, folds, and coils. Many proteins consist of more than one polypeptide chain. Some have other organic compounds bonded to them; examples are glycoproteins, which have oligosaccharides attached, and lipoproteins, which have lipids attached.

A protein becomes denatured, losing its ability to function normally, when some factor changes its normal three-dimensional shape.

 Learn more about amino acids and how peptide bonds form a polypeptide chain.

Section 2.13 Nucleic acids such as DNA and RNA are built of nucleotides. A nucleotide consists of a sugar (such as deoxyribose, the sugar in DNA), one or more phosphate groups, and a base. The nucleotide

ATP transfers energy that powers chemical reactions in cells; coenzymes are nucleotides that assist in energy transfers.

 Explore the structure of DNA.

Review Questions

1. Distinguish between an element, an atom, and a molecule.

2. Explain the difference between an ionic bond and a covalent bond.

3. Ionic and covalent bonds join atoms into molecules. What do hydrogen bonds do?

4. Name three vital properties of water in living cells.

5. Which small organic molecules make up carbohydrates, lipids, proteins, and nucleic acids?

6. Which of the following is the carbohydrate, the fatty acid, the amino acid, and the polypeptide?
   a. $^+NH_3$—CHR—COO$^-$ c. (glycine)$_{20}$
   b. $C_6H_{12}O_6$ d. $CH_3(CH_2)_{16}COOH$

7. Describe the four levels of protein structure. How do a protein's side groups influence its interactions with other substances? What is denaturation?

8. Distinguish among the following:
   a. monosaccharide, polysaccharide, disaccharide
   b. peptide bond, polypeptide
   c. glycerol, fatty acid
   d. nucleotide, nucleic acid

Self-Quiz

Answers in Appendix V

1. The backbone of organic compounds forms when _____ atoms are covalently bonded.

2. Each carbon atom can form up to _____ bonds with other atoms.
   a. four   c. eight
   b. six   d. sixteen

3. All of the following except _____ are small organic molecules that serve as the main building blocks or energy sources in cells.
   a. fatty acids   d. amino acids
   b. simple sugars   e. nucleotides
   c. lipids

4. Which of the following is not a carbohydrate?
   a. glucose molecule   c. margarine molecule
   b. simple sugar   d. polysaccharide

5. _____, a class of proteins, make metabolic reactions proceed much faster than they would on their own.
   a. Nucleic acids   c. Fatty acids
   b. Amino acids   d. Enzymes

6. Examples of nucleic acids are _____.
   a. polysaccharides   c. proteins
   b. DNA and RNA   d. simple sugars

7. Which phrase best describes what a functional group does?
   a. assembles large organic compounds
   b. influences the behavior of organic compounds
   c. splits molecules into two or more parts
   d. speeds up metabolic reactions

8. In _____ reactions, small molecules are linked by covalent bonds, and water can also form.
   a. hydrophilic   c. condensation
   b. hydrolysis   d. ionic

9. Match each type of molecule with its description.
   _____ chain of amino acids   a. carbohydrate
   _____ energy carrier   b. phospholipid
   _____ glycerol, fatty acids,   c. protein
         phosphate   d. DNA
   _____ chain of nucleotides   e. ATP
   _____ one or more sugar units

10. What kinds of bonds often control the shape (or tertiary form) of large molecules such as proteins?
    a. hydrogen   d. inert
    b. ionic   e. single
    c. covalent

Critical Thinking

1. Black coffee has a pH of 5, and milk of magnesia has a pH of 10. Is coffee twice as acidic as milk of magnesia?

2. Your cotton shirt has stains from whipped cream and strawberry syrup, and your dry cleaner says that two separate cleaning agents will be needed to remove the stains. Explain why two agents are needed and what different chemical characteristic each would have.

3. A store clerk tells you that vitamin C extracted from rose hips is better for you than synthetic vitamin C. Based on what you know of the structure of organic compounds, does this claim seem credible? Why or why not?

4. Use the Internet to find three examples of acid rain damage and efforts to combat the problem. You might start with the United States Environmental Protection Agency's acid rain home page.

5. Carbonated drinks get that way when pressurized carbon dioxide gas is forced into flavored water. A chemical reaction between water molecules and some of the CO_2 molecules creates hydrogen ions (H^+) and bicarbonate, which is a buffer. In your opinion, is this reaction likely to raise the pH of a soda above 7, or lower it? Give your reasoning.

Explore on Your Own

It's easy to demonstrate the practical consequences of differences between hydrophilic and hydrophobic molecules. Just try this little kitchen experiment. Take two identical clean plates. Smear one with grease (such as margarine) and pour syrup over the other.

Next run moderately warm water over both plates for thirty seconds and observe the results. Which plate got cleaner, and why? The companies that make dishwashing detergents manipulate them chemically so that their molecules have both hydrophobic and hydrophilic regions. Given what you know about the ability of water by itself to dissolve hydrophilic and hydrophobic substances, why might this be?

When Mitochondria Spin Their Wheels

In the early 1960s, Swedish physician Rolf Luft was treating a young patient who felt weak and too hot all the time. Even on the coldest winter days, she couldn't stop sweating, and her skin was flushed. She was thin in spite of a huge appetite.

Luft inferred that his patient's symptoms pointed to a metabolic disorder. Her cells seemed to be spinning their wheels—a lot of their activity was being dissipated as metabolic heat. Working body cells use oxygen, and tests showed that her cells were consuming oxygen at the highest rate ever recorded.

A cell's energy powerhouses are called mitochondria, which are pictured in the microscope image at right. Mitochondria are specialized compartments, or organelles, that make the cellular fuel ATP. A sample of Dr. Luft's patient's muscle contained too many mitochondria, and their shape was abnormal. In addition, too little ATP was forming inside them.

What is now known as *Luft's syndrome* was the first human disease to be linked directly to a defective cell organelle. A person with this disorder is like a city with half of its power plants shut down.

Skeletal and heart muscles, the brain, and other hardworking body parts with the greatest energy needs are hurt the most.

More than a hundred other mitochondrial disorders are now known. Some of them run in families. These diseases generally affect only a small number of people. While this is good news from one standpoint, it also means that there is relatively little demand for drugs that might help save the lives of affected persons. There isn't much financial incentive for pharmaceutical companies to develop such so-called orphan drugs.

In this chapter we look at how cells are built and operate—bringing in some substances, releasing or keeping out others, and conducting their activities with the proverbial "Swiss watch" precision. As you read, keep in mind that all the body's living cells are adapted to function in the internal steady state called homeostasis.

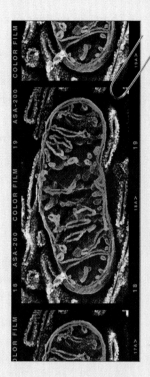

 How Would You Vote? Should pharmaceutical companies receive financial incentives (such as tax breaks) to search for cures for diseases that affect only a small number of people? Cast your vote online at www.thomsonedu.com/biology/starr/humanbio.

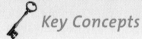 *Key Concepts*

BASIC CELL FEATURES
Cells have an outer plasma membrane, a semifluid interior called cytoplasm, and an inner region of DNA. Most cells are too small to be visible without the aid of a microscope.

CELLS AND THEIR PARTS
All cells except bacteria are eukaryotic: Their cytoplasm contains a nucleus and other organelles, which are compartments that have specialized functions. The DNA is in the nucleus.

HOW CELLS GAIN ENERGY
Cells use organic compounds to make a chemical called ATP, which fuels life processes. The major pathway for making ATP starts in the cytoplasm and ends in organelles called mitochondria.

Links to Earlier Concepts

The living cell is one of the earliest levels of organization in nature (1.3).

In this chapter, you will learn how lipids are organized to form cell membranes (2.10). You will also see where DNA and RNA are found in cells (2.13) and which cell structures use amino acids and carbohydrates as building blocks for other molecules, such as proteins (2.9, 2.11, 2.12).

The chapter explains principles that govern the movement of water and solutes into and out of cells (2.6). It also considers how cells make and use the nucleotide ATP to fuel their activities (2.13).

3.1 What Is a Cell?

LINKS TO
SECTIONS
1.1, 1.3, AND 2.10

From its size and shape to the structure of its parts, a cell is built to allow it to carry out life functions efficiently.

There are trillions of cells in your body, and each one is a highly organized bit of life. A desire to understand cells led early biologists to develop the **cell theory**:

1. **Every organism is composed of one or more cells.**

2. **The cell is the smallest unit having the properties of life.**

3. **All cells come from pre-existing cells.**

These basic ideas still hold true, and they provide the foundation for everything that modern researchers have learned about cells.

ALL CELLS ARE ALIKE IN THREE WAYS

All living cells have three things in common. They have an outer **plasma membrane**, they contain DNA, and they contain cytoplasm.

THE PLASMA MEMBRANE This thin outer covering encloses the cell's internal parts, so that the cell's life-sustaining activities can go on apart from events that may be taking place outside the cell. The plasma membrane does not completely isolate the cell interior. Substances still can move across it, as you will read later in this chapter.

DNA A cell has DNA somewhere inside it, along with molecules that can copy or read the inherited genetic instructions DNA carries.

CYTOPLASM Cytoplasm (SY-toe-plasm) is everything between the plasma membrane and the region of DNA. It consists of a thick, jellylike fluid, the **cytosol**, and various other components.

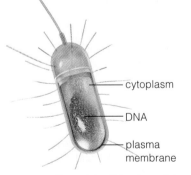

- cytoplasm
- DNA
- plasma membrane

a Bacterial cell (prokaryotic)

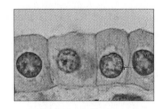

b Animal cell (eukaryotic)

Figure 3.1 The two basic types of cells. (**a**) A prokaryotic cell. (**b**) Eukaryotic cells. These are kidney cells and are shown here in cross section. A dye makes the nucleus look reddish.

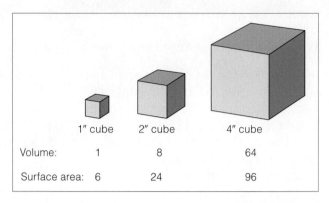

	1" cube	2" cube	4" cube
Volume:	1	8	64
Surface area:	6	24	96

Figure 3.2 The relationship between surface area and volume. Here, the "cells" are boxes, which have six sides. If the linear dimensions of a box double, the volume increases 8 times but the surface area increases only 4 times. As in the text example, if the linear dimensions increase by 4 times, the volume is 64 times greater but the surface area is only 16 times larger.

THERE ARE TWO BASIC KINDS OF CELLS

Cells are classified into two basic kinds, depending on how they are organized internally (Table 3.1). In a **prokaryotic cell** (prokaryotic means "before the nucleus") nothing separates the cell's DNA from other internal cell parts. Bacteria, like the one diagrammed in Figure 3.1*a*, are the only prokaryotic cells.

By contrast, all other cells are **eukaryotic cells** ("true nucleus"). In their cytoplasm are tiny compartments and sacs called **organelles** ("little organs"). One organelle, the nucleus, contains the DNA of a eukaryotic cell. Nuclei are clearly visible in the cells pictured in Figure 3.1*b*.

WHY ARE CELLS SMALL?

You may be wondering how small cells really are. Can any be observed with the unaided eye? There are a few, including the yolks of bird eggs and the fish eggs we call caviar. These cells can be large because when they are fully developed their metabolism is fairly sluggish. They are mainly storehouses for substances that will nourish developing young. Cells that are more active typically are

Table 3.1	Eukaryotic and Prokaryotic Cells Compared		
		Eukaryotic	Prokaryotic
Plasma membrane		yes	yes
DNA-containing region		yes	yes
Cytoplasm		yes	yes
Nucleus inside a membrane		yes	no

BASIC CELL FEATURES

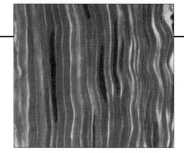

a Smooth muscle cell

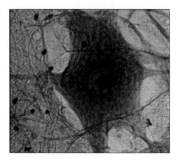

b Motor neuron

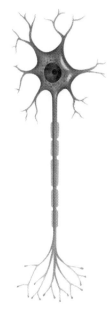

Figure 3.3 Two types of cells in the human body. (**a**) Smooth muscle cells, found in the walls of hollow organs such as the stomach, have an elongated shape. (**b**) A motor neuron, a type of nerve cell, has threadlike extensions.

so small that they can only be seen with a microscope. For instance, a human red blood cell is about 8 millionths of a meter across, so tiny that you could line up 2,000 of them across your thumbnail.

The **surface-to-volume ratio** is responsible for the small size of cells. This ratio is a physical relationship. It dictates that as the linear dimensions of a three-dimensional object increase, the volume of the object increases faster than its surface area does (Figure 3.2). For instance, if a round cell grew like an inflating balloon so that its diameter increased to 4 times the starting girth, the volume inside the cell would be 64 times more than before, but the cell's surface would be just 16 times larger. The cell would not have enough surface area to allow nutrients to flow inward rapidly, or for wastes or cell products to move rapidly outward. In short order the cell would die.

A large, round cell also would have trouble moving materials through its cytoplasm. In small cells, though, random, tiny motions of molecules easily distribute materials. If a cell isn't small, it probably is long and thin or has folds that increase its surface area relative to its volume. The smaller or narrower or more frilly the cell, the more efficiently materials can cross its surface and disperse inside it. Figure 3.3 shows two of the many shapes of cells in your own body. Part *a* depicts long, slender cells in a type of muscle called smooth muscle. The cells of muscle that attaches to the skeleton also are

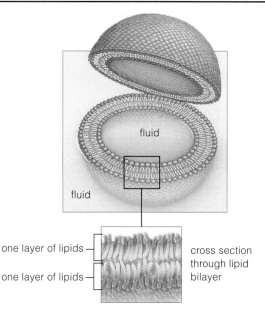

Figure 3.4 *Animated!* Phospholipids arranged in a lipid bilayer.

slender. In the biceps of your upper arm they are many inches long—as long as the muscle itself.

MEMBRANES ENCLOSE CELLS AND ORGANELLES

A eukaryotic cell and its organelles are enclosed by membranes. Most of the molecules in cell membranes are phospholipids, which were introduced in Section 2.10. You may remember that a phospholipid has a hydrophilic (water-loving) head and two fatty acid tails, which are hydrophobic (water-dreading). When a large number of phospholipids are immersed in water, they interact with the water molecules and with one another. They may spontaneously organize into two layers with all the hydrophobic tails sandwiched between all the heads (Figure 3.4). This heads-out, tails-in arrangement is called a **lipid bilayer**. All cell membranes have the lipid bilayer structure. The hydrophilic heads of the phospholipids are dissolved in the watery fluids inside and outside cells.

A cell has an outer plasma membrane, which encloses its jellylike cytoplasm and its DNA. The cells of humans, like those of other complex organisms, are eukaryotic: Their DNA is contained in an organelle, the nucleus.

In prokaryotic cells, the bacteria, the DNA is not contained inside a nucleus.

Most cells are small and have a large surface area relative to their volume. These features permit the efficient inward flow of nutrients and outward flow of wastes.

A cell's membranes are built mostly of phospholipids arranged in a bilayer.

3.2 The Parts of a Eukaryotic Cell

The interior of a cell is divided into organelles, each of which has one or more special functions.

In every eukaryotic cell, at any given moment, a vast number of chemical reactions are going on. Many of the reactions would conflict if they occurred in the same cell compartment. For example, a molecule of fat can be built by some reactions and taken apart by others, but a cell gains nothing if both sets of reactions proceed at the same time on the same fat molecule.

In eukaryotic cells, including those of the human body, organelles (Table 3.2) solve this problem. Their outer membrane physically separates the inside of an organelle from the rest of the cytoplasm. It also controls the types and amounts of substances that enter or leave the organelle. For example, organelles called lysosomes contain enzymes that break down various unwanted substances. If the enzymes escaped from the organelle, they could destroy the entire cell.

Organelles also may serve as "way stations" for operations that occur in steps. Proteins are assembled and modified in steps involving several organelles.

Figure 3.5 shows where organelles and some other structures might be located in a body cell. This is only a general picture of cells. There are major differences in the structures and functions of cells in different tissues.

Table 3.2 Common Features of Eukaryotic Cells	
ORGANELLES AND THEIR MAIN FUNCTIONS:	
Nucleus	Contains the cell's DNA
Endoplasmic reticulum (ER)	Routes and modifies newly formed polypeptide chains; also, where lipids are assembled
Golgi body	Modifies polypeptide chains into mature proteins; sorts and ships proteins and lipids for secretion or for use inside cell
Various vesicles	Transport or store a variety of substances; break down substances and cell structures in the cell; other functions
Mitochondria	Produce ATP
OTHER STRUCTURES AND THEIR FUNCTIONS:	
Ribosomes	Assemble polypeptide chains
Cytoskeleton	Gives overall shape and internal organization to cell; moves the cell and its internal parts

Organelles physically isolate chemical reactions inside cells. They also provide separate locations for activities that occur in a sequence of steps.

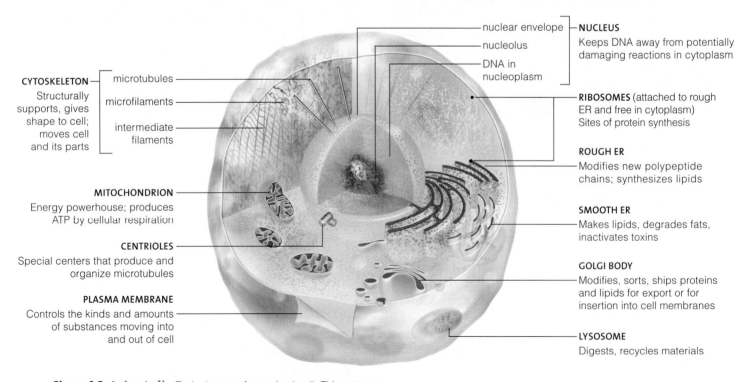

CYTOSKELETON Structurally supports, gives shape to cell; moves cell and its parts
— microtubules
— microfilaments
— intermediate filaments

MITOCHONDRION Energy powerhouse; produces ATP by cellular respiration

CENTRIOLES Special centers that produce and organize microtubules

PLASMA MEMBRANE Controls the kinds and amounts of substances moving into and out of cell

nuclear envelope
nucleolus
DNA in nucleoplasm
— **NUCLEUS** Keeps DNA away from potentially damaging reactions in cytoplasm

RIBOSOMES (attached to rough ER and free in cytoplasm) Sites of protein synthesis

ROUGH ER Modifies new polypeptide chains; synthesizes lipids

SMOOTH ER Makes lipids, degrades fats, inactivates toxins

GOLGI BODY Modifies, sorts, ships proteins and lipids for export or for insertion into cell membranes

LYSOSOME Digests, recycles materials

Figure 3.5 *Animated!* Typical parts of an animal cell. This cutaway diagram corresponds roughly to the micrograph in Figure 3.6.

CELLS AND THEIR PARTS

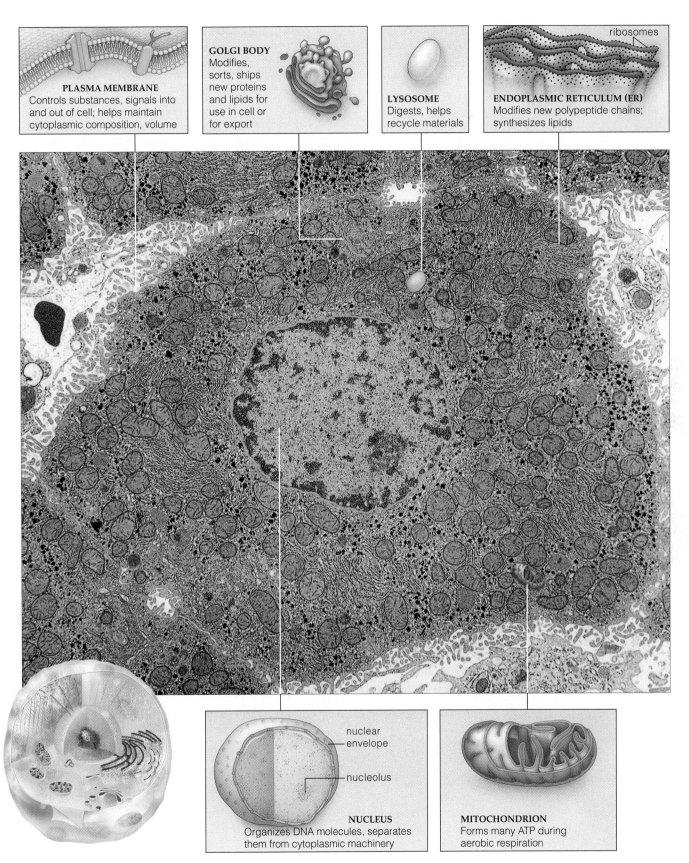

PLASMA MEMBRANE
Controls substances, signals into and out of cell; helps maintain cytoplasmic composition, volume

GOLGI BODY
Modifies, sorts, ships new proteins and lipids for use in cell or for export

LYSOSOME
Digests, helps recycle materials

ENDOPLASMIC RETICULUM (ER)
Modifies new polypeptide chains; synthesizes lipids

ribosomes

nuclear envelope

nucleolus

NUCLEUS
Organizes DNA molecules, separates them from cytoplasmic machinery

MITOCHONDRION
Forms many ATP during aerobic respiration

Figure 3.6 Transmission electron micrograph of a liver cell in cross section. Although this is a rat liver cell, liver cells in humans are not much different.

3.3 The Plasma Membrane: A Double Layer of Lipids

LINKS TO SECTIONS 2.8 AND 2.10

The plasma membrane isn't a solid, rigid wall between a cell's cytoplasm and the fluid outside. If it were, needed substances couldn't enter the cell and wastes couldn't leave it. Instead, the plasma membrane has an oily, fluid quality, something like cooking oil. The membrane also is extremely thin. A thousand stacked like pancakes would be about as thick as this page.

THE PLASMA MEMBRANE IS A MIX OF LIPIDS AND PROTEINS

In Figure 3.4 you've already seen a simple picture of a plasma membrane lipid bilayer with its "sandwich" of phospholipids. Figure 3.7 below gives a more complete idea of how biologists visualize this structure. It is often described as a "mosaic" of proteins and various lipids: phospholipids, glycolipids, and, in human cells (and those of other animals), the lipid cholesterol. The proteins are embedded in the bilayer or attach to its outer or inner surface, where they have specific functions.

What makes the membrane fluid? For one thing, the molecules that make it up move. Most phospholipids can spin on their long axis like a chicken on a rotisserie. They also move sideways and flex their tails—movements that help keep neighboring molecules from packing into a solid layer. The short or kinked hydrophobic tails of lipids in the bilayer also give the membrane a fluid quality.

PROTEINS PERFORM MOST OF THE FUNCTIONS OF CELL MEMBRANES

The proteins that are embedded in or attached to a lipid bilayer carry out most of a cell membrane's functions. Many of these proteins are enzymes; you may recall from Chapter 2 that enzymes speed chemical reactions in cells. Other membrane proteins serve a range of functions. Some are channels through the membrane, while others are transporters that move substances across it. Still others are receptors; they are like docks for signaling molecules, such as hormones, that trigger changes in cell activities. Recognition proteins that wave like flags on the surface of a cell are "fingerprints" that identify the cell as being of a specific type. You will read more about membrane proteins in upcoming chapters.

Before we examine the internal parts of cells, read the *Science Comes to Life* feature on the facing page, which describes how biologists use microscopes to study these tiny bits of life.

The plasma membrane is a lipid bilayer. It is a mix of various lipids and proteins and has a fluid quality. Proteins of the bilayer carry out most of the membrane's functions, such as moving substances across the membrane, binding substances, and serving as identity tags.

Figure 3.7 *Animated!* Cutaway view of a plasma membrane, based on the fluid mosaic model of membrane structure. Some of the proteins in the plasma membrane are drawn as ribbons to suggest the intricate folding of long polypeptide chains. Section 2.12 discusses how the chains form.

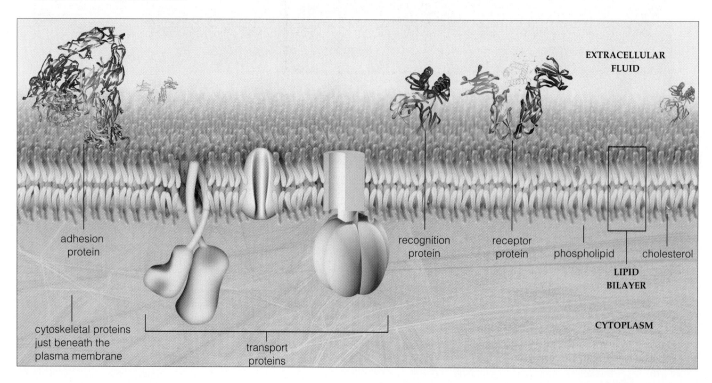

adhesion protein

cytoskeletal proteins just beneath the plasma membrane

transport proteins

recognition protein

receptor protein

phospholipid

cholesterol

EXTRACELLULAR FLUID

LIPID BILAYER

CYTOPLASM

CELLS AND THEIR PARTS

3.4 How Do We See Cells?

Microscopy has allowed us to learn a great deal about cells in the human body. A photograph formed using a microscope is called a micrograph.

The micrographs in Figure 3.8 compare the sorts of detail different types of microscopes can reveal. For example, the red blood cells in Figure 3.8*a* were viewed using a compound light microscope, in which two or more glass lenses bend (refract) incoming light rays to form an enlarged image of a specimen. With this method, the cell must be small or thin enough for light to pass through, and its parts must differ in color or optical density from their surroundings. Unfortunately, most cell parts are nearly colorless and they have about the same density. For this reason, before viewing cells through a light microscope, researchers expose the cells to dyes that react with some cell parts but not with others. Even with the best glass lens system, however, light microscopes only provide sharp images when the diameter of the object being viewed is magnified by 2,000 times or less.

Electron microscopes use magnetic lenses to bend beams of electrons. They reveal smaller details than even the best light microscopes can. There are several types, with new innovations occurring often.

A transmission electron microscope uses a magnetic field as the "lens" that bends a stream of electrons and focuses it into an image, which then is magnified. With a scanning electron microscope, a beam of electrons is directed back and forth across a specimen thinly coated with metal. The metal emits some of its own electrons, and then the electron energy is converted into an image of the specimen's surface on a television screen. Most of the images have fantastic depth (Figure 3.8*b*, right).

A scanning tunneling microscope magnifies objects up to 100 million times (Figure 3.8*c*). The scope's needlelike probe has a single atom at its tip. As an electrical current passes between the tip and a specimen's surface, electrons "tunnel" from the probe to the specimen. A computer analyzes the tunneling motion and makes a 3-D view of the surface.

Figure 3.8 *Animated!* Human red blood cells viewed with different types of microscopes. (**a**) Red blood cells inside a small blood vessel, as revealed by a light microscope. (**b**) Electron micrographs. *Top:* This transmission electron micrograph (TEM) shows the inside of mature red blood cells, which are packed with hemoglobin. *Bottom:* A scanning electron micrograph (SEM) with color added shows the "doughnut without a hole" shape of red blood cells. (**c**) The green-colored image is a micrograph of DNA obtained with a scanning tunneling microscope.

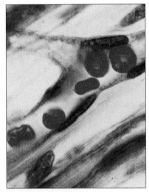

Compound light microscope   **a**

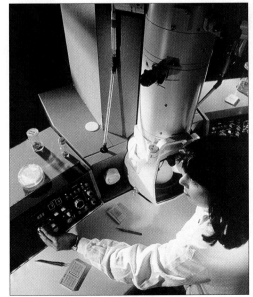

Transmission electron microscope

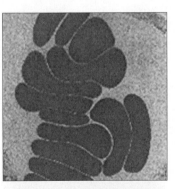
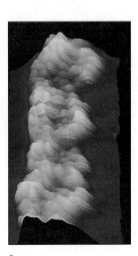
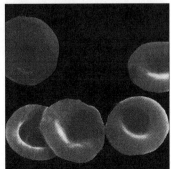

c

b

3.5 The Nucleus

The nucleus is often described as a cell's master control center. It also is a protective "isolation chamber" for the genetic material DNA.

The **nucleus** encloses the DNA of a eukaryotic cell. DNA contains instructions for building a cell's proteins and, through those proteins, for determining a cell's structure and function. In a human cell there are forty-six DNA molecules that would stretch more than 6 feet if they were stretched out end to end.

Figure 3.9 shows the structure of the nucleus, and Table 3.3 lists its five main components. The nucleus has several key functions. First, it prevents DNA from getting entangled with structures in the cytoplasm. When a cell divides, its DNA molecules must be copied so that each new cell receives a full set. Keeping the DNA separate makes it easier to copy and organize these hereditary instructions. In addition, outer membranes of the nucleus are a boundary where cells control the movement of substances to and from the cytoplasm.

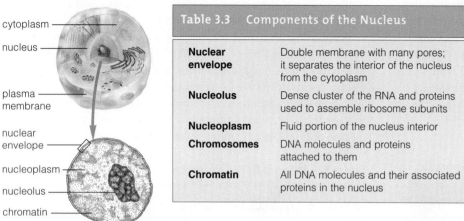

Table 3.3	Components of the Nucleus
Nuclear envelope	Double membrane with many pores; it separates the interior of the nucleus from the cytoplasm
Nucleolus	Dense cluster of the RNA and proteins used to assemble ribosome subunits
Nucleoplasm	Fluid portion of the nucleus interior
Chromosomes	DNA molecules and proteins attached to them
Chromatin	All DNA molecules and their associated proteins in the nucleus

A NUCLEAR ENVELOPE ENCLOSES THE NUCLEUS

Unlike the cell itself, a nucleus has two outer lipid bilayers, one pressed against the other. This double-membrane system is called a **nuclear envelope** (Figure 3.10). The bilayers surround the fluid part of the nucleus (the nucleoplasm), and many proteins are embedded in them. The outer section of the nuclear envelope merges with the membrane of ER, an organelle in the cytoplasm, which you will read more about in Section 3.6.

Threadlike bits of protein attach to the inner surface of the nuclear envelope. They anchor DNA molecules to the envelope and help keep them organized.

Membrane proteins that span both bilayers have a wide variety of functions. Some are receptors or transporters. Others form pores, as you can see in Figure 3.10*b*. The pores are passageways for small ions and for molecules dissolved in the watery fluid inside and outside the nucleus to cross the nuclear membrane.

Figure 3.9 The nucleus of an animal cell. The small arrows on this micrograph (a photograph taken with an electron microscope) point to pores where substances can move through the nuclear envelope in controlled ways.

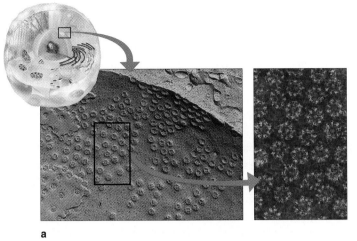

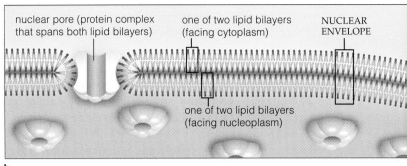

nuclear pore (protein complex that spans both lipid bilayers)

one of two lipid bilayers (facing cytoplasm)

NUCLEAR ENVELOPE

one of two lipid bilayers (facing nucleoplasm)

a

b

Figure 3.10 *Animated!* (**a**) Part of the outer surface of a nuclear envelope. *Left:* This specimen was prepared so as to show the layering of its two lipid bilayers. *Right:* Nuclear pores. Each pore across the envelope is a cluster of membrane proteins. It permits the selective transport of substances into and out of the nucleus. (**b**) Sketch of the nuclear envelope's structure.

THE NUCLEOLUS IS WHERE CELLS MAKE THE UNITS OF RIBOSOMES

As a cell grows, one or more dense masses appear inside its nucleus. Each mass is a **nucleolus** (noo-KLEE-uh-luhs; plural: nucleoli), a construction site where certain RNAs and proteins are combined to make the various parts of ribosomes. These subunits eventually will cross through nuclear pores to the cytoplasm. There, they join briefly to form ribosomes, the "workbenches" where amino acids are assembled into proteins.

DNA IS ORGANIZED IN CHROMOSOMES

When a eukaryotic cell is not dividing, you cannot see individual DNA molecules, nor can you see that each consists of two strands twisted together. The nucleus just looks grainy, as in Figure 3.9. When a cell is preparing to divide, however, it copies all of its DNA so that each new cell will get all the required hereditary instructions. Soon the duplicated DNA molecules are visible as long threads. They then fold and twist into a compact structure:

Early microscopists named the seemingly grainy substance in the nucleus *chromatin*, and they called the compact structures *chromosomes* ("colored bodies"). Today we define **chromatin** as the cell's DNA along with the proteins associated with it. We also understand that chromatin makes up each **chromosome**—a double-stranded DNA molecule that carries genetic information. A key point to remember is that a chromosome can look different at different times, being grainy or compact depending on whether the cell is dividing or is in some other part of its life cycle.

EVENTS THAT BEGIN IN THE NUCLEUS CONTINUE TO UNFOLD IN THE CELL CYTOPLASM

Outside the nucleus, new polypeptide chains for proteins are assembled on ribosomes. Many of them are used at once or stockpiled in the cytoplasm. Others enter the endomembrane system. As you'll read in the next section, this system includes various structures. It is where many proteins get their final form and where lipids are assembled and packaged.

The nucleus, an organelle with two outer membranes, keeps a cell's DNA molecules separated from the cytoplasm.

The separation makes it easier to organize the DNA and to copy it before a cell divides.

Pores across the nuclear envelope control the passage of many substances between the nucleus and the cytoplasm.

one chromosome (one dispersed DNA molecule + proteins; not duplicated)

one chromosome (threadlike and now duplicated; two DNA molecules + proteins)

one chromosome (duplicated and also condensed tightly)

3.6 The Endomembrane System

Some organelles assemble lipids, modify new polypeptide chains into final proteins, and then sort and ship these products to various destinations. These organelles are part of the cell's endomembrane system.

ER IS A PROTEIN AND LIPID ASSEMBLY LINE

The functions of the **endomembrane system** begin with **endoplasmic reticulum**, or **ER**. The ER is a flattened channel that starts at the nuclear envelope and snakes through the cytoplasm (Figure 3.11). At various points inside the channel, lipids are assembled and "raw" polypeptide chains are modified into final proteins. In different places the ER looks rough or smooth, depending mainly on whether the organelles called **ribosomes** are attached to the side of the membrane that faces the cytoplasm. Like a workbench, a ribosome is a platform for building a cell's proteins.

Rough ER is studded with ribosomes (Figure 3.11*b*). Newly forming polypeptide chains that have a built-in signal (a string of amino acids) can enter the space inside rough ER or be incorporated into ER membranes. Once the chains are in rough ER, enzymes in the channel may attach side chains to them. Body cells that secrete finished proteins have extensive rough ER. For example, in your pancreas, ER-rich gland cells make and secrete enzymes that end up in your small intestine and help you digest your meals.

Smooth ER has no ribosomes and curves through the cytoplasm like flat connecting pipes (Figure 3.11*c*). Many cells assemble most lipids inside these pipes. In liver cells, smooth ER inactivates certain drugs and harmful by-products of metabolism. In skeletal muscle cells a type of smooth ER called sarcoplasmic reticulum stores and releases calcium ions essential for muscles to contract.

GOLGI BODIES "FINISH, PACK, AND SHIP"

A **Golgi body** is a series of flattened sacs that often resemble a stack of pancakes (Figure 3.11*d*). Enzymes in the sacs put the finishing touches on proteins and lipids, then sort and package the completed molecules in vesicles

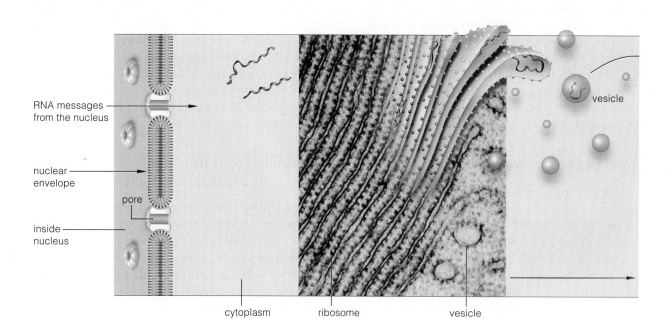

RNA messages from the nucleus

nuclear envelope

pore

inside nucleus

cytoplasm   ribosome   vesicle

vesicle

the cell nucleus

(a) RNA messages are translated into polypeptide chains on ribosomes. Many chains are stockpiled in the cytoplasm or used at once. Others enter the rough ER.

rough ER

(b) Flattened sacs of rough ER form one continuous channel between the nucleus and smooth ER. Polypeptide chains that enter the channel are modified. They will be inserted into organelle membranes or will be secreted from the cell.

Figure 3.11 *Animated!* The endomembrane system. With this system's components, many proteins are processed, lipids are assembled, and both products are sorted and shipped to destinations in the cell or to the plasma membrane to be exported out of the cell.

for shipment to specific locations. A **vesicle** is a tiny sac that moves through the cytoplasm or takes up positions in it. For example, an enzyme in one Golgi region might attach a phosphate group to a new protein and then "pack" the protein into a vesicle, thereby giving it a "mailing tag" to its proper destination. The top pancake of a Golgi body is the organelle's "shipping gate" for molecules to be exported. Here, vesicles form as patches of the membrane bulge out and then break away into the cell's cytoplasm.

A VARIETY OF VESICLES MOVE SUBSTANCES INTO AND THROUGH CELLS

Many kinds of vesicles shuttle substances around cells. A common type, the lysosome, buds from the membranes of Golgi bodies. A **lysosome** is specialized for digestion: It contains a potent stew of enzymes that speed the breakdown of proteins, complex sugars, nucleic acids, and some lipids. Lysosomes may even digest whole cells or cell parts. Often, lysosomes fuse with vesicles that

have formed at a cell's plasma membrane. The vesicles usually contain molecules, bacteria, or other items that attached to the plasma membrane. White blood cells of the immune system take in foreign material in vesicles and dispose of it.

Peroxisomes, another type of vesicle, are tiny sacs of enzymes that break down fatty acids and amino acids. The reactions produce hydrogen peroxide, a potentially harmful substance. But before hydrogen peroxide can injure the cell, another enzyme in peroxisomes converts it to water and oxygen or uses it to break down alcohol. After someone drinks alcohol, nearly half of it is broken down in peroxisomes of liver and kidney cells.

> *In the ER and Golgi bodies of the cytomembrane system, many proteins take on final form, and lipids are assembled.*
>
> *Lipids, proteins, and other substances are packaged in vesicles to be exported, stored, used to build membranes, and carry out other cell activities. Unwanted materials may be broken down in lysosomes and peroxisomes.*

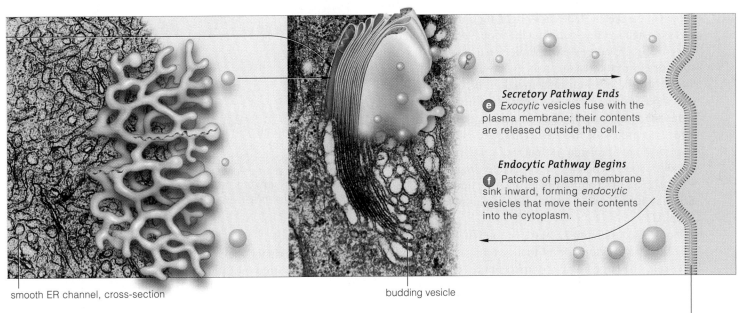

smooth ER channel, cross-section

budding vesicle

Secretory Pathway Ends
(e) *Exocytic* vesicles fuse with the plasma membrane; their contents are released outside the cell.

Endocytic Pathway Begins
(f) Patches of plasma membrane sink inward, forming *endocytic* vesicles that move their contents into the cytoplasm.

smooth ER
(c) Some proteins in the channel continue on, to smooth ER. Many become smooth ER enzymes or membrane proteins. The enzymes make lipids, inactivate toxins, and mediate other tasks.

Golgi body
(d) A Golgi body receives, processes, and then packages substances that arrive in vesicles from the ER. Other vesicles transport the substances to the plasma membrane or other parts of the cell.

plasma membrane
(g) Exocytic vesicles release cell products and wastes to the outside. Endocytic vesicles move nutrients, water, and other substances into the cytoplasm from outside.

3.7 Mitochondria: The Cell's Energy Factories

LINK TO
SECTION
2.13

For as long as a cell lives, it hums with chemical reactions. Nearly all those reactions require energy, which is provided by ATP—a life-sustaining molecule that is made in the cell's sausage-shaped mitochondria.

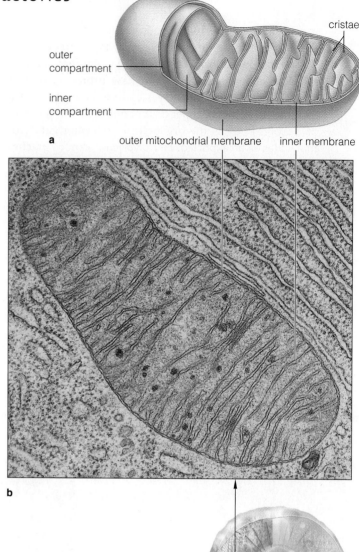

a

MITOCHONDRIA MAKE ATP

Section 2.13 introduced the main energy carrier in cells, ATP. Because ATP can deliver energy to nearly all the reaction sites in a cell, ATP drives nearly all of a cell's activities. ATP forms during reactions that break down organic compounds to carbon dioxide and water in a **mitochondrion** (plural: mitochondria).

Only eukaryotic cells contain mitochondria. The one shown in Figure 3.12 gives you an idea of their structure. The kind of ATP-forming reactions that occur in mitochondria extract far more energy from organic compounds than can be obtained by any other means. The reactions cannot be completed without an ample supply of oxygen. Every time you inhale, you are taking in oxygen mainly for mitochondria in your cells.

ATP FORMS IN AN INNER COMPARTMENT OF THE MITOCHONDRION

A mitochondrion has a double-membrane system. As shown in the sketch at the upper right, the outer membrane faces the cell's cytoplasm. The inner one generally folds back on itself, accordion-fashion. Each fold is a crista (KRIS-tuh; plural: cristae). This membrane system is the key to the mitochondrion's function, because it forms two separate compartments inside the organelle. In the outer one, enzymes and other proteins stockpile hydrogen ions. This process is fueled by energy from electrons. As electrons are depleted of energy, oxygen binds and removes them. When the stockpiled hydrogen ions later flow out of the compartment, energy inherent in the flow (as in a flowing river) powers the reactions that form ATP.

Look back at Figure 3.6 and you can see mitochondria all through the cytoplasm of that one thin slice from one liver cell. The large number of mitochondria indicate that the liver is a highly active, energy-demanding organ.

Mitochondria have intrigued biologists because they are about the same size as bacteria and function like them in many ways as well. Mitochondria even have their own DNA and some ribosomes, and they divide independently of the cell they are in. Many biologists believe mitochondria evolved from ancient bacteria that were consumed by another ancient cell, yet did not die. Perhaps they were able to reproduce inside the predatory cell and its descendants. If they became permanent,

Figure 3.12 *Animated!*
(**a**) Sketch and (**b**) transmission electron micrograph of a thin slice through a typical mitochondrion. Reactions inside this organelle produce ATP, the major energy carrier in cells.

protected residents, they might have lost structures and functions required for independent life while they were becoming mitochondria, the ATP-producing organelles without which we humans could not survive.

The organelles called mitochondria are the ATP-producing powerhouses of eukaryotic cells.

ATP is produced by reactions that take place in the inner compartment formed by a mitochondrion's double-membrane system. These reactions require oxygen.

3.8 The Cell's Skeleton

A cell's internal framework is called the cytoskeleton. It is not permanently rigid, however. Its elements assemble and disassemble as needed for cell activities.

The **cytoskeleton** is a system of interconnected fibers, threads, and lattices in the cytosol (Figure 3.13). It gives cells their shape and internal organization, as well as their ability to move. **Microtubules** are the largest cytoskeleton elements. Their main function is to spatially organize the interior of the cell, although microtubules also help move cell parts.

Microfilaments often reinforce some part of a cell, such as the plasma membrane. Some membrane proteins are anchored in place by microfilaments.

Some kinds of cells also have **intermediate filaments** that add strength much as steel rods strengthen concrete pillars. Intermediate filaments also anchor the filaments of two other proteins, called actin and myosin, which interact in muscle cells and enable the muscle to contract. Chapter 6 looks at this process.

Some types of cells move about by **flagella** (singular: flagellum) or **cilia** (singular: cilium). In both structures nine pairs of microtubules ring a central pair; a system of spokes and links holds this "9 + 2 array" together (Figure 3.14). The flagellum or cilium bends when microtubules in the ring slide over each other. Whiplike flagella propel human sperm (Figure 3.13*b*).

Cilia are shorter than flagella, and there may be more of them per cell. In your respiratory tract, thousands of

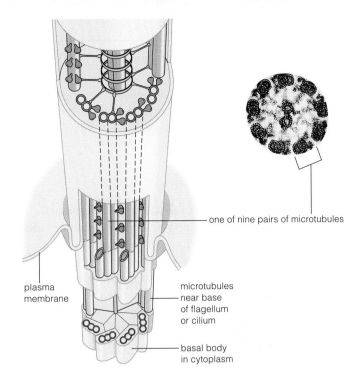

Figure 3.14 *Animated!* How microtubules are arranged inside cilia and flagella.

one of nine pairs of microtubules

plasma membrane

microtubules near base of flagellum or cilium

basal body in cytoplasm

ciliated cells whisk out mucus laden with dust or other undesirable material. The microtubules of cilia and flagella arise from **centrioles**, which remain at the base of the completed structure as a "basal body." As you will read in a later chapter, centrioles have an important role when a cell divides.

The cytoskeleton gives each cell its shape, internal structure, and capacity for movement. Its main elements are microtubules, microfilaments, and intermediate filaments.

Certain types of cells move their bodies or parts by way of flagella or cilia.

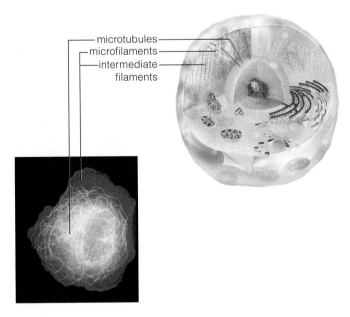

microtubules
microfilaments
intermediate filaments

Figure 3.13 (**a**) The cytoskeleton of a human pancreas cell. The blue region is DNA. (**b**) The "tails" of sperm cells are whiplike flagella.

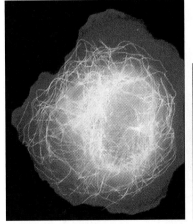

a

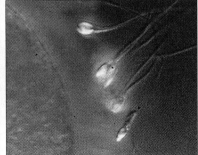

b

3.9 How Diffusion and Osmosis Move Substances across Membranes

LINKS TO
SECTIONS
2.10 AND 2.12

There is fluid on both sides of a cell's plasma membrane. The kinds and amounts of dissolved substances in the fluid are not the same on the two sides. A cell must maintain those differences—a task of the plasma membrane.

THE PLASMA MEMBRANE IS "SELECTIVE"

As you already know, a cell's plasma membrane is a bilayer containing lipids and proteins. These molecules give the membrane **selective permeability**. They allow some substances but not others to enter and leave a cell (Figure 3.15). They also control *when* a substance can cross and how much crosses at a given time. Lipids in the bilayer are mostly nonpolar, so they let small, nonpolar molecules such as carbon dioxide and oxygen slip across. Water molecules are polar, but some can slip through gaps that briefly open up in the bilayer. Ions, and large polar molecules such as glucose, cross the bilayer through the interior of its transport proteins. Why does a solute move one way or another at any given time? The answer starts with concentration gradients.

IN DIFFUSION, A SOLUTE MOVES DOWN A CONCENTRATION GRADIENT

"Concentration" refers to the number of molecules of a substance in a certain volume of fluid. "Gradient" means that the number of molecules in one region is not the same as in another. Therefore, a **concentration gradient** is a difference in the number of molecules or ions of a given substance in two neighboring regions. Molecules are always randomly moving between the two regions, but on balance, unless other forces come into play, they tend to move into the region where they are less concentrated.

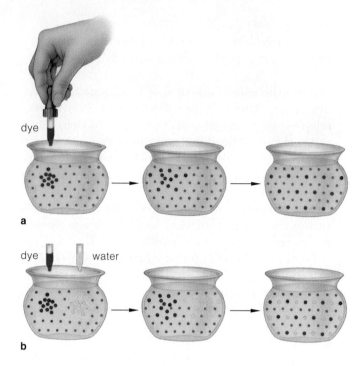

a

b

Figure 3.16 *Animated!* Two examples of diffusion. (**a**) A drop of dye enters a bowl of water. Gradually the dye molecules become evenly dispersed through the molecules of water. (**b**) The same thing happens with the water molecules. Here, red dye and yellow dye are added to the same bowl. Each substance will move down its own concentration gradient.

The net movement of like molecules or ions down a concentration gradient is called **diffusion**. In living organisms, the diffusion of a substance across a cell membrane is called **passive transport**, "passive" because a cell does not have to draw energy from ATP, the cell's chemical fuel, to make it happen. Diffusion moves substances to and from cells, and into and out of the fluids bathing them. Diffusion also moves substances through the cytoplasm of a cell.

If a solution contains more than one kind of solute, each kind diffuses down its own concentration gradient. For example, if you put a drop of dye in one side of a bowl of water, the dye molecules diffuse to the region where they are less concentrated. Likewise, the water molecules move in the opposite direction, to the region where *they* are less concentrated (Figure 3.16).

Molecules diffuse faster when the gradient is steep. Where molecules are most concentrated, more of them move outward, compared to the number that are moving in. As the gradient smooths out, there is less difference in the number of molecules moving either way. Even when the gradient disappears, molecules are still moving, but the total number going one way or the other during a given interval is about the same. For charged molecules,

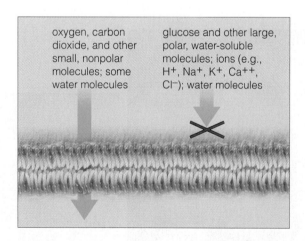

| oxygen, carbon dioxide, and other small, nonpolar molecules; some water molecules | glucose and other large, polar, water-soluble molecules; ions (e.g., H⁺, Na⁺, K⁺, Ca⁺⁺, Cl⁻); water molecules |

Figure 3.15 *Animated!* The selective permeability of cell membranes.

CELLS AND THEIR PARTS

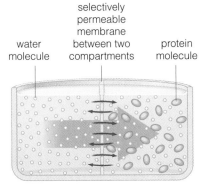

water
molecule

selectively
permeable
membrane
between two
compartments

protein
molecule

Figure 3.17 *Animated!* How a solute concentration gradient affects osmotic movement of water.

Start with a container divided by a membrane that water but not proteins can cross. Pour water into the left side and pour the same volume of a protein-rich solution into the right side. There, proteins occupy some of the space. The net diffusion of water in this example is from left to right (large gray arrow).

transport is influenced by both the concentration gradient and the *electric gradient*—a difference in electric charge across the cell membrane. As you will read in a later chapter, nerve impulses depend on electric gradients.

WATER CROSSES MEMBRANES BY OSMOSIS

Because the plasma membrane is selectively permeable, the concentration of a solute can increase on one side of the membrane but not on the other. For example, the cytoplasm of most cells usually contains solutes (such as proteins) that cannot diffuse across the plasma membrane. When solutes become more concentrated on one side of the plasma membrane, the resulting solute concentration gradients affect how water diffuses across the membrane. **Osmosis** is the name for the diffusion of water across a selectively permeable membrane in response to solute concentration gradients (Figure 3.17).

Tonicity is the ability of a solution to draw water into or out of a cell. When solute concentrations in the fluids on either side of a cell membrane are the same, the fluids are *isotonic* (*iso-* means same) and there is no net flow of water in either direction across the membrane. When the solute concentrations are not equal, one fluid is *hypotonic*— it has fewer solutes. The other has more solutes and it is *hypertonic*. Figure 3.18 shows how the tonicity of a fluid affects red blood cells. A key point to remember: Water always tends to move from a hypotonic solution to a hypertonic one because it moves down its concentration gradient.

If too much water enters a cell by osmosis, in theory the cell will swell up until it bursts. This is not a danger for most body cells because they can selectively move solutes out—and as solutes leave, so does water. Also, the cytoplasm exerts pressure against the plasma membrane. When this pressure counterbalances the tendency of water to follow its concentration gradient, osmosis stops.

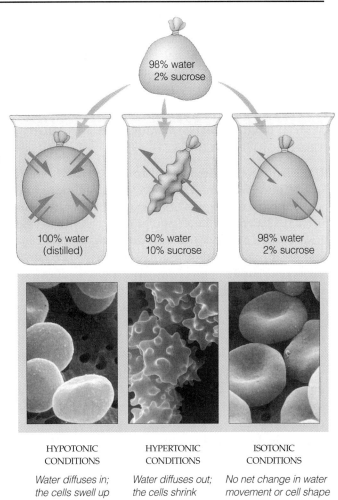

98% water
2% sucrose

100% water
(distilled)

90% water
10% sucrose

98% water
2% sucrose

HYPOTONIC
CONDITIONS

Water diffuses in; the cells swell up

HYPERTONIC
CONDITIONS

Water diffuses out; the cells shrink

ISOTONIC
CONDITIONS

No net change in water movement or cell shape

Figure 3.18 *Animated!* Tonicity and the diffusion of water. In the sketches, membrane-like bags through which water but not sucrose can move are placed in hypotonic, hypertonic, and isotonic solutions. In each container, arrow width represents the relative amount of water movement. The sketches show what happens when red blood cells—which cannot actively take in or expel water—are placed in similar solutions.

Moment to moment, cell activities and other events change the factors that affect the solute concentrations of body fluids and water movements between them. Cells that are not equipped to adjust to such differences shrivel or burst, as Figure 3.18 illustrates. In a later chapter we will discuss ways in which osmotic water movements help maintain the body's proper water balance.

The net movement of like molecules (or ions) from a region of higher concentration to a region of lower concentration is called diffusion.

Osmosis is the net diffusion of water across a selectively permeable membrane. Most body cells have mechanisms for adjusting the movement of water and solutes into and out of the cell.

3.10 Other Ways Substances Cross Cell Membranes

Substances also cross cell membranes by mechanisms called facilitated diffusion, active transport, exocytosis, and endocytosis.

MANY SOLUTES CROSS MEMBRANES THROUGH TRANSPORT PROTEINS

Diffusion directly through a plasma membrane is just one of three ways by which substances can move into and out of a cell (Figure 3.19). You may remember that Section 3.3 mentioned transport proteins, which span the lipid bilayer. Many of them provide a channel for ions and other solutes to diffuse across the membrane down their concentration gradients. The process does not require ATP energy, so it is a form of passive transport (Figure 3.20). It is called **facilitated diffusion** because the transport proteins provide a route for the solute that is crossing the cell membrane.

Two features allow a transport protein to fulfill its role. First, its interior can open to both sides of a cell membrane. Second, when the protein interacts with a solute, its shape changes, then changes back again. The changes move the solute through the protein, from one side of the lipid bilayer to the other. Transport proteins are "choosy" about which solutes pass through them. For example, the protein that transports amino acids will not carry glucose.

As cells use and produce substances, the concentrations of solutes on either side of their membranes are constantly changing. A cell also must actively move certain solutes in, out, and through its cytoplasm. Action requires energy, and so cells have mechanisms called "membrane pumps" that move substances across membranes *against*

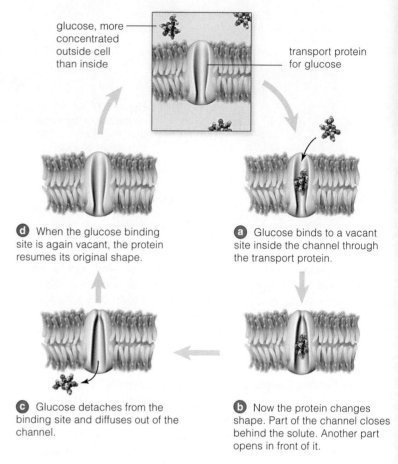

d When the glucose binding site is again vacant, the protein resumes its original shape.

a Glucose binds to a vacant site inside the channel through the transport protein.

c Glucose detaches from the binding site and diffuses out of the channel.

b Now the protein changes shape. Part of the channel closes behind the solute. Another part opens in front of it.

Figure 3.20 *Animated!* Facilitated diffusion, a form of passive transport. In this mechanism, a solute can move in both directions through transport proteins. The solute moves down its concentration gradient. In this example the solute is the sugar glucose.

concentration gradients. This pumping is called **active transport** (Figure 3.21). ATP provides most of the energy for active transport, and membrane pumps can continue working until the solute is *more* concentrated on the side of the membrane where it is being pumped. For example, a calcium pump helps keep the concentration of calcium inside cells at least a thousand times lower than it is outside. This difference lays the chemical foundation for muscle contraction, which we will discuss in Chapter 6.

VESICLES TRANSPORT LARGE SOLUTES

Transport proteins can only move small molecules and ions into or out of cells. To bring in or expel larger molecules or particles, cells use vesicles that form through exocytosis and endocytosis.

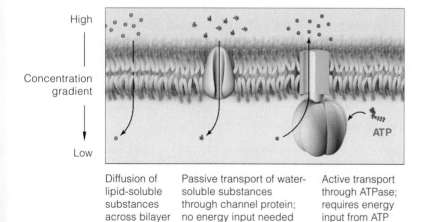

High

Concentration gradient

Low

| Diffusion of lipid-soluble substances across bilayer | Passive transport of water-soluble substances through channel protein; no energy input needed | Active transport through ATPase; requires energy input from ATP |

Figure 3.19 Membrane-crossing mechanisms.

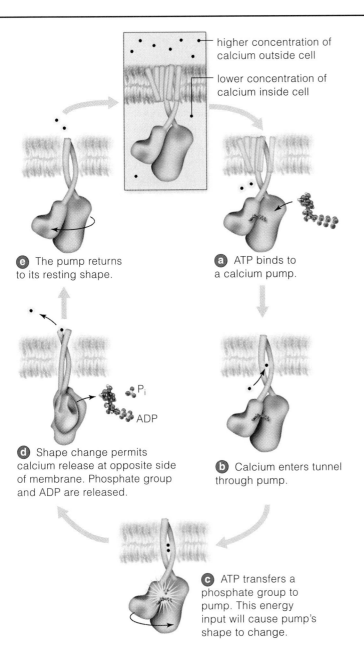

higher concentration of calcium outside cell

lower concentration of calcium inside cell

e The pump returns to its resting shape.

a ATP binds to a calcium pump.

Pᵢ
ADP

d Shape change permits calcium release at opposite side of membrane. Phosphate group and ADP are released.

b Calcium enters tunnel through pump.

c ATP transfers a phosphate group to pump. This energy input will cause pump's shape to change.

Figure 3.21 *Animated!* Active transport across a cell membrane.

---

plasma membrane

exocytic vesicle leaving cytoplasm

a

endocytic vesicle forming

b

Figure 3.22 (**a**) Exocytosis. Cells can release substances when a vesicle's membrane fuses with the plasma membrane (Figure 3.22*a*). Its contents are then released to the outside. (**b**) Endocytosis. A bit of plasma membrane balloons inward beneath water and solutes outside, then pinches off as a vesicle that moves into the cytoplasm.

In **exocytosis** ("moving out of a cell"), a vesicle moves to the cell surface and the protein-studded lipid bilayer of its membrane fuses with the plasma membrane (Figure 3.22*a*). Its contents are then released to the outside.

In **endocytosis** ("coming inside a cell"), a cell takes in substances next to its surface. A small indentation forms at the plasma membrane, balloons inward, and pinches off. The resulting vesicle transports its contents or stores them in the cytoplasm (Figure 3.22*b*). When endocytosis brings organic matter into the cell, the process is called **phagocytosis**, or "cell eating."

Transport proteins carry many solutes across cell membranes. In passive transport, a solute diffuses down its concentration gradient. In active transport, membrane pumps move solutes against their gradient. ATP provides much of the needed energy. Exocytosis and endocytosis move large molecules or particles across the membrane.

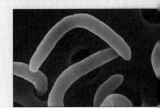

Vibrio cholerae, the cause of cholera

In places where there is little or no proper public sanitation, people run the risk of getting cholera because drinking water and some foods are contaminated by human sewage. Cholera's main symptom is sudden, massive diarrhea that can literally drain a person's body of water in less than 24 hours. The cause is a poison, cholera exotoxin, produced by the bacterium *Vibrio cholerae.* The toxin causes cells to pump out chloride ions, and other solutes follow. As solutes leave, cells lose their water by osmosis. Cholera is a common, deadly threat in parts of Africa, Asia, and South America, and after Hurricane Katrina, officials feared it would strike the U.S. Gulf Coast as well. In developed nations cholera can be treated with antibiotics. Elsewhere patients may recover if they are quickly rehydrated.

3.11 Metabolism: Doing Cellular Work

LINKS TO
SECTIONS
2.5, 2.8, AND 2.13

The 65 trillion living cells in your body need energy for their operations. The raw energy for this cellular work comes from organic compounds in food, which a cell's mitochondria convert to ATP—a chemical form the cell can use.

ATP IS THE CELL'S ENERGY CURRENCY

The chemical reactions in cells are called **metabolism**. Some reactions release energy and others require it. ATP links the two kinds of reactions, carrying energy from one reaction to another. You may remember that ATP is short for adenosine triphosphate, one of the nucleotides. A molecule of ATP consists of the five-carbon sugar ribose to which adenine (a nucleotide base) and three phosphate groups are attached (Figure 3.23a). ATP's stored energy is contained in the bond between the second and third phosphate groups.

Enzymes can break the bond between the second and third phosphate groups of the ATP molecule. The enzymes then can attach the released phosphate group to another molecule. When a phosphate group is moved from one molecule to another, stored energy goes with it.

Cells use ATP constantly, so they must renew their ATP supply. In many metabolic processes, phosphate (symbolized by P_i) or a phosphate group that has been split off from some substance, is attached to ADP, adenosine diphosphate (the prefix *di-* indicates that *two* phosphate groups are present). Now the molecule, with three phosphates, is ATP. And when ATP transfers a phosphate group elsewhere, it reverts to ADP. In this way it completes the **ATP/ADP cycle** (Figure 3.23b).

Like money earned at a job and then spent to pay the rent, ATP is earned in reactions that produce energy and spent in reactions that require it. That is why textbooks often use a cartoon coin to symbolize ATP.

THERE ARE TWO MAIN TYPES OF METABOLIC PATHWAYS

At this moment thousands of reactions are transforming thousands of substances inside each of your cells. Most of these reactions are part of **metabolic pathways**, steps in which reactions take place one after another. There are two main types of metabolic pathways, called anabolism and catabolism.

In **anabolism**, small molecules are put together into larger ones. In these larger molecules, the chemical bonds hold more energy. Anabolic pathways assemble complex carbohydrates, proteins, and other large molecules. The energy stored in their bonds is a major reason why we can use these substances as food.

In **catabolism**, large molecules are broken down to simpler ones. Catabolic reactions disassemble complex carbohydrates, proteins, and similar molecules, releasing their components for use by cells. For example, when a complex carbohydrate is catabolized, the reactions release the simple sugar glucose, the main fuel for cells.

Any substance that is part of a metabolic reaction is called a *reactant*. A substance that forms between the beginning and the end of a metabolic pathway is an *intermediate*. Substances present at the end of a reaction or a pathway are the *end products*.

Many metabolic pathways advance step-by-step from reactants to end products:

enzyme   enzyme   enzyme

A ⟶ B ⟶ C ⟶ D

end product

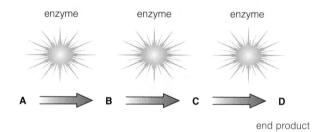

a

base

three phosphate groups

sugar

reactions that release energy

ATP

ADP + P_i

reactions that require energy

ATP

cellular work
(e.g., synthesis, breakdown, or rearrangement of substances; contraction of muscle cells; active transport across a cell membrane)

b

Figure 3.23 *Animated!* (**a**) Structure of the energy carrier ATP. (**b**) ATP connects energy-releasing reactions with energy-requiring ones. In the ATP/ADP cycle, the transfer of a phosphate group turns ATP into ADP, then back again to ATP.

HOW CELLS GAIN ENERGY

In other pathways the steps occur in a cycle, with the end products serving as reactants to start things over.

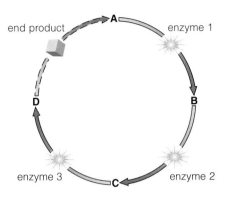

ENZYMES PLAY A VITAL ROLE IN METABOLISM

The metabolic reactions that keep all of us alive require **enzymes**, which you first read about in Section 2.8. Most enzymes are proteins, and they have several key features. Most importantly enzymes are catalysts: they speed up chemical reactions. In fact, enzymes usually make reactions occur hundreds to millions of times faster than would be possible otherwise. Enzymes are not used up in reactions, so a given enzyme molecule can be used over and over.

Each kind of enzyme can only interact with specific kinds of molecules, which are called its **substrates**. The enzyme can chemically recognize a substrate, bind it,

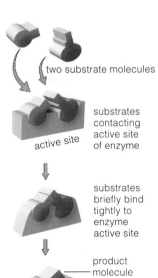

two substrate molecules

substrates contacting active site of enzyme

active site

substrates briefly bind tightly to enzyme active site

product molecule

enzyme unchanged by the reaction

Figure 3.24 *Animated!* How enzymes and substrates fit together. When substrate molecules contact an enzyme's active site, they bind to the site for a brief time and a product molecule forms. When the product molecule is released, the enzyme goes back to its previous shape. It is not changed by the reaction it catalyzed.

and change it in some way. An example is thrombin, one of the enzymes required to clot blood. It only recognizes a side-by-side alignment of two particular amino acids in a protein. When thrombin "sees" this arrangement, it breaks the peptide bond between the amino acids.

An enzyme and its substrate interact at a surface crevice on the enzyme, called an **active site**. Figure 3.24 shows how enzyme action can combine two substrate molecules into a new, larger product molecule.

Powerful as they are, enzymes only work well within a certain temperature range. For example, if a person's body temperature rises too high, the increased heat energy breaks bonds holding an enzyme in its three-dimensional shape. The shape changes, substrates can't bind to the active site as usual, and chemical reactions are disrupted. People usually die if their internal temperature reaches 44°C (112°F).

Enzymes also function best within a certain pH range—in the body, from pH 7.35 to 7.4. Above or below this range most enzymes cannot operate normally.

Organic molecules called **coenzymes** assist with many reactions. Lots of coenzymes, including **NAD$^+$** (nicotinamide adenine dinucleotide) and **FAD** (flavin adenine dinucleotide), are derived from vitamins, which is one reason why vitamins are important in the diet.

To maintain homeostasis the body must have ways of controlling the activity of enzymes. Some of these controls boost or slow the action of existing enzymes. Others adjust how fast enzyme molecules are made, and thus how many are available for a metabolic pathway. For example, when you eat, food arriving in your stomach causes gland cells there to secrete the hormone gastrin into your bloodstream. Stomach cells with receptors for gastrin respond in several ways, such as secreting the ingredients of "gastric juice"—including enzymes that break down food proteins.

Most chemical reactions in cells are organized in the orderly steps of metabolic pathways.

Enzymes speed the rate of chemical reactions, but each one acts only on specific substrates. Enzymes function best within certain ranges of temperature and pH.

3.12 How Cells Make ATP

LINK TO
SECTION
2.9

The chemical reactions that sustain the body depend on energy that cells capture when they produce ATP.

CELLULAR RESPIRATION MAKES ATP

To make ATP, cells break apart carbohydrates, especially glucose, as well as lipids and proteins. The reactions remove electrons from intermediate compounds, then energy associated with the electrons powers the formation of ATP. Human cells typically form ATP by **cellular respiration**. In large, complex organisms like ourselves, this process usually is aerobic, which means that it uses oxygen. Glucose is the most common raw material for cellular respiration, so it will be our example here.

STEP 1: GLYCOLYSIS BREAKS GLUCOSE DOWN TO PYRUVATE

Cellular respiration starts in the cell's cytoplasm, in a set of reactions called **glycolysis**—literally, "splitting sugar." You may recall that glucose is a simple sugar. Each glucose

molecule consists of six carbon atoms, twelve hydrogens, and six oxygens, all joined by covalent bonds. During glycolysis, a glucose molecule is broken into two **pyruvate** molecules, each with three carbons (Figure 3.25).

When glycolysis begins, two ATPs each transfer a phosphate group to glucose, donating energy to it. This kind of transfer is called **phosphorylation**. It adds enough energy to glucose to begin the energy-releasing steps of glycolysis.

The first energy-releasing step breaks the glucose into two molecules of **PGAL** (for phosphoglyceraldehyde), which are converted to intermediates. These molecules then each donate a phosphate group to ADP, forming ATP. The same thing happens with the next intermediate in the sequence, and the end result is two molecules of pyruvate and four ATP. However, because two ATP were invested to start the reactions, the *net* energy yield is only two ATP.

Notice that glycolysis does not use oxygen. If oxygen is not available for the following aerobic steps of cellular respiration, for a short time a cell can still form a small amount of ATP by a process of fermentation, which also does not use oxygen. You will read more about this "back-up" process for forming ATP later in the chapter.

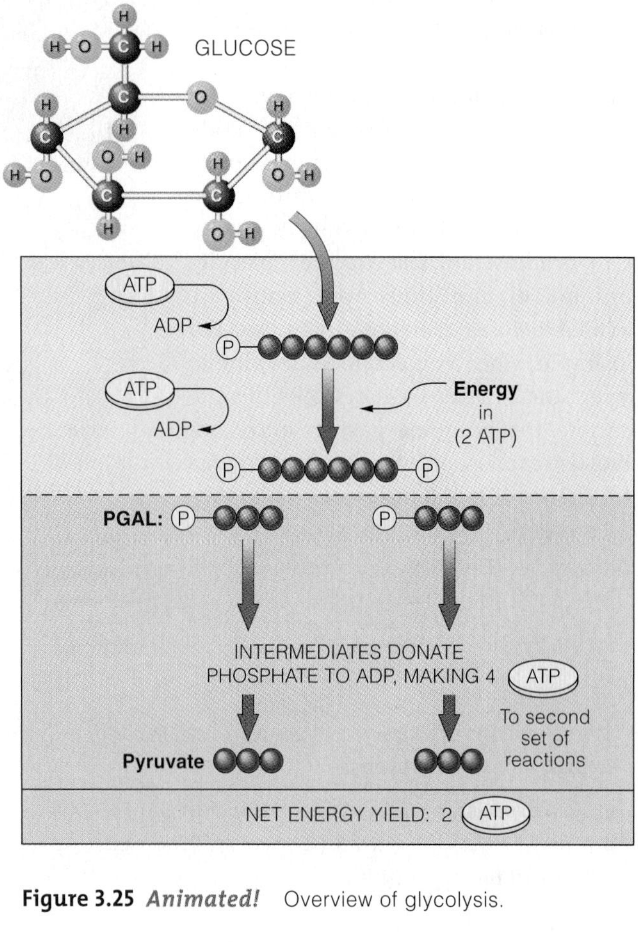

STEP 2: THE KREBS CYCLE PRODUCES ENERGY-RICH TRANSPORT MOLECULES

The pyruvate molecules formed by glycolysis move into a mitochondrion. There the oxygen-requiring phase of cellular respiration will be completed. Enzymes catalyze each reaction, and the intermediate molecules formed at one step become substrates for the next.

In preparatory steps, an enzyme removes a carbon atom from each pyruvate molecule. A coenzyme called coenzyme A combines with the remaining two-carbon fragment and becomes a compound called **acetyl-CoA**. This substance enters the **Krebs cycle**. For each turn of the cycle, six carbons, three from each pyruvate, enter and six also leave, in the form of carbon dioxide. The bloodstream then transports this CO_2 to the lungs where it is exhaled.

Reactions in mitochondria before and during the Krebs cycle have three important functions. First, they produce two molecules of ATP. Second, they regenerate intermediate compounds required to keep the Krebs cycle going. And in a third, crucial step, a large number of the coenzymes called NAD^+ and FAD pick up H^+ and electrons, in the process becoming NADH and $FADH_2$. Loaded with energy, NADH and $FADH_2$ will now move to the site of the third and final stage of reactions that make ATP.

GLUCOSE

ATP
ADP

ATP
ADP

Energy
in
(2 ATP)

PGAL:

INTERMEDIATES DONATE
PHOSPHATE TO ADP, MAKING 4 (ATP)

To second
set of
reactions

Pyruvate

NET ENERGY YIELD: 2 (ATP)

Figure 3.25 *Animated!* Overview of glycolysis.

STEP 3: ELECTRON TRANSPORT PRODUCES MANY ATP MOLECULES

ATP production goes into high gear during the final stage of cellular respiration. In the production "assembly line" chains of reactions capture and use energy released by electrons. Each chain is called an **electron transport system**. It includes enzymes inside the membrane that divides the mitochondrion into two compartments (Figure 3.26). As electrons flow through the system, each step transfers a bit of energy to a molecule that briefly stores it. This gradual releasing of energy reduces the amount of energy that is lost (as heat) while a cell is generating ATP.

As you can see at the bottom left of Figure 3.26, an electron transport system uses electrons and hydrogen ions delivered by NADH and FADH$_2$. The electrons are transferred from one molecule of the transport system to the next in line. When molecules in the chain accept and then donate electrons, they also pick up hydrogen ions in

the inner compartment, then release them to the outer compartment. The blue arrows in Figure 3.26 represent this process. At the end of an electron transport system, oxygen accepts electrons in a reaction that forms water.

As the system moves hydrogen ions into the outer compartment, an H$^+$ concentration gradient develops. As the ions become more concentrated in the outer compartment, they follow the gradient back into the inner compartment, crossing the inner membrane through the interior of enzymes that can catalyze the formation of ATP from ADP and phosphate (P$_i$). This step is shown at the far right of Figure 3.26.

In glycolysis, a carbohydrate such as glucose is broken down to two molecules of pyruvate. The net energy yield of glycolysis is two ATP molecules.

When the two pyruvate molecules from glycolysis enter a mitochondrion, each gives up a carbon atom and the rest of the molecule enters the Krebs cycle. The carbon atoms end up in carbon dioxide. The Krebs cycle and its preparatory steps yield two more ATP molecules.

In the final stage of cellular respiration, electrons and H$^+$ move through transport systems inside mitochondria. ATP forms when hydrogen ions flow through membrane enzymes that add a phosphate group to ADP.

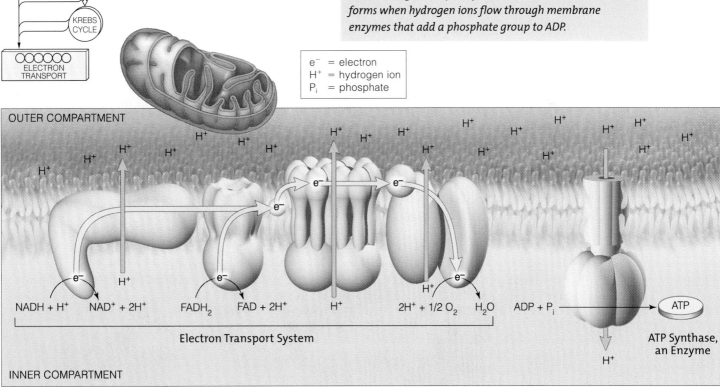

e$^-$ = electron
H$^+$ = hydrogen ion
P$_i$ = phosphate

(a) At the inner mitochondrial membrane, NADH and FADH$_2$ give up electrons to transfer chains. When electrons are transferred through the chains, hydrogen (H$^+$) is shuttled across the membrane to the outer compartment.

(b) Oxygen accepts electrons at the end of the transfer chain.

(c) There are now H$^+$ concentration and electric gradients across the membrane. H$^+$ follows the gradients through the interior of enzymes, to the inner compartment. This flow drives the formation of ATP from ADP and phosphate (P$_i$).

Figure 3.26 *Animated!* How electron transport forms ATP.

3.13 Summary of Cellular Respiration

Figure 3.27 below reviews the steps and ATP yield from cellular respiration. Only this aerobic pathway delivers enough energy to build and maintain a large, active, multicellular organism such as a human. In many types of cells, the third stage of reactions forms thirty-two ATP. When we add these to the final yield from the preceding stages, the total harvest is thirty-six ATP from one glucose molecule. This is a very efficient use of our cellular resources!

While aerobic cellular respiration typically yields thirty-six ATP, the actual amount may vary, depending on conditions in a cell at a given moment—for instance, if a cell requires a particular intermediate elsewhere and pulls it out of the reaction sequence. If you would like to learn more about these metabolic events, see Appendix I at the back of this book.

Cellular respiration begins with glycolysis in the cytoplasm and ends with electron transport systems in mitochondria. From start to finish this aerobic process typically has a net yield of thirty-six ATP for every glucose molecule.

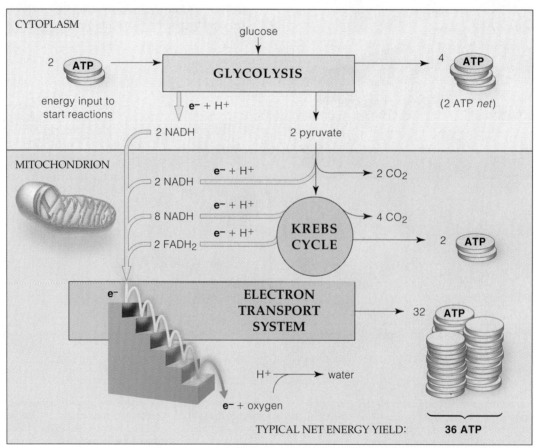

Figure 3.27 *Animated!*   Summary of aerobic cellular respiration.

3.14 Alternative Energy Sources in the Body

Glucose from complex carbohydrates is the body's main energy source. When conditions warrant, however, other substances can supply needed fuel for making ATP.

HOW THE BODY USES CARBOHYDRATES AS FUEL

When glucose from food moves into your bloodstream, a rise in the glucose level in blood prompts an organ, the pancreas, to release insulin. This hormone makes cells take up glucose faster.

If you consume more glucose than your cells need for the moment, one of the intermediates of glycolysis is diverted into an anabolic pathway that makes a storage sugar called glycogen. The detour halts glycolysis, so for the time being no more ATP forms. This switch occurs quite often in muscle and liver cells, which store most of the body's glycogen. Other kinds of cells tend to store excess glucose as fat.

Sudden, intense exercise, such as weightlifting or a sprint, may call on cells in skeletal muscles (which attach to our bones) that use a different kind of ATP-forming mechanism, a process called *lactate fermentation* (Figure 3.28). The process converts pyruvate from glycolysis to lactic acid. It does not use oxygen and produces ATP quickly but not for very long. Muscles feel sore when lactic acid builds up in them.

Between meals, glucose is not moving into your bloodstream and its level in the blood falls. The decline must be offset because nerve cells in the brain use glucose as their preferred energy source. Accordingly, the pancreas responds to falling blood glucose by secreting a hormone that makes liver cells convert glycogen back to glucose and release it to the blood. Thus, hormones control whether the body's cells use glucose as an energy source or store it for future use.

Only about 1 percent of the body's total energy reserves consists of glycogen, however. Of the total energy stores in a typical adult American, 78 percent is in body fat and 21 percent in proteins.

FATS AND PROTEINS ALSO PROVIDE ENERGY

Most of the body's stored fat consists of triglycerides, which accumulate inside the fat cells in certain tissues (called *adipose* tissues) of the buttocks and other locations beneath the skin.

Between meals or during exercise, the body may tap triglycerides as energy alternatives to glucose. Enzymes in fat cells break apart triglycerides into glycerol and fatty acids, which enter the bloodstream. When glycerol reaches the liver, enzymes convert it to PGAL, the intermediate of glycolysis mentioned in Section 3.12. Most body cells take up the circulating fatty acids. Enzymes convert them to acetyl-CoA, which can enter the Krebs cycle. Each fatty acid tail has many more carbon-bound hydrogen atoms than glucose does, so breaking down a fatty acid yields much more ATP. In fact, this pathway can supply about half the ATP required by your muscle, liver, and kidney cells.

The body stores excess fats but not proteins. Enzymes dismantle unneeded proteins into amino acids. Then they remove the molecule's amino group ($—NH_3^+$) and ammonia (NH_3) forms. The cell's metabolic machinery may use leftover carbons to make fats or carbohydrates. Or the carbons may enter the Krebs cycle, where coenzymes can pick up hydrogen as well as electrons removed from the carbon atoms. These can be used to make ATP in electron transport systems in mitochondria. The ammonia is converted to urea, a waste that is excreted in urine.

IMPACTS, ISSUES

Due to an inherited defect in the functioning of their mitochondria, some people have *lactic acidosis*—their cells make far too much lactic acid. High levels of lactate make the blood dangerously acidic and cause muscle weakness, abdominal pain, nausea, fatigue, and, in extreme cases, death.

Figure 3.28 Sprinters, drawing on muscle cells that use lactate fermentation to generate ATP.

Complex carbohydrates, fats, and proteins all can serve as energy sources in the human body.

Certain muscle cells can make a small amount of ATP by the process of lactate fermentation.

Summary

Section 3.1 A living cell has a plasma membrane surrounding an inner region of cytoplasm. The cytoplasm consists of a jellylike cytosol and other components. In a eukaryotic cell, including human cells, membranes divide the cell into functional compartments called organelles. Organelle membranes separate metabolic reactions in the cytoplasm and allow different kinds to proceed in an orderly way.

 Investigate the physical limits on cell size and learn how different types of microscopes function.

Section 3.3 Cell membranes consist mostly of lipids and proteins. The lipids are mainly phospholipids, arranged as a double layer called a lipid bilayer. Various kinds of proteins in the bilayer, or at one of its surfaces, perform most membrane functions.

Some membrane proteins are transport proteins that allow water-soluble substances to cross the membrane. Others are receptors for hormones or other substances. Still other proteins have carbohydrate chains that serve as a cell's identity tags. Adhesion proteins help cells stay together in their proper tissues.

 Learn more about the functions of receptor proteins.

Section 3.5 The largest organelle is the nucleus, where the genetic material DNA is located. The nucleus is surrounded by a double membrane, the nuclear envelope. Pores in the envelope help control the movement of substances into and out of the nucleus. A nucleolus is a dense mass inside the nucleus; nucleoli are where the subunits of ribosomes are constructed.

A cell's DNA and proteins associated with it are called chromatin. Each chromosome in the nucleus is one DNA molecule with its associated proteins.

 Introduce yourself to the major types of organelles and take a close-up look at the nuclear membrane.

Section 3.6 The endomembrane system includes the endoplasmic reticulum, Golgi bodies, and different kinds of vesicles. In this system new proteins are modified into final form and lipids are assembled. Unwanted materials may be broken down in vesicles called lysosomes and peroxisomes.

 Follow a path through the endomembrane system.

Section 3.7 Mitochondria carry out the oxygen-requiring reactions that make ATP, the cell's energy currency. These reactions take place in the inner compartment formed by a mitochondrion's double-membrane system.

Section 3.8 The cytoskeleton consists mainly of microtubules and microfilaments; some types of cells also have intermediate filaments. The cytoskeleton gives a cell its shape and internal structure. In some kinds of cells microtubules are the framework for cilia or flagella, which are used in movement. Cilia and flagella develop from structures called centrioles.

 Learn more about elements of the cytoskeleton and what they do.

Section 3.9 A cell's plasma membrane is selectively permeable—the membrane allows some substances but not others to cross it, according to the cell's needs. Substances cross cell membranes by several transport mechanisms. Diffusion is the random movement of solutes down a concentration gradient. Osmosis is the movement of water across a selectively permeable membrane in response to concentration gradients, a pressure gradient, or both.

 Investigate how substances diffuse across membranes and how water crosses by osmosis.

Section 3.10 In passive transport, a solute moves down its concentration gradient through a membrane transport protein. In active transport, a solute is pumped through a membrane protein *against* its concentration gradient. Active transport requires an energy boost, as from ATP.

Cells use vesicles to take in or expel large molecules or particles. In exocytosis, a vesicle moves to the cell surface and fuses with the plasma membrane. In endocytosis, a vesicle forms at the surface and moves inward. When an endocytic vesicle brings organic matter into a cell, the process is called phagocytosis (cell eating).

 Compare the processes of passive and active transport, and see how vesicles move substances into and out of cells.

Section 3.11 The chemical reactions in a cell are collectively called its metabolism. A metabolic pathway is a stepwise sequence of chemical reactions catalyzed by enzymes—catalytic molecules that speed up the rate of metabolic reactions. Each enzyme interacts only with a specific substrate, linking with it at one or more active sites.

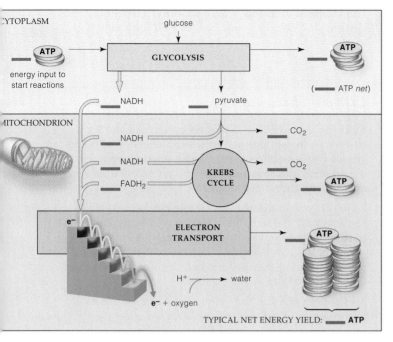

Table 3.4	Summary of Energy Sources in the Human Body	
Starting Molecule	Subunit	Entry Point into the Aerobic Pathway
Complex carbohydrate	Simple sugars (e.g., glucose)	Glycolysis
Fat	Fatty acids	Preparatory reactions for Krebs cycle
	Glycerol	Raw material for key intermediate in glycolysis (PGAL)
Protein	Amino acids	Carbon backbones enter Krebs cycle or preparatory reactions

Anabolism builds large, energy-rich organic compounds from smaller molecules. Catabolism breaks down molecules to smaller ones. Various cofactors, such as the coenzymes NAD^+ and FAD, assist enzymes or carry electrons, hydrogen, or functional groups from a substrate to other sites.

 Investigate how enzymes facilitate chemical reactions.

Section 3.12 The reactions of most anabolic pathways run on energy from ATP. In body cells, cellular respiration produces most ATP molecules. This pathway releases chemical energy from glucose and other organic compounds. ATP is replenished by way of the ATP/ADP cycle.

Section 3.13 In aerobic cellular respiration, oxygen is the final acceptor of electrons removed from glucose. The pathway has three stages: glycolysis (in the cytoplasm), the Krebs cycle, and electron transport, which generates a large amount of ATP in mitochondria. The typical net energy yield of cellular respiration is thirty-six ATP.

 Take a step-by-step journey through glycolysis and cellular respiration.

Section 3.14 The body can extract energy from carbohydrates, fats, and proteins. Complex carbohydrates are broken down to the simple sugar glucose, the body's main metabolic fuel. Alternatives to glucose include fatty acids and glycerol from triglycerides and, in certain circumstances, amino acids from proteins (Table 3.4).

 Learn more about how cells can use different kinds of organic molecules as energy sources.

Review Questions

1. Describe the general functions of the following in a eukaryotic cell: the plasma membrane, cytoplasm, DNA, ribosomes, organelles, and cytoskeleton.

2. Which organelles are part of the cytomembrane system?

3. Distinguish between the following pairs of terms:
   a. diffusion; osmosis
   b. passive transport; active transport
   c. endocytosis; exocytosis

4. What is an enzyme? Describe the role of enzymes in metabolic reactions.

5. In aerobic cellular respiration, which reactions occur only in the cytoplasm? Which ones occur only in a cell's mitochondria?

6. For the diagram of the aerobic pathway shown above, fill in the number of molecules of pyruvate and the net ATP formed at each stage.

1. The plasma membrane _____ .
   a. surrounds the cytoplasm
   b. separates the nucleus from the cytoplasm
   c. separates the cell interior from the environment
   d. both a and c are correct

2. The _____ is responsible for a eukaryotic cell's shape, internal organization, and cell movement.

3. Cell membranes consist mainly of a _____ .
   a. carbohydrate bilayer and proteins
   b. protein bilayer and phospholipids
   c. phospholipid bilayer and proteins

4. _____ carry out most membrane functions.
   a. Proteins   c. Nucleic acids
   b. Phospholipids   d. Hormones

5. The passive movement of a solute through a membrane protein down its concentration gradient is an example of _____ .
   a. osmosis   c. endocytosis
   b. active transport   d. diffusion

6. Match each organelle with its correct function.
   ___ protein synthesis   a. mitochondrion
   ___ movement   b. ribosome
   ___ intracellular digestion   c. smooth ER
   ___ modification of proteins   d. rough ER
   ___ lipid synthesis   e. nucleolus
   ___ ATP formation   f. lysosome
   ___ ribosome assembly   g. flagellum

7. Which of the following statements is *not* true? Metabolic pathways _____ .
   a. occur in stepwise series of chemical reactions
   b. are speeded up by enzymes
   c. may break down or assemble molecules
   d. always produce energy (such as ATP)

8. Enzymes _____ .
   a. enhance reaction   c. act on specific
      rates   substrates
   b. are affected by pH   d. all of the above
                                are correct

9. Match each substance with its correct description.
   ___ a coenzyme or metal ion   a. reactant
   ___ formed at end of a   b. enzyme
       metabolic pathway   c. cofactor
   ___ mainly ATP   d. energy carrier
   ___ enters a reaction   e. end product
   ___ catalytic protein

10. Cellular respiration is completed in the _____ .
    a. nucleus   c. plasma membrane
    b. mitochondrion   d. cytoplasm

11. Match each type of metabolic reaction with its function:
    ___ glycolysis   a. many ATP, NADH, FADH$_2$,
    ___ Krebs cycle   and CO$_2$ form
    ___ electron   b. glucose to two pyruvate
        transport   molecules and some ATP
                           c. H$^+$ flows through channel
                              proteins, ATP forms

12. In a mitochondrion, where are the electron transport systems and enzymes required for ATP formation located?

Critical Thinking

1. Using Section 3.4 as a reference, suppose you want to observe the surface of a microscopic section of bone. Would you benefit most from using a compound light microscope, a transmission electron microscope, or a scanning electron microscope?

2. Jogging is considered aerobic exercise because the cardiovascular system (heart and blood vessels) can adjust to supply the oxygen needs of working cells. In contrast, sprinting the 100-meter dash might be called "anaerobic" (lacking oxygen) exercise, and golf "nonaerobic" exercise. Explain these last two observations.

3. The cells of your body never use nucleic acids as an energy source. Can you suggest a reason why?

Explore on Your Own

In this chapter you learned that an enzyme can only act on certain substrates. Because your saliva contains enzymes that can use some substances as substrates but not others, you can easily gain some insight into practical impacts of this concept (Figure 3.29). Start by holding a bite of plain cracker in your mouth for thirty seconds, without chewing it. What happens to the cracker, which is mostly starch (carbohydrate)? Repeat the test with a dab of butter or margarine (lipid), then with a piece of meat, fish, or even scrambled egg (protein). Based on your results, what type of biological molecules do your salivary enzymes act upon?

Figure 3.29 Putting enzymes to work digesting the different kinds of biological molecules in foods.

4 TISSUES, ORGANS, AND ORGAN SYSTEMS

Stem Cells

Each year tens of thousands of people severely injure their spinal cords. In 1995 actor Christopher Reeve, the movies' Superman, suffered a fall from a horse that left him paralyzed. Until his death in 2004 he was a strong supporter of stem cell research.

Stem cells are the first to form when a fertilized egg starts dividing, and they are plentiful in developing embryos and fetuses. Adults also have them in some tissues, including bone marrow and adipose (fat) tissue. Stem cells are like blank slates. Under the proper conditions, they can give rise to a range of different cell types, including blood cells, cartilage, muscle, and nerve cells.

Stem cells from adults have shown quite a bit of promise for regenerating some kinds of tissues, such as missing cartilage and heart muscle damaged by a heart attack.

Embryonic stem cells are more controversial. In theory, embryonic stem cells can give rise to every kind of cell in the body. Many scientists feel they are the best alternative for research leading to therapies that can replace the damaged or dead nerve cells responsible for paralysis or disorders such as Alzheimer's disease. Other people believe it's unethical to use embryonic cells for any reason, because doing so destroys the embryo. Currently, the United States government funds only research that uses existing embryonic stem cell lines.

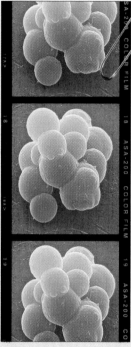

Stem cells start us thinking about how the human body and its parts are put together (anatomy) and how the body functions (physiology). This chapter is an overview of body tissues, organs, and organ systems. A **tissue** is a group of similar cells that perform a particular function. Various tissues combine in certain proportions and patterns to form an organ, such as the heart. An organ system is two or more organs that work together in performing a common task.

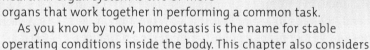

As you know by now, homeostasis is the name for stable operating conditions inside the body. This chapter also considers the controls that help maintain this stability.

 How Would You Vote? Should researchers be allowed to start embryonic stem cell lines from human embryos that are not used for in vitro fertilization? Cast your vote online at www.thomsonedu.com/biology/starr/humanbio.

Key Concepts

FOUR TYPES OF TISSUES
Four types of tissues make up the body. Epithelial tissues line body surfaces. Connective tissues bind, support, strengthen, and insulate other tissues. Muscle tissues contract and move the body. Nervous tissue detects stimuli, integrates the information, and governs responses to it.

ORGANS AND ORGAN SYSTEMS
Organs consist of different tissues organized in specific proportions and patterns. Organs are the components of the body's organ systems. Each organ system has specialized functions, but all interact and contribute to the survival of the body as a whole.

Links to Earlier Concepts

In this chapter we reach the tissue, organ, and organ system levels of biological organization (1.3). As you learn about the differences among different types of tissues, you will also get a look at some of the many variations on basic cell structure (3.1–3.8) that occur in your body. The variations are a reminder that cells which perform different functions must be built to carry out those specialized tasks.

4.1 Epithelium: The Body's Covering and Linings

LINK TO
SECTION
3.8

Like other complex animals, we humans are built of just four basic types of tissues. Epithelial tissues cover the body surface or line its cavities and tubes.

Your skin. The rosy lining of your mouth. Each of these is an example of **epithelium** (plural: epithelia). Epithelium is a sheetlike tissue with one surface that faces the outer environment or a body fluid (Figure 4.1*a*). The tissue's other surface rests on a **basement membrane** that is sandwiched between it and the tissue below (Figure 4.1*a*). A basement membrane has no cells but is packed with proteins and polysaccharides.

The cells in epithelium nestle closely together and they are arranged in one or more layers. The cells also are linked by junctions that perform specific structural or functional tasks. Cells in some epithelia are specialized to absorb substances, others to secrete them.

THERE ARE TWO BASIC TYPES OF EPITHELIA

Epithelium may be "simple," with just one layer of cells, or it may be "stratified" and have several layers. Simple epithelium lines the body's cavities, ducts, and tubes—for example, the chest cavity, tear ducts, and the tubes in the kidneys where urine is formed (Figure 4.1*b–d*). In general, the cells in a simple epithelium function in the diffusion, secretion, absorption, or filtering of substances across the layer.

Some simple epithelia have a single cell layer that looks like several layers when it is viewed from the side because the nuclei of neighboring cells don't line up evenly. Most of the cells also have cilia. This type of simple epithelium is termed *pseudostratified* (*pseudo-* means false). It lines the throat, nasal passages, reproductive tract, and other sites in the body where cilia sweep mucus or some other fluid across the surface of the tissue.

Stratified epithelium has two or more layers of cells, and its typical function is protection. For example, this is the tissue at the surface of your skin, which is exposed to nicks, bumps, scrapes, and so forth.

The two basic types of epithelium are subdivided into categories depending on the shape of cells at the tissue's free surface. A *squamous epithelium* has flattened cells, a *cuboidal epithelium* has cube-shaped cells, and a *columnar epithelium* has tall, elongated cells. The different shapes correlate with different functions. For instance, oxygen and carbon dioxide easily diffuse across the thin simple squamous epithelium that makes up the walls of fine blood vessels, as in Figure 4.1*b*. The plumper cells of cuboidal and columnar epithelia secrete substances. Table 4.1 summarizes the types of epithelium and their roles.

GLANDS DEVELOP FROM EPITHELIUM

A **gland** is a structure that makes and releases specific products, such as saliva or mucus. Some glands consist of a single cell, while others are multicellular. Regardless, each gland develops from epithelial tissue and often stays connected to it. Mucus-secreting goblet cells, for instance, are embedded in epithelium that lines the trachea (your windpipe) and other tubes leading to the lungs. The stomach's epithelial lining contains gland cells that secrete protective mucus and digestive juices.

Glands often are classified according to how their secretions reach the place where they are used. **Exocrine glands** release substances through ducts or tubes. Mucus, saliva, earwax, oil, milk, and digestive enzymes all are exocrine secretions. Many exocrine glands simply release the substance they are specialized to make; salivary glands and most sweat glands are like this. In other cases, a gland's secretions include bits of the gland cells. For instance, milk secreted from a nursing woman's mammary glands contains bits of the glandular epithelial tissue. In still other cases, such as sebaceous (oil) glands in your skin, whole cells full of the material to be secreted are actually shed into the duct, where they burst and their contents spill out.

In contrast to exocrine glands, **endocrine glands** do not release substances through tubes or ducts. They make hormones that directly enter the extracellular fluid bathing the glands. Typically, the bloodstream picks up hormones and carries them to target cells somewhere else in the body. Examples of endocrine glands include the pituitary gland, the thyroid, and other glands that are discussed in Chapter 15.

Table 4.1	Major Types of Epithelium	
Type	**Shape**	**Typical Locations**
Simple (one layer)	Squamous	Linings of blood vessels, lung alveoli (sites of gas exchange)
	Cuboidal	Glands and their ducts, surface of ovaries, pigmented epithelium of eye
	Columnar	Stomach, intestines, uterus
Pseudostratified	Columnar	Throat, nasal passages, sinuses, trachea, male genital ducts
Stratified (two or more layers)	Squamous	Skin (keratinized), mouth, throat, esophagus, vagina (nonkeratinized)
	Cuboidal	Ducts of sweat glands
	Columnar	Male urethra, ducts of salivary glands

free surface
of epithelium

simple
squamous
epithelium

basement
membrane

connective
tissue

a

Figure 4.1 *Animated!* Some basic characteristics of epithelium. (**a**) All epithelia have a free surface. A basement membrane is sandwiched between the opposite surface and underlying connective tissue. The diagram shows simple epithelium, a single layer of cells. The micrograph shows the upper portion of stratified squamous epithelium, which has more than one cell layer. The cells are more flattened toward the surface. (**b–d**) Examples of simple epithelium, showing the three basic cell shapes in this type of tissue.

Epithelia are sheetlike tissues with one free surface. Simple epithelium lines body cavities, ducts, and tubes. Stratified epithelium, like that at the skin surface, typically protects the underlying tissues.

Glands make and secrete substances. They are derived from epithelium, and often remain connected to it. Some glands are single cells, while others are multicellular.

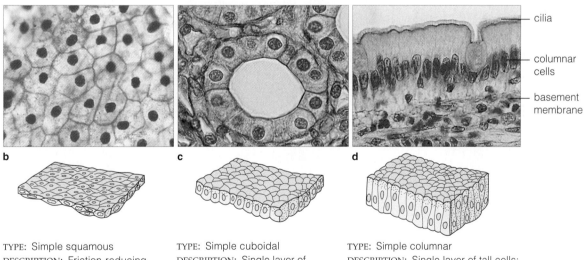

cilia

columnar
cells

basement
membrane

b

TYPE: Simple squamous
DESCRIPTION: Friction-reducing slick, single layer of flattened cells
COMMON LOCATIONS: Lining of blood and lymph vessels, heart; air sacs of lungs; peritoneum
FUNCTION: Diffusion; filtration; secretion of lubricants

c

TYPE: Simple cuboidal
DESCRIPTION: Single layer of squarish cells
COMMON LOCATIONS: Ducts, secretory part of small glands; retina; kidney tubules; ovaries, testes; bronchioles
FUNCTION: Secretion; absorption

d

TYPE: Simple columnar
DESCRIPTION: Single layer of tall cells; free surface may have cilia, mucus-secreting glandular cells, microvilli
COMMON LOCATIONS: Glands, ducts; gut; parts of uterus; small bronchi
FUNCTION: Secretion; absorption; ciliated types move substances

4.2 Connective Tissue: Binding, Support, and Other Roles

Connective tissue binds together, supports, and anchors body parts. In some cases, it provides metabolic support as well.

Connective tissue is the most abundant body tissue. There are two general groups: fibrous connective tissues and specialized types, which include cartilage, bone, blood, and adipose (fat) tissue (Table 4.2). In most kinds of connective tissues cells secrete fiberlike structural proteins and a "ground substance" of polysaccharides. Together these ingredients make up a **matrix** around the cell. The matrix can range from hard to liquid, and it gives each kind of connective tissue its specialized properties.

FIBROUS CONNECTIVE TISSUES ARE STRONG AND STRETCHY

Fibrous connective tissue is subdivided into several categories, with several forms within each category. All the different kinds have cells, fibers, and a matrix, but in different proportions that make each one well-suited to perform its special function.

For example, the various forms of **loose connective tissue** have few fibers and cells, and they are loosely arranged in a jellylike ground substance, as pictured in Figure 4.2*a*. This structure makes loose connective tissue flexible. The example in Figure 4.2*a* wraps many organs and helps support the skin. A "reticular" (netlike) form of loose connective tissue is the framework for soft organs such as the liver, spleen, and lymph nodes.

Dense connective tissue is packed with more collagen fibers than we see in loose connective tissue, so it is less flexible but much stronger. It also comes in several forms. The one pictured in Figure 4.2*b* helps support the skin's lower layer, the dermis. It also wraps around muscles and organs that do not need to stretch much, such as kidneys. In another version of this connective tissue, large bundles of collagen fibers are aligned in the same plane (Figure 4.2*c*). This tissue occurs in tendons, which attach skeletal muscle to bones, and in ligaments, which attach bones to one another. Its structure allows a tendon to resist being torn, and in ligaments the tissue's elastic fibers allow the ligament to stretch so bones can move at joints such as the knee.

Elastic connective tissue is a form of dense connective tissue in which most of the fibers are the protein elastin. As a result, this tissue is elastic and is found in organs that must stretch, such as the lungs, which expand and recoil as air moves in and out.

CARTILAGE, BONE, ADIPOSE TISSUE, AND BLOOD ARE SPECIALIZED CONNECTIVE TISSUES

Like rubber, **cartilage** is both solid and pliable and is not easily compressed. Its matrix is a blend of collagen and

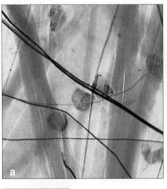
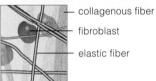

TYPE: Loose connective tissue
DESCRIPTION: Fibroblasts, other cells, plus fibers loosely arranged in semifluid matrix
COMMON LOCATIONS: Under the skin and most epithelia
FUNCTION: Elasticity, diffusion

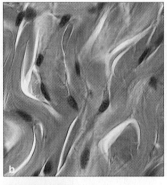

TYPE: Dense, irregular connective tissue
DESCRIPTION: Collagenous fibers, fibroblasts, less matrix
COMMON LOCATIONS: In skin and capsules around some organs
FUNCTION: Support

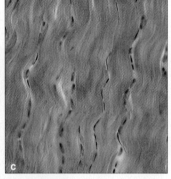
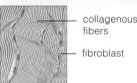

TYPE: Dense, regular connective tissue
DESCRIPTION: Collagen fibers in parallel bundles, long rows of fibroblasts, little matrix
COMMON LOCATIONS: Tendons, ligaments
FUNCTION: Strength, elasticity

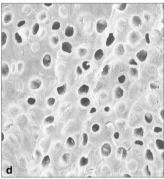
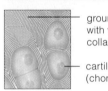

TYPE: Cartilage
DESCRIPTION: Cells embedded in pliable, solid matrix
COMMON LOCATIONS: Ends of long bones, nose, parts of airways, skeleton of embryos
FUNCTION: Support, flexibility, low-friction surface for joint movement

Figure 4.2 *Animated!* Characteristics of connective tissues.

Table 4.2 Connective Tissues at a Glance

Fibrous Connective Tissues

Loose	Collagen and elastin loosely arranged in ground substance; quite flexible and fairly strong
Dense	Mainly collagen; somewhat flexible and quite strong. Collagen fibers are aligned in parallel in the dense connective tissue of tendons and ligaments
Elastic	Mainly elastin; easily stretches and recoils

Special Connective Tissues

Cartilage	Mainly collagen in a watery matrix; resists compression
Bone	Mineral-hardened matrix; very strong
Adipose tissue	Mainly cells filled with fat; soft matrix
Blood	Matrix is the fluid blood plasma, which contains blood cells and other substances

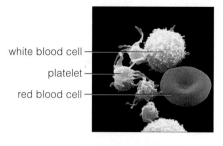

white blood cell
platelet
red blood cell

Figure 4.3 Some components of human blood. This tissue's straw-colored, liquid matrix (plasma) is mostly water in which a variety of substances are dissolved.

elastin fibers in a rubbery ground substance, and the end result is a tissue that can withstand a great deal of physical stress. The collagen-producing cells become trapped inside small cavities in the matrix (Figure 4.2d). Lacking blood vessels, injured cartilage heals slowly.

Most cartilage in the body is whitish, glistening *hyaline cartilage* (hyalin = "glassy"). Hyaline cartilage at the ends of bones reduces friction in movable joints. It also makes up parts of your nose, windpipe (trachea), and ribs. An early embryo's skeleton consists of hyaline cartilage.

Elastic cartilage has both collagen and elastin fibers and it occurs in places where a flexible yet rigid structure is required, such as the flexible outer flaps of your ears. Sturdy and resilient *fibrocartilage* is packed with thick bundles of collagen fibers. It can withstand tremendous pressure, and it forms the cartilage "cushions" in joints such as the knee and in the disks between the vertebrae in the spinal column.

Bone tissue is the main tissue in bones. It is hard because its matrix includes not only collagen fibers and ground substance but also calcium salts (Figure 4.2e). As part of the skeleton our bones serve the body in many ways that you will learn about in Chapter 5.

Adipose tissue stores fat—the way the body deals with carbohydrates and proteins that are not immediately used for metabolism. It is mostly cells packed with fat droplets, with just a little matrix between them (Figure 4.2f). Most of our adipose tissue is located just beneath the skin, where it provides insulation and cushioning.

Blood is classified as connective tissue even though it does not "connect" or bind other body parts. Instead blood's role is transport. Its matrix is the fluid plasma, which contains proteins (blood's "fibers") as well as a variety of blood cells and cell fragments called platelets (Figure 4.3). Chapter 8 discusses this complex tissue.

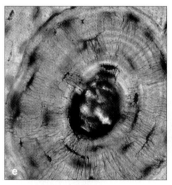

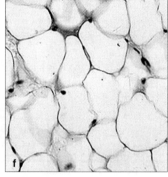

compact bone tissue
blood vessel
bone cell (osteocyte)

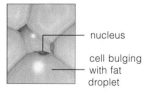

nucleus
cell bulging with fat droplet

TYPE: Bone tissue

DESCRIPTION: Collagen fibers, matrix hardened with calcium

COMMON LOCATIONS: Bones of skeleton

FUNCTION: Movement, support, protection

TYPE: Adipose tissue

DESCRIPTION: Large, tightly packed fat cells occupying most of matrix

COMMON LOCATIONS: Under skin, around heart, kidneys

FUNCTION: Energy reserves, insulation, padding

Connective tissue binds together, supports, strengthens, protects, and insulates other body tissues. All connective tissues consist of cells in a matrix that contains protein fibers and a ground substance.

Fibrous connective tissues include loose, dense, and elastic types, depending on the amount and arrangements of collagen and elastin fibers.

Cartilage, bone, blood, and adipose tissue are specialized connective tissues. Cartilage and bone are both structural materials. Blood is specialized to transport substances. Adipose tissue is a reservoir of stored energy.

4.3 Muscle Tissue: Movement

Muscle tissue, another of the four basic tissue types, has contractile cells that are specialized for moving body parts.

The cells in **muscle tissue** contract, or shorten, when they are stimulated by an outside signal; then they relax and lengthen. Muscle tissue has long, cylindrical cells lined up in parallel. This shape is why muscle cells are often called "muscle fibers." Muscle layers—and muscular organs—contract and relax in a coordinated way. This is how the action of muscles maintains and changes the positions of body parts, movements that range from walking to blinking your eyes. The three types of muscle tissue are skeletal, smooth, and cardiac muscle tissues.

Skeletal muscle is located in muscles that attach to your bones (Figure 4.4*a*). In a typical muscle, such as the biceps, skeletal muscle cells are bundled closely together, in parallel. This arrangement makes them look striped, or *striated*. The bundles, called fascicles, are enclosed by a sheath of dense connective tissue. This arrangement of muscle and connective tissue makes up the organs we call "muscles." The structure and function of skeletal muscle tissue are topics we will consider in Chapter 6.

Smooth muscle cells taper at both ends (Figure 4.4*b*). Junctions hold the cells together (Section 4.5), and they are bundled inside a connective tissue sheath. This type of muscle tissue is specialized for steady, controlled contraction. It is found in the walls of internal organs—including blood vessels, the stomach, and the intestines. The contraction of smooth muscle is said to be "involuntary" because we usually cannot make it contract just by thinking about it (as we can with skeletal muscle).

Cardiac muscle (Figure 4.4*c*) is found only in the wall of the heart and its sole function is to pump blood. As you will read in Chapter 9, special junctions fuse the plasma membranes of cardiac muscle cells. In places, communication junctions allow the cells to contract as a unit. When one cardiac muscle cell is signaled to contract, the cells around it contract, too.

Muscle tissue can contract (shorten) in response to stimulation. It helps move the body and its parts.

Skeletal muscle is attached to bones. Smooth muscle is found in internal organs. Cardiac muscle makes up the walls of the heart.

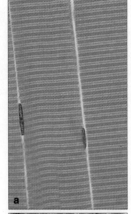

TYPE: Skeletal muscle
DESCRIPTION: Bundles of cylindrical, long, striated contractile cells; many mitochondria; often reflex-activated but can be consciously controlled
LOCATIONS: Partner of skeletal bones, against which it exerts great force
FUNCTION: Locomotion, posture; head, limb movements

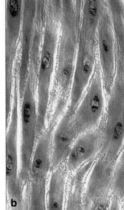

TYPE: Smooth muscle
DESCRIPTION: Contractile cells tapered at both ends; not striated
LOCATIONS: Wall of arteries, sphincters, stomach, intestines, urinary bladder, many other soft internal organs
FUNCTION: Controlled constriction; motility (as in gut); arterial blood flow

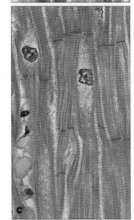

TYPE: Cardiac muscle
DESCRIPTION: Unevenly striated, fused-together cylindrical cells that contract as a unit owing to signals at gap junctions between them
LOCATIONS: Heart wall
FUNCTION: Pump blood forcefully through circulatory system

Figure 4.4 *Animated!* Characteristics and examples of skeletal muscle, smooth muscle, and cardiac muscle tissues.

FOUR TYPES OF TISSUES

4.4 Nervous Tissue: Communication

Of the four types of tissues in the body, nervous tissue has the most control over how the body responds to changing conditions, both internal and external.

The body's **nervous tissue** consists mostly of cells. They include **neurons**, the "nerve cells," as well as other cells that serve various support functions. There are tens of thousands of neurons in the brain and spinal cord, and millions more are present throughout the body. Because neurons transmit signals called nerve impulses, they make up the body's communication lines.

NEURONS CARRY MESSAGES

Like other kinds of cells, a neuron has a cell body that contains the nucleus and cytoplasm. It also has two types of extensions, or cell "processes." Branched processes called dendrites pick up incoming chemical messages. Outgoing messages are conducted by an axon. Depending on the type of neuron, its axon may be very short, or it may be as long as three or four feet. Figure 4.5*a* shows cell processes of a motor neuron, which carries signals to muscles and glands.

A cluster of processes from several neurons forms a **nerve**. Nerves conduct messages from the central nervous system (the brain and spinal cord) to muscles and glands. They also carry messages from specialized sensory receptors back to the central nervous system.

NEUROGLIA ARE SUPPORT CELLS

About 90 percent of the cells in the nervous system are **glial cells** (also called **neuroglia**). The word *glia* means glue, and glial cells were once thought to simply be the "mortar" that physically supported neurons. Today we know that they also have other functions. In the central nervous system, glia called astrocytes ("astro" because they are star-shaped; Figure 4.5*b*) help bring nutrients to neurons and provide physical support. Another type removes debris, microorganisms, or other foreign matter. Outside the brain and spinal cord glia called Schwann cells provide insulation—an extremely important function that helps speed nerve impulses through the body, as described in Chapter 13.

Neurons are the basic units of communication in nervous tissue. Different kinds detect specific stimuli, integrate information, and issue or relay commands for response.

Neuroglia lend structural support to neurons, help nourish and protect them, or provide insulation.

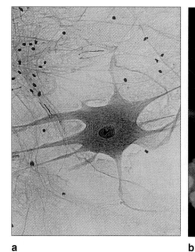 

a   b

Figure 4.5 A sampling of the millions of cells in nerve tissue. (**a**) A motor neuron. This type of nerve cell relays signals from the brain or spinal cord to muscles and glands. (**b**) Astrocytes, a type of neuroglia. These and other kinds of neuroglia make up well over half the volume of nerve tissue and provide vital support and other services for neurons.

4.5 Cell Junctions: Holding Tissues Together

LINKS TO
SECTIONS
3.3, 3.7, AND 3.8

Junctions between the cells in a tissue have various functions, from stopping leaks, holding a tissue's cells firmly together, and serving as channels for chemical communication.

Our tissues and organs would fall into disarray if there were not some way for individual cells to "stick together" and to communicate. Cell junctions meet these needs, and they can be found in all tissues. These cell-to-cell contacts are particularly common where substances must not leak from one body compartment to another.

Figure 4.6 shows some examples of cell junctions. **Tight junctions** (Figure 4.6*a*) are strands of protein that help stop substances from leaking across a tissue. The strands form gasketlike seals that prevent molecules from moving easily across the junction. In epithelium, for example, tight junctions allow the epithelial cells to control what enters the body. For instance, while food is being digested, various types of nutrient molecules can diffuse into epithelial cells or enter them selectively by active transport, but tight junctions keep those needed molecules from slipping *between* cells. Tight junctions also prevent the highly acidic gastric fluid in your stomach from leaking out and digesting proteins of your own body instead of those you consume in food. Actually, that kind of leakage is what happens in people who have peptic ulcers (Section 7.3).

Adhering junctions (Figure 4.6*b*) cement cells together. One type, sometimes called desmosomes, are like spot welds at the plasma membranes of two adjacent cells. They are anchored to the cytoskeleton in each cell and help hold cells together in tissues that are subject to stretching, such as epithelium of the skin, the lungs, and the stomach. Another type of adhering junction (also called zonula adherens junctions) forms a tight collar around epithelial cells.

Gap junctions (Figure 4.6*c*) are channels that connect the cytoplasm of neighboring cells. They help cells communicate by promoting the rapid transfer of ions and small molecules between them. Gap junctions are most plentiful in smooth muscle and cardiac muscle. As you will read in Chapter 9, ions moving through them from muscle cell to muscle cell play a key role in contraction of whole muscles. In other kinds of tissues gap junctions are the conduits for many kinds of signaling molecules.

Tight junctions between cells help stop leaks in a tissue. Adhering junctions cement cells in a tissue together. Gap junctions serve as channels through which ions and small molecules can move from cell to cell.

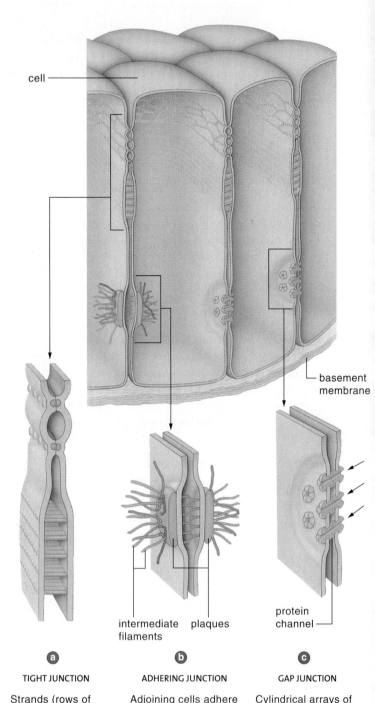

a
TIGHT JUNCTION
Strands (rows of proteins) running parallel with the free surface of the tissue; they block leaking between adjoining cells.

b
ADHERING JUNCTION
Adjoining cells adhere at a mass of proteins (a plaque) anchored beneath their plasma membrane by many intermediate filaments of the cytoskeleton.

c
GAP JUNCTION
Cylindrical arrays of proteins span the plasma membrane of adjoining cells. They pair up as open channels for signals between cells.

Figure 4.6 *Animated!* Cell junctions.

FOUR TYPES OF TISSUES

4.6 Tissue Membranes: Thin, Sheetlike Covers

The body's surfaces and cavities are covered by different kinds of thin, sheetlike membranes. Some mainly provide protection, while others both protect and lubricate organs.

A membrane is assigned to one of two categories, depending on its structure. In one group are *epithelial membranes*, while in the second group are *connective tissue membranes*. Here we'll consider some examples of each.

EPITHELIAL MEMBRANES PAIR WITH CONNECTIVE TISSUE

Epithelial membranes consist of a sheet of epithelium atop connective tissue. For instance, consider the body's *mucous membranes*, sometimes called mucosae (singular: mucosa). These are the pink, moist membranes lining the tubes and cavities of your digestive, respiratory, urinary, and reproductive systems (Figure 4.7*a*). Most mucous membranes are specialized to absorb substances, secrete them, or both. And as you might guess, most mucous membranes, like the lining of the stomach, contain glands, including mucous glands that secrete mucus. Not all do, though. For instance, the mucous membrane lining the urinary tract (including the tubes that carry urine out) has no glands. Later chapters will provide many examples of how mucous membranes protect other tissues and secrete or absorb substances.

Serous membranes are another type of epithelial membrane. These membranes occur in paired sheets; imagine one paper sack inside another, with a narrow space between them, and you'll get the idea. Serous membranes don't have glands, but the layers do secrete a fluid that fills the space between them. Examples include the membranes that line the chest (thoracic) cavity and enclose the heart and lungs. Among other functions, serous membranes help anchor internal organs in place and provide lubricated smooth surfaces that prevent chafing between adjacent organs or between organs and the body wall.

A third type of epithelial membrane is the *cutaneous membrane* (Figure 4.7*c*). You know this hardy, dry membrane as your skin. Its tissues also are part of one of the body's major organ systems, the integumentary system, which we examine in some detail in Section 4.8.

MEMBRANES IN JOINTS CONSIST OF CONNECTIVE TISSUE

A few membranes in the body have no epithelial cells, only connective tissue. These *synovial membranes* (Figure 4.7*d*) line the sheaths of tendons and the capsules around certain joints. Cells in the membranes secrete fluid that lubricates the ends of moving bones or prevents friction between a moving tendon and the bone it is attached to.

> *Epithelial membranes consist of epithelium overlying connective tissue. They line body surfaces, cavities, ducts, and tubes. Different types include mucous and serous membranes and the cutaneous membrane of skin. Most epithelial membranes contain glands and epithelial cells specialized for secretion, absorption, or both.*
>
> *Connective tissue membranes consist only of connective tissue. They line joint cavities.*

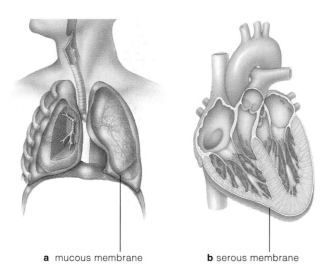

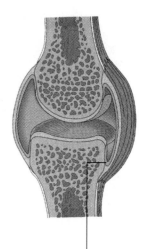

a mucous membrane **b** serous membrane **c** cutaneous membrane (skin) **d** synovial membrane

Figure 4.7 Examples of membranes in the human body.

4.7 Organs and Organ Systems

LINK TO
SECTION
1.3

Tissues begin to develop in the tiny embryo that arises after conception. With time, development produces organs in eleven organ systems.

An **organ** is a combination of two or more kinds of tissue which together perform one or more functions. The heart, for example, contains all four of the tissue types you have read about in previous sections. Much of the heart's wall is cardiac muscle, and several nerves help regulate the heart's life-sustaining beat. Tough dense connective tissue provides a sturdy outer wrapping, while the heart's chambers are lined with connective tissue and epithelium.

The heart and many other major organs are located inside body cavities shown in Figure 4.8. The figure also lists some terms biologists and medical professionals use to describe the positions of various organs. The **cranial cavity** and **spinal cavity** house your brain and spinal cord—the central nervous system. Your heart and lungs reside in the **thoracic cavity**—essentially, inside your chest. The diaphragm muscle separates the thoracic cavity from the **abdominal cavity**, which holds your stomach, liver, most of the intestine, and other organs. Reproductive organs, the bladder, and the rectum are located in the **pelvic cavity**.

Two or more organs combine to make up each of the body's eleven **organ systems**. Each organ system in turn contributes to the survival of all living cells in the body (Figure 4.9). Does this statement seem like a stretch? After all, how could, say, bones and muscles help each microscopically small cell to stay alive? Yet, interactions between your skeletal and muscular systems allow you to move about—toward sources of nutrients and water, for example. Parts of those systems help keep your blood circulating to cells, as when contractions of leg muscles help move blood in veins back to your heart. Blood inside the circulatory system rapidly carries nutrients and other substances to cells and transports products and wastes away from them. Your respiratory system swiftly delivers oxygen from air to your circulatory system and takes up carbon dioxide wastes from it, skeletal muscles assist the respiratory system—and so it goes, throughout the entire body.

> *The body's organ systems each serve a specialized function that contributes to the survival of all living body cells.*

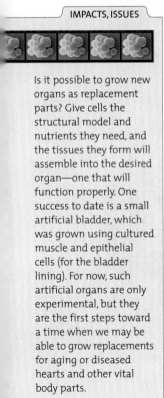

IMPACTS, ISSUES

Is it possible to grow new organs as replacement parts? Give cells the structural model and nutrients they need, and the tissues they form will assemble into the desired organ—one that will function properly. One success to date is a small artificial bladder, which was grown using cultured muscle and epithelial cells (for the bladder lining). For now, such artificial organs are only experimental, but they are the first steps toward a time when we may be able to grow replacements for aging or diseased hearts and other vital body parts.

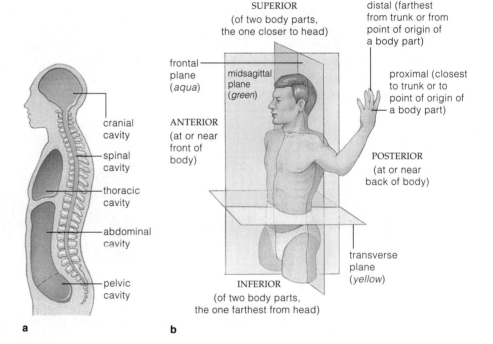

Figure 4.8 *Animated!* (**a**) The major cavities in the human body. (**b**) Directional terms and planes of symmetry for the human body. Notice how the midsagittal plane divides the body into right and left halves. The transverse plane divides it into superior (top) and inferior (bottom) parts. The frontal plane divides it into anterior (front) and posterior (back) parts.

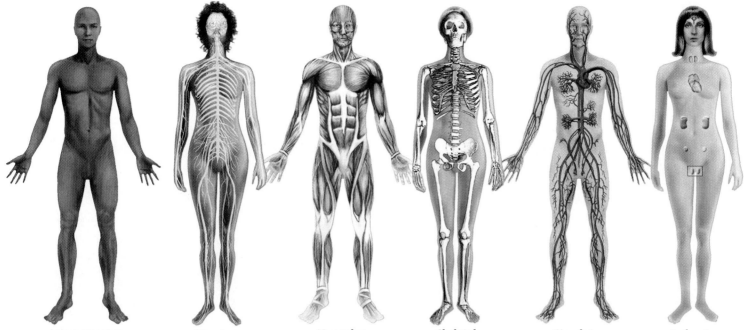

Integumentary System
Protects body from injury, dehydration, and some microbes; controls body temperature; excretes some wastes; receives some sensory information.

Nervous System
Detects external and internal stimuli; controls and coordinates the responses to stimuli; integrates all organ system activities.

Muscular System
Moves body and its parts; maintains posture; generates heat by increasing metabolic activity.

Skeletal System
Supports and protects body parts; provides muscle attachment sites; produces red blood cells; stores calcium, phosphorus.

Circulatory System
Rapidly transports many materials to and from cells; helps stabilize internal pH and temperature.

Endocrine System
Hormonally controls body functioning; works with nervous system to integrate short-term and long-term activities.

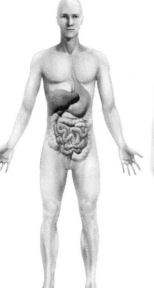

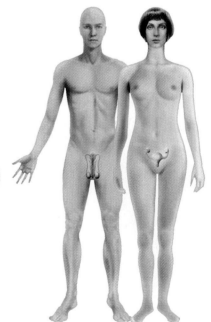

Lymphatic System
Collects and returns tissue fluid to the blood; defends the body against infection and tissue damage.

Respiratory System
Delivers oxygen to all living cells; removes carbon dioxide wastes of cells; helps regulate pH.

Digestive System
Ingests food and water; mechanically, chemically breaks down food and absorbs small molecules into internal environment; eliminates food residues.

Urinary System
Maintains the volume and composition of blood and tissue fluid; excretes excess fluid and blood-borne wastes.

Reproductive System
Female: Produces eggs; after fertilization, affords a protected, nutritive environment for the development of a fetus. *Male:* Produces and transfers sperm to the female. Hormones of both systems also influence other organ systems.

Figure 4.9 *Animated!* Overview of human organ systems and their functions.

4.8 The Integument—Example of an Organ System

Our integument includes the skin and several other components. The skin itself is the body's largest organ, weighing about 9 pounds in an average-sized adult.

The organ system called the **integument** (from Latin *integere*, "to cover") consists of the skin, oil and sweat glands, hair, and nails. As coverings go, skin is pretty amazing. It holds its shape after years of washing and being stretched, blocks harmful solar radiation, is a barrier to many microbes, holds in moisture, and fixes small cuts and burns. Skin also helps regulate body temperature, and signals from sensory receptors in skin help the brain assess what's going on in the outside world. Yet except for places subjected to regular abrasion (such as the palms of the hands and soles of the feet), your skin is generally not much thicker than a piece of construction paper. It is even thinner in some places, such as the eyelids.

Human skin also makes cholecalciferol, a precursor of vitamin D. Vitamin D is a generic name for steroid-like compounds that help the body absorb calcium from food. When vitamin D made in skin is released into the bloodstream in a hormone-like action, skin exposed to sunlight acts like an endocrine gland.

Figure 4.10 *Animated!* (**a**) The structure of human skin. The dark spots in the epidermis are cells to which melanocytes have passed pigment. (**b**, right): A section through human skin.

EPIDERMIS AND DERMIS ARE THE TWO LAYERS OF SKIN

Skin has an outer **epidermis** and an underlying **dermis**. Sweat glands, oil glands, hair follicles, and toenails and fingernails develop from the epidermal tissue (Figure 4.10). The dermis is mainly dense connective tissue, so it contains elastin fibers that make skin resilient and collagen fibers that make it strong. Together, the epidermis and dermis form the cutaneous membrane you read about in Section 4.6. Below the dermis is a subcutaneous ("under the skin") layer, the hypodermis. This is a loose connective tissue that anchors the skin while allowing it to move a bit. Fat stored in the hypodermis helps insulate the body and cushions some of its parts.

The outer part of the epidermis is a stratified squamous epithelium. Its cells arise in deeper layers and are pushed toward the skin's surface as new cells arise beneath them. (This efficient replacement is one reason why the skin

can mend minor damage so quickly.) Due to pressure from the ever-growing cell mass and from normal wear and tear at the surface, older cells are dead and flattened by the time they reach the outer layers. There, they are rubbed off or flake away.

Most cells of the epidermis are **keratinocytes**. These cells make keratin, a tough, water-insoluble protein. By the time they reach the skin surface and have died, all that remain are the keratin fibers inside plasma membranes. This helps make the skin's outermost layer—the stratum corneum—tough and waterproof.

In the deepest layer of epidermis, cells called **melanocytes** produce a brown-black pigment called melanin. The pigment is transferred to keratinocytes and helps give skin its color. A yellow-orange pigment in the dermis, called carotene, also contributes some color. In general, all humans have the same number of melanocytes, but skin color varies due to differences in the distribution and activity of those cells. For example, the pale skin of Caucasians contains only a little melanin, so the pigment hemoglobin inside red blood cells shows through thin-walled blood vessels and the epidermis itself, both of which are transparent. There is more melanin in naturally brown or black skin.

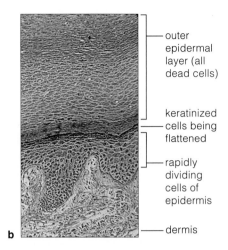

- outer epidermal layer (all dead cells)
- keratinized cells being flattened
- rapidly dividing cells of epidermis
- dermis

b

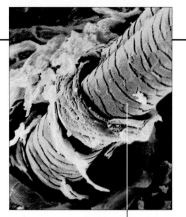

dead, flattened cells of a shaft of hair

Figure 4.11 Close-up of a hair. Dead, flattened hair cells form a tubelike cuticle around the hair shaft.

The epidermis also contains two cell types that help protect the body. *Langerhans cells* are phagocytes ("cell eaters") that engulf bacteria or virus particles, a process that mobilizes the immune system. *Granstein cells* may help control immune responses in the skin.

The dense connective tissue of the dermis makes it quite tough, but this protection has limits. For example, ongoing abrasion—as might happen if you wear a too-tight shoe—separates the epidermis from the dermis, the gap fills with a watery fluid, and you get a blister.

The dermis is laced with small blood vessels and sensitive nerve endings, and hair follicles, sweat glands, and oil glands are embedded in it. On the palms of the hands and soles of the feet it also has ridges that push up corresponding ridges on the epidermis. These ridges loop and curve in the intricate patterns we call fingerprints. The pattern is determined by a person's genes and is different for each of us, even identical twins.

SWEAT GLANDS AND OTHER STRUCTURES ARE DERIVED FROM EPIDERMIS

The body has about 2.5 million sweat glands. The fluid they secrete is 99 percent water; it also contains dissolved salts, traces of ammonia and other wastes, vitamin C, and other substances. A type of sweat gland that is plentiful in the palms of the hands, soles of the feet, forehead, and armpits functions mainly in temperature regulation. Another type is abundant in skin around the genitals. Stress, pain, and sexual foreplay all can increase the amount of sweat they secrete.

Oil glands (or *sebaceous glands*) are everywhere except on the palms and the soles of the feet. The oily substance they release softens and lubricates the hair and skin. Other secretions kill many harmful bacteria. *Acne* is a skin inflammation that develops after bacteria infect the ducts of oil glands.

A **hair** is a flexible structure of mostly keratinized cells, rooted in skin with a shaft above its surface. As cells divide near the base of the root, older cells are pushed upward, then flatten and die. The outermost layer of the shaft consists of flattened cells that overlap like roof shingles (Figure 4.11). These dead cells are what frizz out as "split ends." An average human scalp has about 100,000 hairs. However, the growth and the density of a person's hair are influenced by genes, nutrition, hormones, and stress.

SUNLIGHT PERMANENTLY DAMAGES THE SKIN

Ultraviolet (UV) radiation stimulates the melanin-producing cells of the epidermis. With prolonged sun exposure, melanin levels increase and light-skinned people become tanned. Tanning gives some protection against UV radiation, but over the years, it causes elastin fibers in the dermis to clump together. The skin loses its resiliency and begins to look leathery.

Ultraviolet radiation from sunlight or from the lamps of tanning salons also can activate proto-oncogenes in skin cells (Section 23.2). These genes can trigger cancer, like the squamous cell carcinoma, a common form of skin cancer, shown at right.

Until recently, ozone in the stratosphere intercepted much of the potentially damaging UV radiation that reaches Earth. Today, however—due in large part to human activities described in Chapter 25—the ozone layer over ever-larger regions of the globe is being destroyed faster than natural processes can replace it. The rate of skin cancers now is rapidly increasing.

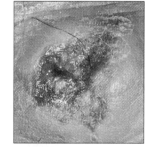

Squamous cell carcinoma

With its layers of keratinized and melanin-shielded epidermal cells, skin helps the body conserve water, limit damage from ultraviolet radiation, and resist mechanical stress. Hairs, oil glands, sweat glands, and other structures associated with skin are derived from epidermis.

4.9 Homeostasis: The Body in Balance

LINKS TO
SECTIONS
1.1 AND 2.7

Cells, tissues, and organs can function properly only when conditions inside the body are stable. Maintaining this stable internal state—homeostasis—demands finely tuned controls.

THE INTERNAL ENVIRONMENT IS A POOL OF EXTRACELLULAR FLUID

The trillions of cells in your body all are bathed in fluid—about 15 liters, or a little less than four gallons. This fluid, called **extracellular** ("outside the cell") **fluid**, is what we mean by the term "internal environment." Much of the extracellular fluid is *interstitial*, meaning that it fills spaces between cells and tissues. The rest is blood plasma, the fluid portion of blood. Substances constantly enter and leave interstitial fluid as cells draw nutrients from it and dump their metabolic waste products into it. Those substances can include ions, compounds such as water, and other materials.

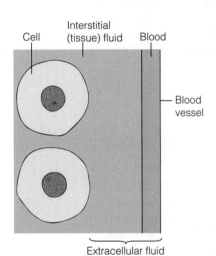

Cell

Interstitial (tissue) fluid

Blood

Blood vessel

Extracellular fluid

All this chemical traffic means that the chemical makeup and volume of extracellular fluid change from moment to moment. If the changes are drastic, they can have drastic effects on cell activities. The number and type of ions in extracellular fluid (such as H^+) are especially crucial, because they must be kept at levels that allow metabolism to continue normally. As you read in Chapter 1, **homeostasis** means "staying the same." The mechanisms of homeostasis operate to maintain stability in the volume and chemical makeup of extracellular fluid.

In maintaining homeostasis, all components of the body work together in the following general way:

- Each cell engages in metabolic activities that ensure its own survival.

- Tissues, which consist of cells, perform one or more activities that contribute to the survival of the whole body.

- Together, the operations of individual cells, tissues, organs, and organ systems help keep the extracellular fluid in a stable state—a state of homeostasis that allows cells to survive.

HOMEOSTASIS REQUIRES THE INTERACTION OF SENSORS, INTEGRATORS, AND EFFECTORS

Three "partners" must interact to maintain homeostasis. They are sensory receptors, integrators, and effectors. **Sensory receptors** are cells or cell parts that can detect a **stimulus**—a specific change in the environment. For a simple example, if someone taps you on the shoulder, there is a change in pressure on your skin. Receptors in the skin translate the stimulus into a signal, which can be sent to the brain. Your brain is an **integrator**, a control point where different bits of information are pulled together in the selection of a response. It can send signals to muscles, glands, or both. Your muscles and glands are **effectors**—they carry out the response, which in this case might include turning your head to see if someone is there. Of course, you cannot keep your head turned indefinitely, because eventually you must eat, use the bathroom, and perform other tasks that maintain body operating conditions.

So how does the brain deal with physiological change? Receptors inform it about how things *are* operating, but the brain also maintains information about how things *should be* operating—that is, information from "set points." When conditions deviate sharply from a set point, the brain brings them back within proper operating range. It does this by sending signals that cause specific muscles and glands to step up or reduce their activity. Set points are important in many physiological mechanisms, including those that influence eating, breathing, thirst, and urination, to name a few.

FEEDBACK MECHANISMS ARE IMPORTANT HOMEOSTATIC CONTROLS

Mechanisms for feedback help keep physical and chemical aspects of the body within tolerable ranges. In

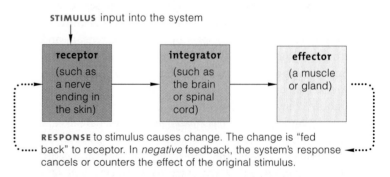

STIMULUS input into the system

receptor (such as a nerve ending in the skin)

integrator (such as the brain or spinal cord)

effector (a muscle or gland)

RESPONSE to stimulus causes change. The change is "fed back" to receptor. In *negative* feedback, the system's response cancels or counters the effect of the original stimulus.

Figure 4.12 *Animated!* Three basic components of negative feedback at the organ level.

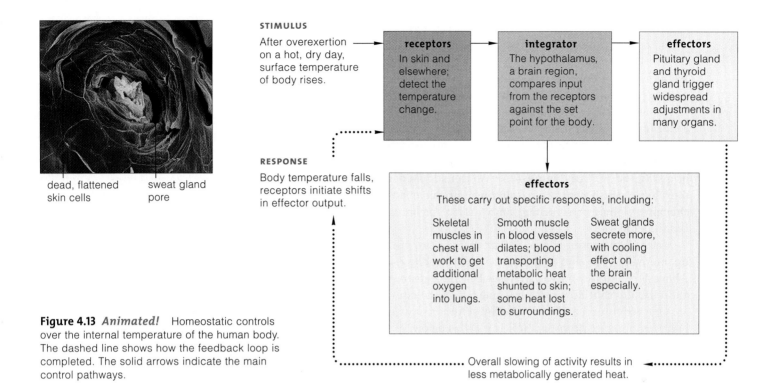

STIMULUS

After overexertion on a hot, dry day, surface temperature of body rises.

receptors	integrator	effectors
In skin and elsewhere; detect the temperature change.	The hypothalamus, a brain region, compares input from the receptors against the set point for the body.	Pituitary gland and thyroid gland trigger widespread adjustments in many organs.

RESPONSE

Body temperature falls, receptors initiate shifts in effector output.

effectors

These carry out specific responses, including:

Skeletal muscles in chest wall work to get additional oxygen into lungs.	Smooth muscle in blood vessels dilates; blood transporting metabolic heat shunted to skin; some heat lost to surroundings.	Sweat glands secrete more, with cooling effect on the brain especially.

Overall slowing of activity results in less metabolically generated heat.

dead, flattened skin cells

sweat gland pore

Figure 4.13 *Animated!* Homeostatic controls over the internal temperature of the human body. The dashed line shows how the feedback loop is completed. The solid arrows indicate the main control pathways.

negative feedback, an activity alters a condition in the internal environment, and this triggers a response that reverses the altered condition (Figure 4.12). By analogy, think of a furnace with a thermostat. The thermostat senses the air temperature and mechanically compares it to a preset point on a thermometer built into the furnace control system. When the temperature falls below the preset point, the thermostat signals a switch that turns on the heating unit. When the air warms enough to match the preset level, the thermostat signals the switch to shut off the heating unit.

In a similar way, negative feedback helps keep body temperature within a normal range (Figure 4.13). For example, when sensors indicate that the skin is getting too hot while you work outside in the sun, mechanisms kick in that slow both the metabolic activity of cells *and* overall activity levels. You may move less and look for shade. At the same time, blood flow to the skin increases and your sweat glands secrete more sweat. As water in sweat evaporates, your body loses more heat. These and other changes curb the body's heat-producing activities and release excess heat to the surroundings.

In a few situations **positive feedback** operates. In this type of mechanism, a chain of events *intensify* a change from an original condition—and after a limited time, the intensifying feedback reverses the change. There are not many instances of positive feedback in

body functions, but one familiar example is childbirth. During labor a fetus exerts pressure on the walls of its mother's uterus. The pressure stimulates the production and secretion of a hormone (oxytocin) that causes the mother's uterine muscles to contract and exert pressure on the fetus, which exerts more pressure on the uterine wall, and so on until the fetus is expelled.

As the body monitors and responds to information about the external world and the internal environment, its organ systems must operate in a coordinated way. In upcoming chapters we will be asking four important questions about how organ systems function:

1. What physical or chemical aspect of the internal environment is each organ system working to maintain as conditions change?

2. How is each organ system kept informed of changes?

3. How does each system process incoming information?

4. What are the responses?

As you will see, all organ systems operate under precise controls of the nervous system and the endocrine system.

Homeostatic control mechanisms maintain the physical and chemical characteristics of the internal environment within ranges that are favorable for cell operations.

Controls over the body's core temperature provide good examples of negative feedback loops.

We humans are **endotherms**, which means "heat from within." The body's **core temperature**—the temperature of the head and torso—is about 37°C, or 98.6°F. It is controlled mainly by metabolic activity, which produces heat, and by negative feedback loops. These homeostatic controls adjust physiological responses for conserving or getting rid of heat (Figure 4.14). We can supplement the physiological controls by altering our behavior—changing clothes or switching on a furnace or an air-conditioner.

Metabolism produces heat. If that heat were to build up internally, your core temperature would steadily rise. Above 41°C (105.8°F), some enzymes become denatured and virtually shut down. By the same token, the rate of enzyme activity generally *decreases* by at least half when body temperature drops by 10°F. If it drops below 35°C (95°F), you are courting danger. As enzymes lose their ability to function, your heart will not beat as often or as effectively, and heat-generating mechanisms such as shivering stop. At this low core temperature breathing

slows, so you may lose consciousness. Below 80°F the human heart may stop beating entirely. Given these stark physiological facts, humans require mechanisms that help maintain the core body temperature within narrow limits.

RESPONSES TO COLD STRESS

Table 4.3 summarizes the major responses to cold stress. They are governed by the **hypothalamus**, a centrally located structure in the brain that includes both neurons and endocrine cells. When the outside temperature drops, thermoreceptors (*thermo-* means heat) at the body surface detect the decrease. When their signals reach the hypothalamus, neurons command smooth muscle in the walls of arterioles in the skin to contract. The resulting **peripheral vasoconstriction** reduces blood flow to capillaries near the body surface, so your body retains heat. For example, when your fingers or toes get cold, as much as 99 percent of the blood that would otherwise flow to your skin is diverted.

In the **pilomotor response** to a drop in outside temperature, your body hair can "stand on end." This

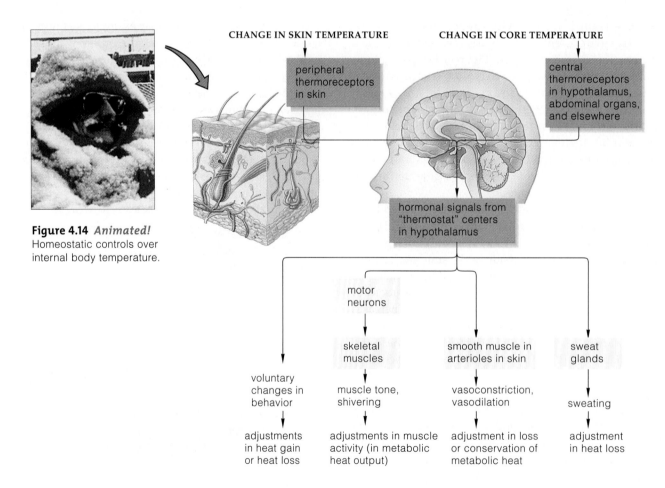

Figure 4.14 *Animated!*
Homeostatic controls over internal body temperature.

ORGANS AND ORGAN SYSTEMS

Table 4.3	Summary of Human Responses to Cold Stress and to Heat Stress	
Environmental Stimulus	**Main Responses**	**Outcome**
Drop in temperature	Vasoconstriction of blood vessels in skin; pilomotor response; behavior changes (e.g., putting on a sweater)	Heat is conserved
	Increased muscle activity; shivering; nonshivering heat production	More heat is produced
Rise in temperature	Vasodilation of blood vessels in skin; sweating; changes in behavior; heavy breathing	Heat is dissipated from body
	Reduced muscle activity	Less heat is produced

happens because smooth muscle controlling the erection of body hair is stimulated to contract. This creates a layer of still air close to the skin that reduces heat losses. (This response is most effective in mammals with more body hair than humans!) Heat loss can be restricted even more by behaviors that reduce the amount of body surface exposed for heat exchange, as when you put on a sweater or hold your arms tightly against your body.

When other responses can't counteract cold stress, signals from the hypothalamus step up skeletal muscle contractions, similar to the low-level contractions that produce muscle tone. The result? You start shivering. Your skeletal muscles contract ten to twenty times per second, boosting heat production throughout the body.

Prolonged or severe exposure to cold can lead to a hormonal response that elevates the rate of metabolism in cells. This *nonshivering heat production* is especially notable in a specialized type of adipose tissue called "brown fat." Heat is generated as the lipid molecules are broken down. Babies (who can't shiver) have this tissue in the neck and armpits and near their kidneys; adults have little brown fat unless they are cold-adapted.

In *hypothermia*, body core temperature falls below the normal range. A drop of only a few degrees leads to mental confusion; further cooling can lead to coma and death. Some victims of extreme hypothermia, mainly children, have survived prolonged immersion in ice-cold water. One reason is that mammals, including humans, have a dive reflex. When the body is submerged, the heart rate slows and blood is shunted to the brain and other vital organs.

Freezing often destroys tissues, a condition we call *frostbite*. Frozen cells may be saved if thawing is precisely controlled. This sometimes can be done in a hospital.

RESPONSES TO HEAT STRESS

Table 4.3 also summarizes the main responses to heat stress. When core temperature rises above a set point, the hypothalamus again orders key responses. In **peripheral vasodilation**, its signals cause blood vessels in the skin to dilate. More blood flows to the skin, where the excess heat that the blood carries is dissipated.

Evaporative heat loss also can be influenced by the hypothalamus, which can activate sweat glands. There are roughly 2.5 million sweat glands in skin, and lots of heat is dissipated when the water in sweat evaporates. With prolonged heavy sweating the body also loses key salts, especially sodium chloride. Losing too many electrolytes can make you feel woozy. People who exercise heavily may consume "sports drinks" that replenish electrolytes.

Sometimes peripheral blood flow and evaporative heat loss can't adequately counter heat stress. The result is *hyperthermia*, in which the core temperature rises above normal. If the increase isn't too great, a person can suffer *heat exhaustion*, in which blood pressure drops due to vasodilation and water losses from heavy sweating. The skin feels cold and clammy, and the person may collapse.

When heat stress is severe enough to completely break down the body's temperature controls, *heat stroke* occurs. Sweating stops, the skin becomes dry, and the core body temperature rapidly rises to a level that can be lethal.

When someone has a fever, the hypothalamus has reset the "thermostat" that dictates what the body's core temperature will be. The normal response mechanisms are brought into play, but they are carried out to maintain a higher temperature.

When a fever starts, heat production increases, heat loss drops, and the person feels chilled. When a fever "breaks," peripheral vasodilation and sweating increase as the body tries to restore the normal core temperature; then you feel warm. The controlled increase in core temperature during a fever seems to enhance the body's immune response, so using fever-reducing drugs such as aspirin or ibuprofen may actually interfere with fever's beneficial effects. A severe fever, however, requires medical supervision because of the dangers it poses.

The hypothalamus governs responses that regulate the body's core temperature.

Physiological responses to cold stress include constriction of blood vessels near the body surface, the pilomotor response, shivering, and sometimes nonshivering heat production.

Responses to heat stress include dilation of peripheral blood vessels and evaporative heat loss.

Summary

Introduction   A tissue is a group of similar cells that perform a common function (Table 4.4). Different tissues combine in certain proportions and patterns to form an organ. In an organ system, two or more organs interact in ways that contribute to the body's survival.

Section 4.1   Epithelial tissues cover external body surfaces and line internal cavities and tubes. Each kind of epithelium has one surface exposed to body fluids or the external environment; the opposite surface rests on a basement membrane sandwiched between it and an underlying connective tissue.

Glands are derived from epithelium. A gland is a cell or a multicellular structure that makes and secretes a specific substance such as saliva. Exocrine glands release substances onto the surface of an epithelium through ducts or tubes. This contrasts with endocrine glands, which secrete substances (hormones) directly into extracellular fluid.

Section 4.2   Connective tissues bind together, support, strengthen, protect, and insulate other tissues. Most have fibers of structural proteins (especially collagen), fibroblasts, and other cells within a matrix. They include fibrous connective tissue and specialized connective tissues such as cartilage, bone, adipose tissue, and blood.

Section 4.3   Muscle tissue contracts (shortens) and then returns to the resting position. It helps move the body or its parts. The three types of muscle tissue are skeletal muscle, smooth muscle, and cardiac muscle.

Section 4.4   Nervous tissue receives and integrates information from inside and ouside the body and sends signals for responses. Neurons are the basic units of the nervous system; they have extensions (axons) that form nerves. Neuroglia (glial cells) are accessory cells of the nervous system. They physically support and help nourish neurons, among other functions.

Section 4.5   Tight junctions help prevent substances from leaking across a tissue. Adhering junctions bind cells together in tissues. Gap junctions link the cytoplasm of neighboring cells.

 Compare the structure and functions of the main types of cell junctions.

Section 4.6   Membranes cover body surfaces and cavities. Those made of epithelium include mucous and serous membranes. Most mucous membranes have glands that secrete mucus; serous membranes help anchor organs and secrete lubricating fluid onto their surfaces. Connective tissue membranes include the synovial membranes of certain joints. The skin is a cutaneous membrane.

Section 4.7   Body organs are located in five major cavities: the cranial cavity (brain); spinal cavity (spinal cord); thoracic cavity (heart and lungs); abdominal cavity (stomach, liver, most of the intestine, other organs); and pelvic cavity (reproductive organs, bladder, rectum). The various organs in the body are arranged into eleven organ systems. Each system performs a specific function, such as transporting blood (cardiovascular system) or reproduction.

 Investigate the function of organ systems and learn about terms used to describe their locations.

Section 4.8   An example of an organ system is the integument, or skin. Skin has an outer epidermis and an underlying dermis. Most epidermal cells are keratinocytes, which make the protein keratin. Keratin makes the skin's outer layer tough and waterproof. Melanocytes in the epidermis produce pigment that gives skin its color. Hair, nails, sweat glands, and oil glands are derived from the epidermis.

Skin protects the rest of the body from abrasion, invading bacteria, ultraviolet radiation, and dehydration. It helps control internal temperature, contains cells that synthesize vitamin D, and serves as a blood reservoir for the rest of the body. Receptors in skin are essential for detecting environmental stimuli.

 Explore the structure of skin and hair.

Section 4.9   Extracellular fluid, which consists of blood and tissue fluid, is the body's internal environment. Tissues, organs, and organ systems work together to maintain the stable state of homeostasis in this fluid environment. Maintaining homeostasis requires sensory receptors, which can detect a stimulus, integrators, and effectors.

Feedback controls help maintain internal conditions. In negative feedback, a change in a condition triggers a response that reverses the change. In positive feedback, a response reverses a change in a condition by intensifying the change for a limited time.

Section 4.10   Physiological responses that govern temperature rely on negative feedback controls that respond to heat stress and cold stress.

Thomson NOW!   *See how negative feedback helps regulate body temperature.*

Table 4.4 Summary of Basic Tissue Types in the Human Body

Tissue	Function	Characteristics
Epithelium	Covers body surface; lines internal cavities and tubes	One free surface; opposite surface rests on basement membrane supported by connective tissue
Connective tissue	Binds, supports, adds strength; some provide protection or insulation	Cells surrounded by a matrix (ground substance) containing structural proteins except in blood
FIBROUS CONNECTIVE TISSUES		
Loose	Elasticity, diffusion	Cells and fibers loosely arranged
Dense	Support. elasticity	Several forms. One has collagen fibers in various orientations in the matrix; it occurs in skin and as capsules around some organs Another form has collagen fibers in parallel bundles; it occurs in ligaments, tendons
Elastic	Elasticity	Mainly elastin fibers; occurs in organs and that must stretch
SPECIALIZED CONNECTIVE TISSUES		
Cartilage	Support, flexibility, low-friction surface	Matrix solid but pliable; no blood supply
Bone	Support, protection, movement	Matrix hardened by minerals
Adipose tissue	Insulation, padding, energy storage	Soft matrix around large, fat-filled cells
Blood	Transport	Liquid matrix (plasma) containing blood cells, many other substances
Muscle tissue	Movement of the body and its parts	Made up of arrays of contractile cells
Nervous tissue	Communication between body parts; coordination, regulation of cell activity	Made up of neurons and support cells (neuroglia)

Review Questions

1. List the general characteristics of epithelium, and then describe the various types of epithelial tissues in terms of specific characteristics and functions.

2. List the major types of connective tissues; add the names and characteristics of their specific types.

3. Identify and describe the tissues shown below.

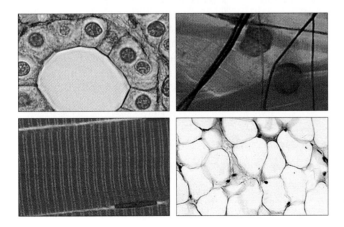

4. List the types of cell junctions and their functions.

5. List the basic types of membranes in the body.

6. What is the difference between a tissue, an organ, and an organ system? From memory, try to make a list of the eleven major organ systems of the human body.

7. What are some functions of skin?

8. Define homeostasis.

9. What is extracellular fluid, and how does the concept of homeostasis pertain to it?

10. What is the difference between negative feedback and positive feedback? What role do these mechanisms play in maintaining homeostasis?

Self-Quiz   *Answers in Appendix V*

1. _____ tissues have closely linked cells and one free surface.
   a. Muscle   c. Connective
   b. Nerve   d. Epithelial

2. Most _____ has collagen and elastin fibers.
   a. muscle tissue   c. connective tissue
   b. nervous tissue   d. epithelial tissue

3. _____, a specialized connective tissue, is mostly plasma with cellular components and various dissolved substances.
   a. Irregular connective tissue   c. Cartilage
   b. Blood   d. Bone

4. _____ tissue detects and integrates information about changes and controls responses to changes.
   a. Muscle
   b. Nervous
   c. Connective
   d. Epithelial

5. _____ can shorten (contract).
   a. Muscle tissue
   b. Nervous tissue
   c. Connective tissue
   d. Epithelial tissue

6. After you eat too many carbohydrates and proteins, your body converts the excess to storage fats, which accumulate in _____.
   a. loose connective tissue
   b. dense connective tissue
   c. adipose tissue
   d. both b and c

7. In _____, physical and chemical aspects of the body are being kept within tolerable ranges by controlling mechanisms.
   a. positive feedback
   b. negative feedback
   c. homeostasis
   d. metastasis

8. Fill in the blanks: _____ detect specific environmental changes, an _____ pulls different bits of information together in the selection of a response, and _____ carry out the response.

9. Match the concepts:
   ____ muscles and glands
   ____ positive feedback
   ____ sites of body receptors
   ____ negative feedback
   ____ brain

   a. integrating center
   b. reverses an altered condition
   c. eyes and ears
   d. effectors
   e. intensifies the original condition

Critical Thinking

1. In people who have the genetic disorder anhidrotic ectodermal dysplasia, patches of tissue have no sweat glands. What kind of tissue are we talking about?

2. The disease called scurvy results from a deficiency of vitamin C, which the body uses to synthesize collagen. Explain why scurvy sufferers tend to lose teeth, and why any wounds heal much more slowly than normal, if at all.

3. The man pictured in Figure 4.15 wears several dozen ornaments in his skin, nearly all of them applied by piercing. Among the skin's many functions, it serves as a

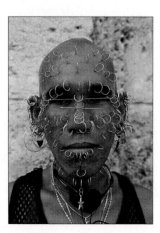

Figure 4.15 An example of heavy body piercing.

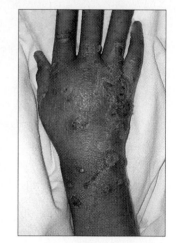

Figure 4.16 Ulcers and blisters that form on the skin of a person affected by porphyria after exposure to sunlight.

barrier to potentially dangerous bacteria, and some people object to extensive body piercing on the grounds that it opens the door to infections. Explain why you do or don't agree with this objection.

4. Porphyria, a genetic disorder, occurs in about 1 in 25,000 individuals. Affected people lack enzymes of a metabolic pathway that forms heme, the iron-containing group in hemoglobin. Accumulating porphyrins, which are intermediates in the pathway, cause awful symptoms, especially after exposure to sunlight. Lesions and scars form on the skin (Figure 4.16). Thick hair grows on the face and hands. The gums retreat and the canine teeth can begin to look like fangs. Symptoms worsen if the affected person consumes alcohol or garlic. Individuals with porphyria can avoid sunlight and aggravating substances. They also can get injections of heme from normal red blood cells. If you are familiar with vampire stories, which date from even before the Middle Ages, can you think of a reason why they may have evolved among people who knew nothing of the cause of porphyria?

Explore on Your Own

As epithelium, your skin contains fibers of collagen and elastin. These structural proteins have different properties that you can see in action when you pull on a patch of skin, such as on the back of your hand. Notice that even if you pull firmly, the skin doesn't tear; which type of protein fiber gives it that tensile strength? Which type returns the skin to its original shape when you let go?

Creaky Joints

Whether you're 18 or 80, you probably have or will develop some degree of osteoarthritis—a disorder in which joints become painfully stiff because their cartilage lining is breaking down or bone spurs have formed there. Disease, sports injuries, obesity, and simple aging cause creaky joints, and common remedies range from nonprescription pain relievers and cartilage-building supplements to injections of steroid drugs. Severely affected joints often are replaced with high-tech artificial ones.

A lot of arthritis sufferers also are exploring less conventional "treatments." Gullible ones eat ground-up cartilage from sharks or baby chicks. Botanicals— herbs and exotic plant extracts—also are finding customers eager to find relief for their symptoms.

There is a long menu of nontraditional, plant-based arthritis remedies, including evening primrose oil, ginger, devil's claw, feverfew, and an Indian herb called ashwaghanda. Do such substances work? Well, in 1998 researchers at a meeting of the American College of Rheumatology reported the results of a carefully designed study of 90 people with

osteoarthritis. Of patients who used botanicals suggested by Ayurveda, the traditional medicine of India, half improved, compared to only one-fifth of patients who received a placebo. Findings were similar in a study that tested the same herbal formula on patients with rheumatoid arthritis.

Critics pointed out that this research was sponsored by a company that sells the herbs. Also, the two types of arthritis involved have different causes and symptoms. How could one "therapy" treat both? In general, no authoritative scientific basis has been established for the purported health effects of many herbal remedies.

Arthritis research introduces this chapter's topic, the **skeletal system**—the skeleton along with cartilages, joints, and straplike ligaments that hold our bones together. As you will learn, your bones are not just a sturdy framework for your soft flesh. They partner with muscles to bring about movement and have an essential role in maintaining the body's calcium balance.

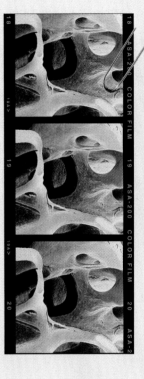

 How Would You Vote? Should claims about "medicinal" exotic plant extracts have to be backed up by independent scientific testing? Cast your vote online at www.thomsonedu.com/biology/starr/humanbio.

Key Concepts

THE STRUCTURE AND FUNCTIONS OF BONES
Bones are built of bone tissue. They store minerals, protect and support soft organs, and function in body movement. Some bones contain marrow where blood cells develop.

THE SKELETON
The skeleton's key function is to serve as the body's strong internal framework. Its 206 bones are organized into two parts, the axial skeleton and the appendicular skeleton.

JOINTS
At joints, different bones touch or are in close contact. Some of these connections permit adjoining bones to move in ways that in turn move body parts, such as the limbs.

BONES UNDER SIEGE
Disorders that affect our bones can upset homeostasis by impairing normal bone functions, such as aiding movement.

Links to Earlier Concepts

With this chapter we begin a survey of the body's eleven organ systems. As you study the skeletal system, you will learn more about the structure and functions of bone tissue, cartilage, and some other connective tissues (4.2) that are major components of the system.

Chapter 1 introduced the concept of homeostasis, and Section 4.9 gave you an overview of mechanisms that help maintain this internal stability. Although courses in human biology usually consider each organ system in turn, it is important to keep in mind that at every moment all of your organ systems are contributing to the survival of the amazingly complex whole that is you.

5.1 Bone—Mineralized Connective Tissue

LINK TO
SECTION
4.2

We begin our look at the skeletal system with bone tissue, the hardened connective tissue that makes up our bones.

Bone is a connective tissue, so it is a blend of living cells and a matrix that contains fibers. Bones are covered by a sturdy two-layer membrane called the periosteum (meaning "around the bone"). The membrane's outer layer is dense connective tissue and the inner layer contains bone cells called **osteoblasts** ("bone formers"). As bone develops, the osteoblasts secrete collagen and some elastin, as well as carbohydrates and other proteins. With time, this matrix around osteoblasts hardens when salts of the mineral calcium are deposited in it. The osteoblasts are trapped in spaces, or lacunae, in the matrix (*lacuna* = hole). At this point their bone-forming function ends and they are called **osteocytes** (*osteo* = bone; *cyte* = cell).

The minerals in bone tissue make it hard, but it is the collagen that gives our bones the strength to withstand the mechanical stresses associated with activities such as standing, lifting, and tugging.

THERE ARE TWO KINDS OF BONE TISSUE

Bones contain two kinds of tissue, compact bone and spongy bone. Figure 5.1 shows where these tissues are in a long bone such as the femur (thighbone). As its name suggests, **compact bone** is a dense tissue that looks solid and smooth. In a long bone, it forms the bone's shaft and the outer part of its two ends. A cavity inside the shaft contains bone marrow.

Compact bone tissue forms in thin, circular layers around small central canals. Each set of layers is called an **osteon** (or sometimes a *Haversian system*). The canals connect with each other and serve as channels for blood vessels and nerves that transport substances to and from osteocytes. Osteocytes also extend slender cell processes into narrow channels called canaliculi that run between lacunae. These "little canals" allow nutrients to move through the hard matrix from osteocyte to osteocyte. Wastes can be removed the same way.

The bone tissue *inside* a long bone's shaft and at its ends looks like a sponge. Tiny, flattened struts are fused together to make up this **spongy bone** tissue, which looks lacy and delicate but actually is quite firm and strong.

A BONE DEVELOPS ON A CARTILAGE MODEL

An early embryo has a rubbery skeleton that consists of cartilage and membranes. Yet, after only about two months of life in the womb, this flexible framework is transformed into a bony skeleton. Once again, we can look at the development of a long bone as an example.

As you can see at the top of Figure 5.2, a cartilage "model" provides the pattern for each long bone. Once the outer membrane is in place on the model, the bone-forming osteoblasts become active and a bony "collar" forms around the cartilage shaft. Then the cartilage inside the shaft calcifies, and blood vessels, nerves, and elements including osteoblasts begin to infiltrate the forming bone. Soon, the marrow cavity forms and osteoblasts produce the matrix that will become mineralized with calcium.

Each end of a long bone is called an epiphysis (e-PIF-uh-sis). As long as a person is growing, each epiphysis is separated from the bone shaft by an *epiphyseal plate* of cartilage. Human growth hormone (GH) prevents the plates from calcifying, so the bone can lengthen. When

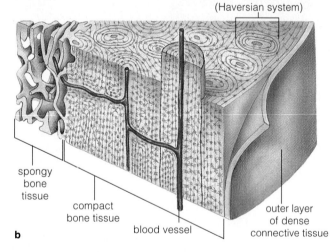

Figure 5.1 *Animated!* (**a**) Spongy and compact bone tissue in a femur. (**b**) Thin, dense layers of compact bone tissue form interconnected arrays around canals that contain blood vessels and nerves. Each array is an osteon (Haversian system). The blood vessel threading through it transports substances to and from osteocytes, living bone cells in small spaces (lacunae) in the bone tissue. Small tunnels called canaliculi connect neighboring spaces.

THE STRUCTURE AND FUNCTIONS OF BONES

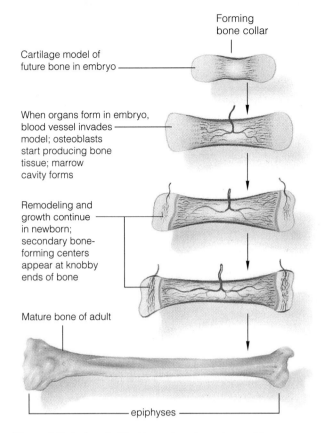

Forming bone collar

Cartilage model of future bone in embryo

When organs form in embryo, blood vessel invades model; osteoblasts start producing bone tissue; marrow cavity forms

Remodeling and growth continue in newborn; secondary bone-forming centers appear at knobby ends of bone

Mature bone of adult

epiphyses

Figure 5.2 *Animated!* How a long bone forms. First, osteoblasts begin to function in a cartilage model in the embryo. The bone-forming cells are active first in the shaft, then at the knobby ends. In time, cartilage is left only in the epiphyses at the ends of the shaft.

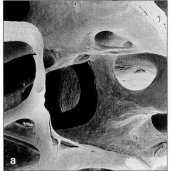
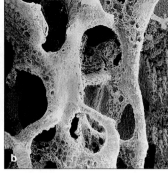

Figure 5.3 Osteoporosis. (**a**) Normal bone tissue. (**b**) After the onset of osteoporosis, replacements of mineral ions lag behind withdrawals. In time the tissue erodes, and the bone becomes hollow and brittle.

growth stops, usually in the late teens or early twenties, bone replaces the cartilage plates.

BONE TISSUE IS CONSTANTLY "REMODELED"

Calcium is constantly entering and leaving a person's bones. Calcium is deposited when osteoblasts form bone, and it is withdrawn when "bone breaker" cells called **osteoclasts** break down the matrix of bone tissue. This ongoing calcium recycling is called **bone remodeling**, and it has several important functions.

Regularly breaking down "old" bone and replacing it with fresh tissue helps keep bone resilient, so it is less likely to become brittle and break. When a bone is subjected to mechanical stress, such as load-bearing exercise, the remodeling process is adjusted so that more bone is deposited than removed. That is why the bones of regular exercisers are denser and stronger than the bones of couch potatoes. On the other hand, when the body must heal a broken bone, osteoclasts release more calcium than usual from bone matrix. Osteoblasts then use the calcium to repair the injured bone tissue.

A child's body requires lots of calcium to meet the combined demands of bone growth and other needs for the calcium stored in bones. Along with dietary calcium, remodeling helps meet the demand. For example, the diameter of a growing child's thighbones increases as

osteoblasts form bone at the surface of each shaft. At the same time, however, osteoclasts break down a small amount of bone tissue *inside* the shaft. Thus the child's thighbones become thicker and stronger to support the increasing body weight, but they don't get too heavy.

Bone remodeling also plays a key role in maintaining homeostasis of the blood level of calcium. Neither our nervous system nor our muscles can function properly unless the blood level of calcium stays within a narrow range. When the level falls below this range, a hormone called PTH stimulates osteoclasts to break down bone and release calcium to the blood. If the level rises too high, another hormone, calcitonin, stimulates osteoblasts to *deposit* calcium in bone tissue. Notice that this control mechanism is an example of negative feedback. You will read more about it in Chapter 15, when we take a closer look at hormones.

As we age, bone tissue may break down faster than it is renewed. This progressive deterioration is called *osteoporosis* (Figure 5.3). When it occurs, the backbone, pelvis (hip bones), and other bones lose mass. Osteoporosis is most common in women past menopause, although men can be affected, too. Deficiencies of calcium and sex hormones, smoking, and a sedentary lifestyle all may contribute to osteoporosis. On the other hand, getting lots of exercise (to stimulate bone deposits) and taking in plenty of calcium can help minimize bone loss. Drug treatments can slow or even help reverse the bone loss.

Bone tissue, including both compact bone and spongy bone, consists of living cells and a nonliving mineralized matrix.

Bones grow, become strong, and are repaired through the process of bone remodeling.

5.2 The Skeleton: The Body's Bony Framework

LINKS TO
SECTIONS
4.3 AND 5.1

We associate body movements with muscles, but muscles can't do the job alone. The force of contraction must be applied against something else. The skeleton fills the bill.

The skeleton consists mainly of bones. From ear bones no larger than a watch battery to massive thighbones, bones vary in size and shape. Some bones, like the thighbone in Figure 5.4, are long and slender. Other bones, like the ankle bones, are short. Still other bones, including the sternum (breastbone), are flat, and still others, such as spinal vertebrae, are "irregular." All bones are alike in some ways, however. They all contain bone tissue and other connective tissue that lines their surfaces and internal cavities. At joints there is cartilage where one bone meets or "articulates" with another. Other tissues associated with bones include nervous tissue and epithelium, which occurs in the walls of blood vessels that carry substances to and from bones.

Long bones and some other bones have cavities that contain **bone marrow**, a connective tissue where blood cells are formed. With time, the red marrow in most long bones is replaced by fatty yellow marrow. For this reason, most of an adult's blood cells form in red bone marrow in irregular bones, such as the hip bone, and in flat bones, such as the sternum. If a person loses a great deal of blood, yellow marrow in long bones can convert to red marrow, which makes red blood cells.

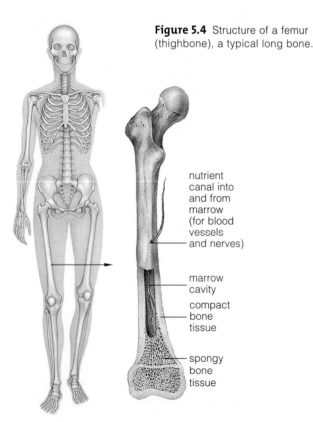

Figure 5.4 Structure of a femur (thighbone), a typical long bone.

nutrient canal into and from marrow (for blood vessels and nerves)

marrow cavity

compact bone tissue

spongy bone tissue

Table 5.1	Functions of Bone

1. **Movement**. Bones interact with skeletal muscles to maintain or change the position of body parts.

2. **Support**. Bones support and anchor muscles.

3. **Protection**. Many bones form hard compartments that enclose and protect soft internal organs.

4. **Mineral storage**. Bones are a reservoir for calcium and phosphorus. Deposits and withdrawals of these mineral ions help to maintain their proper concentrations in body fluids.

5. **Blood cell formation**. Some bones contain marrow where blood cells are produced.

THE SKELETON: A PREVIEW

A fully formed human skeleton has 206 bones, which grow by way of remodeling until a person is about twenty. The bones are organized into an **axial skeleton** and an **appendicular skeleton** (Figure 5.5). The bones of the axial skeleton form the body's vertical, head-to-toe axis. The appendicular ("hanging") skeleton includes bones of the limbs, shoulders, and hips. **Ligaments** connect bones at joints. Ligaments are composed of elastic connective tissue, so they are stretchy and resilient like thick rubber bands. **Tendons** are cords or straps that attach muscles to bones or to other muscles. They are built of connective tissue packed with collagen fibers, which make tendons strong.

BONE FUNCTIONS ARE VITAL IN MAINTAINING HOMEOSTASIS

It can be easy to take our bones for granted simply as girders that support our soft, rather squishy flesh, but in fact bones are complex organs that contribute to homeostasis in many ways (Table 5.1). For instance, bones that support and anchor skeletal muscles help maintain or change the positions of our body parts. Some form hard compartments that enclose and protect other organs; for example, the skull encloses and protects the brain, and the rib cage protects the lungs. As noted in Section 5.1, bones also serve as a "pantry" where the body can store calcium. Because the calcium in bone is in the form of the compound calcium phosphate, bone also is a storage depot for phosphorus.

The fully formed human skeleton consists of 206 bones, in axial and appendicular divisions.

Bones contribute to homeostasis by providing body support, enabling movement, storing minerals, and in some cases, producing blood cells in marrow.

Axial Skeleton

Appendicular Skeleton

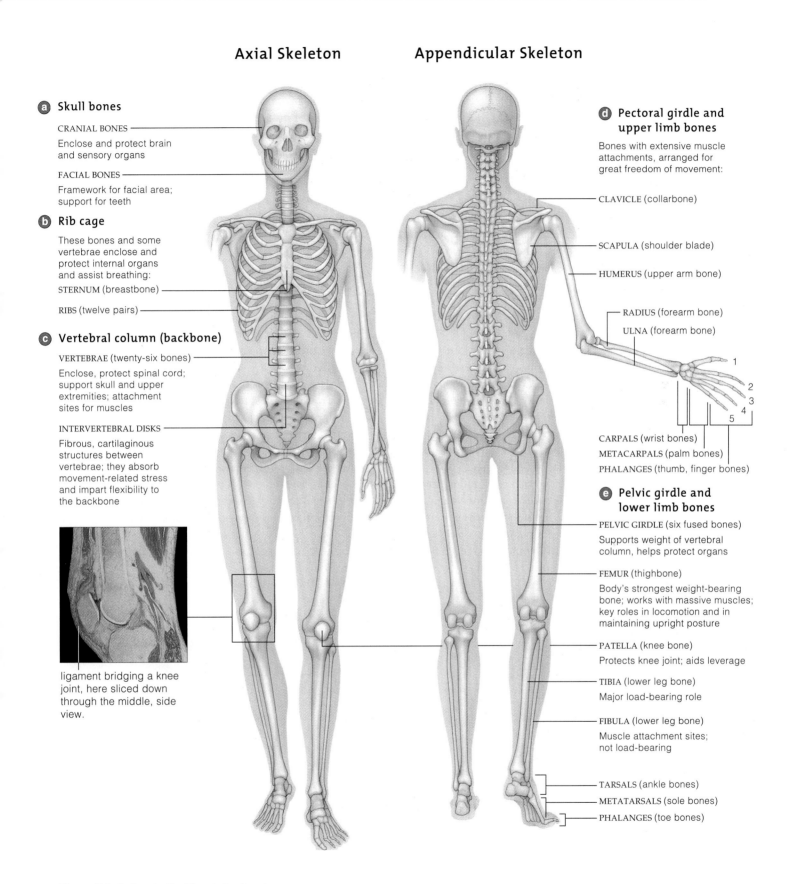

a Skull bones

CRANIAL BONES

Enclose and protect brain and sensory organs

FACIAL BONES

Framework for facial area; support for teeth

b Rib cage

These bones and some vertebrae enclose and protect internal organs and assist breathing:

STERNUM (breastbone)

RIBS (twelve pairs)

c Vertebral column (backbone)

VERTEBRAE (twenty-six bones)

Enclose, protect spinal cord; support skull and upper extremities; attachment sites for muscles

INTERVERTEBRAL DISKS

Fibrous, cartilaginous structures between vertebrae; they absorb movement-related stress and impart flexibility to the backbone

ligament bridging a knee joint, here sliced down through the middle, side view.

d Pectoral girdle and upper limb bones

Bones with extensive muscle attachments, arranged for great freedom of movement:

CLAVICLE (collarbone)

SCAPULA (shoulder blade)

HUMERUS (upper arm bone)

RADIUS (forearm bone)

ULNA (forearm bone)

1
2
3
4
5

CARPALS (wrist bones)

METACARPALS (palm bones)

PHALANGES (thumb, finger bones)

e Pelvic girdle and lower limb bones

PELVIC GIRDLE (six fused bones)

Supports weight of vertebral column, helps protect organs

FEMUR (thighbone)

Body's strongest weight-bearing bone; works with massive muscles; key roles in locomotion and in maintaining upright posture

PATELLA (knee bone)

Protects knee joint; aids leverage

TIBIA (lower leg bone)

Major load-bearing role

FIBULA (lower leg bone)

Muscle attachment sites; not load-bearing

TARSALS (ankle bones)

METATARSALS (sole bones)

PHALANGES (toe bones)

Figure 5.5 *Animated!* The skeletal system. The blue-tinged areas are cartilage.

5.3 The Axial Skeleton

The axial skeleton supports much of our body weight and protects many internal organs.

We begin our tour of the skeleton with bones of the axial skeleton—the skull, vertebral column (backbone), ribs, and sternum (the breastbone).

THE SKULL PROTECTS THE BRAIN

Did you know that your skull consists of more than two dozen bones? These bones are divided into several groups, and while by tradition many of them have names derived from Latin, their roles are simple to grasp. For example, one grouping, the "cranial vault," or **brain case**, includes eight bones that together surround and protect your brain. As Figure 5.6a shows, the *frontal bone* makes up the forehead and upper ridges of the eye sockets. It contains **sinuses**, which are air spaces lined with mucous membrane. Sinuses make the skull lighter, which translates into less weight for the spine and neck muscles to support. But passages link them to the upper respiratory tract—and their ability to produce mucus can mean misery for anyone who has a head cold or pollen allergies. Bacterial infections in the nasal passages can spread to the sinuses, causing *sinusitis*. Figure 5.6c shows sinuses in the cranial and facial bones.

Temporal bones form the lower sides of the cranium and surround the ear canals, which are tunnels that lead to the middle and inner ear. Inside the middle ear are tiny bones that function in hearing (Chapter 14). On either side of your head, in front of each temporal bone, a *sphenoid bone* extends inward to form part of the inner eye socket.

The *ethmoid bone* also contributes to the inner socket and helps support the nose. Two *parietal bones* above and behind the temporal bones form a large part of the skull; they sweep upward and meet at the top of the head. An *occipital bone* forms the back and base of the skull and also encloses a large opening, the *foramen magnum* ("large hole"). Here, the spinal cord emerges from the base of the brain and enters the spinal column (Figure 5.6b). Several passageways run through and between various skull bones for nerves and blood vessels, especially at the base of the skull. For instance, the jugular veins, which carry blood leaving the brain, pass through openings between the occipital bone and each temporal bone.

FACIAL BONES SUPPORT AND SHAPE THE FACE

Figure 5.6 also shows facial bones, many of which you can easily feel with your fingers. The largest is your lower jaw, or **mandible**. The upper jaw consists of two *maxillary bones*. Two *zygomatic bones* form the middle of the hard bumps we call "cheekbones" and the outer parts of the eye sockets. A small, flattened *lacrimal bone* fills out the inner eye socket. Tear ducts pass between this bone and the maxillary bones and drain into the nasal cavity—one reason why your nose runs when you cry. Tooth sockets in the upper and lower jaws also contain the teeth.

Palatine bones make up part of the floor and side wall of the nasal cavity. (Extensions of these bones, together with the maxillary bones, form the back of the hard palate, the "roof" of your mouth.) A *vomer bone* forms part of the nasal septum, a thin "wall" that divides the nasal cavity into two sections.

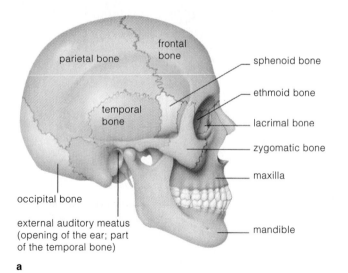

a

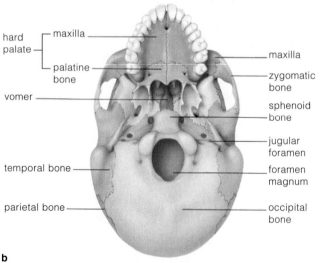

b

Figure 5.6 The skull. (**a**) The irregular junctions between different bones are called sutures. (**b**) An "inferior," or bottom-up, view of the skull. The large foramen magnum is situated atop the uppermost cervical vertebra. (**c**) Sinuses in bones in the skull and face.

THE VERTEBRAL COLUMN IS THE BACKBONE

The flexible, curved human vertebral column—your backbone or spine—extends from the base of the skull to the hip bones (pelvic girdle). This arrangement transmits the weight of a person's torso to the lower limbs. As a result, people who gain a large amount of excess weight may develop problems with their knees and ankles because those joints are not designed to bear such a heavy load. The **vertebrae** are stacked one on top of the other. They have bony projections that form a protected channel for the delicate spinal cord. As sketched in Figure 5.7, humans have seven *cervical* vertebrae in the neck, twelve thoracic vertebrae in the chest area, and five *lumbar* vertebrae in the lower back. During the course of human evolution, five other vertebrae have become fused to form the sacrum, and another four have become fused to form the coccyx, or "tailbone." Counting these, there are thirty-three vertebrae in all.

Roughly a quarter of your spine's length consists of **intervertebral disks**—compressible pads of fibrocartilage sandwiched between vertebrae. The disks serve as shock absorbers and flex points. They are thickest between cervical vertebrae and between lumbar vertebrae. Severe or rapid shocks, as well as changes due to aging, can cause a disk to *herniate* or *"slip."* If the slipped disk ruptures, its jellylike core may squeeze out, making matters worse. And if the changes compress neighboring nerves or the spinal cord, the result can be excruciating pain and the loss of mobility that often comes with pain. Depending on the situation, treatment can range from bed rest and use of painkilling drugs to surgery.

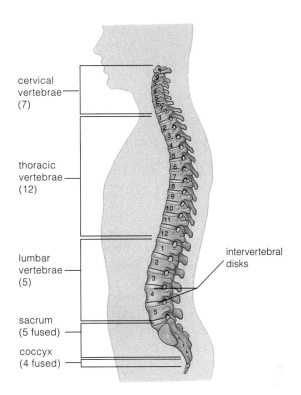

Figure 5.7 Side view of the vertebral column or backbone. The cranium balances on the column's top vertebra.

cervical vertebrae (7)

thoracic vertebrae (12)

lumbar vertebrae (5)

intervertebral disks

sacrum (5 fused)

coccyx (4 fused)

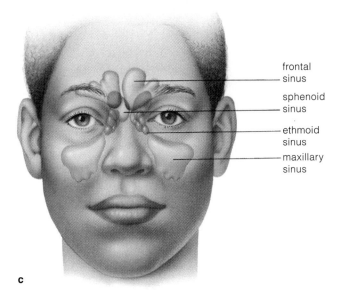

frontal sinus

sphenoid sinus

ethmoid sinus

maxillary sinus

c

THE RIBS AND STERNUM SUPPORT AND HELP PROTECT INTERNAL ORGANS

In addition to protecting the spinal cord, absorbing shocks, and providing flexibility, the vertebral column also serves as an attachment point for twelve pairs of **ribs**, which in turn function as a scaffolding for the body cavity of the upper torso. The upper ribs also attach to the paddle-shaped **sternum** (see Figure 5.5). As you will read in later chapters, this rib cage helps protect the lungs, heart, and other internal organs and is vitally important in breathing.

While the axial skeleton provides basic body support and helps protect internal organs, many movements depend on interactions of skeletal muscles with the bones of the appendicular skeleton—the skeletal component we turn to next.

Bones of the axial skeleton make up the body's vertical axis. They include the skull (and facial bones), the vertebral column, and the ribs and sternum.

Intervertebral disks between the vertebrae absorb shocks and serve as flex points.

5.4 The Appendicular Skeleton

"Append" means to hang, and the appendicular skeleton includes the bones of body parts that we sometimes think of as dangling from the main body frame.

The bones of your arms, hands, legs, and feet are all parts of the appendicular skeleton. It also includes a pectoral girdle at each shoulder and the pelvic girdle at the hips.

THE PECTORAL GIRDLE AND UPPER LIMBS PROVIDE FLEXIBILITY

Each **pectoral girdle** (Figure 5.8) has a large, flat shoulder blade—a **scapula**—and a long, slender collarbone, or **clavicle,** that connects to the breastbone (sternum). The rounded shoulder end of the **humerus,** the long bone of the upper arm, fits into an open socket in the scapula. Your arms can move in a great many ways; they can swing in wide circles and back and forth, lift objects, or tug on a rope. Such freedom of movement is possible because muscles only loosely attach the pectoral girdles and upper limbs to the rest of the body. Although the arrangement is sturdy enough under normal conditions, it is vulnerable to strong blows. Fall on an outstretched arm and you might fracture your clavicle or dislocate your shoulder. The collarbone is the bone most frequently broken.

Each of your upper limbs includes thirty separate bones. The humerus connects with two bones of the forearm—the **radius** (on the thumb side) and the **ulna** (on the "pinky finger" side). The upper end of the ulna joins the lower end of the humerus to form the elbow joint. The bony bump sometimes (mistakenly) called the "wrist bone" is the lower end of the ulna.

The radius and ulna join the hand at the wrist joint, where they meet eight small, curved *carpal* bones. Ligaments attach these bones to the long bones. Blood vessels, nerves, and tendons pass in sheaths over the wrist; when a blow, constant pressure, or repetitive movement (such as typing) damages these tendons, the result can be a painful disorder called *carpal tunnel syndrome* (Section 5.6). The bones of the hand, the five *metacarpals,* end at the knuckles. *Phalanges* are the bones of the fingers.

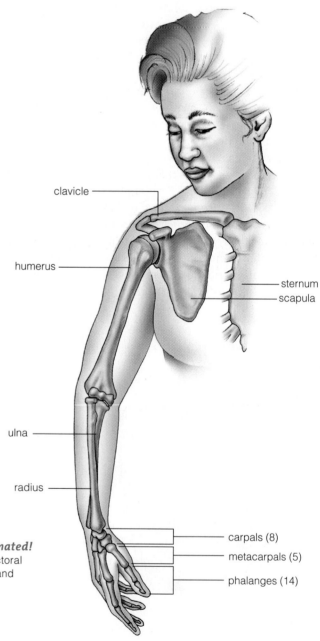

Figure 5.8 *Animated!*
Bones of the pectoral girdle, the arm, and the hand.

clavicle

humerus

sternum

scapula

ulna

radius

carpals (8)

metacarpals (5)

phalanges (14)

THE PELVIC GIRDLE AND LOWER LIMBS SUPPORT BODY WEIGHT

For most of us, our shoulders and arms are much more flexible than our hips and legs. Why? Although there are similarities in the basic "design" of both girdles, this lower part of the appendicular skeleton is adapted to bear the body's entire weight when we are standing. The **pelvic girdle** (Figure 5.9) is much more massive than the combined pectoral girdles, and it is attached to the axial skeleton by extremely strong ligaments. It forms an open basin: A pair of *coxal bones* attach to the lower spine (sacrum) in back, then curve forward and meet at the *pubic arch*. ("Hipbones" are actually the upper *iliac* regions of the coxal bones.) This combined structure is the *pelvis*. In females the pelvis is broader than in males, and it shows other structural differences that are evolutionary adaptations for childbearing. A forensic scientist or paleontologist examining skeletal remains can easily establish the sex of the deceased if a pelvis is present.

The legs contain the body's largest bones. In terms of length, the thighbone, or **femur**, ranks number one. It is also extremely strong. When you run or jump, your femurs routinely withstand stresses of several tons per square inch (aided by contracting leg muscles). The femur's ball-like upper end fits snugly into a deep socket in the coxal (hip) bone. The other end connects with one of the bones of the lower leg, the thick, load-bearing *tibia* on the inner (big toe) side. A slender *fibula* parallels the tibia on the outer (little toe) side. The tibia is your shinbone. A triangular kneecap, the patella, helps protect the knee joint. In spite of this protection, knees are among the joints most often damaged by athletes, both amateur and professional.

The ankle and foot bones correspond closely to those of the wrist and hand. *Tarsal* bones make up the ankle and heel, and the foot contains five long bones, the *metatarsals*. The largest metatarsal, leading to the big toe, is thicker and stronger than the others to support a great deal of body weight. Like fingers, the toes contain phalanges.

> The appendicular skeleton includes bones of the limbs, a pectoral girdle at the shoulders, and a pelvic girdle at the hips.
>
> The thighbone (femur) is the largest bone in the body and one of the strongest. The wrists and hands and ankles and feet have corresponding sets of bones known respectively as carpals and metacarpals and tarsals and metatarsals.

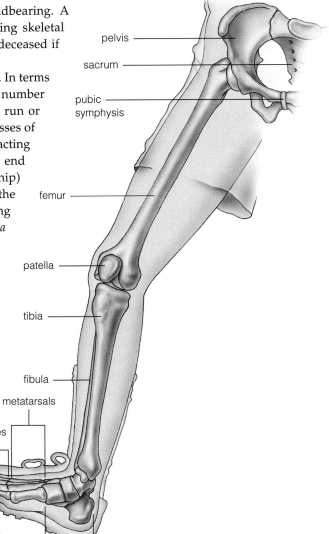

pelvis
sacrum
pubic symphysis
femur
patella
tibia
fibula
metatarsals
phalanges
tarsals

Figure 5.9 *Animated!*
Bones of the pelvic girdle, the leg, and the foot.

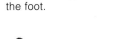

5.5 Joints—Connections between Bones

Joints are areas of contact or near contact between bones. Each of the various types of joints has some form of connective tissue that bridges the gap between bones.

SYNOVIAL JOINTS MOVE FREELY

In the most common type of joint, called a **synovial joint**, adjoining bones are separated by a cavity (Figure 5.10). The articulating ends of the bones are covered with a cushioning layer of cartilage, and they are stabilized by ligaments. A capsule of dense connective tissue surrounds the bones of a synovial joint. Cells that line the inner surface of the capsule secrete a lubricating *synovial fluid* into the joint cavity.

Synovial joints are built to allow movement. In hinge-like synovial joints such as the knee and elbow, the motion is limited to simple flexing and extending (straightening). The ball-and-socket joints at the hips are capable of a wider range of movements: They can rotate and move in different planes—for instance, up-down or side-to-side. Figure 5.11 shows these and some other ways body parts can move at joints.

OTHER JOINTS MOVE LITTLE OR NOT AT ALL

In a **cartilaginous joint**, cartilage fills the space between bones, so only slight movement is possible. Such joints occur between vertebrae and between the breastbone and some of the ribs.

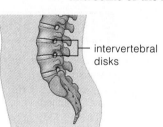

There is no cavity in a **fibrous joint**, and fibrous connective tissue unites the bones. An adult's fibrous joints generally don't allow movement. Examples are the fibrous joints that hold your teeth in their sockets. In a fetus, fibrous joints loosely connect the flat skull bones. During childbirth, these loose connections allow the bones to slide over each other, preventing skull fractures. A newborn baby's skull still has fibrous joints and soft areas called fontanels. With time the joints harden into *sutures*. Much later in life the skull bones may fuse completely.

> A joint connects one bone to another. In all joints, connective tissue bridges the gap between bones.
>
> Freely movable (synovial) joints include the hinge-like knee joint and the ball-and-socket joints at the hips.
>
> Cartilaginous joints have cartilage in the space between bones. They allow only slight movement. In fibrous joints fibrous connective tissue joins the bones.

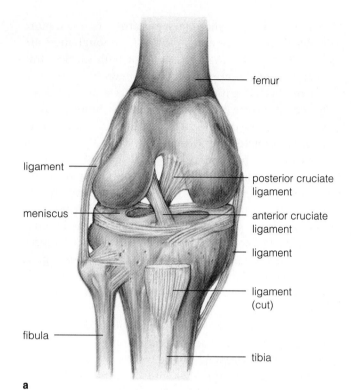

a

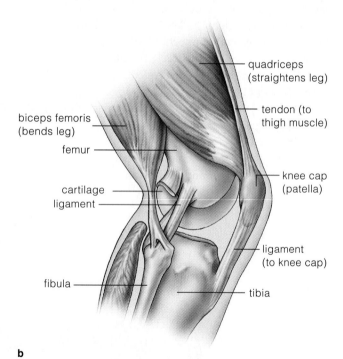

b

Figure 5.10 The knee joint, an example of a synovial joint. The knee is the largest and most complex joint in the body. Part (**a**) shows the joint with muscles stripped away; in (**b**) you can see where muscles such as the quadriceps attach.

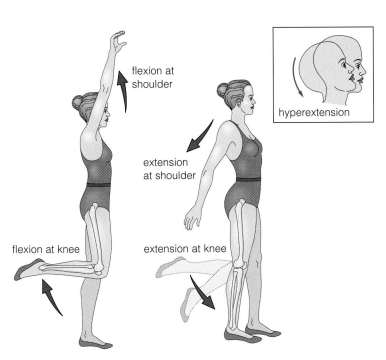

flexion at shoulder

extension at shoulder

hyperextension

flexion at knee

extension at knee

a **flexion and extension**
Flexion reduces the angle between two bones, while extension increases it. Hyperextension, as when you tip your head back, increases the angle beyond 180°.

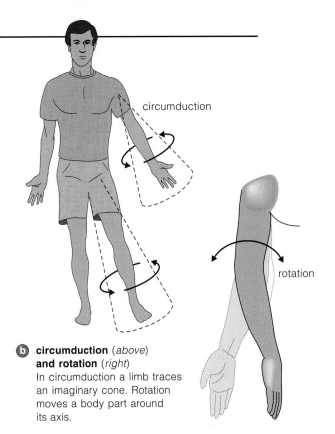

circumduction

rotation

b **circumduction** (*above*) **and rotation** (*right*)
In circumduction a limb traces an imaginary cone. Rotation moves a body part around its axis.

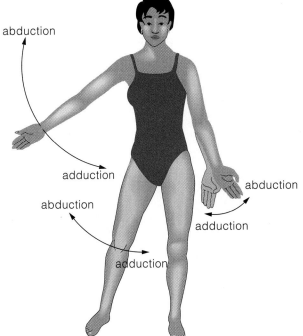

abduction

adduction

abduction

abduction

adduction

abduction

adduction

adduction

c **abduction and adduction**
Abduction moves a limb away from the body's midline; adduction moves a limb toward the midline or beyond it.

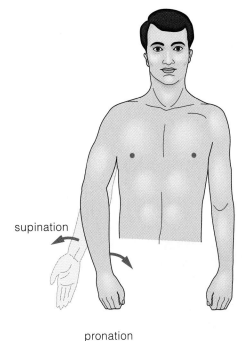

supination

pronation

d **supination and pronation**
In supination forearm bones rotate so that the palms face outward; in pronation the rotation turns the palms to the rear.

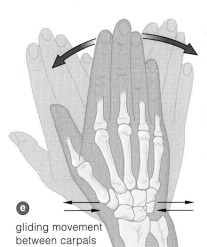

gliding movement between carpals

e

Figure 5.11 (**a–e**) Ways body parts move at synovial joints. The synovial joint at the shoulder permits the greatest range of movement.

 JOINTS

5.6 Disorders of the Skeleton

When a part of the skeleton is injured or affected by disease, impaired movement is the likely result.

INFLAMMATION IS A FACTOR IN SOME SKELETAL DISORDERS

Excessive wear on a joint is the hallmark of *osteoarthritis*. This kind of wear happens when years of use, mechanical stress, or disease wears away the cartilage covering the bone ends of freely movable joints. Often, the arthritic joint is painfully inflamed, and surgeons now routinely replace seriously arthritic hips, knees, and shoulders. Another degenerative joint condition, *rheumatoid arthritis*, results when a person's immune system malfunctions and mounts an attack against tissues in the affected joint. Then, the synovial membrane becomes inflamed and thickens, cartilage is eroded away, and the bones fall out of proper alignment (Figure 5.12). With time the bone ends may even fuse together.

Repetitive movements also can cause inflammation when they damage the soft tissue associated with joints. *Tendinitis*, the underlying cause of conditions such as "tennis elbow," develops when tendons and synovial membranes around joints such as the elbow, shoulders, and fingers become inflamed.

Today one of the most common repetitive motion injuries is *carpal tunnel syndrome*. The "carpal tunnel" is a slight hollow between a wrist ligament and the underside of the wrist's eight carpal bones (see Figure 5.8). Squeezed into this tunnel are several tendons and a nerve that services parts of the hand. Chronic overuse, such as long hours typing at a computer keyboard, can inflame the tendons. When the swollen tendons press on the nerve, the result can be pain, numbness, and tingling in fingers. Simply avoiding the offending motion can help relieve carpal tunnel syndrome. In more serious cases injections of an anti-inflammatory drug are helpful. Sometimes, however, the wrist ligament must be surgically cut to relieve the pressure.

JOINTS ALSO ARE VULNERABLE TO STRAINS, SPRAINS, AND DISLOCATIONS

Synovial joints such as our knees, hips, and shoulders get a lot of use, so it's not surprising that they are vulnerable to mechanical stresses. Stretch or twist a joint suddenly and too far, and you *strain* it. Do something that makes a small tear in its ligaments or tendons and you will have a *sprain*. In fact, a sprained ankle is the most common joint injury. Sprains hurt mainly because of swelling and bleeding from broken small blood vessels. Applying cold (such as an ice pack, 30 minutes on, then 30 minutes off) for the first 24 hours will minimize these effects; after that, doctors usually advise applying heat, such as a hot

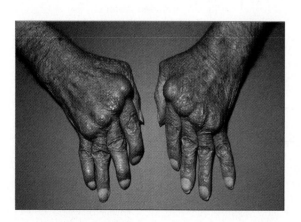

Figure 5.12 Hands affected by rheumatoid arthritis.

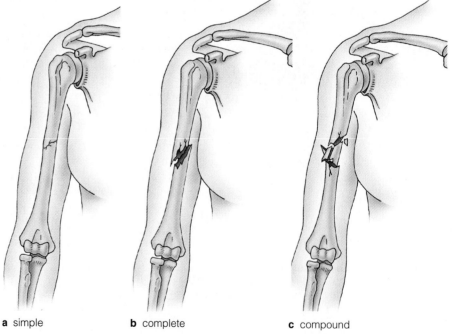

a simple   **b** complete   **c** compound

Figure 5.13 Three types of bone fractures.

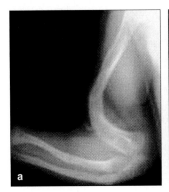

Figure 5.14 Effects of *osteogenesis imperfecta* (OI). (**a**) An X-ray of an arm bone deformed by OI. (**b**) Tiffany has OI. She was born with multiple fractures in her arms and legs. By age six, she had undergone surgery to correct more than 200 bone fractures and to place steel rods in her legs. Every three months she receives intravenous infusions of an experimental drug that may help strengthen her bones.

pad. The warmth speeds tissue repair by increasing blood circulation to the injured tissue.

A blow can *dislocate* a joint—that is, the two bones will no longer be in contact. During collision sports such as football, a blow to a knee often tears a ligament. If the torn part is not reattached within ten days, phagocytic cells in the knee joint's synovial fluid will attack and destroy the damaged tissue.

IN FRACTURES, BONES BREAK

Injuries severe enough to dislocate a joint also often break one or both of the bones involved. Regardless of the cause, most breaks can be classed as either a simple fracture, a complete fracture, or a compound fracture. As you can probably tell from the drawings in Figure 5.13, a simple fracture is the least serious injury; it is really only a crack in a bone. A complete fracture, in which the bone separates into two pieces and the surrounding soft tissue is damaged, is more serious. Even worse is a compound fracture, in part because broken ends or shards of bone puncture the skin, creating an open wound and the chance of infection. It also may be difficult or impossible for a surgeon to reattach all the pieces of a bone that has been shattered in this way.

When a bone breaks into pieces, the situation demands prompt medical attention. Unless the pieces are soon reset into their normal alignment, it's unlikely that the bone will heal properly. Its functioning may be impaired for the rest of a person's life. Today, in addition to the pins and casts that may be used to hold healing bones in place, the injured area may be stimulated with electricity, a procedure that speeds healing.

Overall, injuries to joints and bones tend to heal faster when we're younger. Aging, and bad habits such as smoking cigarettes, slow the body's ability to repair itself.

OTHER BONE DISORDERS INCLUDE GENETIC DISEASES, INFECTIONS, AND CANCER

Some skeleton disorders are inherited, and a few cause lifelong difficulties for affected people. An example is *osteogenesis imperfecta* or OI (Figure 5.14). In this disease the collagen in bone tissue is defective. As a result, the bones are exceptionally brittle and break easily. Children with OI often have stunted growth and must endure repeated hospitalizations to have fractures set. In some cases where the disease has not been detected early, an affected child's parents have been wrongly suspected of child abuse. Unfortunately, there is no cure for OI, but researchers are looking for ways to improve bone strength in affected individuals.

Our bones and bone marrow also can become infected by bacteria when an infection elsewhere spreads (via the bloodstream) or when the microbe enters an open wound. A heavy dose of antibiotics usually can cure the problem, although severe cases may require surgery to clean out the affected bone tissue.

Bone cancer, or *osteosarcoma*, can strike people young and old. It usually occurs in a long bone in an arm or leg, or in a joint such as the hip or knee. Unfortunately, bone cancer spreads rapidly to other organs. The first sign may be a painful swelling, and the most common treatment is to remove the affected bone tissue by amputating the limb involved. Like many other cancers, however, if caught early bone cancer often is curable.

IMPACTS, ISSUES

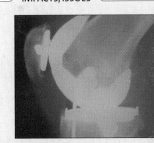

In this replacement knee joint a projection of the joint has been fitted into the end of the patient's femur (*center*) and another projection has been fitted into the tibia below. The hatlike disk at the upper left attaches to the patella—the kneecap. With technological advances, today it may take only about 2 hours to replace a knee joint, even less for a hip. After surgery, walking and standing put stress on the new joint, so the patient's osteoblasts generate new bone that grows into pits on the prosthesis. With normal use and proper care, a new knee or hip may last decades.

Summary

Section 5.1 Bones are organs. In addition to bone tissue, they include other types of connective tissue, nervous tissue, and epithelial tissue (in the walls of blood vessels).

A bone develops as osteoblasts secrete collagen fibers and a matrix of protein and carbohydrate. The matrix eventually surrounds each osteoblast and hardens (mineralizes) as calcium salts are deposited in it. The mature living bone cells, osteocytes, are located inside spaces (lacunae) in the bone tissue.

Bone tissue has both compact bone and spongy bone. Denser compact bone is organized as thin, circular layers called osteons. Small canals through the layers are channels for nerves and blood vessels. In spongy bone, needlelike struts are fused together in a latticework.

A cartilage model provides the pattern for a developing bone. Long bones lengthen at their ends (epiphyses). This process stops by early adulthood. Until then, a cartilage plate separates each epiphysis from the bone shaft. The plates are replaced by bone when growth ends.

Bones grow, gain strength, and are repaired by bone remodeling. In this process, osteoblasts deposit bone and osteoclasts break it down. Hormones largely control bone remodeling.

 Look inside a human femur.

Section 5.2 Bones are the main elements of the skeleton and function in movement by interacting with skeletal muscles. Bones also store minerals and help protect and support other body parts. Ligaments connect bones at joints; tendons attach muscles to bones or to other muscles.

Some bones, including the sternum, hip bones, and femur, contain bone marrow. Blood cells are produced in red bone marrow.

Section 5.3 The skeleton is divided into an axial portion and an appendicular portion (Table 5.2). The axial skeleton forms the body's vertical axis and is a central support structure. In the spine, intervertebral disks of fibrocartilage are shock pads and flex points.

Eight skull bones form the brain case, which protects the brain. The skull's frontal bone contains sinuses—air spaces that make the skull lighter in weight. The mandible (lower jaw) is the largest facial bone.

 Explore the parts of the human skeleton.

Section 5.4 The appendicular skeleton—including limb bones, the pelvic girdle, and pectoral girdles—

Table 5.2 Review of the Skeleton
FUNCTIONS OF BONE:
1. *Movement.* Interact with skeletal muscles to maintain or change the position of body parts.
2. *Support.* Support and anchor muscles.
3. *Protection.* Many bones form hard compartments that enclose and protect soft internal organs.
4. *Mineral storage.* Reservoir for mineral ions, which are deposited or withdrawn and so help maintain ion concentrations in body fluids.
5. *Blood cell formation.* Some bones contain red marrow, where blood cells are produced.
PARTS OF THE SKELETON:
Appendicular portion
Pectoral girdles: clavicle and scapula
Arm: humerus, radius, ulna
Wrist and hand: carpals, metacarpals, phalanges (of fingers)
Pelvic girdle (6 fused bones at the hip)
Leg: femur (thighbone), patella, tibia, fibula
Ankle and foot: tarsals, metatarsals, phalanges (of toes)
Axial portion
Skull: cranial bones and facial bones
Rib cage: sternum (breastbone) and ribs (12 pairs)
Vertebral column: vertebrae (26)

provides support for upright posture and interacts with skeletal muscles in most movements. Major bones of each pectoral girdle are a scapula (shoulder blade) and a clavicle (collarbone). The long bone of the upper arm, the humerus, connects at its top end with the scapula. The other end of the humerus connects with the two major bones of the lower arm, the radius and ulna.

At the hips, bones of the pelvic girdle form the pelvis. Each coxal bone of the pelvis articulates with the upper end of the femur, the longest bone in the body. Major bones of the lower leg are the tibia and fibula. The wrists and hands and ankles and feet have corresponding sets of bones.

Section 5.5 Together with skeletal muscles, the skeleton works like a system of levers in which rigid rods (bones) move about at fixed points (joints). In a synovial joint, a fluid-filled cavity separates adjoining bones. Such joints are freely movable. In cartilaginous joints, cartilage fills the space between bones and allows only slight movements. In fibrous joints, there is no cavity, and fibrous connective tissue unites the bones.

Section 5.6 Diseases and disorders that affect the skeletal system can impair movement and hamper other functions of bones that help maintain homeostasis.

Review Questions

1. Describe the basic elements of bone tissue.

2. What are the two types of bone tissue, and how are they different?

3. Describe how bone first develops.

4. Explain why bone remodeling is important, and give its steps.

5. How do bones contribute to homeostasis?

6. Name the two main divisions of the skeleton.

7. How does a tendon differ from a ligament?

8. What is the function of intervertebral disks? What are they made of?

9. What is a joint?

10. Name at least two different synovial joints. What is the defining feature of this type of joint?

Self-Quiz

Answers in Appendix V

1. The _____ and _____ systems work together to move the body and specific body parts.

2. Bone tissue contains _____.
   a. living cells   d. all of these
   b. collagen fibers   e. only a and b
   c. calcium and phosphorus

3. _____ are shock pads and flex points.
   a. Vertebrae   c. Lumbar bones
   b. Cervical bones   d. Intervertebral disks

4. The hollow center of an osteon (Haversian system) provides space for what vital part of compact bone tissue?
   a. marrow   c. a blood vessel
   b. collagen fibers   d. osteocytes

5. _____ is a type of connective tissue; _____ form(s) in it.
   a. An osteon; collagen
   b. Bone marrow; blood cells
   c. Bone; an osteocyte
   d. A sinus; bone marrow

6. Mineralization of bone tissue requires _____.
   a. calcium ions   c. elastin
   b. osteoclasts   d. all of the above

7. The axial skeleton consists of the _____, while the appendicular skeleton consists of the _____.

8. Match the terms and definitions.
   ___ bone   a. in certain skull bones
   ___ collagen   b. all in the hands
   ___ synovial fluid   c. blood cell production
   ___ osteocyte   d. a fibrous protein
   ___ marrow   e. mature bone cell
   ___ metacarpals   f. lubrication
   ___ mandible   g. mineralized connective
   ___ sinuses   tissue
                          h. the lower jaw

Critical Thinking

1. Hormones and vitamins play important roles in the formation of bones. In particular, vitamin D is needed for bone tissue to mineralize properly. Children who develop the disorder called rickets (Figure 5.15) have suffered from a deficiency of vitamin D, so their bones are abnormally soft and become deformed. Based on your reading in this chapter, which cells in bone tissue are affected by the vitamin deficiency? Do you think bone remodeling might be affected?

2. Growth hormone, or GH, is used clinically to spur growth in children who are unusually short because they have a GH deficiency. However, it is useless for a short but otherwise normal 25-year-old to request GH treatment from a physician. Why?

3. If bleached human bones found lying in the desert were carefully examined, which of the following would not be present? Haversian canals, a marrow cavity, osteocytes, calcium.

4. For young women, the recommended daily allowance (RDA) of calcium is 800 milligrams. During Hilde's pregnancy, the RDA is 1,200 milligrams a day. What might happen to a pregnant woman's bones without the larger amount, and why?

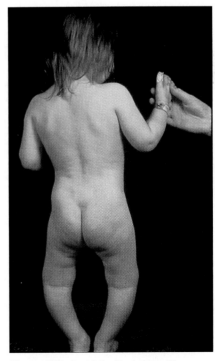

Figure 5.15 A child with rickets. The medical name for this disease is *osteomalacia.*

Explore on Your Own

When it comes to the skeleton and joints, your body can be a great learning tool.

- Feel along the back of your neck beginning at your hairline. Can you feel any lumps made by the bony processes of your spinal vertebrae? In the figure below, locate the C7 vertebra, which in most people is the most prominent. Can you feel it at the base of your neck?

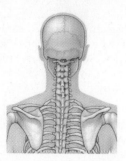

- While seated, feel your kneecap—the patella—move as you flex and extend your lower leg. Just below the patella you should also be able to feel a ligament that attaches it to your tibia.

    Can you find the upper protuberance of your tibia? Moving your fingers around to outside of the joint, can you feel the knobby upper part of the fibula?

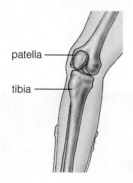

- Using the diagram below as a guide, see if you can locate the ridges of your frontal bone above your eyebrows; the arching part of your zygomatic bone, which forms your "cheekbones"; and the joint where your lower jaw articulates with the temporal bone.

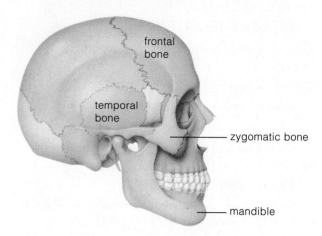

A series of horizontal cartilage rings support the trachea, your windpipe. Rest your fingers on this area and swallow. Can you feel the rings? What happens to them when you swallow? As described in Chapter 7, this movement is part of a mechanism that normally prevents swallowed food or drink from "going down the wrong way" into your lungs.

Pumping Up Muscles

Want to "bulk up" your muscles and be stronger, with more endurance? Just swallow a pill. That is the message to bodybuilders and other athletes from the sellers of substances like "andro"—androstenedione—and THG (tetrahydrogestrinone). Andro made news when Mark McGwire said he used it during his successful effort to break major league baseball's home-run record. An international track star admitted using THG, as did some professional football players. Several American professional baseball players were called before a grand jury investigating a nutritional supplement company that makes THG. Tests have shown that this "food supplement" actually is a chemical cousin of two anabolic (tissue-building) steroids prohibited in sports competitions. The federal Food and Drug Administration has forced THG off the market.

Androstenedione occurs naturally when the body synthesizes the sex hormone testosterone. Studies suggest, however, that andro is not an effective muscle-builder, because it only raises the testosterone level for a few hours. On the other hand, andro does have side effects, including a risk of liver damage. In 2004 the FDA issued an advisory warning of these problems, and companies were ordered to stop distributing the drug.

Creatine is an easy-to-obtain performance enhancer. The body normally produces this substance and gets more from food. Muscle cells use it as a quick energy source when they must contract hard and fast. Controlled studies show that creatine supplements do improve performance during brief, high-intensity exercise. Long-term effects are not known, although there is evidence that in large amounts creatine puts a strain on the kidneys. No regulatory agency checks to see how much creatine is actually present in any commercial product.

With this chapter we look at why we have muscles in the first place. We will begin by reviewing the three types of muscle tissue in the body, and then focus on skeletal muscles, which make up the **muscular system**. As you will read, their coordinated interactions with the skeleton underlie the movements and position changes required for the full range of a person's daily activities.

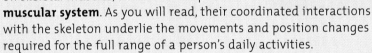 **How Would You Vote?** *Dietary supplements are largely unregulated. Should they be subject to more stringent testing for effectiveness and safety? Cast your vote online at www.thomsonedu.com/biology/starr/humanbio.*

 ## Key Concepts

WHAT SKELETAL MUSCLES DO
Like all muscle tissue, the tissue of skeletal muscles contains long muscle cells, often called muscle fibers, lined up in parallel arrays. Skeletal muscles attach to and pull on bones to move the body and its parts. Skeletal muscles also work with one another.

HOW MUSCLES WORK
A muscle cell is divided into units called sarcomeres. A sarcomere contracts—that is, it shortens—due to sliding interactions between filaments of two proteins, actin and myosin. When sarcomeres shorten, muscle cells shorten, and so does the whole muscle.

 ## Links to Earlier Concepts

Building on what you learned about the skeleton in Chapter 5, this chapter explains how muscles partner with bones to bring about movements of the body and its parts.

You will learn how two proteins, actin and myosin, work together in ways that allow muscle cells to contract (3.8). Our discussion also will draw on your knowledge of how ATP fuels cell activities (3.7), and how active transport moves substances into and out of cells (3.10).

6.1 The Body's Three Kinds of Muscle

LINKS TO
SECTIONS
4.3 AND 4.6

A full 50 percent of the body is muscle tissue. Three different kinds of muscle perform different functions, but in all of them groups of cells contract to produce movement.

THE THREE KINDS OF MUSCLE ARE BUILT AND FUNCTION IN DIFFERENT WAYS

In Chapter 4 we introduced the three basic kinds of muscle tissue—skeletal muscle, smooth muscle, and cardiac muscle. In all of them, cells specialized to contract bring about some type of movement.

Most of the body's muscle tissue is **skeletal muscle**, which interacts with the skeleton to move body parts. Its long, thin cells are often called muscle "fibers" (Figure 6.1*a*). And unlike other body cells, skeletal muscle fibers have more than one nucleus. As you will read later on, their internal structure gives them a striated, or striped, appearance, and bundles of them form skeletal muscles.

Smooth muscle is found in the walls of hollow organs and of tubes, such as blood vessels (Figure 6.1*b*). Its cells are smaller than skeletal muscle cells, and they do not look striped—hence the "smooth" name for this muscle tissue. Junctions link smooth muscle cells, which often are organized into sheets.

You may recall that **cardiac muscle** is found only in the heart (Figure 6.1*c*). It looks striated, like skeletal muscle. Unlike skeletal and smooth muscle, however, cardiac muscle can contract on its own, without stimulation by the nervous system. Special junctions between its cells allow contraction signals to pass between them so fast that for all intents and purposes the cells contract as a single unit.

We do not have conscious control over the contractions of cardiac muscle and smooth muscle, so they are considered to be "involuntary" muscles. We *can* control many of our skeletal muscles, so they are "voluntary" muscles. Figure 6.2 shows the major skeletal muscles in the body. Some are close to the surface, others deep in the body wall. Some, such as facial muscles, attach to the skin. The trunk has muscles of the thorax (chest), spine, abdominal wall, and pelvic cavity. And of course, other muscle groups attach to limb bones.

When we speak of the body's "muscular system," we're talking about skeletal muscle, and so it is the focus of the rest of this chapter. Only skeletal muscle interacts with the skeleton to move the body, its limbs, or other parts. Those movements range from delicate adjustments that help us keep our balance to the cool moves you might execute on a dance floor. Our skeletal muscles also help stabilize joints between bones, and while we won't focus on it in this chapter, muscles also generate body heat.

> The body's three kinds of muscle—skeletal, smooth, and cardiac—each have a specific function in the body. Skeletal muscle makes up the muscular system, which partners with the skeleton to produce movement.

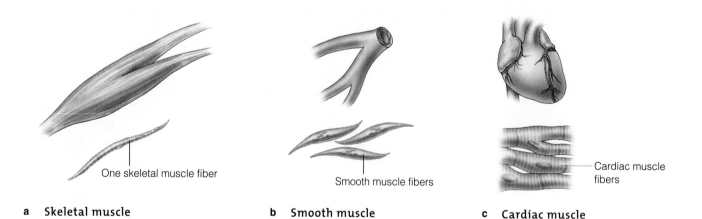

a Skeletal muscle   One skeletal muscle fiber

b Smooth muscle   Smooth muscle fibers

c Cardiac muscle   Cardiac muscle fibers

Figure 6.1 The three kinds of muscle in the body and where each type is found.

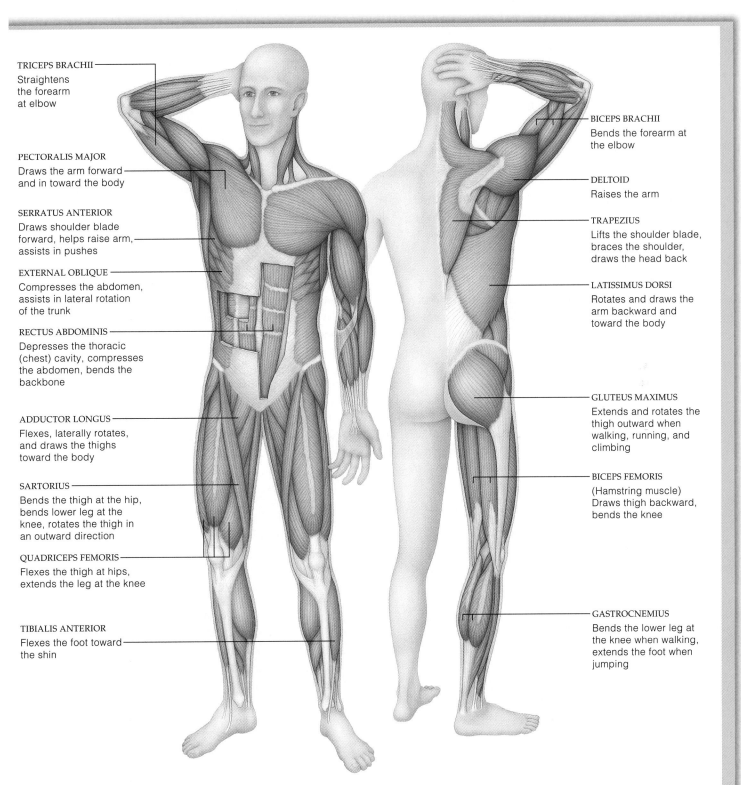

TRICEPS BRACHII
Straightens the forearm at elbow

PECTORALIS MAJOR
Draws the arm forward and in toward the body

SERRATUS ANTERIOR
Draws shoulder blade forward, helps raise arm, assists in pushes

EXTERNAL OBLIQUE
Compresses the abdomen, assists in lateral rotation of the trunk

RECTUS ABDOMINIS
Depresses the thoracic (chest) cavity, compresses the abdomen, bends the backbone

ADDUCTOR LONGUS
Flexes, laterally rotates, and draws the thighs toward the body

SARTORIUS
Bends the thigh at the hip, bends lower leg at the knee, rotates the thigh in an outward direction

QUADRICEPS FEMORIS
Flexes the thigh at hips, extends the leg at the knee

TIBIALIS ANTERIOR
Flexes the foot toward the shin

BICEPS BRACHII
Bends the forearm at the elbow

DELTOID
Raises the arm

TRAPEZIUS
Lifts the shoulder blade, braces the shoulder, draws the head back

LATISSIMUS DORSI
Rotates and draws the arm backward and toward the body

GLUTEUS MAXIMUS
Extends and rotates the thigh outward when walking, running, and climbing

BICEPS FEMORIS
(Hamstring muscle) Draws thigh backward, bends the knee

GASTROCNEMIUS
Bends the lower leg at the knee when walking, extends the foot when jumping

Figure 6.2 *Animated!* Some of the major muscles of the muscular system.

6.2 The Structure and Function of Skeletal Muscles

LINKS TO
SECTIONS
4.2 AND 4.3

Muscle cells generate force by contracting. After a muscle contracts, it can relax and lengthen. As their name suggests, skeletal muscles attach to and interact with bones.

A SKELETAL MUSCLE IS BUILT OF BUNDLED MUSCLE CELLS

A skeletal muscle contains bundles of muscle cells, which look like long, striped fibers (Figure 6.3). Inside each cell are threadlike **myofibrils**, which you'll read more about in Section 6.3. There may be hundreds, even thousands, of cells in a muscle, all bundled together by connective tissue that extends past them to form tendons. A **tendon** is a strap of dense connective tissue that attaches a muscle to bone. Tendons make joints more stable by helping keep the adjoining bones properly aligned. Tendons often rub against bones, but they slide inside fluid-filled sheaths that help reduce the friction (Figure 6.4). Your knees, wrists, and finger joints all have tendon sheaths.

BONES AND SKELETAL MUSCLES WORK LIKE A SYSTEM OF LEVERS

You have more than 600 skeletal muscles, and each one helps produce some kind of body movement. In general,

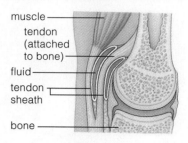

Figure 6.4 A tendon sheath. Notice the lubricating fluid inside each of the sheaths sketched here.

one end of a muscle, called the **origin**, is attached to a bone that stays relatively motionless during a movement. The other end of the muscle, called the **insertion**, is attached to the bone that moves most (Figure 6.5). In effect, the skeleton and the muscles attached to it are like a system of levers in which bones (rigid rods) move near joints (fixed points). When a skeletal muscle contracts, it pulls on the bones it attaches to. Because muscles attach very close to most joints, a muscle only has to contract a short distance to produce a major movement.

MANY MUSCLES ARE ARRANGED AS PAIRS OR IN GROUPS

Many muscles are arranged as pairs or groups. Some work in opposition (that is, antagonistically) so that the action of one opposes or reverses the action of the other. Figure 6.5 shows an antagonistic muscle pair, the biceps and triceps of the arm. Try extending your right arm in front of you, then place your left hand over the biceps in the upper arm and slowly "bend the elbow." Can you feel the biceps contract? When the biceps relaxes and its partner (the triceps) contracts, your arm straightens. This kind of coordinated action comes partly from *reciprocal innervation* by nerves from the spinal cord. When one muscle group is stimulated, no signals are sent to the opposing group, so it does not contract.

Other muscles work in a synergistic, or support, role. Their contraction adds force or helps stabilize another contracting muscle. If you make a fist while keeping your wrist straight, synergist muscles are stabilizing your wrist joint while muscles in your hand are doing the "heavy lifting" of closing your fingers.

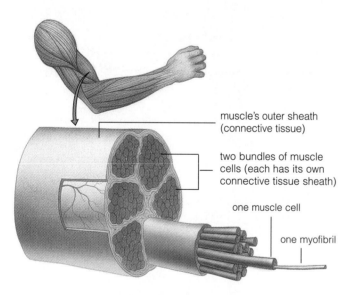

muscle's outer sheath (connective tissue)

two bundles of muscle cells (each has its own connective tissue sheath)

one muscle cell

one myofibril

Figure 6.3 Structure of skeletal muscle. The muscle's cells are bundled together inside a wrapping of connective tissue.

WHAT SKELETAL MUSCLES DO

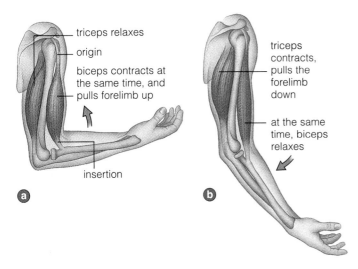

Figure 6.5 *Animated!* Two opposing muscle groups in human arms. (**a**) When the triceps relaxes and its opposing partner (biceps) contracts, the elbow joint flexes and the forearm bends up. (**b**) When the triceps contracts and the biceps relaxes, the forearm is extended down.

"FAST" AND "SLOW" MUSCLE

Your body has two basic types of skeletal muscle (Figure 6.6a). "Slow" or "red" muscle appears crimson because its cells are packed with myoglobin, a reddish protein that binds oxygen for the cell's use in making ATP. Red muscle also is served by larger numbers of the tiny blood vessels called capillaries. (Red muscle is the dark meat in chicken and turkey.) Red muscle contracts fairly slowly, but because its cells are so well equipped to make lots of ATP, the contractions can be sustained for a long time. For example, some muscles of the back and legs—called postural muscles because they aid body support—must contract for long periods when a person is standing. They have a high proportion of red muscle cells. By contrast, the muscles of your hand have fewer capillaries and relatively more "fast" or "white" muscle cells, in which there are fewer mitochondria and less myoglobin. Fast muscle can contract rapidly and powerfully for short periods, but it can't sustain contractions for long periods. This is why you get writer's cramp if you write long-hand for an hour or two.

When an athlete trains rigorously, one goal is to increase the relative size and contractile strength of fast or slow fibers in muscles. The type of sport determines which type of fiber is targeted. A sprinter will benefit

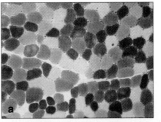

Figure 6.6 Fast and slow skeletal muscle. (**a**) This micrograph shows a cross section of the different kinds of cells in a skeletal muscle. The lighter, "white fibers" are fast muscle. They have little myoglobin and fewer mitochondria than the dark red fibers, which are slow muscle. (**b**) A distance swimmer can work her shoulder muscles for extended periods due to the many well-developed slow muscle cells they contain.

from larger, stronger fast muscle cells in the thighs, while a distance swimmer (Figure 6.6b) will train to increase the number of mitochondria in his or her shoulder muscle cells.

A skeletal muscle consists of hundreds or thousands of muscle cells bundled together by connective tissue. When a skeletal muscle contracts, it pulls on a bone to produce movement.

Tendons strap skeletal muscles to bone.

In many movements, the action of one muscle opposes or reverses the action of another.

The cells in red or "slow" skeletal muscle have features that support slow, long-lasting contractions. The cells in white or "fast" skeletal muscle are specialized for rapid, strong bursts of contraction.

6.3 How Muscles Contract

Bones move—they are pulled in some direction—when the skeletal muscles attached to them contract. In turn, a skeletal muscle contracts when the individual muscle cells in it shorten.

A MUSCLE CONTRACTS WHEN ITS CELLS SHORTEN

When a muscle cell shortens, many units of contraction inside that cell are shortening. Each of these basic units of contraction is a **sarcomere**.

Figure 6.7a is a reminder of how bundles of cells in a skeletal muscle run parallel with the muscle. In Figure 6.7b you can see how each myofibril in a muscle cell is divided into bands, which show up as an alternating light–dark pattern when they are stained and viewed under a microscope. The bands of the myofibrils in the cells line up rather closely; this is why a skeletal muscle cell looks striped.

The dark bands, or Z bands, mark the ends of each sarcomere. Inside a sarcomere are many filaments, some thick, others thin. Each thin filament is like two strands of pearls, twisted together, with one end attached to a

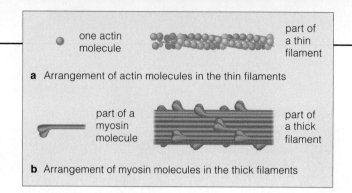

a Arrangement of actin molecules in the thin filaments

b Arrangement of myosin molecules in the thick filaments

Figure 6.8 Actin and myosin filaments.

Z band. The "pearls" are molecules of **actin** (Figure 6.8a), a globular protein that can contract. Other proteins are found near grooves on actin's surface.

Each thick filament is made of molecules of the protein **myosin**. Each myosin molecule has a tail and a double head. In a thick filament many of them are bundled together so that all the heads project outward (Figure 6.8b), away from the sarcomere's center.

As you can see in Figure 6.7, muscle bundles, muscle cells, myofibrils, and their filaments all run in the same direction. Why? This arrangement focuses the force of a contracting muscle onto a bone in a specific direction.

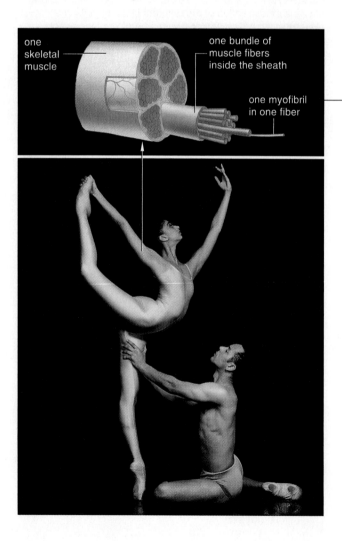

one skeletal muscle

one bundle of muscle fibers inside the sheath

one myofibril in one fiber

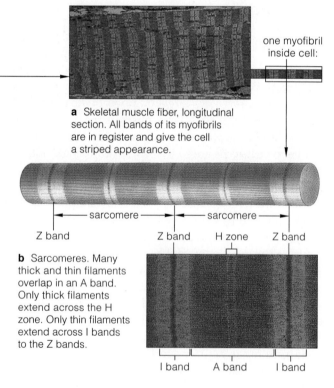

one myofibril inside cell:

a Skeletal muscle fiber, longitudinal section. All bands of its myofibrils are in register and give the cell a striped appearance.

sarcomere — sarcomere —

Z band   Z band   H zone   Z band

b Sarcomeres. Many thick and thin filaments overlap in an A band. Only thick filaments extend across the H zone. Only thin filaments extend across I bands to the Z bands.

I band   A band   I band

Figure 6.7 *Animated!* Zooming down through skeletal muscle from a biceps to filaments of the proteins actin and myosin. These proteins can contract.

MUSCLE CELLS SHORTEN WHEN ACTIN FILAMENTS SLIDE OVER MYOSIN

Years of research have produced an explanation for how sarcomeres shorten and contract a muscle. According to the **sliding filament model**, all of the myosin filaments stay in place. They use short "power strokes" to slide the sets of actin filaments over them, toward the sarcomere's center. It's like two pocket doors closing—but pulling with them both walls to which they are attached. Pulling both sets shrinks the length of the sarcomere (Figure 6.9). Each power stroke is driven by energy from ATP.

Each myosin head connects repeatedly to binding sites on a nearby actin filament. The head is an ATPase, a type of enzyme. It binds ATP and catalyzes a phosphate-group transfer that powers the reaction.

A change in the concentration of calcium ions causes the myosin head to form a **cross-bridge** to the actin (Figure 6.9b). This link tilts the myosin head and pulls the actin filament toward the center of the sarcomere. Next, with the help of energy from ATP, the myosin head's grip on actin is broken and the head returns to its starting position (Figure 6.9c–e). Each contraction of a sarcomere requires hundreds of myosin heads making a series of short strokes down the length of actin filaments.

When a person dies, body cells stop making ATP. In muscle fibers this means that the myosin cross-bridges with actin can't break apart after a power stroke. As a result skeletal muscles "lock up," a stiffening called **rigor mortis** ("stiffness of death"). Rigor mortis lasts for 24 to 60 hours, or until the natural decomposition of dead tissues gets under way. Crime investigators can use this kind of information to help pinpoint the time when a suspicious death occurred.

A skeletal muscle cell contracts when its sarcomeres shorten. Thus sarcomeres are the basic units of muscle contraction.

Powered by ATP, interactions between myosin and actin filaments shorten the sarcomeres of a muscle cell.

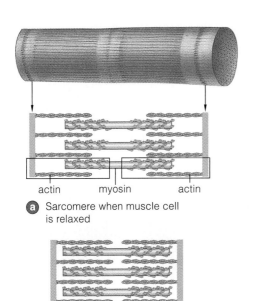

actin   myosin   actin

a Sarcomere when muscle cell is relaxed

b Same sarcomere, contracted

Figure 6.9 *Animated!* (**a**) Actin and myosin filaments in a sarcomere. Interactions between the two kinds of filaments shorten the sarcomere. (**b–e**) Sliding filament model of contraction in the sarcomeres of muscle cells.

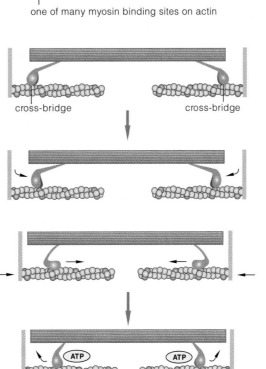
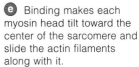

myosin head

one of many myosin binding sites on actin

c A myosin filament in a resting muscle. All the myosin heads were energized earlier by the binding of ATP.

cross-bridge   cross-bridge

d Calcium released from a cellular storage system allows cross-bridges to form; myosin binds to actin filaments.

e Binding makes each myosin head tilt toward the center of the sarcomere and slide the actin filaments along with it.

f Using energy from ATP, the myosin heads drag the actin filaments inward, pulling the Z lines closer together.

ATP   ATP

g New ATP binds to the myosin heads and they detach from actin. The myosin heads return to their original orientation, ready to act again.

6.4 How the Nervous System Controls Muscle Contraction

LINK TO
SECTION
3.6

In response to signals from the nervous system, contracting skeletal muscles move the body and parts of it at certain times, in certain ways.

CALCIUM IONS ARE THE KEY TO CONTRACTION

The nervous system controls the contraction of skeletal muscle cells. Its "orders" reach the muscles by way of motor neurons that stimulate or inhibit contraction of the sarcomeres in muscle cells (Figure 6.10).

section from
spinal cord

motor
neuron

a Signals from the nervous system travel along spinal cord, down motor neuron.

b Endings of motor neuron terminate next to muscle cells.

section from a skeletal muscle

part of one muscle cell

c Signals travel along muscle cell's plasma membrane to sarcoplasmic reticulum around myofibrils.

Figure 6.10 *Animated!* Pathway for signals from the nervous system that stimulate contraction of skeletal muscle.

When neural signals arrive at a muscle cell, they spread rapidly and eventually reach small extensions of the cell's plasma membrane. These **T tubules** connect with a membrane system that laces around the cell's myofibrils (Figure 6.10*d*). The system, called the **sarcoplasmic reticulum** (SR), is a modified version of the endoplasmic reticulum described in Chapter 3. SR takes up and releases calcium ions (Ca^{++}). An incoming nerve impulse triggers the release of calcium ions from the SR. The ions diffuse into myofibrils, and when they reach actin filaments the stage is set for contraction.

Two proteins, troponin and tropomyosin, are found along the surface of actin filaments (Figure 6.11). When incoming calcium binds to troponin, the binding site on the actin filament is uncovered. This allows myosin cross-bridges to attach to the site, and the cycle described in Section 6.3 continues. When nervous system signals shut off, calcium is actively transported back into the SR. Now the binding site on actin is covered up again, myosin can't bind to actin, and the muscle cell relaxes.

NEURONS ACT ON MUSCLE CELLS AT NEUROMUSCULAR JUNCTIONS

The nerve impulses that stimulate a skeletal muscle cell arrive at **neuromuscular junctions**. A motor neuron has

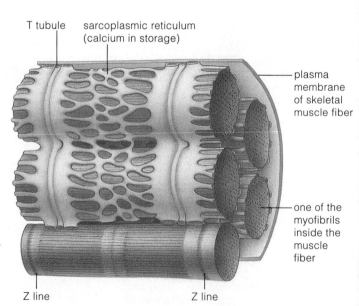

T tubule   sarcoplasmic reticulum (calcium in storage)

plasma membrane of skeletal muscle fiber

one of the myofibrils inside the muscle fiber

Z line   Z line

d Signals trigger the release of calcium ions from sarcoplasmic reticulum threading among the myofibrils. The calcium allows actin and myosin filaments in the myofibrils to interact and bring about contraction.

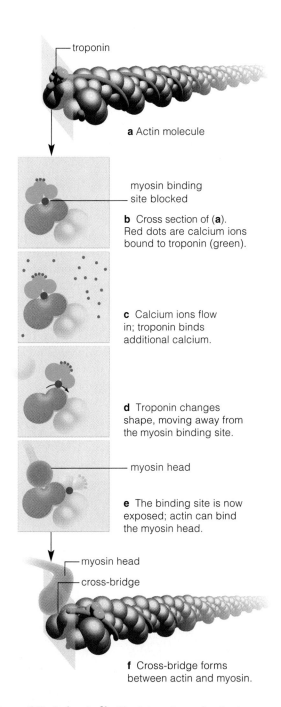

troponin

a Actin molecule

myosin binding
site blocked

b Cross section of (**a**).
Red dots are calcium ions
bound to troponin (green).

c Calcium ions flow
in; troponin binds
additional calcium.

d Troponin changes
shape, moving away from
the myosin binding site.

myosin head

e The binding site is now
exposed; actin can bind
the myosin head.

myosin head

cross-bridge

f Cross-bridge forms
between actin and myosin.

Figure 6.11 *Animated!* The interactions of actin, tropomyosin, and troponins in a skeletal muscle cell.

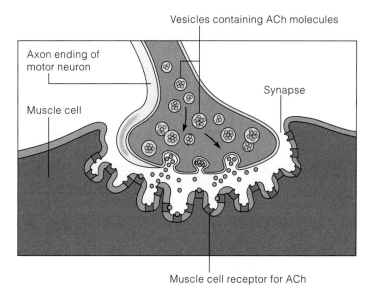

Vesicles containing ACh molecules

Axon ending of
motor neuron

Synapse

Muscle cell

Muscle cell receptor for ACh

Figure 6.12 How a chemical messenger called a neurotransmitter carries a signal across a neuromuscular junction.

of chemical messenger, a *neurotransmitter* called ACh (for acetylcholine) carries the signal from a motor neuron across the gap.

This signaling between a neuron and a muscle cell takes place in several steps. When the motor neuron is stimulated, calcium channels open in the plasma membrane of the neuron's axon endings that are in the neuromuscular junction. Then, calcium ions from the extracellular fluid flow inside the axon endings, and vesicles in each ending release ACh. When ACh binds to receptors on the muscle cell membrane, it may set in motion the events that cause the muscle cell to contract. ACh can excite or inhibit muscle and gland cells, as well as some cells in the brain and spinal cord.

The nervous system controls the contraction of muscle cells by way of signals that spark the release of calcium ions from a membrane system around a muscle cell's myofibrils. Nerve impulses pass from a neuron to a muscle cell across neuromuscular junctions.

extensions called *axons*; neuromuscular junctions are places where the branched endings of axons abut the muscle cell membranes, as you can see in Figure 6.10*b* and Figure 6.12. The neuron endings don't touch a muscle cell; between them there is a gap called a *synapse*. A type

6.5 How Muscle Cells Get Energy

LINKS TO
SECTIONS
3.11 AND 3.14

In resting muscle cells the demand for ATP can skyrocket quickly when the muscle is ordered to contract.

When a resting muscle cell is stimulated to contract, ATP must become available twenty to one hundred times faster than before. A muscle cell has only a little ATP when contraction starts, so a fast reaction forms more of it. An enzyme transfers phosphate from **creatine phosphate** to ADP. Because a cell has about five times as much creatine phosphate as ATP, this reaction can fuel contractions until a slower ATP-forming pathway can kick in (Figure 6.13).

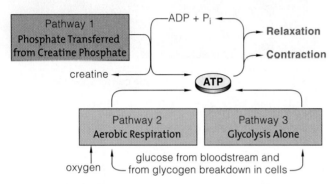

Figure 6.13 *Animated!* Three metabolic pathways by which ATP forms in muscles in response to the demands of physical exercise.

Normally, most of the ATP for muscle contraction comes from the oxygen-using reactions of cellular respiration. It's the same with the first five to ten minutes of moderate exercise. For the next half hour or so of steady activity, that muscle cell depends on glucose and fatty acids delivered by the blood. Beyond that time, fatty acids are the main fuel source (Section 3.14).

If you exercise hard, your respiratory and circulatory systems may not deliver enough oxygen for aerobic cellular respiration in some muscles. Then, glycolysis (which does not use oxygen) will contribute more of the ATP being formed. Although glycolysis doesn't yield much ATP, muscle cells use this metabolic backup as long as stored glycogen can provide glucose—or at least until **muscle fatigue** sets in. This is a state in which a muscle cannot contract, even if it is being stimulated. Fatigue probably is due to an **oxygen debt** that results when your muscles use more ATP than cellular respiration can deliver. The switch to glycolysis produces lactic acid. Along with the already low ATP supply, the rising acidity hampers the contraction of muscle cells. Deep, rapid breathing helps repay the oxygen debt.

> *How much ATP is available inside muscle cells affects whether the fiber can contract, and for how long.*

6.6 Properties of Whole Muscles

A muscle may contract weakly, strongly, or somewhere in between. A contraction may last a long time or only a few thousandths of a second. The properties of whole muscles relate to how individual muscle cells contract.

SEVERAL FACTORS DETERMINE THE CHARACTERISTICS OF A MUSCLE CONTRACTION

A **motor neuron** supplies a number of cells in a muscle. A motor neuron and the muscle cells it synapses with form a **motor unit** (Figure 6.14). The number of cells in a motor unit depends on how precise the muscle control must be. For instance, where precise control is required, as in the tiny muscles that move the eye, motor units have only four or five muscle cells. By contrast, motor units in some large leg muscles include hundreds of cells.

When a motor neuron fires, all the cells in its motor unit contract for a fleeting moment. This response is a **muscle twitch** (Figure 6.15*a*). If a new nerve impulse arrives before a twitch ends, the muscle twitches again. Repeating the stimulation of a motor unit in a short period of time makes all the twitches run together (Figure 6.15*b*). The result is a sustained contraction called **tetanus**

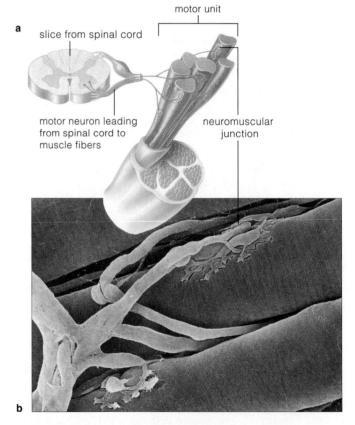

Figure 6.14 (a) Example of motor units present in muscles. (b) The micrograph shows the axon endings of a motor neuron that acts on individual muscle cells in the muscle.

(Figure 6.15c). Our muscles normally contract in this way, which generates three or four times the force of a single twitch. In the disease *tetanus*, muscles stay contracted, sometimes fatally (Figure 6.15d).

A skeletal muscle contains a large number of muscle cells, but not all of them contract at the same time. If a muscle is contracting only weakly—say, as your forearm muscles do when you pick up a pencil—it is because the nervous system is activating only a few of the muscle's motor units. In stronger contractions (when you heft a stack of books) more motor units are stimulated. Even when a muscle is relaxed, however, some of its motor units are contracted. This steady, low-level contracted state is called **muscle tone**. It helps maintain muscles in general good health and is important in stabilizing the skeleton's movable joints.

Muscle tension is the force that a contracting muscle exerts on an object, such as a bone. Opposing this force is a load, either the weight of an object or gravity's pull on the muscle. A stimulated muscle shortens only when muscle tension exceeds the opposing forces.

Isotonically contracting muscles shorten and move a load (Figure 6.16a). *Isometrically* contracting muscles develop tension but don't shorten. This is what happens when you attempt to lift an object that is too heavy (Figure 6.16b).

TIRED MUSCLES CAN'T GENERATE MUCH FORCE

When steady, strong stimulation keeps a muscle in a state of tetanus, the muscle eventually becomes fatigued. Muscle fatigue is a decrease in the muscle's ability to

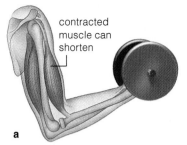
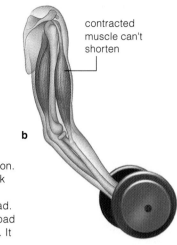

contracted muscle can shorten

a

contracted muscle can't shorten

b

Figure 6.16 (**a**) An isotonic contraction. The load is less than a muscle's peak capacity to contract, so the muscle can contract, shorten, and lift the load. (**b**) In an isometric contraction, the load exceeds the muscle's peak capacity. It contracts, but can't shorten.

generate force (that is, to develop tension). After a few minutes of rest, a fatigued muscle will be able to contract again. How long this recovery takes depends in part on how long and how often the muscle was stimulated before. Muscles trained by a pattern of brief, intense exercise fatigue and recover rapidly. This is what happens during weight lifting. Muscles used in prolonged, moderate exercise fatigue slowly but take longer to recover, often up to a day. Exactly what causes muscle fatigue is unknown, but one factor is depletion of glycogen, the form in which muscles hold glucose in reserve for energy. The build-up of lactic acid, which makes overused muscles sore, also contributes to fatigue.

A motor unit consists of a motor neuron and the muscle fibers it serves. The number of motor units in a muscle correlates with how precisely the nervous system must control a muscle's activity.

A muscle twitch is a brief muscle contraction following a stimulating nerve impulse. Body muscles normally contract in a sustained way called tetanus. Muscle tension is the force a contracting muscle exerts.

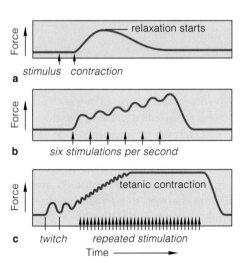

a   stimulus   contraction

relaxation starts

b   six stimulations per second

c   twitch   repeated stimulation

tetanic contraction

Time ⟶

d

Figure 6.15 *Animated!* Recordings of twitches in muscles artificially stimulated in different ways. (**a**) A single twitch. (**b**) Six per second cause a summation of twitches, and (**c**) about 20 per second cause tetanic contraction. (**d**) Painting of a soldier dying of the disease tetanus in a military hospital in the 1800s after the bacterium *Clostridium tetani* infected a battlefield wound.

6.7 Muscle Disorders

LINK TO
SECTION
3.12

Every movement depends on skeletal muscles, so when an accident, disease, or disuse prevents them from functioning normally, nearly every aspect of life is affected.

If you have ever torn a muscle or known someone with a muscle-wasting disease, you are very well aware that any problem that impairs the ability of skeletal muscles to produce movement has a serious impact on activities that most of us take for granted. In general, ills that can befall our skeletal muscles fall into three categories: injuries, disease, and disuse.

STRAINS AND TEARS ARE MUSCLE INJURIES

Given that our muscular system gets almost constant use, it's not surprising that the most common disorders of skeletal muscles are injuries. Lots of people, and athletes especially, strain a muscle at some point in their lives (Figure 6.17). The injury happens when a movement stretches or tears muscle fibers. Usually, there is some bleeding into the damaged area, which causes swelling and a painful muscle spasm. The usual first aid is an ice pack, followed by resting the affected muscle and using anti-inflammatory drugs such as ibuprofen.

When a whole muscle is torn, the aftereffects can last a lifetime. If scar tissue develops while the tear mends, the healed muscle may be shorter than before. As a result it may not function as effectively.

Figure 6.17 For athletes, muscle strains and tears often are "part of the game."

Figure 6.18 A child with Duchenne muscular dystrophy.

SOMETIMES A SKELETAL MUSCLE WILL CONTRACT ABNORMALLY

In a muscle *spasm*, a muscle suddenly and involuntarily contracts. A muscle *cramp* is a painful muscle spasm that doesn't immediately release. Any skeletal muscle can cramp, but the usual "victims" are calf and thigh muscles. In some cases the real culprit is a deficiency of potassium, which is needed for the proper transmission of nerve impulses to muscles and other tissues. Gentle stretching and massage may coax a cramped muscle to release.

Most people experience occasional muscle *tics*. These minor, involuntary twitches are common in muscles of the face and eyelids and may be triggered by anxiety or some other psycho-emotional cause.

MUSCULAR DYSTROPHIES DESTROY MUSCLE FIBERS

Muscular dystrophies are genetic diseases in which muscle fibers break down and the affected muscles progressively weaken and shrivel. *Duchenne muscular dystrophy* (DMD) is the most common form in children (Figure 6.18). It is caused by a single mutant gene that interferes with the ability of sarcomeres in muscle cells to contract. Affected youngsters usually are confined to a wheelchair by their teens, and die by their early twenties.

Myotonic muscular dystrophy is usually seen in adults. It generally affects only the hands and feet and is not life-threatening. "Myo" means muscle, and the name of this disorder indicates that affected muscles contract strongly but do not relax in the normal way.

Figure 6.19 Muscle-building alternatives. (**a**) Aerobic exercise builds endurance and improves overall muscle function. (**b**) Strength training builds larger, stronger muscles but does not improve endurance.

EXERCISE MAKES THE MOST OF MUSCLES

Muscle cells adapt to the activity demanded of them. When severe nerve damage or prolonged bed rest prevents a muscle from being used, the muscle will rapidly begin to waste away, or *atrophy*. Over time, affected muscles can lose up to three-fourths of their mass, with a corresponding loss of strength. More commonly, the skeletal muscles of a sedentary person stay basically healthy but cannot respond to physical demands in the same way that well-worked muscles can.

The best way to maintain or improve the work capacity of your muscles is to exercise them—that is, to increase the demands on muscle fibers to contract. To increase muscle endurance, nothing beats regular aerobic exercise—activities such as walking, biking, jogging, swimming, and aerobics classes (Figure 6.19*a*). Aerobic exercise works muscles at a rate at which the body can keep them supplied with oxygen. It affects muscle fibers in several ways:

1. There is an increase in the number and the size of mitochondria, the organelles that make ATP.

2. The number of blood capillaries supplying muscle tissue increases. This increased blood supply brings more oxygen and nutrients to the muscle tissue and removes metabolic wastes more efficiently.

3. Muscle tissues contain more of the oxygen-binding pigment myoglobin.

Together, these changes produce muscles that are more efficient metabolically and can work longer without becoming fatigued.

By contrast, strength training involves intense, short-duration exercise, such as weight lifting. It affects fast muscle fibers, which form more myofibrils and make more of the enzymes used in glycolysis (which forms some ATP). These changes translate into whole muscles that are larger and stronger (Figure 6.19*b*), but such bulging muscles fatigue rapidly so they don't have much endurance. Fitness experts recommend a workout plan that combines strength training and aerobic workouts.

Starting at about age 30, the tension, or physical force, a person's muscles can muster begins to decrease. This means that, once you enter your fourth decade of life, you may exercise just as long and intensely as a younger person but your muscles cannot adapt to the workouts to the same extent. Even so, being physically active is extremely beneficial. Aerobic exercise improves endurance and blood circulation, and even modest strength training slows the loss of skeletal muscle tissue that is an inevitable part of aging.

Summary

Section 6.1 Muscle tissue includes skeletal, smooth, and cardiac muscle. Although the cells of these different kinds of muscle tissue are structurally specialized for their particular functions, all muscle cells are specialized to generate force—that is, to contract or shorten.

Section 6.2 Skeletal muscle makes up the muscular system. There are more than 600 skeletal muscles, which transmit force to bones and move body limbs or other parts (Table 6.1). Skeletal muscles also help to stabilize joints and generate body heat. Each one contains bundles of muscle fibers (muscle cells) that are wrapped in connective tissue.

Tendons connect skeletal muscle to bones. The origin end of a skeletal muscle is attached to a bone that stays relatively motionless during a movement. The insertion end is attached to the bone that moves most. Some muscles work antagonistically, so the action of one opposes or reverses the action of the other. Other muscles are synergists that facilitate movements.

 Learn about the location and action of skeletal muscles.

Section 6.3 Bones move when they are pulled by the shortening, or contraction, of skeletal muscles. In turn, this shortening occurs because individual muscle fibers are shortening. Skeletal (and cardiac) muscle fibers contain threadlike myofibrils, which are divided lengthwise into sarcomeres, the basic units of contraction. Each sarcomere consists of an array of filaments of the proteins actin (thin) and myosin (thick):

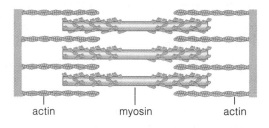

actin   myosin   actin

Interacting actin and myosin filaments shorten a sarcomere. The myosin head attaches to a neighboring actin filament (forming a cross-bridge); then the actin slides over the myosin. This movement is powered by ATP and is called the sliding filament mechanism of muscle contraction.

 Get an in-depth look at the structure and function of skeletal muscles.

Section 6.4 Nerve impulses cause skeletal muscle cells to contract. When the signals reach T tubules

Table 6.1	Review of Skeletal Muscle

FUNCTION OF SKELETAL MUSCLE:
Contraction (shortening) that moves the body and its parts.

MAJOR COMPONENTS OF SKELETAL MUSCLE CELLS:
Myofibrils: Strands containing filaments of the contractile proteins actin and myosin.

Sarcomeres: The basic units of muscle contraction.

Other:

Motor unit: A motor neuron and the muscle cells it controls.

Neuromuscular junction: Synapse between a motor neuron and muscle cells.

(extensions of the cell's plasma membrane) they trigger the release of calcium ions from sarcoplasmic reticulum, a membrane system that wraps around myofibrils in the muscle fiber. The binding of calcium affects proteins on the surface of actin filaments so that the heads of myosin molecules can bind to actin.

A neuromuscular junction is a synapse between a motor neuron and a muscle fiber. A nerve impulse triggers the release of a neurotransmitter called ACh into the synapse. This starts the events that cause the fiber to contract.

 See how the nervous system controls muscle contraction.

Section 6.5 The ATP required for muscle contraction can come from cellular respiration, from glycolysis alone, or from the generation of ATP from creatine phosphate. Glycolysis is used when muscles are exercised so intensively that the full aerobic pathway can't keep up with the demand for ATP. When muscles use more ATP than aerobic respiration can provide, an oxygen debt may develop in muscle tissue.

 Compare sources of energy for muscle contraction.

Section 6.6 A motor neuron and the muscle fibers it controls make up a motor unit. When a stimulus activates a certain number of motor units, it produces a brief muscle contraction called a muscle twitch. If a series of twitches occur close together, a sustained contraction called tetanus develops. Skeletal muscles normally operate near or at tetanus. Some of a muscle's motor units are contracted even when it is relaxed. This low-level state of contraction is called muscle tone and helps keep muscles healthy. Muscle tension is the force that a contracting muscle exerts on an object. When muscle fatigue sets in, a muscle cannot contract even if it is being stimulated.

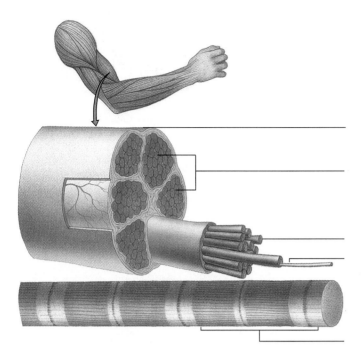

Section 6.7
Injuries are the most common disorders of skeletal muscles, and even healthy muscles may contract abnormally, such as when they cramp. Muscular dystrophies are a set of diseases that destroy muscle fibers and cause skeletal muscles to lose function. A combination of regular aerobic exercise and resistance training helps build muscle and increase its endurance.

Review Questions

1. In a general sense, how do skeletal muscles produce movement?

2. In the diagram above, label the fine structure of a muscle, down to one of its myofibrils. Identify the basic unit of contraction in a myofibril.

3. How do actin and myosin interact in a sarcomere to bring about muscle contraction? What roles do ATP and calcium play?

4. How does a muscle cell incur an oxygen debt? How is this "debt" different from muscle fatigue?

5. What is the function of the sarcoplasmic reticulum in muscle cell contraction?

6. Explain why (a) calcium ions and (b) ACh are vital for muscle contraction.

7. What is a motor unit? Why does a rapid series of muscle twitches yield a stronger overall contraction than a single twitch?

8. What are the structural and functional differences between "slow" and "fast" muscle?

Self-Quiz
Answers in Appendix V

1. The _____ and _____ systems work together to move the body and specific body parts.

2. The three types of muscle tissue are _____, _____, and _____.

3. _____ forms cross-bridges with myosin.
   a. A muscle fiber   c. Myoglobin
   b. A tendon   d. Actin

4. The _____ is the basic unit of muscle contraction.
   a. myofibril   c. muscle fiber
   b. sarcomere   d. myosin filament

5. Skeletal muscle contraction requires _____.
   a. calcium ions   c. arrival of a nerve impulse
   b. ATP   d. all of the above

6. Match the M words with their defining feature.
   ____ muscle   a. actin's partner
   ____ muscle twitch   b. delivers contraction signal
   ____ muscle tension   c. a muscle cannot contract
   ____ myosin   d. motor unit response
   ____ motor neuron   e. force exerted by cross-bridges
   ____ myofibrils   f. muscle cells bundled in connective tissue
   ____ muscle fatigue   g. threadlike parts in a muscle fiber

Critical Thinking

1. You are training athletes for the 100-meter dash. They need muscles specialized for speed and strength, *not* endurance. What muscle characteristics would your training regimen aim to develop? How would you alter it to train marathoners?

2. In 1989, explorer Will Steger and his dogsled team crossed Antarctica, traveling some 3,741 miles. Steger said later that his polar huskies worked the hardest and pulled all the weight. A husky's limb bones and skeletal muscles are suited for long-distance load-pulling. For example, the forelegs move freely, thanks to a deep but not-too-broad rib cage, and the dog also has a well-muscled chest. What kind of muscle fibers and muscle mass would you expect to find in a husky's *hind* legs, which provide much of the brute power to propel a loaded sled?

3. Curare, a poison extracted from a South American shrub, blocks the binding of ACh by muscle cells. What do you suppose would happen to your muscles, including the ones involved in breathing, if a toxic dose of curare entered your bloodstream?

4. At the gym Sean gets on a stair-climbing machine and "climbs" as fast as he can for fifteen minutes. At the end of that time he is breathing hard and his quadriceps and other leg muscles are aching. What is the physiological explanation for these symptoms?

5. In training for a marathon, Lydia plans to take creatine supplements because she heard that they boost an athlete's energy. What is your opinion on this plan?

Explore on Your Own

A good way to improve your understanding of your muscular system is to explore the movements of your own muscles. Try the following quick exercises.

Human hands don't contain many of the muscles that control hand movements. Instead, as you can see in Figure 6.20a, most of those muscles are in the forearm. Tendons extending from one muscle, the flexor digitorum superficialis (the "superficial finger flexor"), bend your fingers. Place one hand on the top of the opposite forearm, and then wiggle your fingers on that side or make a fist several times. Can you feel the "finger flexor" in action?

Place your fingers on the skin above your nose, between your eyebrows. Now frown. The muscle you feel pulling your eyebrows together is the corrugator supercilii. One effect of its contraction is to "corrugate" the skin of your forehead into vertical wrinkles.

A grin calls into action other facial muscles, including the zygomaticus major (Figure 6.20b). On either side of the skull, this muscle originates on the cheekbones and inserts at the corners of the mouth. To feel it contract, place the tips of your index fingers at the corners of your mouth, and then smile.

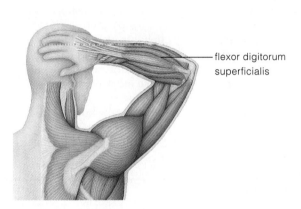

flexor digitorum superficialis

a The flexor digitorum superficialis, a forearm muscle that helps move the fingers.

Figure 6.20 Some muscles to explore.

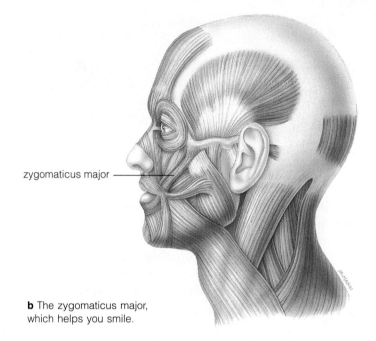

zygomaticus major

b The zygomaticus major, which helps you smile.

Hormones and Hunger

Americans are among the fattest people in the world, with 60 percent of adults overweight or obese. Excess weight is a risk factor for heart disease, diabetes, and some forms of cancer. So what should we do? The seemingly obvious answer is, "Lose the weight." But it's not that simple. Studies confirm what dieters have always thought. Once we put on extra pounds, they're hard to lose.

Like other mammals, we have fat-storing cells concentrated in adipose tissue. This energy warehouse evolved among our early ancestors, who could not always be sure where their next meal was coming from. Stored body fat helped them through the lean times. Once these cells form, however, they are in the body to stay. Take in more calories than you burn, and the cells plump up with fat droplets.

Adipose cells make leptin. This hormone acts on a part of the brain that deals with appetite. Researchers discovered that mutant mice that can't make leptin also eat nonstop. Does a lack of leptin also make overweight people eat too much? Unfortunately, no. Tests show that, if anything, overweight people have more leptin in their blood. It's possible that their leptin receptors aren't functioning

properly. Or perhaps their body cells just don't have enough of the receptors.

The stomach and brain also "weigh in" on our eating habits. Some of their cells secrete ghrelin, a hormone that sparks feelings of hunger. When a person's stomach is empty, more ghrelin is released. The level falls again after a meal.

In one experiment on ghrelin, obese volunteers ate a low-fat, low-calorie diet for six months. On average they lost 17 percent of their body weight. However, the level of ghrelin circulating in the subjects' blood had climbed 24 percent. The now-thinner dieters were hungrier than ever!

The topics of appetite, eating, and body weight take us into the world of digestion and nutrition. In this chapter you will learn how the digestive system brings into the body the nutrients living body cells need to survive. As it makes this key contribution to homeostasis, the digestive system interacts with other organ systems as well.

 How Would You Vote? *Obesity may soon replace smoking as the main cause of preventable deaths in the United States. Fast food is contributing to the problem. Should fast-food items be required to carry health warnings? Cast your vote online at www.thomsonedu.com/biology/starr/humanbio.*

 ## Key Concepts

THE DIGESTIVE SYSTEM
The digestive system mechanically and chemically breaks down food, absorbs nutrient molecules, and then eliminates the residues.

NUTRITION AND BODY WEIGHT
An ideal diet provides enough nutrients, vitamins, and minerals to support the metabolic activity of living cells. A given body weight can be maintained when the amount of energy used by the body balances energy inputs from food.

Links to Earlier Concepts

This chapter shows how carbohydrates (2.9), proteins (2.11), and lipids (2.10) in food are broken down by digestion. Transport mechanisms, including diffusion and osmosis (3.9) and active transport (3.10) are essential to this process, for they provide the routes by which nutrient molecules enter the bloodstream. You will also build on what you have learned about different sources that can supply the body's energy needs (3.14).

7.1 The Digestive System: An Overview

LINKS TO
SECTIONS
4.1 AND 4.7

Stretched out, the gastrointestinal tract would be 6.5 to 9 meters long in an adult (21 to 30 feet). This long tube is where food processing takes place and nutrition begins.

Our **digestive system** is a tube with two openings and many specialized organs. It extends from the mouth to the anus and is also called the **gastrointestinal (GI) tract**.

An interesting fact about the GI tract is that while food or food residues are in it, technically the material is still *outside* the body. Nutrients don't "officially" enter the body until they move from the space inside the digestive tube—its *lumen*—into the bloodstream.

From beginning to end, epithelium lines the surfaces facing the lumen. The lining is coated with thick, moist mucus that protects the wall of the tube and enhances the diffusion of substances across it.

When we eat, food advances in one direction, from the mouth (the oral cavity) through the pharynx, the esophagus, stomach, small intestine, and large intestine. The large intestine ends in the rectum, anal canal, and anus. Figure 7.1 diagrams an adult's digestive system.

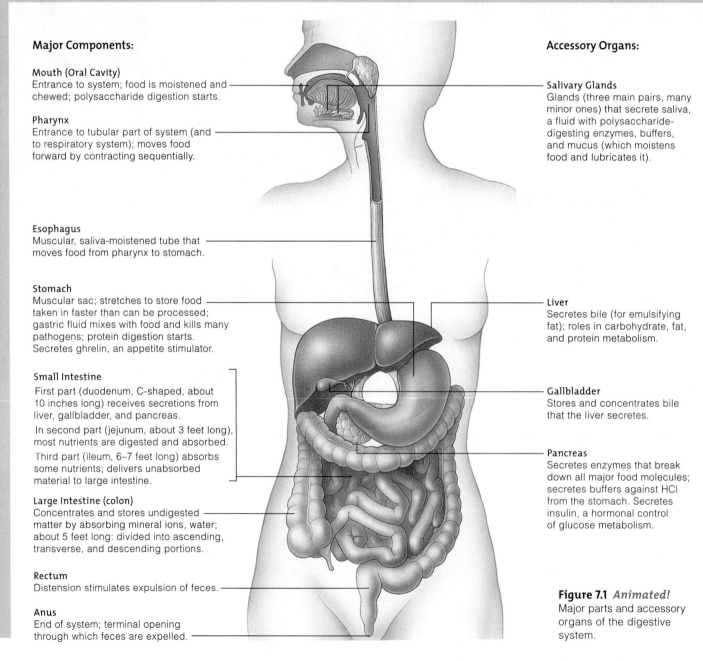

Major Components:

Mouth (Oral Cavity)
Entrance to system; food is moistened and chewed; polysaccharide digestion starts.

Pharynx
Entrance to tubular part of system (and to respiratory system); moves food forward by contracting sequentially.

Esophagus
Muscular, saliva-moistened tube that moves food from pharynx to stomach.

Stomach
Muscular sac; stretches to store food taken in faster than can be processed; gastric fluid mixes with food and kills many pathogens; protein digestion starts. Secretes ghrelin, an appetite stimulator.

Small Intestine
First part (duodenum, C-shaped, about 10 inches long) receives secretions from liver, gallbladder, and pancreas.
In second part (jejunum, about 3 feet long), most nutrients are digested and absorbed.
Third part (ileum, 6–7 feet long) absorbs some nutrients; delivers unabsorbed material to large intestine.

Large Intestine (colon)
Concentrates and stores undigested matter by absorbing mineral ions, water; about 5 feet long; divided into ascending, transverse, and descending portions.

Rectum
Distension stimulates expulsion of feces.

Anus
End of system; terminal opening through which feces are expelled.

Accessory Organs:

Salivary Glands
Glands (three main pairs, many minor ones) that secrete saliva, a fluid with polysaccharide-digesting enzymes, buffers, and mucus (which moistens food and lubricates it).

Liver
Secretes bile (for emulsifying fat); roles in carbohydrate, fat, and protein metabolism.

Gallbladder
Stores and concentrates bile that the liver secretes.

Pancreas
Secretes enzymes that break down all major food molecules; secretes buffers against HCl from the stomach. Secretes insulin, a hormonal control of glucose metabolism.

Figure 7.1 *Animated!* Major parts and accessory organs of the digestive system.

THE DIGESTIVE TUBE HAS FOUR LAYERS

From the esophagus onward, the digestive tube wall has four layers (Figure 7.2). The *mucosa* (the innermost layer of epithelium) faces the lumen—the space through which food passes. The mucosa is surrounded by the *submucosa*, a layer of connective tissue with blood and lymph vessels and nerve cells. The next layer is *smooth muscle*—usually two sublayers, one circling the tube and the other oriented lengthwise. An outer layer, the *serosa*, is a very thin serous membrane (Section 4.7). Circular arrays of smooth muscle in sections of the GI tract are **sphincters**. Contractions of the sphincter muscles can close off a passageway. In your stomach, they help pace the forward movement of food and prevent backflow.

DIGESTIVE SYSTEM OPERATIONS CONTRIBUTE TO HOMEOSTASIS IN KEY WAYS

The following list summarizes the overall operations of the digestive system:

1. **Mechanical processing and motility**. Movements of various parts, such as the teeth, tongue, and muscle layers, break up, mix, and propel food material.

2. **Secretion**. Digestive enzymes and other substances are released into the digestive tube.

3. **Digestion**. Food is chemically broken down into nutrient molecules small enough to be absorbed.

4. **Absorption**. Digested nutrients and fluid pass across the tube wall and into blood or lymph.

5. **Elimination**. Undigested and unabsorbed residues are eliminated from the end of the GI tract.

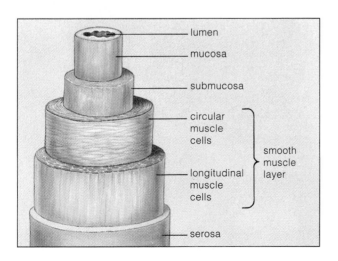

Figure 7.2 The four-layered wall of the gastrointestinal tract. The layers are not drawn to scale.

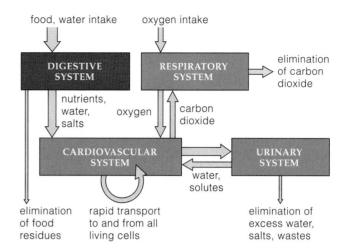

Figure 7.3 Links between the digestive system and other organ systems. As the text describes, it works together with the circulatory system, respiratory system, and urinary system to supply cells with raw materials and to eliminate wastes.

Various accessory structures secrete enzymes and other substances that are essential for different aspects of digestion and absorption. They include glands in the wall of the GI tract, the salivary glands, and the liver, gallbladder, and pancreas.

The "flow diagram" in Figure 7.3 gives a general idea of how the workings of the digestive system fit into the larger picture of homeostasis in the body. For instance, once nutrients from food have entered the bloodstream, the circulating blood carries them throughout the body. The respiratory system keeps all body cells, including those of digestive system tissues and organs, supplied with the oxygen they need for aerobic respiration, and removes carbon dioxide wastes. And although food residues are eliminated by the digestive system itself, the urinary system disposes of many other wastes or unneeded substances (such as excess salt) that enter the blood from the GI tract. Together these adjustments help maintain the proper volume and chemical makeup of the extracellular fluid. With this overview in mind, let's now see how each major component of the digestive system performs its functions.

The digestive tube extends from the mouth to the anus. For most of its length the tube wall consists of four layers, including smooth muscle.

The digestion and absorption of food make a vital contribution to homeostasis as interactions among the digestive, circulatory, respiratory, and urinary systems supply cells with raw materials and dispose of wastes.

7.2 Chewing and Swallowing: Food Processing Begins

LINK TO SECTION 2.9

Food processing begins the moment food enters your mouth, where enzymes begin chemical digestion of starches.

PROCESSING STARTS WITH THE TEETH AND SALIVARY GLAND ENZYMES

In the **oral cavity**, or mouth, the food you eat begins to be broken apart by chewing. Most adults have thirty-two teeth (Figure 7.4*a*); young children have just twenty "primary teeth." A tooth's crown (Figure 7.4*b*) is coated with hardened calcium deposits, the tooth enamel—the hardest substance in the body. It covers a living, bonelike layer called dentin. Dentin and an inner pulp extend into the root. The pulp cavity contains blood vessels and nerves.

The shape of a tooth fits its function. Chisel-shaped incisors bite off chunks of food, and cone-shaped canines (cuspids) tear it. Premolars and molars, with broad crowns and rounded cusps, grind it.

Chewing mixes food with saliva from several **salivary glands** (Figure 7.4*c*). A large *parotid gland* nestles just in front of each ear. *Submandibular glands* lie just below the lower jaw in the floor of the mouth, and *sublingual glands*

are under your tongue. The tongue itself is skeletal muscle covered by a membrane. As described in Chapter 14, its taste receptors respond to dissolved chemicals.

Saliva is mostly water, but it includes other important substances. One, the enzyme **salivary amylase**, breaks down starch; chew on a soda cracker and you can feel it turning to mush as salivary amylase goes to work. A buffer, bicarbonate (HCO_3^-), keeps the pH of your mouth between 6.5 and 7.5, a range within which salivary amylase can function. Saliva also contains *mucins*, proteins that help bind food bits into a lubricated ball called a **bolus** (BOE-lus). Starch digestion continues in the stomach until acids there inactivate salivary amylase.

The roof of the mouth, a bone-reinforced section of the **palate**, provides a hard surface against which the tongue can press food as it mixes it with saliva. Tongue muscle contractions force the bolus into the **pharynx** (FARE-inks; throat). This passageway connects with the windpipe, or *trachea* (Figure 7.5), which leads to the lungs. It also connects with the **esophagus**, which leads to the stomach. Mucus secreted by the membrane lining the pharynx and esophagus lubricates the bolus, helping move food on its way.

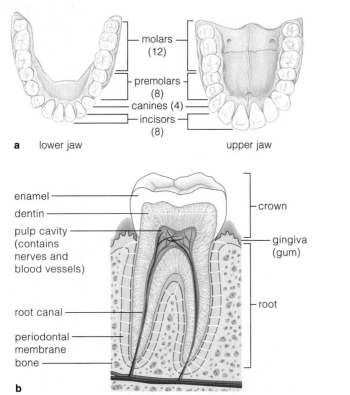

a lower jaw   upper jaw

molars (12)
premolars (8)
canines (4)
incisors (8)

enamel
dentin
pulp cavity (contains nerves and blood vessels)
root canal
periodontal membrane
bone

crown
gingiva (gum)
root

b

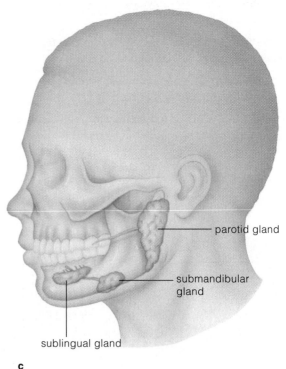

parotid gland
submandibular gland
sublingual gland

c

Figure 7.4 *Animated!* (**a**) Locations of the different types of teeth. (**b**) Anatomy of a human tooth. (**c**) Locations of the salivary glands.

THE DIGESTIVE SYSTEM

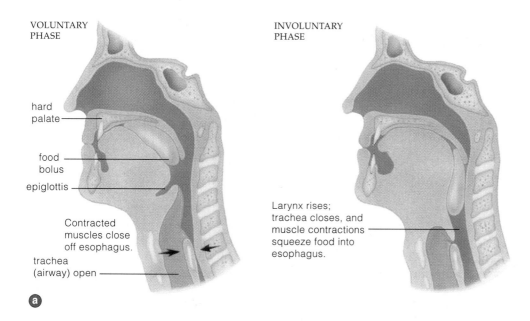

VOLUNTARY
PHASE

hard
palate

food
bolus

epiglottis

Contracted
muscles close
off esophagus.

trachea
(airway) open

a

INVOLUNTARY
PHASE

Larynx rises;
trachea closes, and
muscle contractions
squeeze food into
esophagus.

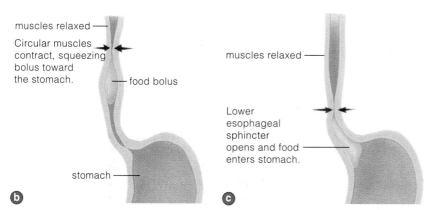

muscles relaxed

Circular muscles
contract, squeezing
bolus toward
the stomach.

food bolus

stomach

b

muscles relaxed

Lower
esophageal
sphincter
opens and food
enters stomach.

c

Figure 7.5 *Animated!*
Swallowing and peristalsis.
(**a**) Contractions of the tongue
push the food bolus into the
pharynx. Next, the vocal cords
seal off the larynx, and the
epiglottis bends downward,
helping to keep the trachea
closed. Contractions of throat
muscles then squeeze the
food bolus into the esophagus.
(**b,c**) Finally, peristalsis in the
esophagus moves the bolus
through a sphincter, and food
enters the stomach.

SWALLOWING HAS VOLUNTARY AND INVOLUNTARY PHASES

Swallowing a mouthful of food might seem simple, but
it involves a sequence of events. Swallowing begins
when voluntary skeletal muscle contractions push a
bolus into the pharynx, stimulating sensory receptors in
the pharynx wall. The receptors trigger a swallowing
reflex in which involuntary muscle contractions keep
food from moving up into your nose and down into the
trachea. As this reflex gets under way, the vocal cords are
stretched tight across the entrance to the larynx (your
"voice box"). Then, the flaplike epiglottis is pressed down
over the vocal cords as a secondary seal. For a moment,

breathing stops as food moves into the esophagus, so
you normally don't choke when you swallow. When
swallowed food reaches the lower end of the esophagus,
it passes through a sphincter into the stomach (Figure
7.5b,c). Waves of muscle contractions called **peristalsis**
(pair-uh-STALL-sis) help push the food bolus along.

*As food is chewed, the teeth and tongue start breaking it up
mechanically. Enzymes in saliva begin the chemical
digestion of starches.*

*Swallowed food moves down the esophagus, through the
lower esophageal sphincter, and into the stomach.*

7.3 The Stomach: Food Storage, Digestion, and More

LINKS TO
SECTIONS
2.11, 4.6, AND 4.7

You may think of your stomach as simply a bag for holding food, but it really is a complex organ with multiple functions in processing the food you eat.

The **stomach** is a muscular, stretchable sac (Figure 7.6a) that has three functions:

1. It mixes and stores ingested food.

2. It produces secretions that help dissolve and break down food particles, especially proteins.

3. It helps control the passage of food into the small intestine.

The stomach wall surface facing the lumen is lined with glandular epithelium. Each day, gland cells in the lining release about two liters of hydrochloric acid (HCl), mucus, and other substances. These include pepsinogens, precursors of digestive enzymes called **pepsins**. Other gland cells secrete *intrinsic factor*, a protein required for vitamin B$_{12}$ to be absorbed in the small intestine. Along with water, these substances make up the stomach's strongly acidic **gastric juice**. Combined with stomach contractions, the acidity converts swallowed boluses into a thick mixture called **chyme** (KIME). The acidity kills most microbes in food. It also can cause "heartburn" when gastric fluid backs up into the esophagus.

The digestion of proteins starts when the high acidity denatures proteins and exposes their peptide bonds. The acid also converts pepsinogens to active pepsins, which break the bonds, "chopping" the protein into fragments. Meanwhile, gland cells secrete the hormone *gastrin*, which stimulates cells that secrete HCl and pepsinogen.

Why don't HCl and pepsin break down the stomach lining? Usually, mucus and bicarbonate protect the lining. These protections, which form the "gastric mucosal barrier," can go awry, however. A common cause is the bacterium *Helicobacter pylori*. It produces a toxin that inflames the lining. Tight junctions that normally prevent HCl from passing between cells of the stomach lining break down, so hydrogen ions and pepsins diffuse into the lining—and that launches further damage. The resulting open sore is a *peptic ulcer*. Antibiotics can cure peptic ulcers caused by *H. pylori*, but about 20 percent of ulcers are related to factors such as chronic emotional stress, smoking, and excessive use of aspirin and alcohol.

Waves of peristalsis empty the stomach of food. These waves mix chyme and build force as they approach the pyloric sphincter at the stomach's base (Figure 7.6b). When a strong contraction arrives, the sphincter closes, squeezing most of the chyme back. Only a small amount moves into the small intestine at a given time. In this way the stomach regulates the rate at which food moves onward, so that food is not passed along faster than it can be processed. Depending mainly on the fat content and acidity of chyme, it can take from two to six hours for a full stomach to empty. When the stomach is empty, its walls crumple into folds called *rugae*.

Water and alcohol are two of a few substances that begin to be absorbed across the stomach wall. Liquids imbibed on an empty stomach pass rapidly to the small intestine, where further absorption occurs. When food is in the stomach, gastric emptying slows. This is why a person feels the effects of alcohol more slowly when drinking accompanies a meal.

> The stomach's functions are storing food, initial digestion of proteins, and regulating passage of food (in the form of chyme) into the small intestine.

IMPACTS, ISSUES

Gastric bypass surgery—"stomach stapling"—is one option for treating extreme obesity. The surgery closes off most of the stomach and part of the small intestine. It reduces the amount of food the patient can comfortably eat and the nutrients he or she can absorb. Gastric bypass also reduces ghrelin secretion. This reduces hunger and helps patients keep off weight. Some patients have lost up to 300 pounds. The surgery has risks, however, and can lead to side effects such as osteoporosis.

Figure 7.6 *Animated!*
(**a**) Structure of the stomach.
(**b**) A peristaltic wave down the stomach, produced by alternating contraction and relaxation of muscles in the stomach wall.

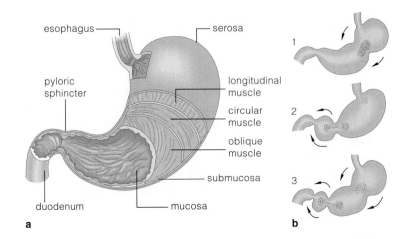

esophagus — serosa

pyloric sphincter

longitudinal muscle

circular muscle

oblique muscle

submucosa

duodenum — mucosa

a

1

2

3

b

THE DIGESTIVE SYSTEM

7.4 The Small Intestine: A Huge Surface Area for Digestion and Absorption

Your small intestine—about an inch and a half in diameter and 6 meters (20 feet) long—absorbs most nutrients.

The key to the **small intestine**'s ability to absorb nutrients is the structure of its wall. Figure 7.7 shows how densely folded the mucosa is, and how the folds all project into the lumen. Each one of the folds has an amazing number of ever tinier projections. And the epithelial cells at the surface of these tiny projections have a brushlike crown of still *smaller* projections, all exposed to the lumen.

What is the benefit of so many folds and projections? Together, all the projections from the intestinal mucosa greatly increase the surface area for absorbing nutrients from chyme. Without that huge surface area, absorption would take place too slowly to sustain life.

Figure 7.7c shows a **villus** (plural: villi). Millions of villi, each one about a millimeter long, are the absorptive "fingers" on the folds of the intestinal mucosa. Their density makes the mucosa look velvety. Small blood vessels (an arteriole and a vein) in each villus and a lymph vessel move substances to and from the bloodstream (Figure 7.7d).

Most cells in the epithelium covering each villus bear microvilli (singular: microvillus). A **microvillus** is a threadlike projection of the epithelial cell's plasma membrane. Each epithelial cell has about 1,700 of them. This dense array gives the epithelium of villi its common name, the "brush border." Gland cells in the lining release digestive enzymes, and defensive cells called phagocytes ("cell eaters") patrol and help protect the lining.

A folded mucosa, millions of villi, and hundreds of millions of microvilli give the small intestine a vast surface area for absorbing nutrients.

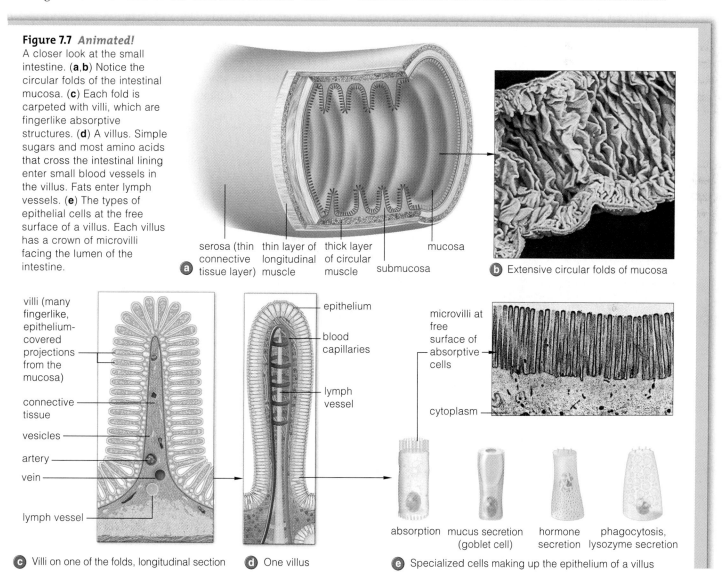

Figure 7.7 *Animated!*
A closer look at the small intestine. (**a**,**b**) Notice the circular folds of the intestinal mucosa. (**c**) Each fold is carpeted with villi, which are fingerlike absorptive structures. (**d**) A villus. Simple sugars and most amino acids that cross the intestinal lining enter small blood vessels in the villus. Fats enter lymph vessels. (**e**) The types of epithelial cells at the free surface of a villus. Each villus has a crown of microvilli facing the lumen of the intestine.

a serosa (thin connective tissue layer) — thin layer of longitudinal muscle — thick layer of circular muscle — submucosa — mucosa

b Extensive circular folds of mucosa

c Villi on one of the folds, longitudinal section
villi (many fingerlike, epithelium-covered projections from the mucosa) — connective tissue — vesicles — artery — vein — lymph vessel

d One villus
epithelium — blood capillaries — lymph vessel

microvilli at free surface of absorptive cells — cytoplasm

e Specialized cells making up the epithelium of a villus
absorption — mucus secretion (goblet cell) — hormone secretion — phagocytosis, lysozyme secretion

7.5 Accessory Organs: The Pancreas, Gallbladder, and Liver

LINKS TO
SECTIONS
2.7 AND 4.1

Before absorbed nutrients start moving throughout the body in the bloodstream, they make a key stop—the liver.

Figure 7.1 listed three important accessory organs of digestion: the pancreas, gallbladder, and liver. These organs are "accessory" because they assist digestion in some way but are outside the digestive tube.

THE PANCREAS PRODUCES A VARIETY OF DIGESTIVE ENZYMES

The long, slender **pancreas** nestles behind and below the stomach. It contains two kinds of gland cells: exocrine cells that make digestive enzymes and release them into the first section of the small intestine, and endocrine cells that make and release hormones into the bloodstream (Figure 7.8). The hormones help regulate blood sugar, and you will learn more about them in Chapter 15.

The pancreas produces four basic kinds of digestive enzymes, which can chemically dismantle the four major categories of food—complex carbohydrates, proteins, lipids, and nucleic acids. These enzymes work best when the pH is neutral or slightly alkaline, and so the pancreas also secretes fluid containing bicarbonate ($NaHCO_3^-$),

which neutralizes the acid in chyme arriving from the stomach. Depending on how often and what you eat, your pancreas may make as much as two quarts of this fluid each day!

THE GALLBLADDER STORES BILE

When the digestive system is processing food, along with pancreatic enzymes and bicarbonate-rich fluid, a fluid called *bile* also is released into the upper small intestine. The liver makes bile, as you will read shortly, and this yellowish fluid is stored in the **gallbladder**, a sausage-shaped, green-colored sac tucked behind the liver. As needed, the gallbladder contracts and empties bile into the small intestine where it aids in the digestion and absorption of fats. When there's no food moving through the GI tract, a sphincter closes off the main bile duct, and bile backs up into the gallbladder.

The gallbladder is one of our more "dispensible" organs. If it is surgically removed—usually due to the presence of gallstones, described in a moment—the duct that connects it to the small intestine enlarges and takes on the role of bile storage. This is why many millions of people today are walking around minus their gallbladder, with no ill effects.

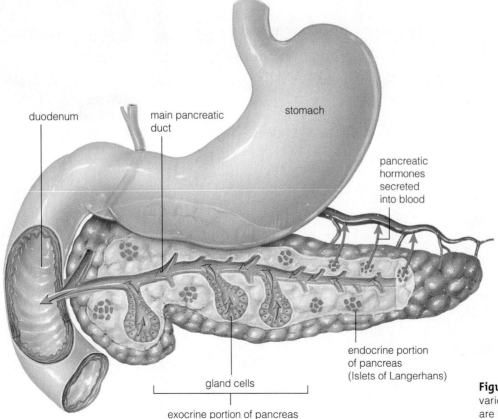

duodenum

main pancreatic duct

stomach

pancreatic hormones secreted into blood

endocrine portion of pancreas (Islets of Langerhans)

gland cells

exocrine portion of pancreas enzymes secreted into duodenum

Figure 7.8 The pancreas. The various regions of the pancreas are shown here much larger than their actual size.

THE LIVER IS A MULTIPURPOSE ORGAN

The **liver** is one of the body's largest organs, and it serves a range of important functions (Table 7.1). Its role in digestion is to secrete bile—as much as 1,500 ml, or 1.5 quarts, every day. Bile contains bile salts, which the liver synthesizes from cholesterol. Bile salts help to emulsify fats in chyme—that is, to break up large fat globules into smaller bits.

The liver not only uses cholesterol to make bile salts, but also helps manage the level of this lipid in the body. Liver cells secrete cholesterol into bile, which bile salts emulsify along with other lipids in chyme. Some of this cholesterol becomes part of the food residues that eventually are excreted in feces. When there is chronically more cholesterol in bile than available bile salts can dissolve, the excess may separate out. Hard *gallstones*, which are mostly lumps of cholesterol, can develop in the gallbladder. They cause severe pain if they become lodged in bile ducts.

The liver also processes nutrient-bearing blood from the small intestine. This blood flows to the **hepatic portal vein**, which carries the blood through vessels in the liver. In the liver, excess glucose is taken up before a hepatic vein returns the blood to the general circulation (Figure 7.9). The liver converts and stores much of this glucose as glycogen.

Besides its digestive functions, the liver helps maintain homeostasis in other ways (Table 7.1). For instance, it processes incoming nutrient molecules into substances the body requires (such as blood plasma proteins) and

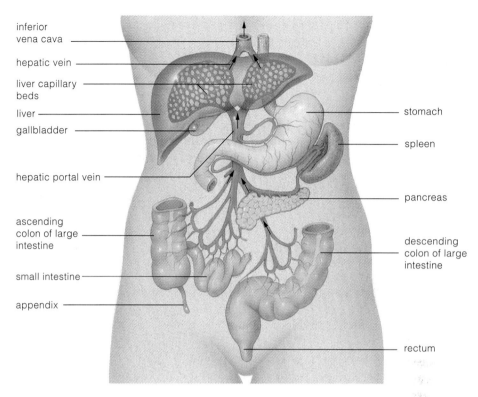

Figure 7.9 Hepatic portal system. Arrows show the direction of blood flow.

removes toxins ingested in food or already circulating in the bloodstream. It also inactivates many hormones and sends them to the kidneys for excretion (in urine). Ammonia (NH_3) that is produced when cells break down amino acids can be dangerously toxic to cells, especially in the nervous system. The circulatory system carries ammonia to the liver, where it is converted to a much less toxic waste product, urea, which also is excreted in urine.

The pancreas, gallbladder, and liver are accessory organs of the digestive system.

The pancreas produces enzymes that can dismantle complex carbohydrates, protein, lipids, and nucleic acids.

The gallbladder stores bile, which is produced in the liver.

The liver also processes nutrient-bearing blood, storing excess glucose as glycogen, removing toxins, and other functions.

Table 7.1	Ways the Liver Contributes to Homeostasis

1. Plays a role in carbohydrate metabolism
2. Partially controls synthesis of proteins in blood; assembles and disassembles certain other proteins
3. Forms urea from nitrogen-containing wastes
4. Assembles and stores some fats; forms bile to aid in fat digestion
5. Inactivates many chemicals (hormones, some drugs)
6. Detoxifies many poisons
7. Breaks down worn-out red blood cells
8. Aids immune response (removes some foreign particles)
9. Absorbs, stores factors needed for red blood cell formation

7.6 Digestion and Absorption in the Small Intestine

The small intestine's vast surface area is the stage for nutrients to be taken into the internal environment—tissue fluid and the bloodstream. Different kinds of nutrients are absorbed by way of several mechanisms.

On average, each day about 9 liters of fluid (roughly 10 quarts) enters the first section of the small intestine, the duodenum (doo-oh-DEE-num). This fluid includes chyme as well as digestive juices and other substances from the pancreas, liver, and gallbladder. All these chemicals have key roles in digestion.

NUTRIENTS ARE RELEASED BY CHEMICAL AND MECHANICAL MEANS

Chyme entering the duodenum triggers hormone signals that stimulate a brief flood of digestive enzymes from the pancreas. As part of "pancreatic juice," these enzymes act on carbohydrates, fats, proteins, and nucleic acids (Table 7.2). For example, like pepsin in the stomach, the pancreatic enzymes trypsin and chymotrypsin digest the polypeptide chains of proteins into peptide fragments. The fragments are then broken down to amino acids by different peptidases (present on the surface of the intestinal mucosa). As noted in Section 7.5, the pancreas also secretes bicarbonate, which buffers stomach acid and so maintains a chemical environment in which pancreatic enzymes can function.

Besides enzymes, fat digestion requires the bile salts in bile secreted by the liver and delivered via the gallbladder. You have already read that bile salts speed up fat digestion by breaking up large units of fat into smaller ones. How does this emulsification process work? Most fats in the average diet are triglycerides, which do not dissolve in water. Accordingly, in chyme they tend to clump into big fat globules. When peristalsis mixes chyme, the globules break up into droplets that become coated with bile salts (see Figure 7.10c). The salts bear a negative charge, so the coated droplets repel each other and stay separated. The droplets give fat-digesting enzymes a much greater surface area to act on. So, because triglycerides are emulsified, they can be broken down much more rapidly to monoglycerides and fatty acids, molecules that are small enough to be absorbed.

When a nutrient, water, or some other substance is absorbed, it crosses the intestine lining into the bloodstream. The vast absorptive surface area of the small intestine, with its villi and microvilli, helps make this process extremely efficient. **Segmentation** helps, too. In this process, rings of smooth muscle in the wall repeatedly contract and relax. This creates a back-and-forth movement that mixes digested material and forces it against the wall:

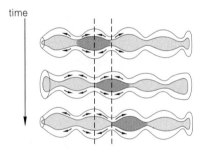

time

Enzyme	Released by:	Active in:	Breaks down:	Resulting Products
DIGESTING CARBOHYDRATES:				
Salivary amylase	Salivary glands	Mouth, stomach	Polysaccharides	Disaccharides, oligosaccharides
Pancreatic amylase	Pancreas	Small intestine	Polysaccharides	Disaccharides, monosaccharides
Disaccharidases	Intestinal lining	Small intestine	Disaccharides	MONOSACCHARIDES* (e.g., glucose)
DIGESTING PROTEINS:				
Pepsins	Stomach lining	Stomach	Proteins	Protein fragments
Trypsin and chymotrypsin	Pancreas	Small intestine	Proteins	Protein fragments
Carboxypeptidase	Pancreas	Small intestine	Peptides	AMINO ACIDS*
Aminopeptidase	Intestinal lining	Small intestine	Peptides	AMINO ACIDS*
DIGESTING FATS:				
Lipase	Pancreas	Small intestine	Triglycerides	FREE FATTY ACIDS, MONOGLYCERIDES*
DIGESTING UCLEIC ACIDS:				
Pancreatic nucleases	Pancreas	Small intestine	DNA, RNA	NUCLEOTIDES*
Intestinal nucleases	Intestinal lining	Small intestine	Nucleotides	NUCLEOTIDE BASES, MONOSACCHARIDES*

Table 7.2   Major Enzymes of Digestion and What They Do

*Products small enough to be absorbed into the internal environment.

THE DIGESTIVE SYSTEM

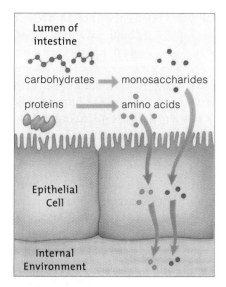

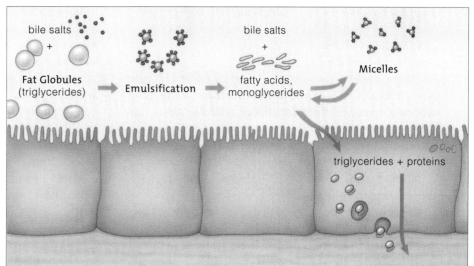

(a) Enzymes from the pancreas and from cells of the intestine lining complete the digestion of carbohydrates to monosaccharides and proteins to amino acids.

(b) Monosaccharides and amino acids are actively transported into epithelial cells, then move out of the cells and into the bloodstream.

(c) The constant movement of the intestine wall breaks up fat globules into small droplets (emulsification). Bile salts prevent the globules from re-forming. Pancreatic enzymes digest the droplets to fatty acids and monoglycerides.

(d) Micelles form as bile salts combine with nutrient molecules and phospholipids. Nutrients can easily slip into and out of the micelles.

(e) Concentrating monoglycerides and fatty acids in micelles creates steeper concentration gradients. Both substances then diffuse into cells of the intestine lining.

(f) Triglycerides re-form in cells of the intestine lining. They become coated with proteins, then are moved (by exocytosis) into the internal environment. They move into lacteals and then into the blood.

Figure 7.10 *Animated!* Digestion and absorption in the small intestine.

DIFFERENT NUTRIENTS ARE ABSORBED BY DIFFERENT MECHANISMS

By the time food is halfway through your small intestine, most of it has been broken apart and digested. Water crosses the intestine lining by osmosis, and cells in the lining also selectively absorb minerals. Figure 7.10 diagrams what happens with other kinds of nutrients.

For instance, transport proteins in the plasma membrane of brush border cells actively move some nutrients, such as the monosaccharide glucose and amino acids, across the lining. After glucose and amino acids are absorbed, they move directly into blood vessels.

By contrast, even after fat globules are emulsified, several more steps are required to move fatty acids and monoglycerides into the bloodstream (Figure 7.10d–f). Both of these kinds of molecules from digested lipids have hydrophobic regions, so they don't dissolve in watery chyme. Instead, the molecules clump with bile salts, along with cholesterol and other substances, and form tiny droplets. Each droplet is a *micelle* (my-CELL).

The molecules inside micelles constantly exchange places with those in chyme. However, the micelles concentrate them next to the intestine lining. When they are concentrated enough, nutrient molecules diffuse out of the micelles and into epithelial cells. Inside an epithelial cell, fatty acids and monoglycerides quickly reunite into triglycerides. Then triglycerides combine with proteins into particles that leave the cells by exocytosis and enter tissue fluid.

Unlike glucose and amino acids, when triglycerides are absorbed they do not move immediately into blood vessels. First they cross into lymph vessels called **lacteals**, which drain into the general circulation.

In the small intestine, most large organic molecules are digested to smaller molecules that can be absorbed.

Pancreatic enzymes secreted into the small intestine act on carbohydrates, fats, proteins, and nucleic acids in chyme. Bile salts emulsify fats (triglycerides), allowing them to be more easily digested.

Substances pass through brush border cells that line the surface of each villus by osmosis, active transport, or diffusion across plasma membranes.

7.7 The Large Intestine

Anything not absorbed in the small intestine moves into the large intestine, which absorbs water and some nutrients and eliminates wastes.

The **large intestine** is about 1.2 meters (5 feet) long (Figure 7.11). It begins as a blind pouch called the *cecum*. The cecum merges with the **colon**, which is divided into four regions in an inverted U-shape. The ascending colon travels up the right side of the abdomen, the transverse colon continues across to the left side, and the descending colon then turns downward. The sigmoid colon makes an S-curve and connects with the **rectum**.

Cells in the colon's lining actively transport sodium ions out of the tube. When the ion concentration there falls, water moves out by osmosis. As water is removed and returned to the bloodstream, the material left in the colon is gradually concentrated into *feces*, a mixture of undigested and unabsorbed matter, bacteria, and a little water. It is stored and finally eliminated. The typical brown color of feces comes mainly from bile pigments.

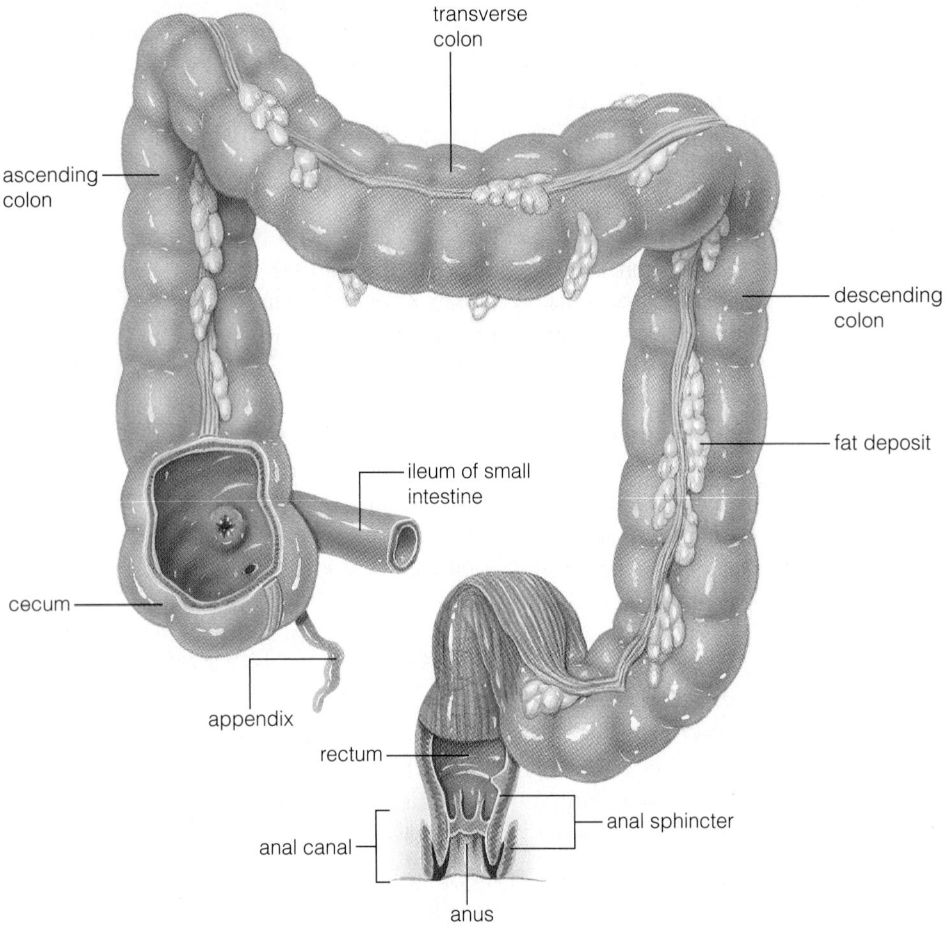

Bacteria make up about 30 percent of the dry weight of feces. Such microorganisms, including *Escherichia coli*, normally inhabit our intestines and are nourished by the food residues there. Their metabolism produces useful fatty acids and some vitamins (such as vitamin K), which are absorbed into the bloodstream. Feces of humans and other animals also can contain disease-causing organisms. Health officials use evidence of such "coliform bacteria," including *E. coli*, in water and food supplies as a measure of fecal contamination.

Your **appendix** projects from the cecum like the little finger of a glove. No one has ever discovered a digestive function for it, but, like the ileum of the small intestine, the appendix is colonized by defensive cells that combat bacteria you may have consumed in food. Feces that become wedged in the appendix can cause **appendicitis**. If an inflamed appendix isn't removed right away, it can burst and spew bacteria into the abdominal cavity where they can cause the life-threatening infection **peritonitis**.

Short, lengthwise bands of smooth muscle in the colon wall are gathered at their ends, like full skirts nipped in at elastic waistbands. As they contract and relax, material in the colon moves back and forth against the wall's absorptive surface. Shortly after you eat, hormone signals and nervous system commands direct large portions of the ascending and transverse colon to contract at the same time. Within a few seconds, residues in the colon may move as much as three-fourths of the colon's length and make way for incoming food. When feces distend the rectal wall, the stretching triggers defecation—elimination of feces from the body. From the rectum feces move into the **anal canal**. The nervous system also controls defecation. It can stimulate or inhibit contractions of sphincter muscles at the **anus**, the terminal opening of the GI tract.

transverse colon

ascending colon

descending colon

fat deposit

ileum of small intestine

cecum

appendix

rectum

anal sphincter

anal canal

anus

Figure 7.11 *Animated!* The large intestine.

> In the large intestine, water and salts are reabsorbed from food residues entering from the small intestine. The remaining concentrated residues are stored and later eliminated as feces.

THE DIGESTIVE SYSTEM

7.8 Managing Digestion and the Processing of Nutrients

As you know, homeostatic controls counter shifts in the internal environment. In digestion, however, controls act in the GI tract, before digested nutrients enter the bloodstream.

NERVES AND HORMONES REGULATE DIGESTION

The nervous system and endocrine system jointly control digestion (Figure 7.12). These controls are sensitive to two factors: the amount of food in the GI tract and the food's chemical makeup.

Food entering the stomach stretches the stomach walls, and then those of the small intestine. This stretching triggers signals from sensory receptors in the walls. Some of the signals give you (by way of processing in your brain) that "full" feeling after you eat. Others can lead to the muscle contractions of peristalsis or the release of digestive enzymes and other substances. Centers in the brain coordinate these activities with factors such as how much blood is flowing to the small intestine, where nutrients are being absorbed.

There are several types of endocrine cells in the GI tract. For example, one type secretes the hormone gastrin into the bloodstream when the stomach contains protein. Gastrin mainly stimulates the release of hydrochloric acid (HCl), which you may recall is a key ingredient in gastric juice. After the stomach empties, the increased acidity there causes another type of endocrine cell to release somatostatin, which shuts down HCl secretion so that conditions in the stomach are less acid. Notice that this is an example of negative feedback.

Endocrine cells in the small intestine also release hormones. One of them, secretin, signals the pancreas to release bicarbonate when acid enters the duodenum. When fat enters the small intestine, a hormone called

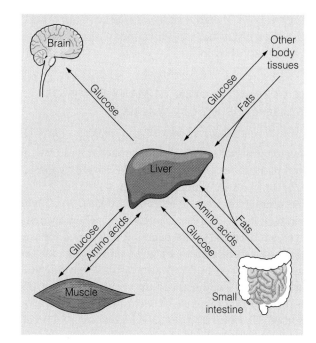

Figure 7.13 The liver's central role in managing nutrients.

CCK (for cholecystokinin) is released. CCK spurs the pancreas to release enzymes and triggers gallbladder contractions that deliver bile into the small intestine. Secretin and CCK also slow the rate at which the stomach empties—the mechanism mentioned in Section 7.3 that prevents food from entering the small intestine faster than it can be processed there. Yet another hormone, GIP (for glucose insulinotropic peptide) is released when fat and glucose are in the small intestine. Its roles include stimulating the release of insulin (from the pancreas), which is required for cells to take up glucose.

After nutrients are absorbed, the blood carries them to the liver, as described in Section 7.5. The liver is like a central shipping, storage, and receiving center. When glucose arrives from the small intestine, it is either shipped back out to the brain and other tissues, or stored as glycogen. Arriving fats may be stored, or used to make lipoproteins and other needed molecules. Liver cells also assemble amino acids into various proteins or process and reship them in a form cells can use to make ATP. Figure 7.13 is a simple visual summary of these activities. They are vital to maintain the body's supply of molecules cells can use as fuel, as building blocks, or in other ways.

Signals from the nervous system and the endocrine system control activity in the digestive system.

When absorbed nutrients reach the liver, they are sent on to the general circulation, stored, or converted to other forms for use in body cells.

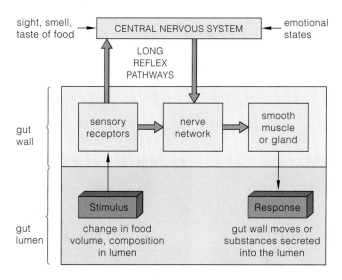

Figure 7.12 Controls over the digestive tract.

7.9 Digestive System Disorders

A disorder of the digestive system is a serious matter if it hampers processes that supply the body with nutrients or other needed substances.

THE GI TRACT IS OPEN TO MANY KINDS OF DISEASE-CAUSING ORGANISMS

The GI tract opens to the outside world, so it is a convenient portal into the body for bacteria, viruses, and other pathogens. Contaminated foods and water also can bring in harmful microorganisms.

A common effect of an intestinal infection is **diarrhea**, or watery feces. Diarrhea can develop when an irritant (such as a bacterial toxin) causes the lining of the small intestine to secrete more water and salts than the large intestine can absorb. It can also result when infections, stress, or other factors speed up peristalsis in the small intestine, so that there isn't time for enough water to be absorbed. Diarrheal diseases are dangerous in part because they dehydrate the body, depleting water and salts nerve and muscle cells need to function properly.

If you have ever had a case of "food poisoning," your stomach or intestines have been colonized by bacteria such as *Salmonella*, which can contaminate meat, poultry, and eggs (Figure 7.14*a*). Humans also are susceptible to several harmful strains of *E. coli* bacteria (Figure 7.14*b*). One of them, called O157:H7, inhabits the intestines of cattle. If a person eats ground beef or some other food that is contaminated with this microbe, it can cause a dangerous form of diarrhea that is coupled with anemia and may lead to kidney failure.

Bacteria that cause tooth decay (dental caries) flourish on food residues in the mouth, especially sugars (Figure 7.14*c*). Daily brushing and flossing are the best way to avoid a bacterial infection of the gums, which can lead to **gingivitis** (jin-juh-VY-tus). This inflammation can spread to the periodontal membrane that helps anchor a tooth in the jaw. Untreated periodontal disease can slowly destroy a tooth's bony socket, which can lead to loss of the tooth and other complications.

Section 7.3 mentioned **peptic ulcers**, open sores in the wall of the stomach or small intestine (Figure 7.14*d*) that are caused by the bacterium *Helicobacter pylori*. This microbe also is responsible for some cases of **gastritis** (inflammation of the GI tract), and even stomach cancer.

COLON DISORDERS RANGE FROM INCONVENIENT TO LIFE-THREATENING

It is normal to "move the bowels," or defecate, anywhere from three times a day to once a week. In **constipation**, food residues remain in the colon for too long, too much water is reabsorbed, and the feces become dry, hard, and difficult to eliminate. Constipation is uncomfortable, and it is a common cause of the enlarged rectal blood vessels known as hemorrhoids.

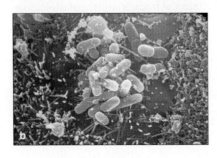

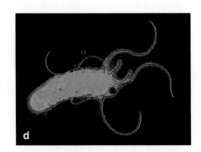

Figure 7.14 Some bacteria that infect the GI tract.
(**a**) Raw poultry may harbor *Salmonella*. Kitchen tools, and the cook's hands, should be thoroughly washed after handling raw poultry.
(**b**) A disease-causing strain of *E. coli* on an intestinal cell.
(**c**) Various bacteria on a human tooth.
(**d**) *Helicobacter pylori*.

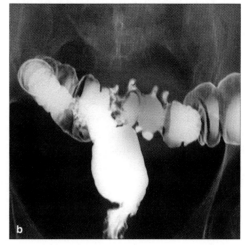

Figure 7.15 Fruits like grapes (**a**) are good sources of both soluble and insoluble fiber, which provides bulk in the diet and helps keep feces moving through the colon. (**b**) X-ray showing a colon in which knoblike diverticula (green areas) have developed.

Constipation is often due to a lack of bulk in the diet. *Bulk* is the volume of fiber (mainly cellulose from plant foods) and other undigested food material that is not decreased by absorption in the colon. Much of it consists of *insoluble fiber* such as cellulose and other plant compounds that humans cannot digest (we lack the required enzymes) and that does not easily dissolve in water. Wheat bran and the edible skins of fruits such as apples, plums, and grapes are just a few examples (Figure 7.15*a*). (*Soluble* fiber consists of plant carbohydrates such as fruit pectins that swell or dissolve in water.) If you chronically eat too little fiber, you are much more likely to be in the 50 percent of the U.S. population in whom the colon has formed diverticula—knoblike sacs where the inner colon lining protrudes through the intestinal wall. Inflammation of a diverticulum is called **diverticulitis**, and it can have quite serious complications, including peritonitis, if an inflamed diverticulum ruptures. A more common form of this disorder is **diverticulosis** (Figure 7.15*b*), in which diverticula are there but have not (yet) become inflamed.

Have you ever heard of someone having a "spastic colon"? This problematical condition also is known as **irritable bowel syndrome** (IBS), and it is the most common intestinal disorder. IBS often begins in early to mid-adulthood, and it affects twice as many women as men. Although the symptoms—abdominal pain and alternating diarrhea and constipation—are distressing, a medical examination rarely turns up signs of disease. While the direct cause of IBS symptoms is a disturbance in the smooth muscle contractions that move material through the colon, the reason for the change is not known.

Colon cancer is the #2 cancer diagnosis in the United States, second only to lung cancer, and it accounts for about 20 percent of all cancer deaths. Colon cancer often gets started when a growth called a polyp develops on the colon wall and becomes malignant. Fortunately, many cases of colon cancer and precancerous polyps can easily be detected by colonoscopy. After the patient is mildly sedated, a physician inserts a viewing tube into the colon and can examine it for polyps and other signs of disease.

The tendency to develop polyps, and colon cancer, can run in families, but usually there is no obvious genetic link. Because colon cancer is much more common in Western societies, some experts have proposed that the typical high-fat, low-fiber Western diet may be a factor, and there is a lot of active research on the issue. Chapter 23 looks in more detail at this and other forms of cancer.

MALABSORPTION DISORDERS PREVENT SOME NUTRIENTS FROM BEING ABSORBED

Anything that interferes with the ability of the small intestine to take up nutrients can lead to a *malabsorption disorder*. Many adults develop **lactose intolerance**, a mild disorder that results from deficiency of the enzyme lactase. It prevents normal breakdown and absorption of lactose, the sugar found in milk and many other milk products.

More serious malabsorption disorders are associated with some diseases that affect the pancreas, including the genetic condition **cystic fibrosis** (CF). Patients with CF don't make the necessary pancreatic enzymes for normal digestion and absorption of fats and other nutrients. (CF also affects the lungs, as you will read in later chapters.) **Crohn's disease** is an inflammatory disorder that can so severely damage the intestinal lining that much of the intestine must be removed.

7.10 The Body's Nutritional Requirements

LINKS TO
SECTIONS
2.9–2.11

Diet definitely has a profound effect on body functions. So, what is an average person "supposed" to eat?

What happens to the nutrients we absorb? As you've read, they are burned as fuel to provide energy and used as building blocks to build and replace tissues. In this section we focus on the three main classes of nutrients—carbohydrates, lipids, and proteins—and we take a look at new guidelines for what makes up a healthy diet.

COMPLEX CARBOHYDRATES ARE BEST

There are many views on the definition of a "proper" diet, but on one point just about all nutritionists can agree: The healthiest carbohydrates are the "complex" ones such as starch—the type of carbohydrate in fleshy fruits, cereal grains, and legumes, including peas and beans.

Complex carbohydrates are easily broken down to glucose, the body's chief energy source. Foods rich in complex carbohydrates also usually are high in fiber, including the insoluble fiber that adds needed bulk to feces and helps prevent constipation (see Section 7.9). By contrast, simple sugars don't have much fiber, and they lack the vitamins and minerals of whole foods.

The average American eats up to two pounds of refined sugars per week. In packaged foods, these sugars often are disguised as corn syrup, corn sweeteners, and dextrose. Refined sugars represent "empty calories" because they add to our caloric intake, but meet no other nutritional needs. Highly refined carbohydrates also have a high *glycemic index*. This means that within minutes of being absorbed, refined carbohydrates cause a surge in the blood levels of sugar and insulin.

Circulating insulin makes cells take up glucose quickly, and it also prevents cells from using stored fat as fuel. At the same time, glucose that is not needed as fuel for cells is stored as fat. When blood sugar levels later fall, we feel hungry. So we eat more, secrete more insulin, and keep storing fat, mainly in the form of triglycerides. Over time, high triglyceride levels increase the risk of heart disease and type 2 diabetes.

THERE ARE GOOD FATS AND BAD FATS

The body can't survive without fats and other lipids. The phospholipid lecithin and the sterol cholesterol both are building blocks of cell membranes. Fat stored in adipose tissue serves as an energy reserve, cushions organs such as the eyes and kidneys, and provides insulation beneath the

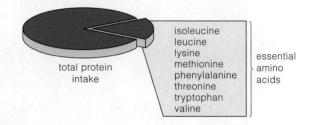

total protein intake

isoleucine
leucine
lysine
methionine
phenylalanine
threonine
tryptophan
valine

essential amino acids

Figure 7.16 The eight essential amino acids.

skin. The brain of a young child won't develop properly without a supply of cholesterol and saturated fat. The body also stores fat-soluble vitamins in adipose tissues.

The liver can manufacture most fats the body needs, including cholesterol, from protein and carbohydrates. The ones it cannot produce are **essential fatty acids**, but whole foods and vegetable oils provide plenty of them. Linoleic acid is an example. You can get enough of it by consuming just one teaspoon a day of corn oil, olive oil, or some other polyunsaturated fat.

Animal fats—the fat in butter, cheese, and fatty meat—are rich in saturated fats and cholesterol. Eating too much of these kinds of foods increases the risk for heart disease and stroke, as well as for certain cancers. The *trans fatty acids*, or "trans fats," are also bad for the cardiovascular system. Food labels are now required to show the amounts of trans fats, saturated fats, and cholesterol per serving. Section 9.8 has more information on "good" and "bad" forms of cholesterol.

PROTEINS ARE VITAL TO LIFE

When the digestive system digests and absorbs proteins, their amino acids become available for protein synthesis in cells. Of the twenty common amino acids, eight are **essential amino acids**. Our cells cannot make them, so we must obtain them from food. The eight are isoleucine, leucine, lysine, methionine, phenylalanine, threonine, tryptophan, and valine (Figure 7.16).

Most animal proteins are *complete*, meaning their ratios of amino acids match human nutritional needs. Nearly all plant proteins are *incomplete*, meaning they lack one or more of the essential amino acids. The proteins of quinoa (KEEN-wah) are a notable exception. In parts of the world where animal protein is a luxury, traditional cuisines include combinations of plant proteins, such as beans with rice, cornbread with chili, tofu with rice, and lentils with wheat bread.

USDA Nutritional Guidelines	
Food Group	Amount Recommended
Vegetables	2.5 cups/day
Dark green vegetables	3 cups/week
Orange vegetables	2 cups/week
Legumes	3 cups/week
Starchy vegetables	3 cups/week
Other vegetables	6.5 cups/week
Fruits	2 cups/day
Milk Products	3 cups/day
Grains	6 ounces/day
Whole grains	3 ounces/day
Other grains	3 ounces/day
Fish, poultry, lean meat	5.5 ounces/day
Oils	24 grams/day

Figure 7.17 From the United States Department of Agriculture, a summary of nutritional guidelines as of 2006. The recommended proportions add up to a daily caloric intake of 2,000 kilocalories for sedentary females between ages ten and thirty. Recommended intake and serving sizes are larger for males and highly active females and less for older females. The USDA recommends varying protein choices (fish, poultry, lean red meats, eggs, beans, nuts, and seeds).

GUIDELINES FOR HEALTHY EATING

The United States Food and Drug Administration has issued a set of nutritional guidelines to replace its earlier "food pyramid." These guidelines are based on nutritional research. They are designed to educate consumers about how a healthy diet can help reduce the risk for chronic diseases, such as diabetes, heart disease, hypertension (high blood pressure), and certain cancers. Figure 7.17 shows the recommended number of servings of various food groups. In comparison with the diet of a typical American, the guidelines call for a reduced intake of refined grains (such as white flour and white rice), trans fats and saturated fats, and refined sugars. The guidelines also suggest eating more whole grains, legumes, dark green and orange vegetables, fruits, and milk products. The full report can be downloaded from www.health.gov.

The new USDA guidelines call for about 55 percent of daily calories to come from complex carbohydrates, and recommends limiting total fat intake to 20 to 30 percent of daily caloric intake. About 4 ounces of lean meat—the rough equivalent of a small hamburger—is enough to meet minimum daily protein requirements.

There are respected alternative diets, however. One of them is the *Mediterranean diet*, which is associated with lower risk of heart disease, among other chronic ills. It emphasizes grains first, and then fruits and vegetables. Its main fat is olive oil, an excellent antioxidant. It limits weekly intakes of animal protein, eggs, and refined sugars, and places red meat at the pyramid's tiny tip.

In recent years, highly promoted diet plans that strictly limit carbohydrates and load up on proteins (and, often, fats) have become wildly popular. Although low-carb diets are controversial and their long-term effects on organs such as the kidneys are not known, millions of dieters swear by them because they can lead to extremely rapid weight loss.

A healthy diet must provide essential nutrients in the proper proportions and amounts.

Complex carbohydrates provide nutrients and fiber without adding "empty" calories.

Fats and other lipids are used for building cell membranes, energy stores, and other needs. Food must provide the essential fatty acids, which the body cannot synthesize.

Proteins are the source of essential amino acids.

7.11 Vitamins and Minerals

Vitamins are organic substances that are essential for growth and survival. No other substances can play their metabolic roles. In the course of evolution, animal cells have lost the ability to synthesize these substances, so we must obtain vitamins from food.

At a minimum, our cells need the vitamins listed in Table 7.3. Each vitamin has specific metabolic functions. Many chemical reactions use several types, and the absence of one affects the functions of others.

Minerals are inorganic substances that are essential for growth and survival. For instance, all of your cells need iron for their electron transport chains. Your red blood cells can't function without iron in hemoglobin, the oxygen-carrying pigment in blood. And neurons stop functioning without sodium and potassium (Table 7.4).

In general, people who are in good health and who eat a balanced diet of whole foods may get most of the vitamins and minerals they need. According to the

Table 7.3 Vitamins: Sources, Functions, and Effects of Deficiencies or Excesses*

Vitamin	Common Sources	Main Functions	Signs of Severe Long-Term Deficiency	Signs of Extreme Excess
FAT-SOLUBLE VITAMINS:				
A	Its precursor comes from beta carotene in yellow fruits, yellow or green leafy vegetables; also in fortified milk, egg yolk, fish liver	Used in synthesis of visual pigments, bone, teeth; maintains epithelia	Dry, scaly skin; lowered resistance to infections; night blindness; permanent blindness	Malformed fetuses; hair loss; changes in skin; liver and bone damage; bone pain
D	D_3 formed in skin and in fish liver oils, egg yolk, fortified milk; converted to active form elsewhere	Promotes bone growth and mineralization; enhances calcium absorption	Bone deformities (rickets) in children; bone softening in adults	Retarded growth; kidney damage; calcium deposits in soft tissues
E	Whole grains, dark green vegetables, vegetable oils	Possibly inhibits effects of free radicals; helps maintain cell membranes; blocks breakdown of vitamins A and C in gut	Lysis of red blood cells; nerve damage	Muscle weakness, fatigue, headaches, nausea
K	Colon bacteria form most of it; also in green leafy vegetables, cabbage	Blood clotting; ATP formation via electron transport	Abnormal blood clotting; severe bleeding (hemorrhaging)	Anemia; liver damage and jaundice
WATER-SOLUBLE VITAMINS:				
B_1 (thiamine)	Whole grains, green leafy vegetables, legumes, lean meats, eggs	Connective tissue formation; folate utilization; coenzyme action	Water retention in tissues; tingling sensations; heart changes; poor coordination	None reported from food; possible shock reaction from repeated injections
B_2 (riboflavin)	Whole grains, poultry, fish, egg white, milk	Coenzyme action	Skin lesions	None reported
Niacin	Green leafy vegetables, potatoes, peanuts, poultry, fish, pork, beef	Coenzyme action	Contributes to pellagra (damage to skin, gut, nervous system, etc.)	Skin flushing; possible liver damage
B_6	Spinach, tomatoes, potatoes, meats	Coenzyme in amino acid metabolism	Skin, muscle, and nerve damage; anemia	Impaired coordination; numbness in feet
Pantothenic acid	In many foods (meats, yeast, egg yolk especially)	Coenzyme in glucose metabolism, fatty acid and steroid synthesis	Fatigue, tingling in hands, headaches, nausea	None reported; may cause diarrhea occasionally
Folate (folic acid)	Dark green vegetables, whole grains, yeast, lean meats; colon bacteria produce some folate	Coenzyme in nucleic acid and amino acid metabolism	A type of anemia; inflamed tongue; diarrhea; impaired growth; mental disorders	Masks vitamin B_{12} deficiency
B_{12}	Poultry, fish, red meat, dairy foods (not butter)	Coenzyme in nucleic acid metabolism	A type of anemia; impaired nerve function	None reported
Biotin	Legumes, egg yolk; colon bacteria produce some	Coenzyme in fat, glycogen formation, and amino acid metabolism	Scaly skin (dermatitis), sore tongue, depression, anemia	None reported
C (ascorbic acid)	Fruits and vegetables, especially citrus, berries, cantaloupe, cabbage, broccoli, green pepper	Collagen synthesis; possibly inhibits effects of free radicals; structural role in bone, cartilage, and teeth; role in carbohydrate metabolism	Scurvy, poor wound healing, impaired immunity	Diarrhea, other digestive upsets; may alter results of some diagnostic tests

*The guidelines for appropriate daily intakes are being worked out by the Food and Drug Administration.

Mineral	Common Sources	Main Functions	Signs of Severe Long-Term Deficiency	Signs of Extreme Excess
Calcium	Dairy products, dark green vegetables, dried legumes	Bone, tooth formation; blood clotting; neural and muscle action	Stunted growth; possibly diminished bone mass (osteoporosis)	Impaired absorption of other minerals; kidney stones in susceptible people
Chloride	Table salt (usually too much in diet)	HCl formation in stomach; contributes to body's acid–base balance; neural action	Muscle cramps; impaired growth; poor appetite	Contributes to high blood pressure in susceptible people
Copper	Nuts, legumes, seafood, drinking water	Used in synthesis of melanin, hemoglobin, and some electron transport chain components	Anemia, changes in bone and blood vessels	Nausea, liver damage
Fluorine	Fluoridated water, tea, seafood	Bone, tooth maintenance	Tooth decay	Digestive upsets; mottled teeth and deformed skeleton in chronic cases
Iodine	Marine fish, shellfish, iodized salt, dairy products	Thyroid hormone formation	Enlarged thyroid (goiter), with metabolic disorders	Goiter
Iron	Whole grains, green leafy vegetables, legumes, nuts, eggs, lean meat, molasses, dried fruit, shellfish	Formation of hemoglobin and cytochrome (electron transport chain component)	Iron-deficiency anemia, impaired immune function	Liver damage, shock, heart failure
Magnesium	Whole grains, legumes, nuts, dairy products	Coenzyme role in ATP-ADP cycle; roles in muscle, nerve function	Weak, sore muscles; impaired neural function	Impaired neural function
Phosphorus	Whole grains, poultry, red meat	Component of bone, teeth, nucleic acids, ATP, phospholipids	Muscular weakness; loss of minerals from bone	Impaired absorption of minerals into bone
Potassium	Diet provides ample amounts	Muscle and neural function; roles in protein synthesis and body's acid–base balance	Muscular weakness	Muscular weakness, paralysis, heart failure
Sodium	Table salt; diet provides ample to excessive amounts	Key role in body's acid–base balance; roles in muscle and neural function	Muscle cramps	High blood pressure in susceptible people
Sulfur	Proteins in diet	Component of body proteins	None reported	None likely
Zinc	Whole grains, legumes, nuts, meats, seafood	Component of digestive enzymes; roles in normal growth, wound healing, sperm formation, and taste and smell	Impaired growth, scaly skin, impaired immune function	Nausea, vomiting, diarrhea; impaired immune function and anemia

*The guidelines for appropriate daily intakes are being worked out by the Food and Drug Administration.

American Medical Association, however, many physicians now recommend that even healthy people can benefit from certain vitamin and mineral supplements, in moderation. For example, vitamins E, C, and A lessen some aging effects and can improve immune function by inactivating free radicals. (A free radical, remember, is an atom or group of atoms that is highly reactive because it has an unpaired electron.) Vitamin K supplements help older women retain calcium and diminish the loss of bone due to osteoporosis.

However, metabolism varies in its details from one person to the next, so no one should take massive doses of any vitamin or mineral supplement except under medical supervision. Also, excessive amounts of many vitamins and minerals can harm anyone. For example, very large doses of the fat-soluble vitamins A and D can accumulate in tissues, especially in the liver, and interfere with normal metabolism. And although sodium has roles in the body's salt–water balance, muscle activity, and nerve function, prolonged, excessive intake of sodium may contribute to high blood pressure in some people.

Severe shortages or self-prescribed, massive excesses of vitamins and minerals can disturb the delicate balances in body function that promote health.

7.12 Calories Count: Food Energy and Body Weight

Attitudes about body weight often are cultural, but excess weight also raises real health issues.

The "fat epidemic" described in this chapter's introduction is spreading around the world. In the United States alone, about 300,000 people die each year due to preventable, weight-related conditions. Lifestyles are becoming more sedentary, and many people simply are eating more: Studies show that since the 1970s portion sizes in most restaurants have doubled. This is one reason why the FDA guidelines noted in Section 7.10 don't say "servings" of food but specify amounts instead.

The scientific standard for body weight is based on the ratio of weight to height (Figure 7.18). A person who is overweight has a higher than desirable weight-for-height. **Obesity** is an excess of body fat—more than 20 percent for males, and 24 percent for females. The World Health Organization has declared obesity a major global health concern, in part because its harmful effects on health are so serious—increasing not only the risk of type 2 diabetes and heart disease, but also osteoarthritis, high blood pressure, kidney stones, and many other ailments.

One indicator of weight-related health risk is the *body mass index* (BMI). It is determined by the formula

$$\text{BMI} = \frac{\text{weight (pounds)} \times 700}{\text{height (inches)}^2}$$

If your BMI value is 27 or higher, the health risk rises dramatically. Other risk factors include smoking, a genetic predisposition for heart disease, and fat stored above the waist (having an "apple shape" or "beer belly").

When someone is overweight, the usual culprit is a chronically unbalanced "energy equation" in which too many food calories are taken in while too few calories are burned. We measure food energy in **kilocalories** (kcal). A kilocalorie is 1,000 calories of heat energy. (Calorie, with a capital "C," is shorthand for a kilocalorie.) A value called **basal metabolic rate** (BMR) measures the amount of energy needed to sustain basic body functions. As a general rule, the younger you are, the higher your BMR. But BMR also varies from person to person, and it is influenced by the amount of muscle tissue in the body, emotions, hormones, and differences in physical activity. Adding BMR to the kcal needed for other demands (such as body movements) gives the total amount of food energy you need to fuel your daily life.

To figure out how many kcal you should take in daily to maintain a desired weight, multiply that weight (in pounds) by 10 if you are sedentary, by 15 if you are fairly active, and by 20 if highly active. From the value you get this way, subtract the following amount:

Age	20–34	Subtract	0
	35–44		100
	45–54		200
	55–64		300
	Over 65		400

For instance, if you want to weigh 120 pounds and are very active, $120 \times 20 = 2,400$ kilocalories. If you are 35 years old and moderately active, then you should take in a total of $1,800 - 100$, or 1,700 kcal a day. Along with

Figure 7.18 How to estimate the "ideal" weight for an adult. The values given are consistent with a long-term Harvard study into the link between excessive weight and increased risk of heart disorders. Depending on certain factors, such as having a small, medium, or large skeletal frame, the "ideal" may vary by plus or minus 10 percent.

Weight Guidelines for Women

Starting with an ideal weight of 100 pounds for a woman who is 5 feet tall, add five additional pounds for each additional inch of height. Examples:

Height (feet)	Weight (pounds)
5' 2"	110
5' 3"	115
5' 4"	120
5' 5"	125
5' 6"	130
5' 7"	135
5' 8"	140
5' 9"	145
5' 10"	150
5' 11"	155
6'	160

Weight Guidelines for Men

Starting with an ideal weight of 106 pounds for a man who is 5 feet tall, add six additional pounds for each additional inch of height. Examples:

Height (feet)	Weight (pounds)
5' 2"	118
5' 3"	124
5' 4"	130
5' 5"	136
5' 6"	142
5' 7"	148
5' 8"	154
5' 9"	160
5' 10"	166
5' 11"	172
6'	178

Table 7.5 Calories Expended in Some Common Activities

Activity	Kcal/hour per pound of body weight	Hours needed to lose 1 lb. fat 120 lbs	155 lbs	185 lbs
Basketball	3.78	7.7	6.0	5.0
Cycling (9 mph)	2.70	10.8	8.4	7.0
Hiking	2.52	11.6	8.9	7.5
Jogging	4.15	7.0	5.4	4.5
Mowing lawn (push mower)	3.06	9.5	7.4	6.2
Racquetball	3.90	7.5	5.8	4.8
Running (9-minute mile)	5.28	5.5	4.3	3.6
Snow skiing (cross-country)	4.43	6.6	5.1	4.3
Swimming (slow crawl)	3.48	8.4	6.5	5.4
Tennis	3.00	9.7	7.5	6.3
Walking (moderate pace)	2.16	13.5	10.4	8.7

To calculate these values for your own body weight, first multiply your weight by the kcal/hour expended for an activity to determine total kcal you use during one hour of the activity. Then divide that number into 3,500 (kcal in a pound of fat) to obtain the number of hours you must perform the activity to burn a pound of body fat.

this rough estimate, factors such as height and gender also must be considered. Males tend to have more muscle and so burn more calories (they have a higher BMR); hence an active woman needs fewer kilocalories than an active man of the same height and weight. Nor does she need as many as another active woman who weighs the same but is several inches taller.

GENES, WEIGHT CONTROL, AND EXERCISE

You've probably noticed that some people have a lot more trouble keeping off excess weight than others do. Although various factors influence body weight, recent research has shown that genes play a major role. As you will read later in this textbook, there are different chemical versions of many genes, and each version may have a slightly different effect. Scientists have identified several dozen genes that govern hormones, such as leptin and ghrelin, that influence appetite, hunger, how the body stores fat, and other weight-related factors. It may be that differences among genes help explain why some people stay slim no matter what and how often they eat, while others wage a lifelong struggle with extra pounds.

Regardless of genes, for most people maintaining a healthy weight over the years requires balancing their "energy budget" so that energy in—calories in food— equals energy used by our cells. Losing a pound of fat requires expending about 3,500 kcal. Although weight-loss diets may accomplish this deficit temporarily, over the long haul keeping off excess weight means pairing a moderate reduction in caloric intake with an increase in physical activity (Table 7.5). Exercise also increases the

mass of skeletal muscles, and even at rest muscle burns more calories than other types of tissues.

Emotions can influence weight gain and loss, sometimes to extremes. People who suffer from **anorexia nervosa** see themselves as fat no matter how thin they become. An anorexic purposely starves and may overexercise as well. Most common among younger women, anorexia can be fatal. Another extreme is the binge–purge disorder called **bulimia**. The term means "having an oxlike appetite." A bulimic might consume as much as 50,000 calories at one sitting and then purposely vomit, take a laxative, or both. Chronic vomiting can erode away the enamel from a person's teeth (due to stomach acid) and rupture the stomach. In severe cases it also can cause chemical imbalances that lead to heart and kidney failure.

IMPACTS, ISSUES

A hormone called PYY is one of the appetite regulators. It is released by glands in the stomach and small intestine after a meal. It acts in the brain to suppress appetite, countering the effects of ghrelin. Volunteers who were given an intravenous dose of PYY before a buffet meal ate less than a control group that got only a saline solution. A nasal spray that delivers the hormone into the bloodstream to suppress appetite is being tested in clinical trials.

To maintain an acceptable body weight, energy input (caloric intake) must be balanced with energy output in the form of metabolic activity and exercise.

Basal metabolic rate, physical activity, age, hormones, and emotions all influence the body's energy use.

Genes govern the hormones that influence appetite, hunger, and how the body stores and uses energy.

Summary

Section 7.1 The digestive system has five main activities.

a. Mechanical processing and motility: Chewing and muscle movements break up, mix, and propel ingested food through the system.

b. Secretion: Digestive enzymes and other substances are released from the salivary glands, pancreas, liver, and glandular epithelium into the digestive tube.

c. Digestion: Food is broken down into particles, then into nutrient molecules small enough to be absorbed.

d. Absorption: Digested organic compounds, fluid, and ions pass into the internal environment.

e. Elimination: Undigested and unabsorbed residues are expelled at the end of the system.

The gastrointestinal tract includes the mouth, pharynx, esophagus, stomach, small intestine, and large intestine. Its associated accessory organs include salivary glands, the liver, the gallbladder, and the pancreas (Table 7.6).

The GI tract is lined with mucous membrane. From the esophagus onward its wall consists of four layers: an innermost mucosa, then the submucosa, then smooth muscle, then the serosa. Sphincters at either end of the stomach and at other locations control the forward movement of ingested material.

 Tour the human digestive system.

Section 7.2 Starch digestion begins in the mouth or oral cavity, where the salivary glands secrete saliva, which contains salivary amylase. Chewed food mixes with saliva to form a bolus that is swallowed. Waves of peristalsis move each bolus down the esophagus to the stomach.

Section 7.3 Protein digestion begins in the stomach, where gastric fluid containing pepsins and other substances is secreted. The stomach contents are reduced to a watery chyme that passes through a sphincter into the small intestine.

Section 7.4 Digestion is completed and most nutrients are absorbed in the small intestine, which has a large surface area for absorption due to its many villi and microvilli.

Section 7.5 Enzymes and some other substances secreted by the pancreas, the liver, and the gallbladder aid digestion. Bile (secreted by the liver and then stored and released into the small intestine by the gallbladder) contains bile salts that speed up the digestion of fats. Micelles aid the absorption of fatty

acids and triglycerides. A hepatic portal vein carries nutrient-laden blood to the liver for processing.

Section 7.6 In the small intestine, a process of segmentation mixes material and forces it close to the absorptive surface. Absorbed glucose and amino acids move into blood vessels in intestinal villi. Triglycerides enter lacteals, then move into blood vessels.

Section 7.7 Peristalsis moves wastes into the large intestine. Water is reabsorbed in the colon; wastes (feces) move on to the rectum and into the anal canal and are eliminated via the anus. The appendix projects from the upper part of the large intestine. It may have a role in immunity.

Section 7.8 The nervous and endocrine systems govern the digestive system. Many controls operate in response to the volume and composition of food in the gut. They cause changes in muscle activity and in the secretion rates of hormones or enzymes.

Section 7.10 Complex carbohydrates are the body's preferred energy source. The diet also must provide

Table 7.6	Summary of the Digestive System
MOUTH (oral cavity)	Start of digestive system, where food is chewed, moistened, polysaccharide digestion begins
PHARYNX	Entrance to tubular parts of digestive and respiratory systems
ESOPHAGUS	Muscular tube, moistened by saliva, that moves food from pharynx to stomach
STOMACH	Sac where food mixes with gastric fluid and protein digestion begins; stretches to store food taken in faster than can be processed; gastric fluid destroys many microbes
SMALL INTESTINE	The first part (duodenum) receives secretions from the liver, gallbladder, and pancreas
	Most nutrients are digested, absorbed in second part (jejunum)
	Some nutrients absorbed in last part (ileum), which delivers unabsorbed material to colon
COLON (large intestine)	Concentrates and stores undigested matter (by absorbing mineral ions and water)
RECTUM	Distension triggers expulsion of feces
ANUS	Terminal opening of digestive system
Accessory Organs:	
SALIVARY GLANDS	Glands (three main pairs, many minor ones) that secrete saliva, a fluid with polysaccharide-digesting enzymes, buffers, and mucus (which moistens and lubricates ingested food)
PANCREAS	Secretes enzymes that digest all major food molecules and buffers against HCl from stomach
LIVER	Secretes bile (used in fat emulsification); role in carbohydrate, fat, and protein metabolism
GALLBLADDER	Stores and concentrates bile from the liver

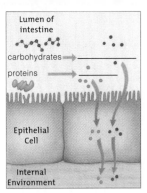
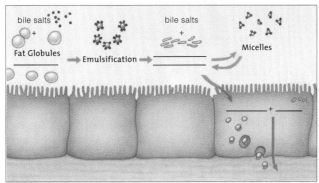

Figure 7.19 Fill in the blanks for substances that cross the lining of the small intestine.

eight essential amino acids, some essential fatty acids, vitamins, and minerals.

Section 7.11 Vitamins and minerals both are essential for normal body growth and functioning. Vitamins are organic substances; minerals are inorganic.

Section 7.12 Food energy is measured in kilocalories. The basal metabolic rate is the amount of kilocalories needed to sustain the body when a person is awake and resting. To maintain acceptable weight and overall health, a person's total energy output must balance caloric intake. Obesity is a health-threatening condition that increases the risk of type 2 diabetes, heart trouble, and some cancers, among other diseases and disorders.

Review Questions

1. What are the main functions of the stomach? The small intestine? The large intestine?

2. Using the sketch below, list the organs and accessory organs of the digestive system. On a separate piece of paper, list the main functions of each.

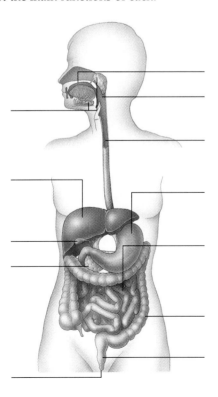

3. Define peristalsis, and list the regions of the GI tract where it occurs. Mention segmentation in your answer.

4. Using the black lines shown in Figure 7.19, name the types of molecules small enough to be absorbed across the small intestine's lining.

Self-Quiz

Answers in Appendix V

1. Different regions of the digestive system specialize in _____ and _____ food and in _____ unabsorbed food residues.

2. Maintaining normal body weight requires that _____ intake be balanced by _____ output.

3. The preferred energy sources for the body are _____.

4. The human body cannot produce its own vitamins or minerals, nor can it produce certain _____ and _____.

5. Which of the following is *not* associated with digestion?
   a. salivary glands   d. gallbladder
   b. thymus gland   e. pancreas
   c. liver

6. Digestion is completed and products are absorbed in the _____.
   a. mouth   c. small intestine
   b. stomach   d. large intestine

7. After absorption, triglycerides, fatty acids, and monoglycerides leave the cell and move into the _____.
   a. bloodstream   c. liver
   b. intestinal cells   d. lacteals

8. Excess carbohydrates and proteins are stored as _____.
   a. amino acids   c. fats
   b. starches   d. monosaccharides

9. Match the digestive system parts and functions.
   ____ liver   a. secrete substances that moisten
   ____ small intestine   food, start polysaccharide
   ____ salivary glands   breakdown
   ____ stomach   b. where protein digestion begins
   ____ large intestine   c. where water is reabsorbed
                           d. where most digestion is
                              completed
                           e. receives blood carrying
                              absorbed nutrients

Critical Thinking

1. A glass of whole milk contains lactose, protein, triglycerides (in butterfat), vitamins, and minerals. Explain what happens to each component when it passes through your digestive tract.

2. Some nutritionists claim that the secret to long life is to be slightly underweight as an adult. If a person's weight is related partly to diet, partly to activity level, and partly to genetics, what underlying factors could be at work to generate statistics that support this claim?

3. As a person ages, the number of body cells steadily decreases and energy needs decline. If you were planning an older person's diet, what kind(s) of nutrients would you emphasize, and why? Which ones would you recommend less of?

4. Along the lines of question 3, formulate a healthy diet for an actively growing seven-year-old.

5. Raw poultry can carry *Salmonella* or *Campylobacter* bacteria, both of which produce toxins that can cause serious diarrhea, among other symptoms. Aside from the discomfort, why does such an infection require immediate medical attention?

6. Dutch cyclist Leontien Zijlaard, the young woman shown in Figure 7.20, won three Olympic gold medals. Four years earlier, she was suffering from anorexia and too weak and malnourished to compete. Many recovered anorexics lead normal—even extraordinary—lives, and researchers are uncovering a wealth of new information about the disorder. Do some research yourself and find answers to the following questions:

Are eating disorders common or rare?

How many people die each year from anorexia and bulemia?

Is there evidence that genes influence these conditions?

Explore on Your Own

This is an exercise you can eat when you're done. All you need is a food item like a hamburger or a salad and paper for jotting notes.

To begin, analyze your meal, noting the various kinds of biological molecules it includes. (For this exercise, ignore nucleic acids.) Then, beginning with your mouth and teeth, write what happens to your meal as it moves through your digestive system. Key questions to consider include: What kinds of enzymes act on the different components of the meal (such as lettuce or meat), and where do they act, as it is digested? What mechanical processes aid digestion? Which ones can you consciously control? Using the tables in Section 7.11, list the vitamins and minerals that your meal likely contains. Finally, analyze your meal in terms of its contribution (or not) to a balanced diet.

Figure 7.20 Dutch cyclist Leontien Zijlaard, who recovered from anorexia and went on to win Olympic gold.

8 BLOOD

Chemical Questions

In 2002 a team of scientists at the Centers for Disease Control in Atlanta reported finding 116 pollutants in the blood and urine of more than 2,500 healthy people who had volunteered to be tested for contaminated body fluids. The volunteers were selected to provide a statistically reliable cross section of the U.S. population. Many of the pollutants that turned up were substances known or strongly suspected to be harmful—toxic metals, chemicals in secondhand cigarette smoke, residues of pesticides and herbicides, and by-products of manufacturing processes.

A similar study by researchers at the Environmental Working Group in Washington, D.C., found a whopping 167 contaminants in the body fluids of volunteers who reported no unusual exposure to polluting chemicals.

Few of the chemicals tracked in the tests even existed when you were born. For the most part they are recent inventions designed to enhance products ranging from lipstick to telephone equipment or to improve agricultural productivity.

Many researchers are concerned that too little is known about the health impacts of many synthetic chemicals. For example, in the CDC study the majority of

subjects, including children, had traces of phthalates in their fluids. These substances, which are used in cosmetics and plastics, are not regulated in the United States. Yet studies using laboratory animals have produced compelling evidence that phthalates cause cancer and abnormalities of the reproductive system.

How serious is the problem? In general, say environmental scientists, children and fetuses are most at risk, because many pollutants affect development. Also, little is known about the effect of long-term exposure to many synthetic chemicals. The metal lead is an example: Levels of lead in blood that were deemed safe in 1970 were later found to pose a serious health threat to children. Ultimately lead was banned for use in paints and some other products.

Our blood can transport substances good and not so good. In this chapter you will learn why blood truly is "the river of life"—and a key player in maintaining homeostasis.

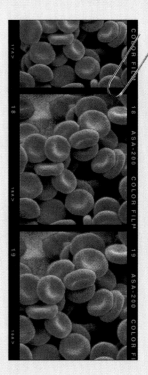

How Would You Vote? Government regulation of substances such as lead seems to be effective: In recent years the levels of several pollutants in the general population have fallen. Should other suspect industrial chemicals be regulated? Cast your vote online at www.thomsonedu.com/biology/starr/humanbio.

 Key Concepts

COMPONENTS AND FUNCTIONS OF BLOOD
Blood consists of watery plasma, red blood cells, white blood cells, and platelets.

Red blood cells transport O_2 and CO_2, white blood cells are part of body defenses, and platelets help clot blood. Blood also helps maintain a stable pH and body temperature.

BLOOD TYPES
Red blood cells have "self" proteins on their surface that establish a person's blood type.

BLOOD CLOTTING
Mechanisms that clot blood help prevent blood loss.

 Links to Earlier Concepts

This chapter is a prelude to our study of the cardiovascular system—the heart and blood vessels— in Chapter 9. It also expands on some topics you have already read about. For example, you will learn more about hemoglobin, the oxygen-carrying protein in red blood cells (2.12), and about the various kinds of blood cells that arise from stem cells in bone marrow (5.2).

This chapter's discussion of blood typing shows a key function of recognition proteins that are embedded in cell plasma membranes (3.3). Section 8.7 on blood clotting provides good examples of how enzymes catalyze chemical reactions that are vital to life.

8.1 Blood: Plasma, Blood Cells, and Platelets

LINKS TO
SECTIONS
2.6, 2.11, 3.9,
AND 7.5

As the old saying goes, human blood is thicker than water. It also flows more slowly, and it is rather sticky. But what exactly is this unusual liquid?

Blood consists of plasma, blood cells, and cell fragments called platelets. If you are an adult woman of average size, your body has about 4 to 5 liters of blood; males have slightly more. In all, blood amounts to about 6 to 8 percent of your body weight.

PLASMA IS THE FLUID PART OF BLOOD

If you fill a test tube with blood, treat it so it doesn't clot, and whirl it in a centrifuge, the tube's contents should look like what you see in Figure 8.1. About 55 percent of whole blood is **plasma**. Plasma is mostly water. It transports blood cells and platelets, and more than a hundred other substances. Most of these "substances" are different plasma proteins, which have a variety of functions.

Plasma proteins determine the fluid volume of the blood—how much of it is water. Two-thirds of plasma proteins are albumin molecules made in the liver. Because there is so much of it—that is, because its concentration is so high—albumin has a major influence on the osmotic movement of water into and out of blood. Albumin also carries many chemicals in blood, from metabolic wastes to therapeutic drugs. Too little albumin can be one cause of *edema*, swelling that occurs when water leaves the blood and enters tissues.

Other plasma proteins include protein hormones and proteins involved in immunity and blood clotting. Lipoproteins carry lipids, and still other plasma proteins transport fat-soluble vitamins.

Plasma also contains ions, glucose and other simple sugars, amino acids, various communication molecules, and dissolved gases—mostly oxygen, carbon dioxide, and nitrogen. The ions (such as Na^+, Cl^-, H^+, and K^+) help maintain the volume and pH of extracellular fluid.

RED BLOOD CELLS CARRY OXYGEN AND CO_2

About 45 percent of whole blood—the bottom portion in your centrifuged test tube—consists of erythrocytes, or **red blood cells**. Each red blood cell is a biconcave disk, like a thick pancake with a dimple on each side. The cell's red color comes from the iron-containing protein hemoglobin. Hemoglobin transports oxygen that the body requires for aerobic respiration. Red blood cells also carry away some carbon dioxide wastes.

red blood cell

white blood cell

platelets

Figure 8.1
Components of blood. In the micrograph the dark red cells are red blood cells. Platelets are pink. The fuzzy gold balls are white blood cells.

Components	Relative Amounts	Functions
Plasma Portion (*50%–60% of total volume*):		
1. Water	91%–92% of plasma volume	Solvent
2. Plasma proteins (albumin, globulins, fibrinogen, etc.)	7%–8%	Defense, clotting, lipid transport, roles in extracellular fluid volume, etc.
3. Ions, sugars, lipids, amino acids, hormones, vitamins, dissolved gases	1%–2%	Roles in extracellular fluid volume, pH, etc.
Cellular Portion (*40%–50% of total volume*):		
1. White blood cells:		
Neutrophils	3,000–6,750	Phagocytosis during inflammation
Lymphocytes	1,000–2,700	Immune responses
Monocytes (macrophages)	150–720	Phagocytosis in all defense responses
Eosinophils	100–360	Defense against parasitic worms
Basophils	25–90	Secrete substances for inflammatory response and for fat removal from blood
2. Platelets	250,000–300,000	Roles in clotting
3. Red blood cells	4,800,000–5,400,000 per microliter	Oxygen, carbon dioxide transport

COMPONENTS AND FUNCTIONS OF BLOOD

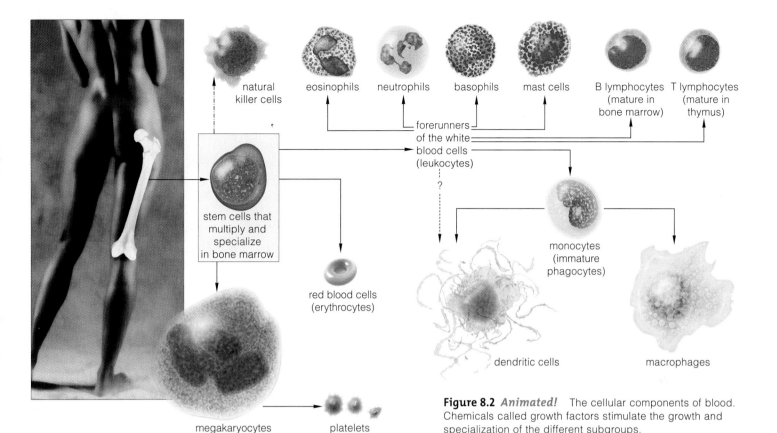

natural killer cells   eosinophils   neutrophils   basophils   mast cells   B lymphocytes (mature in bone marrow)   T lymphocytes (mature in thymus)

forerunners of the white blood cells (leukocytes)

?

stem cells that multiply and specialize in bone marrow

red blood cells (erythrocytes)

monocytes (immature phagocytes)

dendritic cells   macrophages

megakaryocytes   platelets

Figure 8.2 *Animated!* The cellular components of blood. Chemicals called growth factors stimulate the growth and specialization of the different subgroups.

Red blood cells arise from stem cells in bone marrow. You may recall that a **stem cell** stays unspecialized and retains the ability to divide. Some of the daughter cells, however, do become specialized for particular functions, as you can see in Figure 8.2.

WHITE BLOOD CELLS PERFORM DEFENSE AND CLEANUP DUTIES

Leukocytes, or **white blood cells**, make up a tiny fraction of whole blood. (With platelets, they are the thin, pale, middle layer in your test tube.) Leukocytes function in housekeeping and defense. Some scavenge dead or worn-out cells, or material identified as foreign to the body. Others target or destroy disease agents such as bacteria or viruses. Most go to work after they squeeze out of blood vessels and enter tissues. The number of them in the body varies, depending on whether a person is sedentary or highly active, healthy or fighting an infection.

All white blood cells develop from stem cells in bone marrow. In the various kinds of cells, the nucleus varies in its size and shape, and there are other differences as well. **Granulocytes** include neutrophils, eosinophils, and basophils. When this type of cell is stained, various types of granules are visible in its cytoplasm. The majority of leukocytes are neutrophils. They and eosinophils,

basophils, and mast cells have roles in body defenses that you will read more about in Chapter 10.

The leukocytes called **agranulocytes** don't have visible granules in their cytoplasm. One type, called monocytes, develops into macrophages, "big eaters" that engulf and destroy invading microbes and debris. Another type, lymphocytes—B cells, T cells, and natural killer cells—operates in immune responses. Most types of white blood cells live for only a few days or, during a major infection, perhaps a few hours. Others may live for years.

PLATELETS HELP CLOT BLOOD

Some stem cells in bone marrow develop into "giant" cells called megakaryocytes (mega = large). These cells shed bits of cytoplasm that become enclosed in a plasma membrane. The fragments, known as **platelets**, last only about a week, but millions are always circulating in our blood. Platelets release substances that begin the process of blood clotting described in Section 8.7.

Blood consists of plasma, in which proteins and other substances are dissolved; red blood cells; white blood cells; and platelets.

8.2 How Blood Transports Oxygen

A key function of blood is transporting oxygen, and the key to oxygen transport is the protein called hemoglobin.

HEMOGLOBIN IS THE OXYGEN CARRIER

If you were to analyze a liter of blood drawn from an artery, you would find only a quarter teaspoon of oxygen dissolved in the plasma—just 3 milliliters. Yet, like all large, active, warm-bodied animals, humans require a lot of oxygen to maintain the metabolic activity of their cells. Hemoglobin (Hb) meets this need. In addition to the small amount of dissolved oxygen, a liter of arterial blood usually carries around 65 times more O_2 bound to the heme groups of hemoglobin molecules. This oxygen-bearing hemoglobin is called **oxyhemoglobin**.

WHAT DETERMINES HOW MUCH OXYGEN HEMOGLOBIN CAN CARRY?

As conditions change in different tissues and organs, so does the tendency of hemoglobin to bind with and hold on to oxygen. Several factors influence this process. The most important factor is how much oxygen is present relative to the amount of carbon dioxide. Other factors are the temperature and acidity of tissues. Hemoglobin is most likely to bind oxygen in places where blood plasma contains a relatively large amount of oxygen, where the temperature is relatively cool, and where the pH is roughly neutral. This is exactly the environment in our lungs, where the blood must take on oxygen. By contrast, metabolic activity in cells *uses* oxygen. It also increases both the temperature and the acidity (lowers the pH) of tissues. Under those conditions, the oxyhemoglobin of red blood cells arriving in tissue capillaries tends to release oxygen, which then can enter cells. We can summarize these events this way:

LUNGS

TISSUES

more O_2
cooler
less acidic

Hb + O_2 ➡ HbO_2

HbO_2 ➡ Hb + O_2

less O_2
warmer
more acidic

The protein portion of hemoglobin also carries some of the carbon dioxide wastes cells produce, along with hydrogen ions (H^+) that affect the pH of body fluids. You'll read more about hemoglobin in Chapter 11, where we consider the many interacting elements that enable the respiratory system to transport gases efficiently to and from body cells.

You can see the structure of a hemoglobin molecule in Figure 8.3. Notice that it has two parts: the protein globin, and heme groups that contain iron. Globin is built of four linked polypeptide chains, and each chain

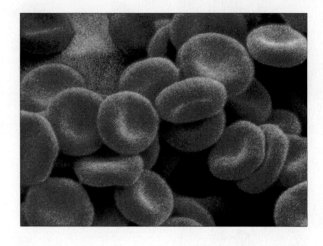

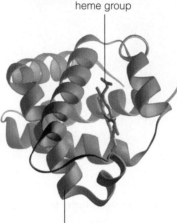

heme group

coiled and twisted
polypeptide chain of
one globin molecule

Figure 8.3 *Animated!*
The structure of hemoglobin. Recall from Chapter 2 that hemoglobin is a globular protein consisting of four polypeptide chains and four iron-containing heme groups. Oxygen binds to the iron in heme groups, which is one reason why humans require iron as a mineral nutrient.

is associated with a heme group. It is the iron molecule at the center of each heme group that binds oxygen.

Oxygen in the lungs diffuses into the blood plasma and then into individual red blood cells. There it binds with the iron in hemoglobin. This oxyhemoglobin is deep red. Hemoglobin that is depleted of oxygen looks scarlet, especially when it is observed through skin and the walls of blood vessels.

Hemoglobin in red blood cells transports oxygen. The oxygen is bound to iron molecules in heme groups in each hemoglobin molecule.

The relative amounts of oxygen and carbon dioxide present in blood, and the temperature and acidity of tissues, affect how much oxygen hemoglobin binds—and therefore the amount of oxygen available to tissues.

8.3 Hormonal Control of Red Blood Cell Production

Red blood cells do not live long. In response to hormones stem cells in bone marrow constantly produce new ones.

Each second, about 3 million new red blood cells enter your bloodstream. They gradually lose their nucleus and other organelles, structures that are unnecessary because red blood cells do not divide or synthesize new proteins.

Red blood cells have enough enzymes and other proteins to function for about 120 days. As they near the end of their life, die, or become damaged or abnormal, phagocytes called macrophages ("big eaters") remove them from the blood. Much of this "cleanup" occurs in the spleen, which is located in the upper left abdomen. As a macrophage dismantles a hemoglobin molecule, amino acids from its proteins return to the bloodstream and the iron in its heme groups returns to red bone marrow, where it may be recycled in new red blood cells. The rest of the heme group is converted to the orangish pigment bilirubin. Liver cells take up this pigment, which is mixed with bile that is released into the small intestine during digestion.

Steady replacements from stem cells in bone marrow keep a person's red blood cell count fairly constant over time. A **cell count** is a tally of the number of cells in a microliter of blood. On average, an adult male's red blood cell count is around 5.4 million. In an adult female the count averages about 4.8 million red blood cells.

Having a stable red blood cell count is important for homeostasis, because body cells need a reliable supply of oxygen. Your kidneys make the hormone erythropoietin, and it stimulates the production of new red blood cells as needed.

The process relies on a negative feedback loop (Figure 8.4). In this loop, the kidneys monitor the level of oxygen in the blood. When it falls below a set point, kidney cells detect the change and in short order they release erythropoietin. It stimulates stem cells in bone marrow to produce more red blood cells. As new red blood cells enter your bloodstream, the blood's capacity for carrying oxygen increases. As the oxygen level rises in your blood and in tissues, this information feeds back to the kidneys. They then make less erythropoietin, and the production of red blood cells in bone marrow drops.

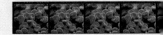
Cells in the kidneys monitor the oxygen-carrying capacity of blood. When more red blood cells are needed to carry oxygen, the kidneys release the hormone erythropoietin, which stimulates the production of new red blood cells by stem cells in bone marrow.

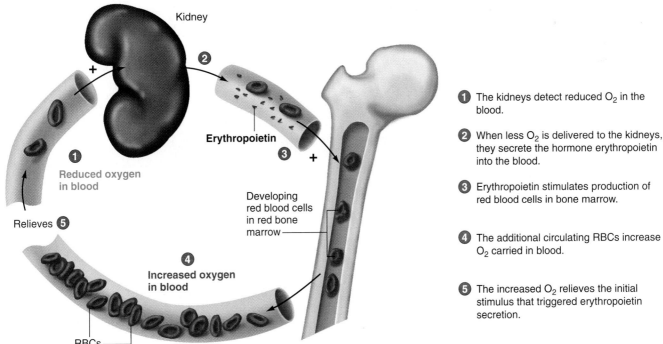

❶ The kidneys detect reduced O_2 in the blood.

❷ When less O_2 is delivered to the kidneys, they secrete the hormone erythropoietin into the blood.

❸ Erythropoietin stimulates production of red blood cells in bone marrow.

❹ The additional circulating RBCs increase O_2 carried in blood.

❺ The increased O_2 relieves the initial stimulus that triggered erythropoietin secretion.

Figure 8.4 The feedback loop that helps maintain a normal red blood cell count.

COMPONENTS AND FUNCTIONS OF BLOOD

8.4 Blood Types—Genetically Different Red Blood Cells

LINK TO
SECTION
3.2

You probably know that there are different human blood types. The differences are due to variations in the surface markers on red blood cells.

Each of your body cells has proteins on its surface that mark the cell as "self." Your genes have determined the chemical characteristics of these self markers, which vary from person to person. The variations are medically important because the markers on cells and substances that are *not* part of an individual's own body are antigens. An **antigen** is a chemical characteristic of a cell, particle, or substance that causes the immune system to mount an immune response. Defensive proteins called *antibodies* identify and attack antigens in a process that is a major topic of Chapter 10.

Our red blood cells bristle with self markers. To date biologists have identified at least 30 common ones, and many more rare ones. Because each kind of marker can have several forms, they are often called "blood groups." Two of them, the Rh blood group and the ABO blood group, are extremely important in situations where the blood of two people mixes. We will look at the Rh blood group in Section 8.5. For now, let's look more closely at the ABO blood group, which is a vital consideration in blood transfusions.

THE ABO GROUP OF BLOOD TYPES INCLUDES KEY SELF MARKERS ON RED BLOOD CELLS

One of our genes carries the instructions for building the ABO self markers on red blood cells. Different versions of this gene carry instructions for different markers, called type A and type B. A third version of the gene does not call for a marker, and red blood cells of someone who has this gene are dubbed type O. Collectively, these markers make up the ABO blood group.

In type A blood, red blood cells bear A markers. Type B blood has B markers, and type AB has both A and B. Type AB blood is quite rare, but a large percentage of people have type O red blood cells—they have neither A nor B markers. Depending on your ABO blood type, your blood plasm also will contain antibodies to other blood types, even if you have never been exposed to them. As you will read shortly, a severe immune response takes place when incompatible blood types are mixed. This is why donated blood must undergo a chemical analysis called **ABO blood typing** (Table 8.1).

MIXING INCOMPATIBLE BLOOD TYPES CAN CAUSE THE CLUMPING CALLED AGGLUTINATION

As you can see in Table 8.1, if you are type A, your body does not have antibodies against A markers but does have them against B markers. If you are type B, you don't have antibodies against B markers, but you do have antibodies against A markers. If you are type AB, you do not have antibodies against either form of the marker. If you are type O, however, you have antibodies against *both* forms of the marker, so you can only receive blood from another type O individual.

In theory, type O people are "universal donors," because they have neither A nor B antigens, and—again, only in theory—type AB people are "universal recipients." In fact, however, as already noted there are *many* markers

Blood Type	Antigens on Plasma Membranes of RBCs	Antibodies in Blood	Safe to Transfuse	
			To	From
A	A	Anti-B	A, AB	A, O
B	B	Anti-A	B, AB	B, O
AB	A + B	none	AB	A, B, AB, O
O	—	Anti-A Anti-B	A, B, AB, O	O

Table 8.1 Summary of ABO Blood Types

BLOOD TYPES

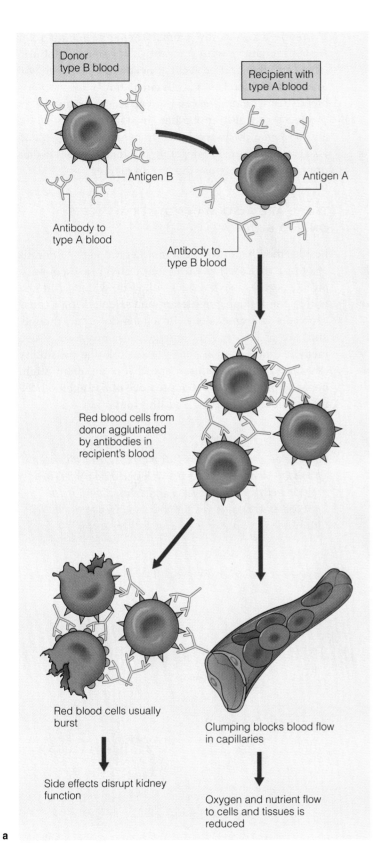

Donor
type B blood

Recipient with
type A blood

Antigen B

Antigen A

Antibody to
type A blood

Antibody to
type B blood

Red blood cells from
donor agglutinated
by antibodies in
recipient's blood

Red blood cells usually
burst

Clumping blocks blood flow
in capillaries

Side effects disrupt kidney
function

Oxygen and nutrient flow
to cells and tissues is
reduced

a

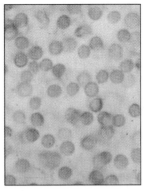

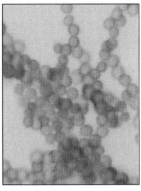

b

compatible blood cells

incompatible blood cells

Figure 8.5 *Animated!* Agglutination in red blood cells. (**a**) Example of an agglutination reaction. This diagram shows what happens when type B blood is transfused into a person who has type A blood. (**b**) What an agglutination reaction looks like. In the micrograph on the left, commingled red blood cells are compatible and have not clumped. The cells on the right are a mix of incompatible ABO types, and they have clumped together. Donated blood is typed in order to avoid an agglutination response when the blood is transfused into another person.

associated with our red blood cells, and any of them can trigger the defense response called **agglutination** (Figure 8.5). When the mixing of incompatible blood causes agglutination, antibodies act against the "foreign" cells and cause them to clump. The clumps can clog small blood vessels, severely damaging tissues throughout the body and sometimes even causing death.

We turn next to the Rh blood group. As you will now read, agglutination is also a danger when mismatched Rh blood types mix.

Like all cells, red blood cells bear genetically determined proteins on their surface. These proteins serve as "self" markers and determine a person's ABO (and Rh) blood type.

When incompatible blood types mix, an agglutination response occurs in which antibodies cause potentially fatal clumping of red blood cells.

8.5 Rh Blood Typing

Another surface marker on red blood cells that can cause agglutination is the "Rh factor," so named because it was first identified in the blood of Rhesus monkeys.

RH BLOOD TYPING LOOKS FOR AN RH MARKER

Rh blood typing determines the presence or absence of an Rh marker. If you are type Rh1, your blood cells bear this marker and you are Rh$^+$ (positive). If you are type Rh$^-$, they don't have the marker and you are Rh$^-$ (negative). When a person's blood type is determined, the ABO blood type and Rh type are usually combined. For instance, if your blood is type A and Rh negative, your blood type will be given as type A$^-$.

Most people don't have antibodies against the Rh marker. But someone who receives a transfusion of Rh1 blood will make antibodies against the marker, and these will continue circulating in the person's bloodstream.

If an Rh$^-$ woman becomes pregnant by an Rh$^+$ man, there is a chance the fetus will be Rh$^+$. During pregnancy or childbirth, some of the fetal red blood cells may leak into the mother's bloodstream. If they do, her body will produce antibodies against Rh (Figure 8.6). If she gets pregnant *again*, Rh antibodies will enter the bloodstream of this new fetus. If its blood is type Rh1, its mother's antibodies will cause its red blood cells to swell and burst. In extreme cases, called *hemolytic disease of the newborn*, so many red blood cells are destroyed that the fetus dies. If the condition is diagnosed before or during a live birth, the baby can survive by having its blood replaced with transfusions free of Rh antibodies.

Currently, a known Rh$^-$ woman can be treated after her first pregnancy with an anti-Rh gamma globulin (RhoGam) that will protect her next fetus. The drug will inactivate Rh1 fetal blood cells circulating in the mother's bloodstream before she can become sensitized and begin producing anti-Rh antibodies. In non-maternity cases, an Rh$^-$ person who receives a transfusion of Rh$^+$ blood also can have a severe negative reaction if he or she has previously been exposed to the Rh marker.

THERE ARE ALSO MANY OTHER MARKERS ON RED BLOOD CELLS

Besides the Rh and AB blood marker proteins, hundreds of others are now known to exist. These markers are a bit like needles in a haystack—they are widely scattered within the human population and usually don't cause problems in transfusions. Reactions do occur, though, and except in extreme emergencies, hospitals use a method called *cross-matching* to exclude the possibility that blood to be transfused and that of a patient might be incompatible due to the presence of a rare blood cell marker outside the ABO and Rh groups.

> *In some people, red blood cells are marked with an Rh protein. If this Rh$^+$ blood mixes with the Rh$^-$ blood of someone else, the Rh$^-$ individual will develop antibodies against it. The antibodies will trigger an immune response against Rh$^+$ red blood cells if the person is exposed to them again.*

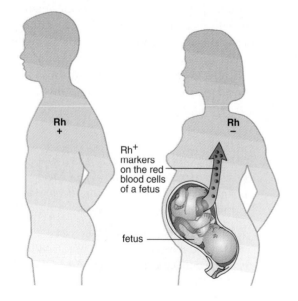

Rh +

Rh −

Rh$^+$ markers on the red blood cells of a fetus

fetus

a

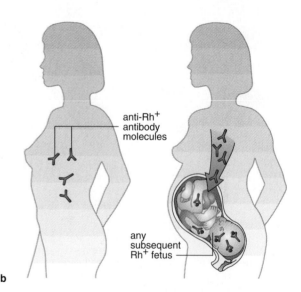

anti-Rh$^+$ antibody molecules

any subsequent Rh$^+$ fetus

b

Figure 8.6 *Animated!* Development of antibodies in response to Rh$^+$ blood. (**a**) Blood cells from an Rh$^+$ fetus leak into the Rh$^-$ mother's bloodstream. (**b**) The mother now develops antibodies against a subsequent Rh$^+$ fetus.

8.6 New Frontiers of Blood Typing

Because blood types are genetically determined, they can be used to help establish a person's genetic heritage.

BLOOD + DNA: INVESTIGATING CRIMES AND IDENTIFYING MOM OR DAD

In addition to helping ensure that a blood transfusion will be safe or that a mother's antibodies will not harm her fetus, the markers on red blood cells have a variety of other uses. For example, investigations of rapes, murders, and sometimes other crimes often compare the blood groups of victims and any possible perpetrators.

Today, blood samples often are used for DNA testing, which provides the most definitive information about a person's genetic heritage. For instance, there is a lot of similarity in the blood types found in and among people of different ethnic backgrounds (Table 8.2; notice that AB is the rarest blood type).

At one time blood typing was also commonly used to help determine the identity of a child's father or mother in cases where parentage was disputed. This is another area in which DNA testing is now the norm.

FOR SAFETY'S SAKE, SOME PEOPLE BANK THEIR OWN BLOOD

A blood transfusion is inherently risky. There is the need for an accurately matched blood type, and the risk of being exposed to blood-borne pathogens such as hepatitis viruses and HIV, the human immunodeficiency virus that causes AIDS. Although in general hospital blood supplies are carefully screened, some people who are slated for elective surgery take the extra precaution of pre-donating blood for an *autologous transfusion* (Figure 8.7). This means they have some of their own blood

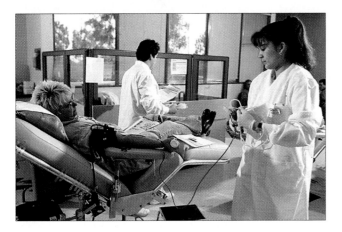

Figure 8.7 Donating blood.

Table 8.2	ABO Blood Groups in the U.S. Population (percentages)			
Blood Group	White	Black	Asian	Native American
AB	4	4	5	<1
B	11	20	27	4
A	40	27	28	16
O	45	49	40	79

removed and stored before the procedure so it can be used during the surgery if a transfusion is necessary.

BLOOD SUBSTITUTES MUST ALSO AVOID SPARKING AN IMMUNE RESPONSE

For years medical researchers have been trying to develop a safe, effective blood substitute that can be used in emergencies when matching a person's blood type isn't feasible, as in an ambulance or on a battlefield. A substitute might also be acceptable to people who refuse blood transfusions on religious grounds. As you've read, however, blood is extremely complex, and red blood cells, which are the crucial oxygen transporters, have many different self markers on their plasma membranes. Under these circumstances, it has been a tall order to find the right recipe for a blood substitute.

To date the most promising approach seems to be a substitute oxygen carrier that will not trigger an immune response. At this writing a product called Oxygent™, pictured at right, is in the final stage of being tested in clinical trials. If this milky-white fluid is approved for general use, it will have to be used with care, because tissues can be damaged if there is too much oxygen in the bloodstream. Inevitably, there will also be negative side effects in some people.

The more we explore options for blood substitutes, the more we understand just what a remarkable substance we have coursing through our arteries and veins!

> *The presence of self markers on red blood cells allows blood to be used to help identify individuals. The markers also make transfusions risky, so some patients opt for autologous transfusions.*
>
> *The need to match blood groups also is a challenge in the development of blood substitutes.*

8.7 Hemostasis and Blood Clotting

Small blood vessels can easily tear or be damaged by a cut or blow. To maintain homeostasis, it is extremely important that small tears are quickly repaired.

HEMOSTASIS PREVENTS BLOOD LOSS

Hemostasis is the name of a process that stops bleeding and so helps prevent the excessive loss of blood. In this process, an affected blood vessel constricts, platelets plug up the tear, and blood coagulates, or clots (Figure 8.8). Although hemostasis can only seal tears or punctures in relatively small blood vessels, most cuts and punctures fall into this category.

When a blood vessel ruptures, smooth muscle in the damaged vessel wall contracts in an automatic response called a spasm. The muscle contraction constricts the blood vessel, so blood flow through it slows or stops. This response can last for up to half an hour, and it is vital in stemming the immediate loss of blood. Then, while the flow of blood slows, platelets arrive and clump together, creating a temporary plug in the damaged wall.

They also release the hormone serotonin and other chemicals that help prolong the spasm and attract more platelets. Lastly, blood coagulates—that is, it converts to a gel—and forms a clot.

FACTORS IN BLOOD ARE ONE TRIGGER FOR BLOOD CLOTTING

Two different mechanisms can cause a blood clot to form. The first is called an "intrinsic" clotting mechanism because it involves substances that are in the blood itself. Figure 8.8 diagrams this process. It gets under way when a protein in the blood plasma, called "factor X," is activated. This triggers reactions that produce thrombin. This is an enzyme that acts on a rod-shaped protein called fibrinogen. The fibrinogen rods stick together, forming long threads of fibrin. The fibrin threads also stick to one another. The result is a net that entangles blood cells and platelets, as you can see in the micrograph in Figure 8.8. The entire mass is a blood clot. With time, the clot becomes more compact, drawing the torn walls of the vessel back together.

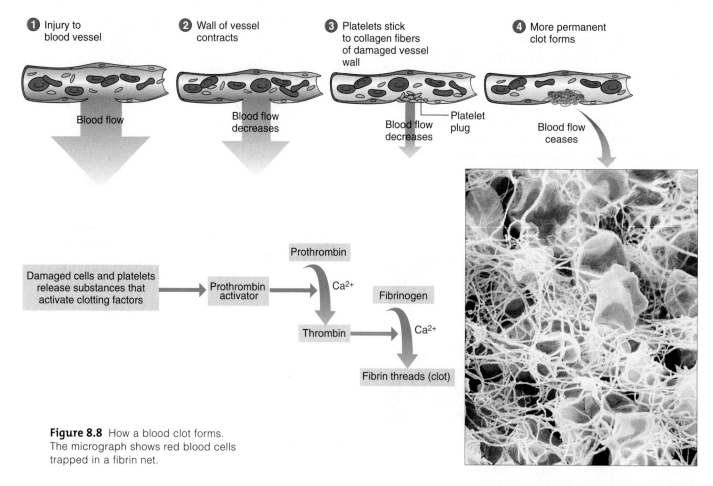

Figure 8.8 How a blood clot forms. The micrograph shows red blood cells trapped in a fibrin net.

FACTORS FROM DAMAGED TISSUE ALSO CAN CAUSE A CLOT TO FORM

Blood also can coagulate through an extrinsic clotting mechanism. "Extrinsic" means that the reactions leading to clotting are triggered by the release of enzymes and other substances *outside* the blood. These chemicals come from damaged blood vessels or from tissue around the damaged area. The substances lead to the formation of thrombin, and the remaining steps are like the steps of the intrinsic pathway.

Because aspirin reduces the aggregation of platelets, it is sometimes prescribed in small doses to help prevent blood clots. A clot that forms in an unbroken blood vessel can be a serious threat because it can block the flow of blood. A clot that stays where it forms is called a *thrombus*, and the condition is called a *thrombosis*.

Even scarier is an *embolus*, a clot that breaks free and circulates through the bloodstream. A person who suffers an *embolism* in the heart, lungs, brain, or some other organ may suddenly die when the roving clot shuts down the organ's blood supply. This is what happens when a person suffers a **stroke**. A blood clot blocks the flow of blood to some part of the brain and the affected brain tissue dies. Strokes can be mild to severe. In serious cases the person may be paralyzed on one side of the body and have trouble speaking. Physical therapy and speech therapy may help minimize the long-term effects.

The disease **hemophilia** is a genetic disorder in which the blood does not contain the usual clotting factors and so does not clot properly. You will read more about this disorder, and its central role in some historic events, in Chapter 21.

THE FORMATION OF A BLOOD CLOT IS A KEY STEP IN HEALING WOUNDED SKIN

When the skin is punctured or torn, blood clotting gets under way immediately to help seal the breach (Figure 8.9). With minor cuts, it usually takes less than 30 minutes for a clot to seal off injured vessels. In a few more hours, phagocytes are at work cleaning up debris and a scab has begun to form. This quick action is vital to minimize blood loss and the chances of infection.

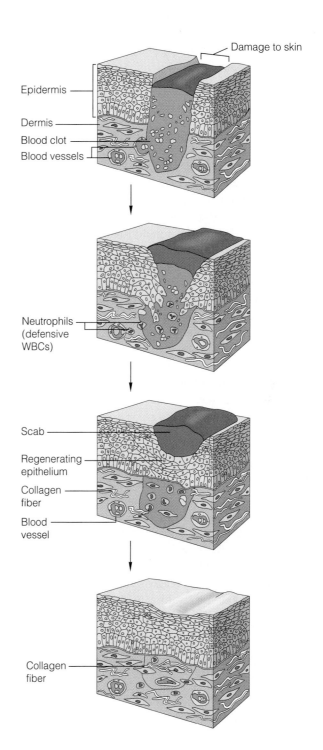

Figure 8.9 How blood clotting helps heal a wound in the skin.

Hemostasis refers to processes that slow or stop the flow of blood from a ruptured vessel.

The mechanisms include spasms that constrict blood vessel walls, the formation of platelet plugs, and blood clotting.

Blood clotting can be triggered by substances in the blood itself (such as thrombin), or by way of reactions involving substances in damaged tissue.

LINK TO SECTION 7.11

Various disorders can hamper the ability of blood cells to function normally.

ANEMIAS ARE RED BLOOD CELL DISORDERS

At least half a dozen **anemias** (meaning "no blood") are signs that red blood cells are not delivering enough oxygen to meet body needs. All anemias result from other, underlying problems. To varying degrees they make a person feel tired and listless, among other symptoms.

Two common types of anemia result from nutrient deficiencies. For example, **iron-deficiency anemia** develops when the body's iron supply is too low to form enough hemoglobin (with its iron-containing heme groups). Folic acid and vitamin B_{12} both are needed for the production of red blood cells in bone marrow. A deficiency of either one can lead to **pernicious anemia**. A balanced diet usually provides both nutrients, but other conditions can prevent them from being absorbed.

The rare malady **aplastic anemia** arises when red bone marrow, including the stem cells that give rise to red and white blood cells and platelets, has been destroyed by radiation, drugs, or toxins.

"Hemolytic" means "blood breaking," and **hemolytic anemias** develop when red blood cells die or are destroyed before the end of their normal useful life. The root cause may be an inherited defect, as in sickle-cell anemia, in which red blood cells take a sickle shape (Figure 8.10*a* and *b*) and can burst. Chapter 20 looks more fully at the genetic trigger for these changes.

Worldwide, **malaria** is a major cause of hemolytic anemia. It is caused by a protozoan that is transmitted by mosquitoes. One life stage of this pathogen multiplies inside red blood cells, leading to disease symptoms such as fever, chills, and trembling. Eventually the red blood cells burst (Figure 8.10*c*). Chapter 18 provides more information about this global scourge.

Like people who have sickle-cell anemia, those with **thalassemia** produce abnormal hemoglobin. Too few red blood cells form, and those that do form are thin and extremely fragile.

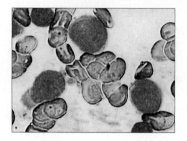

Figure 8.11 Blood from a person with chronic myelogenous leukemia. Abnormal white blood cells (purple) are starting to crowd out normal cells.

CARBON MONOXIDE POISONING PREVENTS HEMOGLOBIN FROM BINDING OXYGEN

Carbon monoxide, or CO, is a colorless odorless gas. It is present in auto exhaust fumes and in smoke from burning wood, coal, charcoal, and tobacco. It binds to hemoglobin at least 200 times more tightly than oxygen does. As a result, breathing even tiny amounts of it can tie up half of the body's hemoglobin and prevent tissues from receiving the oxygen they need. CO poisoning is especially dangerous because an affected person may not realize that the symptoms—headache and feeling "woozy"—are signs of life-threatening distress.

MONONUCLEOSIS AND LEUKEMIAS AFFECT WHITE BLOOD CELLS

Our white blood cells also can be affected by disease. For example, **infectious mononucleosis** is caused by the Epstein-Barr virus, which causes overproduction of lymphocytes. The patient feels achy and tired and runs a low-grade fever for several weeks as the highly contagious disease runs its course.

Far more serious are **leukemias**, which all are the result of cancer in the bone marrow. The word "leukemia" means "white blood," and the hallmark of leukemias (like other cancers) is runaway multiplication of the abnormal cells and destruction of healthy bone marrow.

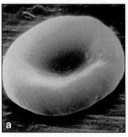 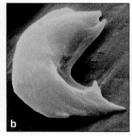 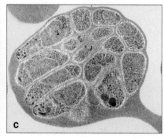

Figure 8.10 Some causes of hemolytic anemia. Scanning electron micrographs of normal (**a**) and sickled (**b**) red blood cells. (**c**) A life stage of the microorganism that causes malaria, about to rupture a red blood cell.

BLOOD TYPES

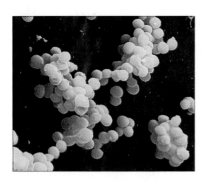

Figure 8.12
The bacterium *Staphylococcus aureus.*

In the most serious forms of leukemia, which tend to strike children, the marrow cavities in bones become choked with cancerous white blood cells. As other types of blood cells (and stem cells) are excluded, leukemia's symptoms develop—fever, weight loss, anemia, internal bleeding, pain, and susceptibility to infections. Modern treatments now save thousands of lives, and there is hope that experimental gene therapies will one day provide more help. Figure 8.11 shows cells of one type of leukemia, called **chronic myelogenous leukemia**.

Viral infections can also hamper or destroy white blood cells. The most notorious culprit is HIV, the human immunodeficiency virus, which causes AIDS. Its ability to kill lymphocytes of the immune system is a major topic in Chapter 18.

TOXINS CAN DESTROY BLOOD CELLS OR POISON THE BLOOD IN OTHER WAYS

A variety of bacteria can release toxins into the blood, a condition called **septicemia**. One of our most fearsome bacterial foes is *Staphylococcus aureus*, or simply "staph A" (Figure 8.12). This microbe produces enzymes that destroy red blood cells and prevent blood clotting. Unfortunately, some strains have become highly resistant to antibiotics, a growing problem that we will discuss again in Chapter 18.

Metabolic poisons in the body cause *toxemia*. For example, the kidneys remove many toxic wastes from the blood. In a person whose kidneys do not function well due to disease or some other cause, the buildup of certain wastes prevents the normal replacement of red blood cells. It also prevents platelets from functioning. Thus the person becomes anemic and blood does not clot properly. Chapter 11 discusses some other extremely serious effects of kidney disease.

Summary

Section 8.1 Blood is a fluid connective tissue that helps maintain homeostasis as it performs the following functions:

a. Transporting oxygen and other substances to and from the extracellular fluid bathing cells.

b. Transporting many proteins and ions. The proteins (such as albumin) help maintain the proper fluid volume of blood. Ions help stabilize the pH of extracellular fluid.

Blood consists of plasma, red and white blood cells, and cell fragments called platelets. Blood cells and platelets arise from stem cells in bone marrow.

Plasma, the liquid part of blood, transports blood cells and platelets. Plasma water is a solvent for proteins, simple sugars, amino acids, mineral ions, vitamins, hormones, and several gases.

Red blood cells carry oxygen from the lungs to body tissues. Major categories of white blood cells are granulocytes and agranulocytes. Granulocytes, such as neutrophils, operate in body defense. Agranulocytes include a type that develops into macrophages, which scavenge dead or worn-out cells and other debris and cleanse tissues of "non-self" material. Still other white blood cells (the lymphocytes) form armies that destroy specific microbes and other disease agents.

Platelets release substances that begin the process of blood clotting.

Section 8.2 Red blood cells contain hemoglobin, an iron-containing pigment molecule that binds reversibly with oxygen, forming oxyhemoglobin. Red blood cells also carry some carbon dioxide (also bound to hemoglobin) from extracellular fluid back to the lungs (to be exhaled).

Section 8.3 Red blood cells live for about 120 days. A cell count measures the number of them in a microliter of blood. Macrophages remove dead or damaged red blood cells while stem cells provide replacements.

Section 8.4 Blood type is determined by certain proteins on the surface of red blood cells. The four main human blood types are A, B, AB, and O. Agglutination is an immune response activated when a person's blood mingles with an incompatible type. Rh blood typing determines the presence or absence of Rh factors (+ or −) on red blood cells. If incompatible Rh types commingle, the immune system will attack and destroy the "foreign" cells.

 Learn about ABO and Rh blood types.

Section 8.7 Processes of hemostasis slow or stop bleeding. They include spasms that constrict blood vessels, the formation of platelet plugs, and blood clotting.

Review Questions

1. What is blood plasma, and what is its function?

2. What are the cellular components of blood? Where do the various kinds come from?

3. Add the missing labels to Figure 8.13 at the right. Then, on a separate sheet of paper, list the factors that affect the tendency of hemoglobin to bind with oxygen.

4. Explain what an agglutination response is, and how it can be avoided when blood is transfused.

5. What is the function of hemostasis? What are the two ways a blood clot can form?

Self-Quiz
Answers in Appendix V

1. Which are *not* components of blood?
    a. plasma
    b. blood cells and platelets
    c. gases and other dissolved substances
    d. all of the above are components of blood

2. The _____ produces red blood cells, which transport _____ and some _____.
    a. liver; oxygen; mineral ions
    b. liver; oxygen; carbon dioxide
    c. bone marrow; oxygen; hormones
    d. bone marrow; oxygen; carbon dioxide

3. The _____ produces white blood cells, which function in _____ and _____.
    a. liver; oxygen transport; defense
    b. lymph glands; oxygen transport; stabilizing pH
    c. bone marrow; day-to-day housekeeping; defense
    d. bone marrow; stabilizing pH; defense

4. In the lungs, the main factor in boosting the tendency of hemoglobin to bind with and hold oxygen is _____.
    a. temperature   c. acidity (pH)
    b. the amount of O_2 d. all are equally important
       relative to the amount
       of CO_2 in plasma

5. Match the blood terms with the best description.
    ___ red blood cell   a. plug leaks
    ___ platelets   b. blood markers
    ___ stem cell   c. blood cell source
    ___ plasma   d. erythrocyte
    ___ A, B, O   e. more than half of whole blood

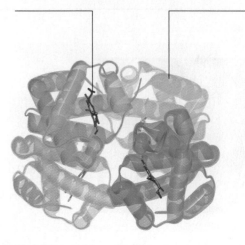

Figure 8.13 A hemoglobin molecule.

Critical Thinking

1. Thrombocytopenia (throm-bo-sye-tow-PEE-ne-ah) is a disorder that develops when certain drugs, bone marrow cancer, or radiation destroys red bone marrow, including stem cells that give rise to platelets. Predict a likely symptom of this disorder.

2. As the text described, when a person's red blood cell count drops, the kidneys receive less oxygen. In response they release erythropoietin, which prompts the bone marrow to make more red blood cells. As the rising number of red blood cells carry *more* oxygen to the kidneys, they stop releasing the hormone. What type of homeostatic control mechanism are we talking about here?

Explore on Your Own

What is your "Blood IQ"?
    To find out how much you know about blood and public blood supplies, visit www.RedCross.org or www.givelife.org. Both are sponsored by the American Red Cross. At the GIVELIFE website, take the ten-question Blood IQ test and see how much you know about blood types and other issues. The websites offer information about blood, blood donation, and even current research on blood substitutes and other topics.

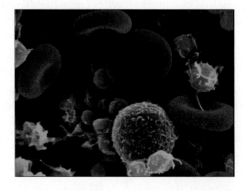

The Breath of Life

Each year in the U.S. 250,000 people have a sudden cardiac arrest, usually outside a hospital. The heart stops beating and blood stops flowing through the vessels. A problem with the electrical signals that stimulate cardiac muscle is the most common cause. The normal heartbeat abruptly shifts into an erratic pattern called ventricular fibrillation (VF).

It happened to Tammy Higgins. The 28-year-old mother collapsed as she was leaving church with her husband Chris and their daughter Lindsay. She had no pulse, and she had stopped breathing. Chris knew that getting oxygen-rich blood to his wife's brain was vital. He immediately began CPR—cardiopulmonary resuscitation. Using this technique, he and others kept Tammy alive until an ambulance arrived.

Emergency medical technicians used a heart-shocking device called a defibrillator to get Tammy's heart restarted. In the hospital, she was diagnosed with a heart rhythm disturbance. She also learned that she was pregnant.

Doctors implanted a tiny defibrillator that constantly monitors the rhythm of Tammy's heart and shocks it back into action if it stops. The device shocked her heart once during her pregnancy and once again in the three years after the

birth of her daughter Nicole. Without the device, either of these cardiac arrests could have been fatal.

Tammy was lucky—her husband and several bystanders knew how to do CPR. But while CPR can keep a person alive for a short while, it can't restore the heart's normal rhythm. That requires the use of a defibrillator. As with CPR, the sooner electrical stimulation begins, the better. For each minute without defibrillation, the odds of survival drop as much as 10 percent.

Automated external defibrillators (AEDs) in public places allow trained bystanders to provide the life-saving shocks before an ambulance arrives. An AED is about the size of a laptop computer. Its voice commands direct the user to perform the appropriate steps. AEDs are now available in many senior centers, shopping malls, hotels, and other public places.

What you learn in this chapter will help you to understand the biology that underlies CPR and the use of an AED. If you would like to learn how to save lives with these methods, the American Heart Association, the American Red Cross, and many community organizations provide training.

 How Would You Vote? Some advocates think that CPR training should be a required mini-course in high schools. People who learn CPR also must be periodically recertified. Would you favor mandatory CPR training in high schools? Cast your vote online at www.thomsonedu.com/biology/starr/humanbio.

 ## Key Concepts

CIRCULATING BLOOD
The cardiovascular system transports oxygen, nutrients, hormones, and other substances swiftly to body cells and carries away wastes and cell products.

PUMPING BLOOD
The heart is a muscular pump. Its contractions provide the force that drives blood through the cardiovascular system's arteries and veins.

BLOOD VESSELS
Various types of blood vessels—arteries, arterioles, capillaries, venules, and veins—are specialized for different blood transport functions.

Links to Earlier Concepts

Building on what you learned about blood in Chapter 8, our focus now shifts to the blood-pumping cardiovascular system—the heart and blood vessels. This chapter looks more closely at cardiac muscle (4.3) and at the specialized cell junctions in this tissue (4.6). You will also see how the dynamic tubelike organs we call blood vessels are built from epithelium, connective tissue, and smooth muscle (4.1–4.3). We also consider links between cardiovascular health and lipoproteins and cholesterol (2.10 and 2.12).

9.1 The Cardiovascular System—Moving Blood through the Body

"Cardiovascular" comes from the Greek kardia (heart) and the Latin vasculum (vessel). The cardiovascular system—also called the circulatory system—is built to rapidly transport blood to every living cell in the body.

THE HEART AND BLOOD VESSELS MAKE UP THE CARDIOVASCULAR SYSTEM

As you can see in Figure 9.1 below, the **cardiovascular system** has two main elements:

- the **heart**, a muscular pump that generates the pressure required to move blood throughout the body
- blood vessels, which are tubes of different diameters that transport blood.

The heart pumps blood into large-diameter **arteries**. From there blood flows into smaller **arterioles**, which branch into even narrower **capillaries**. Blood flows from capillaries into small **venules**, then into large-diameter **veins** that return blood to the heart. Because the heart pumps constantly, the volume of flow through the entire system each minute is equal to the volume of blood returned to the heart each minute.

As you will read later on, the rate and volume of blood flow through the cardiovascular system can be adjusted to suit conditions in the body. For example, blood flows rapidly through arteries, but in capillaries it must flow slowly so that there is enough time for substances moving to and from cells to diffuse into and out of extracellular fluid. This slow flow occurs in *capillary beds*, where blood moves through vast numbers of slender capillaries. By

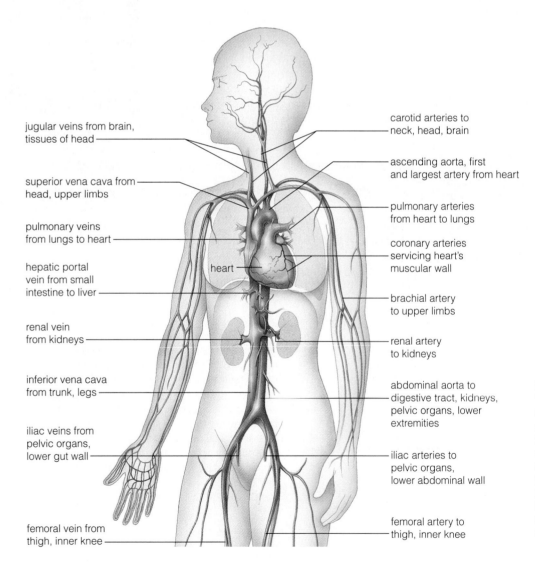

jugular veins from brain, tissues of head

superior vena cava from head, upper limbs

pulmonary veins from lungs to heart

hepatic portal vein from small intestine to liver

renal vein from kidneys

inferior vena cava from trunk, legs

iliac veins from pelvic organs, lower gut wall

femoral vein from thigh, inner knee

heart

carotid arteries to neck, head, brain

ascending aorta, first and largest artery from heart

pulmonary arteries from heart to lungs

coronary arteries servicing heart's muscular wall

brachial artery to upper limbs

renal artery to kidneys

abdominal aorta to digestive tract, kidneys, pelvic organs, lower extremities

iliac arteries to pelvic organs, lower abdominal wall

femoral artery to thigh, inner knee

Figure 9.1 *Animated!* The human cardiovascular system. Arteries, which carry oxygenated blood to tissues, are shaded red. Veins, which carry deoxygenated blood away from tissues, are shaded blue. Notice, however, that for the pulmonary arteries and veins the roles are reversed.

dividing up the blood flow, the capillaries handle the same total volume of flow as the large-diameter vessels, but at a slower pace.

CIRCULATING BLOOD IS VITAL TO MAINTAIN HOMEOSTASIS

Recall from Chapter 8 that blood is aptly called "the river of life." It brings cells such essentials as oxygen, nutrients from food, and secretions, such as hormones. It also takes away the wastes produced by our metabolism, along with excess heat. In fact, cells depend on circulating blood to make constant pickups and deliveries of an amazingly diverse range of substances, including those that move into or out of the digestive system and the respiratory and urinary systems (Figure 9.2).

Homeostasis is one of our constant themes in this book, so it's good to keep in mind that maintaining it would be impossible were it not for our circulating blood. Cells must exchange substances with blood because that is a key way they adjust to changes in the chemical makeup of the extracellular fluid around them—part of the "internal environment" in which they live.

THE CARDIOVASCULAR SYSTEM IS LINKED TO THE LYMPHATIC SYSTEM

The heart's pumping action puts pressure on blood flowing through the cardiovascular system. Partly because of this pressure, small amounts of water and some proteins dissolved in blood are forced out and become part of interstitial fluid (the fluid between cells). An elaborate network of drainage vessels picks up excess extracellular fluid and reclaimable solutes—such as water, proteins, and fatty acids—and returns them to the cardiovascular system. This network is part of the lymphatic system, which we consider in Chapter 10.

The cardiovascular system consists of the heart and the blood vessels. It transports substances to and from the interstitial fluid that bathes all living cells.

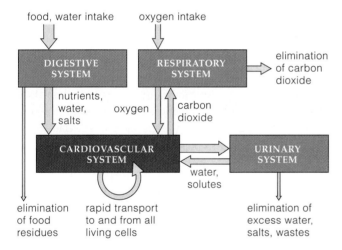

Figure 9.2 Together with the other systems shown here, the cardiovascular system helps maintain favorable operating conditions in the internal environment.

IMPACTS, ISSUES

Recently a large extended family in Iowa provided the first evidence of a gene linked directly to heart attacks. For generations, coronary artery disease has been unusually common in the family. Plaques of cholesterol would build up on the artery walls and block blood flow, leading to a heart attack. Researchers found that affected family members all had the same gene mutation. The faulty gene has been dubbed MEF2A. The next step is to determine whether MEF2A is the culprit in heart disease patients who do not belong to families in which the mutation is inherited.

9.2 The Heart: A Double Pump

LINKS TO
SECTIONS
4.1 AND 4.2

In a lifetime of 70 years, the human heart beats some 2.5 billion times. This durable pump is the centerpiece of the cardiovascular system.

Roughly speaking, your heart is located in the center of your chest (Figure 9.3*a*). Its structure reflects its role as a long-lasting pump. The heart is mostly cardiac muscle tissue, the **myocardium** (Figure 9.3*b*). A tough, fibrous sac, the pericardium (*peri* = around), surrounds, protects, and lubricates it. The heart's chambers have a smooth lining (endocardium) composed of connective tissue and a layer of epithelial cells. The epithelial cell layer, known as endothelium, also lines the inside of blood vessels.

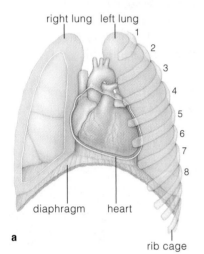

a

THE HEART HAS TWO HALVES AND FOUR CHAMBERS

A thick wall, the **septum**, divides the heart into two halves, right and left. Each half has two chambers: an **atrium** (plural: atria) located above a **ventricle**. Flaps of membrane separate the two chambers and serve as a one-way **atrioventricular valve** (AV valve) between them. The AV valve in the right half of the heart is called a *tricuspid valve* because its three flaps come together in pointed cusps (Figure 9.3*c*). In the heart's left half the AV valve consists of just two flaps; it is called the *bicuspid valve* or *mitral valve*. Tough, collagen-reinforced strands (chordae tendineae, or "heartstrings") connect the AV valve flaps to cone-shaped muscles that extend out from the ventricle wall. When a blood-filled ventricle contracts, this arrangement prevents the flaps from opening backward into the atrium. Each half of the heart also has a half-moon–shaped **semilunar valve** between the ventricle and the arteries leading away from it. During a heartbeat, this valve opens and closes in ways that keep blood moving in one direction through the body.

The heart has its own "coronary circulation." Two **coronary arteries** lead into a capillary bed that services most of the cardiac muscle (Figure 9.4). They branch off the **aorta**, the major artery carrying oxygenated blood away from the heart.

IN A "HEARTBEAT," THE HEART'S CHAMBERS CONTRACT, THEN RELAX

Blood is pumped each time the heart beats. It takes less than a second for a "heartbeat"—one sequence of contraction and relaxation of the heart chambers. The sequence occurs almost simultaneously in both sides of the heart. The contraction phase is called **systole** (SISS-toe-lee), and the relaxation phase is called **diastole** (dye-ASS-toe-lee). This sequence is the **cardiac cycle** diagrammed in Figure 9.5.

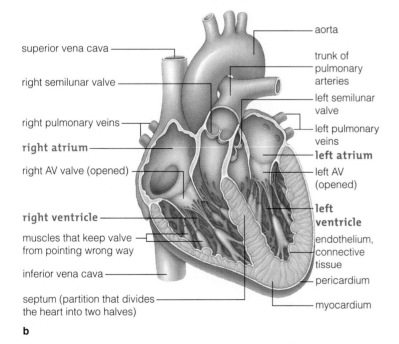

b

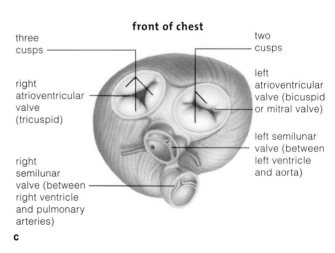

c

Figure 9.3 *Animated!* (**a**) Location of the heart. (**b**) Cutaway view showing the heart's internal organization, and (**c**) valves of the heart. In this drawing, you are looking down at the heart. The atria have been removed so that the atrioventricular (AV) and semilunar valves are visible.

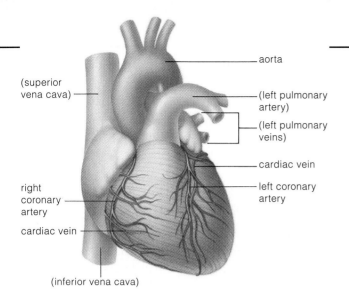

aorta

(superior vena cava)

(left pulmonary artery)

(left pulmonary veins)

cardiac vein

left coronary artery

right coronary artery

cardiac vein

(inferior vena cava)

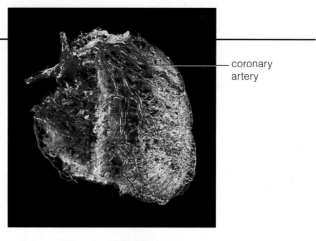

coronary artery

Figure 9.4 Coronary arteries and veins. The photograph shows a resin cast of the heart's arteries and veins.

During the cycle, the ventricles relax before the atria contract, and the ventricles contract when the atria relax. When the relaxed atria are filling with blood, the fluid pressure inside them rises and the AV valves open. Blood flows into the ventricles, which are 80 percent filled by the time the atria contract. As the filled ventricles begin to contract, fluid pressure inside *them* increases, forcing the AV valves shut. The rising pressure then forces the semilunar valves open—and blood flows out of the heart and into the aorta and pulmonary artery. Now the ventricles relax, and the semilunar valves close. For about half a second the atria and ventricles are all in diastole. Then the blood-filled atria contract, and the cycle repeats.

The amount of blood each ventricle pumps in a minute is called the **cardiac output**. On average, every sixty seconds the cardiac output from each ventricle is about 5 liters—nearly all the blood in the body. This means that in a year each half of your heart pumps at least 2.5 million liters of blood. That is more than 600,000 gallons!

The blood and heart movements during the cardiac cycle generate an audible "lub-dup" sound made by the forceful closing of the heart's one-way valves. At each "lub," the AV valves are closing as the two ventricles contract. At each "dup," the semilunar valves are closing as the ventricles relax.

Each half of the heart is divided into an atrium and a ventricle. Valves help control the direction of blood flow.

During a cardiac cycle, contraction of the atria helps fill the ventricles. Contraction of the ventricles provides the force that drives blood away from the heart.

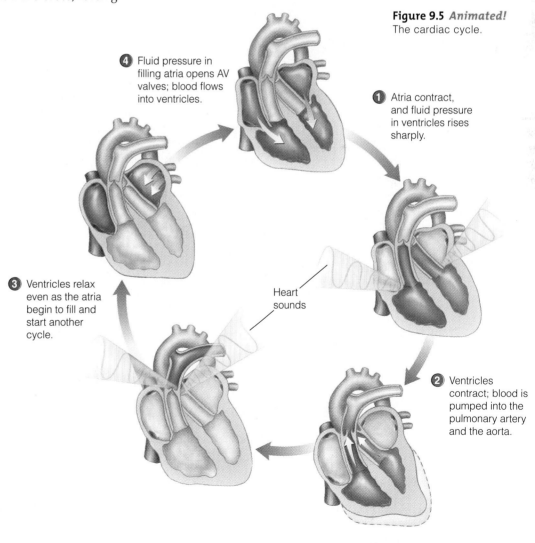

Figure 9.5 *Animated!* The cardiac cycle.

4 Fluid pressure in filling atria opens AV valves; blood flows into ventricles.

1 Atria contract, and fluid pressure in ventricles rises sharply.

3 Ventricles relax even as the atria begin to fill and start another cycle.

Heart sounds

2 Ventricles contract; blood is pumped into the pulmonary artery and the aorta.

9.3 The Two Circuits of Blood Flow

Each half of the heart pumps blood. The two side-by-side pumps are the basis of two cardiovascular circuits through the body, each with its own set of arteries, arterioles, capillaries, venules, and veins.

THE PULMONARY CIRCUIT: BLOOD PICKS UP OXYGEN IN THE LUNGS

The **pulmonary circuit** diagrammed in Figure 9.6*a* receives blood from tissues and circulates it through the lungs for gas exchange. The circuit begins as blood from tissues enters the right atrium, then moves through the AV valve into the right ventricle. As the ventricle fills, the atrium contracts. Blood arriving in the right ventricle is fairly low in oxygen and high in carbon dioxide. When the ventricle contracts, the blood moves through the right semilunar valve into the *main* pulmonary artery, then into the *right* and *left* pulmonary arteries. These arteries carry the blood to the two lungs, where (in capillaries) it picks up oxygen and gives up carbon dioxide that will be exhaled. The freshly oxygenated blood returns through two sets of pulmonary veins to the heart's left atrium, completing the circuit.

(a)

pulmonary circuit for blood flow

right pulmonary artery   left pulmonary artery

capillary bed of right lung

pulmonary trunk

capillary bed of left lung

(to systemic circuit)

(from systemic circuit)

pulmonary veins

heart

(b)

systemic circuit for blood flow

capillary beds of head and upper extremities

(to pulmonary circuit)   aorta

(from pulmonary circuit)

heart

capillary beds of other organs in thoracic cavity

diaphragm (muscular partition between thoracic and abdominal cavities)

capillary bed of liver

hepatic portal vein

capillary beds of intestines

capillary beds of other abdominal organs and lower extremities

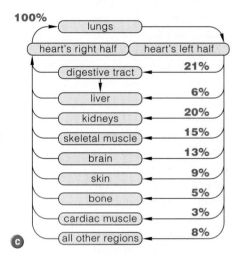

100%

lungs	
heart's right half	heart's left half
digestive tract	21%
liver	6%
kidneys	20%
skeletal muscle	15%
brain	13%
skin	9%
bone	5%
cardiac muscle	3%
all other regions	8%

(c)

Figure 9.6 *Animated!* The (**a**) pulmonary and (**b**) systemic circuits for blood flow in the cardiovascular system. (**c**) Distribution of the heart's output in people napping.

IN THE SYSTEMIC CIRCUIT, BLOOD TRAVELS TO AND FROM TISSUES

In the **systemic circuit** (Figure 9.6b), oxygenated blood pumped by the left half of the heart moves through the body and returns to the right atrium. This circuit begins when the left atrium receives blood from pulmonary veins, and this blood moves through an AV (bicuspid) valve to the left ventricle. This chamber contracts with great force, sending blood coursing through a semilunar valve into the aorta.

As the aorta descends into the torso (see Figure 9.1), major arteries branch off it, funneling blood to organs and tissues where O_2 is used and CO_2 is produced. For example, in a resting person, each minute a fifth of the blood pumped into the systemic circulation enters the kidneys (Figure 9.6c) via *renal arteries*. Deoxygenated blood returns to the right half of the heart, where it enters the pulmonary circuit. Notice that in both the pulmonary and the systemic circuits, blood travels through arteries, arterioles, capillaries, and venules, finally returning to the heart in veins. Blood from the head, arms, and chest arrives through the *superior vena cava*. The *inferior vena cava* collects blood from the lower part of the body.

BLOOD FROM THE DIGESTIVE TRACT IS SHUNTED THROUGH THE LIVER FOR PROCESSING

As you can see near the bottom of Figure 9.6, blood passing through capillary beds in the GI tract travels to another capillary bed in the liver. This is the route described in Section 7.5 by which nutrient-laden blood is sent to the liver through the *hepatic portal vein* after a meal. As blood seeps through this second bed, the liver can remove impurities and process absorbed substances. Part of this processing synthesizes cholesterol. The Impacts/Issues box at right explains how, in people who have too much cholesterol in their blood, drugs called statins can reduce the liver's cholesterol output.

Blood leaving the liver's capillary bed enters the general circulation through a *hepatic vein*. The liver receives oxygenated blood via the *hepatic artery*.

A short pulmonary circuit carries blood through the lungs for gas exchange. A long systemic circuit transports blood to and from tissues.

After meals, the blood in capillary beds in the GI tract is diverted to the liver for processing before reentering the general circulation.

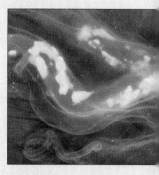

9.4 How Cardiac Muscle Contracts

LINK TO
SECTION
4.3

Unlike skeletal muscle, which contracts only when orders arrive from the nervous system, cardiac muscle contracts— and the heart beats—spontaneously.

ELECTRICAL SIGNALS FROM "PACEMAKER" CELLS DRIVE THE HEART'S CONTRACTIONS

Cardiac muscle cells branch, then link to one another at their endings. Junctions called *intercalated discs* span both plasma membranes of neighboring cells (Figure 9.7). With each heartbeat, signals calling for contraction spread so rapidly across the junctions that cardiac muscle cells contract together, almost as if they were a single unit.

Where do the contraction signals come from? About 1 percent of cardiac muscle cells do not contract, but instead function as the **cardiac conduction system**. Some of these cells are self-exciting "pacemaker" cells—that is, they spontaneously generate and conduct electrical impulses. Those impulses are the signals that stimulate contractions in the heart's contractile cells. Because the cardiac conduction system is independent of the nervous system, the heart will keep right on beating even if all nerves leading to the heart are severed!

Excitation begins with a cluster of cells in the upper wall of the right atrium (Figure 9.8). About 70 times a minute, this **sinoatrial (SA) node** generates waves of excitation. Each wave spreads swiftly over both atria and causes them to contract. It then reaches the **atrioventricular (AV) node** in the septum dividing the two atria.

When a stimulus reaches the AV node, it slows a little, then quickly continues along bundles of conducting fibers

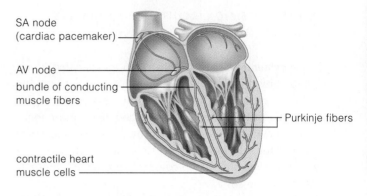

SA node (cardiac pacemaker)

AV node

bundle of conducting muscle fibers

Purkinje fibers

contractile heart muscle cells

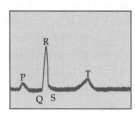

Figure 9.8 *Animated!* The cardiac conduction system. (*left*) Recording of a heartbeat. Letters indicate three waves of electrical activity that were caused by the spread of nerve impulses across cardiac muscle.

that extend to each ventricle. At intervals along each bundle, conducting cells called *Purkinje fibers* pass the signal on to contractile muscle cells in each ventricle. The slow conduction in the AV node is an important part of this sequence. It gives the atria time to finish contracting before the wave of excitation spreads to the ventricles.

Of all cells of the cardiac conduction system, the SA node fires off impulses at the highest frequency and is the first region to respond in each cardiac cycle. It is called the **cardiac pacemaker** because its rhythmic firing is the basis for the normal rate of heartbeat. People whose SA node chronically malfunctions may have an artificial pacemaker implanted to provide a regular stimulus for their heart contractions.

THE NERVOUS SYSTEM ADJUSTS HEART ACTIVITY

The nervous system initiates the contraction of skeletal muscle, but it can only *adjust* the rate and strength of cardiac muscle contraction. Stimulation by one set of nerves increases the force and rate of heart contractions, while stimulation by another set of nerves can slow heart activity. The centers for neural control of heart functions are in the spinal cord and parts of the brain. They are discussed more fully in Chapter 13.

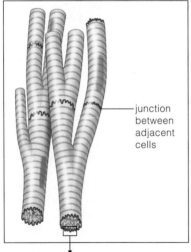

junction between adjacent cells

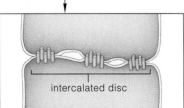

intercalated disc

Figure 9.7 Intercalated discs. The discs contain communication junctions at the ends of abutting cardiac muscle cells. Signals travel rapidly across the junctions and cause cells to contract nearly in unison.

The SA node is the cardiac pacemaker—it establishes a regular heartbeat. Its spontaneous, repeated excitation signals spread along a system of muscle cells that stimulate a rhythmic cycle of contraction in the heart's atria, then the ventricles.

PUMPING BLOOD

9.5 Blood Pressure

Heart contractions generate blood pressure, which changes as blood moves through the cardiovascular system.

BLOOD EXERTS PRESSURE AGAINST THE WALLS OF BLOOD VESSELS

Blood pressure is the fluid pressure that blood exerts against vessel walls. Blood pressure is highest in the aorta; then it drops along the systemic circuit. The pressure typically is measured when a person is at rest (Figure 9.9). For an adult, the National Heart, Lung, and Blood Institute has established blood pressure values under 120/80 as the healthiest (Table 9.1). The first number, *systolic pressure*, is the peak of pressure in the aorta while the left ventricle contracts and pushes blood into the aorta. The second number, *diastolic pressure*, measures the lowest blood pressure in the aorta, when blood is flowing out of it and the heart is relaxed.

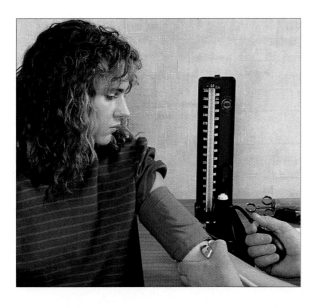

Figure 9.9 *Animated!* Measuring blood pressure. A hollow cuff attached to a pressure gauge is wrapped around the upper arm. The cuff is inflated to a pressure above the highest pressure of the cardiac cycle—at systole, when ventricles contract. Above this pressure, you can't hear sounds through a stethoscope positioned below the cuff and above the brachial artery, because no blood is flowing through the vessel. As air in the cuff is slowly released, some blood flows into the artery. The turbulent flow causes soft tapping sounds. When the tapping starts, the gauge's value is the systolic pressure, measured in millimeters of mercury (Hg). This value measures how far the pressure would force mercury to move upward in a narrow glass column.

More air is released from the cuff. Just after the sounds grow dull and muffled, blood is flowing steadily, so the turbulence and tapping end. The silence corresponds to diastolic pressure at the end of a cardiac cycle, before the heart pumps out blood. A desirable reading is under 80 mm Hg.

Table 9.1	Blood Pressure Values (mm of Hg)	
	Systolic	Diastolic
Normal	100–119	60–79
Hypotension	Less than 100	Less than 60
Prehypertension	120–139	80–139
Hypertension	140 and up	90 and up

Values for systolic and diastolic pressure provide important health information. Chronically elevated blood pressure, or **hypertension**, can be associated with a variety of ills, such as atherosclerosis (Section 9.8). The chart in Figure 9.10 lists some of the major causes and risk factors. Hypertension is a "silent killer" that can lead to a stroke or heart attack. Each year it kills about 180,000 Americans, many of whom may not have had any outward symptoms. Roughly 40 million people in the U.S. are unaware that they have hypertension.

Hypotension is abnormally *low* blood pressure. This condition can develop when for some reason there is not enough water in blood plasma—for instance, if there are not enough proteins in the blood to "pull" water in by osmosis. A large blood loss also can cause blood pressure to plummet. Such a drastic decrease is one sign of a dangerous condition called *circulatory shock*.

> Heart contractions generate blood pressure. Systolic pressure is the peak of pressure in the aorta while blood pumped by the left ventricle is flowing into it. Diastolic pressure measures the lowest blood pressure in the aorta, when blood is flowing out of it.

Risk Factors for Hypertension
1. Smoking
2. Obesity
3. Sedentary lifestyle
4. Chronic stress
5. A diet low in fruits, vegetables, dairy foods, and other sources of potassium and calcium
6. Excessive salt intake (in some individuals)
7. Poor salt management by the kidneys, usually due to disease

Figure 9.10 Some major factors associated with hypertension.

9.6 Structure and Functions of Blood Vessels

LINKS TO
SECTIONS
4.1 AND 4.2

As with all body parts, structure is key to the functions of blood vessels. All our vessels transport blood, but there are important differences in how different kinds "manage" blood flow and blood pressure.

ARTERIES ARE LARGE BLOOD PIPELINES

The wall of an artery has several tissue layers (Figure 9.11a). The outer layer is mainly collagen, which anchors the vessel to the tissue it runs through. A thick middle layer of smooth muscle is sandwiched between thinner layers containing elastin. The innermost layer is a thin sheet of endothelium. Together these layers form a thick, muscular, and elastic wall. In a large artery the wall bulges slightly under the pressure surge caused when a ventricle contracts. In arteries near the body surface, as in the wrist, you can feel the surges as your **pulse**.

The bulging of artery walls helps keep blood flowing on through the system. How? For a moment, some of the blood pumped during the systole phase of each cardiac cycle is stored in the "bulge"; the elastic recoil of the artery then forces that stored blood onward during diastole, when heart chambers are relaxed. In addition to stretchable walls, arteries also have large diameters. For this reason, they present little resistance to blood flow, so blood pressure does not drop much in the large arteries of the systemic and pulmonary circuits (Figure 9.12).

ARTERIOLES ARE CONTROL POINTS FOR BLOOD FLOW

Arteries branch into narrower arterioles, which have a wall built of rings of smooth muscle over a single layer of elastic fibers (Figure 9.11b). Because they are built this way, arterioles can dilate (enlarge in diameter) when the smooth muscle relaxes or constrict (shrink in diameter) when the smooth muscle contracts. Arterioles offer more resistance to blood flow than other vessels do. As the blood flow slows, it can be controlled in ways that adjust how much of the total volume goes to different body regions. For example, you become drowsy after a large meal in part because control signals divert blood away from your brain in favor of your digestive system.

CAPILLARIES ARE SPECIALIZED FOR DIFFUSION

Your body has about 2 miles of arteries and veins but a whopping 62,000 miles of capillaries. Each capillary bed is where substances can diffuse between blood and tissue fluid. This is truly where "the rubber meets the road" when it comes to exchanges of gases (oxygen and carbon dioxide), nutrients, and wastes. As befits its function in diffusion, a capillary has the thinnest wall of any blood vessel—a single layer of flat endothelium (Figure 9.11c).

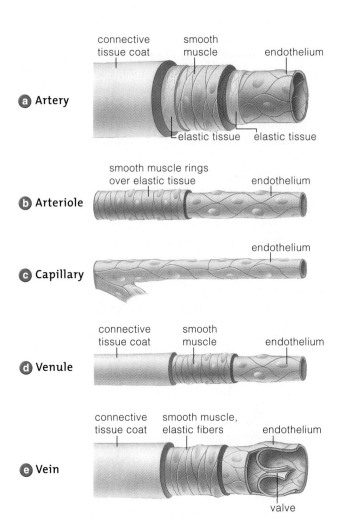

Figure 9.11 *Animated!* Structure of blood vessels.

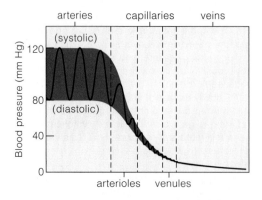

Figure 9.12 Changes in blood pressure in different parts of the cardiovascular system.

BLOOD VESSELS

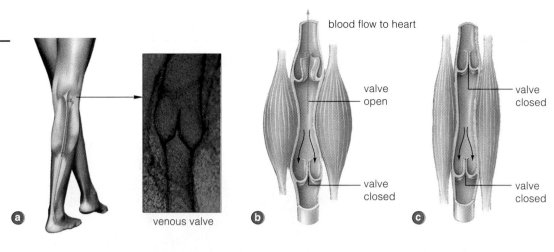

blood flow to heart

valve open

valve closed

a

venous valve

b

valve closed

valve closed

c

Figure 9.13 *Animated!* How contracting skeletal muscles can increase fluid pressure in a vein. (**a**) Valves in medium-sized veins prevent backflow of blood. (**b**) Skeletal muscles next to the vein contract, helping blood flow forward. (**c**) Skeletal muscles relax and valves in the vein shut—preventing backflow.

Blood can't move very fast in capillaries. However, because they are so extensive, capillary beds present less total resistance to flow than do the arterioles leading into them, so overall blood pressure drops more slowly in them. We'll look more closely at how capillaries function in the next section.

VENULES AND VEINS RETURN BLOOD TO THE HEART

Capillaries merge into venules, or "little veins," which in turn merge into large-diameter veins. Venules function a little like capillaries, in that some solutes diffuse across their relatively thin walls (Figure 9.11*d*).

Veins are large-diameter, low-resistance transport tubes to the heart (Figure 9.11*e*). Their valves prevent backflow. When blood starts moving backward due to gravity, it pushes the valves closed. The vein wall can bulge greatly under pressure, more so than an arterial wall. Thus veins are reservoirs for variable volumes of blood. Together, the veins of an adult can hold up to 50 to 60 percent of the total blood volume.

When blood must circulate faster (as during exercise), the smooth muscle in veins contracts. The wall stiffens, the vein bulges less, and venous pressure rises—so more blood flows to the heart (Figure 9.13). Venous pressure also rises when contracting skeletal muscle—especially in the legs and abdomen—bulges against adjacent veins. This muscle activity helps return blood through the venous system.

Obesity, pregnancy, and other factors can weaken venous valves. The walls of a *varicose vein* have become overstretched because, over time, weak valves have allowed blood to pool there.

VESSELS HELP CONTROL BLOOD PRESSURE

Some arteries, all arterioles, and even veins have roles in homeostatic mechanisms that help maintain adequate blood pressure over time. Centers in the brain's medulla monitor resting blood pressure. When blood pressure rises abnormally, they order the heart to contract less often and less forcefully. They also order smooth muscle in arterioles to relax. The result is **vasodilation**—an enlargement (dilation) of the vessel diameter. On the other hand, when the centers detect an abnormal *decrease* in blood pressure, they command the heart to beat faster and contract more forcefully. Neural signals also cause the smooth muscle of arterioles to contract. The result is **vasoconstriction**, a narrowing of the vessel diameter. In some parts of the body arterioles have receptors for hormones that trigger vasoconstriction or vasodilation, thus helping to maintain blood pressure.

Recall that the nervous and endocrine systems also control how blood is allocated to different body regions at different times. In addition, conditions in a particular part of the body can alter blood flow there. For instance, when you run, the amount of oxygen in your skeletal muscle tissue falls, while levels of carbon dioxide, H^+, potassium ions, and other substances rise. These chemical changes cause the smooth muscle in arterioles to relax. The vasodilation results in more blood flowing past the active muscles. At the same time, arterioles in your digestive tract and kidneys constrict.

A **baroreceptor reflex** helps provide short-term control over blood pressure. Baroreceptors are pressure receptors in the **carotid arteries** in the neck, in the arch of the aorta, and elsewhere. They monitor changes in mean arterial pressure ("mean" = the midpoint) and send signals to centers in the brain. As described in Chapter 13, this information is used to coordinate the rate and strength of heartbeats with changes in the diameter of arterioles and veins. The baroreceptor reflex helps keep blood pressure within normal limits in the face of sudden changes—such as when you leap up from a chair.

> *Arteries are the main pipelines for oxygenated blood. Because arterioles can dilate and constrict, they are control points for blood flow (and pressure). Capillary beds are diffusion zones. Blood moves back to the heart through venules and veins. Valves in veins prevent the backflow of blood due to gravity.*

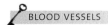

9.7 Capillaries: Where Blood Exchanges Substances with Tissues

LINK TO
SECTION
3.9

Blood enters the systemic circulation moving swiftly in the aorta, but this speed has to slow in order for substances to move into and out of the bloodstream.

A VAST NETWORK OF CAPILLARIES WEAVES CLOSE TO NEARLY ALL LIVING BODY CELLS

Your body comes equipped with one aorta, a few hundred branching arteries and veins, more than half a million arterioles and venules—and as many as 40 billion capillaries! They are so thin that it would take 100 of them to equal the thickness of a human hair. And at least one of these tiny vessels is next to living cells in nearly all body tissues.

In addition to forming a vast network of vessels (Figure 9.14*a*), this branching system also affects the speed at which blood flows through it. The flow is fastest in the aorta, quickly "loses steam" in the more numerous arterioles, and slows to a relative crawl in the narrow capillaries. The flow of blood speeds up again as blood moves into veins for the return trip to the heart.

MANY SUBSTANCES ENTER AND LEAVE CAPILLARIES BY DIFFUSION

Why do we have such an extensive system of capillaries in which blood slows to a snail's pace? Remember from Section 9.6 that capillaries are where all the substances that enter and leave cells are exchanged with the blood, many of them by diffusion. But diffusion is a slow process that is not efficient over long distances. In a large, multicellular organism such as a human, having billions of narrow capillaries solves both these problems. There is a capillary very close to nearly every cell, and in each one the blood is barely moving. As blood "creeps" along in capillaries, there is time for the necessary exchanges of fluid and solutes to take place. In fact, most solutes, including molecules of oxygen and carbon dioxide, diffuse across the capillary wall.

SOME SUBSTANCES PASS THROUGH "PORES" IN CAPILLARY WALLS

Some substances enter and leave capillaries by way of slitlike areas between the cells of capillary walls (Figure 9.14*c*). These "pores" are filled with water. They are passages for substances that cannot diffuse through the lipid bilayer of the cells that make up the capillary wall, but that *can* dissolve in water.

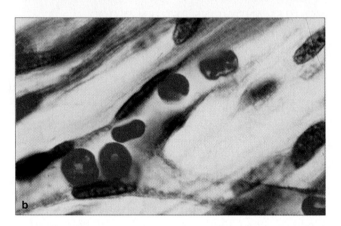

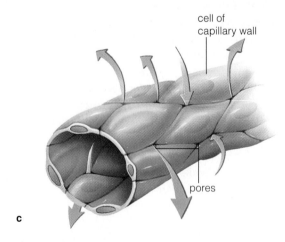

Figure 9.14 Capillaries. (**a**) A resin cast showing a dense network of capillaries. (**b**) Red blood cells moving single file in capillaries. (**c**) How substances pass through slitlike pores in the wall of a capillary.

BLOOD VESSELS

When the blood pressure inside a capillary is greater than pressure from the extracellular fluid outside, water and solutes may be forced out of the vessel—a type of fluid movement called "bulk flow" (Figure 9.15). Various factors affect this process, but on balance, a little more water leaves capillaries than enters them. The lymphatic system, which consists of lymph vessels, lymph nodes, and some other organs, returns the fluid to the blood. This system also plays a major role in body defense, and you will learn more about it in Chapter 10.

Overall, the movements of fluid and solutes into and out of capillaries help maintain blood pressure by adding water to, or subtracting it from, blood plasma. The fluid traffic also helps maintain the proper fluid balance between blood and surrounding tissues.

BLOOD IN CAPILLARIES FLOWS ONWARD TO VENULES

Capillary beds are the "turnaround points" for blood in the cardiovascular system. They receive blood from arterioles, and after the blood flows through the bed it enters channels that converge into venules—the beginning of its return trip to the heart (Figure 9.16).

At the point where a capillary branches into the capillary bed, a wispy ring of smooth muscle wraps around it. This structure, a **precapillary sphincter**, regulates the flow of blood into the capillary. The smooth muscle is sensitive to chemical changes in the capillary bed. It can contract and prevent blood from entering the capillary, or it can relax and let blood flow in.

For example, if you sit quietly and listen to music, only about one-tenth of the capillaries in your skeletal muscles are open. But if you decide to get up and boogie, precapillary sphincters will sense the demand for more blood flow to your muscles to deliver oxygen and carry away carbon dioxide. Many more of the sphincters will relax, allowing a rush of blood into the muscle tissue. The same mechanism brings blood to the surface of your skin when you blush or become flushed with heat.

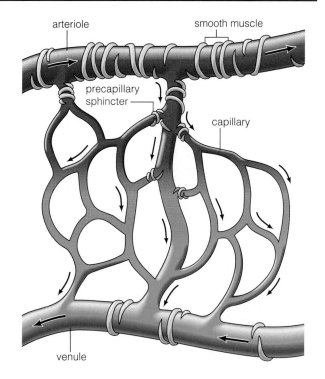

Figure 9.16 The general direction of blood flow through a capillary bed. A precapillary sphincter wraps around the base of each capillary.

The cardiovascular system's extensive network of narrow capillaries ensures that every living cell is only a short distance from a capillary.

In capillary beds, substances move between the blood and extracellular fluid by diffusion, through capillary pores, or by bulk flow. Movements of water and other substances help maintain blood pressure and the proper fluid balance between blood and tissues.

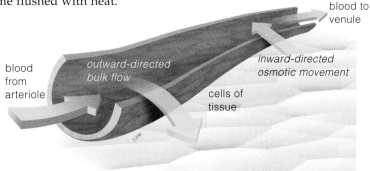

Figure 9.15 *Animated!* The "bulk flow" of fluid into and out of a capillary bed.

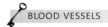

9.8 Cardiovascular Disorders

More than 50 million Americans have cardiovascular disorders, which are the leading cause of death in the United States.

What are your chances of developing a cardiovascular disorder? Some major risk factors include a family history of heart trouble, high levels of blood lipids such as cholesterol and trans fats, hypertension, obesity, smoking, lack of exercise, and simply getting older. Interestingly, however, more than half of people who suffer heart attacks do not have any of these risk factors. What is going on? As Chapter 10 describes, inflammation has powerful, sometimes damaging effects on tissues. Recent studies have implicated this as a trigger for the formation of artery-blocking plaques, which you will read about shortly. Infections by certain viruses and bacteria can trigger inflammation in coronary arteries. Inflamed tissue also leads to the production of *C-reactive protein* (by the liver); this link is why infection-related inflammation and C-reactive protein are listed in Table 9.2.

Another suspect is homocysteine, an amino acid that is released as certain proteins are broken down. Too much of it in the blood also may cause damage that is a first step in a major cardiovascular disorder, atherosclerosis.

Table 9.2	Major Risk Factors for Cardiovascular Disease
1. Inherited predisposition	
2. Elevated blood lipids (cholesterol, trans fats)	
3. Hypertension	
4. Obesity	
5. Smoking	
6. Lack of exercise	
7. Age 50+	
8. Inflammation due to infections by viruses, bacteria	
9. High levels of C-reactive protein in blood	
10. Elevated blood levels of the amino acid homocysteine	

ARTERIES CAN BE CLOGGED OR WEAKENED

In *arteriosclerosis*, or "hardening of the arteries," arteries become thicker and stiffer. In **atherosclerosis**, this condition gets worse as cholesterol and other lipids build up in the artery wall. When this *atherosclerotic plaque* grows large enough to protrude into the artery, there is less room for blood (Figure 9.17).

Coronary arteries and their branches are narrow and vulnerable to clogging by plaques. When the artery is narrowed further to one-quarter of its starting diameter, symptoms can range from mild chest pain, called *angina pectoris*, to a full-scale heart attack.

Having too many lipids in the blood—often, due to a diet high in cholesterol and trans fat—is a major risk factor for atherosclerosis. In the blood, proteins called **LDLs** (*low-density lipoproteins*) bind cholesterol and other fats and carry them to body cells. Proteins called **HDLs** (*high-density lipoproteins*) pick up cholesterol in the blood and carry it back to the liver, where it is mixed into bile and eventually excreted in feces. Because HDLs help remove excess cholesterol from the body, they are called "good cholesterol."

If there are more LDLs in the blood than cells can remove, the surplus increases the risk of atherosclerosis. This is why LDLs are called "bad cholesterol." As LDLs infiltrate artery walls, cholesterol accumulates there. Other changes occur also, and eventually a fibrous net forms over the mass—an atherosclerotic plaque. Blood tests measure the relative amounts of HDLs and LDLs in a person's blood (in milligrams). A total of 200 mg or less per milliliter of blood is considered acceptable (for most people), but experts agree that LDLs should make up only about one-third of this total, or about 70 to 80 mg.

Surgery may be the only answer for a severely blocked coronary artery. In a *coronary bypass*, a section of a large vessel taken from the chest is stitched to the aorta and to

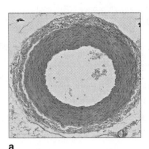

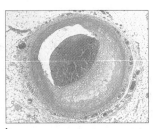

Figure 9.17 Sections from (**a**) a normal artery, and (**b**) one with its lumen narrowed by a plaque. (**c**) Coronary bypasses (green).

a

b

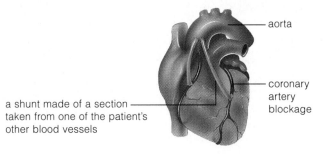

aorta

coronary artery blockage

a shunt made of a section taken from one of the patient's other blood vessels

c

BLOOD VESSELS

the coronary artery below the affected region (Figure 9.17c). In *laser angioplasty*, laser beams vaporize the plaques. In *balloon angioplasty*, a small balloon is inflated inside a blocked artery to flatten a plaque so there is more room in the artery. A small wire cylinder called a stent may then be inserted to help keep the artery open. "Plaque-busting" drugs called statins, which reduce cholesterol in the blood, can help prevent new plaques from forming.

Disease, an injury, or an inborn defect can weaken an artery so that part of its wall balloons outward. This pouchlike weak spot is called an **aneurysm**. Aneurysms can develop in various parts of the cardiovascular system, including vessels in the brain, abdomen, and the aorta. If an aneurysm bursts, it can cause serious and even fatal blood loss. A minor aneurysm may not present any immediate worry, but in the brain, especially, an aneurysm is potentially so dangerous that it requires immediate medical treatment.

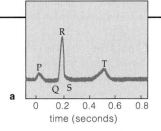

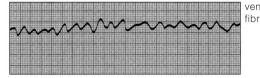

ventricular fibrillation

Figure 9.18 *Animated!* (**a**) ECG of a normal heartbeat. The P wave is generated by electrical signals from the SA node that stimulate contraction of the atria. As the stimulus moves over the ventricles, it is recorded as the QRS wave complex. After the ventricles contract, they rest briefly. The T wave marks electrical activity during this period. (There is also an atrial recovery period "hidden" in the QRS complex.) (**b**) A recording of ventricular fibrillation.

HEART DAMAGE CAN LEAD TO HEART ATTACK AND HEART FAILURE

A **heart attack** is damage to or death of heart muscle. Warning signs of a heart attack include sensations of pain or squeezing behind the breastbone, pain or numbness radiating down the left arm, sweating, and nausea. Women more often experience neck and back pain, fatigue, a sense of indigestion, a fast heartbeat, shortness of breath, and low blood pressure. Risk factors include hypertension, a circulating blood clot (an embolus, described in Section 8.6), and atherosclerosis.

In **heart failure** (HF), the heart is weakened and so does not pump blood as well as it should. Even a basic exertion such as walking can become difficult. Because patients may require repeated hospitalization, HF has become the nation's most costly health problem.

ARRHYTHMIAS ARE ABNORMAL HEART RHYTHMS

An electrocardiogram, or ECG, is a recording of the electrical activity of the cardiac cycle (Figure 9.18a). ECGs reveal **arrhythmias**, or irregular heart rhythms. Some arrhythmias are abnormal, others are not. For example, endurance athletes may have a below-average resting cardiac rate, or *bradycardia*, which is an adaptation to regular strenuous exercise. A cardiac rate above 100 beats per minute, called *tachycardia*, occurs normally during exercise or stressful situations. Serious tachycardia can be triggered by drugs (including caffeine, nicotine, alcohol, and cocaine), excessive thyroid hormones, and other factors.

Ventricular fibrillation is the most dangerous arrythmia. In parts of the ventricles, the cardiac muscle contracts haphazardly, so blood isn't pumped normally. This is what happens in sudden cardiac arrest, as described in the chapter introduction. Ventricular fibrillation is a medical emergency, but with luck, a strong electrical shock to the patient's heart or the use of defibrillating drugs can restore a normal rhythm before the damage is too serious.

A HEART-HEALTHY LIFESTYLE

Everybody ages, and none of us can control the genes we inherit. But as Table 9.2 makes clear, there are some things each of us can do to improve our chances of living free of serious cardiovascular disease. Watching our intake of foods rich in cholesterol and trans fats, getting regular exercise, and not smoking are three strategies, and they provide multiple benefits. A diet that's moderate in fats may also help keep weight under control. Exercise helps with weight control, too. It also relieves stress and helps keep muscles and bones fit and strong. Smoking is bad for just about every body system; you'll get a closer look at the devastating damage it can do to the respiratory system in Chapter 11.

Summary

Section 9.1   The cardiovascular system consists of the heart and blood vessels including arteries, arterioles, capillaries, venules, and veins. The system helps maintain homeostasis by providing rapid internal transport of substances to and from cells.

 Explore the human cardiovascular system.

Section 9.2   The heart muscle is called the myocardium. A partition, the septum, divides the heart into two halves, each with two chambers, an atrium and a ventricle. Valves in each half help control the direction of blood flow. These include a semilunar valve and an atrioventricular valve. Coronary arteries provide much of the heart's blood supply. They branch off the aorta, which carries oxygenated blood away from the heart.

Blood is pumped each time the heart beats, in a cardiac cycle of contraction and relaxation. Systole, the contraction phase, alternates with the relaxation phase, called diastole.

 Learn about the structure and function of the heart.

Section 9.3   The partition between the heart's two halves separates the blood flow into two circuits, one pulmonary and the other systemic.

a. In the pulmonary circuit, deoxygenated blood in the heart's right half is pumped to capillary beds in the lungs. The blood picks up oxygen, then flows to the heart's left atrium.

b. In the systemic circuit, the left half of the heart pumps oxygenated blood to body tissues. There, cells take up oxygen and release carbon dioxide. The blood, now deoxygenated, flows to the heart's right atrium.

Section 9.4   Electrical impulses stimulate heart contractions by way of the heart's cardiac conduction system. In the right atrium, a sinoatrial node—the cardiac pacemaker—generates the impulses and establishes a regular heartbeat. Signals from the SA node pass to the atrioventricular node, a way station for stimulation that triggers contraction of the ventricles. The nervous and endocrine systems can adjust the rate and strength of heart contractions.

Section 9.5   Blood pressure is the fluid pressure blood exerts against vessel walls. It is highest in the aorta, which receives blood pumped by the left ventricle, and drops along the systemic circuit.

 See how blood pressure is measured.

Section 9.6

a. Arteries are strong, elastic pressure reservoirs. They smooth out pressure changes resulting from heartbeats and so smooth out blood flow. When a ventricle contracts, it causes a pressure surge, or pulse, in large arteries.

b. Arterioles are control points for distributing different volumes of blood to different regions.

c. Capillary beds are diffusion zones where blood and extracellular fluid exchange substances.

d. Venules overlap capillaries and veins somewhat in function. Some solutes diffuse across their walls.

e. Veins are blood reservoirs that can be tapped to adjust the volume of flow back to the heart. Valves in some veins, such as those in the limbs, prevent blood that must return to the heart (and lungs) from flowing backward due to gravity.

Blood vessels, especially arterioles, help control blood pressure. Arterioles dilate when monitoring centers in the brain detect an abnormal rise in blood pressure. If blood pressure falls below a set point, the centers trigger vasoconstriction of arterioles. A baroreceptor reflex relies on baroreceptors in carotid arteries. It provides short-term blood pressure control by way of signals that adjust the pressure when sudden changes occur.

Section 9.7   Capillaries are where fluids and solutes move between the bloodstream and body cells. These substances move by diffusion, through pores between cells, and by bulk flow of fluid. The movements help maintain the proper fluid balance between the blood and surrounding tissues, and also help maintain proper blood volume.

Section 9.8   Cardiovascular disorders collectively are the number one cause of death in the U.S. In atherosclerosis, a buildup of cholesterol and other material develops into plaques that narrow the interior space in arteries and reduce blood flow to the heart or other tissues and organs. HDLs (high-density lipoproteins) help transport excess blood cholesterol to the liver for disposal. High levels of LDLs (low-density lipoproteins) and trans fats, smoking, obesity, and inflammation in coronary arteries are some of the major risk factors associated with atherosclerosis.

Disease, injury, or an inborn defect can weaken an artery so that part of its wall balloons outward and forms an aneurysm.

Other serious cardiovascular disorders are heart attack (damage to or death of heart muscle) and heart failure (a weakened heart that cannot pump blood efficiently). An arrhythmia—irregular heart rhythm—can be a sign of heart problems. The most serious arrhythmia is ventricular fibrillation, haphazard contractions of the ventricles that greatly reduce blood pumping.

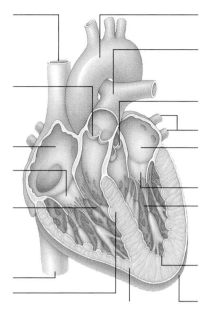

Review Questions

1. List the functions of the cardiovascular system.

2. Define a "heartbeat," giving the sequence of events that make it up.

3. Distinguish between the systemic and pulmonary circuits.

4. Explain the function of (*a*) the sinoatrial node, (*b*) the atrioventricular node, and (*c*) the cardiac pacemaker.

5. State the main function of blood capillaries. Name the main ways substances cross the walls of capillaries.

6. In the diagram above, label the heart's components.

7. State the main functions of venules and veins. What forces work together in returning venous blood to the heart?

Self-Quiz

Answers in Appendix V

1. Cells obtain nutrients from and deposit waste into _____.
   - a. blood
   - b. lymph vessels
   - c. each other
   - d. both a and b

2. The contraction phase of the heartbeat is _____; the relaxation phase is _____.

3. In the pulmonary circuit, the heart's _____ half pumps _____ blood to capillary beds inside the lungs; then _____ blood flows to the heart.
   - a. left; deoxygenated; oxygenated
   - b. right; deoxygenated; oxygenated
   - c. left; oxygenated; deoxygenated
   - d. right; oxygenated; deoxygenated

4. In the systemic circuit, the heart's _____ half pumps _____ blood to all body regions; then _____ blood flows to the heart.
   - a. left; deoxygenated; oxygenated
   - b. right; deoxygenated; oxygenated
   - c. left; oxygenated; deoxygenated
   - d. right; oxygenated; deoxygenated

5. After you eat, blood passing through the GI tract travels through the _____ to a capillary bed in the _____.
   - a. aorta; liver
   - b. hepatic portal vein; liver
   - c. hepatic vein; spleen
   - d. renal arteries; kidneys

6. The cardiac pacemaker _____.
   - a. sets the normal rate of heartbeat
   - b. is the same as the AV node
   - c. establishes resting blood pressure
   - d. all of these are correct

7. Blood pressure is highest in _____ and lowest in _____.
   - a. arteries; veins
   - b. arteries; relaxed atria
   - c. arteries; ventricles
   - d. arterioles; veins

8. _____ contraction drives blood through the systemic and pulmonary circuits; outside the heart, blood pressure is highest in the _____.
   - a. Atrial; ventricles
   - b. Atrial; atria
   - c. Ventricular; arteries
   - d. Ventricular; aorta

9. Match the type of blood vessel with its major function.
   - ____ arteries
   - ____ arterioles
   - ____ capillaries
   - ____ veins
     - a. diffusion
     - b. control of blood distribution
     - c. transport, blood volume reservoirs
     - d. blood transport and pressure regulators

10. Match these three circulation components with their descriptions.
    - ____ capillary beds
    - ____ heart chambers
    - ____ heart contractions
      - a. two atria, two ventricles
      - b. driving force for blood
      - c. zones of diffusion

Critical Thinking

1. A patient suffering from hypertension may receive drugs that decrease the heart's output, dilate arterioles, or increase urine production. In each case, how would the drug treatment help relieve hypertension?

2. Heavy smokers often develop abnormally high blood pressure. The nicotine in tobacco is a potent vasoconstrictor. Explain the connection between these two facts, including what kind of blood vessels are likely affected.

3. Before antibiotics were available, it wasn't uncommon for people in the United States (and elsewhere) to develop *rheumatic fever*. This disease develops after the patient is infected by *Streptococcus pyrogenes*, a hemolytic (red blood cell–destroying) bacterium. The infection can trigger an inflammation that ultimately damages valves in the heart. How must this disease affect the heart's functioning? What kinds of symptoms would arise as a consequence?

4. The highly publicized deaths of several airline travelers led to warnings about "economy-class syndrome." The idea is that economy-class passengers don't have as much leg room as passengers in more costly seating, so they are more likely to sit essentially motionless for long periods on flights—conditions that may allow blood to pool and clots to form in the legs. This condition is called deep-vein thrombosis, or DVT. In addition, low oxygen levels in airplane cabins may increase clotting. If a clot gets large enough to block blood flow or breaks free and is carried to the lungs or brain, it can lethally block an artery.

There could be a time lag between when a clot forms and health problems, so an air traveler who later develops DVT might easily overlook the possible connection with a flight. Studies are now under way to determine whether economy-class travel represents a significant risk of DVT. Given what you know about blood flow in the veins, explain why periodically getting up and moving around in the plane's cabin during a long flight may lower the risk that a clot will form.

Explore on Your Own

As described in Section 9.6, a pulse is the pressure wave created during each cardiac cycle as the body's elastic arteries expand and then recoil. Common pulse points— places where an artery lies close to the body surface— include the inside of the wrist, where the radial artery travels, and the carotid artery at the front of the neck. Monitoring your pulse is an easy way to observe how a change in your posture or activity affects your heart rate.

To take your pulse, simply press your fingers on a pulse point and count the number of "beats" during one minute. For this exercise, take your first measurement after you've been lying down for a few minutes. If you are a healthy adult, it's likely that your resting pulse will be between 65 and 70 beats per minute. Now sit up, and take your pulse again. Did the change in posture correlate with a change in your pulse? Now run in place for 30 seconds and take your pulse rate once again. In a short paragraph, describe what changes in your heart's activity led to the pulse differences.

The Face of AIDS

The photograph below is a snapshot of Chedo Gowero when she was thirteen years old. With both parents dead from AIDS, she left school to support herself, her ten-year-old brother, and her grandmother. She started spending her days gathering firewood and working in the homes and fields of neighbors. To keep her younger brother in school, she proudly assumed adult responsibilities.

Chedo and her brother are among the estimated 12 million African children orphaned because of AIDS. Worldwide, at least 40 million are now infected with HIV, the virus that causes it. Even after

twenty years of high-priority research all over the world, we still do not have an effective vaccine against AIDS.

There are reasons for hope, however. For example, in some African countries, the rates of infection are declining as a result of educational and prevention programs. And most Africans—including Chedo and her brother—are not infected.

HIV disables the immune system, leaving the body vulnerable to all manner of infections and some otherwise rare cancers. And while the ongoing search for a preventive vaccine has been both frustrating and challenging, it is increasing our knowledge of how the body defends itself against threats, the subject of this chapter. We now know a lot about the body's responses to tissue damage, disease-causing agents, and cancer cells. Even so, as AIDS reminds us with every passing day, we still have a great deal to learn.

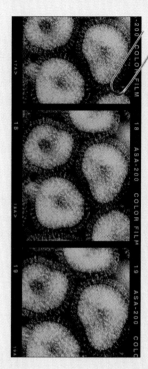

 How Would You Vote? The cost of drugs that extend the life of AIDS patients puts them out of reach of people in most developing countries. Should the federal government offer incentives to companies to discount the drugs for developing countries? What about AIDS patients at home? Who should pay for their drugs? Cast your vote online at www.thomsonedu.com/biology/starr/humanbio.

 ## Key Concepts

THE BODY'S DEFENSES
The human body has a variety of physical, chemical, and cellular defenses against bacteria, viruses, and other disease threats, including cancer. Some of these are innate (inborn), others are acquired as we encounter pathogens.

ACQUIRED IMMUNITY
In acquired immunity white blood cells mount a specific immune response against anything that is chemically recognized as not being a normal part of the body.

IMMUNE SYSTEM DISORDERS
Allergies, autoimmune disorders, and immune deficiencies are the result of faulty or failed immune mechanisms.

Links to Earlier Concepts

In this chapter you will see how several body systems and cells, including the skin (4.8) and white blood cells (8.1), work to fight infection. You will be using what you have learned about proteins (2.11) and processes of endocytosis and phagocytosis (3.10). Finally you will see how circulating blood (9.2) serves as a highway for many defensive cells and how it interacts with the lymphatic system—the body-wide network of vessels and organs where key defensive white blood cells acquire their ability to recognize threats.

10.1 Overview of Body Defenses

LINKS TO
SECTIONS
4.8, 8.1, AND 8.4

Without knowing it, every day we encounter a vast number of health threats. Our defenses include physical barriers and two interacting sets of cells and proteins.

WE ARE BORN WITH SOME GENERAL DEFENSES AND ACQUIRE OTHER, SPECIFIC ONES

Pathogen is the general term for viruses, bacteria, fungi, protozoa, and parasitic worms that cause disease. No matter what we do, we can't really avoid pathogens. They are in the air we breathe, the food we eat, and on everything we touch. This means that our survival depends on having effective defenses against them.

You may remember from Section 8.4 that an **antigen** is something that the body identifies as nonself and that triggers an immune response. Virus particles, foreign cells, toxins, and cancer cells all have antigens on their surface. Most antigens are proteins, lipids, or the large sugar molecules called oligosaccharides.

Immunity is the body's overall ability to resist and combat something that is nonself. The responses involved in immunity all are governed by genes, and they fall into two categories. Each of us is born with some preset responses, which provide **innate immunity**. These responses are launched quickly when tissue is damaged or when the body detects general chemical signals that microbes have invaded. Innate immune responses are carried out by certain white blood cells and proteins in blood plasma.

By contrast, other immune responses develop only after the body detects antigens of specific pathogens, toxins, or abnormal body cells. These responses provide **adaptive immunity**, in which armies of specialized lymphocytes and proteins mount a counterattack against invasion. They take longer to develop, but as you will read in a later section, every adaptive immune response leaves behind cells that "remember" a pathogen and protect against it for a long time, perhaps even for life. Also, some versatile genetic mechanisms underlie adaptive immune responses. They can produce lymphocytes sensitive to billions of different antigens. As a result, adaptive responses can combat billions of potential threats. Table 10.1 summarizes the features of adaptive and innate immunity.

THREE LINES OF DEFENSE PROTECT THE BODY

Biologists often portray the protections of immunity as three "lines of defense." This approach can make it easier to remember what each "line" does, even though in fact all our defenses function as parts of a whole.

The first barrier to invasion is physical. Intact skin and the linings of body cavities and tubes effectively bar most pathogens from entering the body. We'll take a closer look at these barriers in Section 10.3.

The innate immune system is the second line of defense. It swings into action as soon as an antigen has been detected internally. The responses are general; they don't target specific intruders. Still, innate responses can wipe out many pathogens before an infection becomes established. Section 10.4 describes these countermeasures, which include inflammation. When an innate immune response gets under way, it also unleashes the third line of defense, the adaptive immune system.

WHITE BLOOD CELLS AND THEIR CHEMICALS ARE THE DEFENDERS IN IMMUNE RESPONSES

You probably recall from Section 8.1 that stem cells in bone marrow give rise to white blood cells (WBCs). White blood cells, the core of the **immune system**, have crucial roles in both the innate and adaptive immune responses.

Several types of white blood cells are phagocytes, and all of them release chemical signals that help muster or strengthen defense responses. These chemicals include several types of **cytokines**, "cell movers" that promote and regulate many aspects of immunity (Table 10.2). Examples are *interleukins*, which cause inflammation and fever and also stimulate the activity of various kinds of white blood cells. *Interferons* help defend against viruses and activate certain lymphocytes. *Tumor necrosis factor* ("necrosis" means death) triggers inflammation and kills tumor cells. Some white blood cells also secrete enzymes and toxins that kill microbes.

Another chemical weapon is a set of proteins called **complement**. There are about 30 complement proteins. They are carried in the blood and can kill microbes or flag them for phagocytes such as macrophages, which then engulf and destroy the invader.

Table 10.1	Adaptive and Innate Immunity	
	Innate Immunity	**Adaptive Immunity**
Response time:	Immediate	Slower
How antigen is detected:	About 1,000 preset receptors	Billions of different receptors
Triggers:	Damage to tissues; proteins on microbes	Pathogens, toxins, altered body cells
Memory:	None	Long-term

Table 10.2	Chemical Weapons of Immunity
Complement	Directly kills cells; stimulates lymphocytes
Cytokines	Cell–cell and cell–tissue communication:
Interleukins	Cause inflammation and fever, cause T cells and B cells to divide and specialize; stimulate bone marrow stem cells, attract phagocytes, activate NK cells
Interferons	Confer resistance to viruses; activate NK cells
Tumor Necrosis Factor	Causes inflammation; kills tumor cells, causes T cells to accumulate in lymph nodes during infection
Other Chemicals	Various antimicrobial and defensive effects
	Enzymes and peptides; clotting factors; protease inhibitors; toxins; hormones

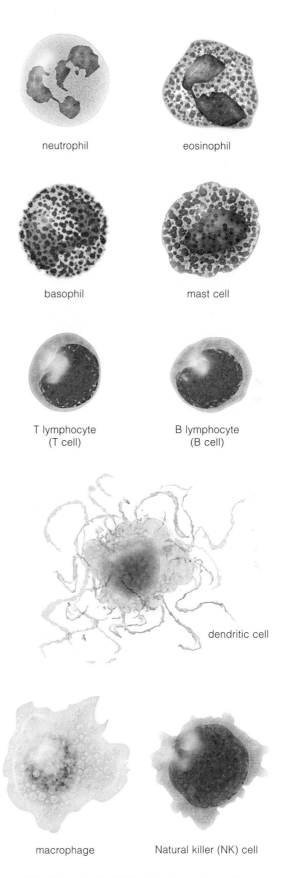

neutrophil

eosinophil

basophil

mast cell

T lymphocyte (T cell)

B lymphocyte (B cell)

dendritic cell

macrophage

Natural killer (NK) cell

Figure 10.1 *Animated!* White blood cells, including lymphocytes, that carry out immune responses.

Many white blood cells also circulate in the blood, as well as in lymph, a pale fluid that circulates in vessels of the lymphatic system. As you will read in the following section, this system works with the cardiovascular system in moving many substances throughout the body and it has major roles in defense.

Figure 10.1 gives a visual summary of white blood cells. About two-thirds of our WBCs are **neutrophils**, which follow chemical trails to infected, inflamed, or damaged tissues. **Basophils** that circulate in blood and **mast cells** in tissues release enzymes and chemicals called histamines when they detect an antigen. **Macrophages** are phagocytes that patrol the bloodstream; each of these "big eaters" can engulf as many as one hundred bacteria. **Eosinophils** target worms, fungi, and other pathogens that are too big for phagocytosis. **Dendritic cells** alert the adaptive immune system when an antigen is present in tissue fluid in the skin and body linings. The **B** and **T lymphocytes**—which we call B and T cells—have the most important roles in adaptive immunity. They are the only defensive cells that are equipped with receptors for specific antigens. **Natural killer cells** (NK cells) also are lymphocytes. They have a role in adaptive immunity, but they act mostly in innate responses that destroy cancer cells and cells that have been infected by a virus.

The body's three lines of defense are physical barriers, innate immunity, and adaptive immunity.

Immune responses are executed by white blood cells and the chemicals they release.

10.2 The Lymphatic System

As you've just read, the **lymphatic system** does several things in the body. It works with the cardiovascular system by picking up fluid that is lost from capillaries and returning it to the bloodstream. The lymphatic system's other key task is defense. As sketched in Figure 10.2, the system consists of drainage vessels, "lymphoid" organs such as the spleen and lymph nodes, and lymphoid tissues. The tissue fluid that has moved into lymph vessels is aptly called **lymph**.

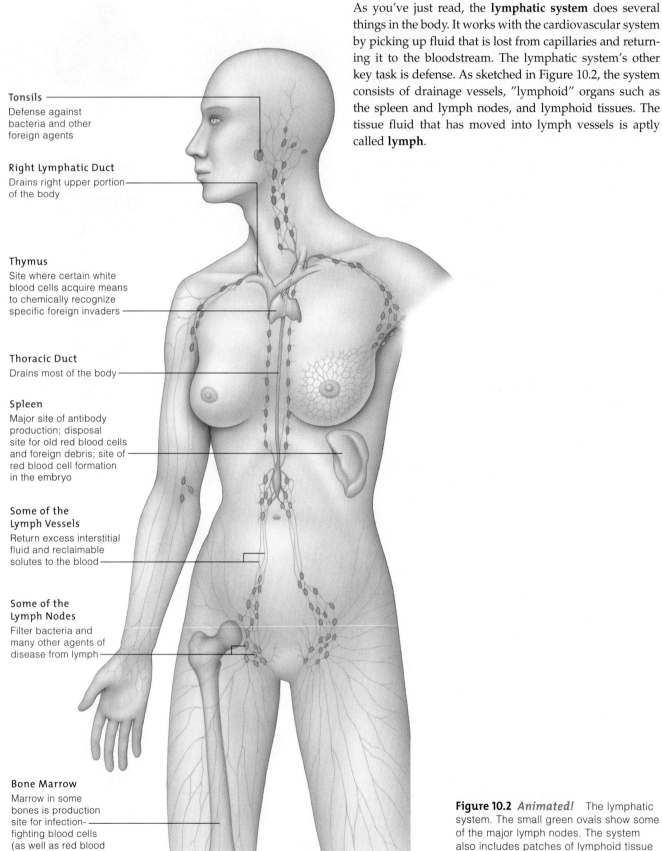

Tonsils
Defense against bacteria and other foreign agents

Right Lymphatic Duct
Drains right upper portion of the body

Thymus
Site where certain white blood cells acquire means to chemically recognize specific foreign invaders

Thoracic Duct
Drains most of the body

Spleen
Major site of antibody production; disposal site for old red blood cells and foreign debris; site of red blood cell formation in the embryo

Some of the Lymph Vessels
Return excess interstitial fluid and reclaimable solutes to the blood

Some of the Lymph Nodes
Filter bacteria and many other agents of disease from lymph

Bone Marrow
Marrow in some bones is production site for infection-fighting blood cells (as well as red blood cells and platelets)

Figure 10.2 *Animated!* The lymphatic system. The small green ovals show some of the major lymph nodes. The system also includes patches of lymphoid tissue in the small intestine and in the appendix.

ACQUIRED IMMUNITY

THE LYMPH VASCULAR SYSTEM FUNCTIONS IN DRAINAGE, DELIVERY, AND DISPOSAL

The **lymph vascular system** consists of lymph capillaries and other vessels that collect water and dissolved substances from tissue fluid and transport them to ducts of the cardiovascular system. The lymph vascular system has three functions, which we could call the "three Ds"—drainage, delivery, and disposal.

To begin with, the system's vessels are drainage channels. They collect water and solutes that have leaked out of the blood in capillary beds (due to fluid pressure there) and return those substances to the bloodstream. The system also picks up fats that the body has absorbed from the small intestine and delivers them to the bloodstream, in the way described in Section 7.5. And finally, lymphatic vessels transport foreign material and cellular debris from body tissues to the lymph vascular system's disposal centers, the lymph nodes.

The lymph vascular system starts at capillary beds (Figure 10.3a), where fluid enters the lymph capillaries. These capillaries don't have an obvious entrance. Instead, water and solutes move into their tips at flaplike "valves." These are areas where endothelial cells overlap (see Figure 9.11c).

Lymph capillaries merge into larger lymph vessels. Like veins, these vessels have smooth muscle in their wall and valves that prevent backflow. They converge into collecting ducts that drain into veins in the lower neck. This is how the lymph fluid is returned to circulating blood. Movements of skeletal muscles and of the rib cage (during breathing) help move fluid through our lymph vessels, just as they do for veins.

LYMPHOID ORGANS AND TISSUES ARE SPECIALIZED FOR BODY DEFENSE

Several elements of the lymphatic system operate in the body's defenses. These parts are the lymph nodes, the spleen, and the thymus, as well as the tonsils and patches of tissue in the small intestine, in the appendix, and in airways leading to the lungs.

The **lymph nodes** are strategically located at intervals along lymph vessels (Figures 10.2 and 10.3b). Before lymph enters the bloodstream, it trickles through at least one of these nodes. A lymph node has several chambers where white blood cells accumulate after they have been produced in bone marrow. During an infection, lymph nodes become battlegrounds where armies of lymphocytes form and where foreign agents are destroyed. Macrophages in the nodes help clear the lymph of bacteria and other unwanted substances.

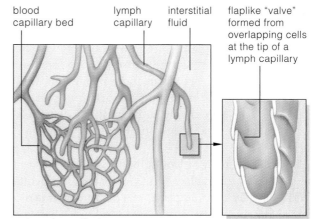

blood capillary bed   lymph capillary   interstitial fluid   flaplike "valve" formed from overlapping cells at the tip of a lymph capillary

a Lymph capillaries

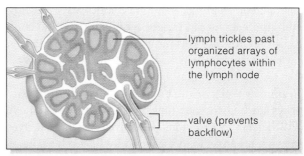

lymph trickles past organized arrays of lymphocytes within the lymph node

valve (prevents backflow)

b A lymph node, cross section

Figure 10.3 *Animated!* (**a**) Some of the lymph capillaries at the start of the drainage network called the lymph vascular system. (**b**) Cutaway diagram of a lymph node. Its inner chambers are packed with highly organized arrays of infection-fighting white blood cells.

The largest lymphoid organ, the **spleen**, is a filtering station for blood and a holding station for lymphocytes. The spleen has inner chambers filled with soft red and white tissue called "pulp." The red pulp is a reservoir of red blood cells and macrophages. (In a developing embryo, the spleen produces red blood cells.) In the white pulp, masses of lymphocytes are arrayed close to blood vessels. If an invader reaches the spleen during an infection, the lymphocytes are mobilized to destroy it, just as in lymph nodes.

The **thymus** is where T cells multiply and become specialized to combat specific foreign antigens. You will soon be learning more about how it functions.

The lymphatic system includes lymph vessels that carry tissue fluid to the blood, transport fats, and carry debris and foreign material to lymph nodes. Lymph nodes, the spleen, and the thymus all function in defense.

10.3 Surface Barriers

LINK TO
SECTION
4.8

Pathogens usually can't get past our skin or the linings of other body surfaces such as the GI tract.

If you showered today, there are probably thousands of microorganisms on every square inch of your skin. They usually are harmless as long as they stay outside the body, and some types grow so densely that they help prevent more harmful species from gaining a foothold (Figure 10.4).

Similarly, normally "friendly" bacteria in the mucosal lining of the GI tract help protect you. In females, lactate produced by *Lactobacillus* bacteria in the vaginal mucosa helps maintain a low vaginal pH that most bacteria and fungi cannot tolerate. Anything that changes the conditions in which these organisms grow can cause an infection. For example, some antibiotics used to cure bacterial infections can trigger a vaginal yeast infection because the drug also kills *Lactobacillus*. If the skin between your toes is chronically moist and warm, these conditions favor the growth of fungi that cause *athlete's foot*.

The inner walls of the branching, tubular respiratory airways leading to your lungs are coated with a sticky mucus. That mucus contains protective substances such as **lysozyme**, an enzyme that chemically attacks and helps destroy many bacteria. Broomlike cilia in the airways sweep out the pathogens.

Lysozyme and some other chemicals in tears, saliva, and gastric fluid offer more protection. Urine's low pH and flushing action help bar pathogens from the urinary tract. In adults, mild diarrhea can rid the lower GI tract of irritating pathogens; blocking it can prolong infection. (In children, however, diarrhea must be controlled to prevent dangerous dehydration.)

Intact skin, mucous membranes, lysozyme, and other physical barriers all help prevent pathogens from entering the internal environment of the body.

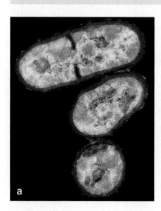
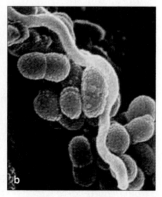

Figure 10.4 Some microbes on body surfaces. (**a**) Overgrowth of this bacterium, *Propionibacterium acnes*, causes acne. (**b**) A variety of *Streptococcus mutans*, which causes dental plaque.

10.4 Innate Immunity

Phagocytosis, inflammation, and fever are the body's "off-the-shelf" mechanisms that act at once to counter threats in general and prevent infection.

Once a pathogen has managed to enter the body, macrophages in tissue fluid are usually the first defenders on the scene. They engulf and destroy virtually anything other than undamaged body cells (Figure 10.5). If they detect an antigen, they release cytokines that, among other effects, attract dendritic cells, neutrophils, and more macrophages.

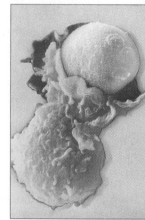

Figure 10.5 A macrophage about to engulf a yeast cell.

Complement is also an important aspect of innate immunity. As they circulate in blood and tissue fluid, complement proteins chemically detect the presence of pathogens. The encounter "activates" a complement protein, which in turn activates more complement molecules, which then activate more, and so on. The cascade of reactions quickly floods a damaged area with complement molecules.

Activated complement molecules attract phagocytes, such as macrophages and neutrophils, to damaged tissues. They also attach themselves to invaders. Phagocytes, in turn, have receptors that bind to the complement proteins. An invader that is coated with complement molecules sticks to the phagocyte, which ingests and kills it. Some complement proteins form *membrane attack complexes* (Figure 10.6). When an attack complex is inserted into the plasma membrane of a pathogen, it forms a pore—that is, a hole—in the membrane and the punctured cell quickly disintegrates.

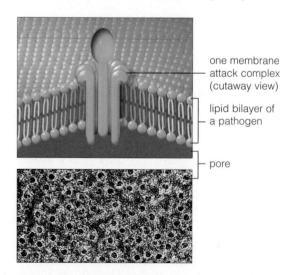

one membrane
attack complex
(cutaway view)

lipid bilayer of
a pathogen

pore

Figure 10.6 *Animated!* Pores formed in the plasma membrane of a bacterium by membrane attack complexes.

ACQUIRED IMMUNITY

Activated complement and cytokines secreted by macrophages both trigger acute (sudden) **inflammation**, a fast, local, general response to tissue invasion (Figure 10.7). Symptoms of inflammation are redness, swelling, warmth, and pain, and they are caused by a series of internal events.

First, mast cells in tissues respond to complement proteins or to an antigen by releasing histamines and cytokines into tissue fluid. Histamines make arterioles in the tissue dilate. As a result, more blood flows through them and the tissue reddens and warms with blood-borne metabolic heat.

Histamine also makes capillaries leaky. The narrow gaps between the cells of the capillary wall become a bit wider, so plasma proteins and phagocytes slip out through them (Figure 10.8). Water flows out as well. As a result of these and other changes, the tissue balloons with fluid. This swelling is called *edema*. The pain that comes with inflammation is due to edema and the effects of inflammatory chemicals.

The plasma proteins leaking into tissue fluid include factors that cause blood to clot. Clots can wall off inflamed areas and delay the spread of microbes into nearby tissues.

A **fever** is a core body temperature above the normal 37°C (98.6°F). Fever develops when cytokines released by macrophages stimulate the brain to release prostaglandins—signaling molecules that can up the set point on the hypothalamic thermostat, which controls core temperature.

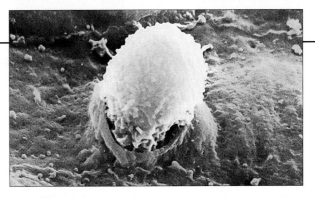

Figure 10.8 A white blood cell squeezing through the wall of a capillary.

Fevers are not usually harmful. A fever of about 39°C (100°F) is actually helpful. Among other benefits, it increases body temperature to a level that is too hot for many pathogens to function normally.

Phagocytosis, inflammation, and fever rid the body of most pathogens before they do major harm. If an infection does take hold, the adaptive immune system takes over. Three defenders we have been discussing—dendritic cells, macrophages, and complement proteins—also take part in adaptive immunity, the topic we turn to next.

Phagocytes such as macrophages, inflammation, and fever are the body's "first strike" weapons against infection. They are the tools of innate immunity, which is a general response to health challenges by pathogens.

LINKS TO SECTIONS 3.1, 3.10, 4.9, AND 8.1

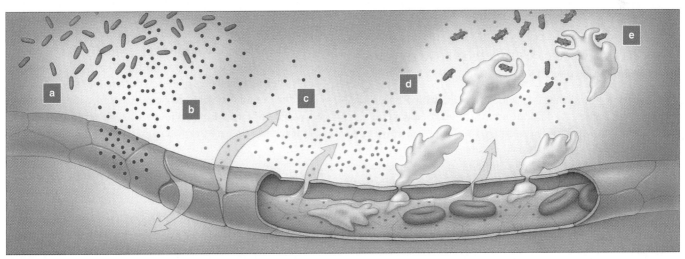

a Bacteria invade a tissue and directly kill cells or release metabolic products that damage tissue.

b Mast cells in tissue release histamine, which then triggers arteriolar vasodilation (hence redness and warmth) as well as increased capillary permeability.

c Fluid and plasma proteins leak out of capillaries; localized edema (tissue swelling) and pain result.

d Plasma proteins attack bacteria. Clotting factors wall off inflamed area.

e Neutrophils, macrophages, and other phagocytes engulf invaders and debris. Activated complement attracts phagocytes and directly kills invaders.

Figure 10.7 *Animated!* Acute inflammation in response to invading bacteria. In addition to combating the attack, the process helps prepare the damaged tissue for repair.

 ACQUIRED IMMUNITY

10.5 Overview of Adaptive Defenses

LINK TO
SECTION
8.4

When physical barriers and inflammation don't deter an invader, the adaptive immune system is mobilized.

ADAPTIVE IMMUNITY HAS THREE KEY FEATURES

Adaptive immunity is the body's third line of defense. It mobilizes B and T cells, which attack cells and substances they recognize as antigens—that is, as foreign intruders. The three defining feaures of adaptive immunity are:

1. The adaptive immune system is *specific*: Each B or T cell makes receptors for only one kind of antigen.

2. A hallmark of adaptive immunity is its *diversity*: B and T cells collectively may have receptors for at least a billion specific threats.

3. The adaptive immune response has *memory*: Some of the B and T cells formed during a first response to an invader are held in reserve for future battles with it.

When a B cell or T cell recognizes an antigen, the meeting stimulates round after round of cell division. In the end, a huge number of identical T cells or B cells form. Each one of them can now counterattack the pathogen.

All the new cells produced by dividing B or T cells are sensitive to the same antigen. Some become **effector cells** that can immediately begin destroying the enemy, while others become **memory cells**. Instead of joining the first battle, memory cells enter a resting phase. If the threat appears again, they will mount a larger, faster response to it. Memory cells are what make you "immune" to a given cold or flu virus once you have recovered from the first infection.

B CELLS AND T CELLS BECOME SPECIALIZED TO ATTACK ANTIGENS IN DIFFERENT WAYS

Chapter 8 noted that lymphocytes arise from stem cells in bone marrow. Cells that are destined to specialize as B cells continue developing in the bone marrow, but cells that will specialize as T cells travel via the blood to the thymus gland. As they complete their development, they split into two groups—helper T cells and effector cells called cytotoxic ("killer") T cells.

When B and T cells are mature, most move into lymph nodes, the spleen, and other lymphoid tissues. Some B and T cells are said to be "naive" until they are activated. Like a defensive "light bulb" blinking on, this activation happens when the cell recognizes an antigen.

When B cells and T cells identify intruders, they attack in different ways. B cells don't directly engage a

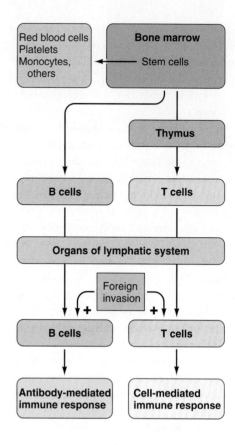

Figure 10.9 A "life history" of B cells and T cells.

pathogen. Instead they produce the defensive proteins called **antibodies**. For this reason their response is called **antibody-mediated immunity**. By contrast, cytotoxic T cells do attack invaders directly, so the T cell response is called **cell-mediated immunity** (Figures 10.9 and 10.10). Helper T cells help launch both responses, and we consider them both in more detail later. For the moment we will complete this overview with a closer look at how T and B cells "learn" they have encountered an antigen.

PROTEINS CALLED MHC MARKERS LABEL BODY CELLS AS SELF

Chapter 8 described the APO self markers on red blood cells. All body cells also have **MHC markers**. These self markers are named after the genes that code for them (*Major Histocompatibility Complex* genes). They are some of the proteins that stick out above the plasma membrane of body cells. T cells have receptors called TCRs (meaning *T Cell Receptor*) that recognize MHC

ACQUIRED IMMUNITY

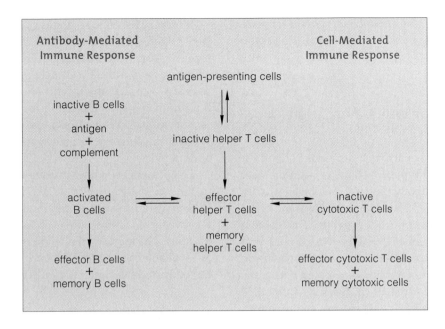

Figure 10.10 *Animated!* Overview of the links between antibody-mediated and cell-mediated immune responses—the two arms of adaptive immunity.

markers and other self tags on body cells. Part of the receptor also can recognize a particular antigen, so the antigen can be linked up with an MHC marker. As it turns out, this is a key step in starting specific immune responses, as you will read in Section 10.6.

ANTIGEN-PRESENTING CELLS INTRODUCE ANTIGENS TO T CELLS AND B CELLS

T cells and B cells can't detect an enemy by themselves. They must meet the threat after it has been "processed" by an **antigen-presenting cell**, an APC. The adaptive immune system allows plenty of opportunities for this to happen, for macrophages, dendritic cells, and B cells all can present antigens. To begin the process, the APC engulfs something bearing an antigen. Then enzymes (made by the APC's lysosomes) break the antigen into pieces. Some of the fragments are joined with MHC markers inside the cell. These *antigen–MHC complexes* move to the plasma membrane and are displayed like "come and get me" flags at the cell's surface.

When a helper T cell binds to an antigen–MHC complex, it releases cytokines. These chemicals are the signals that trigger the repeated rounds of division that produce

huge armies of activated B or T cells. They also stimulate the specialization of activated B and T cells into subgroups of effector and memory cells.

Effector B cells are called **plasma cells**. As you will now read, when called into action they can flood the bloodstream with antibodies at dizzying speed.

Adaptive immune responses are carried out mainly by T and B cells. The responses are specific, they allow for amazing diversity, and they produce memory cells that can mount a faster, stronger response if the same antigen appears in the body again.

Macrophages, dendritic cells, and B cells all can serve as antigen-presenting cells that expose T and B cells to processed antigens—a form that inactive T and B cells can recognize.

Helper T cells release cytokines that stimulate the repeated rounds of cell division by activated T and B cells.

Armies of T and B cells are produced in adaptive immune responses. Their counterattacks are carried out by helper T cells, cytotoxic T cells, and antibodies from plasma (B) cells.

10.6 Antibody-Mediated Immunity: Defending against Threats Outside Cells

LINK TO
SECTION
3.6

Antibodies are crucial for proper functioning of the immune system. This section looks more closely at how antibodies help defend the body and what roles different kinds of antibodies play.

ANTIBODIES DEVELOP WHILE B CELLS ARE IN BONE MARROW

While a B cell is in bone marrow, it develops antibodies. Figure 10.11 shows the typical Y-shape of a simple one. The place where an antibody can bind an antigen usually is near the tip of the two "arms," and its shape and other characteristics are determined by genes. The genetic mechanisms involved ensure that no two B cells will make antibodies that are alike. This is the source of the diversity and specificity of antibody-mediated immunity.

As a B cell matures, it makes copies of its antibodies and they become embedded in its plasma membrane so that the two arms stick out. Before long the B cell bristles with antibodies, each one with receptor sites for one kind of antigen.

ANTIBODIES TARGET PATHOGENS THAT ARE OUTSIDE CELLS

Antibodies can't enter cells and bind to enemies hidden there. Instead, antibodies target pathogens and toxins that are circulating in tissues or body fluids.

To get a more complete picture of how an antibody-mediated response unfolds, let's follow a B cell that has made its antibodies but that has not yet been activated. In this state, a B cell can function as an antigen-presenting cell. When an antigen binds with some of the B cell's antibodies, it links them together. This linking triggers endocytosis, which moves the antigen into the cell. There, antigen–MHC complexes form and are displayed at the B cell surface.

When TCRs of a responding helper T cell bind to the antigen–MHC complex, the T and B cells exchange signals. Then they disengage. This step activates the B cell. Now when the B cell meets an antigen that is *not* part of a complex, the B cell's antibodies bind to it. The binding helps spur the B cell to divide; a boost comes from cytokines secreted from nearby helper T cells. The B cell's descendants become specialized as plasma cells and memory B cells (Figure 10.12).

The plasma cells release huge numbers of antibodies in the bloodstream—up to 2,000 of them each minute. When any of these antibodies binds to an antigen, it flags the invader for destruction by phagocytes and complement proteins. The memory B cells do not engage in battle but are available to respond rapidly to the antigen if it attacks the body another time (Section 10.8).

THERE ARE FIVE CLASSES OF ANTIBODIES, EACH WITH A PARTICULAR FUNCTION

B cells produce five classes of antibodies. Collectively they are called **immunoglobulins**, or Igs. They are the proteins that result from the gene shuffling that takes place while B cells mature (Section 10.5) and while an immune response is under way. We abbreviate them as IgM, IgD, IgG, IgA, and IgE. Each type has antigen-binding sites and other sites with special roles. When B cells secrete them they have roughly these shapes:

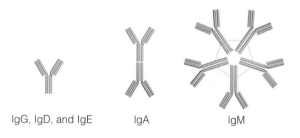

IgG, IgD, and IgE   IgA   IgM

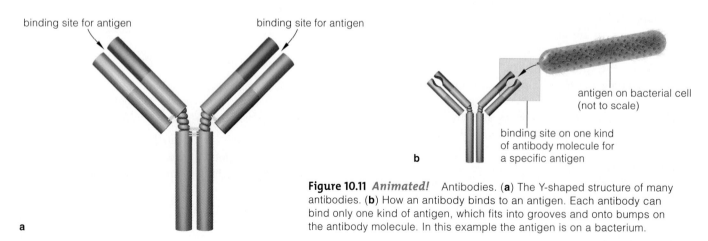

binding site for antigen   binding site for antigen

antigen on bacterial cell (not to scale)

binding site on one kind of antibody molecule for a specific antigen

b

a

Figure 10.11 *Animated!* Antibodies. (**a**) The Y-shaped structure of many antibodies. (**b**) How an antibody binds to an antigen. Each antibody can bind only one kind of antigen, which fits into grooves and onto bumps on the antibody molecule. In this example the antigen is on a bacterium.

a A dendritic cell engulfs and digests a bacterium in interstitial fluid, and then migrates to a lymph node. There, it presents antigen–MHC complexes. Binding to this antigen-presenting cell causes inactive "naïve" helper T cells to multiply, producing effector (active) helper T cells and memory helper T cells.

b Each inactive B cell has more than 100,000 receptors, all specific for the same antigen. This B cell binds to its antigen on the surface of a bacterium in a lymph node. After also binding complement, the bacterium enters the B cell and is digested. Its fragments bind to MHC molecules, and the complexes are displayed at the B cell's surface.

c TCRs of an activated helper T cell bind to the antigen–MHC complexes on the B cell. Binding makes the helper T cell secrete cytokines. These signals cause the B cell to divide, producing a huge population of B cells with identical antigen receptors. The cells specialize into effector B cells and memory B cells.

d Effector B cells begin making huge numbers of IgA, IgG, or IgE antibodies, all of which recognize the same antigen as the original B cell receptor. The new antibodies circulate throughout the body and bind to any remaining bacteria.

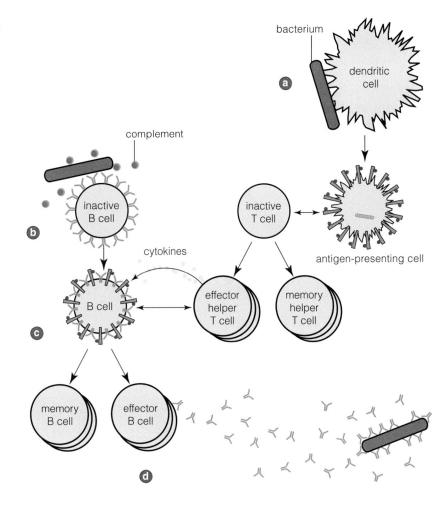

Figure 10.12 *Animated!* Antibody-mediated immune response. This example is a response to a bacterial invasion.

IgM is the first antibody secreted during immune responses and the first one produced by newborns. IgM molecules cluster into a structure with ten antigen-binding sites. This makes it more efficient at binding clumped targets, such as agglutinating red blood cells (Section 8.4) and clumps of virus particles.

Along with IgM, *IgD* is the most common antibody bound to inactive B cells. It may help activate helper T cells.

IgG makes up about 80 percent of the antibodies in blood. It's the most efficient one at turning on complement proteins, and it neutralizes many toxins. This long-lasting antibody easily crosses the placenta. It helps protect the developing fetus with the mother's acquired immunities. IgG secreted into early milk is also absorbed into a suckling newborn's bloodstream.

IgA is the main immunoglobulin in the secretions of exocrine glands, such as tears, saliva, and breast milk. It also is in mucus that coats the respiratory, digestive, and reproductive tracts, areas that are vulnerable to infection. Like IgM, it can form large structures that can bind larger antigens. Bacteria and viruses can't bind to the cells of mucous membranes when IgA is bound to them, and IgA is effective in fighting the pathogens that cause salmonella, cholera, gonorrhea, influenza, and polio.

The *IgE* antibody is involved in allergic reactions, including asthma, hay fever, and hives. IgE also triggers inflammation after attacks by parasitic worms and other pathogens. The tails of IgE antibodies lock onto basophils and mast cells, with their antigen receptors facing outward. When the receptors bind to an antigen, basophils and mast cells release histamine, which promotes inflammation.

B cells secrete five classes of antibodies (immunoglobulins). All help protect the body against diverse threats. They bind to antigens of pathogens or toxins that are outside cells and flag them for destruction by other defenders.

10.7 Cell-Mediated Responses—Defending against Threats Inside Cells

Antibody-mediated responses can't reach threats inside cells. So when cells become infected or altered in harmful ways, other "warrior" cells must come to the defense.

Many pathogens evade antibodies. They hide in body cells, kill them, and often reproduce inside them. They are exposed only briefly after they slip out of one cell and before they infect others. Viruses, bacteria, and some fungi and protozoans all can enter cells. Cell-mediated immune responses are the body's weapons against these dangers as well as abnormal body cells such as cancer cells.

Figure 10.13 gives you an overview of how a cell-mediated immune response takes place. Like an antibody-mediated response, it gets under way when an APC such as a dendritic cell presents an antigen to T cells. The response also leaves cadres of memory cells.

Some of the warriors in cell-mediated immunity are helper T cells and cytotoxic T cells, which respond to

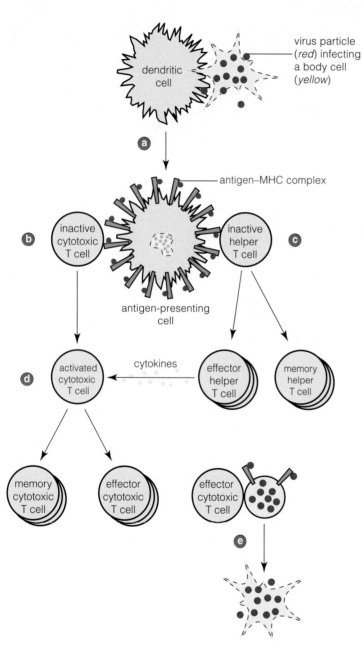

a A dendritic cell patrolling tissues meets and engulfs the remains of a virus-infected cell. The digested antigen fragments bind to MHC molecules, and the complexes are displayed at the cell's surface. The dendritic cell migrates to a lymph node or to the spleen.

b In the lymph node or spleen, TCR receptors on an inactive cytotoxic T cell bind to the complexes on the surface of the antigen-presenting dendritic cell. The interaction activates the cytotoxic T cell.

c In the lymph node, receptors of an inactive helper T cell bind to antigen–MHC complexes on the dendritic cell. The helper T cell is activated and begins dividing. The daughter cells differentiate into effector and memory helper T cells.

d Cytokines released by effector helper T cells stimulate the activated cytotoxic T cell to divide as it returns to circulating blood. Its descendants specialize into effector and memory cytotoxic T cells.

e One of the new circulating cytotoxic T cells encounters antigen–MHC complexes on the surface of a cell infected with the same virus. It touch-kills the cell by injecting it with perforin and enzymes that cause apoptosis.

Figure 10.13 *Animated!* Diagram of a T cell–mediated immune response.

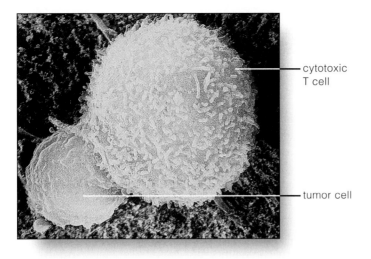

Figure 10.14 A cytotoxic T cell caught in the act of touch-killing a tumor cell.

particular antigens. Others, including NK cells and macrophages, make more general responses.

Helper T cell cytokines stimulate NK cells. These killer cells don't need to have an antigen presented to them. Instead, they simply attack any body cell that has too few or altered MHC markers, or that antibodies have tagged for destruction. They also kill body cells flagged with chemical "stress markers" that develop when a cell is infected or has become cancerous.

Cytotoxic T cells are so sensitive to antigen–MHC complexes and altered body cells that they don't need further signals to start multiplying. They release molecules that can "touch-kill" infected and abnormal body cells (Figure 10.14). Cytotoxic cells also secrete chemicals that cause the genetically programmed death of a target cell. This programmed cell death is called **apoptosis** (a-poh-TOE-sys). The term comes from a Greek word meaning to fall apart, and that's what happens to the cell. Its cytoplasm dribbles out and its DNA and organelles are broken up. After a cytotoxic T cell makes its lethal hit, it disengages from the doomed cell and moves on.

CYTOTOXIC T CELLS CAUSE THE BODY TO REJECT TRANSPLANTED TISSUE

Cytotoxic T cells cause the rejection of tissue and organ transplants. This is partly because features of the MHC markers on donor cells differ enough from the recipient's to be recognized as antigens.

To help prevent rejection, before an organ is transplanted the MHC markers of a potential donor are analyzed to determine how closely they match those of the patient. Because such tissue grafts generally succeed only when the donor and recipient share at least 75 percent of their MHC markers, the best donor is a close relative of the recipient, such as a parent or sibling, who is likely to have a similar genetic makeup.

More commonly, however, the donated organ comes from a fresh cadaver. In addition to having well-matched MHC markers, the donor and recipient also must have compatible blood types (Section 8.4).

After surgery, the organ recipient receives drugs that suppress the immune system. The treatment also may include other therapies designed to fend off an attack by B and T cells. Suppression of the immune system means that the patient must take large doses of antibiotics to control infections. In spite of the difficulties, many organ recipients survive for years beyond the surgery.

Interestingly, not all transplanted tissues provoke a recipient's immune defenses. Two examples are tissues of the eye and the testicles. In simple terms, the plasma membrane of cells of these organs apparently bears receptors that can detect activated lymphocytes in the surroundings. Before such a defender can launch an attack, the protein signals the soon-to-be-besieged cell to secrete a chemical that triggers apoptosis in the approaching lymphocytes—so the attack is averted. Our ability to readily transplant the cornea—the outermost layer of the eye that is vital to clear vision—depends on this mechanism.

Cell-mediated immune responses are mounted against infected or altered body cells. Helper T cells and cytotoxic T cells target antigens. NK cells, macrophages, and various other white blood cells make nonspecific responses.

IMPACTS, ISSUES

Some researchers are actively trying to genetically alter pigs to create varieties with organs bearing common human MHC markers. Why pigs? A major reason is that, anatomically and physiologically, pig organs are quite similar to ours. In theory, such transgenic animals could be sources of readily transplantable organs or cells. Transplantation of organs from one species of animal to another is called xenotransplantation. You can read more about this topic in Chapter 22.

10.8 Immunological Memory

Memory cells produced during an adaptive immune response can provide decades of immunity against pathogens.

Memory cells that form during a primary (first) immune response circulate in the blood for years, even decades. Compared to the B and T cells that initiate a primary response, these patrolling battalions have many more cell "soldiers," so they intercept antigens far sooner. Plasma cells and effector T cells form sooner, in greater numbers, so the infection is ended before the host—you—gets sick (Figure 10.15). Even more memory T and plasma cells form during a secondary adaptive response.

Over time, the kinds of memory T and plasma cells in the body are determined by the antigens you are exposed to. That is why you likely will not have immunity to a bacterium or virus that is not in your usual surroundings.

> *Memory cells enable the adaptive immune system to make a faster, more powerful secondary response to another encounter with a pathogen.*

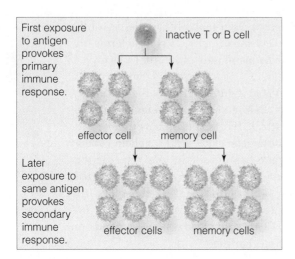

a

First exposure to antigen provokes primary immune response.

inactive T or B cell

effector cell   memory cell

Later exposure to same antigen provokes secondary immune response.

effector cells   memory cells

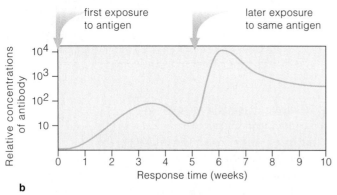

b

first exposure to antigen

later exposure to same antigen

Relative concentrations of antibody

Response time (weeks)

Figure 10.15 *Animated!* Immunological memory. (**a**) In immune responses, some plasma cells and T cells are set aside as memory cells. (**b**) Comparison of a primary and secondary immune response.

10.9 Applications of Immunology

Modern science has developed powerful weapons that can enhance the immune system's functioning or harness it in new ways to treat disease.

IMMUNIZATION GIVES "BORROWED" IMMUNITY

Immunization is a way to increase immunity against specific diseases. In *active immunization*, a **vaccine** is injected into the body or taken orally, sometimes according to a schedule (Figure 10.16). A vaccine is a prepared substance that contains an antigen. The first injection elicits a primary immune response. A "booster shot" given later elicits a secondary response, in which more effector cells and memory cells form and can provide long-lasting disease protection.

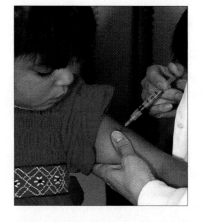

Figure 10.16 From the Centers for Disease Control and Prevention, the 2005 immunization guidelines for children in the United States. Low-cost or free vaccinations are available at many community clinics and health departments.

Recommended Vaccines	Recommended Ages
Hepatitis B	Birth–2 months
Hepatitis B booster 1	1–4 months
Hepatitis B booster 2	6–18 months
DTP Diphtheria, tetanus, and pertussis (whooping cough)	2, 4, and 6 months
DTP booster 1	15–18 months
DTP booster 2	4–6 years
DT	11–12 years
HiB (*Haemophilus influenzae*)	2, 4, and 6 months
HiB booster	12–15 months
Polio	2 and 4 months
Polio booster 1	6–18 months
Polio booster 2	4–6 years
MMR (Measles, Mumps, Rubella)	12–15 months
MMR booster	4–6 years
Pneumococcal	2, 4, and 6 months
Pneumococcal booster 1	12–15 months
Pneumococcal booster 2	2–18 years
Varicella	12–18 months
Hepatitis A series (in some areas)	2–12 years
Influenza	Yearly, 1–18 years

ACQUIRED IMMUNITY

Many vaccines are made from killed or extremely weakened pathogens. For example, weakened poliovirus particles are used for the Sabin polio vaccine. Other vaccines are based on inactivated forms of natural toxins, such as the bacterial toxin that causes tetanus. Today many vaccines are made with harmless genetically engineered viruses (Chapter 22). These "transgenic" viruses incorporate genes from three or more different viruses in their genetic material. After a person is vaccinated with an engineered virus, body cells use the new genes to produce antigens, and immunity is established.

Passive immunization often helps people who are already infected with pathogens, such as those that cause tetanus, measles, hepatitis B, and rabies. A person receives injections of antibodies that have been purified from another source, preferably someone who already has produced a large amount of the antibody. The effects don't last long because the recipient's own B cells are not producing antibodies. However, the injected antibodies may counter the immediate attack.

Vaccines are powerful weapons, but they can fail or have adverse effects. In rare cases, a vaccine can damage the nervous system or result in chronic immunological problems. A physician can explain the risks and benefits.

MONOCLONAL ANTIBODIES ARE USED IN RESEARCH AND MEDICINE

Commercially prepared **monoclonal antibodies** harness antibodies for medical and research uses (Figure 10.17). The term "monoclonal antibody" refers to the fact that the antibodies are made by cells cloned from just a single antibody-producing B cell.

At one time laboratory mice were the "factories" for making monoclonal antibodies. Today most monoclonal antibodies are produced using genetically altered bacteria. Genetically engineered plants such as corn also are being used to make antibodies that may be both cost-effective and safe (few plant pathogens can infect people). The first "plantibody" to be used on human volunteers prevented infection by a bacterium that causes tooth decay.

Monoclonal antibodies have become useful tools in diagnosing health conditions. Because they can recognize and bind to specific antigens, they can detect substances in the body—a bacterial cell, another antibody, or a chemical—even if only a tiny amount is present. Uses include home pregnancy tests and screening for prostate cancer and some sexually transmitted diseases. As you'll read next, monoclonal antibodies also have potential uses as "magic bullets" to deliver drugs used to treat certain forms of cancer.

Figure 10.17
Monoclonal antibody-producing cells stored in liquid nitrogen.

IMMUNOTHERAPIES REINFORCE DEFENSES

Immunotherapy bolsters defenses against infections and cancer cells by manipulating the body's own immune mechanisms. Cytokines that activate B and T cells are being used to treat some cancers. Monoclonal antibodies are another weapon. For example, some aggressive breast cancers have excess HER2 proteins at their surface. The drug Herceptin is a monoclonal antibody that binds to the proteins and draws a response from NK cells. The drug can be a double-edged sword, however, because some healthy body cells also have HER2 proteins, and they are attacked as well.

Monoclonal antibodies also can be bound to poisons to make *immunotoxins*. When these substances bind to an antigen on a cancer cell, they enter it and block processes that allow it to survive and multiply. Some experimental immunotoxins are being tested against HIV, the virus that causes AIDS.

Various body cells, including ones infected by a virus, can make and secrete interferons. When these cytokines reach an uninfected cell, they trigger a chemical attack that prevents the virus from multiplying. *Gamma* interferon, which is made by T cells, has other functions, too. It calls NK cells into action and also boosts the activity of macrophages. Genetically engineered gamma interferon is used to treat hepatitis C, a chronic, potentially lethal viral disease. Some kinds of cells (not lymphocytes) produce *beta* interferon. This protein has recently been approved for the treatment of a type of **multiple sclerosis**, a disease in which the immune system mounts an attack on parts of the nervous system.

Immunization can enhance immunity to specific diseases by stimulating the production of both effector and memory lymphocytes.

Monoclonal antibodies and cytokines such as interferons are important tools in medical research, testing, and the treatment of disease.

10.10 Disorders of the Immune System

Sometimes immune responses are misguided, weakened, or even nonexistent. This section surveys some of the most common immune system difficulties and disorders.

IN ALLERGIES, HARMLESS SUBSTANCES PROVOKE AN IMMUNE ATTACK

Most allergies won't kill you, but they sure can make you miserable. In at least 15 percent of the people in the United States, normally harmless substances can provoke immune responses. These substances are *allergens*, and the response to them is an **allergy**. Common allergens are pollen (Figure 10.18*a*), a variety of foods and drugs, dust mites, fungal spores, insect venom, and ingredients in cosmetics. Some responses start within minutes; others are delayed. Either way, the allergens trigger mild to severe inflammation of mucous membranes and in some cases other tissues as well.

Some people are genetically predisposed to develop allergies. Infections, emotional stress, or changes in air temperature also may cause reactions that otherwise might not occur. When an allergic person first is exposed to certain antigens, IgE antibodies are secreted and bind to mast cells (Figure 10.18*b*). When the IgE binds an allergen, mast cells secrete prostaglandins, histamine, and other substances that fan inflammation. They also cause an affected person's airways to constrict. In **hay fever**, the allergic response is miserably evident in stuffed sinuses, a drippy nose, and sneezing.

Like other allergies, food allergies are skewed responses of the immune system in which a particular food is interpreted as an "invader." The most common culprits are shellfish, eggs, and wheat. Depending on the person and the food involved, symptoms typically include diarrhea, vomiting, and sometimes swelling or tingling of mucous membranes. Some food allergies can be lethal. For example, in people who are allergic to peanuts even a tiny amount can trigger **anaphylactic shock**—a whole-body allergic response in which the person's blood pressure plummets, among other frightening symptoms.

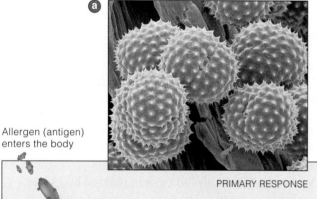

Allergen (antigen) enters the body

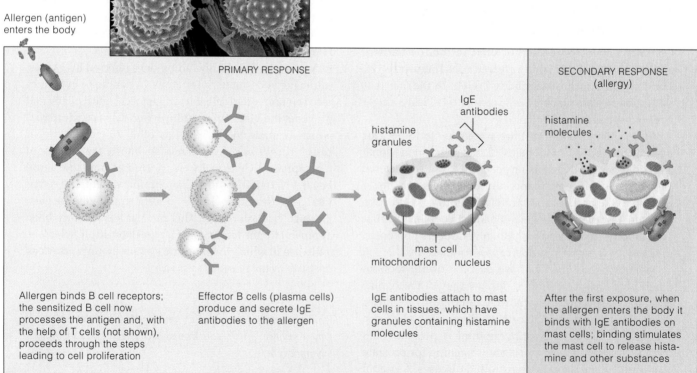

PRIMARY RESPONSE

SECONDARY RESPONSE (allergy)

IgE antibodies

histamine granules

histamine molecules

mast cell

mitochondrion   nucleus

Allergen binds B cell receptors; the sensitized B cell now processes the antigen and, with the help of T cells (not shown), proceeds through the steps leading to cell proliferation

Effector B cells (plasma cells) produce and secrete IgE antibodies to the allergen

IgE antibodies attach to mast cells in tissues, which have granules containing histamine molecules

After the first exposure, when the allergen enters the body it binds with IgE antibodies on mast cells; binding stimulates the mast cell to release histamine and other substances

Figure 10.18 (**a**) Micrograph of ragweed pollen. (**b**) The basic steps leading to an allergic response.

IMMUNE SYSTEM DISORDERS

Anaphylactic shock is also a concern for people who are allergic to wasp or bee venom, for they can die within minutes of a single sting. Air passages to the lungs rapidly constrict, closing almost completely. Fluid gushes from dilated, permeable (and hence extremely leaky) blood vessels all over the body. Blood pressure plummets, which can lead to the complete collapse of the person's cardiovascular system.

The emergency treatment for anaphylactic shock is an injection of the hormone epinephrine. People who know they are at risk (usually because they've already had a bad reaction to an allergen) can carry the necessary medication with them, just in case.

Antihistamines are anti-inflammatory drugs that are often used to relieve short-term allergy symptoms. In some cases a person may opt to undergo a desensitization program. First, skin tests are used to identify offending allergens. Inflammatory responses to some of them can be blocked if the patient's body can be stimulated to make IgG instead of IgE. Gradually, larger and larger doses of specific allergens are administered. Each time, the person's body produces more circulating IgG molecules and IgG memory cells. The IgG will bind with an allergen and block its attachment to IgE. As a result, inflammation is blocked, too.

AUTOIMMUNE DISORDERS ATTACK "SELF"

In an **autoimmune response**, the immune system's powerful weapons are unleashed against normal body cells or proteins. An example is **rheumatoid arthritis**. People with this disorder are genetically predisposed to it. Their macrophages and T and B cells become activated by antigens associated with the joints. Immune responses are mounted against their body's collagen molecules and also apparently against antibodies that have bound to an (as yet unknown) antigen. Tissues in the joints are damaged even more by the inflammation and the complement

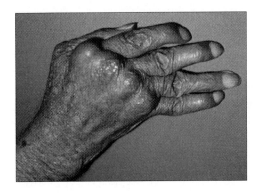

Figure 10.19 Hand of a person affected by rheumatoid arthritis.

system (Figure 10.19). Abnormal repair mechanisms compound the problem. Eventually the affected joints fill with synovial membrane cells and become immobile.

Another common autoimmune disease is **type 1 diabetes**. This is a type of *diabetes mellitus*, in which the pancreas does not secrete enough of the hormone insulin for proper absorption of glucose from the blood. For reasons that are still being investigated, the immune system attacks and destroys the insulin-secreting cells. A viral infection may trigger the response. Chapter 15 looks at the various forms of diabetes in more detail.

Systemic lupus erythematosus (SLE) primarily affects younger women, but males may also develop it. A characteristic symptom is a rash on the face that extends from cheek to cheek, roughly in the shape of a butterfly. The rash is one sign that the affected person has developed antibodies to her or his own DNA and other "self" components. Antigen–antibody complexes accumulate in joints, blood vessel walls, the skin, and the kidneys. In addition to the "butterfly" rash, symptoms include fatigue, painful arthritis, and in some cases a near-total breakdown of kidney function. Therapeutic drugs can help relieve many SLE symptoms, but there is no cure.

Immune responses tend to be stronger in women than in men. And autoimmunity is far more frequent in women. We know that the receptor for estrogen is involved in certain genetic controls. Is the receptor also implicated in autoimmune responses? It's a question that researchers are exploring.

IMMUNE RESPONSES CAN BE DEFICIENT

The term **immunodeficiency** applies when a person's immune system is weakened or lacking altogether. When the body has too few properly functioning lymphocytes, its immune responses are not effective. Both T and B cells are in short supply in **severe combined immune deficiency** (SCID). This disorder usually is inherited, and infants born with it may die early in life. Lacking adequate immune responses, they are extremely vulnerable to infections that are not life-threatening to other people. One type of SCID is now being treated by gene therapy (Chapter 22).

Infection by the human immunodeficiency virus (HIV) causes **AIDS** (acquired immunodeficiency syndrome). HIV is transmitted when body fluids of an infected person enter another person's tissues. The virus cripples the immune system by attacking macrophages and helper T cells. This leaves the body dangerously susceptible to infections and to some otherwise rare forms of cancer. We return to this topic in Chapter 18, which considers the growing global threat of infectious diseases.

Summary

Section 10.1 The body protects itself from pathogens with general and specific responses of white blood cells and chemicals they release. Preset, inborn responses provide innate immunity. Responses that develop after the body detects antigens of specific pathogens provide adaptive immunity (Table 10.3). An antigen is a protein or other type of molecule that triggers an immune response against itself.

The chemical signals that help organize or strengthen immune responses are called cytokines. They include interleukins, interferons, and complement proteins.

Section 10.2 The lymphatic system has a key role in defense. T and B lymphocytes are stationed in lymph nodes, the spleen, the tonsils, and other parts of the system. Lymph vessels also recover water and dissolved substances that have escaped from blood capillaries and return them to the general circulation.

 Learn more about how the lymphatic system functions.

Section 10.3 Physical barriers to infection, such as intact skin and mucous membranes lining body surfaces, are the first line of defense against pathogens. Lysozyme in mucus attacks many bacteria. Other chemical barriers include tears, saliva, and gastric juice. Urine and diarrhea can help flush pathogens from the urinary tract and GI tract.

Table 10.3	The Human Body's Three Lines of Defense against Pathogens

BARRIERS AT BODY SURFACES (*nonspecific* targets)

Intact skin; mucous membranes at other body surfaces

Infection-fighting chemicals in tears, saliva, gastric fluid

Normally harmless bacteria on body surfaces, which outcompete pathogens

Flushing effect of tears, saliva, urination, diarrhea, sneezing, and coughing

INNATE IMMUNE RESPONSES (*nonspecific* targets)

Inflammation

1. Fast-acting white blood cells (neutrophils, eosinophils, and basophils)
2. Macrophages (also take part in immune responses)
3. Complement proteins, blood-clotting proteins, and other infection-fighting cytokines

Organs with pathogen-killing functions (such as lymph nodes)

Some cytotoxic cells (e.g., NK cells) with a range of targets

ADAPTIVE IMMUNE RESPONSES (*specific* targets only)

1. White blood cells (T cells, B cells, and macrophages that interact with them)
2. Communication signals (e.g., interleukins) and chemical weapons (e.g., antibodies, perforins)

Section 10.4 Innate immune responses counter threats in a general way that does not require the detection of specific antigens. They often can prevent an infection from becoming established.

Macrophages are "first responders" that engulf and digest foreign agents and clean up damaged tissue. Complement proteins bind to pathogens and lethally puncture them by inserting membrane attack complexes into the invader's plasma membrane. They also attract phagocytes.

Activated complement and cytokines from macrophages trigger inflammation, a fast, local response to tissue damage. In this response, mast cells in the tissue release histamine, which dilates arterioles. As more blood flows through the arterioles, the tissue reddens and becomes warmer. Fluid leaking from permeable blood capillaries leads to edema, causing swelling and pain. Blood-clotting proteins help repair damaged blood vessels.

Chemical signals triggered by an infection can lead to an increase in the body temperature set point. The result is a fever.

Section 10.5 Adaptive defenses are specific, can combat a great diversity of antigens, and generate memory T and B cells that provide extended immunity to conquered pathogens. B cells and T cells are activated when they recognize an antigen. They then multiply and form large populations of identical cells.

Activated B cells produce antibodies. These are receptors for specific antigens that bind antigens and flag them to be destroyed by phagocytes or other defender cells. Effector B cells, called plasma cells, can release floods of antibodies.

The response by T cells provides cell-mediated immunity. T cells recognize combinations of antigen fragments and MHC self markers. These complexes are produced by antigen-presenting cells (dendritic cells, macrophages, B cells). Cytotoxic T cells are the effectors that attack intruders directly. Helper T cells release cytokines that mobilize and strengthen defense responses.

Section 10.6 Antibodies target pathogens that are outside body cells. In the antibody-mediated response, plasma cells secrete large numbers of antibodies that circulate in the bloodstream. Antibodies are proteins, often Y-shaped, and each has binding sites for one kind of antigen. The five classes of antibodies—IgG, IgD, IgE, IgA, and IgM—are immunoglobulins.

 See how antibodies combat pathogens in the blood and lymphatic system.

Section 10.7 Cell-mediated responses destroy infected cells, tumor cells, and cells of tissue or organ transplants. Cytotoxic T cells secrete chemicals that

can trigger apoptosis, or programmed cell death, in an invading cell.

 See how a cell-mediated immune response unfolds.

Section 10.8 Memory cells that form during a primary adaptive response circulate in the blood for years. They can mount a stronger, more rapid secondary response to a pathogen if it invades the body again.

Section 10.9 In active immunization, a vaccine provokes an immune response, including the production of memory cells. In passive immunization, injections of purified antibodies help patients combat an infection. Monoclonal antibodies are tools in medical research, testing, and the treatment of various diseases.

Section 10.10 An allergy is an immune response to a generally harmless substance. An autoimmune response is an attack by lymphocytes on normal body cells. Immunodeficiency is a weakened or nonexistent capacity to mount an immune response. A prime example is AIDS, caused by the HIV virus.

Review Questions

1. While you're jogging in the surf, your toes land on a jellyfish. Soon the bottoms of your toes are swollen, red, and warm to the touch. Using the diagram at the upper right as a guide, describe how these signs of inflammation came about.

2. Distinguish between:
   a. neutrophil and macrophage
   b. cytotoxic T cell and natural killer cell
   c. effector cell and memory cell
   d. antigen and antibody

3. What is the difference between innate immunity and adaptive immunity?

4. How does a macrophage or a dendritic cell become an antigen-presenting cell?

5. What is the difference between an allergy and an autoimmune response?

Self-Quiz

Answers in Appendix V

1. _____ are barriers to pathogens at body surfaces.
   a. Intact skin and mucous membranes
   b. Tears, saliva, and gastric fluid
   c. Resident bacteria
   d. all are correct

2. Complement proteins function in defense by _____.
   a. neutralizing toxins
   b. enhancing resident bacteria
   c. promoting inflammation
   d. forming pores that cause pathogens to disintegrate
   e. both a and b are correct
   f. both c and d are correct

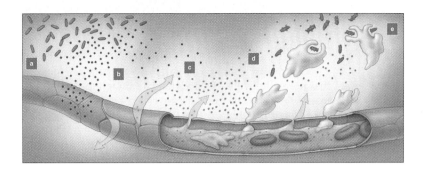

3. _____ are molecules that lymphocytes recognize as foreign and that elicit an immune response.
   a. Interleukins   d. Antigens
   b. Antibodies   e. Histamines
   c. Immunoglobulins

4. Another term for antibodies is _____; there are _____ classes of these molecules.
   a. B cells; three
   b. immunoglobulins; three
   c. B cells; five
   d. immunoglobulins; five

5. Antibody-mediated responses work best against _____.
   a. pathogens inside cells   d. both b and c
   b. pathogens outside cells   e. all are correct
   c. toxins

6. The most common antigens are _____.
   a. nucleotides   c. steroids
   b. triglycerides   d. proteins

7. The ability to develop a secondary immune response is based on _____.
   a. memory cells   d. effector cytotoxic T cells
   b. circulating antibodies   e. mast cells
   c. plasma cells

8. Tears are part of the body's defensive arsenal. What defense category do they fall into, and why?

9. Match the immunity concepts:
   ___ inflammation   a. neutrophil
   ___ antibody secretion   b. plasma cell
   ___ phagocyte   c. nonspecific response
   ___ immunological memory   d. purposely causing
   ___ vaccination   memory cell
   ___ allergy   production
                                 e. basis of secondary
                                    immune response
                                 f. nonprotective immune
                                    response

Critical Thinking

1. New research suggests a link between some microbes that normally live in the body and seemingly unrelated major illnesses. The gum disease called periodontitis itself is not life-threatening, for instance, but it is a fairly good predictor for heart attacks. Bacteria that cause gum disease can trigger inflammation. Thinking back to your reading in Chapter 9, how do you suppose that this response also may be harmful to the heart?

2. Given what you now know about how foreign invaders trigger immune responses, explain why mutated forms of viruses, which have altered surface proteins, pose a monitoring problem for a person's memory cells.

3. Researchers have been trying to develop a way to get the immune system to accept foreign tissue as "self." Can you think of some clinical applications for such a development?

4. Elena developed chicken pox when she was in kindergarten. Later in life, when her children developed chicken pox, she remained healthy even though she was exposed to countless virus particles each day. Explain why.

5. Quickly review Section 4.9 on homeostasis. Then write a short essay on how the immune response contributes to stability in the internal environment.

6. In 1796 the English physician Edward Jenner (below) injected a young boy with material from cowpox scabs. Cowpox is a mild, smallpox-like disease that is not nearly so dangerous as smallpox. Jenner had noticed that people who got cowpox never later got smallpox, which was a major killer at the time. He suspected—rightly as it turned out—that having cowpox somehow made a person immune to smallpox. The boy was his test subject for this hypothesis. After the boy's bout of cowpox was over, Jenner injected him with pus from a smallpox sore. The boy stayed healthy, and the episode led to the discovery of vaccination—a term that literally means "encowment." What do you think would happen if a physician tried this experiment today?

Explore on Your Own

The photograph in Figure 10.20 below shows a reaction to a skin test for tuberculosis. For this test, a health care worker scratches a bit of TB antigen into a small patch of a patient's skin. In people who have a positive reaction to the test, a red swelling develops at the scratch site, usually within a day or two. Even in a person with no medical history of the disease, this response is visible evidence of immunological memory. It shows that there has been an immune response against the tuberculosis bacterium, which the person's immune system must have encountered at some time in the past. Tests for allergies work the same way.

In many communities, a TB test is required for people who are applying for jobs that involve public contact, such as teaching in the public schools. To learn more about this public health measure, find out if the test is required in your community, where it is available, and why public health authorities believe it is important.

Figure 10.20 A positive reaction to a tuberculosis skin test.

Down in Smoke

Each day, about 3,000 teenagers—most younger than 15—join the ranks of smokers in the United States. The first time someone lights up, they typically cough and choke on the irritants in smoke, and they may feel dizzy and nauseated.

So why do smoking "recruits" ignore the threat signals their body is sending and keep on lighting up? Research tells us that teens take up the habit in order to fit in socially. At the time, the threat that tobacco use poses to their health and survival seems remote. And of course, the nicotine in cigarette smoke is extremely addictive.

Tobacco smoke is really bad news for the respiratory system, our topic in this chapter. Cilia that line the airways to the lungs normally sweep away airborne pollutants and microbes. Unfortunately, smoke from a single cigarette immobilizes them for hours. Smoke also kills white blood cells that patrol and defend the respiratory tract. Microbes may start living there, leading to more colds, bronchitis, even asthma attacks.

You're probably aware that smoking also is a major risk factor for lung cancer. It also is linked with other cancers. For example, females who start smoking in their teens are about 70 percent more likely to develop breast cancer than those who don't smoke. Other "bad-efits" of smoking include increased blood pressure, higher levels of "bad" cholesterol (LDL), and lower levels of "good" cholesterol (HDL).

The respiratory system has one basic job—to help maintain homeostasis by bringing in oxygen for body cells and to carry away cells' carbon dioxide wastes. Structures such as lungs and functions such as breathing allow the respiratory system to perform this vital gas exchange.

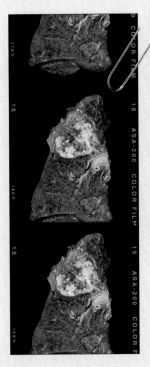

 How Would You Vote? Tobacco is both a worldwide threat to health and a profitable product for American companies. As tobacco use by its citizens declines, should the United States encourage international efforts to reduce tobacco use? Cast your vote online at www.thomsonedu.com/biology/starr/humanbio.

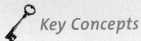 Key Concepts

THE RESPIRATORY SYSTEM
Respiration provides the body with oxygen for aerobic respiration in cells and removes carbon dioxide. These gases enter and leave the body by way of the respiratory system.

GAS EXCHANGE
Oxygen and carbon dioxide are exchanged across the thin walls of alveoli, tiny sacs in the lungs. The cardiovascular system carries gases to and from the lungs.

BREATHING CONTROLS
Various controls regulate respiration. The nervous system controls the rate, depth, and rhythmic pattern of breathing.

Links to Earlier Concepts

In this chapter you will once again use your knowledge of concentration gradients and diffusion (3.9) to help you understand the mechanisms that move oxygen into and carbon dioxode out of the body. You will see how the respiratory system works together with the cardiovascular system (9.1) to supply oxygen and remove carbon dioxide, and you will learn exactly how hemoglobin and red blood cells function in gas exchange (8.1, 9.3, 9.6).

11.1 The Respiratory System—Built for Gas Exchange

Getting oxygen from air and releasing carbon dioxide wastes are the vital functions of the respiratory system.

AIRWAYS ARE PATHWAYS FOR OXYGEN AND CARBON DIOXIDE

Our body cells require oxygen for aerobic respiration and must get rid of carbon dioxide. The **respiratory system** handles these key tasks (Figure 11.1).

When a person breathes quietly, air typically enters and leaves the respiratory system by way of the nose. Hairs at the entrance to the nasal cavity and in its ciliated epithelial lining filter out large particles, such as dust, from incoming air. The air also is warmed in the nose and picks up moisture from mucus. A septum (wall) of bone and cartilage separates the nasal cavity's two chambers. Channels link the cavity with paranasal sinuses above and behind it (which is why nasal sprays for colds or allergies can relieve mucus-clogged sinuses). Tear glands produce moisture that drains into the nasal cavity. Crying increases the flow, which is why your nose "runs" when you cry.

From the nasal cavity, air moves into the **pharynx**. This is the entrance to both the **larynx** (an airway) and

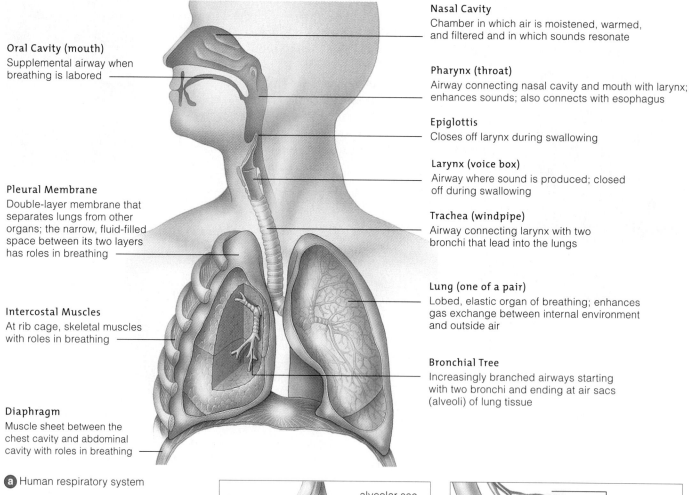

Oral Cavity (mouth)
Supplemental airway when breathing is labored

Pleural Membrane
Double-layer membrane that separates lungs from other organs; the narrow, fluid-filled space between its two layers has roles in breathing

Intercostal Muscles
At rib cage, skeletal muscles with roles in breathing

Diaphragm
Muscle sheet between the chest cavity and abdominal cavity with roles in breathing

Nasal Cavity
Chamber in which air is moistened, warmed, and filtered and in which sounds resonate

Pharynx (throat)
Airway connecting nasal cavity and mouth with larynx; enhances sounds; also connects with esophagus

Epiglottis
Closes off larynx during swallowing

Larynx (voice box)
Airway where sound is produced; closed off during swallowing

Trachea (windpipe)
Airway connecting larynx with two bronchi that lead into the lungs

Lung (one of a pair)
Lobed, elastic organ of breathing; enhances gas exchange between internal environment and outside air

Bronchial Tree
Increasingly branched airways starting with two bronchi and ending at air sacs (alveoli) of lung tissue

a Human respiratory system

Figure 11.1 *Animated!*
Components of the human respiratory system and their functions. Also shown are the diaphragm and other structures with secondary roles in respiration.

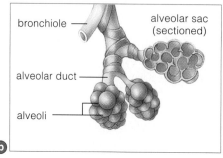

bronchiole

alveolar sac (sectioned)

alveolar duct

alveoli

b

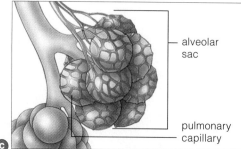

alveolar sac

pulmonary capillary

c

THE RESPIRATORY SYSTEM

Figure 11.2 Color-enhanced scanning electron micrograph of cilia and mucus-secreting cells (orange) in the respiratory tract.

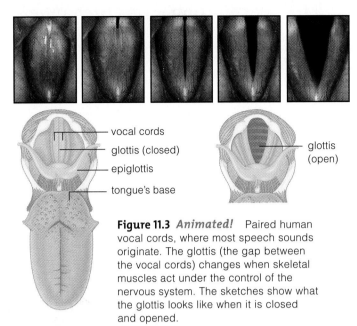

vocal cords
glottis (closed)
epiglottis
tongue's base
glottis (open)

Figure 11.3 *Animated!* Paired human vocal cords, where most speech sounds originate. The glottis (the gap between the vocal cords) changes when skeletal muscles act under the control of the nervous system. The sketches show what the glottis looks like when it is closed and opened.

the esophagus (which leads to the stomach). Nine pieces of cartilage form the larynx. One of these, the thyroid cartilage, is the "Adam's apple."

Luckily, our airways are nearly always open. The flaplike *epiglottis*, attached to the larynx, points up during breathing. However, recall from Chapter 7 that when you swallow, the larynx moves up so that the epiglottis partly covers the opening of the larynx. This helps prevent food from entering the respiratory tract and causing choking.

From the larynx, air moves into the **trachea** (TRAY-key-uh), or windpipe. Press gently at the lower front of your neck, and you can feel some of the bands of cartilage that ring the tube, adding strength and helping to keep it open. The trachea branches into two airways, one leading to each lung. Each airway is a **bronchus** (BRAWN-kus; plural: bronchi). The epithelial lining of bronchi includes mucus-secreting cells and cilia (Figure 11.2). Bacteria and airborne particles stick in the mucus; then the upward-beating cilia sweep the debris-laden mucus toward the mouth.

Near the entrance to the larynx, part of a mucous membrane forms horizontal folds that are the **vocal cords** (Figure 11.3). When you exhale, air is forced through the *glottis*, a gap between the vocal cords that is the opening to the larynx. Air moving through it makes the cords vibrate. By controlling the vibrations we can make sounds. Using our lips, teeth, tongue, and the soft roof over the tongue (the soft palate), we can modify these sounds into speech, song, and other vocalizations.

LUNGS ARE ELASTIC AND PROVIDE A LARGE SURFACE AREA FOR GAS EXCHANGE

Your **lungs** are elastic cone-shaped organs separated from each other by the heart. The left lung has two lobes, the right lung three. The lungs are located inside the rib cage above the **diaphragm**, a sheet of muscle between the thoracic (chest) and abdominal cavities. The lungs are soft and spongy, and they don't attach directly to the chest cavity wall. Instead, each lung is enclosed by a pair of thin membranes called **pleurae** (singular: pleura). You can visualize this arrangement if you think of pushing your closed fist into a fluid-filled balloon. A lung occupies the same kind of position as your fist, and the pleural membrane folds back on itself (as the balloon does) to form a closed *pleural sac*. An extremely narrow *intrapleural space* (*intra-* means between) separates the membrane's two facing surfaces. A thin film of lubricating *intrapleural fluid* in the space reduces chafing between the membranes. Inside each lung, the bronchi narrow as they branch and form "bronchial trees." These narrowing airways are **bronchioles** and their endings are called **respiratory bronchioles**. Tiny air sacs bulge out from their walls. Each sac is an **alveolus** (plural: alveoli), and each lung has about 150 million of them. Usually alveoli are clustered as an alveolar sac. Alveoli are where gases diffuse between the lungs and lung capillaries (Figure 11.1*b* and 11.1*c*).

Collectively the alveoli provide a huge surface area for the diffusion of gases. If they were stretched out as a single layer, they would cover the body several times over—or the floor of a racquetball court!

Taking in oxygen and removing carbon dioxide are the major functions of the respiratory system.

In alveoli inside the lungs, oxygen enters lung capillaries and carbon dioxide leaves them to be exhaled.

11.2 Respiration = Gas Exchange

LINKS TO
SECTIONS
3.7 AND 3.11

All living cells in the body rely on respiration to supply them with oxygen and dispose of carbon dioxide.

Chapter 3 discussed how aerobic respiration inside cells uses oxygen and produces carbon dioxide wastes that must be removed from the body. **Respiration**, in contrast, is the overall exchange of oxygen inhaled from the air for waste carbon dioxide, which is exhaled:

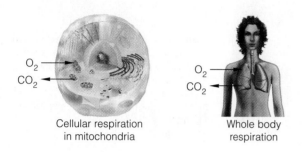

Cellular respiration
in mitochondria

Whole body
respiration

As you just read in Section 11.1, gas exchange takes place in alveoli in the lungs. The alveoli are where the respiratory system's role in respiration ends. From there, the cardiovascular system takes over the task of moving gases (Figure 11.4), which it transports along with other substances arriving from the digestive system and moving into and out of the urinary system.

> *Respiration is the exchange of gases—oxygen and carbon dioxide—with the outside world. The respiratory system brings in oxygen from air and expels carbon dioxide.*
>
> *The cardiovascular system transports O_2 and CO_2 between the lungs and tissues.*

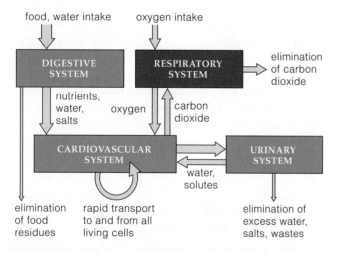

Figure 11.4 Links between the respiratory system, the cardiovascular system, and other organ systems.

11.3 The "Rules" of Gas Exchange

Gas exchange depends on the diffusion of oxygen and carbon dioxide down a concentration gradient.

Gas exchange in the body relies on the tendency of oxygen and carbon dioxide to diffuse down their respective concentration gradients—or, as we say for gases, their *pressure gradients*. When molecules of either gas are more concentrated outside the body, they tend to move into the body, and vice versa.

At sea level the air is about 78 percent nitrogen, 21 percent oxygen, 0.04 percent carbon dioxide, and 0.96 percent other gases. Atmospheric pressure at sea level is about 760 mm Hg, as measured by a mercury barometer (Figure 11.5). Each gas accounts for only *part* of the total pressure exerted by the whole mix of gases. Oxygen's partial pressure is 21 percent of 760, about 160 mm Hg. Carbon dioxide's partial pressure is about 0.3 mm Hg.

Gas exchange must be efficient in order to meet the metabolic needs of a large, active animal such as a human. Various factors influence the process. To begin with, gases enter and leave the body by crossing a thin **respiratory surface** of epithelium. The respiratory surface must be moist, because gases can't diffuse across it unless they are dissolved in fluid. Two factors affect how many gas molecules can move across the respiratory surface in a given period of time. The first is surface area, and the second is the partial pressure gradient across it. The larger the surface area and the steeper the partial pressure gradient, the faster diffusion takes place. The millions of thin-walled alveoli in your lungs provide a huge surface area for gas exchange.

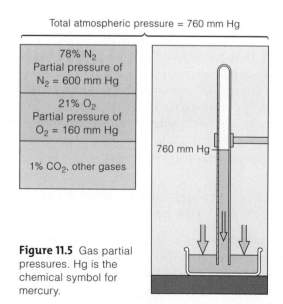

Total atmospheric pressure = 760 mm Hg

78% N_2
Partial pressure of
N_2 = 600 mm Hg

21% O_2
Partial pressure of
O_2 = 160 mm Hg

1% CO_2, other gases

760 mm Hg

Figure 11.5 Gas partial pressures. Hg is the chemical symbol for mercury.

Figure 11.6 Breathing challenges. (**a**) A climber approaches the summit of Chomolungma (Mt. Everest), where the air contains only a small amount of oxygen. (**b**) The human body is unable to extract oxygen dissolved in water.

WHEN HEMOGLOBIN BINDS OXYGEN, IT HELPS MAINTAIN THE PRESSURE GRADIENT

Gas exchange also gets a boost from the hemoglobin in red blood cells. Each hemoglobin molecule binds with as many as four oxygen molecules in the lungs, where the oxygen concentration is high. When blood carries red blood cells into tissues where the oxygen concentration is low, hemoglobin *releases* oxygen. Thus, by carrying oxygen away from the respiratory surface, hemoglobin helps maintain the pressure gradient that helps draw oxygen into the lungs—and into the blood in lung capillaries. Later in this chapter you will learn more about the way oxygen binds to and is released from hemoglobin.

GAS EXCHANGE "RULES" CHANGE WHEN OXYGEN IS SCARCE

In environments where there is less oxygen than normal, the rules of gas exchange change. For instance, the partial pressure of oxygen falls the higher you go (Figure 11.6a). A person who isn't acclimatized to the thinner air at high altitude can become *hypoxic*—meaning that tissues are chronically short of oxygen. Above 2,400 meters (about 8,000 feet), the brain's respiratory centers trigger the response known as *hyperventilation*—faster, deeper breathing—to compensate for the oxygen deficiency.

When you swim or dive, there may be ample oxygen dissolved in the water but the body has no way to extract it. (Gills do this for a fish.) Humans trained to dive without oxygen tanks can stay submerged only for about three minutes (Figure 11.6b).

Deep divers risk "raptures of the deep" or nitrogen narcosis. This condition develops because water pressure increases the deeper you go, and at about 45 meters (150 feet) dangerous levels of nitrogen gas (N_2) start to become dissolved in tissue fluid and move into cells. In brain cells the nitrogen interferes with nerve impulses, and the diver becomes euphoric and drowsy. If a diver ascends from depth too quickly, the falling pressure causes N_2 to enter the blood faster than it can be exhaled, so nitrogen bubbles may form in blood and tissues. The resulting pain (especially in joints) is called *decompression sickness*, or "the bends."

Gas exchange depends on steep partial pressure gradients between the outside and inside of the body. The larger the respiratory surface and the larger the partial pressure gradient, the faster gases diffuse.

When hemoglobin in red blood cells binds oxygen, it helps maintain the pressure gradient that draws air into the lungs.

 GAS EXCHANGE

11.4 Breathing—Air In, Air Out

LINK TO
SECTION
2.6

You will take about 500 million breaths by age 75—even more if you consider that young children breathe faster than adults do. But what does "taking a breath" mean?

WHEN YOU BREATHE, AIR PRESSURE GRADIENTS REVERSE IN A CYCLE

Breathing ventilates the lungs in a continuous, in/out pattern called a **respiratory cycle**. Ventilation has two phases. First, **inspiration**—or inhalation—draws a breath of air into the airways. Then, in the phase of **expiration** or exhalation, a breath moves out.

In each respiratory cycle, the volume of the chest cavity increases, then decreases (Figure 11.7). At the same time, pressure gradients between the lungs and the air outside the body are *reversed*. To understand how this shift affects breathing, it helps to remember that air in your airways (oxygen, carbon dioxide, and the other atmospheric gases) is at the same pressure as the outside atmosphere.

Before you inhale, the pressure inside all your alveoli (called *intrapulmonary pressure*) is also the same as that of outside air.

THE BASIC RESPIRATORY CYCLE As you start to inhale, the diaphragm contracts and flattens, and external intercostal muscle movements lift the rib cage up and out (Figure 11.7a). As the chest cavity expands, the lungs expand too. At that time, the air pressure in alveolar sacs is lower than the atmospheric pressure. Fresh air follows this gradient and flows down the airways, then into the alveoli. If you take a deep breath, the volume of the chest cavity increases even more because contracting neck muscles raise the sternum and the first two ribs.

During normal, quiet breathing, expiration is passive. The muscles that contracted to bring about inspiration simply relax and the lungs recoil, like a stretched rubber band. As the lung volume shrinks, the air in the alveoli is compressed. Because pressure in the sacs now is greater than the atmospheric pressure, air follows the gradient, out of the lungs (Figure 11.7b).

If your lungs must rapidly expel more air—for instance, when you huff and puff while working out—expiration becomes active. Muscles in the wall of the abdomen contract, pushing your diaphragm upward, and other muscle movements reduce the volume of the chest cavity even more. Add to these changes the natural recoil of the lungs, and a great deal of air in the lungs is pushed outward.

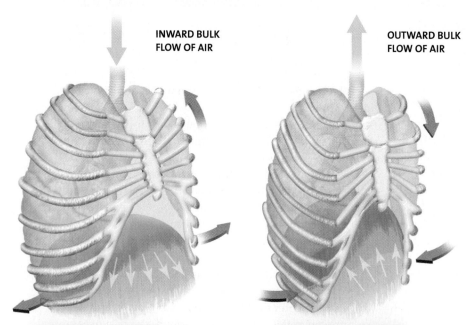

INWARD BULK FLOW OF AIR

OUTWARD BULK FLOW OF AIR

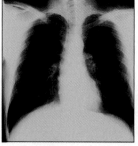

a Inhalation. Diaphragm contracts and moves down. The external intercostal muscles contract and lift the rib cage upward and outward. The lung volume expands.

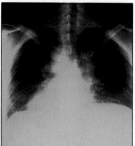

b Exhalation. Diaphragm and external intercostal muscles return to the resting positions. Rib cage moves down. Lungs recoil passively.

Figure 11.7 *Animated!* Changes in the size of the chest (thoracic) cavity during a respiratory cycle. The X-ray image in (**a**) shows how taking a deep breath changes the volume of the thoracic cavity. Part (**b**) shows how the volume shrinks after exhalation.

GAS EXCHANGE

ANOTHER PRESSURE GRADIENT AIDS THE PROCESS

A negative pressure gradient *outside* the lungs contributes to the respiratory cycle. Atmospheric pressure is a bit higher than the pressure in the pleural sac that encloses the lungs. The pressure difference is enough to make the lungs stretch and fill the expanded chest cavity. It keeps the lungs snug against the chest wall even when air is being exhaled, when the lung volume is much smaller than the space inside the chest cavity. As a result, when the chest cavity expands with the next breath, so do the lungs.

You may recall from Chapter 2 that the hydrogen bonds between water molecules prevent them from being easily pulled apart. This cohesiveness of water molecules in the fluid in the pleural sac also helps your lungs hug the chest wall, in much the same way that two wet panes of glass resist being pulled apart.

A "collapsed lung"—medically, called *pneumothorax*—is caused by an injury or illness that allows air to enter the pleural cavity. The lungs can't expand normally and breathing becomes difficult and painful.

HOW MUCH AIR IS IN A "BREATH"?

About 500 milliliters (two cupfuls) of air enters or leaves your lungs in a normal breath. This volume of air is called **tidal volume**. You can increase the amount of air you inhale or exhale, however. In addition to air taken in as part of the tidal volume, a person can forcibly inhale roughly 3,100 milliliters of air, called the *inspiratory reserve volume*. By forcibly exhaling, you can expel an additional

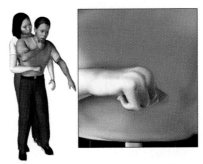
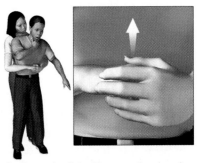

a Place a fist just above the choking person's navel, with the flat of your thumb against the abdomen.

b Cover the fist with your other hand. Thrust both fists up and in with enough force to lift the person off his or her feet.

Figure 11.9 *Animated!* The Heimlich maneuver.

expiratory reserve volume of about 1,200 milliliters of air. **Vital capacity** is the maximum volume of air that can move out of the lungs after you inhale as deeply as possible. It is about 4,800 milliliters for a healthy young man and about 3,800 milliliters for a healthy young woman. As a practical matter, people rarely take in more than half their vital capacity, even when they breathe deeply during strenuous exercise. At the end of your deepest exhalation, your lungs still are not completely emptied of air; another roughly 1,200 milliliters of *residual volume* remains (Figure 11.8).

How much of the 500 milliliters of inspired air is actually available for gas exchange? Between breaths, about 150 milliliters of exhaled "dead" air remains in the airways and never reaches the alveoli. Thus only about 350 (500 − 150) milliliters of fresh air reaches the alveoli each time you inhale. An adult typically breathes at least twelve times per minute. This rate of ventilation supplies the alveoli with 4,200 (350 × 12) milliliters of fresh air every 60 seconds. This is about the volume of soda pop in four 1-liter bottles.

When food "goes down the wrong way" and enters the trachea (instead of the esophagus), air can't be inhaled or exhaled normally. A choking person can suffocate in just a few minutes. The emergency procedure called the Heimlich maneuver can dislodge food from the trachea by elevating the diaphragm muscle. This reduces the chest volume, forcing air up the trachea. Figure 11.9 shows how to perform the maneuver. With luck, the air will rush out with enough force to eject the obstruction.

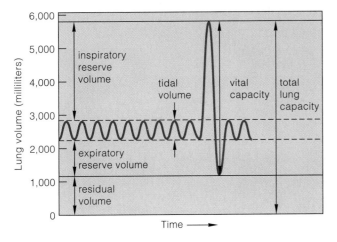

Figure 11.8 *Animated!* Lung volume during quiet breathing and during forced inspiration and expiration ("spikes" above and below the normal tidal volume).

In the respiratory cycle, the air movements of breathing occur as the volume of the chest cavity expands and shrinks. These changes alter the pressure gradients between the lungs and outside air.

11.5 How Gases Are Exchanged and Transported

LINKS TO
SECTIONS
2.6, 2.7, 3.9, 3.11,
8.1, AND 8.2

Ventilation moves gases into and out of the lungs. But it is respiration *that provides body cells with oxygen for cellular respiration and picks up their carbon dioxide wastes.*

Physiologists divide respiration into "external" and "internal" phases. *External* respiration moves oxygen from alveoli into the blood, and moves carbon dioxide in the opposite direction. During *internal* respiration, oxygen moves from the blood into tissues, and carbon dioxide moves from tissues into the blood.

ALVEOLI ARE MASTERS OF GAS EXCHANGE

The alveoli in your lungs are ideally constructed for their function of gas exchange. The wall of each alveolus is a single layer of epithelial cells, supported by a gossamer-thin basement membrane. Hugging the alveoli are lung capillaries (Figure 11.10*a*). They, too, have an extremely thin basement membrane around their wall. In between the two basement membranes is a film of fluid. It may seem like a lot of layers, but the **respiratory membrane** they form is far narrower than even a fine baby hair. This is why oxygen and carbon dioxide can diffuse rapidly across it—oxygen moving in and carbon dioxide moving out (Figure 11.10*b,c*).

Some cells in the epithelium of alveoli secrete *pulmonary surfactant.* This substance reduces the surface tension of the watery film between alveoli. Without it, the force of surface tension can collapse the delicate alveoli. This can happen to premature babies whose underdeveloped lungs do not yet have working surfactant-secreting cells. The result is the dangerous disorder called **infant respiratory distress syndrome**.

HEMOGLOBIN IS THE OXYGEN CARRIER

Blood plasma can carry only so much dissolved oxygen and carbon dioxide. To meet the body's requirements, the gas transport must be improved. The hemoglobin in red blood cells binds and transports both O_2 and CO_2. This pigment enables blood to carry some 70 times more oxygen than it otherwise would and to transport 17 times more carbon dioxide away from tissues.

Air inhaled into your alveoli contains plenty of oxygen and relatively little carbon dioxide. Just the opposite is true of blood arriving from tissues—which, remember, enters lung capillaries at the "end" of the pulmonary circuit (Section 9.3). Thus, in the lungs, oxygen diffuses down its pressure gradient into the blood plasma and then into red blood cells, where up to four oxygen molecules rapidly form a weak, reversible bond with each molecule of hemoglobin. Hemoglobin with oxygen bound to it is called **oxyhemoglobin**, or HbO_2.

The amount of HbO_2 that forms depends on several factors. One is the partial pressure of oxygen—that is, the relative amount of oxygen in blood plasma. In general, the higher its partial pressure, the more oxygen will be picked up by hemoglobin, until oxygen is attached to all hemoglobin binding sites. HbO_2 will give up its oxygen in tissues where the partial pressure of oxygen is lower than in the blood. Figure 11.11 will give you an idea of the pressure gradients in different areas of the body.

In tissues with high metabolic activity—and therefore a greater demand for oxygen—the chemical conditions loosen hemoglobin's "grip" on oxygen. For example, the binding of oxygen weakens as temperature rises or as acidity increases and pH falls. Several events contribute to a falling pH. The reaction that forms HbO_2 releases

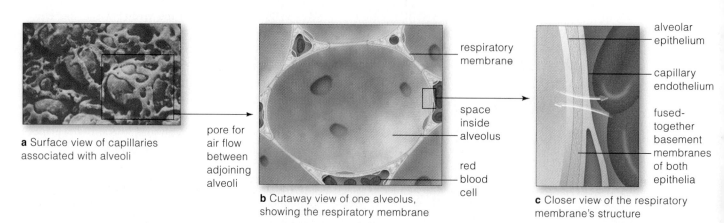

a Surface view of capillaries associated with alveoli

pore for
air flow
between
adjoining
alveoli

respiratory membrane

space
inside
alveolus

red
blood
cell

b Cutaway view of one alveolus, showing the respiratory membrane

alveolar epithelium

capillary endothelium

fused-together basement membranes of both epithelia

c Closer view of the respiratory membrane's structure

Figure 11.10 *Animated!* Gas exchange between blood in pulmonary capillaries and air in alveoli.

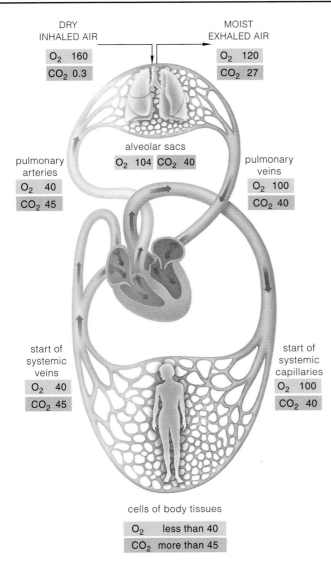

DRY
INHALED AIR

| O_2 | 160 |
| CO_2 | 0.3 |

MOIST
EXHALED AIR

| O_2 | 120 |
| CO_2 | 27 |

alveolar sacs

| O_2 | 104 | CO_2 | 40 |

pulmonary
arteries

| O_2 | 40 |
| CO_2 | 45 |

pulmonary
veins

| O_2 | 100 |
| CO_2 | 40 |

start of
systemic
veins

| O_2 | 40 |
| CO_2 | 45 |

start of
systemic
capillaries

| O_2 | 100 |
| CO_2 | 40 |

cells of body tissues

| O_2 | less than 40 |
| CO_2 | more than 45 |

Figure 11.11 *Animated!* Partial pressure gradients for oxygen and carbon dioxide through the respiratory tract. Remember that *each gas moves from regions of higher to lower partial pressure.*

hydrogen ions (H^+), making the blood more acidic. Blood pH also falls as the level of carbon dioxide given off by active cells increases.

When tissues chronically receive too little oxygen, red blood cells increase their production of a compound called 2,3-diphosphoglycerate, DPG for short. DPG reversibly binds hemoglobin. The more of it that is bound to hemoglobin, the *more loosely* hemoglobin binds oxygen—and thus the more oxygen is available to tissues.

HEMOGLOBIN AND BLOOD PLASMA CARRY CARBON DIOXIDE

As you know, aerobic respiration in cells produces carbon dioxide as a waste. For this reason, there is more carbon dioxide in metabolically active tissues than in the blood in

the nearby capillaries. So, following its pressure gradient, carbon dioxide diffuses into these capillaries. It will be carried toward the lungs in three ways. About 7 percent stays dissolved in plasma. About another 23 percent binds with hemoglobin in red blood cells, forming the compound *carbaminohemoglobin* ($HbCO_2$). Most of the carbon dioxide, about 70 percent, combines with water to form bicarbonate (HCO_3^-). The reaction has two steps. First carbonic acid (H_2CO_3) forms; then it dissociates (separates) into bicarbonate ions and hydrogen ions:

$$CO_2 + H_2O \rightleftharpoons \underset{\text{carbonic acid}}{H_2CO_3} \rightleftharpoons \underset{\text{bicarbonate}}{HCO_3^-} + H^+$$

This reaction takes place in the blood plasma and in red blood cells. However, it is faster in red blood cells, which contain the enzyme carbonic anhydrase. This enzyme increases the reaction rate by at least 250 times. Newly formed bicarbonate in red blood cells diffuses into the plasma, which will carry it to the lungs. The reactions rapidly reduce the amount of carbon dioxide in the blood. This "sopping up" of CO_2 in turn helps maintain the gradient that keeps carbon dioxide diffusing from tissue fluid into the bloodstream.

The reactions that make bicarbonate are reversed in alveoli, where the partial pressure of carbon dioxide is *lower* than it is in surrounding capillaries. The CO_2 that forms as the reactions go in reverse diffuses into the alveoli and is exhaled.

If you look again at the chemical reactions outlined in the pink shaded area above, you can see that the steps that form bicarbonate also produce some H^+, which makes blood more acid. What happens to these hydrogen ions? Hemoglobin binds some of them and thus acts as a buffer (Chapter 2). Certain proteins in blood plasma also bind H^+. These buffering mechanisms are extremely important in homeostasis, because they help prevent an abnormal decline in blood pH.

Driven by its partial pressure gradient, oxygen diffuses from alveoli, through tissue fluid, and into lung capillaries. Carbon dioxide diffuses in the opposite direction, driven by its partial pressure gradient.

Hemoglobin in red blood cells greatly enhances the oxygen-carrying capacity of the blood.

Hemoglobin and blood plasma also carry carbon dioxide.

In plasma, most carbon dioxide is carried in the form of bicarbonate. Buffers help prevent the blood from becoming too acid due to H^+ that is released when bicarbonate forms.

GAS EXCHANGE

11.6 Homeostasis Depends on Controls over Breathing

LINK TO
SECTION
9.4

Your nervous system controls the muscle movements that lead to the normal rhythm of inspiration and expiration. It also controls how often and how deeply you breathe.

A RESPIRATORY PACEMAKER CONTROLS THE RHYTHM OF BREATHING

You normally take about 12 to 15 breaths a minute. If you had to remember to inhale and exhale each time, could you do it, even when you sleep? Fortunately, none of us has to take on that responsibility, because automatic mechanisms ensure that a regular cycle of ventilation moves air into and out of the lungs. Clustered nerve cells in the medulla in the brain stem provide this service. Like the SA node in the heart, these neurons fire impulses spontaneously. They are the pacemaker for respiration.

The diaphragm muscle and the muscles that move the rib cage are under the control of neurons in a system of nerve cells (neurons) running through the brain stem, at the lower rear of the brain. For the moment, we are concerned with two small clusters of cells in the brain stem. One cluster coordinates signals that call for inspiration, "telling" you to take a breath. The other coordinates the signals calling for expiration. The resulting rhythmic contractions of the breathing muscles are fine-tuned by centers in a nearby part of the brain stem (the pons).

As Figure 11.12 suggests, when you inhale, signals from the brain stem travel nerve pathways to the diaphragm and chest. These signals stimulate the rib muscles and diaphragm to contract. As you read in Section 11.4, this causes the rib cage to expand, and air moves into the lungs. When the diaphragm and chest muscles relax, elastic recoil returns the rib cage to its unexpanded state, and air in the lungs moves out. When you breathe rapidly or deeply, stretch receptors in airways send signals to the brain control centers, which then inhibit contraction of the diaphragm and rib cage muscles—so you exhale.

CO₂ IS THE TRIGGER FOR CONTROLS OVER THE RATE AND DEPTH OF BREATHING

You might suppose that body controls over breathing mainly involve monitoring the level of oxygen in blood. However, the nervous system is *more* sensitive to changes in the level of carbon dioxide. Both gases are monitored in blood flowing through arteries. When the conditions warrant, nervous system signals adjust contractions of the diaphragm and muscles in the chest wall and so adjust the rate and depth of your breathing.

Sensory receptors in the medulla of the brain (another part of the brain stem) can detect rising carbon dioxide levels. How? The mechanism is indirect, but (luckily!) extremely sensitive. The receptors detect hydrogen ions that are produced when dissolved CO_2 leaves the bloodstream and enters fluid—called *cerebrospinal fluid*—that bathes the medulla. In cerebrospinal fluid, the drop in

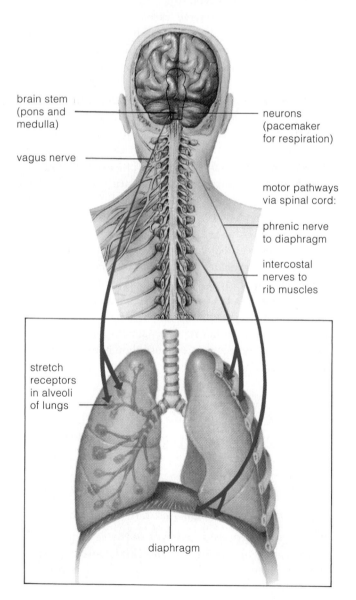

brain stem (pons and medulla)

vagus nerve

neurons (pacemaker for respiration)

motor pathways via spinal cord:

phrenic nerve to diaphragm

intercostal nerves to rib muscles

stretch receptors in alveoli of lungs

diaphragm

Figure 11.12 Controls over breathing. In quiet breathing, centers in the brain stem coordinate signals to the diaphragm and muscles that move the rib cage, triggering inhalation. When a person breathes deeply or rapidly, another center receives signals from stretch receptors in the lungs and coordinates signals for exhalation.

BREATHING CONTROLS

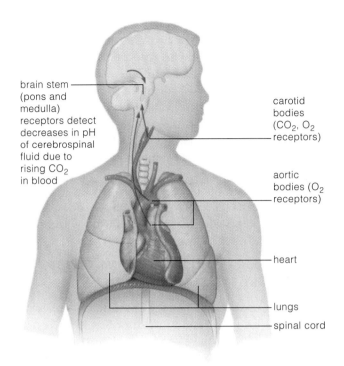

brain stem (pons and medulla) receptors detect decreases in pH of cerebrospinal fluid due to rising CO_2 in blood

carotid bodies (CO_2, O_2 receptors)

aortic bodies (O_2 receptors)

heart

lungs

spinal cord

Figure 11.13 Sensory receptors that detect changes in the concentrations of carbon dioxide and oxygen in the blood.

pH that goes along with increasing H^+ stimulates receptors that signal the change to the brain's respiratory centers (Figure 11.13). In response, breathing becomes more rapid and deeper, and soon the blood level of CO_2 falls. Notice that this is another example of a negative feedback loop helping to maintain homeostasis.

Our brain also receives input from other sensory receptors, including **carotid bodies**, where the carotid arteries branch to the brain, and **aortic bodies** in artery walls near the heart. Both types of receptors detect changes in levels of CO_2 and of oxygen in the blood. They also detect changes in blood pH. The brain responds by increasing the ventilation rate, so more oxygen can be delivered to tissues.

CHEMICAL CONTROLS IN ALVEOLI HELP MATCH AIR FLOW TO BLOOD FLOW

Chemical controls over air flow operate in the lungs, in the millions of alveoli. For example, if you go for a job interview and get nervous, your heart may start pumping hard and fast but your lungs may not be ventilating at a corresponding pace. Then, the flow of blood in alveolar capillaries outpaces the movement of CO_2-laden air out of the lungs. The rising blood level of carbon dioxide affects smooth muscle in the walls of bronchioles. The bronchioles then dilate, so more air flows through them.

On the other hand, a *decrease* in carbon dioxide levels causes the bronchiole walls to constrict, so less air flows through them.

Similar local controls also work on lung capillaries. When too much air is flowing into the lungs relative to the amount of blood traveling through capillaries, the oxygen level rises in parts of the lungs. As with carbon dioxide, this increase in oxygen also affects the smooth muscle in blood vessel walls. The vessels dilate, so more blood flows through them. Conversely, if there is too little air entering the lungs with respect to the volume of blood flowing in the capillaries, the vessels constrict and less blood moves through them.

APNEA IS A CONDITION IN WHICH BREATHING CONTROLS MALFUNCTION

You can voluntarily hold your breath, but not for long. As CO_2 builds up in your blood, "orders" from the nervous system force you to take a breath. The mechanisms by which the nervous system regulates the respiratory cycle normally operate under involuntary control, as you've just read. In some situations, however, a person can fail to breathe when the arterial CO_2 level falls below a set point. Breathing that stops briefly and then resumes spontaneously is called *apnea*. During certain times in the normal sleep cycle (see Section 13.10), breathing may stop for one or two seconds or even minutes—in extreme cases, as often as 500 times a night. Apnea may be a contributing factor in heavy snoring.

Aging also takes its toll on the respiratory system. **Sleep apnea** is a common problem in the elderly, because the mechanisms for sensing a change in oxygen and carbon dioxide levels gradually become less effective over the years. Also, as we age, our lungs lose some of their elasticity. This along with other changes reduces the efficiency of ventilation. Even so, staying physically fit, and maintaining a "lung-healthy" lifestyle in other ways, can go a long way toward keeping the respiratory system functioning well throughout life.

Respiratory centers in the brain stem control the rhythmic pattern of breathing.

Brain centers that adjust the rate and depth of breathing receive information mainly from sensors that monitor blood levels of carbon dioxide.

These controls contribute to homeostasis by helping to maintain proper levels of carbon dioxide, oxygen, and hydrogen ions in arterial blood.

11.7 Disorders of the Respiratory System

A variety of infections and other disorders can prevent the respiratory system from functioning properly. Some of these problems develop when we inadvertently inhale a pathogen or noxious substances, while others we bring on ourselves.

TOBACCO IS A MAJOR THREAT

People who start smoking tobacco begin wreaking havoc on their lungs. Smoke from a single cigarette can prevent cilia in bronchioles from beating for hours. Toxic particles smoke contains can stimulate mucus secretion and kill the infection-fighting phagocytes that normally patrol the respiratory epithelium.

Today we know that cigarette smoke, including "secondhand smoke" inhaled by a nonsmoker, causes **lung cancer** and contributes to other ills. In the body, some compounds in coal tar and cigarette smoke are converted to carcinogens (cancer-causing substances); they trigger genetic damage leading to lung cancer. Susceptibility to lung cancer is related to the number of cigarettes smoked per day and how often and how deeply the smoke is inhaled. In all, cigarette smoking causes at least 80 percent of all lung cancer deaths. Figure 11.14 lists the known health risks associated with tobacco smoking, as well as the benefits of quitting.

A VARIETY OF PATHOGENS CAN INFECT THE RESPIRATORY SYSTEM

Inhaling viruses, bacteria, fungi, or even toxic fumes can cause some common respiratory disorders. A dry cough, chest pain, and shortness of breath all are symptoms of **pneumonia**. The infection causes inflammation in lung tissue, and then fluid (from edema) builds up in the lungs and makes breathing difficult. Bacterial pneumonia can be treated with antibiotics. Sometimes the trigger for pneumonia is **influenza**, in which an infection that began in the nose or throat spreads to the lungs.

RISKS ASSOCIATED WITH SMOKING	REDUCTION IN RISKS BY QUITTING
SHORTENED LIFE EXPECTANCY: Nonsmokers live 8.3 years longer on average than those who smoke two packs daily from the mid-twenties on.	Cumulative risk reduction; after 10 to 15 years, life expectancy of ex-smokers approaches that of nonsmokers.
CHRONIC BRONCHITIS, EMPHYSEMA: Smokers have 4–25 times more risk of dying from these diseases than do nonsmokers.	Greater chance of improving lung function and slowing down rate of deterioration.
LUNG CANCER: Cigarette smoking is a major contributing factor.	After 10 to 15 years, risk approaches that of nonsmokers.
CANCER OF MOUTH: 3–10 times greater risk among smokers.	After 10 to 15 years, risk is reduced to that of nonsmokers.
CANCER OF LARYNX: 2.9–17.7 times more frequent among smokers.	After 10 years, risk is reduced to that of nonsmokers.
CANCER OF ESOPHAGUS: 2–9 times greater risk of dying from this.	Risk proportional to amount smoked; quitting should reduce it.
CANCER OF PANCREAS: 2–5 times greater risk of dying from this.	Risk proportional to amount smoked; quitting should reduce it.
CANCER OF BLADDER: 7–10 times greater risk for smokers.	Risk decreases gradually over 7 years to that of nonsmokers.
CORONARY HEART DISEASE: Cigarette smoking is a major contributing factor.	Risk drops sharply after a year; after 10 years, risk reduced to that of nonsmokers.
EFFECTS ON OFFSPRING: Women who smoke during pregnancy have more stillbirths, and weight of liveborns averages less (hence, babies are more vulnerable to disease, death).	When smoking stops before fourth month of pregnancy, risk of stillbirth and lower birthweight eliminated.
IMPAIRED IMMUNE SYSTEM FUNCTION: Increase in allergic responses, destruction of defensive cells (macrophages) in respiratory tract.	Avoidable by not smoking.
BONE HEALING: Evidence suggests that surgically cut or broken bones require up to 30 percent longer to heal in smokers, possibly because smoking depletes the body of vitamin C and reduces the amount of oxygen reaching the tissues. Reduced vitamin C and reduced oxygen interfere with production of collagen fibers, a key component of bone. Research in this area is continuing.	Avoidable by not smoking.

Figure 11.14 From the American Cancer Society, a list of the risks incurred by smoking and the benefits of quitting. The photograph shows swirls of cigarette smoke at the entrance to the two bronchi that lead into the lungs.

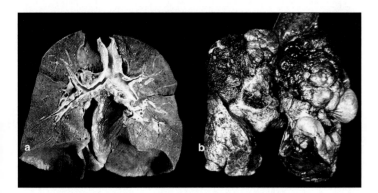

Figure 11.15 (**a**) Normal human lungs (this lung tissue looks darker than normal because it has been chemically preserved). (**b**) Lungs from a person with emphysema.

Tuberculosis (TB) is a serious lung infection caused by the bacterium *Mycobacterium tuberculosis*. It starts with flulike symptoms but eventually can destroy patches of lung tissue and can spread to other parts of the body. Although TB is curable with antibiotics, newer drug-resistant strains of *M. tuberculosis* have made treatment more challenging. Untreated it can be fatal.

A microscopic fungus causes **histoplasmosis**. The symptoms include cough, fever, and inflammation in the lungs and airways. Antifungal drugs can cure "histo," but the infection may spread to the retina of the eye, leading to permanently impaired vision or blindness.

IRRITANTS CAUSE OTHER DISORDERS

In cities, in certain occupations, and anywhere near a smoker, airborne particles and irritating gases put extra workloads on the lungs.

Bronchitis can be brought on when air pollution increases mucus secretions and interferes with ciliary action in the lungs. Ciliated epithelium in the bronchioles is especially sensitive to cigarette smoke. Mucus and the particles it traps—including bacteria—accumulate in airways, coughing starts, and the bronchial walls become inflamed. Bacteria or chemical agents start destroying the wall tissue. Cilia in the lining die, and mucus-secreting cells multiply as the body attempts to get rid of the accumulating debris. Eventually scar tissue forms and can block parts of the respiratory tract.

In an otherwise healthy person, even acute bronchitis is easily treated with antibiotics. When inflammation continues, however, scar tissue builds up and the bronchi become chronically clogged with mucus. Also, the walls of some alveoli break down and become surrounded by stiffer fibrous tissue. The result is **emphysema**, in which the lungs are so distended and inelastic that gases cannot be exchanged efficiently (Figure 11.15). Running, walking,

Figure 11.16 An asthma sufferer using an aerosol inhaler.

even exhaling can be difficult. About 1.3 million people in the United States have emphysema.

Smoking, frequent colds, and other respiratory ailments sometimes make a person susceptible to emphysema. Many emphysema sufferers lack a normal gene coding for a protein that inhibits tissue-destroying enzymes made by bacteria. Emphysema can develop over 20 or 30 years. Unfortunately, by the time the disease is detected, the lungs are permanently damaged.

Millions of people suffer from **asthma**, a disorder in which the bronchioles suddenly narrow when the smooth muscle in their walls contracts in strong spasms. At the same time, mucus gushes from the bronchial epithelium, clogging the constricted passages even more. Breathing can become extremely difficult so quickly that the victim may feel in imminent danger of suffocating. The triggers include allergens such as pollen, dairy products, shellfish, pet hairs, flavorings, or even the dung of tiny mites in house dust. In susceptible people, attacks also can be triggered by noxious fumes, stress, strenuous exercise, or a respiratory infection. While the reasons aren't fully understood, the incidence of asthma in the United States has grown rapidly in the last several decades. Some researchers believe that increased air pollution is at least partly to blame.

Many asthma sufferers rely on aerosol inhalers, which squirt a fine mist into the airways (Figure 11.16). A drug in the mist dilates bronchial passages and helps restore free breathing. Some devices contain powerful steroids that can harm the immune system, so inhalers should be used only with medical supervision.

Summary

Section 11.1 The respiratory system brings air, which contains oxygen, into the body and disposes of carbon dioxide.

Airways include the nasal cavity, pharynx, larynx, trachea, bronchi, and bronchioles. Gas exchange occurs in millions of saclike alveoli located at the end of the terminal respiratory bronchioles. The vocal cords are located near the entrance to the larynx.

Airways lead to the lungs, which are elastic organs located in the rib cage above the diaphragm. Membranes called pleurae enclose the lungs.

 Explore the respiratory system's parts and their functions.

Section 11.2 Respiration is the process of bringing oxygen from air into the blood and removing carbon dioxide from blood. Both these processes occur in the lungs. The cardiovascular system partners with the respiratory system as it circulates blood throughout the body. Aerobic cellular respiration is the process in cells (in mitochondria) that uses oxygen to make ATP and produces carbon dioxide.

Section 11.3 Air is a mixture of oxygen, carbon dioxide, and other gases. Each gas exerts a partial pressure, and each tends to move (diffuse) from areas of higher to lower partial pressure. Following pressure gradients, oxygen tends to diffuse into deoxygenated blood in the lungs, and carbon dioxide tends to diffuse out of the blood and into the lungs to be exhaled.

In respiration, oxygen and carbon dioxide diffuse across a respiratory surface—a moist, thin layer of epithelium in the alveoli of the lungs. Airways carry gases to and from one side of the respiratory surface, and blood vessels carry gases to and away from the other side.

 Investigate the effects of partial pressure gradients in the body.

Section 11.4 Breathing ventilates the lungs in a respiratory cycle. During inspiration (inhalation), the chest cavity expands, pressure in the lungs falls below atmospheric pressure, and air flows into the lungs. During normal expiration (exhalation), these steps are reversed.

The volume of air in a normal breath, called the tidal volume, is about 500 milliliters. Vital capacity is the maximum volume of air that can move out of the lungs after you inhale as deeply as possible.

 Learn more about the respiratory cycle.

Section 11.5 Driven by its partial pressure gradient, oxygen in the lungs diffuses from alveoli into pulmonary capillaries. Then it diffuses into red blood cells and binds with hemoglobin, forming oxyhemoglobin. In tissues where cells are metabolically active, hemoglobin gives up oxygen, which diffuses out of the capillaries, across tissue fluid, and into cells.

Hemoglobin binds with or releases oxygen in response to shifts in oxygen levels, carbon dioxide levels, pH, and temperature.

Driven by its partial pressure gradient, carbon dioxide diffuses from cells across tissue fluid and into the bloodstream. Most CO_2 reacts with water to form bicarbonate; the reactions are speeded by the enzyme carbonic anhydrase. They are reversed in the lungs, where carbon dioxide diffuses from lung capillaries into the air spaces of the alveoli, then is exhaled.

Section 11.6 Gas exchange is regulated by the nervous system and by chemical controls in the lungs. A respiratory pacemaker in the medulla (part of the brain stem) sets the normal, automatic rhythm of breathing in and out (ventilation).

The nervous system monitors the levels of oxygen and carbon dioxide in arterial blood by way of sensory receptors. These include carotid bodies (at branches of carotid arteries leading to the brain), aortic bodies (in an arterial wall near the heart), and receptors in the medulla of the brain. Blood levels of carbon dioxide are most important in triggering nervous system commands that adjust the rate and depth of breathing.

Section 11.7 Infections, toxins in tobacco smoke and polluted air, and accumulated damage from inflammation cause respiratory disorders including cancer, bronchitis, emphysema, pneumonia, tuberculosis, and asthma.

Review Questions

1. In the diagram on page 209, label the components of the respiratory system and the structures that enclose some of its parts.

2. What is the difference between respiration and aerobic cellular respiration?

3. Explain what a partial pressure gradient is and how such gradients figure in gas exchange.

4. What is oxyhemoglobin? Where does it form?

5. What drives oxygen from the air spaces in alveoli, through tissue fluid, and across capillary epithelium? What drives carbon dioxide in the opposite direction?

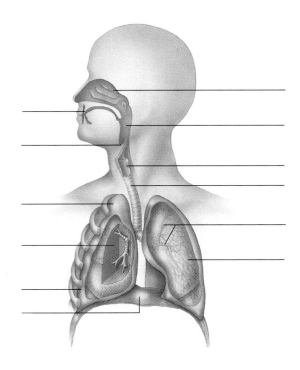

6. How does hemoglobin help maintain the oxygen partial pressure gradient during gas transport in the body?

7. What reactions enhance the transport of carbon dioxide throughout the body? How is carbon dioxide moved out of the body?

8. How do nerve impulses from the brain regulate ventilation of the lungs? How are the rate and depth of breathing controlled?

9. Why does your breathing rate increase when you exercise? What happens to your heart rate at the same time—and why?

Self-Quiz

Answers in Appendix V

1. A partial pressure gradient of oxygen exists between
————.
   a. air and lungs
   b. lungs and metabolically active tissues
   c. air at sea level and air at high altitudes
   d. all of the above

2. The ———— is an airway that connects the nose and mouth with the ————.
   a. oral cavity; larynx
   b. pharynx; trachea
   c. trachea; pharynx
   d. pharynx; larynx

3. Oxygen in air must diffuse across ———— to enter the blood.
   a. pleural sacs
   b. alveolar sacs
   c. a moist respiratory surface
   d. both b and c

4. Each lung encloses a ————.
   a. diaphragm
   c. pleural sac
   b. bronchial tree
   d. both b and c

5. Gas exchange occurs at the ————.
   a. two bronchi
   c. alveoli
   b. pleural sacs
   d. both b and c

6. Breathing ————.
   a. ventilates the lungs
   b. draws air into airways
   c. expels air from airways
   d. causes reversals in pressure gradients
   e. all of the above

7. After oxygen diffuses into lung capillaries it also diffuses into ———— and binds with ————.
   a. tissue fluid; red blood cells
   b. tissue fluid; carbon dioxide
   c. red blood cells; hemoglobin
   d. red blood cells; carbon dioxide

8. Due to its partial pressure gradient, carbon dioxide diffuses from cells into interstitial fluid and into the ————; in the lungs, carbon dioxide diffuses into the ————.
   a. alveoli; bronchioles
   b. bloodstream; bronchioles
   c. alveoli; bloodstream
   d. bloodstream; alveoli

9. Hemoglobin performs which of the following respiratory functions?
   a. transports oxygen
   b. transports some carbon dioxide
   c. acts as a buffer to help maintain blood pH
   d. all of the above

10. Most carbon dioxide in the blood is in the form of ————.
   a. carbon dioxide
   c. carbonic acid
   b. carbon monoxide
   d. bicarbonate

Critical Thinking

1. People occasionally poison themselves with carbon monoxide by building a charcoal fire in an enclosed area. Assuming help arrives in time, what would be the *most* effective treatment: placing the victim outdoors in fresh air, or administering pure oxygen? Explain your answer.

2. Skin divers and swimmers sometimes purposely hyperventilate. Doing so doesn't increase the oxygen available to tissues. It does increase blood pH (making it more alkaline), and it decreases the blood level of carbon dioxide. Based on your reading in this chapter, what effect is hyperventilation likely to have on the neural controls over breathing?

3. Underwater, we humans can't compete with whales and other air-breathing marine mammals, which can stay submerged for extended periods. At the beach one day

you meet a surfer who tells you that special training could allow her to swim underwater without breathing for an entire hour. From what you know of respiratory physiology, explain why she is mistaken.

4. When you sneeze or cough, abdominal muscles contract suddenly, pushing your diaphragm upward. After reviewing the discussion of the respiratory cycle in Section 11.4, explain why this change forcefully expels air out your nose and mouth.

5. Physiologists have discovered that the nicotine in tobacco is as addictive as heroin. The cigarette-smoking child in Figure 11.17 probably is already addicted, and for sure has already begun to endanger her health. Based on the discussion in Section 11.7, what negative health effects might beset her in the coming years?

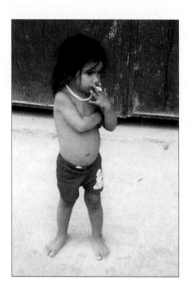

Figure 11.17 A child in Mexico City who is already adept at smoking cigarettes—a behavior that, if continued, will one day endanger her capacity to breathe.

Explore on Your Own

Air pollution is a serious problem in many parts of the world. Even if you don't live near a large urban area, you may be breathing the kinds of air pollutants shown in the chart in Figure 11.18. The ultrafine particulates can stay in the air for weeks or months before they settle to earth or are washed down by rain, and all of them are known to cause respiratory problems, especially in people who have asthma or emphysema.

Explore this health issue by finding out if your community monitors its air quality. If so, what do authorities consider to be the greatest threats to the health of you and your fellow citizens? Where do these pollutants come from?

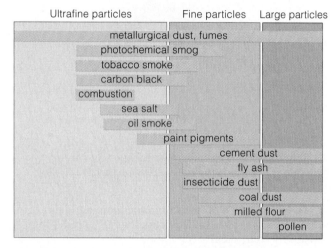

Figure 11.18 Examples of the kinds of particles that may be present in the air you breathe.

Truth in a Test Tube

Light or dark? Clear or cloudy? A lot or a little? Today physicians routinely check the chemical composition of our urine. Acidic urine can signal metabolic problems. Alkaline urine can signal a bacterial infection. Too much protein dissolved in urine might mean the kidneys are not functioning properly. Specialized urine tests can detect chemicals produced by cancers of the kidney, bladder, and prostate gland.

Do-it-yourself urine tests have now become popular for monitoring a woman's fertile period or early signs she may be pregnant. A test for older women may reveal declining hormone levels that signal the onset of menopause.

Not everyone is anxious to have their urine tested. Olympic athletes can be stripped of their medals when mandatory urine tests reveal they use prohibited drugs. Major League Baseball players agreed to urine tests only after repeated allegations that some star players had taken prohibited steroids. Each year the National Collegiate Athletic Association (NCAA) tests urine samples from about 3,300 student athletes for performance-enhancing substances and also for street drugs.

If you use marijuana, cocaine, Ecstasy, or other kinds of illegal drugs, urine tells the tale. For example, after the active ingredient of marijuana enters the blood, the liver converts it to another compound. The kidneys filter the blood and they add this telltale compound to the newly forming urine. It can take up to ten days for all of it to be metabolized and removed from the body. Until then, urine tests can detect it.

That urine is such a remarkable indicator of health, hormonal status, and drug use is a tribute to the urinary system. Each day, your two kidneys filter all of the blood in your body not once but thirty times. When all goes well, the kidneys eliminate excess water and excess or harmful solutes, including many metabolic by-products, toxins, hormones, and drugs.

In the last several chapters you have considered organ systems that work to keep cells supplied with oxygen, nutrients, water, and other substances. Now we turn to the urinary system, which helps maintain the proper chemical composition and volume of body fluids— the internal environment.

 How Would You Vote? Many companies use urine testing to screen prospective employees for drug and alcohol use. Some people say this is an invasion of privacy. Do you think employers should be allowed to require a person to undergo urine testing before being hired? Cast your vote online at www.thomsonedu.com/biology/starr/humanbio.

Key Concepts

MAINTAINING THE EXTRACELLULAR FLUID
The body must eliminate chemical wastes from extracellular fluid and manage the levels of water and solutes in it. The urinary system partners with other organ systems in this task.

THE URINARY SYSTEM
The urinary system consists of the kidneys, ureters, bladder, and urethra. In the kidneys, structures called nephrons filter water and solutes from the blood, returning needed substances to it and eliminating the rest in urine.

WHAT THE KIDNEYS DO
The kidneys form urine in steps called filtration, reabsorption, and secretion. Hormones and a thirst mechanism adjust the chemical makeup of urine.

Links to Earlier Concepts

This chapter explores how operations of the urinary system help maintain homeostasis in the extracellular fluid—that is, blood and tissue fluid. As you study these processes, you will tap your understanding of pH and buffer systems (2.7) and of osmosis and transport mechanisms (3.9, 3.10). You will also use what you have learned about the functions of blood, blood circulation by the cardiovascular system, and the movement of substances into and out of blood capillaries (8.1, 9.1, 9.7).

12.1 The Challenge: Shifts in Extracellular Fluid

LINKS TO
SECTIONS
2.11, 3.12, 7.5,
7.6, 7.7, 8.1, 9.6,
9.7, AND 11.5

If you are an adult female in good health, by weight your body is about 50 percent fluid. If you are an adult male, the ratio is about 60 percent. The chemical makeup of this fluid changes constantly as water and solutes enter and leave it.

Fluid is vital both in our anatomy (body structures) and in our physiology (body functions). This fluid occurs both outside our cells and inside them. Recall that tissue fluid fills the spaces between cells and other components of our body tissues. Blood—which is mostly watery plasma—circulates in blood vessels. Together, tissue fluid, blood plasma, and the relatively small amounts of other fluids (such as lymph) outside cells are the body's **extracellular fluid**, or ECF.

The fluid *inside* our cells is **intracellular fluid**. From previous chapters you know that a variety of gases and other substances move constantly between intracellular and extracellular fluid. Those exchanges are crucial for keeping cells functioning smoothly. They cannot occur properly unless the volume and composition of the ECF are stable.

Yet the ECF is always changing, because gases, cell secretions, ions, and other materials enter or leave it. To maintain stable conditions in the ECF, especially the concentrations of water and vital ions such as sodium (Na^+) and potassium (K^+), there must be mechanisms that remove substances as they enter the extracellular fluid or add needed ones as they leave it. This task is the job of the **urinary system**. Before examining the parts of this system, however, we will look more closely at the traffic of substances into and out of extracellular fluid.

THE BODY GAINS WATER FROM FOOD AND METABOLIC PROCESSES

Ordinarily, each day you take in about as much water as your body loses (Table 12.1). Some of the water is absorbed from foods and liquids you consume. The rest is a by-product of metabolic reactions, including cellular respiration and condensation reactions, that produce water.

Thirst influences how much water we take in. When there is a water deficit in body tissues, our brain "urges" us to seek out water—for example, from a water fountain or a cold drink from the refrigerator. We will be looking at the thirst mechanism later in the chapter.

THE BODY LOSES WATER IN URINE, SWEAT, FECES, AND BY EVAPORATION

Water leaves the body by four routes: Excretion in urine, evaporation from the lungs and skin, sweating, and in feces. Of these four routes of water loss, **urinary excretion** is the one over which the body can exert the most control. Urinary excretion eliminates excess water, as well as excess or harmful solutes, in the form of **urine**. Some water also evaporates from our skin and from the respiratory surfaces of the lungs. These are sometimes called "insensible" water losses, because a person is not always aware they are taking place. As noted in Chapter 7, normally very little water that enters the GI tract is lost; most is absorbed and only a little is eliminated in feces.

Table 12.1	Normal Daily Balance between Water Gain and Water Loss in Adult Humans		
Water Gain (milliliters)		**Water Loss (milliliters)**	
Ingested in solids:	850	Urine:	1,500
Ingested as liquids:	1,400	Feces:	200
Metabolically derived:	350	Evaporation:	900
	2,600		2,600

SOLUTES ENTER EXTRACELLULAR FLUID FROM FOOD, METABOLISM, AND OTHER WAYS

Solutes enter the body's extracellular fluid mainly by four routes. When we eat, a variety of nutrients (including glucose) and mineral ions (such as potassium and sodium ions) are absorbed from the GI tract. So are drugs and food additives. Living cells also continually secrete substances into tissue fluid and circulating blood. The respiratory system brings oxygen into the blood, and respiring cells add carbon dioxide to it (Figure 12.1).

SOLUTES LEAVE THE ECF BY URINARY EXCRETION, IN SWEAT, AND DURING BREATHING

Solutes including mineral ions and metabolic wastes leave extracellular fluid in several ways. Carbon dioxide is the most abundant metabolic waste. We get rid of it by exhaling it from our lungs. All other major wastes—and metabolism produces more than 200 of them—leave in urine. One, uric acid, is formed when cells break down nucleic acids. If it builds up in the ECF, it can crystallize and collect in the joints, causing the painful condition called *gout*.

Other major metabolic wastes include by-products of processes that break down proteins. One of these wastes, ammonia, is formed in "deamination" reactions, which remove the nitrogen-containing amino groups from amino acids. Ammonia is highly toxic if it accumulates in the body. Reactions in the liver combine ammonia with carbon dioxide, producing the much less toxic **urea**. Accordingly, urea is the main waste product when cells break down proteins. About half of the urea filtered from blood in the kidneys is reabsorbed. The rest is excreted. Protein breakdown also produces creatine, phosphoric acid, sulfuric acid, and small amounts of other, nitrogen-containing compounds, some of which are toxic. These also are excreted.

Although sweat carries away a small percentage of urea and uric acid, by far the most nitrogen-containing wastes are removed by our kidneys while they filter other substances from the blood. In addition to removing wastes and excess water, the kidneys also help maintain the balance of important ions such as sodium, potassium, and calcium. These ions are sometimes called **electrolytes** because a solution in which they are dissolved will carry an electric current. Chapter 13 describes the extremely important roles electrolytes have in the functioning of the nervous system.

Normally only a little of the water and solutes that enter the kidneys leaves as urine. In fact, except when you drink lots of fluid (without exercise), all but about 1

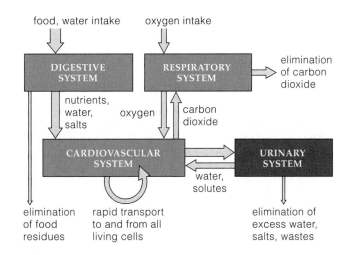

Figure 12.1 How activities of the urinary system coordinate with those of some other organ systems.

percent of the water is returned to the blood. However, the chemical composition of the fluid that is returned has been adjusted in vital ways. Just how this happens will be our focus as we turn our attention to the urinary system.

The kidneys adjust the volume and chemical composition of the blood. In this way they help maintain homeostasis in the extracellular fluid.

Each day the body gains water consumed in liquids and solid foods and from metabolism. It loses an approximately equal amount of water through urinary excretion, evaporation, sweating, and elimination in feces.

The body gains solutes from digested food, secretion by cells, metabolism, and respiration. Solutes are removed by urinary excretion, respiration, and sweating.

12.2 The Urinary System—Built for Filtering and Waste Disposal

LINKS TO
SECTIONS
3.11, 8.3, AND 9.7

The urinary system consists of filtering organs—the kidneys—and structures that carry and store urine.

Each **kidney** is a bean-shaped organ about the size of a rolled-up pair of socks (Figure 12.2). It has several internal lobes. In each lobe, an outer *cortex* wraps around a central region, the *medulla*, as you can see sketched in Figure 12.2c. The whole kidney is wrapped in a tough coat of connective tissue, the *renal capsule* (from the Latin *renes*, meaning kidneys). The kidney's central cavity is called the *renal pelvis*.

Our kidneys have several functions. They produce the hormone erythropoietin, which stimulates the production of red blood cells (Section 8.3). They also convert vitamin D to a form that stimulates the small intestine to absorb calcium in food. In addition, kidneys make the enzyme renin, which helps regulate blood pressure, as you will read later in this chapter. The main function of kidneys, however, is to remove metabolic wastes from the blood and adjust fluid balance in the body.

In addition to the two kidneys, the urinary system includes "plumbing" that transports or stores urine. Once urine has formed in a kidney, it flows into a tubelike **ureter**, then on into the **urinary bladder**, where it is stored until you urinate. It leaves the bladder through the **urethra**, a muscular tube that opens at the body surface.

NEPHRONS ARE THE KIDNEY FILTERS

Each kidney lobe contains blood vessels and more than a million slender tubes called **nephrons**. Nephrons are the structures that filter water and solutes from blood.

A nephron is shaped a little like the piping under a sink (Figure 12.3a). Its wall is a single layer of epithelial cells, but the cells and junctions between them vary in different parts of the tube. Water and solutes pass easily through some parts, but other parts completely block solutes unless they are moved across by active transport (Section 3.11).

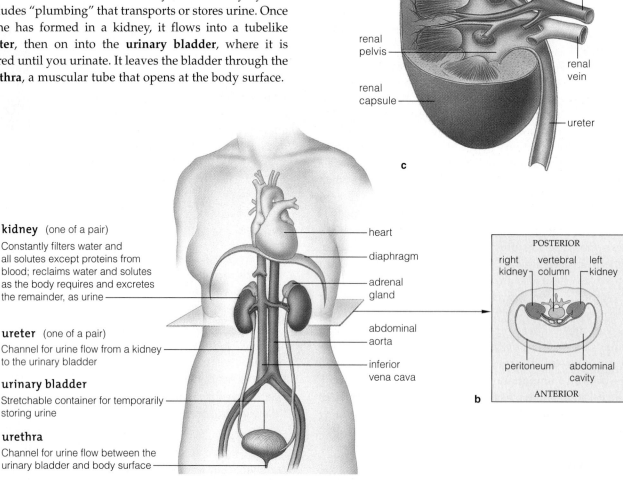

Figure 12.2 *Animated!* (**a**) The urinary system and its functions. (**b**) The two kidneys, ureters, and urinary bladder are located between the abdominal cavity's wall and its lining, the peritoneum. (**c**) Internal structure of a kidney.

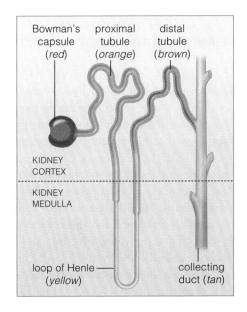

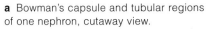

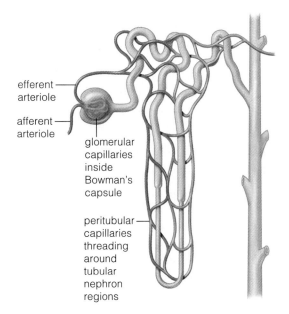

a Bowman's capsule and tubular regions of one nephron, cutaway view.

b Blood vessels associated with the nephron.

Figure 12.3 *Animated!* (**a**) Diagram of a nephron. Interacting with two sets of capillaries, nephrons are a kidney's blood-filtering units. (**b**) The arterioles and capillaries associated with a nephron.

As sketched in Figure 12.3*b*, the nephron wall balloons around a tiny cluster of blood capillaries called the **glomerulus** (plural: glomeruli). The cuplike wall region, called the **Bowman's** (glomerular) **capsule**, receives the substances that are filtered from blood. The rest of the nephron is a winding tubule ("little tube"). Filtrate flows from the cup into the **proximal tubule** (proximal means "next to"), then through a hairpin-shaped **loop of Henle** and into the **distal tubule** ("most distant" from the glomerular capsule). This part of the nephron tubule empties into a collecting duct.

SPECIAL VESSELS TRANSPORT BLOOD TO, IN, AND AWAY FROM NEPHRONS

Each hour, about 75 gallons of blood course through your kidneys, delivered by the renal artery. An **afferent arteriole** brings blood to each nephron (afferent means "carrying toward"). The blood flows into the glomerular capillaries inside Bowman's capsule. These capillaries are not like capillaries in other parts of the body, however. Slitlike pores between the cells of their walls make them much more permeable than other capillaries. Thus it is much easier for water and solutes to move across the wall.

The glomerular capillaries also do not channel blood to venules, as other capillaries do (Section 9.7). Instead,

the glomerular capillaries merge to form an **efferent** ("carrying away from") **arteriole**. This arteriole branches into yet another set of capillaries, called **peritubular** ("around the tubule") **capillaries**. As you can see in Figure 12.3*b*, the peritubular capillaries weave around a nephron's tubules. They merge into venules, which carry filtered blood out of the kidneys.

As we see next, the elaborate network of capillaries that feed and drain blood from nephrons is a key factor in the kidneys' ability to fine-tune the chemical makeup of the blood.

The urinary system consists of two kidneys, two ureters, the urinary bladder, and the urethra.

Kidney nephrons filter water and solutes from blood. A tiny cluster of capillaries called a glomerulus is the nephron's blood-filtering unit.

The capillaries in a glomerulus have pores in their walls that make the vessels much more permeable than usual.

Afferent arterioles deliver blood to nephrons and efferent arterioles carry it away. Peritubular capillaries weave around nephron tubules and deliver filtered blood back to the general circulation.

The processes that form urine normally ensure that only nonessential substances are excreted from the body.

The fluid we call urine forms in a sequence of three steps called filtration, reabsorption, and secretion. Figure 12.4 gives you an overview of these steps.

FILTRATION REMOVES A LARGE AMOUNT OF FLUID AND SOLUTES FROM THE BLOOD

Blood pressure generated by the heart's contractions is the driving force for **filtration**, the first step in forming urine. Efferent arterioles have small diameters, so they deliver blood to the glomerulus under high pressure. This pressure forces about 20 percent of the blood plasma into Bowman's capsule. Blood cells, platelets, proteins, and other large solutes remain in the blood. Everything else—water and some small solutes such as glucose, amino acids, vitamins, sodium, and urea—can filter out of the glomerular capillaries and into Bowman's capsule. From there the filtrate flows into the proximal tubule (Figure 12.5a), where reabsorption can begin.

NEXT, REABSORPTION RETURNS USEFUL SUBSTANCES TO THE BLOOD

The body cannot afford to lose the huge amounts of water and valuable solutes such as glucose, amino acids, and electrolytes that are filtered from the blood by the kidneys. Fortunately, most of the filtrate is recovered by **reabsorption**. In this process, substances leak or are pumped out of the nephron tubule and then enter peritubular capillaries and so return to the bloodstream.

Most reabsorption takes place across the walls of proximal tubules. As in all parts of the tubule, the walls in this area are only one cell thick. Figure 12.5b shows what happens with water, glucose, and salt (ions of sodium and chloride, Na^+ and Cl^-). All these substances can diffuse from the filtrate in a tubule into and through the cells of the tubule wall. On the outer side of the cells, active transport (through proteins in the cells' plasma membranes) moves glucose and Na^+ into the tissue fluid. Sodium ions (Na^+) are positively charged, and negatively charged ions, including chloride (Cl^-), follow the sodium.

As the concentration of solutes rises in the fluid, water moves out of the tubule cells by osmosis. In a final step, solutes are actively transported into peritubular capillaries and water again follows by osmosis. These substances now have been reabsorbed. The solutes and water that remain in the tubule become part of urine.

Reabsorption usually returns almost 99 percent of the filtrate's water, all of the glucose and most amino acids, all but about 0.5 percent of the salt (sodium and chloride ions), and 50 percent of the urea to the blood (Table 12.2).

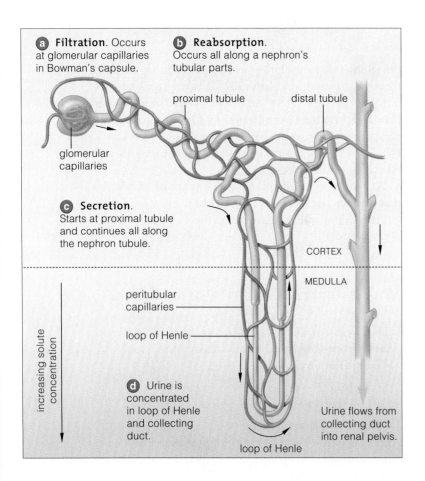

a Filtration. Occurs at glomerular capillaries in Bowman's capsule.

b Reabsorption. Occurs all along a nephron's tubular parts.

proximal tubule   distal tubule

glomerular capillaries

c Secretion. Starts at proximal tubule and continues all along the nephron tubule.

CORTEX

MEDULLA

peritubular capillaries

loop of Henle

increasing solute concentration

d Urine is concentrated in loop of Henle and collecting duct.

Urine flows from collecting duct into renal pelvis.

loop of Henle

Figure 12.4 *Animated!* Overview of the steps that form urine.

Table 12.2	Average Daily Reabsorption Values for a Few Substances		
	Amount Filtered	Percentage Excreted	Percentage Reabsorbed
Water	180 liters	1	99
Glucose	180 grams	0	100
Amino acids	2 grams	5	95
Sodium ions	630 grams	0.5	99.5
Urea	54 grams	50	50

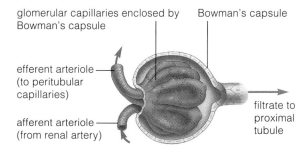

glomerular capillaries enclosed by Bowman's capsule

Bowman's capsule

efferent arteriole (to peritubular capillaries)

afferent arteriole (from renal artery)

filtrate to proximal tubule

(a) Filtration. Water and solutes forced out across the glomerular capillary wall collect in Bowman's capsule, which drains into the proximal tubule.

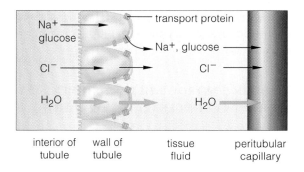

Na+
glucose

transport protein

Na+, glucose

Cl⁻

Cl⁻

H₂O

H₂O

| interior of tubule | wall of tubule | tissue fluid | peritubular capillary |

(b) Reabsorption. As filtrate flows through the proximal tubule, ions and some nutrients are actively and passively transported outward, into tissue fluid. Water follows, by osmosis. Cells of peritubular capillaries transport them into blood. Water again follows by osmosis.

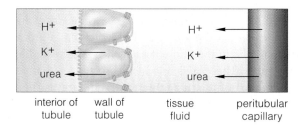

H⁺

H⁺

K⁺

K⁺

urea

urea

| interior of tubule | wall of tubule | tissue fluid | peritubular capillary |

(c) Secretion. Transport proteins move H⁺, K⁺, urea, and wastes out of peritubular capillaries. Transporters in the nephron tubule move them into the filtrate.

Figure 12.5 *Animated!* Filtration, reabsorption, and secretion in a nephron.

SECRETION RIDS THE BODY OF EXCESS HYDROGEN IONS AND SOME OTHER SUBSTANCES

Secretion takes up unwanted substances that have been transported out of peritubular capillaries and adds them to the urine that is forming in nephron tubules (Figure 12.5c). Among other functions, this highly controlled process rids the body of urea and of excess hydrogen ions (H^+) and potassium ions (K^+).

Secretion is crucial to maintaining the body's acid–base balance, which you will read about in a later section. It also helps ensure that some wastes (such as uric acid and some breakdown products of hemoglobin) and foreign substances (such as antibiotics and some pesticides) do not build up in the blood. The drug testing noted in the chapter introduction relies on the use of urinalysis to detect drug residues that have been secreted into urine.

Homeostasis requires that the total volume of fluid in the blood and tissues stay fairly stable. Blood and tissue fluid are mostly water, and while your kidneys are removing impurities from your blood they are also adjusting the amount of water that is excreted in urine or returned to the bloodstream.

URINATION IS A CONTROLLABLE REFLEX

Urination is urine flow from the body. It is a reflex response. As the bladder fills, tension increases in the smooth muscle of its strong walls. Where the bladder joins the urethra, an *internal urethral sphincter* built of smooth muscle helps prevent urine from flowing into the urethra. As tension in the bladder wall increases, though, the sphincter relaxes; at the same time, the bladder walls contract and force urine through the urethra.

Skeletal muscle forms an *external urethral sphincter* closer to the urethral opening. Learning to control it is the basis of urinary "toilet training" in young children.

Urine consists of water and solutes not needed to maintain the chemical balance of extracellular fluid, as well as water-soluble wastes.

Urine forms through the sequence of steps called filtration, reabsorption, and secretion.

In filtration, water and other small molecules are filtered from the blood and into the nephron.

Reabsorption recaptures needed water and solutes.

Secretion adds unwanted substances into urine, including hydrogen ions and foreign substances such as antibiotics.

12.4 How Kidneys Help Manage Fluid Balance and Blood Pressure

LINKS TO
SECTIONS
2.3 AND 9.5

In addition to removing wastes from blood, the kidneys concentrate urine by way of mechanisms that also help regulate blood volume and blood pressure.

Overall, the total volume of your body fluids, including blood plasma, doesn't vary much. This is because during reabsorption, the kidneys adjust how much water and salt (sodium + chloride ions) the body conserves or excretes in urine. As you know, blood and tissue fluid are mostly water. In general, when the volume of blood increases or decreases, so does blood pressure. The kidneys help ensure that the volume of extracellular fluid, and blood in particular, stays within a normal range.

WATER FOLLOWS SALT AS URINE FORMS

Although about two-thirds of filtered salt and water is reabsorbed in the proximal tubule, the filtrate usually still contains more of both than the body can afford to lose in urine. This situation is addressed as the filtrate enters the loop of Henle, which descends into the kidney medulla (Figure 12.6). There the loop is surrounded by extremely salty tissue fluid. Water can pass through the thin wall of the loop's descending limb, so more water moves out by osmosis and is reabsorbed. As the water leaves, the salt concentration in the fluid still inside the descending limb increases until it matches that in the fluid outside.

Now the filtrate "rounds the turn" of the loop and enters the ascending limb. The wall of this part of the nephron tubule does not allow water to pass through. This is an important variation in the tubule's structure, because here sodium is actively transported out of the ascending limb—but water cannot move with it.

The filtrate now moves into the distal tubule. Its cells continue to remove salt, but also do not let water escape. Hence, a dilute urine moves on into the collecting duct.

Naturally, as salt leaves the filtrate moving through a nephron tubule, the concentration of solutes rises outside the tubule and falls inside it. This steep gradient helps drive the reabsorption of valuable solutes, which move into peritubular capillaries. It also draws water out of the descending limb by osmosis.

Urea boosts the gradient. As water is reabsorbed, the urea left behind in the filtrate becomes concentrated. Some of it will be excreted in urine, but when filtrate enters the final portion of the collecting duct, some urea also will diffuse out—so the concentration of solutes in the inner medulla rises even more.

Drink a large glass of water and the next time you "go" your urine may be pale and dilute. If you sleep eight hours without a break, your urine will be concentrated and darker yellow. As described next, hormonal controls allow the kidneys to vary the amount of water in urine. These controls also adjust blood pressure.

HORMONES CONTROL WHETHER KIDNEYS MAKE URINE THAT IS CONCENTRATED OR DILUTE

When you do not take in as much water as your body loses, the salt concentration in your blood rises. In the brain, receptors sense this change and trigger the release of antidiuretic hormone, or **ADH**. This communication molecule acts on the cells in distal tubules and collecting ducts so that more water moves out of them and is reabsorbed (Figure 12.7). As a result, the urine becomes more concentrated. Gradually the increased water in blood reduces the salt concentration there. It also increases the blood volume and blood pressure, and less ADH is released. Figure 12.8 shows this negative feedback loop.

A decrease in the volume of extracellular fluid also activates cells in the efferent arterioles that bring blood to nephrons. These cells release an enzyme called renin. They are part of the **juxtaglomerular apparatus** (Figure 12.9a). *Juxta-* means "next to," and this "apparatus" is an area where arterioles of the glomerulus come into contact with a nephron's distal tubule.

Renin triggers reactions that produce a protein called angiotensin I and then convert it to angiotensin II. Among other effects, angiotensin II stimulates cells of the adrenal cortex, the outer portion of a gland perched on top of each kidney, to secrete the hormone **aldosterone** (Figures 12.7 and 12.9b). Aldosterone causes cells of the distal tubules and collecting ducts to reabsorb sodium faster, so less sodium and less water is excreted in urine.

What must the kidneys do to make dilute urine? Not much. Urine is automatically dilute as long as ADH levels are low, so little of the hormone acts on the distal tubules and collecting ducts.

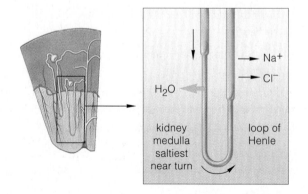

Figure 12.6 Reabsorption of water and salt in the loop of Henle.

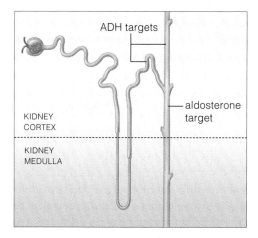

Figure 12.7 Where ADH and aldosterone act in kidney nephrons.

A *diuretic* is any substance that promotes the loss of water in urine. For example, caffeine reduces the reabsorption of sodium along nephron tubules, so more water is excreted.

A THIRST CENTER MONITORS SODIUM

When you don't drink enough, after a while you realize you're thirsty. Why? The concentration of salt in your blood has risen, and this change reduces the amount of saliva your salivary glands produce. A drier mouth stimulates nerve endings that signal a **thirst center** in the brain. The center also receives signals from the same sensors that stimulate the release of ADH. In this case the signals are relayed to a part of the brain that "tells" you to find and drink fluid.

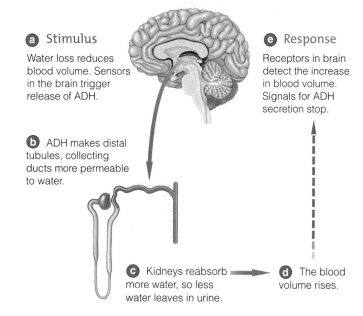

a Stimulus Water loss reduces blood volume. Sensors in the brain trigger release of ADH.

b ADH makes distal tubules, collecting ducts more permeable to water.

c Kidneys reabsorb more water, so less water leaves in urine.

d The blood volume rises.

e Response Receptors in brain detect the increase in blood volume. Signals for ADH secretion stop.

Figure 12.8 A negative feedback loop from the kidneys to the brain that helps adjust the fluid volume of the blood.

In a nephron tubule, water and salt can be reabsorbed or excreted as required to maintain the volume of the extracellular fluid, including blood.

ADH stimulates the kidneys to conserve water. It acts on distal tubules and collecting ducts. Aldosterone promotes the reabsorption of sodium, which indirectly increases the amount of water the body retains.

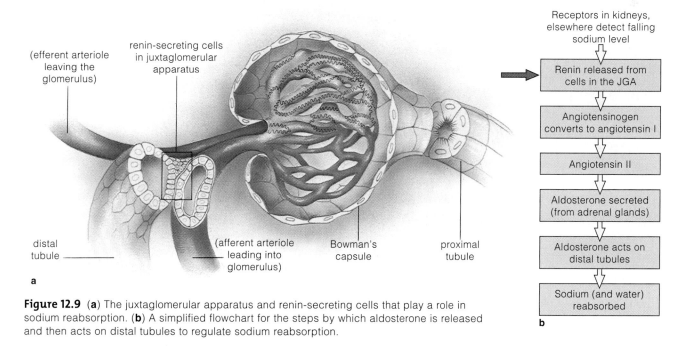

Figure 12.9 (**a**) The juxtaglomerular apparatus and renin-secreting cells that play a role in sodium reabsorption. (**b**) A simplified flowchart for the steps by which aldosterone is released and then acts on distal tubules to regulate sodium reabsorption.

12.5 Removing Excess Acids and Other Substances in Urine

LINKS TO
SECTIONS
2.7, 7.6,
AND 11.5

As the kidneys form urine, nephrons make adjustments that also help keep the extracellular fluid from becoming too acidic or too basic.

You may recall from Chapter 2 that normal pH in the blood and other body fluids is between 7.37 and 7.43. Because acids lower pH and bases raise it, pH reflects the body's **acid–base balance**—the relative amounts of acidic and basic substances in extracellular fluid. Remember also that a buffer system involves substances that reversibly bind and release H^+ and OH^- ions. Buffers minimize pH changes as acidic or basic molecules enter or leave body fluids.

Chapter 11 described how bicarbonate (HCO_3^-) serves as a buffer in the lungs. It forms when carbon dioxide combines with water. The bicarbonate then reacts with H^+ to form carbonic acid, and enzyme action converts carbonic acid into water and carbon dioxide. The CO_2 is exhaled, while the hydrogen ions are now a part of water molecules. H^+ is not eliminated permanently, however. Only the kidneys can do that. They also restore the buffer bicarbonate.

Depending on changes in the acid–base balance of the blood that enters nephrons, the kidneys can either excrete bicarbonate or form new bicarbonate and add it to the blood. The necessary chemical reactions go on in the cells of nephron tubule walls. For example, when the blood is too acid (a too high concentration of H^+), water and carbon dioxide combine with the help of an enzyme. They form carbonic acid that then can be broken into bicarbonate and H^+. Figure 12.10 summarizes these steps.

As you can see, bicarbonate produced in the reactions moves into peritubular capillaries. It ends up circulating in the blood, where it buffers excess H^+. When the blood is too basic (alkaline), chemical adjustments in the kidneys normally ensure that less bicarbonate is reabsorbed into the bloodstream.

The H^+ that is formed in the tubule cells is secreted into the filtrate in the tubule. There the excess H^+ may combine with phosphate ions, ammonia (NH_3), or even bicarbonate. In this way the excess H^+ is excreted.

When the kidneys have finished processing fluid and solutes from the blood, the result is urine that contains a wide variety of substances. In addition to revealing traces of drugs, urinalysis provides a chemical snapshot of many physiological processes in the body, and it can be extremely helpful in diagnosing illness or disease. For example, glucose in urine may be a sign of diabetes. White blood cells (pus) frequently indicate a urinary tract infection. Red blood cells in urine can reveal bleeding due to infection, cancer, or an injury. Abnormally high levels of albumin or other proteins in urine may indicate kidney disease, severe hypertension, or other disorders. Bile pigments enter the urine when liver functions are impaired by cirrhosis and hepatitis.

It's relatively easy to keep your urinary tract healthy. Drink plenty of water and practice careful hygiene to minimize the chances for bacteria to invade the urethra. People who are susceptible to bladder infections may also want to limit their intake of alcohol, caffeine, and spicy foods, all of which can irritate the bladder.

Along with buffering systems and the respiratory system, the kidneys help keep the extracellular fluid from becoming to acidic or too basic.

The urinary system eliminates excess hydrogen ions and also replenishes bicarbonate used in buffering reactions.

Figure 12.10 How the kidneys remove H^+ from the body, preventing the blood from becoming too acidic.

12.6 Kidney Disorders

From the preceding sections, you can sense that good health depends on normal kidney function. Disorders or injuries that interfere with it can be mild to severe.

For example, **kidney stones** are deposits of uric acid, calcium salts, and other substances that have settled out of urine and collected in the renal pelvis. Smaller kidney stones usually are eliminated naturally during urination. Larger ones can become lodged in the renal pelvis or ureter or even in the bladder or urethra. The blockage can partially dam urine flow and cause intense pain and kidney damage. Large kidney stones must be removed medically or surgically. A procedure called *lithotripsy* uses high-energy sound waves to blast the stone to fragments that are small enough to pass out in the urine.

Urinary tract infections routinely plague millions of people. Women especially are susceptible to bladder infections because of their urinary anatomy: The female urethra is short, just a little over an inch long. (An adult male's urethra is about 9 inches long.) The outer opening of a female's urethra also is close to the anus, so it is fairly easy for bacteria from outside the body to make their way to a female's bladder and trigger an inflammation called **cystitis**—or even all the way to the kidneys to cause **pyelonephritis**. In both sexes, urinary tract infections sometimes result from sexually transmitted microbes, including the microorganisms that cause *chlamydia*. Chapter 16 provides more information on this topic.

Polycystic kidney disease is an inherited disorder in which cysts (semisolid masses) form in the kidneys and in many cases gradually destroy normal kidney tissue. Frequent urinary tract infections are a common early symptom; in severe cases, kidney dialysis (Figure 12.11 and described below) or a kidney transplant is the only real option for treatment.

Nephritis is an inflammation of the kidneys. It can be caused by various factors, including bacterial infections. As you may remember from Chapter 10, inflamed tissue tends to swell as fluid accumulates in it. However, because a kidney is "trapped" inside the tough renal capsule, it can't increase in size. As a result, hydrostatic pressure builds up in or around glomerular capillaries, blocking them and hampering or preventing the passage of blood. Then, of course, blood filtering becomes difficult or impossible.

Glomerulonephritis is an umbrella term for several disorders (often involving faulty immune responses) that can severely damage the kidneys. Hypertension and diabetes can disrupt blood circulation to and within the kidneys, sometimes virtually blocking the flow of blood through the glomeruli. In roughly 13 million people in the United States kidney nephrons have become so

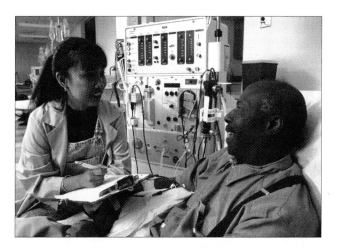

Figure 12.11 *Animated!* Patient undergoing hemodialysis.

impaired that the filtering of blood and formation of urine are woefully inefficient. Control of the volume and composition of the extracellular fluid is disturbed, and toxic by-products of protein breakdown can accumulate in the bloodstream. Patients can suffer nausea, fatigue, and memory loss. In advanced cases, death may result. A kidney dialysis machine can restore the proper solute balances. Like the kidneys, the machine helps maintain extracellular fluid by selectively removing and adding solutes to the patient's bloodstream.

"Dialysis" refers to the exchange of substances across a membrane between chemically different solutions. In *hemodialysis*, the dialysis machine is connected to an artery or a vein, and then blood is pumped through tubes made of a material similar to cellophane. The tubes are submerged in a warm-water bath. The precise mix of salts, glucose, and other substances in the bath sets up the correct gradients with the blood. Dialyzed blood is returned to the body.

Hemodialysis generally must be performed three times a week. It is a temporary measure for patients with reversible kidney disorders. In chronic cases, the procedure must be used for the rest of the patient's life or until a healthy kidney can be transplanted.

Although chronic kidney disease can impose some inconveniences, with proper treatment and a controlled diet, many people are able to pursue a surprisingly active, close-to-normal lifestyle.

Summary

Section 12.1   The fluid inside cells is intracellular fluid. By contrast, body cells are bathed by extracellular fluid (ECF)—various types and amounts of substances dissolved in water. The ECF fills tissue spaces and (in the form of blood plasma) fills blood vessels. Its volume and composition are maintained only when the daily intake of water and solutes is in balance. The following processes maintain the balance:

   a. The body takes in water by absorbing it from the GI tract and by metabolism. Water is lost by urinary excretion, evaporation from the lungs and skin, sweating, and elimination in feces.

   b. Solutes are gained by absorption from the GI tract, secretion, respiration, and metabolism. They are lost by excretion, respiration, and sweating. The solutes include important electrolytes such as ions of sodium, potassium, and calcium.

   c. Losses of water and solutes are controlled mainly by adjusting the volume and composition of urine.

Section 12.2   The urinary system consists of two kidneys, two ureters, a urinary bladder, and a urethra. In the kidneys, blood is filtered and urine forms in nephrons.

   a. A nephron starts as a cup-shaped capsule that is followed by three tubelike regions: the proximal tubule, loop of Henle, and distal tubule, which empties into a collecting duct.

   b. The Bowman's (glomerular) capsule surrounds a set of highly permeable capillaries. Together, they are a blood-filtering unit, the glomerulus.

 Explore the anatomy of the urinary system and kidneys.

Section 12.3   Urine forms through a sequence of steps: filtration, reabsorption, and secretion.

   a. Filtration of blood at the glomerulus of a nephron, which puts water and small solutes into the nephron.

   b. Reabsorption. Water and solutes to be retained leave the nephron's tubular parts and enter the peritubular capillaries that thread around them. Many solutes are reabsorbed passively, following their concentration gradients back into the bloodstream. In other instances, active transport is required. Sodium is reabsorbed by active transport. The reabsorption of water is always passive, by osmosis occurring along water's concentration gradient. A small amount of water and solutes remains in the nephron.

   c. Secretion. Some ions and a few other substances leave the peritubular capillaries and enter the nephron, for disposal in urine.

 Learn more about the processes that form urine.

Section 12.4   During reabsorption in kidney nephrons, water and salt are reabsorbed or excreted as required to conserve or eliminate water. The mechanisms that concentrate urine also help regulate blood volume and blood pressure.

   Urine becomes more or less concentrated by the action of two hormones, ADH and aldosterone, on cells of distal tubules and collecting ducts as follows:

   a. ADH is secreted when the body must conserve water; it enhances reabsorption from the distal nephron tubule and collecting ducts. Inhibition of ADH allows more water to be excreted.

   b. Aldosterone conserves sodium by enhancing its reabsorption in the distal tubule. It is secreted when cells in the juxtaglomerular apparatus (next to the distal tubule) secrete renin, an enzyme that triggers reactions that lead to aldosterone secretion. By contrast, inhibition of aldosterone allows more sodium to be excreted. Because "water follows salt," aldosterone indirectly influences water reabsorption.

Section 12.5   Together with the respiratory system and other mechanisms, the kidneys also help maintain the body's overall acid–base balance. They help regulate pH by eliminating excess hydrogen ions and replenishing the supply of bicarbonate, which acts as a buffer elsewhere in the body.

Section 12.6   Urinary tract infections can develop when bacteria enter through the urethra. Disorders and diseases that damage kidney tissues can interfere seriously with the excretion of wastes in urine and with the kidneys' ability to help regulate blood volume and pressure.

Review Questions

1. Label the parts of this kidney and nephron:

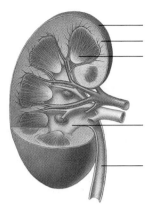 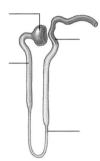

2. How does the formation of urine help maintain the body's internal environment?

3. Explain what is meant when we talk about filtration, reabsorption, and secretion in the kidneys.

4. Which hormone or hormones promote (a) water conservation, (b) sodium conservation, and (c) thirst behavior?

5. Explain how the kidneys help to maintain the balance of acids and bases in extracellular fluid.

Self-Quiz

Answers in Appendix V

1. The body gains water by _____.
   a. absorption in the gut   c. responding to thirst
   b. metabolism   d. all of the above

2. The body loses water by way of the _____.
   a. skin   d. urinary system
   b. lungs   e. c and d
   c. digestive system   f. a through d

3. Water and small solutes enter nephrons during _____.
   a. filtration   c. secretion
   b. reabsorption   d. both a and b

4. Kidneys return water and small solutes to blood by _____.
   a. filtration   c. secretion
   b. reabsorption   d. both a and c

5. Some substances move out of the peritubular capillaries and are moved into the nephron during _____.
   a. filtration   c. secretion
   b. reabsorption   d. both a and c

6. Reabsorption depends on _____.
   a. osmosis across the nephron wall
   b. active transport of sodium across the nephron wall
   c. a steep solute concentration gradient
   d. all of the above

7. _____ directly promotes water conservation.
   a. ADH   c. Aldosterone
   b. Renin   d. both b and c

8. _____ enhances sodium reabsorption.
   a. ADH   c. Aldosterone
   b. Renin   d. both b and c

9. Match the following salt–water balance concepts:
   ____ aldosterone   a. blood filter of a nephron
   ____ nephron   b. controls sodium reabsorption
   ____ thirst mechanism   c. occurs at nephron tubules
   ____ reabsorption   d. site of urine formation
   ____ glomerulus   e. controls water gain

Critical Thinking

1. A urinalysis reveals that the patient's urine contains glucose, hemoglobin, and white blood cells (pus). Are any of these substances abnormal in urine? Explain.

2. As a person ages, nephron tubules lose some of their ability to concentrate urine. What is the effect of this change?

3. Fatty tissue holds the kidneys in place. Extremely rapid weight loss may cause this tissue to shrink so that the kidneys slip from their normal position. On rare occasions, the slippage can put a kink in one or both ureters and block urine flow. Suggest what might then happen to the kidneys.

4. Licorice is used as a remedy in Chinese traditional medicine and also is a flavoring for candy. When licorice is eaten, one of its components triggers the formation of a compound that mimics aldosterone and binds to receptors for it. Based on this information, explain why people who have high blood pressure are advised to avoid eating much licorice.

5. Drinking too much water can be a bad thing. If someone sweats heavily and drinks lots of water, their sodium levels drop. The resulting "water intoxication" can be fatal. Why is the sodium balance so important?

6. As the text noted, two-thirds of the water and solutes that the body reclaims by reabsorption in nephrons occurs in the proximal tubule. Proximal tubule cells have large numbers of mitochondria and demand a great deal of oxygen. Explain why.

Explore on Your Own

The rider and horse shown at the left are living examples of the mammalian body's ability to cool itself by producing sweat. Since sweat is mostly water, how is heavy sweating likely to affect the concentration of urine, especially if the athlete— in this case, a polo player— doesn't remember to drink fluid during the match? (You may well have observed this effect in your own body after exercise.)

Drink one quart of water in one hour. What changes might you expect (and can you observe) in your kidney function and the nature of your urine?

13 THE NERVOUS SYSTEM

In Pursuit of Ecstasy

"Ecstasy" is an illegal but popular drug that relieves anxiety and sharpens the senses. Users say they feel socially accepted and get a mild high. Ecstasy also can leave you dying in a hospital, foaming at the mouth, and bleeding from every orifice as your temperature skyrockets. When Lorna Spinks was nineteen years old, her life ended that way.

Her anguished parents released the photograph at right, taken just minutes after her death. They wanted others to know what Lorna did not: Ecstasy can kill.

Ecstasy's active ingredient, MDMA, is related to amphetamine, or "speed." It interferes with the function of serotonin, a signaling molecule that works in the brain. MDMA causes neurons to release too much serotonin. Instead of being cleared away, as normally happens, serotonin molecules simply saturate receptors on the target cells, which can't be released from overstimulation.

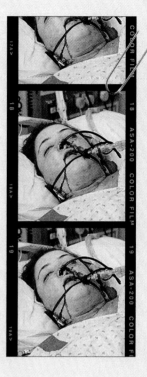

An overdose killed Lorna. Waves of panic and seizures set in. Her heart pounded, her blood pressure soared, and her temperature rose so high that organ systems shut down one by one.

An MDMA overdose does not often end in death, but other problems are common. For example, neurons can't rebound quickly when the brain's serotonin stores are depleted. Below-normal levels of serotonin can contribute to loss of concentration, depression, and memory problems. As studies of Ecstasy users reveal, the more often you use it, the worse memory loss becomes. If you stop using the drug, it can be many months before your brain function returns to normal.

How we function as individuals depends largely on our nervous system—and whether we nurture or abuse it. In this chapter we start by looking at the structure and function of neurons. Then we'll examine how neurons interact in the nervous system and how the brain serves as the body's master control center.

 How Would You Vote? *Would you support legislation that forces nonviolent drug offenders to enter drug rehab programs as an alternative to jail? Cast your vote online at www.thomsonedu.com/biology/starr/humanbio.*

 ## Key Concepts

HOW NEURONS WORK
Messages that travel through the nervous system are based on electrical charges across the plasma membrane of neurons and on signals sent to the next cell in line.

THE NERVOUS SYSTEM
Different parts of the nervous system detect information about conditions inside and outside the body, process it, and then select or control muscles and glands that carry out responses.

THE BRAIN
The brain receives, integrates, stores, and retrieves sensory information. It also orders and coordinates responses by adjusting body activities.

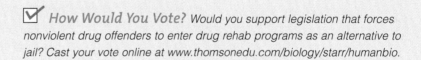 ## Links to Earlier Concepts

In this chapter you will draw on your knowledge of the structure of cell plasma membranes (3.3), concentration gradients and diffusion (3.9), and mechanisms of active and passive transport (3.10). You will also gain a deeper understanding of neuromuscular junctions and how nervous system signals lead to muscle contractions (6.4).

Above all, you will learn a great deal more about neurons and glial cells that make up the body's nervous tissue (4.4).

13.1 Neurons—The Communication Specialists

LINKS TO
SECTIONS
3.3 AND 4.4

Sensing and responding to sounds, sights, hunger, fear—it all begins with cells of the nervous system.

What does your **nervous system** do? Its role is to detect and integrate information about external and internal conditions, then select or control muscles and glands that carry out responses. The system's communication cells are **neurons**, and there are three basic kinds:

Sensory neurons collect and relay information about stimuli to the spinal cord and brain. A stimulus is a form of energy, such as light, detected by a specific receptor.

Interneurons in the spinal cord and brain receive and process sensory input, and send signals to other neurons.

Motor neurons relay signals from interneurons to the body's effectors—muscles and glands—that carry out the specified responses.

NEURONS HAVE SEVERAL FUNCTIONAL ZONES

Every neuron has a cell body, where its nucleus and organelles are. The cell body and slender extensions called **dendrites** are *input zones* for information. Nearby is a **trigger zone**. In motor neurons and interneurons the trigger zone is called the *axon hillock* ("little hill"). At this patch of the neuron's plasma membrane, information travels along a slender and often long extension called an **axon**, the neuron's *conducting zone*. As you can see in the diagram of a motor neuron in Figure 13.1, dendrites tend to be shorter than axons and their number and length vary, depending on the type of neuron. The axon's endings are *output zones* where messages are sent to other cells.

Only about 10 percent of your nervous system consists of neurons. The other 90 percent consists of **neuroglia**, or simply "glia." There are various types of glia, including the star-shaped astroglia shown in Figure 13.2. They help maintain proper concentrations of vital ions in the fluid around neurons, among other tasks. Some glia physically support and protect neurons, as you will read later, and some provide insulation that allows signals to move rapidly along sensory and motor neurons. Others do "clean-up" duty in the central nervous system.

Neurons function so well in communication in part because they are *excitable*—that is, a neuron can respond to certain stimuli by producing an electrical signal. Let's look at the factors that make this possible.

PROPERTIES OF A NEURON'S PLASMA MEMBRANE ALLOW IT TO CARRY SIGNALS

Chapter 3 noted that the plasma membrane's lipid bilayer prevents charged substances—such as ions of potassium (K^+) and sodium (Na^+)—from freely crossing it. Even so, ions can move through the interior of channel proteins that span the bilayer (Figure 13.3). Some channels are

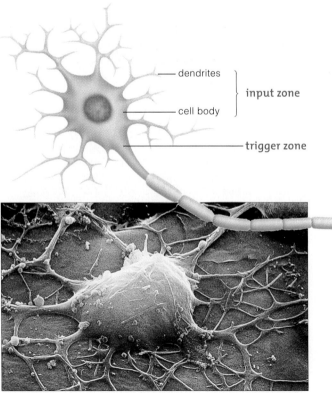

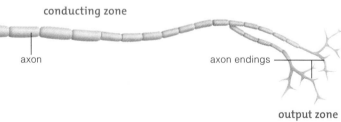

dendrites — input zone
cell body —

trigger zone

conducting zone

axon

axon endings

output zone

Figure 13.1 *Animated!* The functional zones of a motor neuron. The micrograph shows a motor neuron with its plump cell body and branching dendrites.

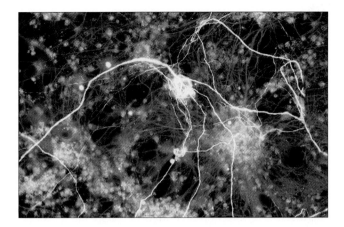

Figure 13.2 In this light micrograph of brain tissue, astrocytes appear orange and a neuron is yellow.

Because the ion concentrations on either side of a neuron's plasma membrane differ, the cytoplasm next to the membrane is negatively charged, compared to the fluid just outside the membrane. Electrical charges may be measured in millivolts, and for many neurons, the steady charge difference across the plasma membrane is about −70 millivolts (indicating that the cytoplasm side of the membrane is more negative). This difference is called the **resting membrane potential**. The name means that the charge difference has the potential to do physiological work in the body.

Various kinds of signals occur in the nervous system, but not all of them spark nerve impulses. Only a signal that is strong enough when it arrives at a resting neuron's input zone may spread to a trigger zone. When a strong enough signal does arrive, however, it can cause the voltage difference across the plasma membrane to reverse, just for an instant. As we see next, these reversals launch nervous system signals.

always open, so that ions can steadily "leak" (diffuse) in or out. Other channels have "gates" that open only under certain circumstances. These controls over ion movements mean that the concentrations of different ions can be different on either side of the plasma membrane.

For example, in a resting neuron, the gated sodium channels are closed. The cell's plasma membrane also does not allow much sodium to leak inward but is more permeable to K⁺. As a result, each ion has a concentration gradient across the membrane (Figure 13.3a). Following the rules of diffusion, sodium tends to move in and potassium tends to move out.

Sensory neurons, interneurons, and motor neurons make up the communication lines of the nervous system.

In a resting neuron, differences in the concentrations of Na⁺ and K⁺ across the plasma membrane produce a resting membrane potential—a difference in electrical charge across the plasma membrane. The difference has the potential to do physiological work.

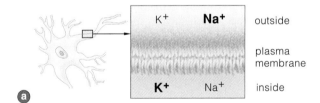

Figure 13.3 *Animated!* Ions and a neuron's plasma membrane. (**a**) Gradients of sodium (Na⁺) and potassium (K⁺) ions across a neuron's plasma membrane. (**b**) How ions cross the plasma membrane of a neuron. They are selectively allowed to cross at protein channels and pumps that span the membrane.

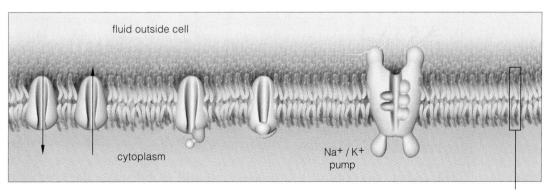

Passive transporters with open channels let ions steadily leak across the membrane.

Other passive transporters have voltage-sensitive gated channels that open and shut. They assist diffusion of Na⁺ and K⁺ across the membrane as the ions follow concentration gradients.

Active transporters pump Na⁺ and K⁺ across the membrane, against their concentration gradients. They counter ion leaks and restore resting membrane conditions.

lipid bilayer of neuron membrane

13.2 Action Potentials = Nerve Impulses

LINK TO
SECTION
3.9

It is easy to understand how a nerve impulse develops if you remember that ions can only cross cell membranes through "tunnels" in transport proteins.

As you've just read, when a strong enough signal arrives at a resting neuron's input zone, it can trigger the reversal of the voltage difference across the plasma membrane. The signal has this effect when it causes sodium gates in the membrane to open, so that Na$^+$ rushes into the neuron.

As the positively charged sodium flows inward, the cytoplasm next to the plasma membrane becomes less negative (Figure 13.4 *a,b*). Then, more gates open, more sodium enters, and so on (note that this is an example of positive feedback). When the voltage difference across the neuron's plasma membrane shifts by a certain minimum amount—the **threshold** level of stimulation—the result is an **action potential**, the "nerve impulse" that is a neuron's communication signal.

The threshold for an action potential can be reached at any patch of plasma membrane where there are voltage-sensitive gated channels for sodium ions. Because of the positive feedback, when the threshold level is reached, the opening of more sodium gates doesn't depend any longer on the strength of the stimulus. The gates open on their own.

It's important to remember that an action potential occurs only if the stimulus to a neuron is strong enough. A weak stimulus—say, mechanical pressure from a tiny insect walking on your skin—that arrives at an input zone may not upset the ion balance enough to cause an action potential. This is because input zones don't have gated sodium channels, so sodium can't flood in there. On the other hand, a neuron's trigger zone is riddled with sodium channels. If a stimulus at an input zone is strong enough to spread to the trigger zone, an action potential may "fire."

ACTION POTENTIALS SPREAD BY THEMSELVES

To transmit messages throughout the body, action potentials must spread to other neurons or to cells in muscles or glands. Each action potential propagates itself—it moves away from its starting point. This self-propagation is possible in part because the changes in membrane potential leading up to an action potential don't lose strength. When the change spreads from one patch of a neuron's plasma membrane to another patch, approximately the same number of gated channels open (Figure 13.4 *c,d*). Action potentials never spread back into the trigger zone. They always propagate *away* from it, for reasons described next.

A NEURON CAN'T "FIRE" AGAIN UNTIL ION PUMPS RESTORE ITS RESTING POTENTIAL

When an incoming signal causes an action potential in a neuron's trigger zone, that area of the cell's plasma membrane can't receive another signal until its resting membrane potential is restored.

To understand how this happens, remember that a neuron's resting membrane potential is due in part to the differing concentrations of Na$^+$ and K$^+$ across the plasma membrane. There also is an *electric* gradient across the membrane. The inside of the cell is a little more negative than the outside, partly because there are many negatively charged proteins in the cytoplasm. In addition, K$^+$ can diffuse out of the neuron, moving down its concentration gradient. Together, these factors mean that in a resting motor neuron, some Na$^+$ is always leaking *into* the cell, down its electrochemical gradient, and K$^+$ is leaking *out* down *its* concentration gradient.

Figure 13.4 *Animated!* (**a**,**b**) Steps leading to an action potential. (**c**,**d**) How an action potential propagates.

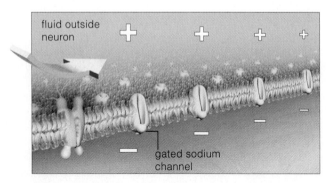

a In a membrane at rest, the inside of the neuron is negative relative to the outside. An electrical disturbance (yellow arrow) spreads from an input zone to an adjacent trigger zone of the membrane, which has a large number of gated sodium channels.

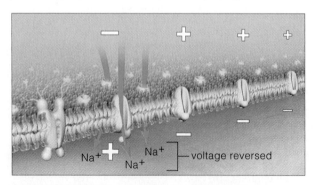

b A strong disturbance initiates an action potential. Sodium gates open. Sodium flows in, reducing the negativity inside the neuron. The change causes more gates to open, and so on until threshold is reached and the voltage difference across the membrane reverses.

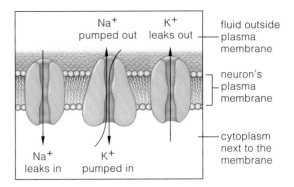

Figure 13.5 Pumping and leaking processes that maintain the distribution of sodium and potassium ions across a resting neuron's plasma membrane. The inward and outward movements of each kind of ion are balanced.

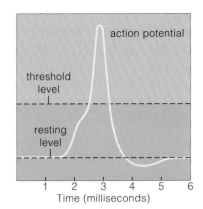

Figure 13.6 *Animated!* A recording of an action potential. You can see clearly how the membrane potential spikes once threshold is reached.

A neuron can't respond to an incoming signal unless the concentration and electric gradients across its plasma membrane are in place. Yet the Na⁺ and K⁺ leaks never stop. You might expect that the net amount of K⁺ in the cell would continue to fall while the amount of Na⁺ would slowly, surely increase. This imbalance doesn't develop, however, because a resting neuron expends energy on an active transport mechanism that maintains the gradients. This mechanism takes the form of carrier proteins called **sodium–potassium pumps** that span the membrane (Figure 13.5). With energy from ATP, they actively transport potassium *in* and sodium *out*.

ACTION POTENTIALS ARE "ALL-OR-NOTHING"

There is no such thing as a "small" or "large" action potential. Every action potential in a neuron spikes to the same level above threshold as an all-or-nothing event. That is, once the positive-feedback cycle of opening sodium gates starts, nothing will stop the full spiking. If threshold is not reached, the disturbance to the plasma membrane will fade away as soon as the stimulus is removed. Figure 13.6 shows a recording of the voltage difference across a neuron's plasma membrane before, during, and after an action potential.

Each spike lasts for about a millisecond. At the place on the membrane where the charge reversed, the gated sodium channels close and the influx of sodium stops. About halfway through the action potential, potassium channels open, so potassium ions flow out and restore the original voltage difference across the membrane. And sodium–potassium pumps restore ion gradients, as you've read. Later, after the resting membrane potential has been restored, most potassium gates are closed and sodium gates are in their initial state, ready to be opened again when a suitable stimulus arrives.

> An action potential occurs when a neuron's resting membrane potential briefly reverses. Action potentials self-propagate and always move away from the trigger zone.
>
> After an action potential, sodium–potassium pumps restore the neuron's resting potential.
>
> An action potential is an all-or-nothing event. Once the spiking starts, nothing can stop it.

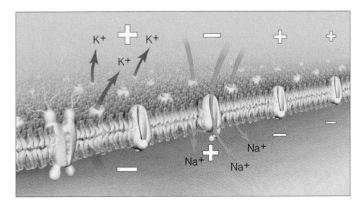

c At the next patch of membrane, another group of gated sodium channels open. In the previous patch, some K⁺ moves out through other gated channels. That region becomes negative again.

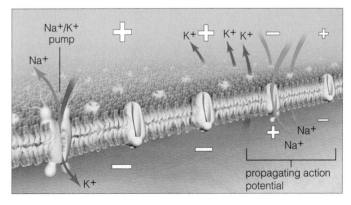

d After each action potential, the sodium and potassium concentration gradients in a patch of membrane are not yet fully restored. Active transport at sodium–potassium pumps restores them.

13.3 Chemical Synapses: Communication Junctions

LINK TO
SECTION
6.6

When action potentials reach a neuron's output zone, they usually stop there. But they may prompt the neuron to release signaling molecules that diffuse to a receiving cell. This is one way that information flows from cell to cell.

Action potentials can trigger the release of one or more **neurotransmitters**. These are signaling molecules that diffuse across a **chemical synapse**, a narrow gap between a neuron's output zone and the input zone of a neighboring cell (Figure 13.7). Some chemical synapses occur between neurons, others between a neuron and a muscle cell or gland cell.

At a chemical synapse, one of the two cells stores neurotransmitter molecules in synaptic vesicles in its cytoplasm. This is the *pre*synaptic cell. Gated channels for calcium ions span the cell's plasma membrane, and they open when an action potential arrives. There are more calcium ions outside the cell, and when they flow in (down their gradient), synaptic vesicles fuse with the plasma membrane. Neurotransmitter molecules in the vesicles now pour into the synapse, diffuse across it, and bind with receptor proteins on the plasma membrane of the *post*synaptic, or receiving, cell. Binding changes the shape of these proteins, so that a channel opens up through them. Ions then diffuse through the channels and enter the receiving cell (Figure 13.7*b*).

NEUROTRANSMITTERS CAN EXCITE OR INHIBIT A RECEIVING CELL

How a receiving cell responds to a neurotransmitter depends on several factors—the type and amount of neurotransmitter, the kinds of receptors the cell has, and the types of channels at its input zone. *Excitatory* signals help drive the membrane toward an action potential. *Inhibitory* signals have the opposite effect.

One neurotransmitter, **acetylcholine (ACh)**, can excite *or* inhibit different target cells in the brain, spinal cord, glands, and muscles. Figure 13.8 shows a chemical synapse between a motor neuron and a muscle cell. ACh released from the neuron diffuses across the gap and binds to receptors on the muscle cell membrane. It excites this kind of cell, triggering action potentials that cause muscle contraction.

Serotonin is a neurotransmitter that acts on brain cells that govern sleeping, sensory perception, regulation of body temperature, and emotional states. Some neurons secrete *nitric oxide* (NO), a gas that controls blood vessel dilation. It is not stored in synaptic vesicles but instead is manufactured as needed. As an example, a sexually aroused male has an erection when NO calls on blood vessels in his penis to dilate, allowing blood to rush in. Section 13.10 looks at the role of some neurotransmitters in disorders of the nervous system.

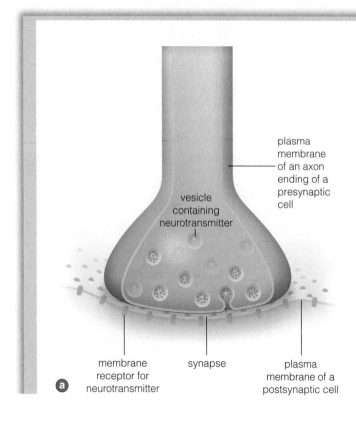

membrane receptor for neurotransmitter

synapse

plasma membrane of a postsynaptic cell

(a)

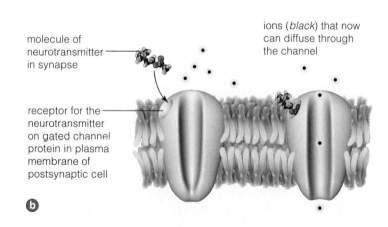

ions (*black*) that now can diffuse through the channel

molecule of neurotransmitter in synapse

receptor for the neurotransmitter on gated channel protein in plasma membrane of postsynaptic cell

plasma membrane of an axon ending of a presynaptic cell

vesicle containing neurotransmitter

(b)

Figure 13.7 *Animated!* Example of a chemical synapse. (**a**) Only a narrow gap separates a presynaptic cell from a postsynaptic one. (**b**) A neurotransmitter carries signals from the presynaptic neuron to the receiving cell.

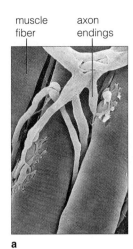

muscle fiber   axon endings

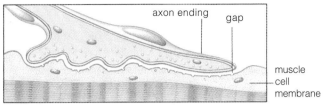

axon ending   gap

muscle cell membrane

Motor end plate (troughs in muscle cell membrane)

b

Figure 13.8 One kind of chemical synapse—a neuromuscular junction. (**a**) Micrograph of a neuromuscular junction. It forms between axon endings of motor neurons and muscle cells. (**b**) The axon's myelin sheath stops at the neuromuscular junction, so the membranes of the two cells are exposed. There are troughs in the muscle cell membrane where the axon endings are positioned.

a

Neuromodulators can magnify or impede the effects of a neurotransmitter. These substances include natural painkillers called *endorphins*. Endorphins inhibit nerves from releasing substance P, which conveys information about pain. In athletes who push themselves beyond normal fatigue, endorphins can produce a euphoric "high." Endorphins also may have roles in functions such as memory, learning, and sexual behavior.

COMPETING SIGNALS ARE "SUMMED UP"

Between 1,000 and 10,000 communication lines form synapses with a typical neuron in your brain. And your brain contains at least *100 billion* neurons, humming with messages about doing what it takes to be a human.

At any moment, many signals are washing over the input zones of a receiving neuron. All of them are graded potentials (their magnitude can be large or small), and they compete for control of the membrane potential at the trigger zone. The ones called EPSPs (for excitatory postsynaptic potentials) *depolarize* the membrane—they bring it closer to threshold. On the other hand, IPSPs (inhibitory postsynaptic potentials) may *hyperpolarize* the membrane (drive it away from threshold) or help keep the membrane at its resting level.

Synaptic integration tallies up the competing signals that reach an input zone of a neuron at the same time—a little like adding up the pros and cons of a certain course of action. This process, called *summation*, is how signals arriving at a neuron are suppressed, reinforced, or sent onward to other cells in the body.

Integration occurs when neurotransmitter molecules from more than one presynaptic cell reach a neuron's input zone at the same time. Signals also are integrated after a neurotransmitter is released repeatedly, over a short time period, from a neuron that is responding to a rapid series of action potentials.

NEUROTRANSMITTER MOLECULES MUST BE REMOVED FROM THE SYNAPSE

The flow of signals through the nervous system depends on the rapid, controlled removal of neurotransmitter molecules from synapses. Some of the neurotransmitter molecules diffuse out of the gap. Enzymes cleave others in the synapse, as when acetylcholinesterase breaks down ACh. Also, membrane transport proteins actively pump the neurotransmitter molecules back into presynaptic cells or into neighboring neuroglia.

Some drugs can block the reuptake of certain neurotransmitters. For example, some antidepressant drugs (such as Prozac) alter a person's mood by blocking the reuptake of serotonin.

Neurotransmitters are signaling molecules that bridge the synapse (a narrow gap) between two neurons or between a neuron and a muscle cell or gland cell. They may excite or inhibit the activity of different kinds of target cells.

In synaptic integration, incoming signals that excite or inhibit the same postsynaptic cell are summed. In this way messages traveling through the nervous system can be reinforced or downplayed, sent onward or suppressed.

IMPACTS, ISSUES

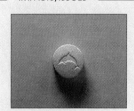

The MDMA in Ecstasy acts mainly at axon endings that release serotonin. It stimulates the release of *all* stored serotonin, which floods the chemical synapses and saturates receptors on postsynaptic cells. Ecstasy also slows the uptake of serotonin by the same axon endings. Repeated use of Ecstasy may cause the axon endings to wither and die. If the endings grow back, they may be abnormal or in the wrong places.

13.4 Information Pathways

Once a message is "sent" in the nervous system, what determines where it will go? That depends on how neurons are organized in the body.

NERVES ARE LONG-DISTANCE LINES

Nerves are communication lines between the brain or spinal cord and the rest of the body. A **nerve** consists of nerve fibers, which are the long axons of sensory neurons, motor neurons, or both. Connective tissue encloses most of the axons like electrical cords inside a tube (Table 13.1 and Figure 13.9*a*). In the central nervous system (the brain and spinal cord) nerves are called **nerve tracts**.

Each axon has an insulating **myelin sheath**, which allows action potentials to propagate faster than they would otherwise. The sheath consists of glia called **Schwann cells**, which wrap around the long axons like jelly rolls. As you can see in Figure 13.9*b*, an exposed node, or gap, separates each cell from the next one. There, voltage-sensitive, gated sodium channels pepper the plasma membrane. In a manner of speaking, action potentials jump from node to node (a phenomenon also called saltatory conduction, after a Latin word meaning

Table 13.1	Basic Components of the Nervous System
Neuron	Nervous system cell specialized for communication
Nerve fiber	Long axon of one neuron
Nerves	Long axons of several neurons enclosed by connective tissue

"to jump"). The sheathed areas between nodes hamper the movement of ions across the plasma membrane, so stimulation tends to travel along the membrane until the next node in line. At each node, however, the flow of ions can produce a new action potential. In large sheathed axons, action potentials propagate at a remarkable 120 meters (nearly 400 feet) per second!

There are no Schwann cells in the central nervous system. There, glia called oligodendrocytes sheath all myelinated axons.

REFLEX ARCS ARE THE SIMPLEST NERVE PATHWAYS

Sensory and motor neurons of certain nerves take part in automatic responses called reflexes. A **reflex** is a simple, stereotyped movement (it is always the same) in response to a stimulus. In the simplest **reflex arcs**, sensory neurons synapse directly on motor neurons.

The *stretch reflex* contracts a muscle after gravity or some other load has stretched the muscle. Suppose you hold out a bowl and keep it stationary as someone loads peaches into it, adding weight to the bowl. When your hand starts to drop, a muscle in your arm (the biceps) is stretched.

In the muscle, stretching activates receptor endings that are a part of muscle spindles. These are sensory organs in which specialized cells are enclosed in a sheath that runs parallel with the muscle. The receptor endings are the input zones of sensory neurons whose axons synapse with motor neurons in the spinal cord (Figure 13.10). Axons of the motor neurons lead back to the stretched muscle. Action potentials that reach the axon endings trigger the release of ACh, which triggers contraction. As long as receptors continue to send messages, the motor neurons are excited. This allows them to maintain your hand's position.

In most reflex pathways, the sensory neurons also interact with several interneurons. These excite or inhibit motor neurons as needed for a coordinated response.

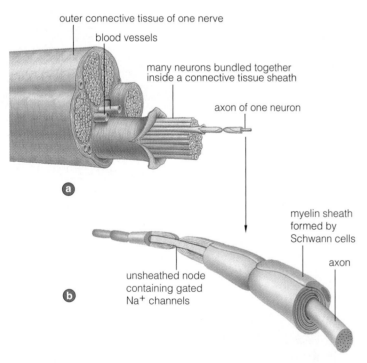

outer connective tissue of one nerve

blood vessels

many neurons bundled together inside a connective tissue sheath

axon of one neuron

(a)

myelin sheath formed by Schwann cells

axon

unsheathed node containing gated Na$^+$ channels

(b)

Figure 13.9 *Animated!* (**a**) Structure of a nerve. (**b**) Structure of a sheathed axon. A myelin sheath (a series of Schwann cells wrapped like a jelly roll around the axon) blocks the flow of ions except at nodes between Schwann cells.

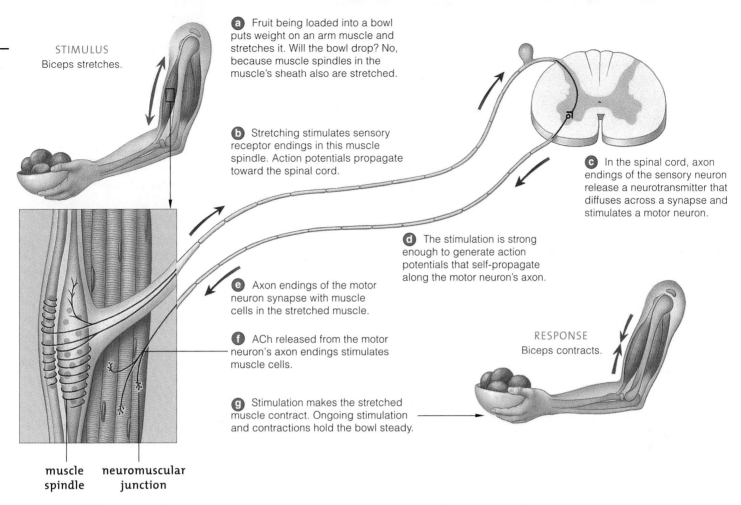

STIMULUS
Biceps stretches.

a Fruit being loaded into a bowl puts weight on an arm muscle and stretches it. Will the bowl drop? No, because muscle spindles in the muscle's sheath also are stretched.

b Stretching stimulates sensory receptor endings in this muscle spindle. Action potentials propagate toward the spinal cord.

c In the spinal cord, axon endings of the sensory neuron release a neurotransmitter that diffuses across a synapse and stimulates a motor neuron.

d The stimulation is strong enough to generate action potentials that self-propagate along the motor neuron's axon.

e Axon endings of the motor neuron synapse with muscle cells in the stretched muscle.

f ACh released from the motor neuron's axon endings stimulates muscle cells.

RESPONSE
Biceps contracts.

g Stimulation makes the stretched muscle contract. Ongoing stimulation and contractions hold the bowl steady.

muscle spindle   neuromuscular junction

Figure 13.10 *Animated!* How nerves are organized in a reflex arc that deals with muscle stretching. In a skeletal muscle, stretch-sensitive receptors of a sensory neuron are located in muscle spindles. The stretching generates action potentials, which reach axon endings in the spinal cord. These synapse with a motor neuron that carries signals to contract from the spinal cord back to the stretching muscle.

IN THE BRAIN AND SPINAL CORD, NEURONS INTERACT IN CIRCUITS

The diagram below shows the overall direction of the flow of information in your nervous system. Sensory nerves relay information into the spinal cord, where they form chemical synapses with interneurons. The spinal cord and brain contain only interneurons, which integrate the signals. Many interneurons synapse with motor neurons, which carry signals away from the spinal cord and brain.

In the brain and spinal cord, interneurons are grouped into blocks of hundreds or thousands. The blocks in turn are parts of circuits. Each block receives signals—some that excite, others that inhibit—and then integrates the messages and responds with new ones. For example, in some brain regions the circuits diverge—the processes of

neurons in one block fan out to form connections with other blocks. Elsewhere signals from many neurons are funneled to just a few. And in still other places in the brain, neurons synapse back on themselves, repeating signals among themselves. These "reverberating" circuits include the ones that make your eye muscles twitch rhythmically as you sleep.

Nerves containing the long axons of sensory neurons, motor neurons, or both, connect the brain and spinal cord with the rest of the body.

In a reflex arc, sensory neurons synapse directly on motor neurons. This is the simplest path of information flow.

Interneurons in the brain and spinal cord are organized in information-processing blocks.

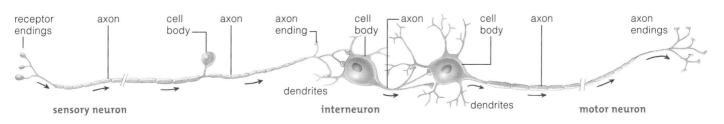

receptor endings | axon | cell body | axon | axon ending | cell body | axon | cell body | axon | axon endings

dendrites

sensory neuron   **interneuron**   dendrites   **motor neuron**

13.5 Overview of the Nervous System

Up to this point we have focused on the signals that travel through the nervous system. This section gives an overview of how the system as a whole is organized.

Humans have the most intricately wired nervous system in the animal world (Figure 13.11). We can simplify its complexity by dividing it according to function into two main regions. The brain and spinal cord make up the **central nervous system** (CNS). All of the nervous system's interneurons are in this system. The **peripheral nervous system** (PNS) consists mainly of nerves that thread through the rest of the body and carry signals into and out of the central nervous system.

Nerves that carry sensory input to the central nervous system sometimes are called *afferent* ("bringing to") nerves. Nerves that carry motor messages away from the central nervous system to muscles and glands may be termed *efferent* ("carrying outward") nerves.

As Figure 13.12 shows, the peripheral nervous system is organized into *somatic* and *autonomic* subdivisions, and the autonomic nerves are subdivided yet again. We'll consider the roles of those nerves in Section 13.6.

The peripheral nervous system consists of thirty-one pairs of spinal nerves that carry signals to and from the spinal cord and twelve pairs of cranial nerves that carry signals to and from the brain (Figure 13.13). At some places in the PNS, cell bodies of several neurons occur in

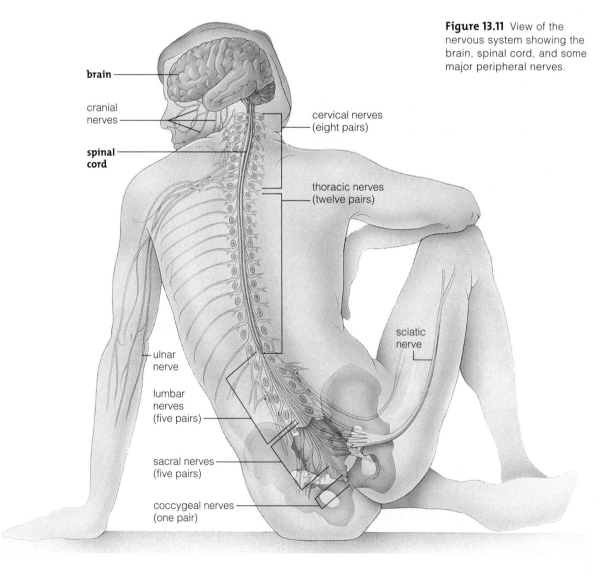

brain

cranial nerves

spinal cord

cervical nerves (eight pairs)

thoracic nerves (twelve pairs)

ulnar nerve

sciatic nerve

lumbar nerves (five pairs)

sacral nerves (five pairs)

coccygeal nerves (one pair)

Figure 13.11 View of the nervous system showing the brain, spinal cord, and some major peripheral nerves.

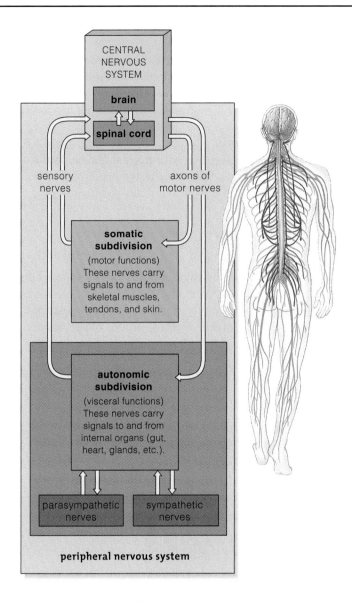

Figure 13.12 *Animated!* Divisions of the nervous system. The central nervous system is color-coded blue, somatic nerves are green, and autonomic nerves are red.

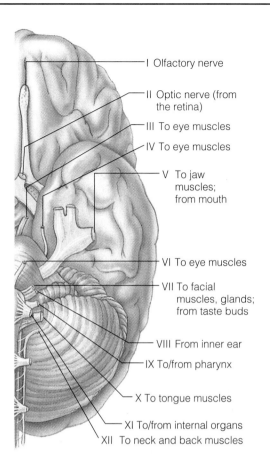

- I Olfactory nerve
- II Optic nerve (from the retina)
- III To eye muscles
- IV To eye muscles
- V To jaw muscles; from mouth
- VI To eye muscles
- VII To facial muscles, glands; from taste buds
- VIII From inner ear
- IX To/from pharynx
- X To tongue muscles
- XI To/from internal organs
- XII To neck and back muscles

Figure 13.13 Cranial nerves. Twelve pairs of cranial nerves extend from different regions of the brain stem. By tradition, Roman numerals are used to designate cranial nerves.

clusters called **ganglia** (singular: ganglion). The central and peripheral nervous systems both also have glia, such as the oligodendrocytes (CNS) and Schwann cells (PNS) described in Sections 13.1 and 13.4.

Throughout our lives, our remarkable nervous system integrates the array of body functions in ways that help maintain homeostasis. Operations of the CNS also give us much of our "humanness," as you will read shortly.

The nervous system is divided into the central nervous system (CNS) and the peripheral nervous system (PNS). The CNS consists of the brain and spinal cord. The PNS consists of nerves that carry signals to and from the CNS.

Many popular food fish (such as albacore tuna) contain high levels of mercury. This metal also enters lakes and waterways in industrial waste. Mercury can damage neurons, and several long-term studies have linked brain damage in children to mercury-tainted fish eaten by their mothers. The FDA now warns pregnant or nursing mothers to avoid eating fish likely to be contaminated.

13.6 Major Expressways: Peripheral Nerves and the Spinal Cord

Peripheral nerves and the spinal cord interconnect as the major expressways for information flow through the body.

THE PERIPHERAL NERVOUS SYSTEM CONSISTS OF SOMATIC AND AUTONOMIC NERVES

Nerves of the PNS are grouped by function. To begin with, cranial and spinal nerves are subdivided into two groups. **Somatic nerves** carry signals related to movements of the head, trunk, and limbs. **Autonomic nerves** carry signals beween internal organs and other structures.

In somatic nerves, sensory axons carry information from receptors in skin, skeletal muscles, and tendons to the central nervous system. Their motor axons deliver commands from the brain and spinal cord to skeletal muscles. In the autonomic category, motor axons of spinal and cranial nerves carry messages to smooth muscle, cardiac (heart) muscle, and glands (Figure 13.14).

Unlike somatic neurons, single autonomic neurons do not extend the entire distance between muscles or glands and the central nervous system. Instead, preganglionic ("before a ganglion") neurons have cell bodies inside the spinal cord or brain stem, but their axons travel through nerves to autonomic system ganglia *outside* the CNS. There, the axons synapse with postganglionic ("after a ganglion") neurons, which make the actual connection with "effectors"—the body's muscles and glands.

AUTONOMIC NERVES ARE DIVIDED INTO PARASYMPATHETIC AND SYMPATHETIC GROUPS

Autonomic nerves are divided into *parasympathetic* and *sympathetic* nerves. Normally these two sets of nerves work antagonistically—the signals from one oppose those of the other. However, both these groups of nerves carry excitatory and inhibitory signals to internal organs. Often

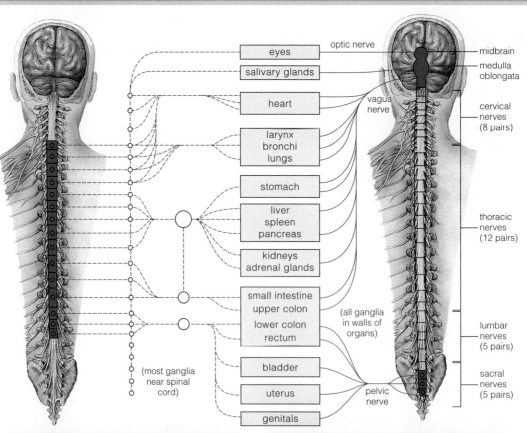

Figure 13.14 *Animated!* The autonomic nervous system. This is a diagram of the major sympathetic and parasympathetic nerves leading out from the central nervous system to some major organs. Remember, there are *pairs* of both kinds of nerves, servicing the right and left halves of the body. The ganglia are simply clusters of cell bodies of the neurons that are bundled together in nerves.

optic nerve
eyes
salivary glands
heart
vagus nerve
larynx
bronchi
lungs
stomach
liver
spleen
pancreas
kidneys
adrenal glands
small intestine
upper colon
lower colon
rectum
(all ganglia in walls of organs)
bladder
uterus
pelvic nerve
genitals

midbrain
medulla oblongata
cervical nerves (8 pairs)
thoracic nerves (12 pairs)
lumbar nerves (5 pairs)
sacral nerves (5 pairs)

(most ganglia near spinal cord)

a sympathetic outflow from the spinal cord

Examples of Responses
Heart rate increases
Pupils of eyes dilate (widen, let in more light)
Glandular secretions decrease in airways to lungs
Salivary gland secretions thicken
Stomach and intestinal movements slow down
Sphincters (rings of muscle) contract

Examples of Responses
Heart rate decreases
Pupils of eyes constrict (keep more light out)
Glandular secretions increase in airways to lungs
Salivary gland secretions become dilute
Stomach and intestinal movements increase
Sphincters (rings of muscle) relax

b parasympathetic outflow from the spinal cord and brain

THE NERVOUS SYSTEM

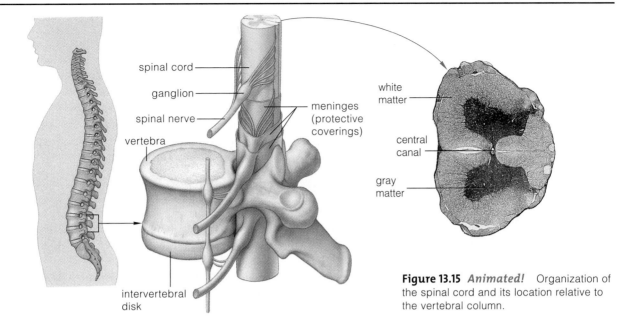

spinal cord

ganglion

spinal nerve

vertebra

meninges (protective coverings)

white matter

central canal

gray matter

intervertebral disk

Figure 13.15 *Animated!* Organization of the spinal cord and its location relative to the vertebral column.

their signals arrive at the same time at muscle or gland cells and compete for control. When that happens, synaptic integration leads to minor adjustments in an organ's activity.

Parasympathetic nerves dominate when the body is not receiving much outside stimulation. They tend to slow down the body overall and divert energy to basic "housekeeping" tasks, such as digestion.

Sympathetic nerves dominate during heightened awareness, excitement, or danger. They tend to shift activity away from housekeeping tasks. For example, as you read this, sympathetic nerves are prompting your heart to beat a bit faster, and parasympathetic nerves are commanding it to beat a little slower. Integration of these opposing signals influences your heart rate. If something scares or excites you, parasympathetic input to your heart drops. At the same time, sympathetic nerves release norepinephrine, a neurotransmitter that makes your heart beat faster and makes you breathe faster and sweat. In this **fight–flight response**, you are primed to fight (or play) hard or to get away fast.

When the stimulus for the fight–flight response stops, sympathetic activity may fall and parasympathetic activity may rise. This "rebound effect" can occur after someone has been mobilized, say, to rush onto a street to save a child from an oncoming car. The person may well collapse as soon as the child has been swept out of danger.

THE SPINAL CORD IS THE PATHWAY BETWEEN THE PNS AND THE BRAIN

The **spinal cord** is a vital expressway for signals between the peripheral nervous system and the brain. It threads through a canal made of bones of the vertebral column (Figure 13.15). Most of the cord consists of nerve tracts

(bundles of myelinated axons). Because the myelin sheaths of these axons are white, the tracts are called *white matter*. The cord also contains dendrites, cell bodies of neurons, interneurons, and neuroglial cells. These form its *gray matter*. In cross section, the cord's gray matter looks a little bit like a butterfly. The cord lies inside a closed channel formed by the bones of the vertebral column. Those bones, and ligaments attached to them, protect the soft nervous tissue of the cord. So do three coverings of connective tissue called *meninges* layered around the spinal cord and brain. They are discussed in Section 13.7.

In addition to carrying signals between the peripheral nervous system and the brain, the spinal cord is a control center for reflexes. Sensory and motor neurons involved in many reflex movements of skeletal muscle make direct connections in the cord; such *spinal reflexes* do not require the brain's input. When you jerk your hand away from a hot stove burner, you are experiencing a spinal reflex in action. Information about the sensory stimulus also reaches higher brain centers and so you become aware of "hot burner!" even as your hand is moving away from it. In addition to all of the above, the spinal cord contributes to some *autonomic reflexes*, which deal with internal organ functions such as bladder emptying.

The peripheral nervous system consists of the nerves to and from the brain and spinal cord.

Its somatic nerves deal with skeletal muscle movements. Its autonomic nerves govern functions of internal organs, such as the heart and glands. Autonomic nerves are divided into parasympathetic nerves (for housekeeping functions) and sympathetic nerves (for aroused states).

The spinal cord carries signals between peripheral nerves and the brain. It also is a control center for certain reflexes.

13.7 The Brain—Command Central

The brain is divided into three main regions, each one containing centers that manage specific biological tasks.

The spinal cord merges with the **brain**. Like the cord, the brain is protected by bones (of the cranium) and by the three **meninges** (meh-NIN-jeez). These are membranes of connective tissue layered between the skull and the brain (Figure 13.16). Meninges cover and protect the fragile CNS neurons and blood vessels that service the tissue. The leathery, outer membrane, the *dura mater,* is folded double around the brain. Its upper surface attaches to the skull. The lower surface is the outer covering of the brain and separates its right and left hemispheres. A thinner middle layer is called the *arachnoid,* and the even more delicate *pia mater* wraps the brain and spinal cord. The meninges also enclose fluid-filled spaces that cushion and nourish the brain.

THE BRAIN IS DIVIDED INTO A HINDBRAIN, MIDBRAIN, AND FOREBRAIN

The brain's three divisions are the *hindbrain*, *midbrain*, and *forebrain* (Figure 13.17). In the hindbrain and midbrain are centers that control many simple, basic reflexes; this tissue is the **brain stem**. As the ancestors of humans evolved, expanded layers of gray matter developed over the brain stem. These changes in brain structure have been correlated with our species' increasing reliance on three major sensory organs: the nose, ears, and eyes. They are topics in Chapter 14.

HINDBRAIN The medulla oblongata, cerebellum, and pons are all components of the hindbrain. The **medulla oblongata** contains reflex centers for a number of vital tasks, such as respiration and blood circulation. It also coordinates motor responses with certain complex reflexes, such as coughing. In addition, the medulla influences brain centers that help you sleep or wake up.

The **cerebellum** integrates sensory signals from the eyes, inner ears, and muscles with motor signals from the forebrain to coordinate movement and balance. It helps control dexterity. Some of its activities may also be crucial in language and other forms of mental "agility."

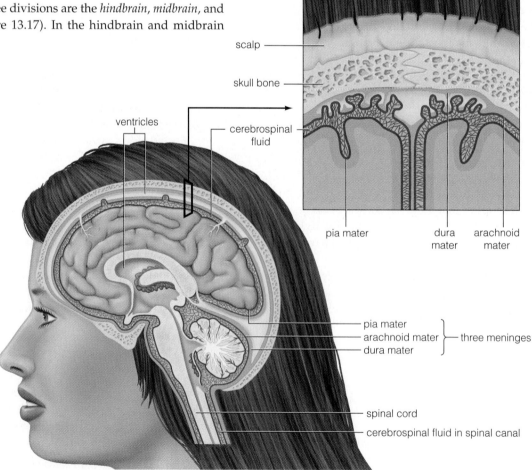

scalp

skull bone

ventricles

cerebrospinal fluid

pia mater

dura mater

arachnoid mater

pia mater
arachnoid mater } three meninges
dura mater

spinal cord

cerebrospinal fluid in spinal canal

Figure 13.16 Location of the three meninges in relation to the brain. Cerebrospinal fluid fills the space between the arachnoid and the pia mater.

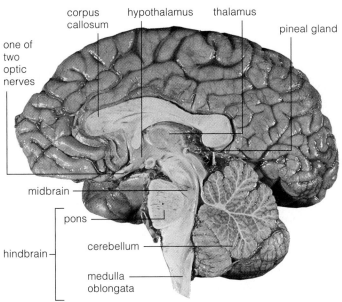

forebrain

corpus callosum

hypothalamus

thalamus

pineal gland

one of two optic nerves

midbrain

pons

hindbrain

cerebellum

medulla oblongata

Figure 13.17 *Animated!* The brain's right hemisphere. Each hemisphere has a cerebral cortex, a layer of gray matter about 2–4 millimeters (1/8 inch) thick. Interneuron cell bodies and dendrites, unmyelinated axons, glia, and blood vessels make up the gray matter.

Bands of many axons extend from the cerebellum into the **pons** ("bridge") in the brain stem. The pons directs the signal traffic between the cerebellum and the higher integrating centers of the forebrain.

MIDBRAIN The midbrain coordinates reflex responses to sights and sounds. It has a roof of gray matter, the *tectum* (Latin for "roof"), where visual and auditory sensory input converges before being sent on to higher brain centers.

FOREBRAIN The forebrain is the most highly developed brain region. It includes the cerebrum, olfactory bulbs, and the thalamus and hypothalamus. In the **cerebrum**, information is processed and sensory input and motor responses are integrated. A pair of *olfactory bulbs* deal with sensory information about smell. The **thalamus** is mainly a sensory relay switchboard. In it, incoming signals in sensory nerve tracts are relayed to clusters of neuron cell bodies (called *nuclei*), then relayed onward. The nuclei also process some outgoing motor information. As you will read in Section 13.10, Parkinson's disease results when functioning of *basal nuclei* in the thalamus is disrupted.

Located below the thalamus, the **hypothalamus** has evolved into the body's "supercenter" for controlling homeostatic adjustments in the activities of internal organs. As noted in previous chapters, for example, it helps govern states such as thirst and hunger. The hypothalamus also has roles in sexual behavior and emotional expression, such as when fear causes a person to break into a sweat.

CEREBROSPINAL FLUID FILLS CAVITIES AND CANALS IN THE BRAIN

Our brain and spinal cord would both be highly vulnerable to damage if they were not protected by bones and meninges. In addition, as you can see in Figure 13.16, both of them are surrounded by **cerebrospinal fluid**, or CSF. This transparent fluid forms from blood plasma and is chemically similar to it. It is secreted from specialized capillaries inside a system of fluid-filled cavities and canals in the brain. The cavities in the brain are called ventricles. They connect with each other and with the central canal of the spinal cord and are filled with cerebrospinal fluid. The fluid also fills the space between the innermost layer of the meninges and the brain itself. Because the enclosed cerebrospinal fluid can't be compressed, it helps cushion the brain and spinal cord from jarring movements. Some diseases, such as **meningitis** (inflammation of the meninges), can be diagnosed by analyzing a sample of the CSF.

Maintaining homeostasis in the fluid that bathes your brain's neurons is vital. Yet as you know, the chemical makeup of extracellular fluid, including levels of ions and other substances, is constantly changing. In the brain this problem is solved by the unusual structure of brain capillaries. Their walls are much less permeable than the walls of capillaries elsewhere in the body, so substances must pass *through* the wall cells, rather than between them, to reach the brain. This **blood–brain barrier** helps control which blood-borne substances are allowed to reach the brain's neurons.

Transport proteins embedded in the plasma membrane of the cells in brain capillary walls allow glucose and other water-soluble substances to cross the barrier. However, lipid-soluble substances are another matter. They quickly diffuse through the lipid bilayer of the plasma membrane. This "lipid loophole" in the blood–brain barrier is one reason why lipid-soluble chemicals such as caffeine, nicotine, alcohol, barbiturates, heroin, and anesthetics can rapidly affect brain function.

The brain's main divisions are the hindbrain, midbrain, and forebrain. Their functions range from reflex controls over basic survival functions (as in the brain stem in the hindbrain) to the complex integration of sensory information and motor responses.

In the brain and spinal cord, cerebrospinal fluid fills a system of cavities and canals. This fluid provides a cushion against jarring movement.

13.8 A Closer Look at the Cerebrum

If you are an average-sized adult, your brain weighs about 1,300 grams (three pounds) and contains at least 100 billion neurons! Its cerebrum is where much of "the action" is.

THERE ARE TWO CEREBRAL HEMISPHERES

The human cerebrum looks somewhat like a much-folded walnut, and a deep fissure divides it into left and right **cerebral hemispheres** (Figure 13.18). Each hemisphere has a thin outer layer of gray matter, the **cerebral cortex**. (The cerebral cortex weighs about a pound. If stretched flat, it would cover a surface area of two and a half square feet.) Below the cortex are the white matter (axons) and the basal nuclei—patches of gray matter in the thalamus.

Each cerebral hemisphere receives, processes, and coordinates responses to sensory input mainly from the opposite side of the body. (For instance, "cold" signals from an ice cube in your left hand travel to your right cerebral hemisphere.) The left hemisphere deals mainly with speech, analytical skills, and mathematics. In most people it dominates the right hemisphere, which deals more with visual–spatial relationships, music, and other creative activities. A band of nerve tracts, the corpus callosum, carries signals back and forth between the hemispheres and coordinates their operations.

Each hemisphere is divided into frontal, occipital, temporal, and parietal lobes, which process different signals (Figure 13.19*a*). EEGs and PET scans (Figure 13.19*b*) can reveal activity in each lobe. EEG, short for

electroencephalogram, is a recording of summed electrical activity in some part of the brain.

THE CEREBRAL CORTEX CONTROLS THOUGHT AND OTHER CONSCIOUS BEHAVIOR

Everything you comprehend, communicate, remember, and voluntarily act upon arises in the cerebral cortex; it governs conscious behavior. Functionally, the cortex is divided into *motor* areas (control of voluntary motor activity), *sensory* areas (perception of the meaning of sensations), and *association* areas (the integration of information that precedes a conscious action). None of these areas functions alone; consciousness arises by way of interactions throughout the cortex.

MOTOR AREAS   In the frontal lobe of each hemisphere, the whole body is spatially mapped out in the primary motor cortex. This area controls coordinated movements of skeletal muscles. Thumb, finger, and tongue muscles get much of the area's attention, indicating how much control is required for voluntary hand movements and verbal expression (Figure 13.20).

Also in the frontal lobe are the premotor cortex, Broca's area, and the frontal eye field. The premotor cortex deals with learned patterns or motor skills. Repetitive motor actions, such as bouncing a ball, are evidence that your motor cortex is coordinating the movements of several muscle groups. Broca's area (usually in the left hemisphere) and a corresponding area in the right hemisphere control the tongue, throat, and lip muscles used in speech. It kicks in when we are about to speak and even when we plan voluntary motor activities other than speaking (so you can talk on the phone and write down a message at the same time). Above Broca's area is the frontal eye field. It controls voluntary eye movements.

SENSORY AREAS   Sensory areas occur in different parts of the cortex. In the parietal lobe, the body is spatially mapped out in the primary somatosensory cortex. This area is the main receiving center for sensory input from the skin and joints. The parietal lobe also has a primary cortical area dealing with perception of taste. At the back of the occipital lobe is the primary visual cortex, which receives sensory inputs from your eyes. Perception of sounds and of odors arises in primary cortical areas in each temporal lobe.

ASSOCIATION AREAS   Association areas occupy all parts of the cortex except the primary motor and sensory regions. Each integrates, analyzes, and responds to many inputs. For instance, the visual association area surrounds

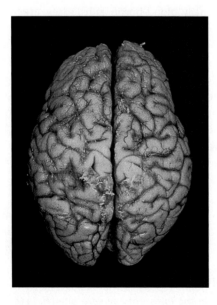

Figure 13.18 A top-down view of the brain's two cerebral hemispheres.

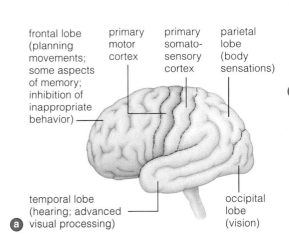

frontal lobe (planning movements; some aspects of memory; inhibition of inappropriate behavior)

primary motor cortex

primary somato-sensory cortex

parietal lobe (body sensations)

temporal lobe (hearing; advanced visual processing)

occipital lobe (vision)

a

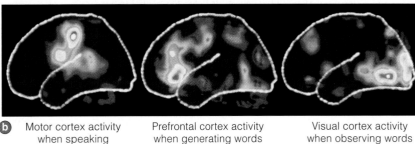

b Motor cortex activity when speaking

Prefrontal cortex activity when generating words

Visual cortex activity when observing words

Figure 13.19 *Animated!* (**a**) Primary receiving and integrating centers for the human cerebral cortex. Primary cortical areas receive signals from receptors on the body's periphery. Association areas coordinate and process sensory input from different receptors. The PET scans (**b**) show which brain regions were active when a person performed three specific tasks: speaking, generating words, and observing words.

the primary visual cortex. It helps us recognize something we see by comparing it with visual memories. Neural activity in the most complex association area—the prefrontal cortex—is the basis for complex learning, intellect, and personality. Without it, we would be incapable of abstract thought, judgment, planning, and concern for others.

THE LIMBIC SYSTEM: EMOTIONS AND MORE

The prefrontal cortex interacts intimately with the **limbic system**, which is located inside the cerebral hemispheres. It governs our emotions and has roles in memory (Figure 13.21). The limbic system includes parts of the thalamus along with the hypothalamus, the amygdala, and the

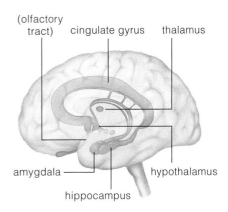

(olfactory tract)

cingulate gyrus

thalamus

amygdala

hypothalamus

hippocampus

Figure 13.21 Key structures in the limbic system, which encircles the upper brain stem. The amygdala and the cingulate gyrus are especially important in emotions. The hypothalamus is a clearinghouse for emotions and visceral activity. Both the hippocampus and the amygdala help convert stimuli into long-term memory (Section 13.9).

hippocampus. It is called our "emotional–visceral brain" because its operations produce "gut" reactions such as rage. The system's links with other brain centers also allow it to correlate self-gratifying behavior, such as eating and sex, with the activities of associated organs. Responses of the limbic system also are part of the reason you may feel "warm and fuzzy" when you recall the cologne of a special person who wore it.

Each cerebral hemisphere receives, processes, and coordinates responses to sensory input mainly from the opposite side of the body. The corpus callosum carries signals between them.

The left hemisphere deals mainly with speech, analytical skills, and mathematics. It usually dominates the right hemisphere, which deals more with creative activity, such as visual–spatial relationships and music.

The cerebral cortex, the outermost layer of gray matter of each hemisphere, contains motor, sensory, and association areas. Communication among these areas governs conscious behavior. The cerebral cortex also interacts with the limbic system, which governs emotions and memory.

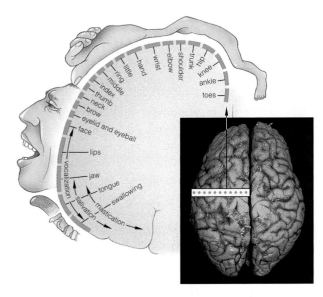

little
ring
middle
index
thumb
neck
brow
eyelid and eyeball
face
lips
vocalization
jaw
tongue
swallowing
salivation
mastication
hand
wrist
elbow
shoulder
trunk
hip
knee
ankle
toes

Figure 13.20 *Animated!* Diagram of a slice through the primary motor cortex of the left cerebral hemisphere. The distortions to the human body draped over the diagram indicate which body parts receive the most precise control.

13.9 Memory and Consciousness

Even before you were born, your brain began to build your memory—to store and retrieve information about your unique experiences.

MEMORY IS HOW THE BRAIN STORES AND RETRIEVES INFORMATION

Learning and adaptive modifications of our behavior would be impossible without **memory**. Information is stored in stages. *Short-term* storage is a stage of neural excitation that lasts a few seconds to a few hours. It is limited to bits of sensory information—numbers, words of a sentence, and so on. In *long-term* storage, seemingly unlimited amounts of information get tucked away more or less permanently, as shown in Figure 13.22.

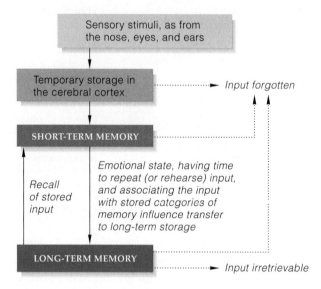

Figure 13.22 Stages of memory processing, starting with the temporary storage of sensory inputs in the cerebral cortex.

Not all of the sensory input bombarding the cerebral cortex ends up in memory storage. Only some is selected for transfer to brain structures involved in short-term memory. Information in these holding bins is processed for relevance, so to speak. If irrelevant, it is forgotten; otherwise it is consolidated with the banks of information in long-term storage structures.

The human brain processes facts separately from skills. Dates, names, faces, words, odors, and other bits of explicit information are *facts*, soon forgotten or filed away in long-term storage, along with the circumstance in which they were learned. Hence you might associate the smell of bread baking, say, with your grandmother's kitchen. By contrast, *skills* are gained by practicing specific motor activities. A skill such as maneuvering a snowboard or playing a piano concerto is best recalled by actually

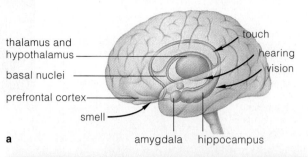

Figure 13.23 *Animated!* (**a**) Possible circuits involved in fact memory. (**b**) A dramatic demonstration of skill memory.

performing it, rather than by recalling the circumstances in which the skill was first learned.

Separate memory circuits handle different kinds of input. A circuit leading to fact memory (Figure 13.23*a*) starts with inputs at the sensory cortex that flow to the amygdala and hippocampus in the limbic system. The amygdala is the gatekeeper, connecting the sensory cortex with parts of the thalamus and with parts of the hippocampus that govern emotional states. Information flows on to the prefrontal cortex, where multiple banks of fact memories are retrieved and used to stimulate or inhibit other parts of the brain. The new input also flows to basal nuclei, which send it back to the cortex in a feedback loop that reinforces the input until it can be consolidated in long-term storage.

Skill memory also starts at the sensory cortex, but this circuit routes sensory input to a region deeper in the brain that promotes motor responses (Figure 13.23*b*). Motor skills entail muscle conditioning. As you might suspect, the circuit extends to the cerebellum, the brain region that coordinates motor activity.

Amnesia is a loss of fact memory. How severe the loss is depends on whether the hippocampus, amygdala, or both are damaged, as by a head blow. Amnesia does not affect a person's capacity to learn new skills.

STATES OF CONSCIOUSNESS INCLUDE ALERTNESS AND SLEEPING

The spectrum of consciousness ranges from being wide awake and fully alert to drowsiness, sleep, and coma. When you are awake and alert, the neural chattering in your brain shows up as wavelike patterns in an EEG. As mentioned earlier, EEGs are recordings of the summed electrical activity of the brain's neurons. PET scans like the ones in Figure 13.19b (Section 13.8) can show the exact location of brain activity as it takes place.

A network of neurons in the brain called the **reticular formation** (Figure 13.24a) helps govern muscle activity associated with maintaining balance, posture, and muscle tone. It also promotes chemical changes that influence whether you stay awake or fall asleep. Serotonin is a neurotransmitter released from one of the network's sleep centers. It inhibits neurons that arouse the brain and maintain wakefulness. High serotonin levels trigger drowsiness and sleep. When substances released from another brain center inhibit its effects, you wake up.

EEGs from electrodes placed on the scalp show up as wave forms. Figure 13.24b shows the patterns for full alertness and for the two major sleep stages: slow-wave, "normal" sleep and REM (*rapid eye movement*) sleep.

Most of the time you spend sleeping is slow-wave sleep. During this stage, your heart rate, breathing, and muscle tone change very little and you can be easily roused. About every 90 minutes, however, you normally enter a period of REM sleep, in which the eyelids flicker and the eyeballs move rapidly back and forth. Dreaming occurs during REM sleep, and it is much harder to wake up during this time. The heart rate and breathing become more erratic, and muscle tone decreases dramatically.

You probably know from personal experience that people who are deprived of sleep feel tired and cranky and have difficulty concentrating. In fact, researchers still do not know exactly why sleep is so important for us. Although neural activity changes during sleep, the brain clearly is not resting. One hypothesis is that sleep is the time when the brain does essential "housekeeping," such as consolidating memories and solidifying connections involved in learning.

Circuits between the cerebral cortex and parts of the limbic system, thalamus, and hypothalamus produce memories as sensory messages are processed through short-term and long-term storage.

The spectrum of consciousness, which includes sleeping and states of arousal, is influenced by the reticular formation.

Sleep has two stages, slow-wave sleep and REM sleep, in which the brain's activity is altered. Sleep may be important in memory and learning.

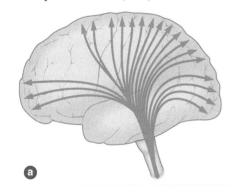

(a)

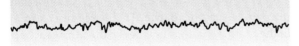

Awake, eyes open

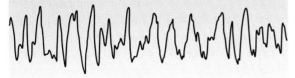

Slow-wave sleep

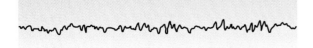

(b) REM sleep

Figure 13.24 States of consciousness. (**a**) The communication pathways of the reticular formation. (**b**) EEG patterns for alertness, slow-wave sleep, and REM (rapid eye movement) sleep. The brain waves of the student pictured below would resemble the pattern in (**a**).

THE BRAIN

13.10 Disorders of the Nervous System

Injuries, viruses, bacteria, even skewed immune responses can seriously damage the nervous system.

In 1817, physician James Parkinson observed troubling symptoms in certain people navigating the streets of London. They walked slowly, taking short, shuffling steps. And their limbs trembled, sometimes violently. Today we know that the culprit is a degenerative brain disorder that now is called **Parkinson's disease**, or PD (Figure 13.25*a*). In PD, neurons in the basal nuclei of the thalamus (Section 13.7) begin to die. Those neurons make neurotransmitters (dopamine and norepinephrine) that are needed for normal muscle function, so PD symptoms include muscle tremors and balance problems, among others. Treatments include drugs that help replace absent neurotransmitters. Surgical treatments are available for advanced cases, but none is a cure.

Like PD, **Alzheimer's disease** involves the progressive degeneration of brain neurons. At the same time, there is an abnormal buildup of amyloid protein, leading to the loss of memory and intellectual functions. Alzheimer's is associated with aging and is considered in more detail in Chapter 17.

Figure 13.25
(**a**) Muhammad Ali during his prime as a champion athlete. After being diagnosed with Parkinson's disease, he established the Muhammad Ali Parkinson's Research Center in Phoenix, Arizona. (**b**) San Francisco 49ers star quarterback Steve Young retired after suffering a series of concussions.

Meningitis is an often fatal disease caused by a bacterial or viral infection. Symptoms include headache, a stiff neck, and vomiting. They develop when the meninges covering the brain and/or spinal cord become inflamed. **Encephalitis** is inflammation of the brain. It is usually caused by a viral infection, such as by HIV or a herpes virus. Like meningitis it can be extremely dangerous; the early symptoms include fever, confusion, and seizures.

In young adults, the most common acquired disease of the nervous system is **multiple sclerosis** (MS). An autoimmune disease that may be triggered by a viral infection in susceptible people, MS involves progressive destruction of myelin sheaths of neurons in the central nervous system. The symptoms, which develop over time, include muscle weakness or stiffness, fatigue, and slurred speech.

A blow to the head or neck can cause a **concussion**. Blurred vision and a brief loss of consciousness result when the blow temporarily upsets the electrical activity of brain neurons (Figure 13.25*b*).

Damage to the spinal cord can lead to lost sensation and muscle weakness or **paralysis** below the site of the injury. Immediate treatment is crucial to limit swelling. Although cord injuries usually have severe consequences, intensive therapy during the first year after an injury can improve the patient's long-term prognosis. Harnessing nerve growth factors or stem cells to repair spinal cord injuries is a major area of medical research.

Brain injury, birth trauma, or other assaults can cause various forms of *epilepsy*, or **seizure disorders**. In some cases the trigger may be an inherited predisposition. Each seizure results when the brain's normal electrical activity suddenly becomes chaotic. Worldwide, many thousands of people develop recurrent seizures either as children or later in life. All but the most intractable cases usually respond well to modern anticonvulsant drugs.

One of the most common of all physical ailments is the pain we call **headache**. There are no sensory nerves in the brain, so it does not "feel pain." Instead, headache pain typically is due to tension (stretching) in muscles or blood vessels of the face, neck, and scalp.

Throbbing *migraine* headaches are infamous for being extremely painful and lasting for up to 72 hours. In the U.S. alone, 28 million people, mainly female, suffer from migraines, which can be triggered by hormonal changes, fluorescent lights, and certain foods (such as chocolate), among other causes. Often, a migraine is accompanied by nausea, vomiting, or sensitivity to light. Migraines may be treated with prescription painkillers and drugs that act as neuromodulators (Section 13.3) to reduce the sensitivity of affected brain neurons to stimuli that trigger the headache in the first place.

THE BRAIN

13.11 The Brain on Drugs

Psychoactive drugs bind to CNS neuron receptors that normally bind to neurotransmitters. As a result, neurons send or receive altered messages.

DRUGS CAN ALTER MIND AND BODY FUNCTIONS

Psychoactive drugs affect parts of the brain that govern consciousness and behavior. Some also alter heart rate, respiration, sensory processing, and muscle coordination. Many affect a pleasure center in the hypothalamus and artificially fan the sense of pleasure we associate with eating, sex, or other activities.

Stimulants include caffeine, nicotine, cocaine, and amphetamines—including Ecstasy (MDMA). They increase alertness and physical activity at first, then depress you. Nicotine mimics ACh, directly stimulating various sensory receptors. It also increases the heart rate and blood pressure. At first amphetamines cause a flood of the neurotransmitters norepinephrine and dopamine, which stimulate the brain's pleasure center. Over time, however, the brain slows its production of dopamine and norepinephrine and depends more on the amphetamine. Chronic users may become psychotic, depressed, and malnourished. They may also develop heart problems.

Cocaine stimulates the pleasure center by *blocking* the reabsorption of dopamine and other neurotransmitters. It also weakens the cardiovascular and immune systems.

Alcohol alters cell functions. It produces a high at first, then *depresses* brain activity. Drinking only an ounce or two diminishes judgment and can produce disorientation and uncoordinated movements. *Blood alcohol concentration* (BAC) measures the percentage of alcohol in the blood. In most states, someone with a BAC of 0.08 per milliliter is considered legally drunk. When the BAC reaches 0.15 to 0.4, a drinker is visibly intoxicated and can't function normally. A BAC greater than 0.4 can kill.

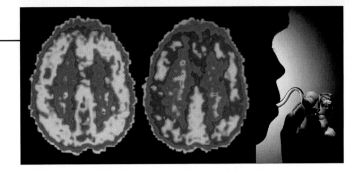

Figure 13.26 Effects of crack cocaine. (**a**) A PET scan of normal brain activity. (**b**) A PET scan showing cocaine's long-term effect. Red areas are most active; yellow, green, and blue indicate the least activity.

Morphine, an *analgesic* (painkiller), is derived from the seed pods of the opium poppy. Like its cousin heroin, it blocks pain signals by binding with certain receptors on neurons in the central nervous system. Both morphine and the synthetic version OxyContin produce euphoria. Thousands of people who obtained OxyContin illegally or by subterfuge have overdosed and died.

Marijuana is a *hallucinogen*. In low doses it is like a depressant. It slows but doesn't impair motor activity, and it elicits a mild euphoria. It can also cause disorientation, anxiety, and hallucinations. Like alcohol, it skews the performance of complex tasks, such as driving a car.

DRUG USE CAN LEAD TO ADDICTION

The body may develop *tolerance* to a drug, meaning that it takes larger or more frequent doses to produce the same effect. Tolerance reflects physical drug dependence. The liver produces enzymes that detoxify drugs in the blood. Tolerance develops when the level of detoxifying liver enzymes increases in response to the ongoing presence of the drug in the bloodstream. In effect, a drug user must increase his or her intake to stay ahead of the liver's growing ability (up to a point) to break down the drug.

In psychological drug dependence, or *habituation*, a user begins to crave the feelings associated with a particular drug. Without a steady supply of it the person can't "feel good" or function normally (Figure 13.26). Table 13.2 lists warning signs of potentially serious drug dependence. Habituation and tolerance both are evidence of addiction.

When different psychoactive drugs are used together, they can interact dangerously. For example, alcohol and barbiturates (such as Seconal and Nembutal) both depress the central nervous system. Used at the same time, they can lethally depress respiratory centers in the brain.

Table 13.2　Warning Signs of Drug Addiction*
1. Tolerance—it takes increasing amounts of the drug to produce the same effect.
2. Habituation—it takes continued drug use over time to maintain self-perception of functioning normally.
3. Inability to stop or curtail use of the drug, even if there is persistent desire to do so.
4. Concealment—not wanting others to know of the drug use.
5. Extreme or dangerous behavior to get and use a drug, as by stealing, asking more than one doctor for prescriptions, or jeopardizing employment by drug use at work.
6. Deteriorating professional and personal relationships.
7. Anger and defensive behavior when someone suggests there may be a problem.
8. Preferring drug use over previously customary activities.

*Having three or more of these signs may be cause for concern.

Psychoactive drugs alter consciousness and behavior. Some also affect physiological events, such as heart rate, respiration, sensory processing, and muscle coordination in harmful, even dangerous, ways.

Summary

Section 13.1 The nervous system detects, interprets, and responds directly to sensory stimuli. Sensory neurons collect information when they respond to external or internal stimuli. Interneurons receive sensory signals, integrate them with other input, and then send signals that influence other neurons. Motor neurons relay messages away from the brain and spinal cord to muscles or glands. Neuroglia provide various kinds of support for neurons.

Extensions of a neuron's cytoplasm form axons and dendrites. Axons carry outgoing signals, and dendrites receive them.

An unstimulated neuron shows a steady voltage difference across its plasma membrane. This difference is called the resting membrane potential. A resting neuron maintains concentration gradients of potassium ions, sodium ions, and other ions across the membrane.

 Review the structure and properties of neurons.

Section 13.2 When the voltage difference across the membrane exceeds a threshold level, gated sodium channels in the membrane open and close rapidly and suddenly reverse the voltage difference. This reversal is a nerve impulse, or action potential. A sodium–potassium pump restores ion gradients after an action potential. Action potentials self-propagate away from the point of stimulation.

Thomson NOW! *View an action potential step by step.*

Section 13.3 Action potentials self-propagate along the neuron membrane until they reach a chemical synapse with another neuron or with a muscle or gland cell. The presynaptic cell releases a neurotransmitter into the synapse. The neurotransmitter excites or inhibits the receiving (postsynaptic) cell. Synaptic integration sums up the various signals acting on a neuron. Neuromodulators boost or reduce the effects of neurotransmitters.

Thomson NOW! *See what happens at a synapse between a motor neuron and a muscle cell.*

Section 13.4 Nerves consist of the long axons of motor neurons, sensory neurons, or both. A myelin sheath formed by Schwann cells insulates each axon, so that action potentials propagate along it much more rapidly. Nerve pathways extend from neurons in one body region to neurons or effectors in different regions.

A reflex is a simple, stereotyped movement. Reflex arcs, in which sensory neurons directly signal motor neurons that act on muscle cells, are the simplest nerve pathways. In more complex reflexes, interneurons coordinate and refine the responses.

Thomson NOW! *Observe what happens during a stretch reflex.*

Section 13.5 The central nervous system consists of the brain and spinal cord. The peripheral nervous system consists of nerves and ganglia in other body regions.

Section 13.6 The peripheral nervous system's somatic nerves deal with skeletal muscles involved in voluntary body movements and sensations arising from skin, muscles, and joints. Its autonomic nerves deal with the functions of internal organs.

Table 13.3	Summary of the Central Nervous System*	
FOREBRAIN	Cerebrum	Localizes, processes sensory inputs; initiates, controls skeletal muscle activity. Governs memory, emotions, abstract thought
	Olfactory lobe	Relays sensory input from nose to olfactory centers of cerebrum
	Thalamus	Has relay stations for conducting sensory signals to and from cerebral cortex; has role in memory
	Hypothalamus	With pituitary gland, a homeostatic control center; adjusts volume, composition, temperature of internal environment. Governs organ-related behaviors (e.g., sex, thirst, hunger) and expression of emotions
	Limbic system	Governs emotions; has roles in memory
	Pituitary gland	With hypothalamus, provides endocrine control of metabolism, growth, development
	Pineal gland	Helps control some circadian rhythms; also has role in reproductive physiology
MIDBRAIN	Roof of midbrain (tectum)	In humans and other mammals, its reflex centers relay sensory input to the forebrain
HINDBRAIN	Pons	Tracts bridge cerebrum and cerebellum; other tracts connect spinal cord with forebrain. With the medulla oblongata, controls rate and depth of respiration
	Cerebellum	Coordinates motor activity for moving limbs and maintaining posture, and for spatial orientation
	Medulla oblongata	Its tracts relay signals between spinal cord and pons; its reflex centers help control heart rate, adjustments in blood vessel diameter, respiratory rate, vomiting, coughing, other vital functions
SPINAL CORD		Makes reflex connections for limb movements. Its tracts connect brain, peripheral nervous system

*The reticular formation extends from the spinal cord to the cerebral cortex.

Autonomic nerves are subdivided into sympathetic and parasympathetic groups. Parasympathetic nerves govern basic tasks such as digestion and tend to slow the pace of other body functions. In situations that demand increased awareness (excitement, danger), sympathetic nerves dominate. Their signals produce the fight–flight response, a state of intense arousal.

Spinal cord nerve tracts carry signals between the brain and the PNS. The cord also is a center for many reflexes.

 Explore the structure of the spinal cord and compare sympathetic and parasympathetic responses.

Section 13.7 The brain is divided into two cerebral hemispheres and has three main divisions (Table 13.3). It and the spinal cord are protected by bones (skull and vertebrae) and by the three meninges. Both are cushioned by cerebrospinal fluid. Specialized capillaries create a blood–brain barrier that prevents some blood-borne substances from reaching brain neurons.

The hindbrain includes the medulla oblongata, pons, and cerebellum. It contains reflex centers for vital functions and muscle coordination.

Midbrain centers coordinate and relay visual and auditory information. The midbrain, medulla oblongata, and pons make up the brain stem.

In the forebrain the thalamus relays sensory information and helps coordinate motor responses. The hypothalamus monitors internal organs and influences behaviors related to their functions (such as thirst). The limbic system has roles in learning, memory, and emotional behavior.

Section 13.8 The cerebral cortex is devoted to receiving and integrating information from sense organs and coordinating motor responses in muscles and glands.

 Review the structure and function of the brain.

Section 13.9 Memory occurs in short-term and long-term stages. Long-term storage depends on chemical or structural changes in the brain. States of consciousness vary between total alertness and deep coma. The levels are governed by the brain's reticular activating system.

Section 13.11 Psychoactive drugs cause brain neurons to send or receive altered signals and can change mental and physical functioning. Drug use can lead to physical tolerance and/or psychological dependence—both of which are elements of addiction.

Review Questions

1. Define sensory neuron, interneuron, and motor neuron.
2. What are the functional zones of a motor neuron?
3. Define an action potential.
4. What is a synapse? Explain the difference between an excitatory and an inhibitory synapse.
5. Explain what happens during synaptic integration.
6. What is a reflex? Describe the events of a stretch reflex.
7. Distinguish between the following:
   a. neurons and nerves
   b. somatic system and autonomic system
   c. parasympathetic and sympathetic nerves

Self-Quiz
Answers in Appendix V

1. The nervous system senses, interprets, and issues commands for responses to _____.
2. A neuron responds to adequate stimulation with _____, a type of self-propagating signal.
3. When action potentials arrive at a synapse between a neuron and another cell, they stimulate the release of molecules of a _____ that diffuse over to that cell.
4. In the simplest kind of reflex, _____ directly signal _____, which act on muscle cells.
   a. sensory neurons; interneurons
   b. interneurons; motor neurons
   c. sensory neurons; motor neurons
   d. motor neurons; sensory neurons
5. The accelerating flow of _____ ions through gated channels across the membrane triggers an action potential.
   a. potassium
   b. sodium
   c. hydrogen
   d. a and b are correct
6. _____ nerves slow down the body overall and divert energy to housekeeping tasks; _____ nerves slow down housekeeping tasks and increase overall activity during times of heightened awareness, excitement, or danger.
   a. Autonomic; somatic
   b. Sympathetic; parasympathetic
   c. Parasympathetic; sympathetic
7. Match each of the following central nervous system regions with some of its functions.
   ____ spinal cord
   ____ medulla oblongata
   ____ hypothalamus
   ____ limbic system
   ____ cerebral cortex

   a. receives sensory input, integrates it with stored information, coordinates motor responses
   b. monitors internal organs and related behavior (e.g., hunger)
   c. governs emotions
   d. coordinates reflexes
   e. makes reflex connections for limb movements, internal organ activity

Critical Thinking

1. In some cases of ADD (attention deficit disorder) the impulsive, erratic behavior can be normalized with drugs that *stimulate* the central nervous system. Explain this finding in terms of neurotransmitter activity in the brain.

2. Meningitis is an inflammation of the membranes that cover the brain and spinal cord. Diagnosis involves making a "spinal tap" (lumbar puncture) and analyzing a sample of cerebrospinal fluid for signs of infection. Why analyze this fluid and not blood?

3. In newborns and premature babies, the blood–brain barrier is not fully developed. Explain why this might be reason enough to pay careful attention to their diet.

4. At one time people deeply feared contracting the disease called *tetanus* (caused by the bacterium *Clostridium tetani*) because it usually meant an agonizing death. The bacterial toxin blocks the release of neurotransmitters that help control motor neurons. The result is continuous contraction of skeletal muscles and, eventually, the heart muscle also. Victims are aware of their plight, for the brain is not affected. What neurons does the toxin affect?

5. Research now demonstrates that new neurons can form in the adult brain, although slowly. Based on your reading in this chapter, name some diseases for which the ability to grow new brain neurons might be helpful.

Explore on Your Own

The "knee jerk" patellar reflex diagrammed in Figure 13.27 is a familiar example of a reflex arc. A doctor will often use a small rubber-tipped instrument to test for this reflex, but you can easily trigger it yourself just by tapping the base of one of your kneecaps (the patella) with a knuckle. The reflex occurs when stretch receptors in a tendon attached to the patella are activated, leading to a contraction of the quadriceps femoris, the extensor muscle of the knee.

Try tapping the base of one of your kneecaps (not too forcefully) and see if you can elicit the patellar reflex. Then write a short paragraph describing the path of the reflex arc, including the location and kind of neurons that integrated the sensory information and ordered a muscle response.

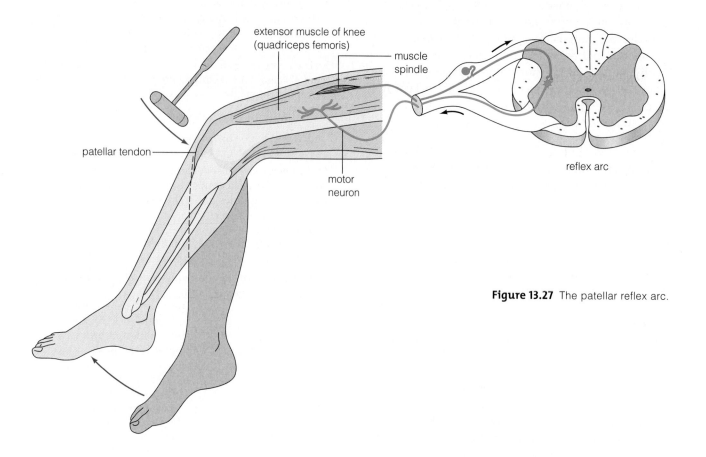

Figure 13.27 The patellar reflex arc.

Private Eyes

Terrorist threats. Identity theft. Today the promise of "foolproof" security seems too good to be true. Yet something close to it is already in place at some airports, high-security chemical plants and office buildings, and even hospitals. You may have heard of this security breakthrough. It's iris scanning.

Iris scanning technology relies on the unique, spokelike arrangement of smooth muscle fibers in the iris of the eye, the surface region of the eyes colored brown or blue or some other hue. Like fingerprints, each person's iris pattern is different from that of every other human on Earth.

For iris scanning to be a reliable identity check, each person's iris pattern must first be recorded digitally and entered into an electronic database. A person who later wants to gain entry to a secure location—or maybe just withdraw money from an ATM—looks into a scanning device that can instantly compare their eyes' iris pattern with the patterns in the stored database. Advocates and manufacturers of iris scanners say that the technology is 99 percent foolproof.

All over the world, millions of people regularly pass through airports and immigration checkpoints. They withdraw money from ATMs, visit offices, and check into and out of hospitals. Where would the "iris databases" come from? There are lots of possibilities. Travelers might be required to provide an "iris print" when applying for a passport or visa, or even when buying a ticket for travel. Your bank could conceivably iris-print you when you open an account. Employers could require potential employees to allow an iris print as part of the job application process. Already, a few hospitals in Europe make iris prints of newborns to prevent "baby switches" when new parents take infants home.

Our eyes are part of the powerful sensory system we call vision. In this chapter we consider vision and other sensory means by which the brain receives signals from the external and internal world. The signals are decoded in ways that give rise to awareness of sounds, sights, odors, and other sensations—including those that enable us to feel physical pain and to maintain our balance.

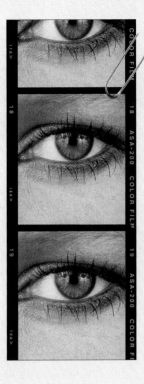

 How Would You Vote? Do you favor laws that allow employers and others to collect the information required for iris scanning—and to be protected from liability if the scans were misused? Cast your vote online at www.thomsonedu.com/biology/starr/humanbio.

 ## Key Concepts

SENSORY RECEPTORS
There are six main categories of sensory receptors, each based on the type of stimulus—such as light, sound, chemicals, or pressure—that the receptor detects.

SENSORY SYSTEMS
Sensory systems receive information about the external and internal world and convey it to the central nervous system. Processing in the CNS leads to vision, hearing, taste, smell, and other senses.

Links to Earlier Concepts

Senses sometimes are called "outposts of the brain" because they are the brain's sources of information about events going on inside and outside the body. Accordingly, this chapter builds directly on the discussion of the nervous system in Chapter 13. You will draw on what you have learned about action potentials, neurotransmitters, synapses between neurons and other cells, and nerves (13.2, 13.3, 13.4). You will also learn more about how the brain processes sensory input of all kinds (13.8).

LINKS TO
SECTIONS
13.2, 13.4,
AND 13.8

Sensory systems are the front doors of the nervous system. They notify the brain and spinal cord of specific changes inside and outside the body.

In a **sensory system**, energy from a stimulus activates receptors, which transduce it—that is, convert it—to a form that travels to the brain and may trigger a sensation or perception:

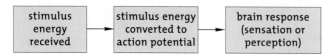

| stimulus energy received | → | stimulus energy converted to action potential | → | brain response (sensation or perception) |

A **stimulus** (plural: stimuli) is a form of energy that activates receptor endings of a sensory neuron. That energy is converted to the electrochemical energy of action potentials—the nerve impulses by which the brain receives information and sends out commands in response. The brain's responses are *sensations*—conscious awareness of a stimulus. By contrast, a *perception* is an understanding of what the sensation means.

Table 14.1 lists the six major categories of sensory receptors, based on the type of stimulus energy that each kind of receptor detects. **Mechanoreceptors** detect forms of mechanical energy: changes in pressure, position, or acceleration. **Thermoreceptors** are sensitive to heat or cold. **Nociceptors** (pain receptors) detect damage to tissues. **Chemoreceptors** detect chemical energy of substances dissolved in the fluid around them. **Osmoreceptors** detect

a

changes in water volume (solute concentration) in some body fluid. **Photoreceptors** detect visible light.

Regardless of their differences, all sensory receptors convert the stimulus energy to action potentials.

Action potentials that move along sensory neurons are all the same. So how does the brain determine the nature of a given stimulus? It assesses *which* nerves are carrying action potentials, the *frequency* of the action potentials on each axon in the nerve, and the *number* of axons that responded.

First, specific sensory areas of the brain can interpret action potentials only in certain ways. That is why you

Table 14.1	Major Categories of Sensory Receptors	
Category	**Examples**	**Stimulus**
MECHANORECEPTORS		
Touch, pressure	Certain free nerve endings and Merkel discs in skin	Mechanical pressure against body surface
Baroreceptors	Carotid sinus (artery)	Pressure changes in fluid (blood) that bathes them
Stretch	Muscle spindle in skeletal muscle	Stretching of muscle
Auditory	Hair cells in organ inside ear	Vibrations (sound waves)
Balance	Hair cells in organ inside ear	Fluid movement
THERMORECEPTORS	Certain free nerve endings	Change in temperature (heating, cooling)
NOCICEPTORS (PAIN RECEPTORS)*	Certain free nerve endings	Tissue damage (e.g., distortions, burns)
CHEMORECEPTORS		
Internal chemical sense	Carotid bodies in blood vessel wall	Substances (O_2, CO_2, etc.) dissolved in extracellular fluid
Taste	Taste receptors of tongue	Substances dissolved in saliva, etc.
Smell	Olfactory receptors of nose	Odors in air, water
OSMORECEPTORS	Hypothalamic osmoreceptors	Change in water volume (solute concentration) of fluid that bathes them
PHOTORECEPTORS		
Visual	Rods, cones of eye	Wavelengths of light

*Extremely intense stimulation of any sensory receptor also may be perceived as pain.

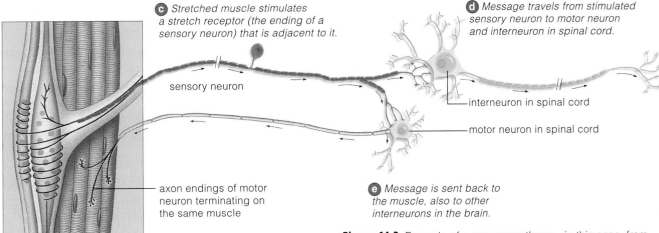

c Stretched muscle stimulates a stretch receptor (the ending of a sensory neuron) that is adjacent to it.

sensory neuron

axon endings of motor neuron terminating on the same muscle

b muscle spindle

d Message travels from stimulated sensory neuron to motor neuron and interneuron in spinal cord.

interneuron in spinal cord

motor neuron in spinal cord

e Message is sent back to the muscle, also to other interneurons in the brain.

Figure 14.2 Example of a sensory pathway—in this case, from receptors called muscle spindles to the spinal cord and brain.

"see stars" when your eye is poked, even in the dark. The mechanical pressure on photoreceptors in the eye generates signals that the associated optic nerve carries to the brain, which always interprets signals from an optic nerve as "light." As described in Section 14.2, the brain has what amounts to a detailed map of the sources of different sensory stimuli.

Second, a strong signal makes receptors fire action potentials more often and longer. The same receptor can detect the sounds of a whisper and a screech. The brain senses the difference through frequency variations in the signals that the receptor sends to it.

Third, a stronger stimulus can recruit more sensory receptors than a weaker stimulus can. Gently tap a spot of skin on your arm and you activate only a few receptors. Press hard on the same spot and you activate more. The increase translates into action potentials in many sensory neurons at once. Your brain interprets the

combined activity as an increase in the intensity of the stimulus. Figure 14.1 charts an example of this effect.

In some cases the frequency of action potentials slows or stops even when the stimulus continues at constant strength. For instance, after you put on a T-shirt, you quickly become only dimly aware of its pressure against your skin. This diminishing response to an ongoing stimulus is called **sensory adaptation**.

Some mechanoreceptors adapt rapidly to a sustained stimulus and only signal when it starts and stops. Other receptors adapt slowly or not at all; they help the brain monitor particular stimuli all the time.

The gymnast in Figure 14.2a is holding his position in response to signals from his skin, skeletal muscles, joints, tendons, and ligaments. For example, how fast and how far a muscle stretches depends on activation of stretch receptors in muscle spindles (Section 13.4). By responding to changes in the length of muscles, his brain helps him maintain his balance and posture.

In the rest of this chapter we explore examples of the body's sensory receptors. Receptors that are found at more than one location in the body contribute to somatic ("of the body") sensations. Other receptors are restricted to sense organs, such as the eyes or ears, and contribute to what are called the "special senses."

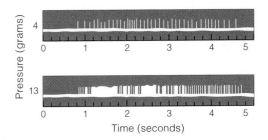

Figure 14.1 *Animated!* Action potentials recorded from a single pressure receptor of the human hand. The recordings correspond to variations in stimulus strength. A thin rod was pressed against the skin with the pressure indicated on the vertical axis of this diagram. Vertical bars above each thick horizontal line represent individual action potentials. The increases in frequency correspond to increases in the strength of the stimulus.

A sensory system has sensory receptors for specific stimuli, nerve pathways that conduct information from receptors to the brain, and brain regions that receive and process the information.

The brain senses a stimulus based on which nerves carry the incoming signals, the frequency of action potentials traveling along each axon, and the number of axons that have been recruited.

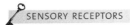

14.2 Somatic Sensations

LINKS TO SECTIONS 4.8, 10.3, AND 13.8

Somatic sensations start with receptors in body surface tissues, skeletal muscles, and the walls of internal organs.

Receptors for somatic senses are located in different parts of the body. **Somatic sensations** come about when receptor signals reach the **somatosensory cortex**, the cerebrum's outermost layer of gray matter. In this area, interneurons are organized like maps of individual parts of the body surface, just as they are for the motor cortex. The largest areas of the map correspond to body parts where there is the greatest density of sensory receptors. These body parts, including the fingers, thumbs, and lips, have the sharpest sensory acuity and require the most intricate control (Figure 14.3).

RECEPTORS NEAR THE BODY SURFACE SENSE TOUCH, PRESSURE, AND MORE

There are thousands of sensory receptors in your skin, providing information about touch, pressure, cold, warmth, and pain (Figure 14.4). Logically enough, places with the most sensory receptors, such as the fingertips and the tip of the tongue, are the most sensitive to stimulation. Less sensitive areas, such as the back of the hand, don't have nearly as many receptors.

Several types of **free nerve endings** in the epidermis and many connective tissues detect touch, pressure, heat, cold, or pain. These nerve endings are simple structures.

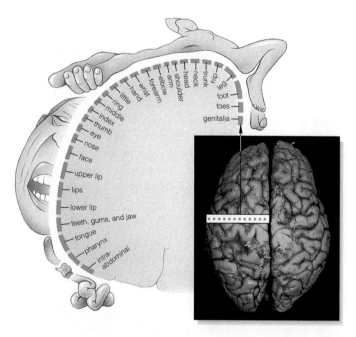

Figure 14.3 *Animated!* The somatosensory cortex "map" of body parts. This strip of cerebral cortex is a little wider than 2.5 centimeters (an inch), from the top of the head to just above the ear.

Basically, they are thinly myelinated or unmyelinated ("naked") dendrites of sensory neurons. One type coils around hair follicles and detects the movement of the hair inside. That might be how, for instance, you become aware that a spider is gingerly making its way across your arm. Free nerve endings sensitive to chemicals such as histamine may be responsible for the sensation of itching.

Encapsulated receptors are more complex. They are enclosed in a capsule of epithelial or connective tissue and are named for the biologist who discovered them. One type, Merkel's discs, adapt slowly and are the most important receptors for steady touch. In lips, fingertips, eyelids, nipples, and genitals there are many Meissner's corpuscles, which are sensitive to light touching. Deep in the dermis and in joint capsules are Ruffini endings, which respond to steady touching and pressure.

The Pacinian corpuscles widely scattered in the skin's dermis are sensitive to deep pressure and vibrations. They also are located near freely movable joints (like shoulder and hip joints) and in some internal organs. Onionlike layers of membrane alternating with fluid-filled spaces enclose this sensory receptor's ending. This arrangement enhances the receptor's ability to detect rapid pressure changes associated with vibrations.

Sensing limb motions and changes in body position relies on mechanoreceptors in skin, skeletal muscles, joints, tendons, and ligaments. Examples include the stretch receptors of muscle spindles described in Section 14.1.

PAIN IS THE PERCEPTION OF BODILY INJURY

Pain is perceived injury to some body region. The most important pain receptors are free nerve endings called *nociceptors* (from the Latin word *nocere*, "to do harm"). Several million of them are distributed throughout the skin and in internal tissues, except for the brain.

Somatic pain starts with nociceptors in skin, skeletal muscles, joints, and tendons. One group is the source of prickling pain, like the jab of a pin when you stick your finger. Another contributes to itching or the feeling of warmth caused by chemicals such as histamine. Sensations of *visceral pain*, which is associated with internal organs, are related to muscle spasms, muscle fatigue, too little blood flow to organs, and other abnormal conditions.

When cells are damaged, they release chemicals that activate neighboring pain receptors. The most potent are bradykinins. They open the floodgates for histamine, prostaglandins, and other substances associated with inflammation (Section 10.3).

When signals from pain receptors reach interneurons in the spinal cord, the interneurons release a chemical

SENSORY RECEPTORS

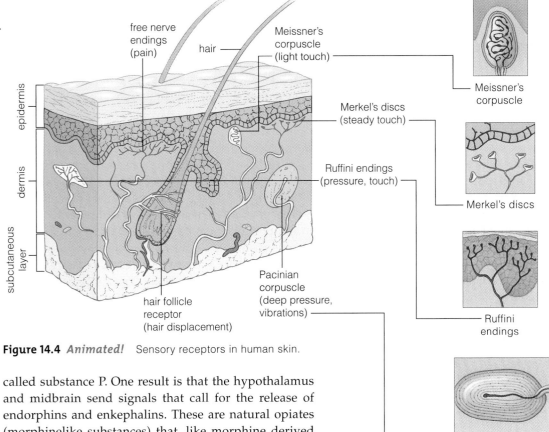

Figure 14.4 *Animated!* Sensory receptors in human skin.

called substance P. One result is that the hypothalamus and midbrain send signals that call for the release of endorphins and enkephalins. These are natural opiates (morphinelike substances) that, like morphine derived from opium poppies, reduce our ability to perceive pain. Morphine, hypnosis, and natural childbirth techniques may also stimulate the release of these natural opiates.

REFERRED PAIN IS A MATTER OF PERCEPTION

A person's perception of pain often depends on the brain's ability to identify the affected tissue. Get hit in the face with a snowball and you "feel" the contact on facial skin. However, sensations of pain from some internal organs may be wrongly projected to part of the skin surface. This response, called *referred pain*, is related to the way the nervous system is built. Sensory information from the skin and from certain internal organs may enter the spinal cord along the same nerve pathways, so the brain can't accurately identify their source. For example, as shown in Figure 14.5, a heart attack can be felt as pain in skin above the heart and along the left shoulder and arm.

Referred pain is not the same as the *phantom pain* reported by amputees. Often they sense the presence of a missing body part, as if it were still there. In some undetermined way, sensory nerves that were severed during the amputation continue to respond to the trauma. The brain projects the pain back to the missing part, past the healed region.

> *Various kinds of free nerve endings and encapsulated receptors detect touch, pressure, heat and cold, pain, limb motions, and changes in body position.*

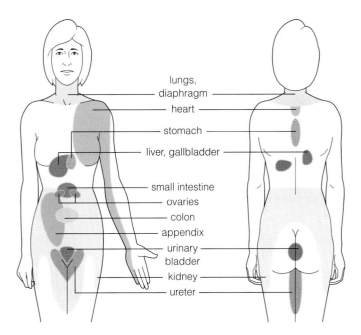

Figure 14.5 *Animated!* Referred pain. When receptors in some internal organs detect painful stimuli, the brain projects the sensation to certain skin areas instead of localizing the pain at the organs.

14.3 Taste and Smell: Chemical Senses

LINKS TO
SECTIONS
5.3, 13.7,
AND 13.8

Some of life's simple pleasures—a delicious meal, the perfume of a rose—are available to us through the chemical "magic" of two special senses, taste and smell.

Taste and smell are *chemical* senses. They begin at chemoreceptors, which are activated when they bind a chemical substance that is dissolved in fluid around them. Although these receptors wear out, new ones replace them. In both cases, sensory information travels from the receptors through the thalamus and on to the cerebral cortex, where perceptions of the stimulus take shape and are fine-tuned. The input also travels to the limbic system, which can integrate it with emotional states and stored memories.

GUSTATION IS THE SENSE OF TASTE

The technical term for taste is *gustation* (as in gusto!). Sensory organs called taste buds hold our **taste receptors** (Figure 14.6). You have about 10,000 taste buds scattered over your tongue, the roof of your mouth (the palate), and your throat.

A taste bud has a pore through which fluids in the mouth (including, of course, saliva) contact the surface of receptor cells. The stimulated receptor in turn stimulates a sensory neuron, which conveys the message to centers in the brain where the stimulus is interpreted. Every perceived taste is some combination of five primary tastes: sweet, sour, salty, bitter, and *umami* (the brothy or savory taste we associate with aged cheese and meats).

Strictly speaking, the flavors of most foods are some combination of the five basic tastes, plus sensory input from olfactory receptors in the nose. Simple as this sounds, research has shown that our taste sense encompasses a complex set of molecular mechanisms. *Science Comes to Life* on the facing page examines some of these findings.

The olfactory element of taste is extremely important. In addition to odor molecules in inhaled air, molecules of volatile chemicals are released as you chew food. These waft up into the nasal passages. There, as described next, the "smell" inputs contribute to the perception of a smorgasbord of complex flavors. This is why anything that dulls your sense of smell—such as a head cold— also seems to diminish food's flavor.

OLFACTION IS THE SENSE OF SMELL

Olfactory receptors (Figure 14.7) detect water-soluble or volatile (easily vaporized) substances. We now know that when odor molecules bind to receptors on olfactory neurons in cells of the nose's olfactory epithelium, the resulting action potential travels directly to olfactory bulbs in the frontal area of the brain. There, other neurons forward the message to a center in the cerebral cortex, which interprets it as "fresh bread," "pine tree," or some other substance.

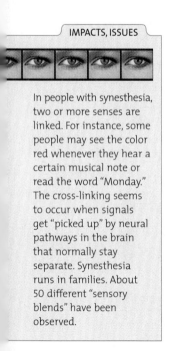

IMPACTS, ISSUES

In people with synesthesia, two or more senses are linked. For instance, some people may see the color red whenever they hear a certain musical note or read the word "Monday." The cross-linking seems to occur when signals get "picked up" by neural pathways in the brain that normally stay separate. Synesthesia runs in families. About 50 different "sensory blends" have been observed.

Figure 14.6 *Animated!* Location of taste receptors in the tongue.

a

tonsil

bitter
sour
salty
sweet

b

c

taste bud

hairlike ending of taste receptor

d

sensory nerve

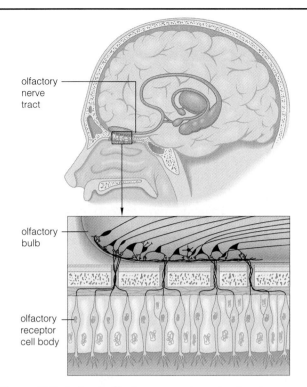

Figure 14.7 *Animated!* Sensory pathway leading from olfactory receptors in the nose to primary receiving centers in the brain.

From an evolutionary perspective, olfaction is an ancient sense—and for good reason. Food, potential mates, and predators give off substances that can diffuse through air (or water) and so give clues or warnings of their whereabouts. Even with our rather insensitive sense of smell, we humans have about 10 million olfactory receptors in patches of olfactory epithelium in the upper nasal passages. (A bloodhound has more than 200 million.)

Just inside your nose, next to the vomer bone (Section 5.3), is a *vomeronasal organ*, or "sexual nose." (Some other mammals also have one.) Its receptors detect pheromones, which are signaling chemicals with roles in social interactions of many animal species. Pheromones can affect the behavior—and maybe the physiology—of other individuals. For instance, one or more pheromones in the sweat of females may account for the common observation that women of reproductive age who are in regular, close contact with one another often come to have their menstrual periods on a similar schedule.

Taste depends on receptors in sensory organs called taste buds in the tongue. The receptors bind molecules dissolved in fluid. Sensory neurons then relay the message to the brain. The five primary tastes are sweet, sour, salty, bitter, and umami.

Olfaction (smell) relies on receptors in patches of epithelium in the upper nasal passages. Neural signals along olfactory neurons travel directly to the olfactory bulbs in the brain.

14.4 A Tasty Morsel of Sensory Science

Taste buds make eating not only a means of securing nutrients, but also one of life's pleasures. In ways that have been poorly understood until recently, the sensory receptors in taste buds distinguish the five main taste categories you've just read about.

Each category is associated with particular "tastant" molecules. When you eat food, however, which taste category (or combination of them) you ultimately perceive depends on the chemical nature of the signal and on how it is processed by the receptor. In each case, some event causes the receptor cell to release a neurotransmitter that triggers action potentials in a nearby sensory neuron.

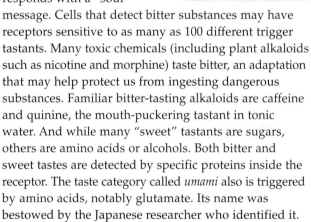

For example, when you taste "salt," the receptor cell's response is due to the influx of Na^+ through sodium ion channels in its plasma membrane. When tastant molecules are acidic—that is, they release hydrogen ions that block membrane potassium ion channels—a receptor responds with a "sour" message. Cells that detect bitter substances may have receptors sensitive to as many as 100 different trigger tastants. Many toxic chemicals (including plant alkaloids such as nicotine and morphine) taste bitter, an adaptation that may help protect us from ingesting dangerous substances. Familiar bitter-tasting alkaloids are caffeine and quinine, the mouth-puckering tastant in tonic water. And while many "sweet" tastants are sugars, others are amino acids or alcohols. Both bitter and sweet tastes are detected by specific proteins inside the receptor. The taste category called *umami* also is triggered by amino acids, notably glutamate. Its name was bestowed by the Japanese researcher who identified it.

Each taste bud has receptors that can respond to tastants in at least two—and in some cases all five—of the taste classes. Various tastants commingle (together with odors) into our perceptions of countless flavors.

Not all taste receptors are equally sensitive. "Bitter" ones tend to be extremely sensitive and so can detect tiny amounts of bitter tastants—and thus potential poisons. Sour tastants are needed in higher concentrations before the stimulus registers. Even higher levels of sweet and salty substances must be present for the stimulus to register. So why can relatively small amounts of artificial sweeteners so readily sweeten foods? Their molecular characteristics make them 150 times (aspartame) to more than 600 times (saccharin) as potent as plain sucrose.

14.5 Hearing: Detecting Sound Waves

The sense of hearing depends on structures in the ear that trap and process sounds traveling through air.

Sounds are waves of compressed air. They are a form of mechanical energy. If you clap your hands, you force out air molecules, creating a low-pressure state in the area they vacated. The pressure variations can be depicted as a wave form, and the *amplitude* of its peaks corresponds to loudness. The *frequency* of a sound is the number of wave cycles per second. Each cycle extends from the start of one wave to the start of the next (Figure 14.8).

The sense of hearing starts with vibration-sensitive mechanoreceptors deep in the ear. When sound waves travel down the ear's auditory canal, they reach a membrane and make it vibrate. The vibrations cause a fluid inside the ear to move, the way water in a waterbed sloshes. In your ear, the moving fluid bends the tips of hairs on mechanoreceptors. With enough bending, the end result will be action potentials sent to the brain, where they are interpreted as sound.

THE EAR GATHERS AND SENDS "SOUND SIGNALS" TO THE BRAIN

Each of your ears consists of three regions (Figure 14.9*a*), each with its own role in hearing. The *outer ear* provides a pathway for sound waves to enter the ear, setting up vibrations. The vibrations are amplified in the *middle ear*. In the *inner ear*, vibrations of different sound frequencies are "sorted out" as they stimulate different patches of receptors. Inner ear structures include *semicircular canals*, which are involved in balance—the topic of Section 14.6. It also contains the coiled **cochlea** (KAHK-lee-uh), where key events in hearing take place. As you'll now read, a coordinated sequence of events in the ear's various regions provides the brain with the auditory input it can interpret to give us a hearing sense.

SENSORY HAIR CELLS ARE THE KEY TO HEARING

Hearing begins when the outer ear's fleshy flaps collect and channel sound waves through the auditory canal to the **tympanic membrane** (the eardrum). Sound waves

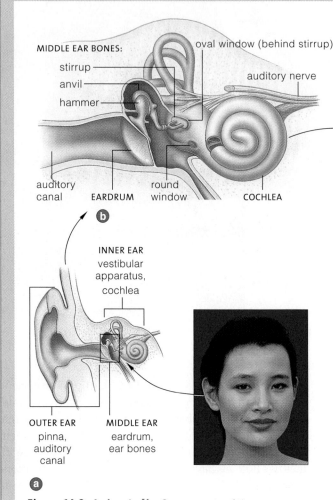

Figure 14.9 *Animated!* Components of the ear.

cause the membrane to vibrate, which in turn causes vibrations in a leverlike array of three tiny bones of the middle ear: the *malleus* ("hammer"), *incus* ("anvil"), and stirrup-shaped *stapes*. The vibrating bones transmit their motion to the *oval window*, an elastic membrane over the entrance to the cochlea. The oval window is much smaller than the tympanic membrane. So, as the middle-ear bones vibrate against its small surface with the full energy that struck the tympanic membrane, the force of the original vibrations is amplified.

Now the action shifts to the cochlea. If we could uncoil the cochlea, we would see that a fluid-filled chamber folds around an inner *cochlear duct* (Figure 14.9*c*). Each "arm" of the outer chamber functions as a separate compartment (the *scala vestibuli* and *scala tympani*, respectively). The amplified vibrations of the oval window create pressure waves in the fluid within the chambers. In turn, these waves are transmitted to the fluid in the cochlear duct. On the floor of the cochlear duct is a *basilar membrane*,

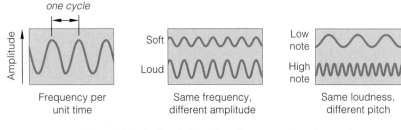

Figure 14.8 *Animated!* Wavelike properties of sound.

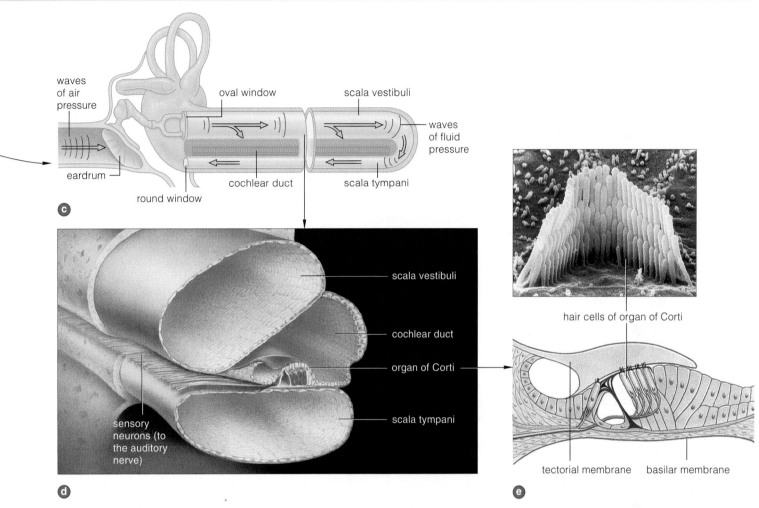

waves of air pressure

oval window

scala vestibuli

waves of fluid pressure

eardrum

cochlear duct

scala tympani

round window

c

scala vestibuli

cochlear duct

organ of Corti

scala tympani

sensory neurons (to the auditory nerve)

d

hair cells of organ of Corti

tectorial membrane   **basilar membrane**

e

and resting on the basilar membrane is a specialized **organ of Corti**, which includes sensory **hair cells**.

Slender projections at the cell tips rest against an overhanging **tectorial** ("rooflike") **membrane**, which is not a membrane at all but a jellylike structure. When pressure waves in the cochlear fluid vibrate the basilar membrane, its movements can press hair cell projections against the tectorial membrane so that the projections bend like brush bristles. Affected hair cells release a neurotransmitter, triggering action potentials in neurons of the auditory nerve, which carries them to the brain.

Different sound frequencies cause different parts of the basilar membrane to vibrate—and, accordingly, to bend different groups of hair cells. Apparently, the total number of hair cells stimulated in a given region determines the loudness of a sound. The perceived tone or "pitch" of a sound depends on the frequency of the vibrations that excite different groups of hair cells. The higher the frequency, the higher the pitch.

Eventually, pressure waves en route through the cochlea push against the *round window*, a membrane at the far end of the cochlea. As the round window bulges outward toward the air-filled middle ear, it serves as a "release valve" for the force of the waves. Air also moves through an opening in the middle ear into the *eustachian tube*. This tube runs from the middle ear to the throat (pharynx), permitting air pressure in the middle ear to be equalized with the pressure of outside air. When you change altitude (say, during a plane trip), this equalizing process makes your ears "pop."

Sounds such as amplified music and the thundering of jet engines are so intense that long-term exposure to them can permanently damage the inner ear (Section 14.7). Sounds produced by these modern technologies exceed the functional range of the evolutionary ancient hair cells in the ear.

> Hearing relies on mechanoreceptors called hair cells, which are attached to membranes inside the cochlea of the inner ear. Pressure waves generated by sound cause membrane vibrations that bend hair cells. The bending produces action potentials in neurons of the auditory nerve.

14.6 Balance: Sensing the Body's Natural Position

Like most other animals, we humans apparently have a sense of the "natural" position for the body (and its parts), given the predictable way we return to it after being tilted or turned upside down. The baseline against which our brain assesses the body's displacement from its natural position is called the "equilibrium position."

Our sense of balance relies partly on messages from receptors in our eyes, skin, and joints. In addition, there are organs of equilibrium located in a part of the inner ear called the **vestibular apparatus**. This "apparatus" is a closed system of sacs and three fluid-filled **semicircular canals** (Figure 14.10). The canals are positioned at right angles to one another, corresponding to the three planes of space. Inside them, some sensory receptors monitor dynamic equilibrium—that is, rotating head movements. Elsewhere in the vestibular apparatus are the receptors that monitor the straight-line movements of acceleration and deceleration.

The receptors attuned to rotation are on a ridge of the swollen base of each semicircular canal (Figure 14.11). As in the cochlea, these receptors are sensory hair cells; their delicate hairs project up into a jellylike *cupula* ("little cap"). When your head rotates horizontally or vertically or tilts diagonally, fluid in a canal corresponding to that direction moves in the opposite direction. As the fluid presses against the cupula, the hairs bend. This bending

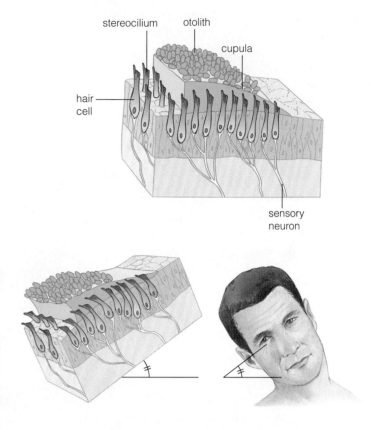

Figure 14.11 How otoliths move in response to gravity when the head tilts.

is the first step leading to action potentials that travel to the brain—in this case, along the vestibular nerve.

The head's position in space tracks *static* equilibrium. The receptors attuned to it are located in two fluid-filled sacs in the vestibular apparatus, the utricle and saccule shown in Figure 14.10. Each sac contains an *otolith* organ, which has hair cells embedded in a jellylike "membrane." The material also contains hard bits of calcium carbonate called *otoliths* ("ear stones"). Movements of the membrane and otoliths signal changes in the head's orientation relative to gravity, as well as straight-line acceleration and deceleration. For example, if you tilt your head, the otoliths slide in that direction, the membrane mass shifts, and tips of the hair cells bend (Figure 14.11). The otoliths also press on hair cells if your head accelerates, as when you start running or are riding in an accelerating vehicle.

Action potentials from the vestibular apparatus travel to reflex centers in the brain stem. As the signals are integrated with information from your muscles and eyes, the brain orders compensating movements that

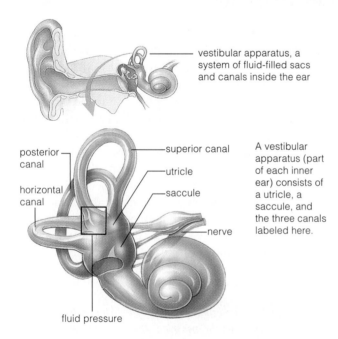

vestibular apparatus, a system of fluid-filled sacs and canals inside the ear

posterior canal
horizontal canal
superior canal
utricle
saccule
nerve
fluid pressure

A vestibular apparatus (part of each inner ear) consists of a utricle, a saccule, and the three canals labeled here.

Figure 14.10 *Animated!* The vestibular apparatus, an organ of equilibrium.

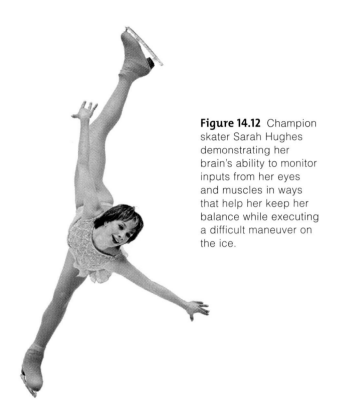

Figure 14.12 Champion skater Sarah Hughes demonstrating her brain's ability to monitor inputs from her eyes and muscles in ways that help her keep her balance while executing a difficult maneuver on the ice.

help you keep your balance when you stand, walk, dance, or move your body in other ways (Figure 14.12).

Motion sickness can result when extreme or continuous motion overstimulates hair cells in the balance organs. It can also be caused by conflicting signals from the ears and eyes about motion or the head's position. As people who are predisposed to motion sickness know all too well, action potentials triggered by the sensory input can reach a brain center that governs the vomiting reflex.

Balance involves a sense of the natural position for the body or its parts. This sense relies heavily on sensory signals from the vestibular apparatus, a closed system of fluid-filled canals and sacs in the inner ear.

The semicircular canals are situated at angles that correspond to the three planes of space. Sensory receptors inside them detect rotation, acceleration, and deceleration of the head.

Otolith organs contain sensory hair cells embedded in a jellylike membrane. Movements of the membrane and otoliths signal changes in the head's orientation relative to gravity, as well as straight-line acceleration and deceleration.

14.7 Disorders of the Ear

The hearing apparatus of our ears is remarkably sturdy, but its functioning also can be damaged by a variety of illnesses and injuries.

Children have short eustachian tubes, so they especially are susceptible to **otitis media**—a painful inflammation of the middle ear that usually is caused by the spread of a respiratory infection such as a cold. The usual treatment is a course of antibiotics. In some cases pus and fluid can build up and cause the eardrum to tear. The rupture usually will heal on its own.

Ear infections, taking lots of aspirin, and other triggers can cause the ringing or buzzing in the ears known as **tinnitus**. Many cases of chronic tinnitus have no detectable cause. While the condition is not a serious health threat, it can be extremely annoying.

Deafness is the partial or complete inability to hear. Some people suffer from congenital (inborn) deafness, and in other cases aging, disease, or environmentally caused damage is the culprit. About one-third of adults in the United States will suffer significant hearing loss by the time they are 65. Researchers believe that most cases of this progressive deafness are due to the long-term effects of living in a noisy world.

The loudness of a sound is measured in decibels. A quiet conversation occurs at about 50 decibels. Rustling papers make noise at a mere 20 decibels. The delicate sensory hair cells in the inner ear (Figure 14.13) begin to be damaged when a person is exposed to sounds louder than about 75–85 decibels over long periods. Some MP3 players can crank out sound at well over 100 decibels. At 130 decibels—typical of a rock concert or shotgun blast—permanent damage can occur much more quickly. Protective earwear is a must for anyone who regularly operates noisy equipment or who works around noisy machinery such as aircraft.

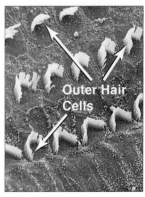
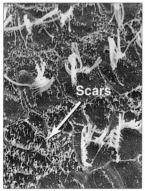

a b

Figure 14.13 (**a**) Healthy sensory hair cells of the inner ear. (**b**) Hair cells damaged by exposure to loud noise.

14.8 Vision: An Overview

All organisms may be sensitive to light. Vision, however, requires (1) a system of photoreceptors and (2) brain centers that can receive and interpret the patterns of action potentials from different parts of the photoreceptor system.

The sense of **vision** is an awareness of the position, shape, brightness, distance, and movement of visual stimuli. Our **eyes** are sensory organs that contain tissue with a dense array of photoreceptors.

THE EYE IS BUILT FOR PHOTORECEPTION

The eye has three layers (Table 14.2), sometimes called "tunics." The outer layer consists of a sclera and a transparent **cornea**. The middle layer consists mainly of a choroid, ciliary body, and iris. The key feature of the inner layer is the retina (Figure 14.14).

The *sclera*—the dense, fibrous "white" of the eye—protects most of the eyeball, except for a "front" region formed by the cornea. Moving inward, the thin, dark-pigmented *choroid* underlies the sclera. It prevents light from scattering inside the eyeball and contains most of the eye's blood vessels.

Behind the transparent cornea is a round, pigmented **iris** (after *irid*, which means "colored circle"). The iris has more than 250 measurable features (such as pigments and fibrous tissues). This is why, as you read earlier, the iris can be used for identification. Look closely at someone's eye, and you will see a "hole" in the center of

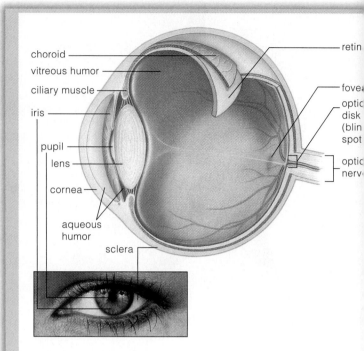

Figure 14.14 *Animated!* Structure of the eye.

the iris. This *pupil* is the entrance for light. When bright light hits the eye, circular muscles in the iris contract and shrink the pupil. In dim light, radial muscles contract and enlarge the pupil.

Behind the iris is a saucer-shaped **lens**, with onionlike layers of transparent proteins. Ligaments attach the lens to smooth muscle of the *ciliary body*; this muscle functions in focusing light, as we will see shortly. The lens focuses incoming light onto a dense layer of photoreceptor cells behind it, in the retina. A clear fluid, *aqueous humor* (body fluids were once called "humors"), bathes both sides of the lens. A jellylike substance (*vitreous humor*) fills the chamber behind the lens.

The **retina** is a thin layer of neural tissue at the back of the eyeball. It has a pigmented basement layer that covers the choroid. Resting on the basement layer are densely packed photoreceptors that are linked with a variety of neurons. Axons from some of these neurons converge to form the optic nerve at the back of the eyeball. The optic nerve is the trunk line to the thalamus— which, recall, sends signals on to the **visual cortex**. The place where the optic nerve exits the eye is a "blind spot" because there are no photoreceptors there.

The surface of the cornea is curved. This means that incoming light rays hit it at different angles and, as they pass through the cornea, their trajectories (paths) bend (Figure 14.15*a*). There, because of the way the rays were

Table 14.2	Parts of the Eye
Wall of eyeball	(three layers)
Sensory Tunic (inner layer)	*Retina*. Absorbs, transduces light energy
	Fovea. Increases visual acuity
Vascular Tunic (middle layer)	*Choroid*. Blood vessels nutritionally support wall cells; pigments prevent light scattering
	Ciliary body. Its muscles control lens shape; its fine fibers hold lens upright
	Iris. Adjusting iris controls incoming light
	Pupil. Serves as entrance for light
	Start of optic nerve. Carries signals to brain
Fibrous Tunic (outer layer)	*Sclera*. Protects eyeball
	Cornea. Focuses light
Interior of eyeball	
Lens	Focuses light on photoreceptors
Aqueous humor	Transmits light, maintains pressure
Vitreous body	Transmits light, supports lens and eyeball

SENSORY SYSTEMS

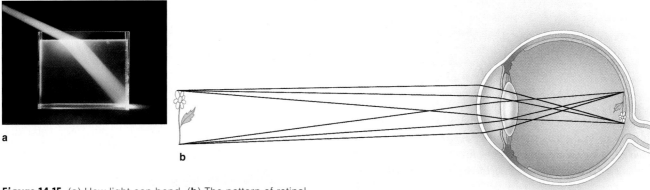

a

b

Figure 14.15 (**a**) How light can bend. (**b**) The pattern of retinal stimulation in the eye. Being curved, the cornea alters the trajectories of light rays as they enter the eye. The pattern is upside-down and reversed left to right, compared to the stimulus.

Figure 14.16 *Animated!* Focusing light on the retina by adjusting the lens (visual accommodation). A ciliary muscle encircling the lens attaches to it. (**a**) When it contracts, the lens bulges. The focal point moves closer and brings close objects into focus. (**b**) When the muscle relaxes, the lens flattens, so the focal point moves farther back. This can bring distant objects into focus.

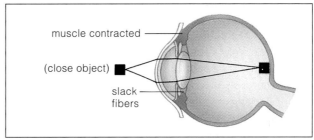

a Accommodation for close objects (lens bulges)

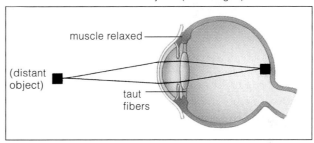

b Accommodation for distant objects (lens flattens)

bent at the curved cornea, the rays converge at the back of the eyeball. They stimulate the retina in a pattern that is upside-down and reversed left to right relative to the original source of the light rays. Figure 14.15*b* gives a simplified diagram of this outcome. The "upside-down and backwards" orientation is corrected in the brain.

EYE MUSCLE MOVEMENTS FINE-TUNE THE FOCUS

Light rays from sources at different distances from the eye strike the cornea at different angles. As a result, they will be focused at different distances behind it, and adjustments must be made so that the light will be focused precisely on the retina. Normally, the lens can be adjusted so that the focal point coincides exactly with the retina. A ciliary muscle adjusts the shape of the lens. As you can see in Figure 14.14 the muscle encircles the lens and attaches to it by ligaments. When the muscle contracts, the lens bulges, so the focal point moves closer. When the muscle relaxes, the lens flattens, so the focal point moves farther back (Figure 14.16). Adjustments like these are called *accommodation*. If they are not made, rays from distant objects will be in focus at a point just in front of the retina, and rays from very close objects will be focused behind it.

Sometimes the lens can't be adjusted enough to place the focal point on the retina. Sometimes also, the eyeball is not shaped quite right. The lens is too close to or too far

away from the retina, so accommodation alone cannot produce a precise match. Eyeglasses or contact lenses can correct these problems, which are called farsightedness and nearsightedness, respectively. Section 14.10 examines a variety of eye disorders.

Eyes are sensory organs specialized for photoreception.

In the outer eye layer, the sclera protects the eyeball and the cornea focuses light.

In the middle layer, the choroid prevents light scattering, the iris controls incoming light, and the ciliary body and lens aid in focusing light on photoreceptors.

Photoreception occurs in the retina of the inner layer.

Adjustments in the position or shape of the lens focus incoming visual stimuli onto the retina.

14.9 From Visual Signals to "Sight"

LINKS TO
SECTIONS
13.2, 13.3,
AND 13.7

Our vision sense is based on the sensory pathway from the retina to the brain. It is an excellent example of how different neurons can interrelate to one another in the nervous system.

"Seeing" something is a multistep process that begins when our eyes receive raw visual information. The information then is transmitted and processed in ways that lead to awareness of light and shadows, of colors, and of near and distant objects in the world around us.

RODS AND CONES ARE THE PHOTORECEPTORS

The flow of information begins as light reaches the retina, at the back of the eyeball. The retina has a basement layer, a pigmented epithelium on top of the choroid. Resting on this layer are 150 million photoreceptors called **rod cells** and **cone cells** (Figure 14.17 and Table 14.3). Rod cells detect very dim light. At night or in dark places, they detect changes in light intensity across the visual field—the start of coarse perception of motion. Cone cells detect bright light—the start of sharp vision and perception of color in the daytime or in brightly lit spaces. (You don't see color in moonlight because the light isn't bright enough to stimulate cones.)

VISUAL PIGMENTS IN RODS AND CONES INTERCEPT LIGHT ENERGY

The outer part of a rod cell contains stacks of several hundred membrane disks, each packed with more than a billion molecules of a visual pigment called **rhodopsin**. The membrane stacks and the large number of pigment molecules hugely increase the odds that rods will detect light energy, which comes packaged as photons. (Each photon is a given amount of light energy.) The action potentials that result when rod pigments absorb even a few photons can allow a person to see objects in dimly lit surroundings such as dark rooms or late at night.

Each rhodopsin molecule consists of a protein (opsin) to which a signal molecule, *cis*-retinal, is bound. The retinal is derived from vitamin A. That's one reason why too little vitamin A in the diet can impair a person's vision, especially at night.

When the photons of blue-green light stimulate rhodopsin, it changes shape. The change is the start of the process that converts light energy to action potentials. It triggers a cascade of reactions that alter the distribution of ions across the rod cell's plasma membrane. As a result, there is a slow down in the ongoing release of a neurotransmitter that inhibits neurons next to the rods. No longer inhibited, the neurons start sending signals about the visual stimulus on toward the brain.

Table 14.3	Rods and Cones Compared	
CELL TYPE	**SENSITIVE TO**	**RELATED PERCEPTION**
Rod	Dim light	Coarse perception of movement
Cone	Bright light	Daytime vision and perception of color

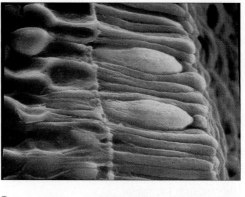

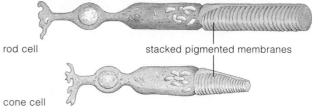

rod cell   stacked pigmented membranes

cone cell

Figure 14.17 Photoreceptors: rods and cones.

Daytime vision and the sense of color start when photons are absorbed by cone cells, each with a different kind of visual pigment—red, green, or blue. Here again, the absorption of photons curtails the release of a neurotransmitter that would otherwise prevent the neurons next to the photoreceptors from firing.

Near the center of the retina is a funnel-shaped depression called the **fovea** (Figure 14.18). Photoreceptors are denser there than anywhere else in the retina. Visual acuity, the ability to discriminate between two objects, also is greatest there. The fovea's cluster of cones enables you to distinguish between neighboring points in space— like the *e* and the period at the end of this sentence.

THE RETINA PROCESSES SIGNALS FROM RODS AND CONES

Neurons in the eye are organized in layers above the rods and cones. As you can see in Figure 14.19, signals flow from rods and cones to *bipolar* interneurons, then to interneurons called *ganglion cells*. The axons of ganglion cells form the two optic nerves to the brain.

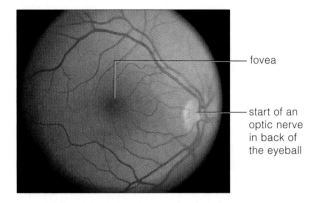

Figure 14.18 Location of the fovea and the start of the optic nerve.

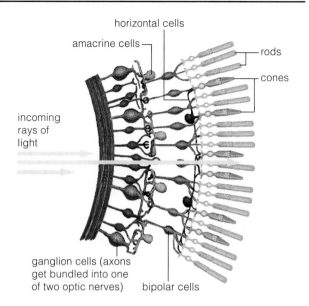

Figure 14.19 *Animated!* Organization of photoreceptors and sensory neurons in the retina.

Before visual signals leave the retina, they converge dramatically. Input from the 150 million photoreceptors funnels onto only 1 million ganglion cells. Signals also flow sideways among *horizontal* cells and *amacrine* cells. These neurons act jointly to dampen or strengthen the signals before they reach ganglion cells. In other words, a great deal of synaptic integration and processing goes on even before visual information is sent to the brain.

RECEPTIVE FIELDS IN THE RETINA   Different neurons in the eye respond to light stimuli in different ways. The retina's surface is organized into "receptive fields," areas that influence the activity of individual sensory neurons. For example, for each ganglion cell, the field is a circle. Some cells respond best to a small spot of light, ringed by dark, in the field's center. Others respond to motion, to a spot of a particular color, or to a rapid change in light intensity—when you switch on the light in a dark room, for instance. In one experiment, certain neurons fired when a hard-edged bar was oriented in some ways but not others. These differences in receptive fields of different neurons help prevent the brain from being bombarded with a confusing array of visual signals.

SIGNALS MOVE ON TO THE VISUAL CORTEX   The part of the outside world that a person actually sees is called the "visual field." The right side of each retina intercepts light from the left half of the visual field and the left side intercepts light from the right half. As you can see in Figure 14.20, signals from each eye "criss-cross." The optic nerve leading out of each eye delivers signals from the left visual field to the right cerebral hemisphere, and signals from the right go to the left hemisphere.

Axons of the optic nerves end in an island of gray matter in the cerebrum (the lateral geniculate nucleus). Its layers each have a map corresponding to receptive fields of the retina. Each map's interneurons deal with one aspect of a visual stimulus—its form, movement,

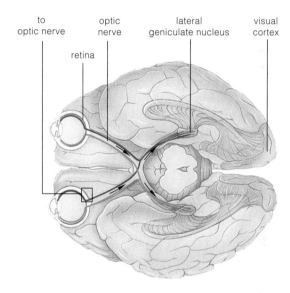

Figure 14.20 *Animated!*   Sensory pathway from the retina to the brain.

depth, color, texture, and so on. After initial processing all the visual signals travel rapidly, at the same time, to different parts of the visual cortex. There, a final round of integration produces the sensation of sight.

> *Rods and cones are the eye's photoreceptors. Rods detect dim light. Cones detect bright light and provide our sense of color.*
>
> *The eye analyzes information on the distance, shape, brightness, position, and movement of a visual stimulus.*
>
> *Visual signals move through layers of neurons in the retina before moving on to the brain.*

14.10 Disorders of the Eye

For us humans, our eyes are the single most important source of information about the outside world. It's no wonder then that we pay a lot of attention to eye disorders.

Problems that disrupt normal eye functions range from injuries and diseases to inherited abnormalities and natural changes associated with aging. The outcomes range from some relatively harmless conditions, such as nearsightedness, to total blindness.

MISSING CONE CELLS CAUSE COLOR BLINDNESS

Occasionally, some or all of the cone cells that selectively respond to light of red, green, or blue are missing. The rare people who have only one of the three kinds of cones are totally color-blind. They see the world only in shades of gray.

Consider a common inherited abnormality, **red-green color blindness**. It shows up most often in males, for reasons you can read about in Chapter 21. The retina lacks some or all of the cone cells with pigments that normally respond to light of red or green wavelengths. Most of the time, color-blind people have trouble distinguishing red from green only in dim light. However, some cannot distinguish between the two even in bright light.

MALFORMED EYE PARTS CAUSE COMMON FOCUSING PROBLEMS

Some inherited vision problems are due to misshapen eye structures that affect the eye's ability to focus light. In **astigmatism**, for example, one or both corneas have an uneven curvature; they cannot bend incoming light rays to the same focal point.

In **myopia**, or nearsightedness, the eyeball is wider than it is high, or the ciliary muscle responsible for adjusting the lens contracts too strongly. Then, images of distant objects are focused in front of the retina instead of on it (Figure 14.21a). **Hyperopia**, farsightedness, is the opposite problem. The eyeball is "taller" than it is wide (or the lens is "lazy"), so close images are focused behind the retina (Figure 14.21b).

THE EYES ALSO ARE VULNERABLE TO INFECTIONS AND CANCER

The eyes are vulnerable to pathogens including viruses, bacteria, and fungi. Health authorities estimate that in the U.S., about 1 in every 50 visits to a doctor's office is for **conjunctivitis**, inflammation of the transparent membrane (the conjunctiva) that lines the inside of the eyelids and

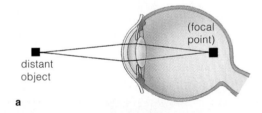

a

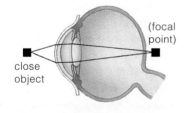

b

Figure 14.21 *Animated!* Examples of nearsighted and farsighted vision.

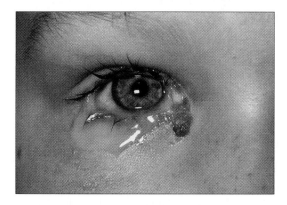

Figure 14.22 A child with conjunctivitis.

covers the sclera (the white of the eye). Symptoms include redness, discomfort, and a discharge. In children, conjunctivitis usually is caused by a bacteria infection; in adults it more often is triggered by allergy (Figure 14.22). Most cases of conjunctivitis are easily treated with antibiotics.

Trachoma is a highly contagious disease that has blinded millions, mostly in North Africa and the Middle East. The culprit is a bacterium that also is responsible for the sexually transmitted disease chlamydia (Chapter 16). Trachoma damages both the eyeball and the conjunctiva. Then, other bacteria can enter the damaged tissues and cause secondary infections. In time the cornea can become so scarred that blindness follows.

Various forms of herpes simplex, a virus that causes cold sores and genital herpes, also can infect the cornea. Because blindness can result from a **herpes infection** in the eyes, a pregnant woman who has a history of genital herpes likely will be delivered by Caesarian section to avoid any chance of exposing her newborn to the virus.

Malignant melanoma is the most common eye cancer. It typically develops in the choroid (the eye's middle layer) and may not trigger noticeable vision problems until it has spread to other parts of the body. About 1 in 20,000 babies is born with **retinoblastoma**, a cancer of the retina. Because it can readily spread along the optic nerve to the brain, the affected eye often is removed surgically. If both eyes are affected, radiation therapy may be used to try to save one of them.

AGING INCREASES THE RISK OF CATARACTS AND SOME OTHER EYE DISORDERS

Cataracts, clouding of the eye's lens, are associated with aging, although an injury or diabetes can also cause them to develop. The underlying change may be an alteration in the structure of transparent proteins that make up the lens. This change in turn may skew the trajectory of incoming light rays. If the lens becomes totally opaque, no light can enter the eye.

Even a normal lens loses some of its natural flexibility as we grow older. This normal stiffening is why people over 40 years old often must start wearing eyeglasses.

Elderly people can suffer **macular degeneration**, in which a portion of the retina breaks down and is replaced by scar tissue. As a result, a "blind spot" develops. Often both eyes are affected. Treatment is difficult unless the problem is detected early.

Glaucoma results when too much aqueous humor builds up inside the eyeball. Blood vessels that service the retina collapse under the increased fluid pressure. An affected person's vision deteriorates as neurons of the retina and optic nerve die. Although chronic glaucoma often is associated with advanced age, the problem really starts in a person's middle years. If detected early, the fluid pressure can be relieved by drugs or surgery before the damage becomes severe.

MEDICAL TECHNOLOGIES CAN REMEDY SOME VISION PROBLEMS AND TREAT EYE INJURIES

Today many different procedures are used to correct eye disorders. In *corneal transplant surgery*, the defective cornea is removed; then an artificial cornea (made of clear plastic) or a natural cornea from a donor is stitched in place. Within a year, the patient is fitted with eyeglasses or contact lenses. Similarly, cataracts sometimes can be surgically corrected by removing the lens and replacing it with an artificial one.

Severely nearsighted people may opt for procedures that eliminate the need for corrective lenses. So-called "lasik" (for laser-assisted in situ keratomilieusis) and "lasek" (for laser-assisted subepithelial keratectomy) use a laser to reshape the cornea. All or part of the surface of the cornea is peeled back and then replaced into position after the defect being treated is corrected. *Conductive keratoplasty* (CK) uses radio waves to reshape the cornea and bring near vision back into focus.

Retinal detachment is the eye injury we read about most often. It may follow a physical blow to the head or an illness that tears the retina. As the jellylike vitreous body oozes through the torn region, the retina is lifted from the underlying choroid. In time it may peel away entirely, leaving its blood supply behind. Early symptoms of the damage include blurred vision, flashes of light that occur in the absence of outside stimulation, and loss of peripheral vision. Without medical help, the person may become totally blind in the damaged eye.

A detached retina may be treatable with *laser coagulation*, a painless technique in which a laser beam seals off leaky blood vessels and "spot welds" the retina to the underlying choroid.

Summary

Section 14.1   A stimulus is a form of energy that the body detects by means of sensory receptors. A sensation is a conscious awareness that stimulation has occurred. Perception is understanding what the sensation means.

Sensory receptors are endings of sensory neurons or specialized cells next to them. They respond to stimuli, which are specific forms of energy, such as mechanical pressure and light.

a. Mechanoreceptors detect mechanical energy that is associated with changes in pressure (e.g., sound waves), changes in position, or acceleration.

b. Thermoreceptors detect the presence of or changes in radiant energy from heat sources.

c. Nociceptors (pain receptors) detect tissue damage. Their signals are perceived as pain.

d. Chemoreceptors detect chemical substances that are dissolved in the body fluids around them.

e. Osmoreceptors detect changes in water volume (hence solute concentrations) in the surrounding fluid.

f. Photoreceptors detect light.

A sensory system has sensory receptors for specific stimuli and nerve pathways from those receptors to receiving and processing centers in the brain. The brain assesses each stimulus based on which nerve pathway is delivering the signals, how often signals are traveling along each axon of the pathway, and the number of axons that were recruited into action. When sensory adaptation occurs, the response to a stimulus decreases.

The special senses include taste, smell, hearing, balance, and vision. The receptors associated with these senses typically are in sense organs or another specific body region.

 See how the intensity of a sensory stimulus affects the frequency of action potentials to the brain.

Section 14.2   Somatic sensations include touch, pressure, pain, temperature, and muscle sense. Receptors associated with these sensations occur in various parts of the body. Their signals are processed in the somatosensory cortex of the brain. The simplest receptors, which include various mechanoreceptors, thermoreceptors, and nociceptors, are free nerve endings in the skin or internal tissues. Some somatic sensations also arise when encapsulated receptors respond to stimuli.

 Learn about many of the sensory receptors in skin.

Section 14.3   Taste and smell are chemical senses. Their sensory pathways travel from chemoreceptors to processing regions in the cerebral cortex and limbic system. Taste buds in the tongue and mouth contain the taste receptors. The sense of smell relies on olfactory receptors in patches of epithelium in the upper nasal passages.

Section 14.5   The sense of hearing requires components of the outer, middle, and inner ear that respectively collect, amplify, and respond to sound waves that vibrate the tympanic membrane (eardrum). The vibrations are transferred to fluid in the cochlea of the inner ear, where they in turn vibrate the tectorial membrane. Ultimately the fluid movements bend sensory hair cells in the organ of Corti. The bending triggers neural signals that are carried to the brain via the auditory nerve.

 Explore the structure and function of the ear.

Section 14.6   Balance organs are located in the vestibular apparatus of the inner ear. Sensory receptors in these semicircular canals (including hair cells) detect gravity, velocity, acceleration, and other factors that affect body positions and movements.

Section 14.8   Eyes are the sensory organs associated with the sense of vision. Key eye structures include the cornea and lens, which focus light; the iris, which adjusts incoming light; and the retina, which contains photoreceptors (rods and cones). The optic nerve at the back of the eyeball transmits visual signals to the visual cortex in the brain.

 Investigate the structure and function of the eye.

Section 14.9   The rod cells and cone cells detect dim and bright light, respectively. Light detection in rods depends on changes in the shape of the visual pigment rhodopsin. The visual pigments in cones respond to colors. Visual signals are processed in the retina before being sent on to the brain. In the retina, abundant receptors in the fovea provide sharp visual acuity.

Learn about the organization of the retina and how visual stimuli are processed.

Review Questions

1. When a receptor cell detects a specific kind of stimulus, what happens to the stimulus energy?

2. Name six categories of sensory receptors and the type of stimulus energy that each type detects.

3. How do somatic sensations differ from special senses?

4. Explain where free nerve endings are located in the body and note some functions of the various kinds.

5. What is pain? Describe one type of pain receptor.

6. What are the stimuli for taste receptors?

7. How do "smell" signals arise and reach the brain?

8. Label the parts of the ear:

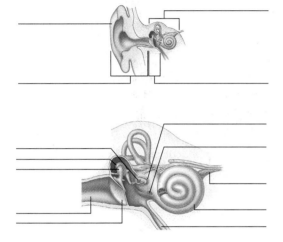

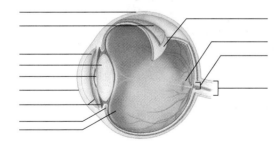

9. In the ear, sound waves cause the tympanic membrane to vibrate. What happens next in the middle ear? In the inner ear?

10. Label the parts of the eye:

11. How does the eye focus the light rays of an image? What do nearsighted and farsighted mean?

Self-Quiz

Answers in Appendix V

1. A _____ is a specific form of energy that can elicit a response from a sensory receptor.

2. Awareness of a stimulus is called a _____ .

3. _____ is understanding what particular sensations mean.

4. A sensory system is composed of _____ .
    a. nerve pathways from specific receptors to the brain
    b. sensory receptors
    c. brain regions that deal with sensory information
    d. all of the above

5. _____ detect energy associated with changes in pressure, body position, or acceleration.
    a. Chemoreceptors   c. Photoreceptors
    b. Mechanoreceptors   d. Thermoreceptors

6. Detecting substances present in the body fluids that bathe them is the function of _____ .
    a. thermoreceptors   c. mechanoreceptors
    b. photoreceptors   d. chemoreceptors

7. Which of the special senses is based on the following events? Membrane vibrations cause fluid movements, which lead to bending of mechanoreceptors and firing of action potentials.
    a. taste   c. hearing
    b. smell   d. vision

8. Rods differ from cones in the following ways:
    a. They detect dim light, not bright light.
    b. They have a different visual pigment.
    c. They are not located in the retina.
    d. all of the above
    e. a and b only

9. The outer layer of the eye includes the _____ .
    a. lens and choroid   c. retina
    b. sclera and cornea   d. both a and c are correct

10. The inner layer of the eye includes the _____ .
    a. lens and choroid   c. retina
    b. sclera and cornea   d. start of optic nerve

11. Your visual field is _____ .
    a. a specific, small area of the retina
    b. what you actually "see"
    c. the area where color vision occurs
    d. where the optic nerve starts

12. Match each of the following terms with the appropriate description.
    ____ somatic senses   a. produced by strong
        (general senses)   stimulation
    ____ special senses   b. endings of sensory
    ____ variations in   neurons or specialized
        stimulus intensity   cells next to them
    ____ action potential   c. taste, smell, hearing,
    ____ sensory receptor   balance, and vision
                               d. frequency and number of
                                   action potentials
                               e. touch, pressure,
                                   temperature, pain, and
                                   muscle sense

Critical Thinking

1. Juanita started having bouts of dizziness. Her doctor asked her whether "dizziness" meant she felt lightheaded as if she were going to faint, or whether it meant she had sensations of *vertigo*—that is, a feeling that she herself or objects near her were spinning around. Why was this clarification important for the diagnosis?

2. Michael, a 3-year-old, experiences chronic middle-ear infection, which is common among youngsters, in part due to an increase in antibiotic-resistant bacteria. This year, despite antibiotic treatment, an infection became so advanced that he had trouble hearing. Then his left eardrum ruptured and a jellylike substance dribbled out. The pediatrician told Michael's parents not to worry, that if the eardrum had not ruptured on its own she would have had to drain it. Suggest a reason why the physician concluded that this procedure would have been needed.

3. Jill is diagnosed with sensorineural deafness, a disorder in which sound waves are transmitted normally to the inner ear but they are not translated into neural signals that travel to the brain. Sometimes the cause is a problem with the auditory nerve, but in Jill's case it has to do with a problem in the inner ear itself. Where in the inner ear is the disruption most likely located?

4. Larry goes to the doctor complaining that he can't see the right side of the visual field with either eye. Where in the visual signal-processing pathway is the problem?

Explore on Your Own

As Section 14.10 described, there are various forms of color blindness. Figure 14.23 shows simple tests, called Ishihara plates, which are standardized tests for different forms of color blindness. For instance, you may have one form of red-green color blindness if you see the numeral "7" instead of "29" in the circle in part *a*. You may have another form if you see a "3" instead of an "8" in the circle in part *b*.

If you do this exercise and have questions about your color vision, visit your doctor to determine whether additional testing is in order.

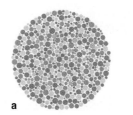
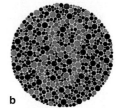

a   b

Figure 14.23 Color blindness tests.

Hormones in the Balance

In 2001, researchers at Dartmouth College discovered what may be another villainous role for the poison arsenic. It appears to be an endocrine disrupter, a chemical that affects health by interfering with normal hormonal processes. Along with some other chemicals, arsenic upsets the action of hormones by blocking them or mimicking them. The problem with a mimic is that, unlike natural hormones, it can't be turned off. As a result, the body loses control of the affected function. The Dartmouth team discovered that arsenic disrupts the action of glucocorticoids, hormones that help regulate blood sugar and that are "on" switches for genes that may protect against cancer.

Arsenic is present in water supplies in many areas. Long-term exposure to it in drinking water is associated with bladder, lung, and skin cancers, birth defects, and other problems. In 2001, around the same time that the Dartmouth study was published, the Environmental Protection Agency cited the risks and lowered the allowable arsenic levels in water supplies. In 2003 some factions lobbied to get the levels increased again.

Other possible endocrine disrupters are also coming under scrutiny. One is atrazine, a widely used herbicide. Golf courses, cornfields, sorghum crops—all have been doused with it to kill weeds. Atrazine now is found in groundwater throughout the American Midwest. Some scientists, including Tyrone Hayes (*right*), suspect it is linked to reproductive problems in several species of frogs.

Other suspects are PCBs, chemicals that were long used as fluid insulation in electrical transformers. Now banned, PCBs linger in the environment. They have been implicated in reproductive disorders in humans.

Research on endocrine disrupters is controversial. Some researchers suspect that they are contributing to an earlier onset of puberty and to low sperm counts. Other scientists dismiss the hypotheses as junk science. Ongoing studies should clarify the picture.

This chapter focuses on hormones—their sources, targets, and interactions. What you learn here may help you evaluate research that may affect human life in a big way.

 How Would You Vote? Some pesticides may disrupt hormone function in humans and other animals. Should they remain in use while researchers study their safety? Cast your vote online at www.thomsonedu .com/biology/starr/humanbio.

 Key Concepts

HOW HORMONES WORK
Hormones control many body functions and influence behavior. Hormones bind to and activate receptors on target cells. Their signals are converted into forms that work inside target cells to bring about a response.

THE ENDOCRINE SYSTEM
Most hormones are released from endocrine glands and tissues. The hypothalamus and pituitary glands control much of this activity. In addition, some hormones are released in response to chemical changes in particular tissues.

Links to Earlier Concepts

This chapter builds on your understanding of the roles of different cell structures (3.3, 3.4). It expands on what you have learned about the roles of two key brain components, the hypothalamus and the pituitary gland (13.7), and provides more examples of how homeostatic feedback loops help regulate body functions (4.9).

You will also see more examples of how proteins in cell plasma membranes function in physiological processes (3.3)—in this case, by serving as receptors for hormone molecules.

LINKS TO
SECTIONS
3.3 AND 14.3

The activities of billions of cells must be integrated in order for the body to function normally. This integration depends on signaling molecules. The nervous system's neurotransmitters are one kind of signaling molecule, and hormones are another.

HORMONES ARE SIGNALING MOLECULES THAT ARE CARRIED IN THE BLOODSTREAM

All the body's signaling molecules—neurotransmitters, pheromones, local signaling molecules, and hormones—act on target cells. A **target cell** is any cell that has receptors for the signaling molecule and that may alter its activities in response. A target may or may not be next to the cell that sends the signal.

Hormones, our main topic here, are secreted by the body's endocrine glands, endocrine cells, and some neurons. They travel the bloodstream to target cells some distance away. Many types of body cells also release "local signaling molecules" that alter conditions within nearby tissues. Prostaglandins are an example. Their targets include smooth muscle cells in the walls of bronchioles, which then constrict or dilate and so alter air flow in the lungs (Section 11.5). Certain exocrine glands secrete **pheromones**, which we touched on briefly in Chapter 14. Unlike other signaling molecules, a pheromone acts on targets *outside* the body. It diffuses through air (or water) to the cells of other animals of the same species. Pheromones help integrate behaviors, such as those related to reproduction. You may recall from Section 14.3 that the vomeronasal organ in the human nose can detect pheromones. Do pheromones act to trigger impressions, such as spontaneous good or bad "feelings" about someone you just met? We don't yet know the answer to that fascinating question.

HORMONE SOURCES: THE ENDOCRINE SYSTEM

The word hormone—from the Greek *hormon*, "to set in motion"—was coined in 1900 by scientists studying food digestion in dogs. They discovered that a certain substance (which they dubbed "secretin") released by gland cells in the canine GI tract could stimulate the pancreas. Later researchers identified other hormones and their sources. There are several major sources of hormones in the human body, as you can see in Figure 15.1.

These hormone-producing glands, organs, and cells became known as the **endocrine system**. The name is misleading, because it implies that there is a separate hormone-based control system for the body. (*Endon* means "within"; *krinein* means "separate.") However, from biochemical studies and electron microscopy, we now know that endocrine sources and the nervous system function in intricately connected ways, as you will see later in this chapter.

HORMONES OFTEN INTERACT

Today we also know that often, two or more hormones affect the same process. There are three common kinds of hormone "partnerships":

1. **Opposing interaction**. The effect of one hormone may oppose the effect of another. Insulin, for example, reduces the level of glucose in the blood, and glucagon increases it.

2. **Synergistic interaction**. The combined action of two or more "cooperating" hormones may be required to trigger a certain effect on target cells. For instance, a woman's mammary glands can't produce and secrete milk without the synergistic interaction of three other hormones: prolactin, oxytocin, and estrogen.

3. **Permissive interaction**. One hormone can exert its effect on a target cell only when a different hormone first "primes" the target cell. For example, even if one of a woman's eggs is fertilized, she can become pregnant only if the lining of her uterus has been exposed to estrogens, then to progesterone.

Hormones are signaling molecules secreted by endocrine glands, endocrine cells, and some neurons. The bloodstream distributes hormones to distant target cells.

Together, the glands and cells that secrete hormones make up the endocrine system.

Hormones may interact in opposition, in cooperation (synergistically), or permissively (a target cell must be primed by exposure to one hormone in order to respond to a second one).

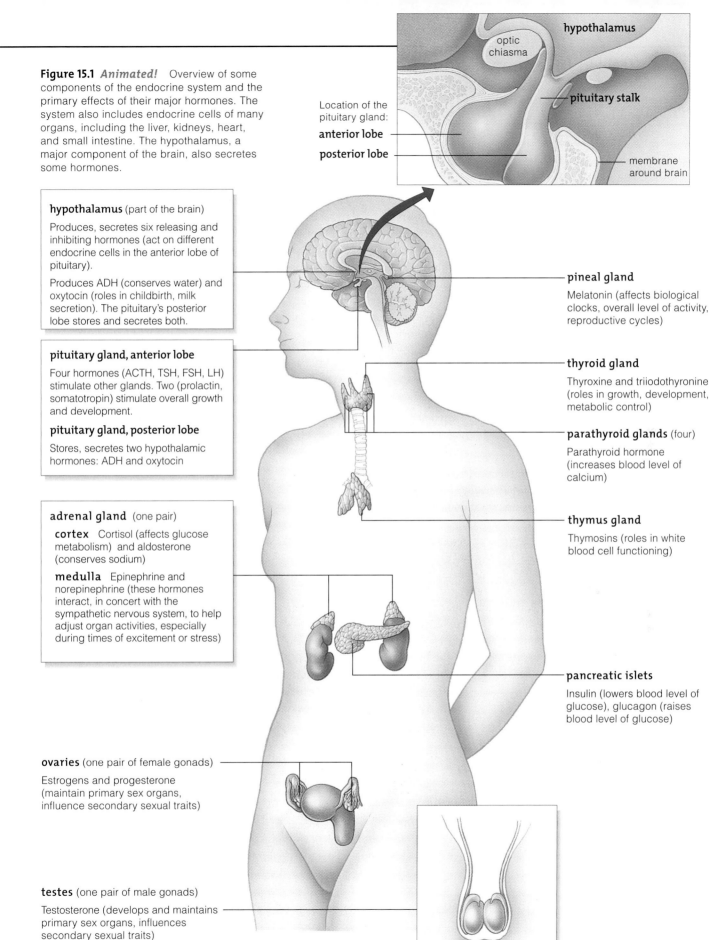

Figure 15.1 *Animated!* Overview of some components of the endocrine system and the primary effects of their major hormones. The system also includes endocrine cells of many organs, including the liver, kidneys, heart, and small intestine. The hypothalamus, a major component of the brain, also secretes some hormones.

hypothalamus

optic chiasma

pituitary stalk

Location of the pituitary gland:

anterior lobe

posterior lobe

membrane around brain

hypothalamus (part of the brain)

Produces, secretes six releasing and inhibiting hormones (act on different endocrine cells in the anterior lobe of pituitary).

Produces ADH (conserves water) and oxytocin (roles in childbirth, milk secretion). The pituitary's posterior lobe stores and secretes both.

pituitary gland, anterior lobe

Four hormones (ACTH, TSH, FSH, LH) stimulate other glands. Two (prolactin, somatotropin) stimulate overall growth and development.

pituitary gland, posterior lobe

Stores, secretes two hypothalamic hormones: ADH and oxytocin

adrenal gland (one pair)

cortex Cortisol (affects glucose metabolism) and aldosterone (conserves sodium)

medulla Epinephrine and norepinephrine (these hormones interact, in concert with the sympathetic nervous system, to help adjust organ activities, especially during times of excitement or stress)

ovaries (one pair of female gonads)

Estrogens and progesterone (maintain primary sex organs, influence secondary sexual traits)

testes (one pair of male gonads)

Testosterone (develops and maintains primary sex organs, influences secondary sexual traits)

pineal gland

Melatonin (affects biological clocks, overall level of activity, reproductive cycles)

thyroid gland

Thyroxine and triiodothyronine (roles in growth, development, metabolic control)

parathyroid glands (four)

Parathyroid hormone (increases blood level of calcium)

thymus gland

Thymosins (roles in white blood cell functioning)

pancreatic islets

Insulin (lowers blood level of glucose), glucagon (raises blood level of glucose)

15.2 Types of Hormones and Their Signals

LINKS TO
SECTIONS
2.10, 2.11, 3.2,
AND 3.3

Like other signaling molecules, hormones bind with protein receptors of target cells. What happens next depends on whether a hormone is a steroid or a peptide hormone.

HORMONES COME IN SEVERAL CHEMICAL FORMS

Hormones vary in their chemical structure, which affects how they function. **Steroid hormones** are lipids derived from cholesterol. **Amine hormones** are modified amino acids. **Peptide hormones** consist of a few amino acids. **Protein hormones** are longer amino acid chains. Table 15.1 lists some examples of each.

Regardless of their chemical makeup, hormones affect cell activities by binding to protein receptors of target cells. The signal is then transduced, or converted, into a form that can work in the cell. Then the cell's activity changes:

Some hormones cause a target cell to increase its uptake of a substance, such as glucose. Others stimulate or inhibit the target in ways that alter its rates of protein synthesis, modify existing proteins or other structures in the cytoplasm, or even change a cell's shape.

Two factors have a major influence on how a target cell responds to hormone signals. To begin with, different hormones activate different kinds of mechanisms in target cells. And second, not all types of cells can respond to a given signal. For instance, many types of cells have receptors for the hormone cortisol, so it has widespread effects in the body. If only a few cell types have receptors for a hormone, its effects in the body are limited to the receptive locations.

Table 15.1	Categories of Hormones and a Few Examples
Steroid hormones	Estrogens, progesterone, testosterone, aldosterone, cortisol
Amines	Melatonin, epinephrine, norepinephrine, thyroid hormone (thyroxine, triiodothyronine)
Peptides	Oxytocin, antidiuretic hormone, calcitonin, parathyroid hormone
Proteins	Growth hormone (somatotropin), insulin, prolactin, follicle-stimulating hormone, luteinizing hormone

STEROID HORMONES INTERACT WITH CELL DNA

Cells in the adrenal glands and in the primary reproductive organs—ovaries and testes—all make steroid hormones. Estrogen made in the ovaries and testosterone made in the testes are familiar examples. Thyroid hormones and vitamin D differ chemically from steroid hormones but behave like them.

Figure 15.2*a* illustrates how a steroid hormone may act. Being lipid-soluble, it may diffuse directly across the lipid bilayer of a target cell's plasma membrane. Once inside the cytoplasm, the hormone molecule usually moves into the nucleus and binds to a receptor. In some cases it binds to a receptor in the cytoplasm, and then the hormone–receptor complex enters the nucleus. There the complex interacts with a particular gene—a segment of the cell's DNA. Genes carry the instructions for making proteins. By turning genes on or off, steroid hormones turn protein synthesis on or off. This change in a target cell's activity is the response to the hormone signal.

Some steroid hormones act in another way. They bind receptors on cell membranes and change the membrane properties in ways that affect the target cell's function.

NONSTEROID HORMONES ACT INDIRECTLY, BY WAY OF SECOND MESSENGERS

Nonsteroid hormones—the amine, peptide, and protein hormones—do not enter a target cell. Their chemical makeup makes them water-soluble, so they can't cross a target cell's lipid-rich plasma membrane. Instead, when this type of hormone binds to receptors in the plasma membrane, the binding sets in motion a series of reactions that activate enzymes. These reactions lead to the target cell's response.

For instance, consider a liver cell that has receptors for glucagon, a peptide hormone. As sketched in Figure 15.2*b*, this type of receptor spans the plasma membrane and extends into the cytoplasm. When a receptor binds glucagon, the cell produces a **second messenger**, a small molecule in the cytoplasm that relays signals from hormone–receptor complexes into the cell. A molecule called **cyclic AMP** (cyclic adenosine monophosphate) is this messenger. (The hormone itself is the "first messenger.")

An activated enzyme (adenylate cyclase) launches a cascade of reactions by converting ATP to cyclic AMP. Molecules of cyclic AMP are signals for the cell to activate molecules of another enzyme called a protein kinase. These act on other enzymes, and so forth, until a final

HOW HORMONES WORK

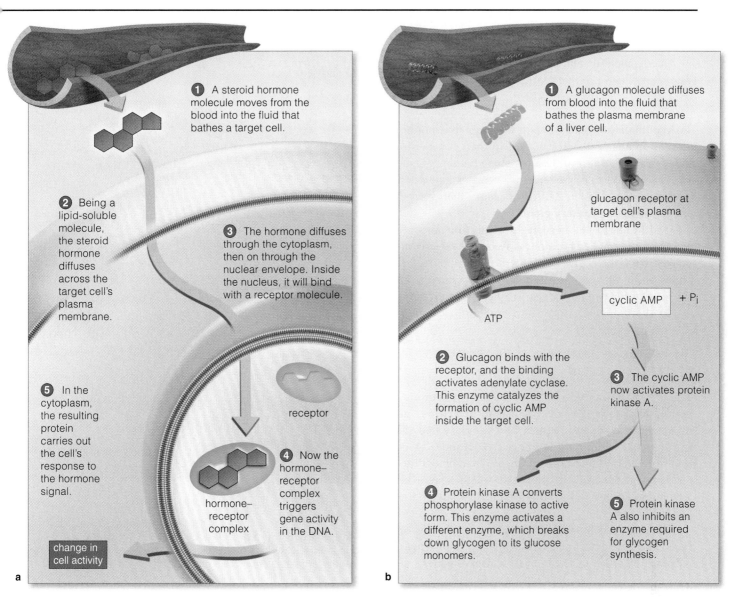

Figure 15.2 (**a**) Example of a mechanism by which a steroid hormone initiates changes in a target cell's activities. (**b**) Example of how a peptide hormone initiates changes in the activity of a target cell. Here, glucagon binds to a receptor and triggers reactions inside the cell. Cyclic AMP, a type of second messenger, relays the signal to the cell's interior.

reaction converts stored glycogen in the cell to glucose. Soon a huge number of molecules are taking part in the final response to the glucagon–receptor complex.

Epinephrine is another amine hormone. Like glucagon, it combines with specific receptors at the surface of the target cell. Binding triggers the release of cyclic AMP as a second messenger that assists in the target cell response.

A slightly different example is a muscle cell that has receptors for insulin, a protein hormone. Among other things, when insulin binds to the receptor, the complex stimulates transporter proteins to insert themselves into the plasma membrane so that the cell can take up glucose faster. The signal also activates enzymes that catalyze reactions allowing the cell to store glucose.

Hormones interact with receptors at the plasma membrane or in the cytoplasm of target cells. Ultimately, the hormone influences protein synthesis in a target cell.

Steroid hormones interact with a target cell's DNA after entering the nucleus or after binding a receptor in the cell's cytoplasm. Some steroid hormones act by altering properties of the plasma membrane itself.

With other types of hormones, the signal for change in a cell's activity comes when the hormone binds to membrane receptors and an enzyme system is activated. Often a second messenger relays the signal to the cell's interior, where the full response unfolds.

15.3 The Hypothalamus and Pituitary Gland: Major Controllers

LINKS TO SECTIONS 5.1, 9.5, AND 13.7

The hypothalamus and pituitary gland interact as a major brain center that controls activities of other organs, many of which also have endocrine functions.

The **hypothalamus** is a part of the forebrain that monitors internal organs and states related to their functioning, such as eating. It also influences some behaviors, such as sexual behavior. It has secretory neurons that extend down into the slender stalk to its base, then into the lobed, pea-sized **pituitary gland**.

In addition to its nervous system functions, the hypothalamus makes hormones. Two of these are later secreted from the pituitary's *posterior* lobe. Others have targets in the *anterior* lobe of the pituitary, which makes and secretes its own hormones. Most of these govern the activity of other endocrine glands (Table 15.2).

THE POSTERIOR PITUITARY LOBE PRODUCES ADH AND OXYTOCIN

Figure 15.3 shows the cell bodies of certain neurons in the hypothalamus. Their axons extend downward into the posterior lobe, ending next to a capillary bed. The neurons make antidiuretic hormone (ADH) and oxytocin, then store them in the axon endings. When one of these hormones is released, it diffuses through tissue fluid and enters capillaries, then travels the bloodstream to its targets.

ADH acts on cells of kidney nephrons and collecting ducts. Recall from Chapter 12 that ADH promotes water reabsorption when the body must conserve water.

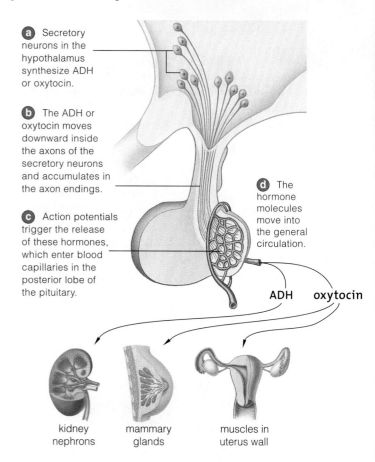

a Secretory neurons in the hypothalamus synthesize ADH or oxytocin.

b The ADH or oxytocin moves downward inside the axons of the secretory neurons and accumulates in the axon endings.

c Action potentials trigger the release of these hormones, which enter blood capillaries in the posterior lobe of the pituitary.

d The hormone molecules move into the general circulation.

ADH   oxytocin

kidney nephrons   mammary glands   muscles in uterus wall

Figure 15.3 *Animated!* Links between the hypothalamus and the posterior lobe of the pituitary. Also shown are main targets of the posterior lobe's hormones.

Pituitary Lobe	Secretions	Designation	Main Targets	Primary Actions
POSTERIOR Nervous tissue (extension of hypothalamus)	Antidiuretic hormone	ADH	Kidneys	Induces water conservation required in control of extracellular fluid volume (and, indirectly, solute concentrations)
	Oxytocin	OT	Mammary glands	Induces milk movement into secretory ducts
			Uterus	Induces uterine contractions
ANTERIOR Mostly glandular tissue	Corticotropin	ACTH	Adrenal cortex	Stimulates release of adrenal steroid hormones
	Thyrotropin	TSH	Thyroid gland	Stimulates release of thyroid hormones
	Gonadotropins: Follicle-stimulating hormone	FSH	Ovaries, testes	In females, stimulates egg formation; in males, helps stimulate sperm formation
	Luteinizing hormone	LH	Ovaries, testes	In females, stimulates ovulation, corpus luteum formation; in males, promotes testosterone secretion, sperm release
	Prolactin	PRL	Mammary glands	Stimulates and sustains milk production
	Growth hormone (also called somatotropin)	GH (STH)	Most cells	Promotes growth in young; induces protein synthesis, cell division; roles in glucose, protein metabolism in adults

Table 15.2   Summary of Hormones Released from the Pituitary Gland

THE ENDOCRINE SYSTEM

Through the gamut of events that can cause a drop in blood pressure—from water lost in sweat to severe blood loss from an injury—the hypothalamus monitors the shifts and releases ADH into the bloodstream when blood pressure falls below a set point. ADH causes the arterioles in some tissues to constrict, and so systemic blood pressure rises. (For this reason, ADH is sometimes called vasopressin.)

Oxytocin has roles in reproduction in both males and females. For example, in pregnant women it triggers muscle contractions in the uterus during labor and causes milk to be released when a mother nurses her infant. New studies suggest that oxytocin is a "cuddle hormone" that helps to stimulate affectionate behavior. In sexually active people, both male and female, it apparently is a chemical trigger for feelings of satisfaction after sexual contact.

THE ANTERIOR PITUITARY LOBE PRODUCES SIX OTHER HORMONES

Inside the pituitary stalk, a capillary bed picks up hormones secreted by the hypothalamus and delivers them to another capillary bed in the anterior lobe. There the hormones leave the bloodstream and act on various target cells. As Figure 15.4 shows, those anterior pituitary cells produce and secrete six hormones:

Corticotropin	ACTH
Thyrotropin	TSH
Follicle-stimulating hormone	FSH
Luteinizing hormone	LH
Prolactin	PRL
Growth hormone (somatotropin)	GH (or STH)

All these hormones have widespread effects. ACTH and TSH orchestrate secretions from the adrenal glands and thyroid gland, respectively. FSH and LH both influence reproduction, as you'll read in Chapter 16. Prolactin is best known for its role in stimulating and sustaining the production of breast milk, after other hormones have primed the tissues. There also is evidence that it promotes the synthesis of the male sex hormone testosterone.

Growth hormone (GH) affects most body tissues. It stimulates protein synthesis and cell division and has a major influence on growth, especially of cartilage and bone. You may recall from Chapter 5 that GH sustains the epiphyseal plates at the ends of growing long bones. GH is equally important as a "metabolic hormone." It stimulates cells to take up amino acids and promotes the breakdown and release of fat stored in adipose tissues, making more fatty acids available to cells. GH also adjusts the rate at which cells take up glucose. In this way it helps to maintain proper blood levels of that cellular fuel.

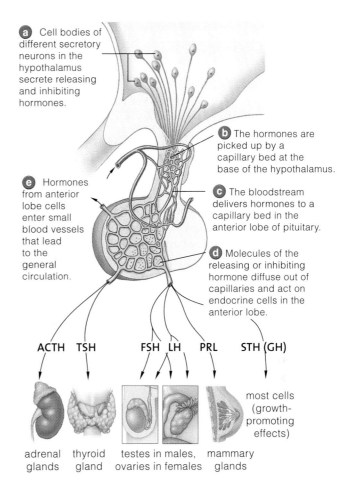

a Cell bodies of different secretory neurons in the hypothalamus secrete releasing and inhibiting hormones.

b The hormones are picked up by a capillary bed at the base of the hypothalamus.

e Hormones from anterior lobe cells enter small blood vessels that lead to the general circulation.

c The bloodstream delivers hormones to a capillary bed in the anterior lobe of pituitary.

d Molecules of the releasing or inhibiting hormone diffuse out of capillaries and act on endocrine cells in the anterior lobe.

ACTH TSH   FSH LH   PRL   STH (GH)

most cells (growth-promoting effects)

adrenal glands   thyroid gland   testes in males, ovaries in females   mammary glands

Figure 15.4 *Animated!* Links between the hypothalamus and the anterior lobe of the pituitary. Also shown are main targets of the anterior lobe's secretions.

Most hypothalamic hormones that act in the anterior pituitary are **releasers** that stimulate target cells to secrete other hormones. For example, GnRH (gonadotropin-releasing hormone) triggers the secretion of FSH and LH, both of which are classified as gonadotropins. Likewise, TRH (thyrotropin-releasing hormone) stimulates the secretion of TSH. Other hypothalamic hormones are **inhibitors**. They *block* secretions from their targets in the anterior pituitary. One of them, somatostatin, inhibits the secretion of growth hormone and thyrotropin.

The hypothalamus and pituitary gland interact to control the secretion of numerous hormones. The pituitary's posterior lobe stores and secretes ADH and oxytocin, both of which target specific types of cells.

The anterior lobe of the pituitary produces and secretes six hormones: ACTH, TSH, FSH, LH, PRL, and GH. These trigger the release of other hormones from other endocrine glands, with wide-ranging effects throughout the body.

15.4 Factors That Influence Hormone Effects

The effects of a hormone will depend on a variety of factors, including how well controls operate, interactions with other hormones, and the status of target cells.

PROBLEMS WITH CONTROL MECHANISMS CAN RESULT IN SKEWED HORMONE SIGNALS

The body does not churn out enormous numbers of hormone molecules. Two researchers, Roger Guilleman and Andrew Schally, realized this when they isolated TRH, the first-known releasing hormone. In four years of work, they dissected 500 tons of brains and 7 metric tons of hypothalamic tissue from sheep and ended up with only a pinhead-sized amount of TRH.

In general, endocrine glands release small quantities of hormones in short bursts. Controls over the frequency of hormone release prevent hormones from being either overproduced or underproduced. If something interferes with the controls, the body's form and functioning may be altered in abnormal ways.

For example, as we noted in Section 15.3, the pituitary gland releases growth hormone. Because this hormone has such widespread effects on bodily growth, if the pituitary secretes too much or too little of it the impact can be profound. For instance, **gigantism** results when growth hormone is overproduced during childhood. Affected adults are proportionally like an average-sized person but much larger (Figure 15.5a). **Pituitary dwarfism** results when the pituitary makes too little growth hormone. Affected adults are proportionally similar to an average person but much smaller.

What happens if too much growth hormone is secreted during adulthood, when long bones such as the femur no longer can lengthen? Bone, cartilage, and other connective tissues in the hands, feet, and jaws thicken abnormally. So do epithelia of the skin, nose, eyelids, lips, and tongue. The result is **acromegaly** (Figure 15.5b).

As another example, ADH secretion may fall or stop if the pituitary's posterior lobe is damaged, as by a blow to the head. This is one cause of **diabetes insipidus**. A person with this disease excretes a great deal of dilute urine—so much that serious dehydration can occur. Hormone replacement therapy is an effective treatment for this form of diabetes.

a

b

Figure 15.5 (**a**) A child affected by pituitary gigantism. This boy is twelve years old and six feet, five inches tall. His mother is standing beside him.

(**b**) Acromegaly, resulting from excessive production of GH during adulthood. Before this person reached maturity, she was symptom-free.

THE ENDOCRINE SYSTEM

HORMONE INTERACTIONS, FEEDBACK, AND OTHER FACTORS ALSO INFLUENCE A HORMONE'S EFFECTS

At least four factors can influence the effects of a given hormone. To begin with, hormones often interact with one another, as noted in Section 15.1. Also, negative feedback mechanisms usually control the secretion of hormones. When the concentration of a hormone rises or falls in some body region, the change triggers events that either slow or boost further secretion. Target cells also may react differently to a hormone at different times, depending on the hormone's concentration and on how the cell's hormone receptors are functioning at that moment. Finally, environmental cues, such as light, can trigger or prevent the release of a hormone. Table 15.3 lists hormones from sources other than the pituitary gland. The rest of this chapter will provide examples of their effects and how they are controlled.

Normally, the release of a hormone is tightly regulated. These controls and a hormone's effects are influenced by hormone interactions, feedback mechanisms, variations in the state of target cells, and sometimes environmental cues.

Table 15.3	Hormone Sources Other Than the Hypothalamus and Pituitary		
Source	**Secretion(s)**	**Main Targets**	**Primary Actions**
PANCREATIC ISLETS	Insulin	Muscle, adipose tissue	Lowers blood-sugar level
	Glucagon	Liver	Raises blood-sugar level
	Somatostatin	Insulin-secreting cells	Influences carbohydrate metabolism
ADRENAL CORTEX	Glucocorticoids (including cortisol)	Most cells	Promote protein breakdown and conversion to glucose
	Mineralocorticoids (including aldosterone)	Kidney	Promote sodium reabsorption; control salt–water balance
ADRENAL MEDULLA	Epinephrine (adrenalin)	Liver, muscle, adipose tissue	Raises blood level of sugar, fatty acids; increases heart rate, force of contraction
	Norepinephrine	Smooth muscle of blood vessels	Promotes constriction or dilation of blood vessel diameter
THYROID	Triiodothyronine, thyroxine	Most cells	Regulate metabolism; have roles in growth, development
	Calcitonin	Bone	Lowers calcium levels in blood
PARATHYROIDS	Parathyroid hormone	Bone, kidney	Elevates levels of calcium and phosphate ions in blood
THYMUS	Thymosins, etc.	Lymphocytes	Have roles in immune responses
GONADS:			
Testes (in males)	Androgens (including testosterone)	General	Required in sperm formation, development of genitals, maintenance of sexual traits; influence growth, development
Ovaries (in females)	Estrogens	General	Required in egg maturation and release; prepare uterine lining for pregnancy; required in development of genitals, maintenance of sexual traits; influence growth, development
	Progesterone	Uterus, breasts	Prepares, maintains uterine lining for pregnancy; stimulates breast development
PINEAL	Melatonin	Hypothalamus	Influences daily biorhythms
ENDOCRINE CELLS OF STOMACH, GUT	Gastrin, secretin, etc.	Stomach, pancreas, gallbladder	Stimulate activity of stomach, pancreas, liver, gallbladder
LIVER	IGFs (Insulin-like growth factors)	Most cells	Stimulate cell growth and development
KIDNEYS	Erythropoietin	Bone marrow	Stimulates red blood cell production
	Angiotensin*	Adrenal cortex, arterioles	Helps control blood pressure, aldosterone secretion
	Vitamin D3*	Bone, gut	Enhances calcium resorption and uptake
HEART	Atrial natriuretic hormone	Kidney, blood vessels	Increases sodium excretion; lowers blood pressure

*These hormones are not produced in the kidneys but are formed when enzymes produced in kidneys activate specific substances in the blood.

LINKS TO
SECTIONS
5.1 AND 10.3

15.5 The Thymus, Thyroid, and Parathyroid Glands

Thymus gland hormones stimulate T cells to mature, while hormones from the thyroid gland are vital to normal development and metabolism. Parathyroid hormone helps regulate calcium levels in the blood.

THYMUS GLAND HORMONES AID IMMUNITY

The thymus gland lies beneath the upper part of the breastbone or sternum (see Figure 15.1). The hormones it releases, called thymosins, help infection-fighting T cells mature. The thymus is quite large in children but it shrinks to a relatively small size as an adolescent matures into adulthood.

THYROID HORMONES AFFECT METABOLISM, GROWTH, AND DEVELOPMENT

The **thyroid gland** is located at the base of the neck in front of the trachea, or windpipe (Figure 15.6). Its main secretions, thyroxine (T_4) and triiodothyronine (T_3), are known jointly as *thyroid hormone* (TH). TH affects the body's overall metabolic rate, growth, and development. The thyroid also makes calcitonin. This hormone helps lower the level of calcium (and of phosphate) in blood. Proper feedback control is essential to the functioning of both these hormones.

For instance, thyroid hormones cannot be synthesized without iodide, a form of iodine. Iodine-deficient diets cause one or both lobes of the thyroid gland to enlarge (Figure 15.7). The enlargement, a **simple goiter**, occurs after low blood levels of thyroid hormones cause the anterior pituitary to secrete TSH (the thyroid-stimulating hormone thyrotropin). The thyroid attempts to make the hormones but cannot do so, which leads to continued secretion of TSH, and so on, in a sustained, but abnormal feedback loop. *Hypothyroidism* is the clinical name for low blood levels of thyroid hormones. Affected adults tend to be overweight, sluggish, intolerant of cold, confused, and depressed. Simple goiter is no longer common in places where people use iodized salt.

Graves disease and other so-called toxic goiters are the result of *hyperthyroidism*, or excess thyroid hormones in the blood. Symptoms include increased heart rate and blood pressure and unusually heavy sweating. Some cases are autoimmune disorders, in which antibodies wrongly stimulate thyroid cells. In other cases the cause can be traced to inflammation or a tumor in the thyroid gland. Some people are genetically predisposed to the disorder.

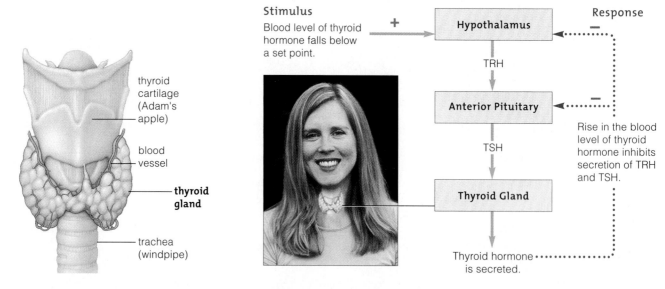

Figure 15.6 *Animated!* The feedback loop that controls the secretion of thyroid hormone.

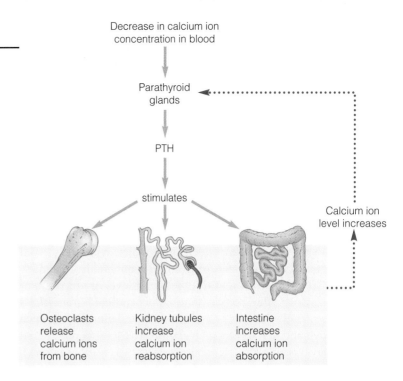
Decrease in calcium ion
concentration in blood

Parathyroid
glands

PTH

stimulates

| Osteoclasts release calcium ions from bone | Kidney tubules increase calcium ion reabsorption | Intestine increases calcium ion absorption |

Calcium ion
level increases

Figure 15.8 How PTH regulates calcium homeostasis.

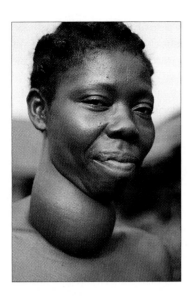

Figure 15.7 A case of goiter caused by a diet low in the micronutrient iodine.

PTH FROM THE PARATHYROIDS IS THE MAIN CALCIUM REGULATOR

We humans have four **parathyroid glands** located on the back of the thyroid gland, as shown in the diagram here. These little glands secrete parathyroid hormone (PTH), the main regulator of the calcium level in blood. Calcium is important for muscle contraction as well as

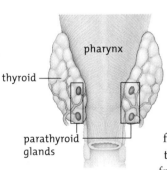

for the activation of enzymes, the formation of bone, blood clotting, and other tasks. The parathyroids secrete more PTH when the blood level of calcium falls below a set point, and they reduce their secretions when the calcium level rises. The hormone calcitonin from the thyroid gland contributes to processes that remove calcium from the blood.

You may recall that Section 5.1 discussed bone remodeling, the process in which bone is deposited or broken down, depending on the level of calcium in the blood. PTH is the hormone in charge of remodeling, and it acts on the skeleton and kidneys. When the blood level of calcium falls below a set point, PTH prompts the bone cells called osteoclasts to secrete enzymes that digest bone tissue (Figure 15.8). This process releases calcium ions (and phosphate) that can be used elsewhere in the body. In the kidneys, PTH also stimulates the

reabsorption of calcium from the filtrate flowing through nephrons. At the same time, PTH helps to activate vitamin D. The activated form, which is also a hormone, improves the absorption of calcium from food in the GI tract.

In children who have vitamin D deficiency, too little calcium and phosphorus are absorbed, so the rapidly growing bones do not develop properly. Children who have the resulting bone disorder, **rickets**, develop bowed legs and other skeletal abnormalities.

Calcium has so many essential roles in the body that disorders related to parathyroid functioning can be quite serious. For example, excess PTH (*hyperparathyroidism*) causes so much calcium to be withdrawn from a person's bones that the bone tissue is dangerously weakened. The excess calcium in the bloodstream may cause kidney stones, and muscles don't function normally. The central nervous system's operations may be so disrupted that the affected person dies.

Thymosins produced in the thymus help T cells mature.

Thyroid hormones affect metabolism, growth, and development. Negative feedback loops control the output of these and other endocrine glands.

Parathyroid glands release PTH, the main regulator of calcium levels in blood.

15.6 Adrenal Glands and Stress Responses

LINKS TO
SECTIONS
7.8, 10.4, 12.6,
AND 13.6

Different parts of the adrenal glands secrete hormones that help regulate blood levels of glucose, influence blood pressure, and regulate blood circulation, among other roles.

THE ADRENAL CORTEX PRODUCES GLUCOCORTICOIDS AND MINERALOCORTICOIDS

We have two adrenal glands, one on top of each kidney. The outer part of each gland is the **adrenal cortex** (Figure 15.9). There, cells secrete two major types of steroid hormones, the glucocorticoids and mineralocorticoids.

Glucocorticoids raise the blood level of glucose. For instance, the body's main glucocorticoid, *cortisol*, is secreted when the body is stressed and glucose is in such demand that its blood level drops to a low set point. That level is an alarm signal; it starts a stress response, which a negative feedback mechanism later cuts off. Among other effects, it promotes protein breakdown in muscle and stimulates the liver to take up amino acids, from which liver cells synthesize glucose in a process called **gluconeogenesis**. Cortisol also reduces how much glucose tissues such as skeletal muscle take up from the blood. This effect is sometimes called "glucose sparing." Glucose sparing is extremely important in homeostasis, for it helps ensure that the blood will carry enough glucose to meet the needs of the brain, which generally cannot use other molecules for fuel. Cortisol also promotes the breakdown of fats and the use of the resulting fatty acids for energy.

Figure 15.9 diagrams how negative feedback governs the release of cortisol. In this example, when the blood level of cortisol rises above a set point, the hypothalamus begins to secrete less of the releasing hormone CRH. The anterior pituitary responds by secreting less ACTH, and the adrenal cortex secretes less cortisol. In a healthy person, daily cortisol secretion is highest when the blood glucose level is lowest, usually in the early morning. Chronic severe *hypoglycemia*, an ongoing low glucose concentration in the blood, can develop when the adrenal cortex makes too little cortisol. Then, the mechanisms that spare glucose and generate new supplies in the liver do not operate properly.

Glucocorticoids also suppress inflammation. The adrenal cortex increases its secretion of these chemicals when a person experiences unusual physical stress such as a painful injury, severe illness, or an allergic reaction. The ample supply of cortisol and other signaling molecules helps speed recovery. That is why doctors prescribe cortisol-like drugs, such as cortisone, for patients with asthma and serious inflammatory disorders. Cortisone is the active ingredient in many over-the-counter products for treating skin irritations.

Long-term use of large doses of glucocorticoids has serious side effects, including suppressing the immune system. Long-term stress has the same effect, as you'll read shortly.

For the most part, **mineralocorticoids** adjust the concentrations of mineral salts, such as potassium and sodium, in the extracellular fluid. The most abundant mineralocorticoid is aldosterone. You may recall from Section 12.6 that aldosterone acts on the distal tubules of kidney nephrons, stimulating them to reabsorb sodium ions and excrete potassium ions. The reabsorption of sodium in turn promotes reabsorption of water from the tubules as urine is forming. A variety of circumstances, such as falling blood pressure or falling blood levels of sodium (falling blood volume as water moves out by osmosis), can trigger the release of aldosterone.

In a developing fetus and early in puberty, the adrenal cortex also secretes large amounts of sex hormones. The main ones are androgens (male sex hormones). Female sex hormones (estrogens and progesterone) are also produced. In adults, however, the reproductive organs generate most sex hormones.

HORMONES FROM THE ADRENAL MEDULLA HELP REGULATE BLOOD CIRCULATION

The **adrenal medulla** is the inner part of the adrenal gland shown in Figure 15.9. It contains neurons that release two substances, epinephrine and norepinephrine. Both act as neurotransmitters when they are secreted by neurons elsewhere in the body. When the adrenal medulla secretes them, however, their hormonelike effects help regulate blood circulation and carbohydrate use when the body is stressed or excited. For example, they increase the heart rate, dilate arterioles in some areas and constrict them in others, and dilate bronchioles. Thus the heart beats faster and harder, more blood is shunted to heart and muscle cells from other regions, and more oxygen flows to energy-demanding cells throughout the body. These are aspects of the fight–flight response noted in Chapter 13.

The operation of the adrenal medulla provides another example of negative feedback control. For example, when the hypothalamus sends the necessary signal (by way of sympathetic nerves) to the adrenal medulla, the neuron axons will start to release norepinephrine into the synapse between the axon endings and the target cells. Soon, norepinephrine molecules collect in the synapse, setting the stage for a localized negative feedback mechanism. As the accumulating norepinephrine binds to receptors on the axon endings, the release of norepinephrine shuts down in short order.

THE ENDOCRINE SYSTEM

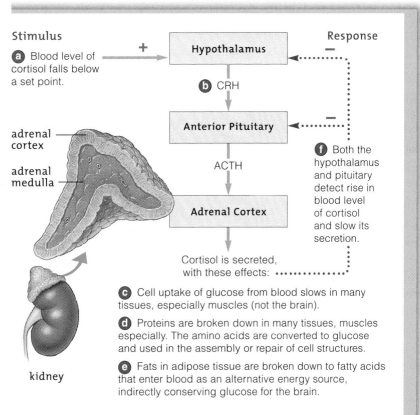

Stimulus

a Blood level of cortisol falls below a set point.

Response

+ → **Hypothalamus** ← **−**

b CRH

adrenal cortex

adrenal medulla

Anterior Pituitary ← **−**

ACTH

f Both the hypothalamus and pituitary detect rise in blood level of cortisol and slow its secretion.

Adrenal Cortex

Cortisol is secreted, with these effects: •••••••••••

c Cell uptake of glucose from blood slows in many tissues, especially muscles (not the brain).

d Proteins are broken down in many tissues, muscles especially. The amino acids are converted to glucose and used in the assembly or repair of cell structures.

e Fats in adipose tissue are broken down to fatty acids that enter blood as an alternative energy source, indirectly conserving glucose for the brain.

kidney

Figure 15.9 *Animated!* Structure of the adrenal gland. One gland rests atop each kidney. The diagram shows a negative feedback loop that governs the secretion of cortisol.

LONG-TERM STRESS CAN DAMAGE HEALTH

As you've just read, when the body is stressed, nervous system commands trigger the fight–flight response and the release of cortisol, epinephrine, and norepinephrine. In daily life, most people also encounter a wide variety of psychosocial stressors—an exam, financial difficulties, a new job or romance, and the like. As you can see from this short list, some stressors are positive, others are negative. Not everyone reacts the same way to life's challenges, but there is ample evidence that being routinely "stressed out" by negative stressors may contribute to hypertension and related cardiovascular disease. And because cortisol suppresses the immune system, people who experience a lot of "bad" stress also may be more susceptible to disease. Chronic negative stress also is linked to insomnia, anxiety, and depression.

Fortunately, research also shows that social connections seem to moderate the effects of stress, as does regular physical exercise. Friends, family, support groups, and counselors can not only make you feel better, they may make you healthier as well.

The adrenal cortex secretes glucocorticoids such as cortisol and mineralocorticoids such as aldosterone.

Cortisol raises blood glucose levels and suppresses inflammation. Aldosterone helps regulate blood pressure by adjusting the reabsorption of potassium and sodium in the kidneys.

The adrenal medulla makes two hormones, epinephrine and norepinephrine, that adjust blood circulation and the use of blood glucose in the fight–flight response to stress.

15.7 The Pancreas: Regulating Blood Sugar

LINK TO
SECTION
7.5

Insulin and glucagon from the pancreas are examples of hormones that work antagonistically, the action of one opposing the action of the other. Controls over when these two hormones are secreted maintain the glucose level in blood.

The pancreas is a gland with both exocrine *and* endocrine functions. As described in Chapter 7, its exocrine cells release digestive enzymes into the small intestine. The endocrine cells of the pancreas are located in roughly 2 million scattered clusters. Each cluster, a **pancreatic islet**, contains three types of hormone-secreting cells:

1. *Alpha cells* secrete *glucagon*. Between meals, cells use the glucose delivered to them by the bloodstream. When the blood glucose level decreases below a set point, secreted glucagon acts on cells in the liver and muscles. It causes glycogen (a storage polysaccharide) and amino acids to be converted to glucose. In this way glucagon raises the glucose level in the blood.

2. *Beta cells* secrete the hormone *insulin*. After meals, when a lot of glucose is circulating in the blood, insulin stimulates muscle and adipose cells to take up glucose. It also promotes synthesis of fats, glycogen, and to a lesser extent, proteins, and inhibits the conversion of proteins to glucose. In this way insulin lowers the glucose level in the blood.

3. *Delta cells* secrete *somatostatin*. This hormone acts on beta cells and alpha cells to inhibit secretion of insulin and glucagon, respectively. Somatostatin is part of several hormone-based control systems. For example, it is released from the hypothalamus to block secretion of growth hormone; it is also secreted by cells of the GI tract, where it acts to inhibit the secretion of various substances involved in digestion.

Figure 15.10 shows how pancreatic hormones help keep blood glucose levels fairly constant, even with all the variations in when and how much we eat. When the body can't produce enough insulin or when target cells can't respond to it, the body does not store glucose in the normal way. Then, body mechanisms for metabolizing carbohydrates, proteins, and fats are disrupted.

Alpha cells of the pancreatic islets secrete glucagon when the blood level of glucose falls below a set point. Islet beta cells secrete insulin when blood levels of glucose rise above the set point.

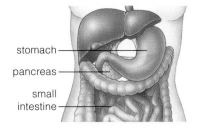

stomach
pancreas
small intestine

Figure 15.10 *Animated!* How cells that secrete insulin and glucagon respond to a change in the level of glucose in blood. These two hormones work antagonistically to maintain the glucose level in its normal range.

(**a**) *After* a meal, glucose enters blood faster than cells can take it up and its level in blood increases. In the pancreas, the increase (**b**) stops alpha cells from secreting glucagon and (**c**) stimulates beta cells to secrete insulin. In response to insulin, (**d**) adipose and muscle cells take up and store glucose, and cells in the liver synthesize more glycogen. As a result, insulin *lowers* the blood level of glucose (**e**).

(**f**) *Between* meals, the glucose level in blood falls. The decrease (**g**) stimulates alpha cells to secrete glucagon and (**h**) slows the insulin secretion by beta cells. (**i**) In the liver, glucagon causes cells to convert glycogen back to glucose, which enters the blood. As a result, glucagon *raises* the blood level of glucose (**j**).

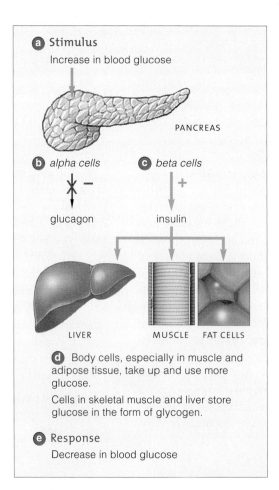

a Stimulus
Increase in blood glucose

PANCREAS

b *alpha cells*

glucagon

c *beta cells*

insulin

LIVER   MUSCLE   FAT CELLS

d Body cells, especially in muscle and adipose tissue, take up and use more glucose.

Cells in skeletal muscle and liver store glucose in the form of glycogen.

e Response
Decrease in blood glucose

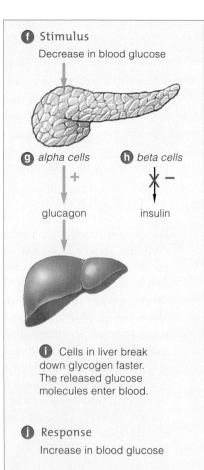

f Stimulus
Decrease in blood glucose

g *alpha cells*

glucagon

h *beta cells*

insulin

i Cells in liver break down glycogen faster. The released glucose molecules enter blood.

j Response
Increase in blood glucose

THE ENDOCRINE SYSTEM

15.8 Disorders of Glucose Homeostasis

Chronically having too much or too little glucose in your blood is a serious, even life-threatening condition.

Too little insulin can lead to **diabetes mellitus**. Because target cells can't take up glucose from blood, glucose builds up in the blood (*mellitus* means "honey" in Greek). The kidneys move excess glucose into the urine, water is also lost, and the body's water–solute balance is upset. Affected people become dehydrated and extremely thirsty. They also lose weight as their glucose-starved cells break down protein and fats for energy. Fat breakdown releases ketones, so these acids build up in the blood and urine. This leads to excess water loss, which in turn can lead to dangerously low blood pressure and heart failure. Another outcome is **metabolic acidosis**, a lower than optimal pH in blood that can disrupt brain function.

In **type 1 diabetes**, an autoimmune response destroys beta cells in the pancreas. Only about 1 in 10 diabetics is type 1, the more immediately dangerous of the two types of diabetes. It may be caused by a viral infection in combination with genetic susceptibility. Symptoms tend to appear early in life so type 1 diabetes also is called "juvenile-onset diabetes." Affected people survive with insulin injections, but their life span may be shortened by associated cardiovascular problems.

TYPE 2 DIABETES IS A GLOBAL HEALTH CRISIS

In **type 2 diabetes**, insulin levels are close to or above normal, but for any of several reasons target cells can't respond properly to the hormone. The beta cells break down and produce less and less insulin. According to the World Health Organization, in the United States and other developed countries type 2 diabetes has reached crisis proportions, right along with its major risk factor—obesity.

Although we don't know exactly why, blood loaded with sugar damages capillaries. Over time, the blood supply to the kidneys, eyes, lower legs, and feet may be so poor that cells and tissues die. This fact explains some of the awful complications of diabetes (Table 15.4), including amputations due to gangrene in blood-starved tissues.

Diabetes also correlates strongly with cardiovascular disease. Even diabetics in their 20s and 30s are at high risk of suffering a stroke or heart attack.

METABOLIC SYNDROME IS A WARNING SIGN

As many as 20 million Americans unknowingly have "prediabetes"—slightly elevated blood sugar that increases the risk of developing type 2 diabetes. Useful diagnostic tools are urinalysis that looks for signs of elevated glucose, and a fasting glucose test that measures the baseline

Figure 15.11 A diabetic checks his blood glucose by placing a blood sample into a glucometer. Compared with Caucasians, Hispanics and African Americans are about 1.5 times more likely to be diabetic. Native Americans and Asians are at even greater risk.

amount of glucose in the blood several hours after a person has eaten.

An early indicator that someone may be at risk for diabetes is a constellation of features that are lumped together as **metabolic syndrome**. These features are:

- A waist measuring more than 35 inches for males and more than 40 inches for females. This characteristic is sometimes called being "apple shaped."

- Blood pressure of 130/85 mm Hg or higher

- Low levels (under 40 mg/dL for males, 50 mg/dL for females) of HDL, the "good" cholesterol

- Fasting glucose of 110 mg/dL or higher

- Fasting triglyceride level of 150 mg/dL or higher

Type 2 diabetes can be controlled by a combination of proper diet, regular exercise, and sometimes by drugs that improve insulin secretion or activity. Many people with diabetes become expert at monitoring their blood glucose and adjusting their meals to match (Figure 15.11).

Table 15.4	Some Complications of Diabetes
Eyes	Changes in lens shape and vision; damage to blood vessels in retina; blindness
Skin	Increased susceptibility to bacterial and fungal infections; patches of discoloration; thickening of skin on the back of hands
Digestive system	Gum disease; delayed stomach emptying that causes heartburn, nausea, vomiting
Kidneys	Increased risk of kidney disease and failure
Heart and blood vessels	Increased risk of heart attack, stroke, high blood pressure, and atherosclerosis
Hands and feet	Impaired ability to sense pain; formation of calluses, foot ulcers; possible amputation of toes, a foot, or leg because of tissue death caused by poor circulation

15.9 Some Final Examples of Integration and Control

LINKS TO
SECTIONS
7.8, 9.2, 10.2,
AND 14.3

Endocrine cells in parts of the brain, the heart, and even the GI tract produce hormones. They and some other signaling molecules have major effects on a range of body functions.

LIGHT/DARK CYCLES INFLUENCE THE PINEAL GLAND, WHICH PRODUCES MELATONIN

Like other animals, humans are physiologically sensitive to light. Many ancient vertebrates had a light-sensitive "third eye" on top of the head. In humans a version of this photosensitive organ still exists, as a lump of tissue in the brain called the **pineal gland**. The pineal gland secretes the hormone *melatonin* into cerebrospinal fluid and the bloodstream. Melatonin influences sleep/wake cycles, and possibly some aspects of human reproduction.

Melatonin is secreted in the dark, so the amount in the bloodstream varies from day to night. It also changes with the seasons, because winter days are shorter than summer days.

The human cycle of sleep and arousal is evidence of an internal **biological clock** that seems to tick in synchrony with day length. Melatonin seems to influence the clock, which can be disturbed by circumstances that alter a person's accustomed exposure to light and dark. Jet lag is an example; some air travelers use melatonin supplements to try to adjust their sleep/wake cycles more quickly.

Seasonal affective disorder, or **SAD**, hits some people in winter. Symptoms typically include depression and an overwhelming desire to sleep. The problem may result from a biological clock that is out of sync with changes in day length during winter, when days are shorter and nights longer. The symptoms get worse when a person with SAD takes melatonin. On the other hand, symptoms improve dramatically when the person is exposed to intense light, which shuts down the pineal gland.

Some researchers have proposed that a drop in melatonin may help trigger the onset of puberty, when the reproductive organs and structures start to mature. For instance, if disease destroys the pineal gland, an affected child can enter puberty prematurely. There is still much more to learn about this possible link.

HORMONES ALSO ARE PRODUCED IN THE HEART AND GI TRACT

Beyond the endocrine glands we have been considering, hormones are also made and secreted by specialized cells elsewhere. For example, the two heart atria secrete *atrial natriuretic peptide*, or ANP. This hormone has a variety of effects, including helping to regulate blood pressure. When your blood pressure rises, ANP acts to inhibit the reabsorption of sodium ions—and hence water—in the kidneys. As a result, more water is excreted, the blood volume decreases, and blood pressure falls.

The digestive tract produces the hormones gastrin and secretin, among others. You may recall that gastrin stimulates the release of stomach acid when proteins are being digested. Secretin stimulates the pancreas to secrete bicarbonate.

Many body cells can detect changes in their chemical environment and alter their activity to match. The "local signaling molecules" they release act on the secreting cell itself or in its immediate vicinity. Cells quickly take up most of the molecules, so that only a few enter the general circulation. Prostaglandins and growth factors, our next topics, are examples.

PROSTAGLANDINS HAVE MANY EFFECTS

Biochemists have identified more than sixteen different kinds of the fatty acids called **prostaglandins**. In fact, the plasma membranes of most cells contain the precursors of prostaglandins. They are released continually, but the rate at which they are synthesized often increases in response to chemical changes in a particular area.

What are some known effects of prostaglandins? With signals from hormones (epinephrine and norepinephrine) as the trigger, at least two prostaglandins cause smooth muscle in the walls of blood vessels to constrict or dilate, helping to adjust blood flow in the area. Similar prostaglandin effects occur in the smooth muscle of the airways. Prostaglandins may aggravate inflammation and allergic responses to dust and pollen.

Prostaglandins have a major impact on reproductive organs. For example, many women experience painful cramping and heavy bleeding when they menstruate, both effects due to prostaglandins acting on smooth muscle of the uterus. (Drugs such as ibuprofen and aspirin block the synthesis of prostaglandins, relieving the discomfort.) Prostaglandins also influence the menstrual cycle in other ways, and along with oxytocin they stimulate contractions of the uterus during labor. We will look much more closely at some of these events in Chapter 17.

GROWTH FACTORS INFLUENCE CELL DIVISION

A doctor treating someone with major spinal cord damage, like the athletes in Figure 15.12, could one day have a powerful chemical arsenal at hand. Researchers are unraveling the workings of hormonelike proteins called **growth factors**—and working to convert this knowledge into treatments for specific kinds of tissue damage.

Figure 15.12 Growth factors that spur the growth of neurons may one day be part of the standard treatment for people who have severe spinal cord injuries. Already, experiments have shown that NGF can help severed nerves regrow.

Growth factors influence growth by regulating the rate at which certain cells divide. *Epidermal growth factor* (EGF) influences the growth of many cell types. So does *insulinlike growth factor* (IGF) made by the liver. *Nerve growth factor* (NGF) is another example. It promotes the survival and growth of neurons in developing embryos. NGF seems to define the direction in which the embryonic neurons grow: Target cells lay down a chemical path that leads the elongating axons to them.

The list of known growth factors is expanding rapidly. For instance, various ones enhance the rate at which wounds heal or stimulate the growth of blood vessels. This latter type might be harnessed to perform that function in cardiac muscle that has been damaged by a heart attack. Promising cancer research is exploring ways to block growth factors that spur the growth of blood vessels that supply malignant tumors.

PHEROMONES MAY BE IMPORTANT COMMUNICATION MOLECULES IN HUMANS

In discussing the sense of smell in Section 14.3, we touched on possible human pheromones. It's not surprising that humans may communicate via pheromones; some of our close primate relatives (such as rhesus monkeys), and bears, dogs, various insects, and many other animals make pheromones that act as territory markers, sex attractants, and other signals to members of their own species. Pheromones are released outside the individual and then travel (through the air, for instance) to reach another individual. The various hypotheses about human pheromones are still tentative, but research on this subject is continuing.

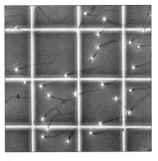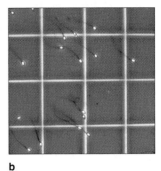

a   b

Figure 15.13 Normal and low sperm counts. The grid lines help clinicians count sperm. (a) Normally, there are about 113 million sperm per milliliter of semen per sample. (b) Low sperm count of 60 to 70 million sperm per milliliter. Human sperm counts have been declining.

ARE ENDOCRINE DISRUPTERS AT WORK?

The chapter introduction discussed the controversy over endocrine disrupters—substances in the environment that some biologists suspect are causing problems with the reproduction or development of various species. Between 1938 and 1990, according to one study, the sperm counts of males in Western countries declined about 40 percent. Approximately fifty studies since then support these findings: Sperm counts *have* gone down (Figure 15.13). Are environmental estrogens the culprits? Possibly. Several pesticides are estrogen mimics; they bind to estrogen receptors on cells. Cells of both males and females have these receptors, and males who work with the pesticide—and estrogen mimic— Kepone have lower than normal sperm counts. Recently a well-controlled study of geographic differences in the quality of semen showed that men from Columbia, Missouri, had fewer healthy sperm than men from New York, Minneapolis, or Los Angeles. The researchers suspect that exposure to some agricultural chemicals may play a role. More studies are now under way.

Endocrine cells in the pineal gland, heart, and GI tract all produce hormones. Other cells in various tissues produce local signaling molecules, such as prostaglandins, which act on the secreting cell itself or in the immediate vicinity.

Growth factors help regulate the rate at which target cells divide.

Summary

Section 15.1 Hormones and other signaling molecules help ensure that the activities of individual body cells mesh in ways that benefit the whole body. Hormones are produced by cells and glands of the endocrine system. They move through the bloodstream to distant target cells.

Other signaling molecules include neurotransmitters, local signaling molecules such as prostaglandins, and pheromones. All are chemical secretions by one cell that adjust the behavior of other, target cells. Any cell with receptors for the signal is the target.

Hormones may interact in opposition, synergistically (in cooperation), or permissively (a target cell must first be primed by one hormone in order to respond to a second one).

Section 15.2 Steroid and nonsteroid hormones exert their effects on target cells by different mechanisms.

Receptors for steroid (and thyroid) hormones are inside target cells. The hormone-receptor complex binds to the DNA, and binding activates genes and the synthesis of proteins.

Amine, peptide, and protein hormones interact with receptors on the plasma membrane of target cells. Responses to them are often mediated by a second messenger, such as cyclic AMP, inside the cell.

Most of these nonsteroid hormones alter the activity of target cell proteins. The resulting target cell responses help maintain homeostasis in extracellular fluid or contribute to normal development or reproductive functioning.

Section 15.3 The hypothalamus and pituitary gland interact to integrate many body activities.

ADH and oxytocin, two hypothalamic hormones, are stored in and released from the posterior lobe of the pituitary. ADH influences extracellular fluid volume. Oxytocin has roles in reproduction.

Six additional hypothalamic hormones are releasers or inhibitors. They control secretions of various cells of the pituitary's anterior lobe.

Of the six hormones produced in the anterior lobe, two (prolactin and growth hormone) have general effects on cells in a variety of body tissues. Four (ACTH, TSH, FSH, and LH) act on specific endocrine glands.

 Study how the hypothalamus and pituitary interact.

Section 15.4 Hormones are produced in small amounts, usually in short bursts. If too much or too little of a hormone is produced, the body's form and functioning may become abnormal.

Body responses to hormones may be influenced by hormone interactions, homeostatic feedback loops to the hypothalamus and pituitary, variations in hormone concentrations, and the number and kinds of receptors on a target cell.

Section 15.5 Thyroid hormone affects overall metabolism, growth, and development. The thyroid also makes calcitonin, which helps lower blood levels of calcium and phosphate. Parathyroid hormone is the main regulator of blood calcium levels.

Section 15.6 The adrenal cortex makes two kinds of steroid hormones, the glucocorticoids and mineralocorticoids. Glucocorticoids, mainly cortisol, raise the blood level of glucose and suppress inflammation. Mineralocorticoids adjust concentrations of minerals such as potassium and sodium in body fluids.

The adrenal medulla releases epinephrine and norepinephrine. Their hormonelike effects include the regulation of blood pressure and the metabolism of carbohydrates. (Certain neurons also release them as neurotransmitters.)

 See how negative feedback maintains cortisol levels.

Section 15.7 Blood levels of glucose are regulated by insulin and glucagon, which are secreted in the pancreatic islets by beta and alpha cells, respectively. Insulin stimulates muscle and adipose cells to take up glucose, while glucagon stimulates glucose-releasing reactions in muscle and the liver. Negative feedback governs both processes. Somatostatin released by islet delta cells can inhibit the release of insulin, glucagon, and some other hormones.

 See how the actions of insulin and glucagon regulate blood sugar.

Section 15.9 The pineal gland in the brain produces melatonin in response to light/dark cycles. Melatonin influences sleep/wake cycles as part of an internal biological clock related to day length.

The heart and GI tract secrete hormones that help adjust blood pressure and operate in digestion. Many cells produce prostaglandins in response to local chemical changes. Effects of prostaglandins range from adjusting blood flow in an area to promoting contraction of smooth muscle. Hormonelike growth factors regulate the rate at which cells divide.

Review Questions

1. Distinguish among hormones, neurotransmitters, local signaling molecules, and pheromones.

2. A hormone molecule binds to a receptor on a cell membrane. It doesn't enter the cell; rather, the binding activates a second messenger inside the cell that triggers an amplified response to the hormonal signal. Is the signaling molecule a steroid or a nonsteroid hormone?

3. Which hormones produced in the posterior and anterior lobes of the pituitary gland have the targets indicated? (*Fill in the blanks using the abbreviations noted in Section 15.3.*)

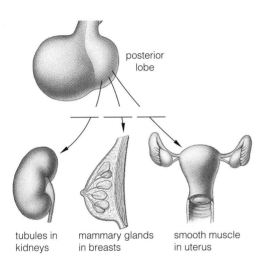

posterior lobe

tubules in kidneys   mammary glands in breasts   smooth muscle in uterus

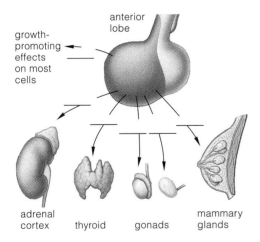

anterior lobe

growth-promoting effects on most cells

adrenal cortex   thyroid   gonads   mammary glands

4. Name the main endocrine glands and state where each is located in the body.

5. Give two examples of feedback control of hormone activity.

Self-Quiz   Answers in Appendix V

1. _____ are molecules released from a signaling cell that have effects on target cells.
   a. Hormones   d. Pheromones
   b. Neurotransmitters   e. a and b
   c. Local signaling molecules   f. All of the above

2. Hormones are produced by _____.
   a. endocrine glands and cells   d. a and b
   b. some neurons   e. a and c
   c. exocrine cells   f. a, b, and c

3. ADH and oxytocin are hypothalamic hormones secreted from the pituitary's _____ lobe.
   a. anterior   c. primary
   b. posterior   d. secondary

4. _____ has effects on body tissues in general.
   a. ACTH   c. LH
   b. TSH   d. Growth hormone

5. Which of the following stimulate the secretion of hormones?
   a. neural signals   d. environment cues
   b. local chemical changes   e. All of the above can
   c. hormonal signals   stimulate hormone
                                  secretion.

6. _____ lowers blood sugar levels; _____ raises the level of blood sugar.
   a. Glucagon; insulin   c. Gastrin; insulin
   b. Insulin; glucagon   d. Gastrin; glucagon

7. The pituitary detects a rising hormone concentration in blood and inhibits the gland that is secreting the hormone. This is a _____ feedback loop.
   a. positive
   b. negative

8. Second messengers assist _____.
   a. steroid hormones   c. only thyroid hormones
   b. nonsteroid hormones   d. both a and b

9. Match the hormone source with the closest description.
   ____ adrenal cortex   a. affected by day length
   ____ adrenal medulla   b. cortisol source
   ____ thyroid gland   c. roles in immunity
   ____ parathyroids   d. adjust(s) blood calcium
   ____ pancreatic islets   level
   ____ pineal gland   e. epinephrine source
   ____ thymus   f. insulin, glucagon
                            g. hormones require iodine

10. Match the endocrine control concepts.
   ____ oxytocin   a. released by the anterior
   ____ ACTH   pituitary and affects the
   ____ ADH   adrenal gland
   ____ growth hormone   b. influences extracellular
   ____ estrogen   fluid volume
                           c. has general effects on
                              growth
                           d. triggers uterine
                              contractions
                           e. a steroid hormone

Critical Thinking

1. Addison's disease develops when the adrenal cortex doesn't secrete enough mineralocorticoids and glucocorticoids. President John F. Kennedy was diagnosed with the disease when he was a young man. Before he started treatment with hormone replacement therapy, he was hypoglycemic and lost weight. Which missing hormone was responsible for his weight loss? How might Addison's disease have affected his blood pressure?

2. A physician sees a patient whose symptoms include sluggishness, depression, and intolerance to cold. After eliminating other possible causes, the doctor diagnoses an endocrine disorder. What disorder fits the symptoms? Why does the doctor suspect that the underlying cause is a malfunctioning anterior pituitary gland?

3. Marianne is affected by type 1 insulin-dependent diabetes. One day, after injecting herself with too much insulin, she starts to shake and feels confused. Following her doctor's suggestion, she drinks a glass of orange juice—a ready source of glucose—and soon her symptoms subside. What caused her symptoms? How would a glucose-rich snack help?

4. Some of the people in a remote village in Pakistan have an inherited form of dwarfism (Figure 15.14). The men are about 130 centimeters (a little over 4 feet) tall and the women about 115 centimeters (3 feet, 6 inches) tall. Northwestern University researchers discovered that the affected villagers have low levels of somatotropin, but a normal gene for this growth hormone. However, the gene coding for a receptor for one of the hypothalamic-releasing hormones is mutated. That receptor occurs on cells of the anterior pituitary. Explain how a faulty receptor may cause the abnormality.

Figure 15.14
Scientist Hiralal Maheshwari, with two men who have a heritable form of pituitary dwarfism.

Explore on Your Own

This Student Stress Scale lists a variety of life events that cause stress for young adults. The score for each event represents its relative impact on stress-related physiological responses. In general, people who score 300 points or more have the highest stress-related health risk. A score of 150–300 points indicates a moderate (50-50) stress-related health risk. A score below 150 indicates the lowest stress-related health risk, about a 1 in 3 chance of a significant, negative change in health status.

Although this test is only a general measure of stress, it can help you decide if you can benefit from adding to or improving your stress management activities, such as getting exercise, including some "down time" in your daily schedule, or seeking counseling.

Event	Points	
Death of a close family member	100	___
Death of a close friend	73	___
Parents' divorce	65	___
Jail term	63	___
Major personal injury or illness	63	___
Marriage	58	___
Being fired from a job	50	___
Failing an important course	47	___
Change in health of family member	45	___
Pregnancy (or causing one)	45	___
Sex problems	44	___
Serious argument with close friend	40	___
Change in financial status	39	___
Change of major	39	___
Trouble with parents	39	___
New romantic interest	38	___
Increased workload at school	37	___
Outstanding personal achievement	36	___
First quarter/semester in college	35	___
Change in living situation	31	___
Serious argument with instructor	30	___
Lower grades than expected	29	___
Change in sleeping habits	29	___
Change in social activities	29	___
Change in eating habits	28	___
Chronic car trouble	26	___
Change in number of family get-togethers	26	___
Too many missed classes	25	___
Change of college	24	___
Dropping more than one class	23	___
Minor traffic violations	20	___

Total _____

Adapted from the Holmes and Rahe Life Event Scale.

Sperm with a Nose for Home?

Human sperm are more than just champion swimmers. It seems that they can sense a variety of chemicals and navigate toward the source. Some researchers say this "nose for home" could be how sperm find their way to a woman's egg.

Back in the 1990s scientists discovered that human sperm have olfactory receptors virtually identical to those in the nose. Then, in 2003, a team of German investigators identified an olfactory receptor they dubbed hOR17-4, and showed that it responds to several chemicals used commercially to add a pleasing floral scent to various products. One of the compounds, called bourgeonal, causes sperm to make a beeline for the odor source.

Since it's unlikely that human eggs manufacture bourgeonal, the search is on for a chemically similar sperm attractant made by human eggs, or by cells in the female reproductive tract.

How does this kind of research apply to the real world? One outcome might be a test that screens a man's sperm for those that are most strongly attracted to the "egg scent." Those sperm could then be used for artificial insemination when a

couple has trouble conceiving naturally. On the flip side, a substance that *blocked* the ability of sperm to "sniff out" their targets might be the basis for a new contraceptive method.

In this chapter we turn our focus to the male and female reproductive systems—the only body system that does not contribute to homeostasis. Instead, its role is to continue our species. For both men and women, the **reproductive system** consists of a pair of primary reproductive organs, or **gonads**, plus accessory glands and ducts. Male gonads are **testes** (singular: testis), and female gonads are **ovaries**. The testes produce sperm and the ovaries produce egg cells. Both release sex hormones that influence reproduction. We'll conclude the chapter with a look at how sexually transmitted diseases can affect reproductive organs.

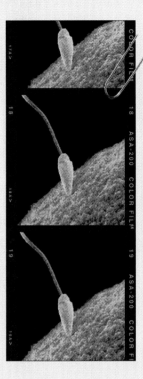

 How Would You Vote? In recent years the rate of teenage pregnancies in the United States has fallen slightly. One contributing factor has been greater access to contraception. Some people feel strongly that teens should not use contraception without parental consent. What's your opinion? Cast your vote online at www.thomsonedu .com/biology/starr/humanbio.

Key Concepts

THE MALE REPRODUCTIVE SYSTEM
A male's reproductive system consists of testes and accessory ducts and glands. Hormones control its functions, including making sperm.

THE FEMALE REPRODUCTIVE SYSTEM
A pair of ovaries are a female's primary reproductive organs. Hormones control their functions, including the development of oocytes (eggs).

SEXUAL INTERCOURSE AND FERTILITY
Sexual intercourse between a male and female is the usual first step toward pregnancy. Various methods exist for limiting or enhancing fertility.

SEXUALLY TRANSMITTED DISEASES
Sexual contact can transmit bacteria, viruses, and other disease-causing pathogens.

Links to Earlier Concepts

This chapter builds on what you have learned about hormones, including the steroid sex hormones estrogen and testosterone (15.1, 15.2). You will see more examples of how negative feedback loops regulate basic body functions (4.9), in this case the production of sperm in males and the menstrual cycle in females.

You will also expand your knowledge of the specialized cell structures called flagella, which propel sperm (3.8), and you will once again encounter chromosomes, the structures that carry genes (3.5).

16.1 The Male Reproductive System

LINKS TO
SECTIONS
2.7, 3.5,
AND 15.9

On any given day, millions of sperm are developing in an adult male's reproductive tract. The testes also secrete hormones that govern reproductive functions and traits associated with maleness.

SPERM (AND EGGS) DEVELOP FROM GERM CELLS

Cells in a male's gonads that give rise to sperm are sometimes called *germ cells*, from a Latin word that means "to sprout." As described in a later section, germ cells in a female's ovaries give rise to eggs. All germ cells are *diploid*—like nearly all other body cells, they have a normal number of chromosomes, which is 46. By contrast, sperm and eggs, which unite during reproduction, can have only 23 chromosomes. This is termed a *haploid* number of chromosomes because it is one-half the normal number. Sperm and eggs are also called *gametes* [GAM-eets], from a Greek word that means "to marry." When two haploid gametes unite at fertilization, the diploid chromosome number is restored. With these principles in mind, let's begin our tour of the male reproductive system.

SPERM FORM IN TESTES

Figure 16.1 shows an adult male's reproductive system and Table 16.1 lists the functions of its organs. In an embryo that is genetically programmed to become male, two testes form on the abdominal cavity wall. Before birth, the testes descend into the scrotum, a pouch of skin below the pelvic region. When a baby boy is born, his testes are fully formed miniatures of the adult organs.

Figure 16.2 shows the position of the scrotum in an adult male. If sperm cells are to develop properly, the temperature inside the scrotum must stay a few degrees cooler than body core temperature. To this end, a control mechanism stimulates or inhibits the contraction of smooth muscles in the scrotum's wall. It helps assure that the scrotum's internal temperature does not stray far from 95°F. When the air just outside the body gets too cold, contractions draw the scrotum closer to the body mass, which is warmer. When the air is warmer outside, the muscles relax and lower the scrotum.

Inside each testis are a large number of small, highly coiled tubes, the **seminiferous tubules**. Sperm begin to form in these tubules, as you'll read in Section 16.2.

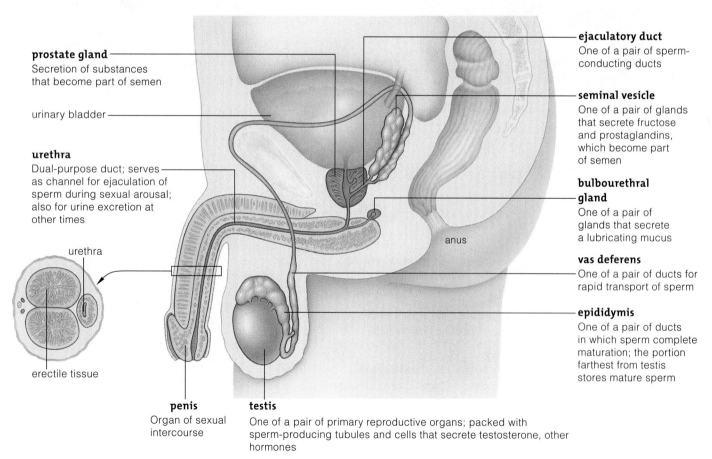

prostate gland
Secretion of substances that become part of semen

urinary bladder

urethra
Dual-purpose duct; serves as channel for ejaculation of sperm during sexual arousal; also for urine excretion at other times

urethra

erectile tissue

penis
Organ of sexual intercourse

testis
One of a pair of primary reproductive organs; packed with sperm-producing tubules and cells that secrete testosterone, other hormones

ejaculatory duct
One of a pair of sperm-conducting ducts

seminal vesicle
One of a pair of glands that secrete fructose and prostaglandins, which become part of semen

bulbourethral gland
One of a pair of glands that secrete a lubricating mucus

vas deferens
One of a pair of ducts for rapid transport of sperm

epididymis
One of a pair of ducts in which sperm complete maturation; the portion farthest from testis stores mature sperm

anus

Figure 16.1 *Animated!* Parts of the male reproductive system and their functions.

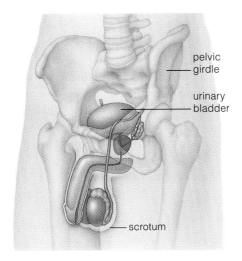

pelvic
girdle

urinary
bladder

scrotum

Figure 16.2 Position of the male reproductive system relative to the pelvic girdle and urinary bladder.

Table 16.1	Organs and Accessory Components of the Human Male Reproductive System	

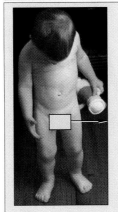

Reproductive Organs

Testis (2)	Sperm, sex hormone production
Epididymis (2)	Sperm maturation site and subsequent storage
Vas deferens (2)	Rapid transport of sperm
Ejaculatory duct (2)	Conduction of sperm to penis
Penis	Organ of sexual intercourse

Accessory Glands

Seminal vesicle (2)	Secretion of large part of semen
Prostate gland	Secretion of part of semen
Bulbourethral gland (2)	Production of mucus that functions in lubrication

SPERM MATURE AND ARE STORED IN THE COILED EPIDIDYMIS

Human sperm are not quite mature when they leave the testes. First they enter a pair of long, coiled ducts, the epididymides (singular: epididymis). Gland cells in the walls of these ducts secrete substances that trigger the finishing touches on sperm. Until sperm leave the body, they are stored in the last stretch of each epididymis.

When a male is sexually aroused, contracting muscle in the walls of his reproductive organs propels mature sperm into and through a pair of thick-walled tubes, the **vas deferentia** (singular: vas deferens). From there, contractions move sperm through a pair of ejaculatory ducts and on through the urethra to the outside. The urethra passes through the **penis**, the male sex organ, and also carries urine.

SUBSTANCES FROM SEMINAL VESICLES AND THE PROSTATE GLAND HELP FORM SEMEN

Glandular secretions become mixed with sperm as they travel through the urethra. The result is **semen**, a thick fluid that is eventually expelled from the penis during sexual activity. As semen is beginning to form, a pair of **seminal vesicles** secrete fructose. The sperm use this sugar for energy. Seminal vesicles also secrete certain kinds of prostaglandins. You may recall from Section 15.9 that these signaling molecules can induce muscle contractions. During sex, the prostaglandins cause contractions in muscles of a female's reproductive tract, and so aid the movement of sperm through it toward the egg.

Substances secreted by the **prostate gland** probably help buffer the acidic environment that sperm encounter in the female reproductive tract. The vaginal pH is about 3.5 to 4.0, but sperm motility improves at pH 6. Two **bulbourethral glands** secrete mucus-rich fluid into the urethra when a male is sexually aroused. This fluid neutralizes acids in any traces of urine in the urethra, and this more alkaline environment creates more hospitable surroundings for the 150 to 350 million sperm that pass through the channel in a typical ejaculation.

The testes and prostate gland both are sites where cancer can develop. At least 5,000 cases of testicular cancer are diagnosed each year in the United States, mostly among young men. This cancer kills about half of its victims. Prostate cancer, which is more common among men over 50, kills 40,000 older men annually—almost the same mortality rate recorded for breast cancer in women. As with other cancers, early detection is the key to survival. Causes and treatments of cancers are the subject of Chapter 23.

Males have a pair of testes—primary reproductive organs that produce sperm and sex hormones—as well as accessory glands and ducts.

Sperm develop mainly in the seminiferous tubules of the testes. When sperm are nearly mature, they leave each testis and enter the long, coiled epididymis, where they remain until ejaculated.

Secretions from the seminal vesicles and the prostate gland mix with sperm to form semen.

THE MALE REPRODUCTIVE SYSTEM

16.2 How Sperm Form

LINKS TO
SECTIONS
3.4, 3.8,
AND 15.3

*Sperm are specialized to unite with eggs in reproduction.
Here, we trace the steps by which sperm develop.*

SPERM FORM IN SEMINIFEROUS TUBULES

A testis is smaller than a golf ball, yet packed inside it are 125 meters—over 400 feet—of seminiferous tubules. As many as 30 wedge-shaped lobes divide the interior, and each holds two or three coiled tubules (Figure 16.3a).

Inside the walls of seminiferous tubules are cells called *spermatogonia* (Figure 16.3b). They are the starting point for several rounds of cell division, including a type called *mitosis* and a type called *meiosis*. You'll read more about mitosis and meiosis in Chapter 19; here the main thing to keep in mind is that meiosis is necessary to form sperm and eggs.

Spermatogonia develop into *primary spermatocytes*, which become *secondary spermatocytes* after a first round of meiosis (meiosis I). A second round of cell division (meiosis II) produces *spermatids*. The spermatids gradually develop into *spermatozoa*, or simply **sperm**, the male gametes. The "tail" of each sperm, a flagellum, arises at the end of the process, which takes nine to ten weeks. All the while, the developing cells are nourished by and receive chemical signals from neighboring **Sertoli cells** that line the seminiferous tubule.

The testes produce sperm from puberty onward. Millions are in different stages of development on any given day. A mature sperm has a tail, a midpiece, and a head (Figure 16.3c). Inside the head, a nucleus contains DNA organized into chromosomes. An enzyme-containing cap, the **acrosome**, covers most of the head. Its enzymes help the sperm penetrate protective material around an egg at fertilization. In the midpiece, mitochondria supply energy for the tail's movements.

HORMONES CONTROL SPERM FORMATION

Male reproductive function depends on testosterone, LH, and FSH. **Leydig cells** (also called interstitial cells), located in tissue between the seminiferous tubules in testes (Figure 16.4a), secrete **testosterone**. This hormone governs the growth, form, and functions of the male reproductive tract. It stimulates sexual behavior, and at puberty it promotes the development of male **secondary sexual traits**, including facial hair and deepening of an adolescent male's voice.

LH (luteinizing hormone) and **FSH** (follicle-stimulating hormone) are released from the anterior lobe of the pituitary gland (Figure 16.4b). These two hormones were named for their effects in females, but are chemically the same in males.

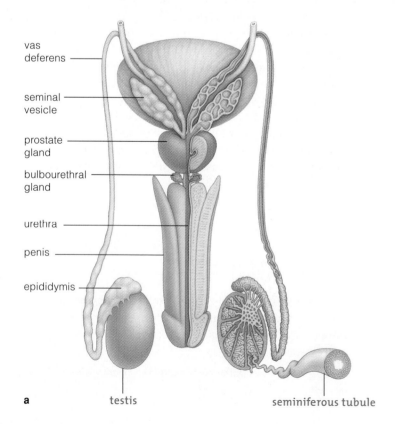

vas deferens

seminal vesicle

prostate gland

bulbourethral gland

urethra

penis

epididymis

a   testis   seminiferous tubule

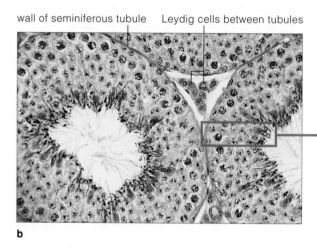

wall of seminiferous tubule   Leydig cells between tubules

b

Figure 16.3 *Animated!* (**a**) The male reproductive tract viewed from behind. (**b**) Leydig cells in tissue spaces between seminiferous tubules. In this light micrograph you can see parts of three of the tubules. (**c**) How sperm form, starting with a diploid germ cell. Cell divisions by mitosis, then meiosis, produce haploid cells that develop into sperm. (**d**) The structure of a mature sperm, the male gamete.

THE MALE REPRODUCTIVE SYSTEM

Since the hypothalamus controls secretions of LH, FSH, and testosterone, it controls the formation of sperm. When the testosterone level in a male's blood falls below a set point, the hypothalamus secretes GnRH. This releasing hormone prompts the pituitary's anterior lobe to release LH and FSH, which have targets in the testes. LH stimulates Leydig cells to secrete testosterone, which stimulates diploid germ cells to become sperm. Sertoli cells have FSH receptors. FSH is crucial to starting sperm formation (called *spermatogenesis*) at puberty, but researchers do not know whether it is essential for the normal functioning of mature testes.

A high level of testosterone in a male's blood inhibits the release of GnRH. Also, when the sperm count is high, Sertoli cells release inhibin, a hormone that acts on the hypothalamus and pituitary to inhibit the release of GnRH and FSH. Accordingly, feedback loops to the hypothalamus begin to operate, causing testosterone secretion and sperm formation to decline.

The formation of sperm depends on the hormones testosterone, LH, and FSH. Testosterone also governs the development of secondary sexual traits. Feedback loops from the testes to the hypothalamus and pituitary gland control the secretion of these hormones.

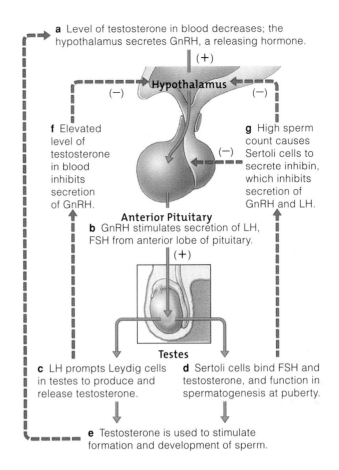

a Level of testosterone in blood decreases; the hypothalamus secretes GnRH, a releasing hormone.

Hypothalamus

f Elevated level of testosterone in blood inhibits secretion of GnRH.

g High sperm count causes Sertoli cells to secrete inhibin, which inhibits secretion of GnRH and LH.

Anterior Pituitary
b GnRH stimulates secretion of LH, FSH from anterior lobe of pituitary.

Testes

c LH prompts Leydig cells in testes to produce and release testosterone.

d Sertoli cells bind FSH and testosterone, and function in spermatogenesis at puberty.

e Testosterone is used to stimulate formation and development of sperm.

Figure 16.4 *Animated!* Negative feedback loops to the hypothalamus and pituitary gland from the testes.

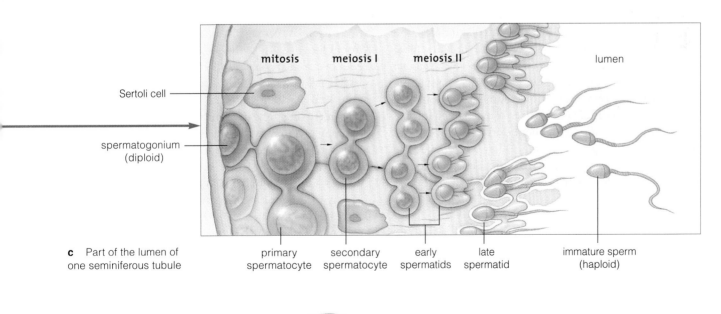

mitosis meiosis I meiosis II lumen

Sertoli cell

spermatogonium (diploid)

c Part of the lumen of one seminiferous tubule

primary spermatocyte

secondary spermatocyte

early spermatids

late spermatid

immature sperm (haploid)

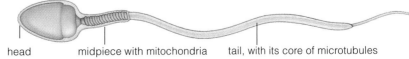

d Sketch of a mature human sperm. The head consists of DNA surrounded by the acrosome.

head   midpiece with mitochondria   tail, with its core of microtubules

16.3 The Female Reproductive System

We now consider the female reproductive system. Its biological function is to nurture developing offspring from conception until birth.

OVARIES ARE A FEMALE'S PRIMARY REPRODUCTIVE ORGANS

Figure 16.5 shows the parts of the female reproductive system and Table 16.2 summarizes their functions. A female's primary reproductive organs are her two **ovaries**. The ovaries release sex hormones, and during a woman's reproductive years they also produce eggs. The ovarian hormones influence the development of female secondary sexual traits. These traits include the "filling out" of breasts, hips, and buttocks as fat deposits accumulate in those areas.

A female's immature eggs are called **oocytes**. When an oocyte is released from an ovary, it moves into the neighboring **oviduct** (sometimes called a *fallopian tube*). The oviducts are where a sperm may fertilize an egg. Fertilized or not, an egg travels down the oviduct into the hollow, pear-shaped **uterus**. In this organ, a baby can grow and develop. The wall of the uterus consists of a thick layer of smooth muscle (the myometrium) and an interior lining, the **endometrium**. The endometrium includes epithelium, connective tissue, glands, and blood vessels. The lower part of the uterus is the *cervix*. The *vagina* leads from the cervix to the body surface. It is a muscular tube that receives the penis and sperm and functions as part of the birth canal.

A female's external genitals are collectively called the *vulva*. Outermost are a pair of fat-padded skin folds, the *labia majora*. They enclose a smaller pair of skin folds, the *labia minora*, that are laced with blood vessels. The labia minora partly enclose the *clitoris*, a small organ sensitive to sexual stimulation. From a developmental standpoint, the female's clitoris is analogous to a male's penis.

A female's urethra opens about midway between her clitoris and her vaginal opening. Whereas in males the urethra carries both urine and sperm, in females it is separate and is not involved in reproduction.

Table 16.2	Female Reproductive Organs
Ovaries	Produce oocytes and sex hormones
Oviducts	Conduct oocytes from ovary to uterus
Uterus	Chamber where new individual develops
Cervix	Secretes mucus that enhances sperm movement into uterus and (after fertilization) reduces the embryo's risk of bacterial infection
Vagina	Organ of sexual intercourse; birth canal

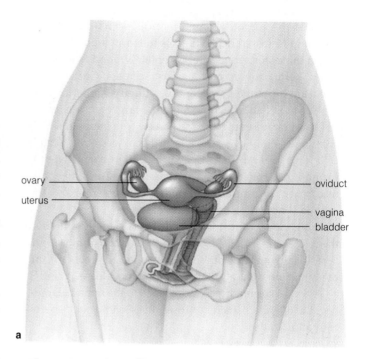

Figure 16.5 *Animated!* (**a**) Position of the female reproductive system relative to the pelvic girdle and urinary bladder. (**b**) Parts of the system and their functions.

DURING THE MENSTRUAL CYCLE, AN OOCYTE IS RELEASED FROM AN OVARY

Like all female primates, a woman has a **menstrual cycle**. It takes about twenty-eight days to complete one cycle, although this can vary from month to month and from woman to woman. During the cycle, an oocyte matures (from a *primary* oocyte to a *secondary* oocyte) and is released from an ovary. All the while, hormones are priming the endometrium to receive and nourish an embryo in case the oocyte is fertilized. If the oocyte is *not* fertilized, a blood-rich fluid starts flowing out through the vaginal canal. This recurring flow is **menstruation**, and it marks the first day of a new cycle. The disintegrating endometrium is being sloughed off, only to be rebuilt once again.

The events just sketched out advance through three phases (Table 16.3). The cycle starts with a *menstrual phase*. This is the time of menstruation, when the endometrium disintegrates. Next comes the *proliferative phase*, when the endometrium begins to thicken again. The end of this phase coincides with ovulation—the release of an oocyte from an ovary. During the cycle's final phase, called the *progestational* ("before pregnancy") *phase*, an endocrine structure called the corpus luteum ("yellow body") forms. It secretes a flood of the sex hormones **progesterone** and **estrogen**, which prime the endometrium for pregnancy.

All three phases are governed by feedback loops to the hypothalamus and pituitary gland from the ovaries.

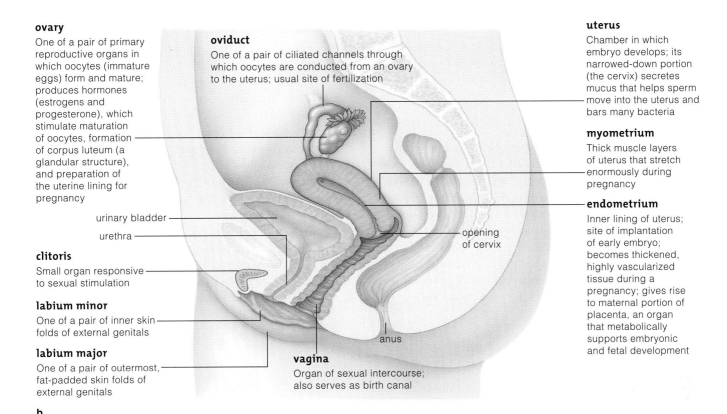

ovary

One of a pair of primary reproductive organs in which oocytes (immature eggs) form and mature; produces hormones (estrogens and progesterone), which stimulate maturation of oocytes, formation of corpus luteum (a glandular structure), and preparation of the uterine lining for pregnancy

urinary bladder

urethra

clitoris

Small organ responsive to sexual stimulation

labium minor

One of a pair of inner skin folds of external genitals

labium major

One of a pair of outermost, fat-padded skin folds of external genitals

b

oviduct

One of a pair of ciliated channels through which oocytes are conducted from an ovary to the uterus; usual site of fertilization

opening of cervix

anus

vagina

Organ of sexual intercourse; also serves as birth canal

uterus

Chamber in which embryo develops; its narrowed-down portion (the cervix) secretes mucus that helps sperm move into the uterus and bars many bacteria

myometrium

Thick muscle layers of uterus that stretch enormously during pregnancy

endometrium

Inner lining of uterus; site of implantation of early embryo; becomes thickened, highly vascularized tissue during a pregnancy; gives rise to maternal portion of placenta, an organ that metabolically supports embryonic and fetal development

Table 16.3	Phases of the Menstrual Cycle	
Phase	Events	Days of the Cycle*
Menstrual phase	Menstruation; endometrium breaks down	1–5
	Follicle matures in ovary; endometrium rebuilds	6–13
Proliferative phase	Endometrium begins to thicken, ovulation occurs	14
Progestational phase	Lining of endometrium develops to receive a possible embryo	15–28

*Assumes a 28-day cycle.

FSH and LH promote cyclic changes in the ovaries, which correspond closely with the menstrual cycle and are described in the following section.

A female's first menstruation, or *menarche*, usually occurs between the ages of ten and sixteen. Menstrual cycles continue until *menopause*, which usually occurs in a woman's late 40s or early 50s. By then, her body's secretion of reproductive hormones has diminished, as has her sensitivity to pituitary reproductive hormones. The decreasing estrogen levels may trigger a range of temporary symptoms, including moodiness and "hot flashes." These sudden bouts of sweating and feeling uncomfortably warm result when blood vessels in the skin dilate. Section 17.13, which covers aging in various body systems, describes other physiological changes associated with menopause. When a woman's menstrual cycles stop altogether, the fertile phase of her life is over.

In *endometriosis*, endometrial tissue spreads and grows outside the uterus. Scar tissue may form on one or both ovaries or oviducts, leading to infertility. In the United States, as many as 10 million women are affected each year. Endometriosis may develop when menstrual flow backs up through the oviducts and spills into the pelvic cavity. Or perhaps some cells became situated in the wrong place when the woman was a developing embryo, then were stimulated to grow during puberty, when her sex hormones became active. Regardless, the symptoms include pain during menstruation, sex, or urination. Treatment ranges from doing nothing in mild cases to surgery to remove the abnormal tissue or sometimes even the whole uterus.

Ovaries, a female's primary reproductive organs, produce oocytes (immature eggs) and sex hormones. Endometrium lines the uterus, where embryos develop.

Sex hormones—estrogens and progesterone—are secreted as part of a menstrual cycle during a female's reproductive years.

16.4 The Ovarian Cycle: Oocytes Develop

LINKS TO
SECTIONS
13.8 AND 15.3

As the menstrual cycle proceeds, a cycle in the ovaries produces an oocyte—the first step toward pregnancy.

IN THE OVARIAN CYCLE, HORMONES GUIDE THE STEPS LEADING TO OVULATION

A newborn girl's ovaries contain about 2 million primary oocytes. All but about 300 are later resorbed, though new research suggests that the ovaries may make fresh oocytes until menopause. In each oocyte, meiosis I begins but then is halted by genetic instructions. Meiosis resumes, usually in one oocyte at a time, with each menstrual cycle.

In the **ovarian cycle**, a primary oocyte develops further and is ovulated (Figure 16.6). Step *a* shows a primary oocyte near an ovary's surface. A layer of *granulosa cells* surrounds and nourishes it. The primary oocyte and the cell layer around it make up a **follicle**. At the start of a menstrual cycle, the hypothalamus is secreting enough GnRH to make the anterior pituitary step up *its* secretion of FSH and LH (Figure 16.7). The blood level of those hormones increases, and *that* causes the follicle to grow. (FSH, recall, is short for follicle-stimulating hormone.) The ovarian cycle is under way.

The oocyte starts to grow, and more layers of cells form around it. Glycoprotein deposits widen the space between these layers and the oocyte. With time the deposits coat the oocyte with a **zona pellucida**.

Both FSH and LH stimulate cells outside the zona pellucida to secrete estrogens. An estrogen-containing fluid builds up in the follicle (now called a secondary or Graafian follicle), and estrogen levels in the blood start to rise. Several hours before being released from the

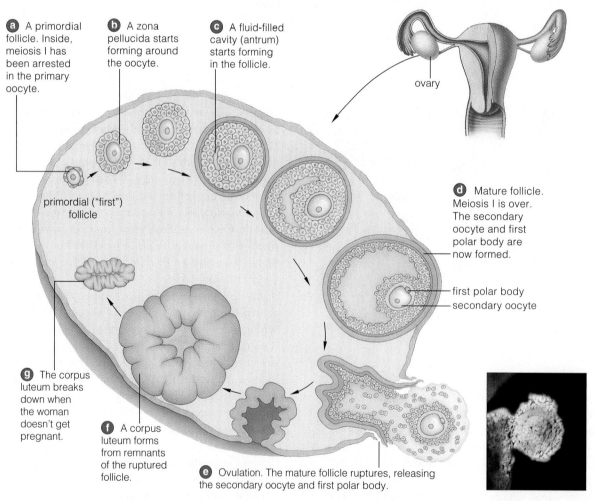

a A primordial follicle. Inside, meiosis I has been arrested in the primary oocyte.

b A zona pellucida starts forming around the oocyte.

c A fluid-filled cavity (antrum) starts forming in the follicle.

ovary

primordial ("first") follicle

d Mature follicle. Meiosis I is over. The secondary oocyte and first polar body are now formed.

first polar body
secondary oocyte

g The corpus luteum breaks down when the woman doesn't get pregnant.

f A corpus luteum forms from remnants of the ruptured follicle.

e Ovulation. The mature follicle ruptures, releasing the secondary oocyte and first polar body.

Figure 16.6 *Animated!* Cyclic changes in the ovary, which is shown in cross section. A follicle stays in the same place in an ovary all through the menstrual cycle. It does not "move around" as in this diagram, which shows the *sequence* of events. In the cycle's first phase, a follicle grows and matures. At ovulation, the second phase, the mature follicle ruptures and releases a secondary oocyte. In the third phase, a corpus luteum forms from the follicle's remnants. It self-destructs if the woman does not become pregnant. The micrograph shows a secondary oocyte being released from an ovary. It will enter an oviduct, the passageway to the uterus.

THE FEMALE REPRODUCTIVE SYSTEM

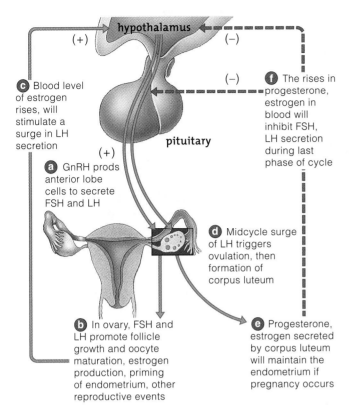

Figure 16.7 *Animated!* Feedback control of hormonal secretion during an ovarian (and menstrual) cycle. A positive feedback loop from an ovary to the hypothalamus causes a surge in LH secretion. This surge triggers ovulation. Afterward, negative feedback loops to the hypothalamus and pituitary inhibit FSH secretion. They prevent another follicle from maturing until the cycle is completed.

(Labels in figure 16.7):

c Blood level of estrogen rises, will stimulate a surge in LH secretion

a GnRH prods anterior lobe cells to secrete FSH and LH

f The rises in progesterone, estrogen in blood will inhibit FSH, LH secretion during last phase of cycle

d Midcycle surge of LH triggers ovulation, then formation of corpus luteum

b In ovary, FSH and LH promote follicle growth and oocyte maturation, estrogen production, priming of endometrium, other reproductive events

e Progesterone, estrogen secreted by corpus luteum will maintain the endometrium if pregnancy occurs

ovary, the oocyte completes the meiotic cell division (meiosis I) that was arrested years before. Now, there are two cells. One, a large, **secondary oocyte**, gets most of the cytoplasm. The other cell is the *first polar body*. (It may divide again.) The secondary oocyte also gets a haploid number of chromosomes—the required number for a gamete. It now begins a second round of meiosis (meiosis II), which again is arrested in preparation for fertilization by a sperm.

About halfway through the ovarian cycle, a woman's pituitary gland detects the rising estrogen level. It releases LH, which causes vascular changes that make the follicle swell. The surge also causes enzymes to break down the bulging follicle wall, which soon ruptures. Fluid escapes, along with the secondary oocyte and polar body (Figure 16.6e). The midcycle surge of LH has triggered **ovulation**—the release of a secondary oocyte from the ovary.

Once released into the abdominal cavity, the secondary oocyte enters an oviduct. Long, ciliated projections from the oviduct (called *fimbriae*) extend like an umbrella over part of the ovary. Movements of the projections and cilia sweep the oocyte into the channel. If fertilization takes

place, it usually occurs while the oocyte is in the oviduct. At fertilization, the oocyte will finish meiosis II and become a mature **ovum**, the *egg*.

THE OVARIAN CYCLE DOVETAILS WITH THE MENSTRUAL CYCLE

As Section 16.3 noted, estrogens released early in the menstrual cycle stimulate growth of the endometrium and its glands, paving the way for a pregnancy. Just before the midcycle LH surge, cells of the follicle wall start secreting estrogens and progesterone. At ovulation, the estrogens act on tissue around the cervical canal, which leads to the vagina. The cervix starts to secrete large amounts of a thin, clear mucus—an ideal medium for sperm to swim through. Blood vessels grow rapidly in the thickened endometrium.

After ovulation, another structure dominates events. Granulosa cells left behind in the follicle differentiate into the **corpus luteum**. The midcycle surge of LH triggers formation of this structure, hence the name *luteinizing* hormone.

Recall that the corpus luteum secretes progesterone as well as some estrogen. The progesterone prepares the woman's reproductive tract for the arrival of an embryo. For example, it causes mucus in the cervix to become thick and sticky, which may prevent bacteria from entering the uterus. Progesterone also maintains the endometrium during a pregnancy.

A corpus luteum lasts for about twelve days. During that time, the hypothalamus signals for a decrease in FSH secretion, which prevents other follicles from developing. If no embryo implants in the endometrium, the corpus luteum secretes prostaglandins that disrupt its own functioning.

After the corpus luteum breaks down, progesterone and estrogen levels fall rapidly, so the endometrium also starts to break down. Deprived of oxygen and nutrients, its blood vessels constrict and its tissues die. Blood escapes from the ruptured walls of weakened capillaries. The blood and sloughed endometrial tissues make up the menstrual flow. As the cycle begins anew a few days later, the rising levels of estrogen stimulate regrowth of the endometrium.

> In the ovarian cycle, a midcycle surge of LH triggers ovulation, the release of the secondary oocyte and the polar body from the ovary.
>
> The cyclic release of hormones helps pave the way for fertilization of an egg and prepares the endometrium and other parts of a female's reproductive tract for pregnancy.

16.5 Visual Summary of the Menstrual and Ovarian Cycles

The menstrual cycle has been compared to a hormonal symphony, with many parts being brought together into a harmonious whole. Before you continue your reading, take a moment to review Figure 16.8. It correlates the cyclic changes in the ovary and uterus with the changes in hormone levels that bring about the events of each menstrual cycle.

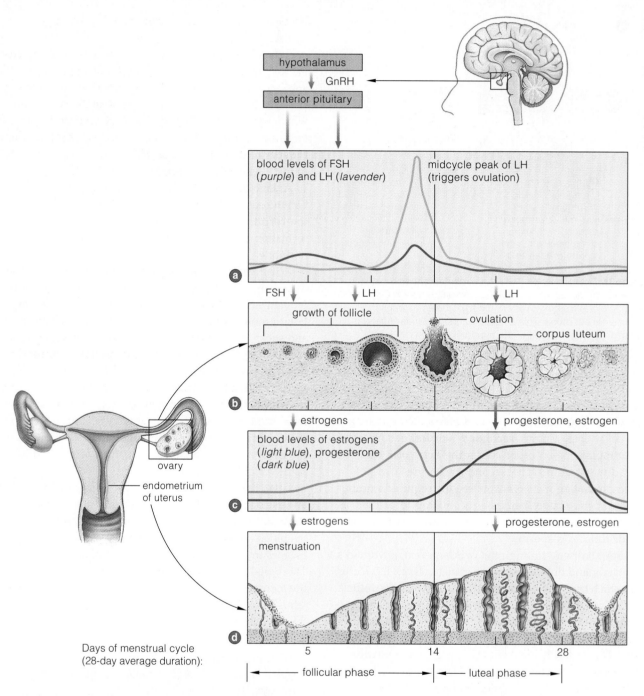

Figure 16.8 *Animated!* Hormones and their effects during one menstrual cycle. (**a**) GnRH, a releasing hormone from the hypothalamus, stimulates the anterior pituitary to secrete FSH and LH. (**b**) FSH and LH stimulate a follicle to grow, an oocyte to mature, and the ovaries to secrete progesterone and estrogens that cause rebuilding of the endometrium. (**c**) A midcycle LH surge triggers ovulation and the formation of a corpus luteum. Progesterone and some estrogens released by the corpus luteum maintain the endometrium, but if no pregnancy occurs, they stop being released and the corpus luteum breaks down.

THE FEMALE REPRODUCTIVE SYSTEM

16.6 Sexual Intercourse

The penis and vagina are mechanically compatible for sexual intercourse, which may lead to pregnancy.

SEXUAL INTERCOURSE INVOLVES PHYSIOLOGICAL CHANGES IN BOTH PARTNERS

Coitus and copulation are both technical terms for sexual intercourse. The male sex act involves an *erection*, in which the limp penis stiffens and lengthens. It also involves *ejaculation*, the forceful expulsion of semen into the urethra and out from the penis. As shown in Figure 16.1, the penis has lengthwise cylinders of spongy tissue. The outer cylinder has a mushroom-shaped tip (the glans penis). Inside it is a dense array of sensory receptors that are activated by friction. In a male who is not sexually aroused, the large blood vessels leading into the cylinders are constricted. In aroused males, these blood vessels vasodilate, so blood flows into the cylinders faster than it flows out. Blood collects in the spongy tissue, and the organ stiffens and lengthens—a mechanism that helps the penis penetrate into the female's vagina.

In a female, arousal includes vasodilation of blood vessels in her genital area. This causes vulvar tissues to engorge with blood and swell. Mucus-rich secretions flow from the cervix, lubricating the vagina.

During coitus, pelvic thrusts stimulate the penis as well as the female's clitoris and vaginal wall. The mechanical stimulation triggers rhythmic, involuntary contractions in smooth muscle in the male reproductive tract, especially the vas deferens and the prostate. The contractions rapidly force sperm out of each epididymis. They also force the contents of seminal vesicles and the prostate gland into the urethra. The resulting mixture, semen, is ejaculated into the vagina.

During ejaculation, a sphincter closes off the neck of the male's bladder and prevents urine from being excreted. Ejaculation is a reflex response; once it begins, it cannot be stopped.

Emotional intensity, heavy breathing, and heart pounding, as well as generalized contractions of skeletal muscles, accompany the rhythmic throbbing of the pelvic muscles. For both partners, **orgasm**—the culmination of the sex act—typically is accompanied by strong sensations of release, warmth, and relaxation.

Some people mistakenly believe that unless a woman experiences orgasm, she cannot become pregnant. This is not true, however. A female can become pregnant from intercourse *regardless* of whether she experiences orgasm, and even if she is not sexually aroused. All that is required is that a sperm meet up with a secondary oocyte that is traveling down one of her oviducts.

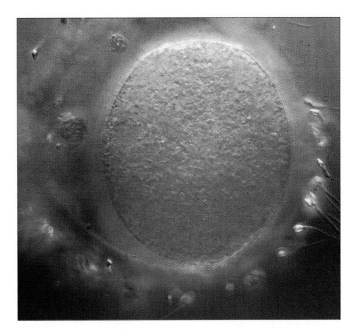

Figure 16.9 A secondary oocyte surrounded by sperm. If fertilization ensues, it will set the stage for the development of a new individual, continuing the human life cycle.

INTERCOURSE CAN PRODUCE A FERTILIZED EGG

If sperm enter the vagina a few days before or after ovulation or anytime between, an ovulated egg may be fertilized. Within thirty minutes after ejaculation, muscle contractions in the uterus move the sperm deeper into the female reproductive tract. Only a few hundred sperm will actually reach the upper portion of the oviduct, which is where fertilization usually takes place. The remarkable micrograph in Figure 16.9 shows living sperm around a secondary oocyte.

As you will read in Chapter 17, the meeting of sperm and secondary oocyte is the first of several intricately orchestrated events that lead to actual **fertilization**—the beginning of a new individual.

Sexual intercourse (coitus) typically involves a sequence of physiological changes in both partners.

During arousal, blood vessels dilate so that more blood flows to the penis (males) and vulva (females). Orgasm involves muscular contractions (including those leading to ejaculation of semen into the vagina) and sensations of release, warmth, and relaxation.

A female may become pregnant through intercourse even if she is not sexually aroused or does not experience orgasm.

Fertilization can occur when a sperm encounters a secondary oocyte, usually in the oviduct.

16.7 Controlling Fertility

Many sexually active people choose to exercise control over whether their activity will produce a child. Here we consider the biological bases of different forms of birth control.

NATURAL BIRTH CONTROL IS CHANCY

The most effective method of birth control is complete *abstinence*—no sexual intercourse whatsoever. A modified form of abstinence is the *rhythm method*, also called the "fertility awareness" or *sympto-thermal method*. The idea is to refrain from intercourse during the woman's fertile period, starting a few days before ovulation and ending a few days after. Her fertile period is identified and tracked by keeping records of the length of her menstrual cycles and sometimes by examining her cervical secretions and taking her temperature each morning when she wakes up. (Core body temperature rises by one-half to one degree just after ovulation.) The method is not very reliable (Figure 16.10). Ovulation can be irregular, and it can be easy to miscalculate. Also, sperm already in the vaginal tract may survive until ovulation.

Withdrawal, removing the penis from the vagina before ejaculation, also is not very effective because fluid released from the penis before ejaculation may contain sperm. *Douching*, or rinsing out the vagina with a chemical right after intercourse, is next to useless. It takes less than 90 seconds for sperm to move past the cervix into the uterus.

SURGICAL SOLUTIONS ARE RELIABLE

Controlling fertility by surgical intervention is less chancy. In *vasectomy*, a physician makes a tiny incision in a man's scrotum, then severs and ties off each vas deferens (Figure 16.11*a*). The procedure takes about twenty minutes and requires only a local anesthetic. Afterward, sperm can't leave the testes and so can't be present in the man's semen. Having a vasectomy does not disrupt male sex hormones. A vasectomy alternative is the Vasclip, a hinged plastic device about the size of a rice grain, which simply closes off the vas deferens.

In *tubal ligation*, a woman's oviducts are cauterized or cut and tied off (Figure 19.11*b*). The procedure most often is performed in a hospital.

PHYSICAL AND CHEMICAL BARRIERS VARY IN EFFECTIVENESS

Spermicidal foam and spermicidal jelly kill sperm. They are packaged in an applicator and placed in the vagina just before intercourse. Neither is reliable unless used with another device, such as a diaphragm or condom.

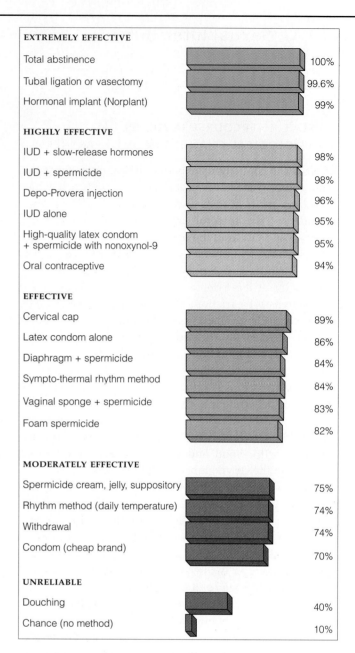

EXTREMELY EFFECTIVE	
Total abstinence	100%
Tubal ligation or vasectomy	99.6%
Hormonal implant (Norplant)	99%
HIGHLY EFFECTIVE	
IUD + slow-release hormones	98%
IUD + spermicide	98%
Depo-Provera injection	96%
IUD alone	95%
High-quality latex condom + spermicide with nonoxynol-9	95%
Oral contraceptive	94%
EFFECTIVE	
Cervical cap	89%
Latex condom alone	86%
Diaphragm + spermicide	84%
Sympto-thermal rhythm method	84%
Vaginal sponge + spermicide	83%
Foam spermicide	82%
MODERATELY EFFECTIVE	
Spermicide cream, jelly, suppository	75%
Rhythm method (daily temperature)	74%
Withdrawal	74%
Condom (cheap brand)	70%
UNRELIABLE	
Douching	40%
Chance (no method)	10%

Figure 16.10 Comparison of the effectiveness of some contraceptive methods in the United States. Percentages shown are based on the number of unplanned pregnancies per 100 couples who used the method as the only form of birth control for one year. For example, "94% effectiveness" for oral contraceptives (the birth control pill) means that, on average, 6 of every 100 women using them will become pregnant.

A *diaphragm* is a flexible, dome-shaped device that is positioned over the cervix before intercourse. It must be fitted by a doctor, used with foam or jelly, and inserted correctly with each use. The *cervical cap* is smaller and can be left in place for up to three days with just a single dose of spermicide. The *contraceptive sponge* is a disposable disk that contains a spermicide and covers the cervix. After being wetted it is inserted up to 24 hours before intercourse. No prescription or special fitting is required.

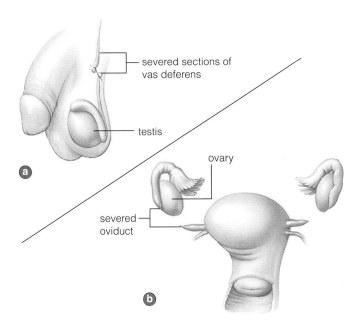

severed sections of vas deferens

testis

ovary

severed oviduct

Figure 16.11 Surgical methods of birth control. (**a**) Vasectomy and (**b**) tubal ligation.

The *intrauterine device*, or IUD, is a plastic or metal device that is placed into the uterus, where it hampers implantation of a fertilized egg. Available by prescription, IUDs have been associated with a variety of complications and any woman considering getting one should discuss the matter fully with her physician.

Condoms are thin, tight-fitting sheaths of latex or animal skin worn over the penis during intercourse. An intact condom is about 85 to 93 percent reliable, and latex condoms help prevent the spread of sexually transmitted diseases. A pouchlike latex "female condom" that is inserted into the vagina has also been developed.

A widely used method of fertility control is the *birth control pill*—any of a number of formulations of synthetic estrogens and progesterones. These hormones block the normal release of anterior pituitary hormones (LH and FSH) that are required for eggs to mature and be ovulated. Oral contraceptives are prescription drugs and must be matched with each patient's needs.

Used as directed, the pill is one of the most reliable methods of controlling fertility. Often, the hormone blend corrects erratic menstrual cycles and reduces cramping. Some users experience (usually temporary) side effects, including nausea and weight gain. Continued use may lead to blood clots in at-risk women. Complications are more likely in women who smoke, and most physicians won't prescribe an oral contraceptive for a smoker.

NuvaRing is a flexible plastic ring that slides into the vagina. It slowly releases the same hormones used in birth control pills and must be replaced once a month.

Products such as injections that contain progestin (Depo-Provera, Lunelle) or hormone-containing implants (Norplant) inhibit ovulation or prevent implantation of the embryo. Both can cause heavier menstrual periods, and Norplant rods can be difficult to remove.

Medically, pregnancy only begins after an embryo implants. *Morning-after pills* such as mifepristone (RU-486) and Preven interfere with the hormones that control events between ovulation and implantation. Contrary to their common name, these chemicals actually work up to seventy-two hours after unprotected intercourse.

Researchers are working to develop new and better methods of fertility control. Examples include implants that biodegrade and so don't require surgical removal. Other efforts are aimed at developing a contraceptive for men that reduces the sperm count.

ABORTION IS HIGHLY CONTROVERSIAL

An induced or surgical *abortion* removes or dislodges an implanted embryo or fetus from the womb. In that sense, the procedure has little to do with fertility, because the pregnancy is already under way. More than 1,500,000 abortions are performed in the United States each year, even while the difficult legal and social conflict over legalized abortion rages on.

During the first trimester (twelve weeks), abortions performed in a clinical setting usually are fast, painless, and free of complications. Even so, polls show that for both medical and moral reasons, the majority of people in the U.S. view sexually responsible behavior as being preferable to an abortion. Aborting a late-term fetus is extremely controversial unless the mother's life is threatened.

This textbook cannot offer any "right" answers to a question about the morality of abortion or any other reproductive decision. It can only offer a serious explanation of how a new individual develops to help you objectively assess the biological basis of human life.

The most effective methods for preventing conception are abstinence, chemical barriers to conception, and surgery or implants that block the vas deferens or oviducts. A frank discussion with a physician is the best starting point for fertility control.

16.8 Options for Coping with Infertility

LINK TO SECTION 15.3

In the United States, about one in every six couples is infertile—unable to conceive a child after a year of trying. Causes run the gamut from hormonal imbalances that prevent ovulation, oviducts blocked by effects of disease, a low sperm count, or sperm that are defective in a way that impairs fertilization.

FERTILITY DRUGS STIMULATE OVULATION

In about one-third of cases, infertility can be traced to poor quality oocytes or to irregular or absent ovulation. These situations are most common in women over the age of 37. A couple's first resort may be fertility drugs, in the hope that one or more ovarian follicles will produce a healthy oocyte. One commonly used drug, clomiphene, stimulates the pituitary gland to release FSH. As noted in Section 15.3, this hormone triggers ovulation. A drug called *human menopausal gonadotropin* (hMG) is basically a highly purified form of FSH. Injected directly into the bloodstream, it stimulates ovulation in 70 to 90 percent of women who receive it.

Although fertility drugs have been used with great success since the 1970s, they can cause undesirable side effects, including the fertilization of several eggs at once. The result is a high-risk pregnancy that can result in babies with neurological and other problems.

ASSISTED REPRODUCTIVE TECHNOLOGIES INCLUDE ARTIFICIAL INSEMINATION AND IVF

Artificial insemination was one of the first methods of *assisted reproductive technology*, or ART. In this approach, semen is placed into a woman's vagina or uterus, usually by syringe, around the time she is ovulating. This procedure may be chosen when a woman's partner has a low sperm count, because his sperm can be concentrated prior to the procedure. In *artificial insemination by donor* (AID), a sperm bank provides sperm from an anonymous donor. AID produces about 20,000 babies in the United States every year.

In vitro fertilization (IVF) is literally "fertilization in glass." If a couple's sperm and oocytes are normal, they can be used. Otherwise, variations of the technology are available that use sperm, oocytes, or both, from donors (Figure 16.12). Sperm and oocytes are placed in a glass laboratory dish in a solution that simulates the fluid in oviducts. If fertilization takes place, about 12 hours later *zygotes* (fertilized eggs in the first stage of development) are transferred to a chemical solution that will support further development. Two to four days later, one or more embryos are transferred to the woman's uterus. An embryo implants in about 20 percent of cases. In vitro fertilization often produces more embryos than

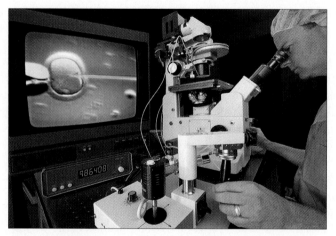

Figure 16.12 Doctor inserting a human sperm into an egg during in vitro fertilization. He is viewing the cell through a microscope. The procedure is magnified on a monitor. The egg, held in place by the tip of a pipette, is being pierced by a microneedle (the thin "line" on the right).

SEXUAL INTERCOURSE AND FERTILITY

Artificial Insemination and Embryo Transfer	In Vitro Fertilization
1. Father is infertile. Mother is inseminated by donor and carries child.	1. Mother is fertile but unable to conceive. Egg from mother and sperm from father are combined in laboratory. Embryo is placed in mother's uterus.
2. Mother is infertile but able to carry child. Egg donor is inseminated with father's sperm. Then embryo is transferred and mother carries child.	2. Mother is infertile but able to carry child. Egg from donor is combined with sperm from father and implanted in mother.
3. Mother is infertile and unable to carry child. Egg donor is inseminated with father's sperm and carries child.	3. Mother is fertile but father is infertile. Egg from mother is combined with sperm from donor.
4. Both parents are infertile, but mother is able to carry child. Egg donor is inseminated by sperm donor. Then embryo is transferred and mother carries child.	4. Both parents are infertile, but mother can carry child. Egg and sperm from donors are combined in laboratory, then embryo is transferred to mother.
	5. Mother is infertile and unable to carry child. Egg of donor is combined with sperm from father. Embryo is transferred to donor who carries child.
	6. Both parents are fertile, but mother is unable to carry child. Egg from mother and sperm from father are combined. Embryo is transferred to surrogate.
	7. Father is infertile. Mother is fertile but unable to carry child. Egg from mother is combined with sperm from donor. Embryo is transferred to surrogate mother.

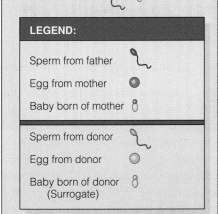

LEGEND:

Sperm from father

Egg from mother

Baby born of mother

Sperm from donor

Egg from donor

Baby born of donor (Surrogate)

In traditional in vitro fertilization, several early cell clusters may be inserted into the uterus to improve the odds that at least one will survive. More than one often does. In one study, 40 percent of IVF patients gave birth to twins. The transfer of a slightly more advanced embryo (called a blastocyst), a more recent technique, allows the cluster to develop longer before transfer into the uterus. Survival chances are better, so fewer have to be implanted. This reduces the risk of multiple births.

Figure 16.13 Some options for assisted reproductive technologies.

can be used in a given procedure. The fate of unused embryos (which are stored frozen) has prompted ethical debates, such as whether such embryos should be used as a source of embryonic stem cells (Chapter 4).

A procedure called ICSI is a variation on IVF. ICSI stands for *i*ntra*c*ytoplasmic *s*perm *i*njection. A single sperm is injected into an egg using a tiny glass needle. Although IVF and ICSI are both in common use, evidence is mounting that babies conceived through any form of in vitro fertilization have a much higher risk of low birth weight and related developmental problems later on.

In *IVF with embryo transfer* (Figure 16.13) a fertile female volunteer is inseminated with sperm from a man whose female partner is infertile. If a pregnancy results, the developing embryo is transferred to the infertile woman's uterus or to a "surrogate mother." This approach is technically difficult and has major legal complications. It isn't a common solution to infertility.

In a technique called GIFT (*g*amete *i*ntrafallopian *t*ransfer) sperm and oocytes are collected and placed into an oviduct (fallopian tube). About 20 percent of the time, the oocyte is fertilized and a normal pregnancy follows. An alternative is ZIFT (*z*ygote *i*ntrafallopian *t*ransfer). First, oocytes and sperm are placed in a laboratory dish. If fertilization occurs, the zygote is placed in a woman's oviducts. GIFT and ZIFT have about the same success rate as in vitro fertilization.

Commonly used fertility drugs include hormones that stimulate ovulation.

A variety of assisted reproductive technologies now exist. In vitro fertilization brings together sperm and oocytes in a laboratory dish, where conception may occur. Other techniques for overcoming infertility include intrafallopian transfers and artificial insemination.

16.9 A Trio of Common Sexually Transmitted Diseases

We turn now to diseases that are transmitted during sexual activity. Three of the most common ones are chlamydia, gonorrhea, and syphilis, all caused by bacteria.

CHLAMYDIAL INFECTIONS AND PID ARE MOST COMMON IN YOUNG PEOPLE

One of the most common **sexually transmitted diseases** (**STDs**) is caused by the bacterium *Chlamydia trachomatis* (Figure 16.14*a*). This infection is often called **chlamydia** for short. Each year an estimated 3 million Americans are infected, about two-thirds of them under age 25. Around the world, *C. trachomatis* infects roughly 90 million people annually. At least 30 percent of newborns who are treated for eye infections and pneumonia were infected with *C. trachomatis* during birth.

The bacterium infects cells of the genital and urinary tract. Infected men may have a discharge from the penis and a burning sensation when they urinate. Women may have a vaginal discharge as well as burning and itching. Often, however, *C. trachomatis* is a "stealth" STD with no outward signs of infection. About 80 percent of infected women and 40 percent of infected men don't have noticeable symptoms—yet they can still pass the bacterium to others.

Once a bout of chlamydia is under way, the bacteria will migrate to the person's lymph nodes, which become enlarged and tender. Impaired lymph drainage can cause swelling in the surrounding tissues.

Chlamydia can be treated with antibiotics. However, because so many people are unaware they're infected, this STD does a lot of damage. Between 20 and 40 percent of women with genital chlamydial infections develop pelvic inflammatory disease (**PID**). PID strikes about 1 million women each year, most often sexually active women in their teens and twenties.

Although PID can arise when microorganisms that normally inhabit the vagina ascend into the pelvic region (typically as a result of excessive douching), it is also a serious complication of both chlamydial infection and gonorrhea. Usually, a woman's uterus, oviducts, and ovaries are affected. Pain may be so severe that infected women often think they are having an attack of acute appendicitis. If the oviducts become scarred, additional complications, such as chronic pelvic pain and even sterility, can result. PID is the leading cause of infertility among young women. An affected woman may also have chronic menstrual problems.

As soon as PID is diagnosed, a woman usually will be prescribed a course of antibiotics. Advanced cases can require hospitalization and hysterectomy (removal of the uterus). A woman's partner should also be treated, even if the partner has no symptoms.

GONORRHEA MAY INITIALLY HAVE NO SYMPTOMS

Like chlamydial infection, **gonorrhea** can be cured if it is diagnosed promptly. Gonorrhea is caused by *Neisseria gonorrhoeae* (Figure 16.14*b*). This bacterium (also called gonococcus) can infect epithelial cells of the genital tract, the rectum, eye membranes, and the throat. Each year in the United States there are about 650,000 new cases reported; there may be up to 10 million unreported cases. Part of the problem is that the initial stages of the disease can be so uneventful that, as with chlamydial infection, a carrier may be unaware of being infected.

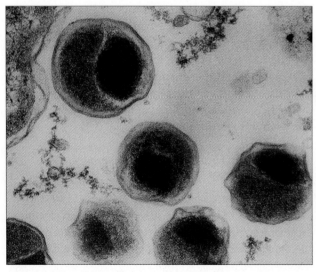

a

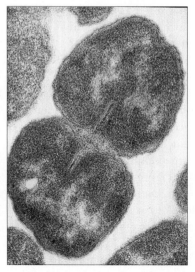

b

Figure 16.14 (**a**) Color-enhanced micrograph of *Chlamydia trachomatis* bacteria. (**b**) *Neisseria gonorrhoeae*, or gonococcus, a bacterium that typically is seen as paired cells, as shown here.

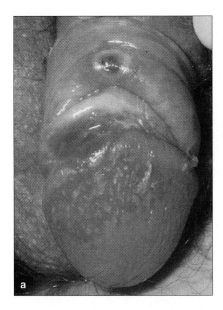

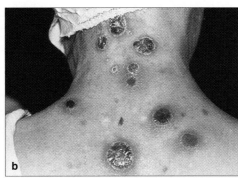

Early on, males have symptoms that usually are easy to detect. Within a week, yellow pus begins to ooze from the penis. Urinating is more frequent and may be painful. A man can become sterile if untreated gonorrhea leads to inflammation of his testicles or scarring of the vas deferens.

The early stages of gonorrhea can be dangerously asymptomatic in females. For example, a woman may not experience burning while urinating, and she may not have a vaginal discharge that seems abnormal. In the absence of worrisome symptoms, a woman's gonorrhea infection may well go untreated and all the while, the bacteria may be spreading into her oviducts. Eventually, she may experience violent cramps, fever, and vomiting. She may even become sterile due to scarring and blocking of her oviducts from pelvic inflammatory disease.

Antibiotics can kill the gonococcus and thus prevent complications of gonorrhea. Penicillin was once the most commonly used drug treatment. Unfortunately, antibiotic-resistant strains of gonococcus have developed. As a result, many doctors now order testing to determine the strain responsible for a particular patient's illness and then treat the infection with an appropriate antibiotic.

Many people believe that once cured of gonorrhea, they can't be reinfected. That is not true, partly because there are at least sixteen different strains of *N. gonorrhoeae*.

SYPHILIS AFFECTS A WIDE RANGE OF ORGANS

Syphilis is caused by the bacterium *Treponema pallidum*. There are an estimated 12 million new cases every year in the United States.

The bacterium, a *treponeme*, is transmitted by sexual contact. Once it reproduces, an ulcer called a chancre ("shanker," Figure 16.15*a*) develops. Usually the chancre is flat rather than bumpy, is not painful, and teems with treponemes. It becomes visible 1 to 8 weeks after infection and is a symptom of the *primary stage* of syphilis. Syphilis can be diagnosed in a cell sample taken from a chancre. By then, however, bacteria have already moved into the person's bloodstream.

The *secondary stage* of syphilis begins a couple of months after the chancre appears. Lesions can develop in mucous membranes, the eyes, bones, and the central nervous system. A blotchy rash breaks out over much of the body (Figure 16.15*b*). After the rash goes away, the infection enters a latent stage that can last for years. In the meantime, the disease does not produce major outward symptoms and can be detected only by laboratory tests.

Usually, the *tertiary stage* of syphilis begins from 5 to 20 years after infection. Lesions may develop in the skin and internal organs, including the liver, bones, and aorta. Scars form; the walls of the aorta can weaken. Treponemes also damage the brain and spinal cord in ways that lead to various forms of insanity and paralysis. Infected women who become pregnant typically have miscarriages, stillbirths, or sickly and syphilitic infants.

Penicillin may cure syphilis during the early stages, although antibiotic-resistant strains have now developed.

Chlamydial infection is the most common STD caused by a bacterium. It is curable with antibiotics, as gonorrhea and syphilis often are. However, there are now some antibiotic-resistant strains of the gonorrhea and syphilis bacteria.

Pelvic inflammatory disease is a dangerous complication of chlamydial infection and gonorrhea.

16.10 The Human Immunodeficiency Virus and AIDS

LINKS TO
SECTIONS
10.2, 10.5, 10.6,
AND 10.7

HIV infection rates continue to skyrocket. Worldwide, roughly 11 of every 1,000 adults between the ages of 15 and 49 are infected. Virtually all will develop AIDS.

AIDS is a group of diseases caused by infection with **HIV**, the **human immunodeficiency virus**. HIV destroys T cells, crippling the immune system. It leaves the body vulnerable to infections and rare forms of cancer (Figure 16.16). Physicians diagnose AIDS if a patient has a severely depressed immune system, tests positive for HIV, and has one or more "indicator diseases," including types of pneumonia, cancer, recurrent yeast infections, and drug-resistant tuberculosis. Worldwide, HIV has infected an estimated 34 million to 46 million people (Table 16.4).

HIV IS TRANSMITTED IN BODY FLUIDS

HIV is transmitted when body fluids, especially blood and semen, of an infected person enter another person's tissues. The virus can enter through any kind of cut or abrasion, anywhere on or in the body. HIV-infected blood also can be present on toothbrushes and razors; on needles used to inject drugs intravenously, pierce ears, do acupuncture, or create tattoos; and on contaminated medical equipment.

The most common mode of transmission is sex with an infected partner. HIV in semen and vaginal secretions enters a partner's body through epithelium lining the penis, vagina, rectum, or (rarely) mouth. Anything that damages the epithelial linings, such as other sexually transmitted diseases, anal intercourse, or rough sex, increases the odds that the virus will be transmitted.

HIV is not effectively transmitted by food, air, water, casual contact,

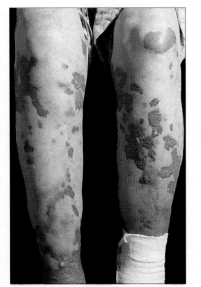

Figure 16.16 The lesions that are a sign of Kaposi's sarcoma.

or insect bites. However, infected mothers can transmit HIV to their babies during pregnancy, birth, and breast-feeding. HIV also travels in tiny amounts of infected blood in syringes that are shared and reused by IV drug abusers and patients in cash-strapped hospitals of developing countries.

Almost half of HIV-infected adults worldwide are women. Some of those infections are due to intravenous drug abuse, but most are the result of sexual contact with infected men. In the United States about 15 percent of new infections in men come from heterosexual contact. Young people are also being hit hard; in recent years, more young adults in the United States have died from AIDS than from any other single cause.

HIV INFECTION BEGINS A TITANIC STRUGGLE

HIV is a retrovirus with a lipid envelope, a bit of plasma membrane acquired as it budded from an infected cell. Various proteins spike from the envelope, span it, or line its inner surface. Inside the envelope, viral coat proteins enclose two RNA strands and several copies of *reverse transcriptase*, a retroviral enzyme. Once inside a host cell, this enzyme uses the RNA as a template to make DNA, which gets inserted into a host chromosome.

HIV can only infect cells that have a certain type of surface receptor. Macrophages, dendritic cells, and helper T cells (Section 10.6) have this receptor. Once the viral enzyme is in a cell, it uses the viral RNA as a template to make DNA, genetic instructions in the form of genes that then are inserted into one of the host cell's chromosomes. In some cells, the inserted genes remain silent but are activated in a later round of infection. Eventually, though, instructions for making new viral particles are read out. A process called *transcription* rewrites the genetic message in DNA as RNA, and these RNA instructions then are "translated" into protein (Chapter 22). These steps are summarized in Figure 16.17.

After HIV successfully infects a person, virus particles begin to circulate in the bloodstream. At this stage, many people have a bout of flulike symptoms. B cells make antibodies to HIV that can be detected by diagnostic tests for HIV infection. Armies of helper T cells and killer T cells also form. During some phases of infection, however, the virus infects an estimated 2 *billion* helper T cells and produces *100 million to 1 billion new HIV particles* each day. They bud from the plasma membrane of the helper T cell or are released when the membrane ruptures.

Region	AIDS Cases	New HIV Cases
Sub-Saharan Africa	25,400,000	3,100,000
South/Southeast Asia	7,100,000	890,000
Latin America	1,700,000	240,000
Central Asia/East Europe	1,400,000	210,000
East Asia	1,100,000	290,000
North America	1,000,000	44,000
Western/Central Europe	610,000	21,000
Middle East/North Africa	540,000	92,000
Caribbean Islands	440,000	53,000
Australia/New Zealand	35,000	5,000

Table 16.4   Global Cases of HIV and AIDS*

*Global estimates as of December 2004, www.unaids.org

SEXUALLY TRANSMITTED DISEASES

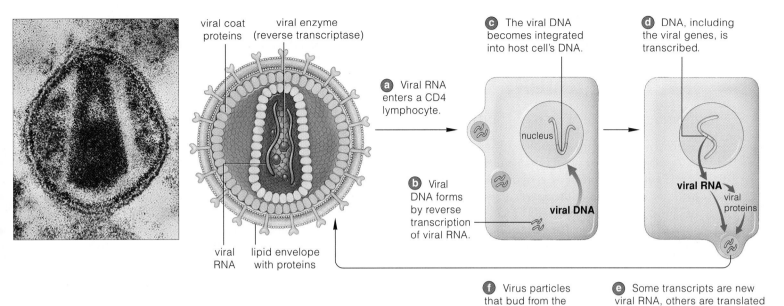

viral coat proteins

viral enzyme (reverse transcriptase)

c The viral DNA becomes integrated into host cell's DNA.

d DNA, including the viral genes, is transcribed.

a Viral RNA enters a CD4 lymphocyte.

nucleus

b Viral DNA forms by reverse transcription of viral RNA.

viral DNA

viral RNA

viral proteins

viral RNA

lipid envelope with proteins

f Virus particles that bud from the infected cell may attack a new one.

e Some transcripts are new viral RNA, others are translated into proteins. Both self-assemble into new virus particles.

Figure 16.17 *Animated!* Replication cycle of HIV.

Over time, billions of HIV particles and masses of infected T cells accumulate in lymph nodes. The number of circulating virus particles also increases and the body produces fewer and fewer helper T cells to replace those it has lost. As the number of healthy helper T cells drops, the person may lose weight and experience symptoms such as fatigue, nausea, heavy night sweats, enlarged lymph nodes, and a series of minor infections. With time, one or more of the typical AIDS indicator diseases appear. These are the diseases that eventually kill the individual.

About 5 percent of HIV-infected people don't develop symptoms of AIDS. Some may have been infected with a less virulent strain of HIV. Others may have a genetic mutation that results in the absence of the necessary receptor on some T cells and macrophages. As a result, the person is more resistant to HIV infection. People who have the mutation are *not* immune to HIV and AIDS. However, the mutation does slow the rate at which HIV infects cells, and thus the onset of AIDS. This delay allows time for a counterattack by anti-HIV treatments.

WHAT ABOUT DRUGS AND VACCINES?

Currently available drugs do not cure HIV-infected people, because we cannot go after HIV genes that are already inserted into someone's DNA. Also, since HIV mutates so rapidly, it can rapidly develop resistance to drugs. Even so, researchers have developed a fairly effective arsenal of anti-HIV drugs. Protease inhibitors block the action of HIV protease, an enzyme required for the assembly of new virus particles. Other drugs inhibit an enzyme that HIV needs to replicate itself.

At present the preferred treatment is a drug "cocktail" that often consists of a protease inhibitor and two anti-HIV drugs. This regimen can sometimes suppress HIV rather effectively, at least for a time. The drug cocktails are costly, however, and they may have serious side effects. The search also is on for compounds that might disrupt the ability of HIV to enter cells. Such "entry inhibitors" are now being tested in humans.

As you read in the chapter introduction, making an effective AIDS vaccine is a tall order. A vaccine works by stimulating the immune system to respond to proteins (antigens) produced by the invading virus. But as already noted, HIV mutates rapidly as it replicates in the immune system. In a single person it can have many different genetic forms, each form presenting the immune system with a different antigen. No single vaccine can keep up with this challenge.

Despite the obstacles, researchers are not giving up. Some efforts revolve around producing a synthetic virus with just enough HIV components to activate the immune system. At this writing, there are a wide variety of experimental vaccines in the pipeline.

AIDS is caused by the human immunodeficiency virus (HIV). HIV is a retrovirus, with RNA instead of DNA. It is transmitted only when blood, semen, or certain other body fluids of an infected person enter another person's tissues.

When HIV infects a cell, its RNA is integrated into the host cell's DNA. Eventually the host cell begins producing new HIV particles.

16.11 A Rogue's Gallery of Other STDs

A variety of viruses, parasites, and fungi cause disorders that can be transmitted by sexual contact. Some of these STDs are merely inconvenient. Others are serious threats to health.

GENITAL HERPES IS A LIFELONG INFECTION

Infections with herpes simplex viruses, or HSV, are extremely contagious. HSV is transmitted by contact with active viruses or sores that contain them (Figure 16.18). Mucous membranes of the mouth or genitals and broken or damaged skin are especially susceptible.

In 2005 the National Institutes of Health estimated that in the United States, one in five people over the age of twelve—roughly 45 million people—have one of the two strains of HSV that cause **genital herpes**. Type 1 strains infect mainly the lips, tongue, mouth, and eyes. Type 2 strains cause most genital infections. Symptoms most often develop within two weeks after infection, although sometimes they are mild or absent. Usually, small, painful blisters erupt on the penis, vulva, cervix, urethra, or anal tissues. The sores can also occur on the buttocks, thighs, or back. The initial flare-up may cause flulike symptoms for several days. Within three weeks the sores crust over and heal.

Every so often the virus may be reactivated. Then it produces new, painful sores at or near the original site of infection. Recurrences can be triggered by stress, sexual intercourse, menstruation, a rise in body temperature, or other infections.

There is no cure for herpes. Between flare-ups, HSV simply is latent in nervous tissue. However, several antiviral drugs inhibit the virus's ability to reproduce. They also reduce the shedding of virus particles from sores, and sores are often less painful and heal faster.

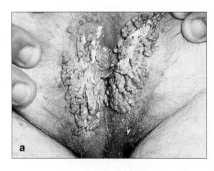

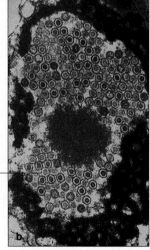

virus particles

Figure 16.18 (**a**) Genital warts caused by the human papillomavirus. (**b**) Particles of herpes virus in an infected cell.

HUMAN PAPILLOMAVIRUS CAN CAUSE CANCER

Genital warts are painless growths caused by infection of epithelium by the **human papillomavirus (HPV)**. The warts can develop months or years after a person is exposed to the virus. Usually they occur in clusters on the penis, the cervix, or around the anus (Figure 16.18*b*). Certain forms of HPV are thought to cause more than 80 percent of cases of invasive cervical cancer, a rare but serious form of cervical cancer. Any woman who has a history of genital warts should tell her physician, who may recommend an annual *Pap smear*, which is a test for abnormal growth of cervix cells.

HEPATITIS CAN BE SEXUALLY TRANSMITTED

Two types of hepatitis can be transmitted through sex. Like HIV, the **hepatitis B** virus (HBV) is transmitted in blood or body fluids such as saliva, vaginal secretions, and semen. However, HBV is far more contagious than HIV. The number of sexually transmitted hepatitis B cases is growing; in the United States, about 750,000 people are living with the disease, and about 80,000 new cases are reported each year. The virus attacks the liver. A key symptom is jaundice, yellowing of the skin and whites of the eyes as the liver loses its ability to process bilirubin pigments produced by the breakdown of hemoglobin from red blood cells. In about 10 percent of cases the HBV infection becomes chronic. Carriers are people who don't have symptoms but who can easily spread infection to their intimate contacts. Chronic hepatitis can lead to liver cirrhosis or cancer. The only treatment is rest. However, people at known risk for getting the disease (such as health care workers and anyone who requires repeated blood transfusions) can be vaccinated against the virus.

In 2005 more than 170 million people worldwide were living with the **hepatitis C** virus (HVC), which causes severe liver cirrhosis and sometimes cancer. It is carried in the blood and can reside in the body for years before symptoms develop. Long associated with IV drug abuse, HVC can be transmitted sexually if contaminated blood enters a sex partner's body through cut or torn skin.

SOME STDS ARE CAUSED BY PARASITES

Several animal parasites can be transmitted by close body contact. One is **pubic lice**, also called crab lice or simply "crabs" (Figure 16.19*a*). These tiny relatives of spiders usually turn up in the pubic hair, although they can make their way to any hairy spot on the body. They cling to hairs and attach their small, whitish eggs ("nits") to the base of the hair shaft. Itching and irritation can be

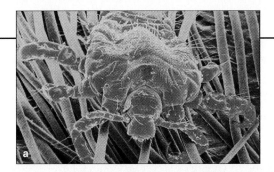

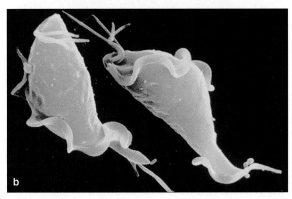

Figure 16.19 (**a**) A crab louse, magnified 120 times. Crab lice are large enough to be visible on the skin, generally as moving brownish dots. (**b**) The protozoan parasite *Trichomonas vaginalis*, which causes trichomoniasis.

intense when the parasites bite into the skin and suck blood. Antiparasitic drugs get rid of pubic lice.

Many microorganisms may live inside the vagina, although its rather acidic pH usually keeps pathogens in check. When certain vaginal infections do occur, they can be transmitted to a sex partner during intercourse. Any factor that alters the vagina's usual chemistry (such as taking an antibiotic) can trigger overgrowth of *Candida albicans*, a type of yeast (a fungus) that often lives in the vagina. A vaginal yeast infection, or **candidiasis**, causes a "cottage cheesy" discharge and itching and irritation of the vulva. A male may notice itching, redness, and flaky skin on his penis. Yeast infections are easily treated by over-the-counter and prescription medications, but both partners may need to be treated to prevent reinfection.

Trichomonas vaginalis, a protozoan parasite (Figure 16.19*b*), can cause the severe vaginal inflammation called **trichomoniasis**. The symptoms include a foul-smelling vaginal discharge and burning and itching of the vulva. An infected male may find urination painful and have a discharge from the penis, both due to an inflamed urethra. Usually both partners are treated with an antibiotic.

Common sexually transmitted diseases include genital herpes, genital warts, type B hepatitis, and infections by certain fungi and parasites.

16.12 Eight Steps to Safer Sex

The only people who are not at risk of STDs are those who are celibate (never have sex) or who are in a long-term, mutually monogamous relationship in which both partners are disease-free. The following guidelines can help you minimize your risk of acquiring or spreading an STD.

1. Use a latex condom during either genital or oral sex to greatly reduce your risk of being exposed to HIV, gonorrhea, herpes, and other diseases. With the condom, use a spermicide that contains nonoxynol-9, which may help kill virus particles. Condoms are available for men and women.

2. Limit yourself to one partner who also has sex only with you.

3. Get to know a prospective partner before you have sex. A friendly but frank discussion of your sexual histories, including any previous exposure to an STD, is very helpful.

4. If you decide to become sexually intimate, be alert to the presence of sores, a discharge, or any other sign of possible trouble in your partner's genital area.

5. Avoid abusing alcohol and drugs. Studies show that alcohol and drug abuse both are correlated with unsafe sex practices.

6. Learn about and be alert for symptoms of STDs. If you have reason to think you have been exposed, abstain from sex until a medical checkup rules out any problems. Self-treatment won't help. See a doctor or visit a clinic.

7. Take all prescribed medication and don't share it with a partner. Unless both of you take a full course of medication, your chances of reinfection will be great. Your partner may need to be treated even if he or she does not have symptoms.

8. If you do become exposed to an STD, avoid sex until medical tests confirm that you are not infected.

Figure 16.20 NBA legend Magic Johnson, a torchbearer of the 2002 Winter Olympics. He contracted HIV through heterosexual sex and credits his survival to AIDS drugs and informed medical care. He continues his campaign to educate others about AIDS.

Summary

Section 16.1 Testes are a male's primary reproductive organs. The male reproductive system also includes accessory ducts and glands.

Sperm develop mostly in the seminiferous tubules and mature in the epididymis. The seminal vesicles, bulbourethral glands, and prostate gland produce fluids that mix with sperm, forming semen.

A vas deferens leading from each testis transports sperm outward when a male ejaculates.

 Learn about the male reproductive system.

Section 16.2 The hormones testosterone, LH (luteinizing hormone), and FSH (follicle-stimulating hormone) control the formation of sperm. They are part of feedback loops among the hypothalamus, anterior pituitary, and testes. Sertoli cells, which line the seminiferous tubules, nourish sperm. Leydig cells in tissue between the tubules secrete testosterone.

A mature sperm cell has a head, midpiece, and tail. Covering much of the head is the acrosome, which contains enzymes that help a sperm penetrate an egg.

In both males and females, gonadotropin-releasing hormone (GnRH) from the hypothalamus stimulates the anterior pituitary to release LH and FSH.

Section 16.3 The paired ovaries, which produce eggs, are a female's primary reproductive organs. Accessory glands and ducts, such as the oviducts, are also part of the female reproductive system. Oviducts open into the uterus, which is lined by the endometrium.

Unless a fertilized egg begins to grow in the uterus, the endometrium proliferates, then is shed in the three-phase menstrual cycle, which averages about 28 days.

 Learn about the female reproductive system.

Sections 16.4, 16.5 The menstrual cycle overlaps with an ovarian cycle. At the end of each menstrual period, a follicle (containing an oocyte) matures in an ovary. Under the influence of hormones, the endometrium starts to rebuild.

A midcycle peak of LH triggers ovulation, the release of a secondary oocyte from the ovary.

A corpus luteum forms from the remainder of the follicle. It secretes progesterone that prepares the endometrium to receive a fertilized egg and helps maintain the endometrium during pregnancy. When no egg is fertilized, the corpus luteum degenerates, and the endometrial lining is shed through menstruation.

The hormones estrogen, progesterone, FSH, and LH control the maturation and release of eggs, as well as changes in the endometrium. They are part of feedback loops involving the hypothalamus, anterior pituitary, and ovaries.

 Observe the cyclic changes in an ovary and the effects of hormones on the menstrual cycle.

Section 16.6 Coitus (sexual intercourse) is the usual way in which egg (a secondary oocyte) and sperm meet for fertilization. It typically involves a sequence of physiological changes in both partners. Orgasm is the culmination of the sex act.

Sections 16.7, 16.8 An increasing variety of physical, chemical, surgical, or behavioral interventions are available for controlling unwanted pregnancies and for helping infertile couples. Inevitably, efforts to control fertility raise important ethical questions.

Sections 16.9–16.12 Sexually transmitted diseases (STDs) are passed by sexual activity. AIDS is a group of diseases caused by infection with the human immunodeficiency virus (HIV). HIV destroys T cells, dendritic cells, and macrophages, crippling the immune system. It is transmitted when blood, semen, or another contaminated body fluid of an infected person enters the body. Sexual contact and IV drug abuse are the most common modes of HIV transmission.

In addition to AIDS, viral STDs include genital herpes, genital warts (HPV), and viral hepatitis. Bacteria cause chlamydia, gonorrhea, and syphilis. Untreated STDs can seriously harm health. Only people who abstain from sexual contact or who are in an infection-free monogamous relationship can be sure of not being exposed to an STD.

Review Questions

1. Distinguish between:
   a. seminiferous tubule and vas deferens
   b. sperm and semen
   c. Leydig cells and Sertoli cells
   d. primary oocyte and secondary oocyte
   e. follicle and corpus luteum
   f. the three phases of the menstrual cycle

2. Which hormones influence the development of sperm?

3. Which hormones influence the menstrual and ovarian cycles?

4. List four events that are triggered by the surge of LH at the midpoint of the menstrual cycle.

5. What changes occur in the endometrium during the ovarian cycle?

6. Label the parts of the male reproductive system and state their functions.

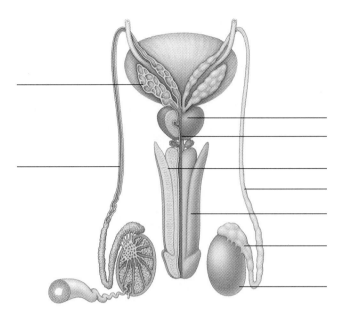

7. Label the parts of the female reproductive system and list their functions.

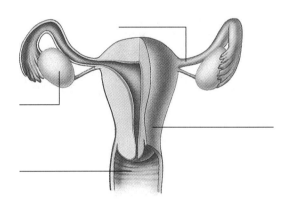

8. Figure 16.21 shows the billowing opening to the oviduct into which an ovulated oocyte is swept. Which oocyte stage is ovulated? What happens to it if (a) it encounters a sperm cell there or (b) it does not meet up with sperm?

Figure 16.21 The entrance to an oviduct, the tubelike channel to the uterus.

Self-Quiz   *Answers in Appendix V*

1. Besides producing gametes (sperm and eggs), the primary male and female reproductive organs also produce sex hormones. The _____ and the pituitary gland control secretion of both.

2. _____ production is continuous from puberty onward in males; _____ production is cyclic and intermittent in females.
   a. Egg; sperm   c. Testosterone; sperm
   b. Sperm; egg   d. Estrogen; egg

3. The secretion of _____ controls the formation of sperm.
   a. testosterone   c. FSH
   b. LH   d. all of the above are correct

4. During the menstrual cycle, a midcycle surge of _____ triggers ovulation.
   a. estrogen   c. LH
   b. progesterone   d. FSH

5. Which is the correct order for one turn of the menstrual cycle?
   a. corpus luteum forms, ovulation, follicle forms
   b. follicle grows, ovulation, corpus luteum forms

6. In order for sexual intercourse to produce a pregnancy, both partners must experience _____ .
   a. orgasm   c. affection
   b. ejaculation   d. none of the above

Critical Thinking

1. Counselors sometimes advise a couple who wish to conceive a child to use an alkaline (basic) douche immediately before intercourse. Speculate about what the doctors' reasoning might be.

2. In the "fertility awareness" method of birth control, a woman gauges her fertile period each month by monitoring changes in the consistency of her vaginal mucus. What kind of specific information does such a method provide? How does it relate to the likelihood of getting pregnant?

3. Some women experience premenstrual syndrome (PMS), which can include a distressing combination of mood swings, fluid retention (edema), anxiety, backache and joint pain, food cravings, and other symptoms. PMS usually develops after ovulation and lasts until just before or just after menstruation begins. A woman's doctor can recommend strategies for managing PMS, which often include diet changes, regular exercise, and use of diuretics or other drugs. Many women find that taking vitamin B6 and vitamin E helps reduce pain and other symptoms. Although the precise cause of PMS is unknown, it seems clearly related to the cyclic production of ovarian hormones. After reviewing Figure 16.6, suggest which hormonal changes may trigger PMS in affected females.

4. Some infertile couples are willing to go to considerable lengths to have a baby (Figure 16.13). From your reading of Section 16.8, which of the variations of reproductive technologies produces a child that is least related (genetically) to the infertile couple? Would you view having a child by that method as preferable to adopting a baby? Why?

5. The absence of menstrual periods, or amenorrhea, is normal in pregnant and postmenopausal women and in girls who have not yet reached puberty. However, in females of reproductive age amenorrhea can result from tumors of the pituitary or adrenals. Based on discussions in this chapter and Chapter 15, speculate about why such tumors might disrupt monthly menstruation.

Explore on Your Own

Public health agencies maintain statistics on the incidence of STDs. They use the numbers to measure the success of public education efforts, to identify increases in reported cases of various STDs, and to monitor the appearance of drug-resistant strains of disease-causing organisms. Table 16.5 below shows the estimated number of new cases of seven STDs in the United States and around the world. Infection by human papillomavirus (HPV) is the most widespread and fastest growing STD in the United States.

To explore how these health concerns are affecting your community or state, go online and find out if your local or state public health department maintains statistics on STDs (most do). Then see which are the most prevalent STDs in your area and whether the numbers have been rising or declining. If someone thinks they may have been exposed, what resources are available for confidential testing?

Table 16.5	Estimated New STD Cases per Year*	
STD	U.S. Cases	Global Cases
HPV infection	5,500,000	20,000,000
Trichomoniasis	5,000,000	174,000,000
Chlamydia	3,000,000	92,000,000
Genital herpes	1,000,000	20,000,000
Gonorrhea	650,000	62,000,000
Syphilis	70,000	12,000,000
AIDS	40,000	4,900,000

*Global data on HPV and genital herpes were last compiled in 1997.

17 DEVELOPMENT AND AGING

IMPACTS, ISSUES

Fertility Factors and Mind-Boggling Births

In December of 1998, Nkem Chukwu of Texas gave birth to octuplets—six girls and two boys. Born prematurely, the babies' combined weight was just over ten pounds. Odera, the smallest, weighed less than a pound (520 grams) and died of heart and lung failure six days later. The other newborns had to remain in the hospital for three months, but now are healthy youngsters.

Chukwu had received a fertility drug, which caused many eggs to mature and be ovulated at the same time. She had the option to reduce the number of embryos but chose to carry all of them to term.

As Chapter 16 noted, multiple births are becoming common, increasing almost 60 percent since the mid-1980s. The incidence of triplets and other higher-order multiple births has quadrupled.

The sharp increase in higher-order multiple births worries some doctors. Carrying more than one embryo increases the risk of miscarriage, premature delivery, and delivery complications that require surgery, such as cesarean section. Compared to single births, newborn weights are lower and mortality rates are higher. The babies are more likely to have development delays and other problems. Also, the parents face more physical, emotional, and financial burdens.

With this chapter, we consider how a human being develops. We start with principles that govern how all the specialized cells and tissues of an adult come into being—a biological journey we all have made.

 How Would You Vote? Should we restrict the use of fertility drugs to conditions that could limit the number of embryos that form? Cast your vote online at www.thomsonedu.com/biology/starr/humanbio.

 Key Concepts

EARLY DEVELOPMENT
A new individual develops in gene-guided steps that begin when gametes—sperm and eggs—form in parents. Each step builds on body structures formed in the preceding one.

PRENATAL DEVELOPMENT
Early development forms a multicellular embryo and its organs. In the fetal phase, organs and other structures grow and mature.

BIRTH AND LATER DEVELOPMENT
Body structures and functioning change throughout life as a person moves from infancy into childhood, adolescence, adulthood, and the later years.

Links to Earlier Concepts

This chapter builds on the principles of reproduction introduced in Chapter 16 and draws on your understanding of hormones that influence the menstrual cycle (16.4).

Here you will also learn the basics of how organ systems such as the digestive system (Chapter 7), nervous system (Chapter 13), and male and female reproductive systems (Chapter 16) start to develop in an embryo. You will use your knowledge of the cardiovascular system as you study some special features of this system in a developing fetus. You will see how a cascade of hormones of the hypothalamus and pituitary set the stage for birth (15.3), and you will learn more about the effects of aging on various organ systems.

17.1 The Six Stages of Early Development: An Overview

LINKS TO SECTIONS 4.8, 10.5, 16.2, AND 16.4

Biologists divide the early development of the human body into six stages, beginning when gametes form.

IN THE FIRST THREE STAGES, GAMETES FORM, AN EGG IS FERTILIZED, AND CLEAVAGE OCCURS

Development begins when sperm or eggs—the **gametes**—form and mature in the male and female who will become a new individual's parents. The next stage is **fertilization**. It begins when a sperm enters a secondary oocyte. After a sequence of steps, fertilization produces a **zygote** (ZYE-goat, "yoked together"), the first cell of the new individual.

Next comes **cleavage**, when cell divisions convert the zygote to a ball of cells (Figure 17.1). This is the point in your existence when you first became a multicellular creature. The first cleavage divides the zygote into two cells. After the third round of cleavage, there are sixteen embryonic cells arranged in a compact ball called a *morula* (MOE-roo-lah) from a Latin word for mulberry.

One of the more interesting results of cleavage is that each new cell—called a *blastomere*—ends up with a particular portion of the egg's cytoplasm. Which bit of cytoplasm a blastomere receives helps determine the developmental fate of cells that arise from it. For instance, one blastomere may receive cytoplasm that contains molecules of a protein that can activate, say, the gene coding for a certain hormone. Later on, *only* the descendants of that blastomere will make the hormone.

IN STAGE FOUR, THREE PRIMARY TISSUES FORM

After cleavage comes *gastrulation* (gas-trew-LAY-shun), a process that rearranges the morula's cells. It lays out the basic organization for the body as cells are arranged into three primary tissues, called **germ layers**. The outer layer is called **ectoderm**, the middle layer **mesoderm**, and the innermost layer **endoderm**. Subgroups of cells in each layer will give rise to the various tissues and organs in the body (Table 17.1).

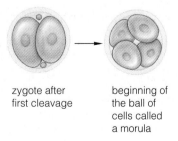

zygote after first cleavage

beginning of the ball of cells called a morula

Figure 17.1 How cleavage in a zygote creates a ball of cells. In the diagram of the first cleavage, the small spheres are polar bodies (Section 16.4).

Table 17.1 Tissues and Organs Derived from the Three Germ Layers in an Embryo

Germ Layer	Body Parts in an Adult
Ectoderm	Nervous system and sense organs Pituitary gland Outer layer of skin (epidermis) and its associated structures, such as hair
Mesoderm	Cartilage, bone, muscle, and various connective tissues Cardiovascular system (including blood) Lymphatic system Urinary system Reproductive system Outer layers of the digestive tube and of structures that develop from it, including parts of the respiratory system
Endoderm	Lining of the digestive tube and of structures that develop from it, such as the lining of the respiratory airways

IN STAGES FIVE AND SIX, ORGANS BEGIN TO FORM, THEN GROW AND BECOME SPECIALIZED

Organogenesis is the name for the overall process by which organs form. During this phase, different sets of cells get their basic biological identity—that is, they come to have a specific structure and function. Their identity assigned, cells can give rise to different tissues (nervous tissue, muscle tissue, and so on), which in turn become arranged in organs. In the final stage of development, *growth and tissue specialization*, our organs grow larger and take on the properties required for them to function in specialized ways—such as pumping blood or filtering wastes. This stage continues into adulthood as, for example, the reproductive organs mature.

Three crucial processes accomplish the changes that mold our specialized tissues and organs. The first of these processes is **cell determination**. It establishes which of several possible developmental paths an embryonic cell can follow—for example, whether its fate is to become the forerunner of some kind of nervous tissue, or of epithelium. It's a little like freshman college students being divided into liberal arts majors, business majors, science majors, and so on. In an early embryo, a given cell's fate (and, eventually, the fate of its descendants) depends on where in the embryo the cell originates—the portion of the egg's cytoplasm that a blastomere receives. As an embryo's development progresses, each cell's fate also is influenced by physical interactions and chemical signaling that goes on between groups of cells.

Next, a gene-guided process of **cell differentiation** takes place. To continue our analogy, think of a group of

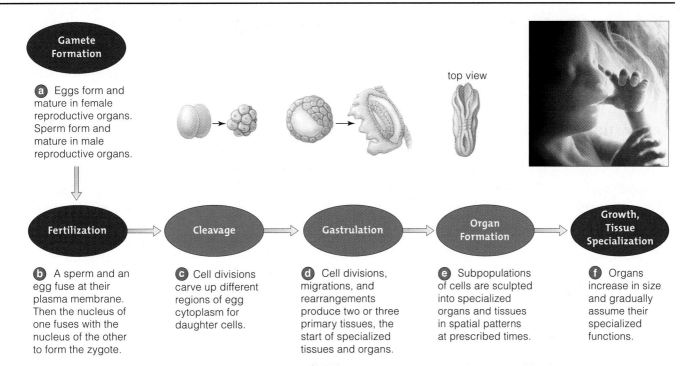

Gamete Formation

a Eggs form and mature in female reproductive organs. Sperm form and mature in male reproductive organs.

top view

Fertilization

b A sperm and an egg fuse at their plasma membrane. Then the nucleus of one fuses with the nucleus of the other to form the zygote.

Cleavage

c Cell divisions carve up different regions of egg cytoplasm for daughter cells.

Gastrulation

d Cell divisions, migrations, and rearrangements produce two or three primary tissues, the start of specialized tissues and organs.

Organ Formation

e Subpopulations of cells are sculpted into specialized organs and tissues in spatial patterns at prescribed times.

Growth, Tissue Specialization

f Organs increase in size and gradually assume their specialized functions.

college science majors, some of whom go on to specialize in biology while others specialize in physics, still others in chemistry, and so forth. As cells differentiate, they come to have specific structures and the ability to make products that are associated with particular functions. For example, you may remember from Chapter 10 how various subsets of T cells, with different functions, differentiate in the thymus.

Morphogenesis ("the beginning of form") produces the shape and structure of each body region. It involves cell division in limited areas and growth and movements of cells and tissues from one place to another. For instance, most of the bones of your face are descended from cells that migrated from the back of your head when you were an early embryo.

As you will read shortly, during morphogenesis sheets of tissue fold and certain cells die on cue. Figure 17.2 summarizes the six stages of development. It is important to remember that by the end of each stage, the embryo has become more complex than it was before. Normal development requires that each stage be completed before the next one begins.

Development begins when gametes form in each parent. The next stages are fertilization, then cleavage of the zygote, then gastrulation, organogenesis, and finally the growth and specialization of tissues.

Each stage of embryonic development builds on structures that were formed during the stage preceding it. For proper development, each stage must be successfully completed before the next begins.

Figure 17.2 *Animated!* The development of an embryo from fertilization to about 6 weeks. (**a**–**f**) For clarity, the membranes surrounding the embryo are not shown. Several stages are shown in cross section.

Conjoined twins form when an embryo partially splits after day 12. The twins remain joined, usually at the chest or abdomen. In 2002, a team at UCLA's Mattel Hospital successfully separated these Guatemalan sisters who had been joined at the head.

17.2 The Beginnings of You—Fertilization to Implantation

LINKS TO
SECTIONS
16.2, 16.4,
AND 16.6

If sperm enter a female's vagina during the fertile period of her menstrual cycle, an oocyte can be fertilized and an embryo can begin to develop.

FERTILIZATION UNITES SPERM AND OOCYTE

As sperm swim through the cervix and uterus and into the oviducts, they are not quite ready to fertilize an oocyte. First, *capacitation* occurs. In this process, chemical changes weaken the membrane over the sperm's acrosome. Only a sperm that is capacitated ("made able") can fertilize an oocyte (Figure 17.3). Of the millions of sperm in the vagina after an ejaculation, just several hundred reach the upper part of an oviduct, which is where fertilization usually occurs. Uterine muscle contractions help move sperm toward the oviducts.

When a capacitated sperm contacts an oocyte, enzymes are released from the now-fragile region of cell membrane covering the acrosome. Many sperm can reach and bind to the oocyte, and acrosome enzymes clear a path through the zona pellucida. Usually, however, only one sperm fuses with the oocyte. Rapid chemical changes in the oocyte cell membrane block more sperm from entering.

Fusion with a sperm stimulates the completion of the cell division process (meiosis II) that began when the oocyte was being formed in an ovary (Section 16.4). The result is a mature egg, or **ovum** (plural: ova), plus another polar body. (Remember that one or, often, two polar bodies are produced when meiosis I gives rise to the secondary oocyte; thus there usually are three tiny polar bodies "packaged" with the ovum.) The nuclei of the sperm and ovum swell up, then fuse. Recall that a sperm or oocyte has only twenty-three chromosomes, *half* the number present in other body cells. Fertilization combines them into a full diploid set of forty-six chromosomes. Thus a zygote has all the DNA required to guide proper development of the embryo.

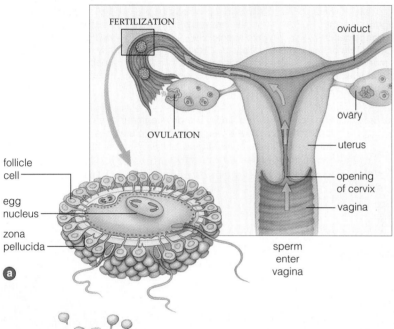

a

follicle cell

egg nucleus

zona pellucida

b

nuclei fuse

fusion of sperm nucleus with egg nucleus

c

d

Figure 17.3 *Animated!* Fertilization. (**a**) Many sperm surround a secondary oocyte. Acrosomal enzymes clear a path through the zona pellucida. (**b**) When a sperm penetrates the secondary oocyte, cortical granules in the oocyte cytoplasm release substances that prevent other sperm from penetrating the zona pellucida. Penetration also stimulates the second meiotic division of the oocyte's nucleus. (**c**) The sperm tail degenerates; its nucleus enlarges and fuses with the oocyte nucleus. (**d**) At fusion, fertilization is completed. The zygote has formed.

CLEAVAGE PRODUCES A MULTICELLULAR EMBRYO

For several days the zygote moves down the oviduct, sustained by nutrients from the ovum or from maternal secretions. On the way, three cleavages convert the single-celled zygote into a morula (Figure 17.4), the ball of cells described in Section 17.1. The morula's cells occupy the same volume as the zygote did, although they are smaller and their shape and activities differ.

When the morula reaches the uterus, a fluid-filled cavity begins to open up in it. This change transforms the morula into a **blastocyst** (*blast-* = bud). It has two tissues: a surface epithelium called the *trophoblast* (*tropho-* = to nourish) and a small clump of

EARLY DEVELOPMENT

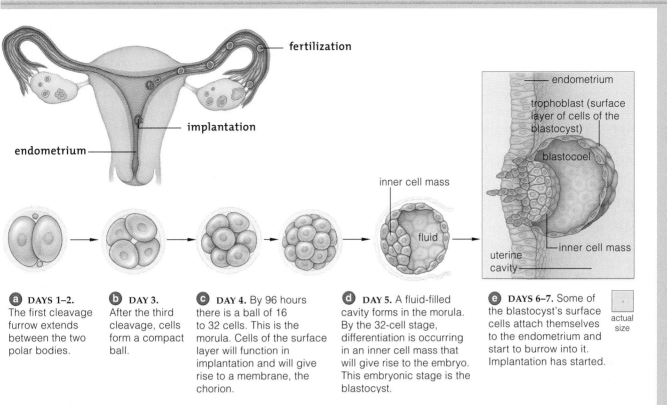

a **DAYS 1–2.** The first cleavage furrow extends between the two polar bodies.

b **DAY 3.** After the third cleavage, cells form a compact ball.

c **DAY 4.** By 96 hours there is a ball of 16 to 32 cells. This is the morula. Cells of the surface layer will function in implantation and will give rise to a membrane, the chorion.

d **DAY 5.** A fluid-filled cavity forms in the morula. By the 32-cell stage, differentiation is occurring in an inner cell mass that will give rise to the embryo. This embryonic stage is the blastocyst.

e **DAYS 6–7.** Some of the blastocyst's surface cells attach themselves to the endometrium and start to burrow into it. Implantation has started.

Figure 17.4 *Animated!* Steps from fertilization through implantation.

cells called the **inner cell mass** (Figure 17.4*e*). The **embryo** develops from the inner cell mass. Sometimes a split separates the two cells produced by the first cleavage, the inner cell mass, or an even later stage. Then, separate embryos develop as *identical twins*, who have the same genetic makeup. *Fraternal twins* result when two eggs are fertilized at roughly the same time by different sperm. Fraternal twins need not be the same sex and they don't necessarily look any more alike than other siblings do.

IMPLANTATION GIVES A FOOTHOLD IN THE UTERUS

About a week after fertilization, **implantation** begins as the blastocyst breaks out of the zona pellucida. Cells of the epithelium then invade the endometrium and cross into the underlying connective tissue. This gives the blastocyst a foothold in the uterus. As time passes it will sink deep into the connective tissue of the uterus, and the endometrium will close over it.

Occasionally a fertilized egg implants in the wrong place—in the oviduct or even in the external surface of the ovary or in the abdominal wall. This *ectopic (tubal) pregnancy* cannot go to full term and must be terminated by surgery. Sometimes it leads to permanent infertility.

Implantation is complete two weeks after the secondary oocyte was ovulated. Menstruation, which would begin at this time if the woman were not pregnant, doesn't occur because the implanted blastocyst secretes HCG (human chorionic gonadotropin). HCG stimulates the corpus luteum to continue secreting both estrogen and progesterone, which prevent the uterus lining from being shed. By the third week of pregnancy, HCG can be detected in the mother's blood or urine. At-home pregnancy tests use chemicals that change color when urine contains HCG.

Fertilization of an egg by a sperm produces a zygote, a single cell with a full set of parental chromosomes. Further development produces a multicellular blastocyst that implants in the endometrium of the uterus.

17.3 How the Early Embryo Takes Shape

The embryonic period begins shortly after fertilization and lasts for eight weeks. During that time, the basic body plan of the embryo develops.

A developing baby is considered an embryo for most of the first trimester, or three months, of the nine months of gestation. When the three germ layers—the ectoderm, mesoderm, and endoderm—are in place, organogenesis begins and the embryo's organ systems start to develop.

FIRST, THE BASIC BODY PLAN IS ESTABLISHED

By the time a woman has missed her first menstrual period, the embryo has implanted and the inner cell mass has been transformed into a pancake-shaped **embryonic disk**. Around day 15, gastrulation has rearranged cells so that a faint "primitive streak" appears at the midline of the disk (Figure 17.5a). Now, ectoderm along the midline thickens to establish the beginnings of a **neural tube**. This tube is the forerunner of the embryo's brain and spinal cord. Some of its cells also give rise to a flexible rod of cells called a *notochord*. The vertebral column will form around this rod.

These events establish the body's long axis and its bilateral symmetry. In other words, the embryonic disk is reshaped in ways that provide the body with the basic form we see in all vertebrates.

Figure 17.5 *Animated!* Hallmarks of the embryonic period of development—the appearance of a primitive streak foreshadowing the brain and spinal cord—and the formation of somites and pharyngeal arches. These are dorsal views (of the embryo's back) except for days 24–25, which is a side view.

On the surface of the embryonic disk near the neural tube, the third primary tissue layer—mesoderm—also has been forming. Toward the end of the third week, some mesoderm gives rise to **somites** (SOE-mites). These are paired blocks of mesoderm; they will be the source of most bones and skeletal muscles of the neck and trunk. The dermis overlying these regions comes from somites as well. Structures called pharyngeal arches start to form; they will contribute to development of the face, neck, mouth, and associated parts. In other mesodermal tissues spaces open up. Eventually, these spaces will merge to form the cavity (called the *coelom*) between the body wall and the gastrointestinal tract.

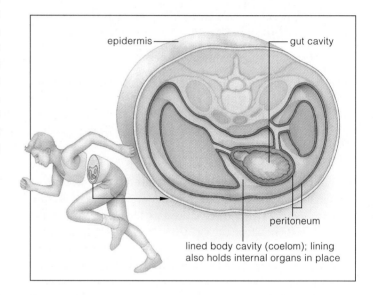

epidermis — gut cavity

peritoneum

lined body cavity (coelom); lining also holds internal organs in place

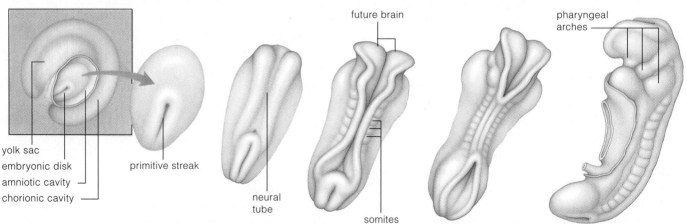

yolk sac
embryonic disk
amniotic cavity
chorionic cavity

primitive streak

neural tube

future brain

somites

pharyngeal arches

a DAY 15. A primitive streak appears along the axis of the embryonic disk. This thickened band of cells marks the onset of gastrulation.

b DAYS 19–23. Cell migrations, tissue folding, and other morphogenetic events lead to the formation of a hollow neural tube and to somites (bumps of mesoderm). The neural tube gives rise to the brain and spinal cord. Somites give rise to most of the axial skeleton, skeletal muscles, and much of the dermis.

c DAYS 24–25. By now, some cells have given rise to pharyngeal arches, which contribute to the face, neck, mouth, nasal cavities, larynx, and pharynx.

NEXT, ORGANS DEVELOP AND TAKE ON THE PROPER SHAPE AND PROPORTIONS

After gastrulation, organs and organ systems begin to form. An example is neurulation, the first stage in the development of the nervous system. Figure 17.6 shows how cells of ectoderm at the embryo's midline elongate and form a neural plate, the first sign that a region of ectoderm is starting to develop into nervous tissue. Next, cells near the middle become wedge-shaped. The changes in cell shape cause the neural plate to fold over and meet at the embryo's midline to form the neural tube.

The folding of sheets of cells is extremely important in morphogenesis. The folding takes place as microtubules lengthen and rings of microfilaments in cells constrict (Figure 17.6a). Morphogenesis also requires cells to move from one place to another. Migrating cells use extensions of their cytoplasm called pseudopodia ("false feet") to move along genetically determined routes. When the cells reach their destination, they come into contact with cells already there. Forerunners of neurons interact this way as your nervous system is forming (Figure 17.6b).

Migrating cells "find their way" in part by following adhesive cues. For instance, as the nervous system is developing, migrating Schwann cells stick to adhesion proteins on the surface of axons but not on blood vessels. Adhesive cues also tell the cells when to stop. Cells migrate to places where the cues are strongest, then stay there once they arrive.

Successful development of an embryo requires that body structures form according to normal patterns, in a certain sequence. **Apoptosis**—genetically programmed cell death—helps sculpt body parts. In this process, inside cells that are destined to die, enzymes switch on and begin digesting cell parts. For instance, morphogenesis at the ends of limb buds first produced paddle-shaped hands at the ends of your arms. Then epithelial cells between the lobes in the paddles died on cue, leaving separate fingers. Figure 17.7 shows what can happen when apoptosis does not occur normally while a human hand is forming.

In the third week after fertilization the new individual's basic body plan is established. Then morphogenesis produces the shape and proportions of body parts.

During morphogenesis, cells divide and migrate, tissues grow and fold, and certain cells die by apoptosis.

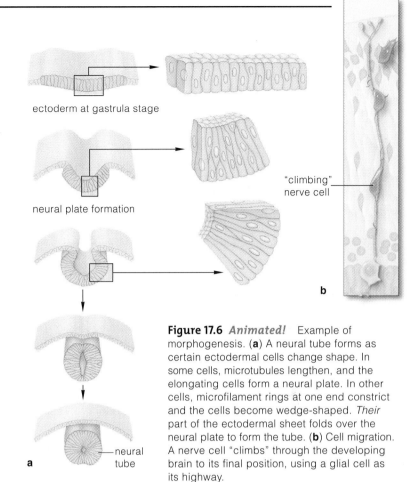

ectoderm at gastrula stage

neural plate formation

"climbing" nerve cell

neural tube

b

a

Figure 17.6 *Animated!* Example of morphogenesis. (**a**) A neural tube forms as certain ectodermal cells change shape. In some cells, microtubules lengthen, and the elongating cells form a neural plate. In other cells, microfilament rings at one end constrict and the cells become wedge-shaped. *Their* part of the ectodermal sheet folds over the neural plate to form the tube. (**b**) Cell migration. A nerve cell "climbs" through the developing brain to its final position, using a glial cell as its highway.

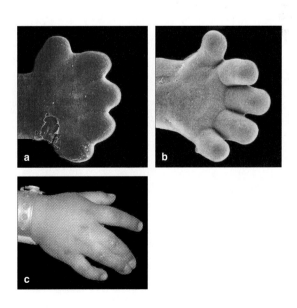

Figure 17.7 *Animated!* How fingers form. (**a**) At first webs of tissue connect the digits. (**b**) Then, cells in the webs die by apoptosis and the digits are separated. (**c**) Fingers that remained attached when embryonic cells did not die on cue.

17.4 Vital Membranes Outside the Embryo

LINKS TO
SECTIONS
15.3 AND 16.3

Implantation and several weeks following it are a dynamic time. In addition to developmental steps that begin shaping the embryo's body, outside the embryo specialized membranes form, including the all-important placenta.

FOUR EXTRAEMBRYONIC MEMBRANES FORM

As you've already read, during implantation, the inner cell mass of the blastocyst is transformed into an embryonic disk (Figure 17.8*a*). Only certain cells of the disk will give rise to the embryo. Others give rise to **extraembryonic membranes**: the yolk sac, the amnion, the allantois, and the chorion.

As the term indicates, extraembryonic membranes are not part of the embryo. One of them, the **yolk sac**, forms below the embryonic disk. Functioning only briefly, it is the source of early blood cells and of germ cells that will become gametes. Parts of it give rise to the embryo's digestive tube.

The **amnion** forms a fluid-filled sac that encloses the embryo. The amniotic fluid insulates the embryo, absorbs shocks, and prevents the embryo from drying out. Just outside it is the **allantois**, which gives rise to blood vessels that will invade the **umbilical cord**. These vessels are the embryo's contribution to circulatory "plumbing" that will link the embryo with its lifeline, the placenta.

The **chorion** wraps around the embryo and the other three membranes (Figure 17.8*c*). It continues the secretion of HCG that began when the embryo implanted. HCG will prevent the lining of the uterus (the endometrium) from breaking down until the placenta can produce enough estrogen and progesterone to maintain the lining.

THE PLACENTA IS A PIPELINE FOR OXYGEN, NUTRIENTS, AND OTHER SUBSTANCES

Three weeks after fertilization, almost a fourth of the inner surface of the uterus has become a spongy tissue, the developing **placenta**. It is the link through which nutrients and oxygen pass from the mother to the embryo and waste products from the embryo pass back to the mother's bloodstream. Said another way, the placenta is a way of sustaining a developing baby while allowing its blood vessels to develop apart from the mother's.

Although the fully developed placenta is considered an organ, it's useful to think of it as a close association of the chorion and the upper cells of the endometrium where the embryo implanted. The mother's side of the placenta is endometrial tissue that contains arterioles and venules. As the chorion develops (from the trophoblast), tiny projections sent out from the implanting blastocyst develop into many *chorionic villi*. Inside each villus are small blood vessels (Figure 17.9).

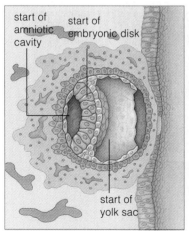

a DAYS 10–11. The yolk sac, embryonic disk, and amniotic cavity have started to form from parts of the blastocyst.   actual size

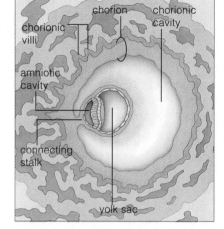

b DAY 12. Blood-filled spaces form in maternal tissue. The chorionic cavity starts to form.   actual size

c DAY 14. A connecting stalk has formed between the embryonic disk and chorion. Chorionic villi, which will be features of a placenta, start to form.   actual size

Figure 17.8 *Animated!*   How extraembryonic membranes (the amnion, chorion, and yolk sac) begin to form.

While chorionic villi are developing, the erosion of the endometrium that began with implantation continues. As capillaries in the endometrium are broken down, spaces in the disintegrating endometrial tissue fill with maternal blood. The chorionic villi extend into these spaces. As the embryo develops, very little of its blood ever mixes with that of its mother. Oxygen and nutrients simply diffuse out of the mother's blood vessels, across the blood-filled spaces in the endometrium, then into the embryo's blood vessels. Carbon dioxide and other wastes diffuse in the opposite direction, leaving the embryo.

Besides nutrients and oxygen, many other substances taken in by the mother—including alcohol, caffeine, drugs, pesticide residues, and toxins in cigarette smoke—can cross the placenta, as can HIV.

> *Extraembryonic membranes—the yolk sac, amnion, allantois, and chorion—begin to form shortly after implantation.*
>
> *The amnion forms a fluid-filled sac around the embryo. The allantois gives rise to blood vessels of the umbilical cord. The chorion helps protect the embryo and secretes HCG.*
>
> *The placenta is a spongy tissue in which maternal and embryonic blood vessels are closely associated. By way of the placenta, the embryo's bloodstream can take up nutrients and oxygen from the mother and also discharge wastes that her bloodstream will transport away.*

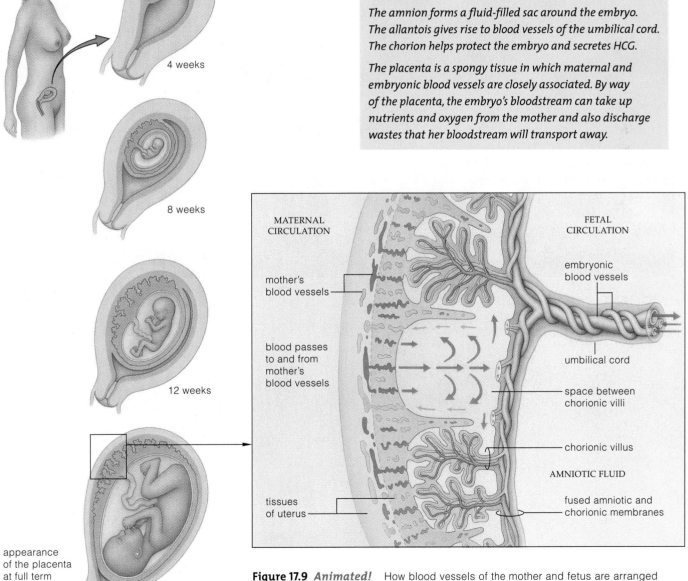

4 weeks

8 weeks

12 weeks

appearance of the placenta at full term

MATERNAL CIRCULATION

mother's blood vessels

blood passes to and from mother's blood vessels

tissues of uterus

FETAL CIRCULATION

embryonic blood vessels

umbilical cord

space between chorionic villi

chorionic villus

AMNIOTIC FLUID

fused amniotic and chorionic membranes

Figure 17.9 *Animated!* How blood vessels of the mother and fetus are arranged in a full-term placenta. Blood vessels from the fetus extend through the umbilical cord and into chorionic villi. Maternal blood spurts into spaces between villi. Oxygen, carbon dioxide, and other small solutes diffuse across the surface of the placental membrane; there is no large-scale mingling of the two bloodstreams.

17.5 The First Eight Weeks—Human Features Emerge

By the end of the fourth week of the embryonic period, the embryo has grown to 500 times its original size. Over the next few weeks recognizable human features will appear.

In an embryo's first few weeks of life, it grows rapidly and its cells begin to specialize. Morphogenesis begins to sculpt limbs, fingers, and toes. The circulatory system becomes more intricate, and the umbilical cord forms. Growth of the all-important head now surpasses that of any other body region (Figure 17.10*a*). The embryonic period ends as the eighth week draws to a close. The embryo is no longer merely "a vertebrate." As you can see from Figure 17.10*c*, its features now clearly define it as a human.

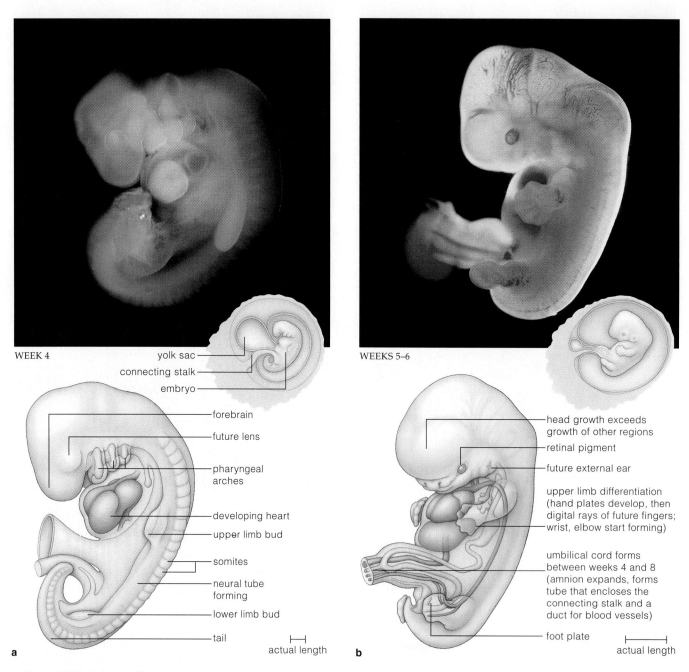

WEEK 4

- yolk sac
- connecting stalk
- embryo

- forebrain
- future lens
- pharyngeal arches
- developing heart
- upper limb bud
- somites
- neural tube forming
- lower limb bud
- tail

actual length

a

WEEKS 5–6

- head growth exceeds growth of other regions
- retinal pigment
- future external ear
- upper limb differentiation (hand plates develop, then digital rays of future fingers; wrist, elbow start forming)
- umbilical cord forms between weeks 4 and 8 (amnion expands, forms tube that encloses the connecting stalk and a duct for blood vessels)
- foot plate

actual length

b

Figure 17.10 *Animated!* (**a**) Human embryo at four weeks. As is true of all vertebrates, it has a tail and pharyngeal arches. (**b**) The embryo at five to six weeks after fertilization. (**c**, *facing page*) An embryo poised at the boundary between the embryonic and fetal periods. It now has features that are distinctly human. It is floating in fluid within the amniotic sac. The chorion, which normally covers the amniotic sac, has been opened and pulled aside.

As the second half of the first trimester gets under way, gonads begin to develop. In an embryo that has inherited X and Y sex chromosomes, a sex-determining region of the Y chromosome now triggers development of testes (Figure 17.11). Sex hormones made by the testes then influence the development of the entire reproductive system. An embryo with XX sex chromosomes will be female, and female reproductive structures begin to form in her body. Notice that no hormones are required to stimulate development of female gonads—all that is necessary is the *absence* of testosterone.

After eight weeks the embryo is just over 1 inch long, its organ systems are formed, and it is designated a **fetus**. As the first trimester ends, a heart monitor can detect the fetal heartbeat. The genitals are well formed, and a doctor often can determine a baby's sex using ultrasound.

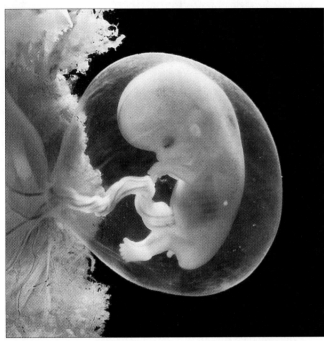

WEEK 8

final week of embryonic period; embryo looks distinctly human compared to other vertebrate embryos

upper and lower limbs well formed; fingers and then toes have separated

early tissues of all internal, external structures now developed

tail has become stubby

c

actual length

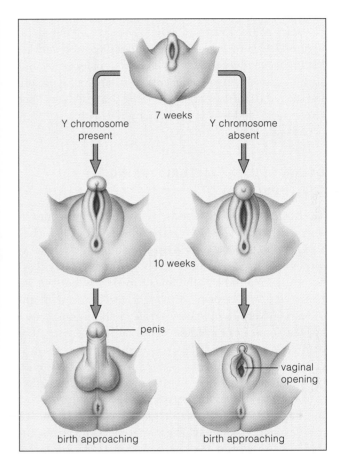

Figure 17.11 Developing genitals of male and female embryos.

MISCARRIAGE

Miscarriage, the spontaneous expulsion of an embryo or fetus, occurs in more than 20 percent of all conceptions, usually during the first trimester. Many factors can trigger a miscarriage (also called spontaneous abortion), but in as many as half the cases the embryo (or the fetus) has one or more genetic disorders that prevent it from developing normally.

During the first eight weeks of life an embryo gradually develops distinctly human body features. At the end of this period the developing individual is termed a fetus.

17.6 Development of the Fetus

LINKS TO
SECTIONS
9.3 AND 11.5

In the second and third trimesters, organs and organ systems gradually mature in preparation for birth.

IN THE SECOND TRIMESTER MOVEMENTS BEGIN

When the fetus is three months old, it is about 4.5 inches long. Soft, fuzzy hair (the lanugo) covers its body. Its reddish skin is wrinkled and protected from abrasion by a thick, cheesy coating called the *vernix caseosa*.

The second trimester of development extends from the start of the fourth month to the end of the sixth. Figure 17.12 shows what the fetus looks like at 16 weeks. Its tiny facial muscles now produce frowns, squints, and sucking movements—evidence of a sucking reflex. Before the second trimester ends, the mother can easily feel her fetus's arms and legs move. During the sixth month, its eyelids and eyelashes form.

ORGAN SYSTEMS MATURE DURING THE THIRD TRIMESTER

The third trimester extends from the seventh month until birth. At seven months the fetus is about 11 inches long, and soon its eyes will open. Although the fetus is growing much larger and rapidly becoming "babylike," not until the middle of the third trimester will it be able to survive on its own. Although development might seem relatively complete by the seventh month, at that age few fetuses can maintain a normal body temperature or breathe normally. However, with intensive medical care, fetuses as young as 23 to 25 weeks have survived early delivery. A baby born before seven months' gestation is at high risk of *respiratory distress syndrome* (described in Chapter 11) because its lungs lack surfactant and so can't expand adequately. The longer the baby can stay in its mother's uterus, the better. By the ninth month, its survival chances are about 95 percent.

THE BLOOD AND CIRCULATORY SYSTEM OF A FETUS HAVE SPECIAL FEATURES

The steady maturation of its organs and organ systems readies the fetus for independent life. For the circulatory system, however, the path toward independence requires a detour. Several temporary bypass vessels form and will function until birth. As Figure 17.13 shows, two umbilical arteries inside the umbilical cord transport deoxygenated blood and metabolic wastes from the fetus to the placenta. There, the fetal blood gives up wastes, takes on nutrients, and exchanges gases with the mother's blood. Fetal hemoglobin is slightly different from adult hemoglobin. It binds oxygen more readily, helping ensure that adequate

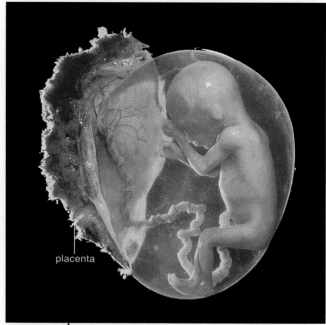

placenta

WEEK 16
Length: 16 centimeters (6.4 inches)
Weight: 200 grams (7 ounces)

WEEK 29
Length: 27.5 centimeters (11 inches)
Weight: 1,300 grams (46 ounces)

WEEK 38 (full term)
Length: 50 centimeters (20 inches)
Weight: 3,400 grams (7.5 pounds)

During fetal period, length measurement extends from crown to heel (for embryos, it is the longest measurable dimension, as from crown to rump).

Figure 17.12 (*top*) The fetus at 16 weeks. During the fetal period, movements begin as soon as nerves establish functional connections with developing muscles. Legs kick, arms wave, fingers grasp, the mouth puckers. These reflex actions will be vital skills in the world outside the uterus. The drawing shows a baby at full term—ready to be born.

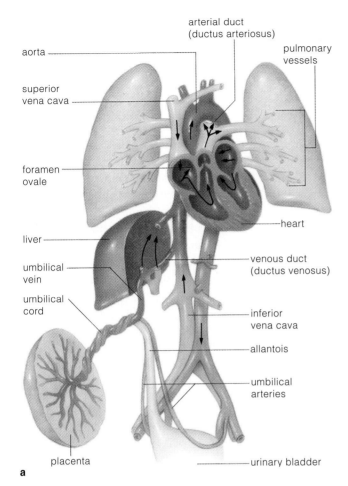

aorta
superior vena cava
foramen ovale
liver
umbilical vein
umbilical cord
placenta

arterial duct (ductus arteriosus)
pulmonary vessels
heart
venous duct (ductus venosus)
inferior vena cava
allantois
umbilical arteries
urinary bladder

a

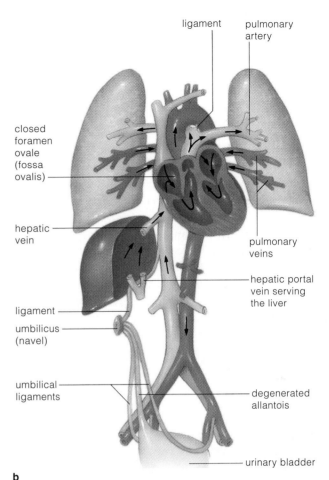

closed foramen ovale (fossa ovalis)
hepatic vein
ligament
umbilicus (navel)
umbilical ligaments

ligament
pulmonary artery
pulmonary veins
hepatic portal vein serving the liver
degenerated allantois
urinary bladder

b

Figure 17.13 Blood circulation in a fetus (*arrows*). (**a**) Umbilical arteries carry deoxygenated blood from fetal tissues to the placenta. There blood picks up oxygen and nutrients from the mother's bloodstream and returns to the fetus via the umbilical vein. Blood mainly bypasses the lungs, moving through the foramen ovale and the arterial duct. It bypasses the liver by moving through the venous duct.

(**b**) At birth the foramen ovale closes, and the pulmonary and systemic circuits of blood flow become completely separate. The arterial duct, venous duct, umbilical vein, and portions of the umbilical arteries become ligaments, and the allantois degenerates.

oxygen will reach developing fetal tissues. The oxygenated blood, enriched with nutrients, returns from the placenta to the fetus in the umbilical vein.

Other temporary vessels divert blood past the lungs and liver. These organs don't develop as rapidly as some others, because (by way of the placenta) the mother's body can perform their functions. The lungs of a fetus are collapsed and won't function for gas exchange until the newborn takes its first breaths after birth. Until then, its lung tissues receive only enough blood to sustain their development. A little of the blood entering the heart's right atrium flows into the right ventricle and moves on to the lungs. Most of it, however, travels through a gap in the interior heart wall (called the *foramen ovale*, or "oval opening") or into an arterial duct (*ductus arteriosus*) that bypasses the nonfunctioning lungs entirely.

Likewise, most blood bypasses the fetal liver because the mother's liver performs most liver functions (such as

nutrient processing) until birth. Nutrient-laden blood from the placenta travels through a venous duct (the *ductus venosus*) past the liver and on to the heart, which pumps it to body tissues. At birth, blood pressure in the heart's left atrium increases. This causes a valvelike flap of tissue to close off the foramen ovale, which then gradually seals and separates the pulmonary and systemic circuits of blood flow (Figure 17.13*b*). The temporary vessels that have formed in a fetus gradually close during the first few weeks after birth.

The organs and organ systems of a fetus mature during the second and third trimesters. Because the fetus exchanges gases and receives nutrients via its mother's bloodstream, its circulatory system develops temporary vessels that bypass the lungs and liver until birth.

17.7 Birth and Beyond

LINKS TO SECTIONS 15.3 AND 16.2

Birth, or parturition, takes place about 39 weeks after fertilization—about 280 days from the start of the woman's last menstrual period.

HORMONES TRIGGER BIRTH

Usually within two weeks of a pregnant woman's "due date," the birth process, "labor," begins when smooth muscle in her uterus starts to contract. These contractions are the indirect result of a cascade of hormones from the fetus's hypothalamus, pituitary, and adrenal glands, which is triggered by an as-yet-unknown signal that says, in effect, it's time to be born. The hormonal flood causes the placenta to produce more estrogen. Rising estrogen in turn calls for a rush of oxytocin and of prostaglandins (also produced by the placenta), which stimulate the uterine contractions. For about the next 2 to 18 hours, the contractions will become stronger, more painful, and more frequent.

LABOR HAS THREE STAGES

Labor is divided into three stages that we can think of loosely as "before, during, and after." In the first stage, uterine contractions push the fetus against its mother's cervix. Initial contractions occur about every 15 to 30 minutes and are relatively mild. As the cervix gradually dilates to a diameter of about 10 centimeters (4 inches, or "5 fingers"), contractions become more frequent and intense. Usually, the amniotic sac ruptures during this stage, which can last 12 hours or more.

The second stage of labor, actual birth of the fetus, typically occurs less than an hour after the cervix is fully dilated. This stage is usually brief—under 2 hours. Strong contractions of the uterus and abdominal muscles occur every 2 or 3 minutes, and the mother feels an urge to push. Her efforts and the intense contractions move the soon-to-be newborn through the cervix and out through the vaginal canal, usually head first (Figure 17.14). Complications can develop if the baby begins to emerge in a "bottom-first" (*breech*) position; the attending physician may use hands or forceps to aid the delivery.

After the baby is expelled, the third stage of labor gets under way. Uterine contractions force fluid, blood, and the placenta (now called the afterbirth) from the mother's body. The umbilical cord—the lifeline to the mother—is now severed. A lasting reminder of this separation is the scar we call the navel—the site where the umbilical cord was attached.

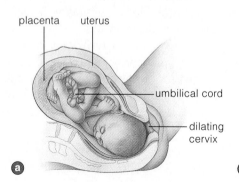

placenta   uterus

umbilical cord

dilating cervix

a

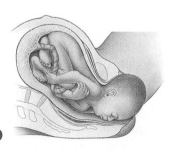

b

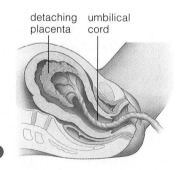

detaching placenta   umbilical cord

c

Figure 17.14 *Animated!* Expulsion of the fetus during birth. The afterbirth—the placenta, fluid, and blood—is expelled shortly afterward.

BIRTH AND LATER DEVELOPMENT

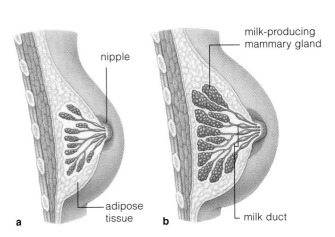

Figure 17.15 *Animated!* (**a**) Breast anatomy. (**b**) Breast of a lactating female.

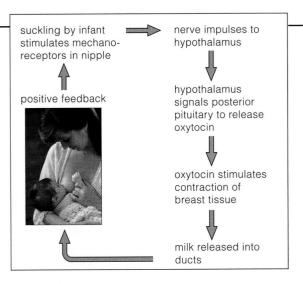

Figure 17.16 How a positive feedback mechanism keeps milk flowing to a suckling infant.

Without the placenta to remove wastes, carbon dioxide builds up in the baby's blood. Together with other factors, including handling by medical personnel, this stimulates control centers in the brain, which respond by triggering inhalation—the newborn's crucial first breath.

As the infant's lungs begin to function, the bypass vessels of the fetal circulation begin to close, soon to shut completely. The fetal heart opening, the foramen ovale, normally closes slowly during the first year of life.

Most full-term pregnancies end in the birth of a healthy infant. Yet babies born prematurely—especially before about eight months of intrauterine life—can suffer complications because their organs have not developed to the point where they can function independently. Then, attempts to sustain the baby's life under conditions that will permit the necessary additional development may require a variety of advanced medical technologies. Even then, the majority of extremely premature infants do not survive.

HORMONES ALSO CONTROL MILK PRODUCTION IN A MOTHER'S MAMMARY GLANDS

Milk production provides an excellent example of how hormones may interact in the body. In this case, a total of four hormones are involved. During pregnancy, estrogen and progesterone stimulate the growth of mammary glands and ducts in the mother's breasts (Figure 17.15). For the first few days after birth, those glands produce colostrum, a pale fluid that is rich in proteins, antibodies, minerals, and vitamin A. Then prolactin secreted by the pituitary stimulates milk production, or **lactation**.

The "let-down" or flow of milk from a nursing mother's mammary glands is a reflex and an example of positive feedback (Figure 17.16). When a newborn nurses, mechanoreceptors in the nipple send nerve impulses to the hypothalamus, which in turn stimulates the mother's pituitary to release oxytocin, which causes the mother's breast tissues to contract. This forces milk into the ducts. This response continues as long as the baby suckles. Oxytocin also triggers contractions of uterine muscle that will help to "shrink" the uterus back to its normal size.

The mother's cervix dilates during the first stage of labor. The baby is born during the second stage. In the third stage, uterine contractions expel the placenta.

Lactation, or milk production, begins a few days after birth. It is stimulated by the hormone prolactin, which is released from the mother's pituitary and acts on her breast (mammary gland) tissues.

Suckling triggers a reflex in which oxytocin released from the pituitary acts to force milk into mammary ducts. The response continues as long as the infant suckles.

17.8 Potential Disorders of Early Development

From fertilization until birth, a woman's future child is at the mercy of her diet, health habits, and lifestyle. This section looks at common concerns with regard to nutrition, the risk of infection, and the use of legal and illegal drugs.

GOOD MATERNAL NUTRITION IS VITAL

A pregnant woman must nourish her unborn child as well as herself. In general, the same balanced diet that is good for her should also provide her developing baby with all the carbohydrates, lipids, and proteins it needs. Vitamins and minerals are a different story, however. Physicians recommend that a pregnant woman take supplemental vitamins and minerals, not only for her own benefit but also to meet the needs of her fetus. This is particularly true for the nutrient folic acid (folate), which is required for the neural tube to develop properly. If too little folic acid is available, a birth defect called

spina bifida ("split spine") may develop, in which the neural tube doesn't close and separate from ectoderm. The infant may be born with part of its spinal cord exposed inside a cyst. Infection is a serious danger, and the resulting neurological problems can include poor bowel and bladder control. To prevent neural tube defects, folic acid now is added to wheat flour and other widely used foods.

A pregnant woman must eat enough to gain between 20 and 35 pounds, on average. If she gains much less than that, she may be putting her fetus at risk. Infants who are severely underweight have more complications after delivery. As birth approaches, the growing fetus demands more and more nutrients from the mother's body. For example, the brain grows the most in the weeks just before and after birth. Poor nutrition during that time, especially protein deficiency, can have repercussions on intelligence and other brain functions later in life.

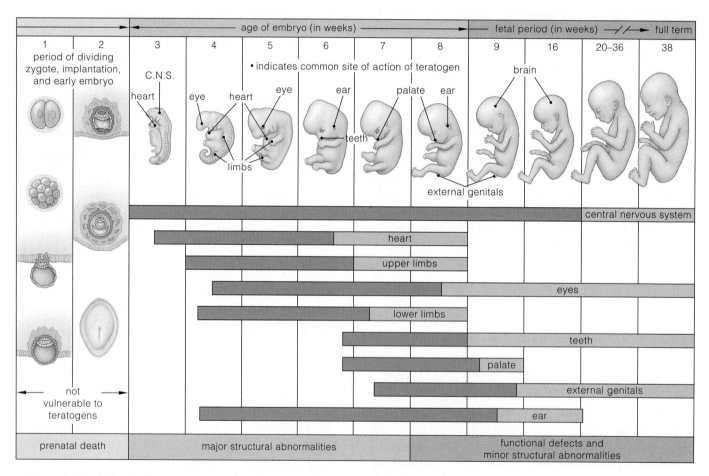

Figure 17.17 Animated! Sensitivity to teratogens. Light blue also indicates periods in which organs are most sensitive to damage from alcohol, viral infection, and so on. Numbers signify the week of development.

INFECTIONS PRESENT SERIOUS RISKS

A pregnant woman's IgG antibodies cross the placenta. They can help protect her developing infant from all but the most severe bacterial infections. Other **teratogens**—agents that can cause birth defects—are more serious threats. Some viral diseases can be dangerous during the first six weeks of pregnancy, when the organs of a fetus are forming (Figure 17.17). For example, if a pregnant woman contracts **rubella** (German measles) during this time, there is a 50 percent chance that some organs of the embryo won't form properly. If she contracts the virus when the embryo's ears are forming, her newborn may be deaf. With time, the risk of damage diminishes, and vaccination before pregnancy can eliminate it entirely.

PRESCRIPTION DRUGS CAN HARM

During its first trimester in the womb, an embryo is extremely sensitive to drugs the mother takes. In the 1960s many women using the tranquilizer thalidomide gave birth to infants with missing or severely deformed arms and legs. Although it wasn't known at the time, thalidomide alters the steps required for normal limbs to develop. When the connection became clear, thalidomide was withdrawn from the market (although it now has other medical uses). However, other commonly used tranquilizers—as well as some sedatives and barbiturates—might cause similar, although less severe, damage. Even certain anti-acne drugs, such as retinoic acid, increase the risk of facial and cranial deformities. The antibiotic streptomycin causes hearing problems and may adversely affect the nervous system; a pregnant woman who uses the antibiotic tetracycline may have a child whose teeth are yellowed.

ALCOHOL AND OTHER DRUGS CAN ALSO HARM

Like many other drugs, alcohol crosses the placenta and affects the fetus. *Fetal alcohol syndrome* (FAS) is a constellation of defects that can result from alcohol use by a pregnant woman. Tragically, FAS is one of the most common causes of mental retardation in the United States. Babies born with it typically have a smaller than normal brain and head, facial deformities, poor motor coordination, and, sometimes, heart defects (Figure 17.18). The symptoms can't be reversed; FAS children never catch up, physically or mentally. Between 60 and 70 percent of alcoholic women give birth to infants with

 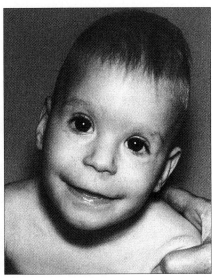

Figure 17.18 An infant with fetal alcohol syndrome (FAS). Obvious symptoms are low and prominent ears, poorly developed cheekbones, and a long, smooth upper lip. The child may have growth problems and abnormalities of the nervous system.

FAS. There may be no "safe" drinking level during pregnancy. Many doctors urge near or total abstinence from alcohol during pregnancy.

A pregnant woman who uses cocaine, especially crack, prevents her child's nervous system from developing normally. As a result, the child may be chronically irritable as well as abnormally small.

Research evidence suggests that tobacco smoke reduces the level of vitamin C in a pregnant woman's blood, and in that of her fetus as well. Cigarette smoke also harms the growth and development of a fetus in other ways. A pregnant woman who smokes daily will give birth to an underweight newborn even if the mother's weight, nutritional status, and all other relevant variables are the same as those of pregnant nonsmokers. A pregnant smoker also has a greater risk of miscarriage, stillbirth, and premature delivery. A long-term study at Toronto's Hospital for Sick Children showed that toxins in tobacco build up even in the fetuses of nonsmokers who are exposed to secondhand smoke at home or work.

Just how cigarette smoke damages a fetus is not known. However, its demonstrated effects are additional evidence that the placenta cannot protect a developing fetus from every danger.

17.9 Prenatal Diagnosis: Detecting Birth Defects

A growing number of options now enable us to detect more than 100 genetic disorders before a child is born.

Amniocentesis samples fluid from within the amnion, the sac that contains the fetus (Figure 17.19). During the fourteenth to sixteenth weeks of pregnancy, the thin needle of a syringe is inserted through the mother's abdominal wall, into the amnion. The physician must take care that the needle doesn't puncture the fetus and that no infection occurs. Amniotic fluid contains sloughed fetal cells; as the syringe withdraws fluid, some of those cells are included. They are then cultured and tested for genetic abnormalities.

Chorionic villus sampling (CVS) uses tissue from the chorionic villi of the placenta. CVS is tricky. Using ultrasound, the physician guides a tube through the vagina, past the cervix, and along the uterine wall, then removes a small sample of chorionic villus cells by suction. The method can be used by the eighth week of pregnancy; results are available within days. Both CVS and amniocentesis involve a small risk of triggering miscarriage. With CVS there also is a slight chance the future child will have missing or underdeveloped fingers or toes.

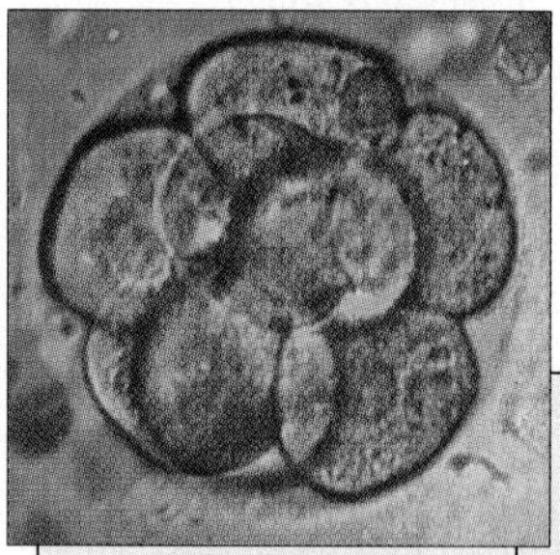

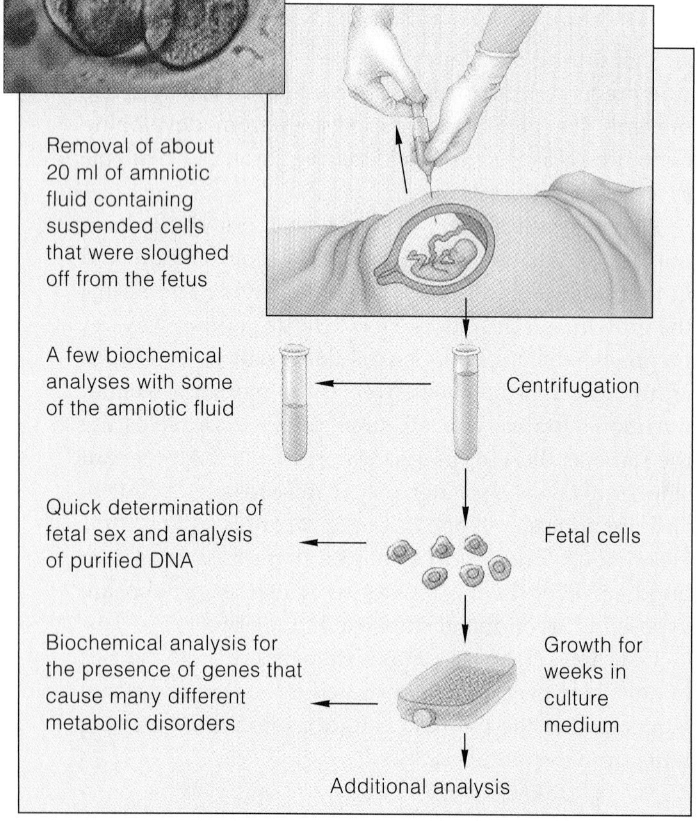

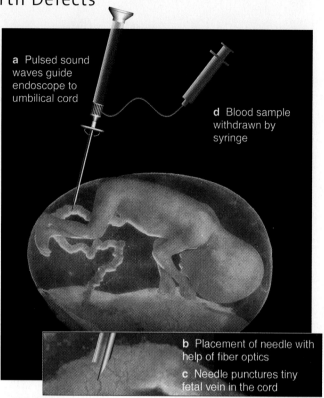

Figure 17.20 Fetoscopy for prenatal diagnosis.

a Pulsed sound waves guide endoscope to umbilical cord

d Blood sample withdrawn by syringe

b Placement of needle with help of fiber optics

c Needle punctures tiny fetal vein in the cord

Removal of about 20 ml of amniotic fluid containing suspended cells that were sloughed off from the fetus

A few biochemical analyses with some of the amniotic fluid

Centrifugation

Quick determination of fetal sex and analysis of purified DNA

Fetal cells

Biochemical analysis for the presence of genes that cause many different metabolic disorders

Growth for weeks in culture medium

Additional analysis

Figure 17.19 *Animated!* Procedure for amniocentesis. *(inset, above)* An eight-cell-stage human embryo.

Physicians also have available several methods of embryo screening. In *preimplantation diagnosis,* an embryo conceived by in vitro fertilization (Section 16.8) is analyzed for genetic defects using recombinant DNA technology. The testing occurs at the eight-cell stage (Figure 17.19, *inset*). According to one view, the tiny, free-floating ball is a *pre*-pregnancy stage. Like the unfertilized eggs discarded monthly during menstruation, the ball is not attached to the uterus. All cells in the ball have the same genes and are not yet committed to giving rise to specialized cells of a heart, lungs, or other organs. Doctors take one of the undifferentiated cells and analyze its genes for suspected disorders. If the cell has no detectable genetic defects, the ball is inserted into the uterus. Embryo screening is designed to help parents who are at high risk of having children with a genetic birth defect. Even so, it raises questions of morality in the minds of some people.

Fetoscopy allows direct visualization of the developing fetus. A fiber-optic device, an endoscope, uses pulsed sound waves to scan the uterus and visually locate parts of the fetus, umbilical cord, or placenta (Figure 17.20). Fetoscopy has been used to diagnose blood cell disorders, such as sickle-cell anemia and hemophilia. Like amniocentesis and CVS, fetoscopy has risks; it increases the risk of a miscarriage by 2 to 10 percent.

17.10 From Birth to Adulthood

After a child enters the world, a gene-dictated course of further growth and development leads to adulthood.

THERE ARE MANY TRANSITIONS FROM BIRTH TO ADULTHOOD

Table 17.2 summarizes the prenatal (before birth) and postnatal (after birth) stages of life. A newborn is called a *neonate*. During infancy, which lasts until about 15 months of age, the child's nervous and sensory systems mature rapidly, and a series of growth spurts makes its body longer. Figure 17.21 shows how body proportions change during childhood and adolescence. In adolescence, **puberty** marks the arrival of sexual maturity as a person's reproductive organs begin to function. Sex hormones trigger the appearance of secondary sex characteristics such as pubic and underarm hair, and behavior changes. A mix of hormones triggers another growth spurt at this time. Boys usually grow most rapidly between the ages of 12 and 15, whereas girls tend to grow most rapidly between the ages of 10 and 13. After several years, the

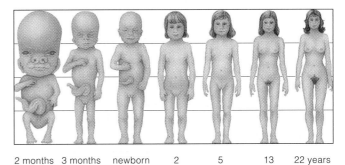

2 months 3 months newborn 2 5 13 22 years

Figure 17.21 *Animated!* Changes in body proportions during prenatal and postnatal growth.

influence of sex hormones causes the cartilaginous plates near the ends of long bones to harden into bone. Humans stop growing by their early twenties.

ADULTHOOD IS ALSO A TIME OF BODILY CHANGE

Although in the United States the average life expectancy is 72 years for males and 79 years for females, we reach the peak of our physical potential in adolescence and early adulthood. A healthy diet, regular exercise, and other beneficial lifestyle habits can go far in keeping a person vigorous for decades of adult life. Even so, after about age 40, body parts begin to undergo structural changes. There is also a gradual loss of efficiency in bodily functions, as well as increased sensitivity to environmentally induced stress. This steady deterioration is built into the life cycle of all organisms in which cells become highly specialized. The process, technically called *senescence*, is what we all know as aging.

Aging leads to many structural changes in the body. Beginning around age 40, there is a gradual decline in bone and muscle mass. Our skin develops more wrinkles, and more fat is deposited. Less obvious are a variety of gradual physiological changes. This chapter's concluding sections explore the phenomenon of aging, beginning with some ideas about its causes.

Table 17.2	Stages of Human Development: A Summary
PRENATAL PERIOD	
1. Zygote	Single cell resulting from union of sperm and egg at fertilization
2. Morula	Solid ball of cells produced by cleavages
3. Blastocyst	Ball of cells with surface layer and inner cell mass
4. Embryo	All developmental stages from 2 weeks after fertilization until end of eighth week
5. Fetus	All developmental stages from the ninth week until birth (about 39 weeks after fertilization)
POSTNATAL PERIOD	
6. Newborn (neonate)	Individual during the first 2 weeks after birth
7. Infant	Individual from 2 weeks to about 15 months after birth
8. Child	Individual from infancy to about 12 or 13 years
9. Pubescent	Individual at puberty, when secondary sexual traits develop; girls between 10 and 16 years, boys between 13 and 16 years
10. Adolescent	Individual from puberty until about 3 or 4 years later; physical, mental, emotional maturation
11. Adult	Early adulthood (between 18 and 25 years); bone formation and growth completed. Changes proceed very slowly afterward.
12. Old age	Aging culminates in general body deterioration

Following birth, development proceeds through childhood and adolescence, which includes the arrival of sexual maturity at puberty. Puberty is the gateway to the adult phase of life. After about age 30, developmental changes associated with aging become increasingly apparent.

17.11 Time's Toll: Everybody Ages

LINKS TO
SECTIONS
2.4, 2.5, 4.2, 5.3,
5.6, 6.5, 16.1,
AND 16.3

Time takes a toll on body tissues and organs. To some extent, our genes determine how long each of us will live.

Aging is a gradual loss of vitality as cells, tissues, and organs function less and less efficiently. At about age 40, the skin begins to noticeably wrinkle and sag, body fat tends to accumulate, and muscles and joints are more easily injured and take longer to heal. Stamina declines, and we become increasingly susceptible to disorders such as heart disease, arthritis, and cancer.

Structural changes in certain proteins may contribute to many changes we associate with aging (Figure 17.22). Remember from Chapter 4 that many connective tissues consist largely of collagen. A collagen molecule is stabilized by molecular bonds that form cross-link segments of the chain. As a person grows older, new cross-links develop, and the protein becomes more and more rigid. Changes in collagen's structure in turn can alter the structure and functioning of organs and blood vessels that contain it. Age-related cross-linking is also thought to affect many enzymes and possibly DNA.

GENES MAY DETERMINE THE MAXIMUM HUMAN LIFE SPAN

Does an internal, biological clock control aging? That's one prominent hypothesis. After all, each species has a maximum life span. For example, we know the maximum is about 20 years for dogs and 12 weeks for butterflies. So far as we can document, no human has lived beyond 122 years. The consistency of life span within species is a sign that genes help govern aging.

One idea is that each type of cell, tissue, and organ is like a clock that ticks at its own genetically set pace. When researchers investigated this possibility, they grew normal human embryonic cells, all of which divided about 50 times, then died.

In the body, human cells divide eighty or ninety times, at most. As discussed in Chapter 19, cells copy their chromosomes before they divide. Capping the ends of chromosomes are repeated stretches of DNA called *telomeres*. A bit of each telomere is lost during each cycle of cell division. When only a nub remains, cells stop dividing and die.

Cancer cells, and the cells in gonads that give rise to both sperm and oocytes, are exceptions to this rule. Both these types of cells make telomerase, an enzyme

Figure 17.22 As we age, structural and functional changes affect virtually all body tissues and organs. Graying hair, "age spots," and skin wrinkles are just three outward signs of these changes.

that causes telomeres to lengthen. Apparently, that is why such cells can divide over and over, without dying.

CUMULATIVE DAMAGE TO DNA MAY ALSO PLAY A ROLE IN AGING

A "cumulative assaults" hypothesis proposes that aging results from mounting damage to DNA combined with a decline in DNA's mechanisms of self-repair. Chapter 2 described how free radicals can damage DNA and other biological molecules. If changes in DNA aren't fixed, they may endanger the synthesis of enzymes and other proteins that are required for normal cell operations.

Some interesting studies of aging have implicated problems in the processes by which DNA replicates and repairs itself. For instance, researchers have correlated Werner's syndrome, a disorder that causes premature aging in young adults, with a harmful mutation in the gene that carries instructions for an enzyme that is thought to be essential for DNA self-repair. With a "bad" form of this enzyme, accumulating DNA damage may progressively undermine cell functions.

Ultimately, it's quite possible that aging involves processes in which genes, free radical damage, a decline in DNA repair mechanisms, and even other factors all come into play.

Aging may result from a combination of factors, including an internal biological clock that ticks out the life spans of cells, and the accumulation of errors in and damage to DNA.

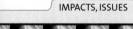

17.12 Aging Skin, Muscle, Bones, and Reproductive Systems

Some obvious signs of aging show up in our skin, muscles, skeleton, and reproductive systems.

CHANGES IN CONNECTIVE TISSUE AFFECT SKIN, MUSCLES, AND BONES

Skin changes often are early obvious signs of aging. The normal replacement of sloughed epidermis through cell division begins to slow. Cells called fibroblasts are a major component of connective tissue, and the number of them in the dermis starts to decrease. Also, the elastin fibers that give skin its flexibility are slowly replaced with more rigid collagen. As a result of these changes, the skin becomes thinner and less elastic, so it sags and wrinkles develop. Wrinkling increases as less fat is stored in the hypodermis, the layer beneath the dermis. The skin also becomes drier as sweat and oil glands begin to break down and are not replaced. The loss of sweat glands and subcutaneous fat is one reason why older people tend to have difficulty regulating their body temperature. As hair follicles die or become less active, there is a general loss of body hair. And as pigment-producing cells die and are not replaced, the remaining body hair begins to appear gray or white.

Fibers in skeletal muscle atrophy, in part because of a corresponding loss of motor neurons that synapse with muscle fibers. In general, aging muscles lose mass and strength, and lost muscle tends to be replaced by fat and, with time, by collagen. Importantly, however, the extent to which this muscle replacement takes place depends on a person's diet and exercise habits. Typically, an 80-year-old has about half the muscle strength he or she had at 30. Staying physically active can help slow all these changes (Figure 17.23).

Our bones become weaker, more porous, and brittle as we age. Over the years bones lose some collagen and elastin, but mainly bones weaken from the loss of calcium and other minerals after about age 40. Older people may be prone to develop osteoporosis. Women naturally have less bone calcium than men do, and they begin to lose it earlier. Without medical intervention, some women lose as much as half of their bone mass by age 70; therapeutic drugs are now available to help retard calcium loss.

With the passing years, intervertebral disks gradually deteriorate, reducing the distance between vertebrae. This is why people tend to get shorter, generally by about a centimeter every ten years from middle age onward.

Ninety percent of people over 40 have some degree of joint breakdown. The cartilage in joints deteriorates from the sheer wear and tear of daily movements, the surfaces of the joined bones begin to wear away, and the joints become more difficult to move. At least 15 percent of adults develop osteoarthritis, which you may recall is a

Figure 17.23 Moderate physical activity that is maintained as a person ages slows age-related deterioration of the muscular and skeletal systems, among other benefits.

chronic inflammation of cartilage, in one or more joints. Although many factors contribute to osteoarthritis, it is most common in older people. This observation suggests that age-related changes in bone are often involved.

REPRODUCTIVE SYSTEMS AND SEXUALITY CHANGE

Levels of most hormones stay steady throughout life. However, the sex hormones are exceptions. Falling levels of estrogens and progesterone trigger menopause in women; in older men, falling levels of testosterone reduce fertility. That said, whereas menopause brings a woman's reproductive period to an end, men can and have fathered children into their 80s. Males and females both retain their capacity for sexual response well into old age.

Menopause usually begins in a woman's late 40s or early 50s. Over a period of several years, her menstrual periods become irregular, then stop altogether as her ovaries become less sensitive to the hormones FSH and LH (Chapter 15) and gradually stop secreting estrogen in response. For many women menopause is relatively free of unpleasant symptoms. However, declining estrogen levels may trigger "hot flashes" (intense sweating and uncomfortable warmth), thinning of the vaginal walls, and some loss of natural lubrication. Postmenopausal women also have increased risk of osteoporosis and heart disease. Hormone replacement therapy (HRT) can counter some side effects of estrogen loss, but over time it also brings an increased risk of certain cancers. For this reason many women opt not to have HRT.

After about age 50, men gradually begin to take longer to achieve an erection due to vascular changes that cause the penis to fill with blood more slowly. In the United States, more than half of men over 50 also have urinary tract problems caused by age-related enlargement of the prostate gland.

Changes in connective tissue contribute to age-related declines in the functioning of muscles, bones, and joints. Declining sex hormones lead to age-related changes in the male and female reproductive systems.

17.13 Age-Related Changes in Some Other Body Systems

LINKS TO
SECTIONS
7.12, 9.2, 10.3,
10.4, 10.9, 11.1,
12.2, 13.3, 13.4,
AND 14.10

Irreversible age-related changes take place in virtually every organ, from the brain on down.

THE NERVOUS SYSTEM AND SENSES DECLINE

You are born with most of the neurons you will ever have. Chances are that as you age, any that are lost will not be replaced. Even in a healthy person, brain neurons die throughout life, and as they do the brain shrinks slightly, losing about 10 percent of its mass after 80 years. On the other hand, the brain has more neurons than it "needs" for various functions. In addition, when certain types of neurons are lost or damaged, other neurons may produce new dendrites and synaptic connections and so take up the slack.

Over time, however, the death of some neurons and structural changes in others apparently do interfere with nervous system functions. For instance, in nearly anyone who lives to old age, tangled clumps of cytoplasmic fibrils develop in the cytoplasm of many neuron cell bodies. These *neurofibrillary tangles* may disrupt normal cell metabolism, although their exact effect is not understood. Clotlike plaques containing protein fragments called *beta amyloid* also develop between neurons.

Symptoms of the dementia called **Alzheimer's disease** (AD) include progressive memory loss and disruptive personality changes. In affected people, the brain tissue contains masses of neurofibrillary tangles and is riddled with beta amyloid plaques (Figure 17.24). It is not clear whether the amyloid plaques are a cause of Alzheimer's or simply one effect of another, unknown disease process.

For instance, the brains of Alzheimer's patients also have lower than normal amounts of the neurotransmitter acetylcholine, and some evidence suggests that this shortage may be related to beta amyloid buildup. Also under investigation is the hypothesis that AD results from chronic inflammation of brain tissue, just as inflammation leads to arthritis and promotes coronary artery disease.

Treatments for AD patients are limited. Drugs can temporarily help alleviate some symptoms or slow the progression of the disease. Researchers have also tested vaccines designed to stimulate the immune system to generate plaque-fighting antibodies. To date, however, the most promising candidates have had unacceptable side effects.

Some cases of AD are inherited. The increased risk is significant for people who inherit one version of a gene that codes for *apolipoprotein E*, a lipid-binding protein. Around 16 percent of the U.S. population has one or two copies of this gene, called apoE-4. Those with two copies have a 90 percent chance of developing AD. Of the AD-related genes discovered thus far, most are associated with early-onset Alzheimer's, which develops before the age of 65. However, many thousands of the 4 million Americans who have the disease first showed symptoms in their late 70s or 80s. Does genetics play a role in those cases, also? Possibly. More and more genes are being found to have forms that affect how much amyloid builds up in the brain. However, many investigators believe that AD probably has many contributing factors, including as-yet-undetermined environmental ones.

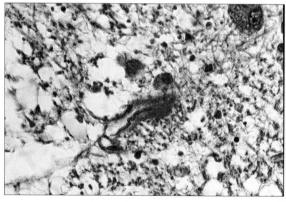

a

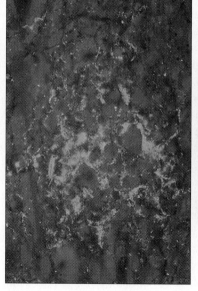

b

c

Figure 17.24 (**a**) A neurofibrillary tangle in brain tissue from a patient with Alzheimer's disease. (**b**) Three beta amyloid plaques. (**c**) Older people who do not have Alzheimer's disease also develop neurofibrillary tangles and beta amyloid plaques, but not nearly as many.

Even otherwise healthy people begin to have some difficulty with short-term memory after about age 60 (the so-called "senior moments") and may find it takes longer to process new information. Perhaps because aging CNS neurons tend to lose some of their insulating myelin sheath, older neurons do not conduct action potentials as efficiently. In addition, neurotransmitters such as acetylcholine may be released more slowly. As a result of such changes, movements and reflexes become slower, and some coordination is lost.

As we age, our sensory organs become less efficient at detecting or responding to stimuli. For example, the taste buds become less sensitive over time. As noted in Chapter 14, people also tend to become farsighted as they grow older because the lens of the eye loses its elasticity and is altered in other ways that prevent it from properly flexing to bend incoming light during focusing.

THE CARDIOVASCULAR AND RESPIRATORY SYSTEMS DETERIORATE

Our heart and lungs also function less efficiently with increasing age. In the lungs, walls of alveoli break down, so there is less total respiratory surface available for gas exchange. If a person does not develop an enlarged heart due to cardiovascular disease, the heart muscle becomes slightly smaller, and its strength and ability to pump blood diminish. As a result, less blood and oxygen are delivered to muscles and other tissues. In fact, decreased blood supply may be a factor in age-related changes throughout the body. Blood transport is also affected by structural changes in aging blood vessels. Elastin fibers in blood vessel walls are replaced with connective tissue containing collagen or become hardened with calcium deposits, and so vessels become stiffer. Cholesterol plaques and fatty deposits often cause further narrowing of arteries and veins (Section 9.14). This is why people often find that their resting blood pressure rises as they get older. However, as with the muscular and skeletal systems, lifestyle choices such as not smoking, eating a healthy diet, and getting regular exercise can help a person maintain vigorous respiratory and cardiovascular systems well past middle age.

THE IMMUNE, DIGESTIVE, AND URINARY SYSTEMS BECOME LESS EFFICIENT

Other organs and organ systems also change as the years go by. In the immune system, the number of T cells falls and B cells become less active. Older people are more likely to develop autoimmune diseases, such as rheumatoid arthritis. Why? One hypothesis holds that, as

Table 17.3		Some Physiological Changes in Aging		
Age	Maximum Heart Rate	Lung Capacity	Muscle Strength	Kidney Function
25	100%	100%	100%	100%
45	94%	82%	90%	88%
65	87%	62%	75%	78%
85	81%	50%	55%	69%

Note: Age 25 is the benchmark for maximal efficiency of physiological functions.

DNA repair mechanisms become less effective, mutations in genes that code for self-markers are not fixed. If the markers change, this could provoke immune responses against the body's own cells.

In the aging digestive tract, glands in the mucous membranes that line the stomach and small and large intestines gradually break down, and the pancreas secretes fewer digestive enzymes. Although it is vital for older people to maintain adequate nutrition, we require fewer food calories as we age. By age 50, a person's basal metabolic rate (BMR) is only 80 to 85 percent of what it was in childhood and will keep declining about 3 percent every decade. This is why people tend to gain weight in middle age, unless they compensate for a falling BMR by consuming fewer calories, increasing their physical activity, or both.

Over time the muscular walls of the large intestine, bladder, and urethra become weaker and less flexible. As the urinary sphincter is affected, many older people experience urine leakage, or **urinary incontinence**. Women who have borne children may have more trouble with urinary incontinence because their pelvic floor muscles are weak. On the other hand, our kidneys may continue to function well, despite the fact that nephrons gradually break down and lose some of their ability to maintain the balance of water and ions in body fluids (Table 17.3). In general, however, even aging kidneys have more than enough nephrons to function well.

In many body systems, such as the nervous, sensory, and immune systems, aging correlates with the death of cells that are not replaced and reduced activity of other cells.

Ultimately, aging involves a steady decline in the finely tuned ebb and flow of substances and chemical reactions that maintain homeostasis.

Summary

Section 17.1   Human development unfolds in six stages:

a. The formation of gametes—eggs and sperm—which mature inside the reproductive organs of parents.

b. Fertilization, in which the DNA (on chromosomes) of a sperm and an egg are brought together in a single cell (the zygote).

c. Cleavage, when the fertilized egg undergoes cell divisions that form the early multicellular embryo. The destiny of various cell lines is established in part by the portion of cytoplasm inherited at this time.

d. Gastrulation, when the organizational framework of the whole body is laid out. Endoderm, ectoderm, and mesoderm form; all the tissues of the adult body will develop from these three germ layers.

e. Organogenesis, when organs start developing.

f. Growth and tissue specialization, when organs enlarge and acquire their specialized chemical and physical properties. Tissues and organs continue to mature as the fetus develops and even after birth.

In cell differentiation cells come to have specific structures and functions; morphogenesis produces the shape and structure of particular body regions. Both these processes are guided by genes.

 See what happens during fertilization, and track the stages in the formation of a human hand.

Section 17.2   When an oocyte fuses with a sperm, its arrested cell division resumes. The result is a mature egg, or ovum. Fertilization now produces a diploid zygote, a single cell that has a full set of the forty-six human chromosomes. During the first week or so after fertilization, cell divisions and other changes transform the zygote into a multicellular blastocyst, which attaches to the mother's uterus during implantation. The blastocyst includes the inner cell mass, a small clump of cells from which the embryo develops.

Section 17.3   Gastrulation and morphogenesis shape the basic body plan. A key step is the formation of the neural tube, the forerunner of the brain and spinal cord, from ectoderm. Mesoderm gives rise to somites that are the source of the skeleton and most muscles. During morphogenesis sheets of cells fold and cells migrate to new locations in the developing embryo.

 Observe the early stages of human development.

Section 17.4   During implantation the inner cell mass is transformed into an embryonic disk. Some of its cells give rise to four extraembryonic membranes that serve key functions (Table 17.4).

a. Yolk sac: contributes to the embryo's digestive tube and helps form blood cells and germ cells.

b. Allantois: its blood vessels become arteries in the umbilical cord and vessels of the placenta; they function in oxygen transport and waste excretion.

c. Amnion: a fluid-filled sac that surrounds and protects the embryo from mechanical shocks and keeps it from drying out.

d. Chorion: a protective membrane around the embryo and the other membranes; a major part of the placenta.

The embryo and its mother exchange nutrients, gases, and wastes by way of the placenta, a spongy organ of endometrium and extraembryonic membranes.

Section 17.5   The first eight weeks of development are the embryonic period; thereafter the developing individual is considered a fetus. By the ninth week of development, the fetus clearly looks human.

Section 17.6   During the last three months of gestation (the third trimester), the fetus grows rapidly and many organs mature. However, because the fetus exchanges gases and receives nourishment via its mother's bloodstream, its own circulatory system routes blood flowing to the lungs and liver through temporary blood vessels.

Section 17.7   Birth takes place approximately 39 weeks after fertilization. Labor advances through three stages; a baby is born at the end of stage two, and the afterbirth (placenta) is expelled in stage three.

Estrogen and progesterone stimulate growth of the mammary glands. At parturition, contractions of the uterus dilate the cervix and expel the fetus and afterbirth. After delivery, nursing causes the secretion of hormones that stimulate lactation—the production and release of milk.

Table 17.4	Summary of Extraembryonic Membranes
Membrane	**Function**
Yolk sac	Source of digestive tube; helps form blood cells and forerunners of gametes
Allantois	Source of umbilical blood vessels and vessels of the placenta
Amnion	Sac of fluid that protects the embryo and keeps it moist
Chorion	Forms part of the placenta; protects the embryo and the other extraembryonic membranes

Section 17.10 Human development can be divided into a prenatal period before birth, followed by the neonate (newborn) stage, childhood, adolescence, and adulthood. The process of aging is called senescence.

Sections 17.11–17.13 Over time the human body shows changes in structure and a decline in its functional efficiency. The precise cause of aging is unknown.

Review Questions

1. Define and describe the main features of the following developmental stages: fertilization, cleavage, gastrulation, and organogenesis.

2. Label the following stages of early development:

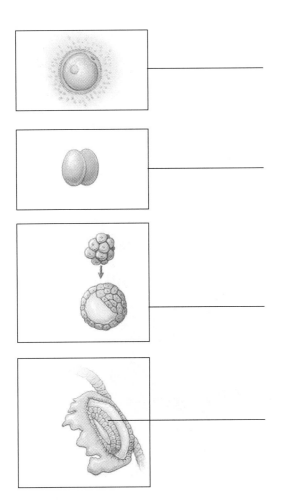

3. Define cell differentiation and morphogenesis, two processes that are critical for development. Which two mechanisms are the basic foundation for cell differentiation and morphogenesis?

4. Summarize the development of an embryo and a fetus. When are body parts such as the heart, nervous system, and skeleton largely formed?

Self-Quiz

Answers in Appendix V

1. Development cannot proceed properly unless each of the following processes is successfully completed before the next begins, starting with _____.
   a. gamete formation
   b. fertilization
   c. cleavage
   d. gastrulation
   e. organ formation
   f. growth, tissue specialization

2. During cleavage, the _____ is converted to a ball of cells, which in turn is transformed into the _____.
   a. zygote; blastocyst
   b. trophoblast; embryonic disk
   c. ovum; embryonic disk
   d. blastocyst; embryonic disk

3. In the week following implantation, cells of the _____ will give rise to the embryo.
   a. blastocyst
   b. trophoblast
   c. embryonic disk
   d. zygote

4. The developmental process called _____ produces the shape and structure of particular body regions.

5. _____ is the gene-guided process by which cells in different locations in the embryo become specialized.
   a. Implantation
   b. Neurulation
   c. Cell differentiation
   d. Morphogenesis

6. In a human zygote, the cell divisions of cleavage produce an embryonic stage known generally as a

   _____.
   a. zona pellucida
   b. gastrula
   c. blastocyst
   d. larva

7. Match each developmental stage with its description.
   ____ cleavage
   ____ gamete formation
   ____ organ formation
   ____ cell differentiation
   ____ gastrulation
   ____ fertilization
   a. egg and sperm mature in parents
   b. sperm, egg nuclei fuse
   c. germ layers form
   d. zygote becomes a ball of cells called a morula
   e. cells come to have specific structures and functions
   f. starts when germ layers split into subgroups of cells

8. Of the four extraembryonic membranes, only the _____ is not needed in order for an embryo to develop properly.
   a. yolk sac
   b. allantois
   c. amnion
   d. chorion
   e. This is a trick question, because all are needed.

9. Of the following, _____ cannot cross the placenta.
   a. alcohol
   b. the mother's antibodies
   c. antibiotics
   d. toxic substances in tobacco smoke
   e. all can cross the placenta

Critical Thinking

1. How accurate is the statement "A pregnant woman must do everything for two"? Give some specifics to support your answer.

2. A renowned developmental biologist, Lewis Wolpert, once observed that birth, death, and marriage are not the most important events in human life—rather, Wolpert said, gastrulation is. In what sense was he correct?

3. One of your best friends tells you that she and her husband think she might be pregnant. She feels she can wait until she's several months along before finding an obstetrician. *You* think she could use some medical advice sooner, and you suggest she discuss her plans with a physician as soon as possible. What kinds of health issues might you be concerned about?

4. In an ectopic pregnancy, an embryo implants outside the mother's uterus, often in an oviduct (Figure 17.25). The complications of ectopic pregnancy are life-threatening for the mother, and in fact each year in the U.S. a few pregnant women die when their situation is not diagnosed in time. Tragically, the only option is to surgically remove the embryo, which was doomed from the beginning. Based on what you know about where an embryo normally develops, explain why an ectopic embryo could not have long survived.

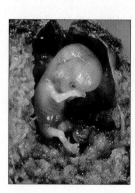

Figure 17.25 A tubal pregnancy.

Explore on Your Own

Housecats can carry the parasite that causes **toxoplasmosis**. In an otherwise healthy person the disease may only produce flulike symptoms, but it is dangerous for a pregnant woman and her fetus. A mother-to-be may suffer a miscarriage, and if the parasite infects a fetus it causes birth defects. An infected cat may not appear to be ill, but its feces will contain infectious cysts. This is why some physicians advise pregnant women to avoid contact with cats and not to clean sandboxes, take care of housecat "accidents," or empty a litterbox.

All that said, toxoplasmosis is not especially common—so is it something a cat-loving expectant mother should take seriously? To explore this health concern, research the kinds of birth defects caused by toxoplasmosis and find out what stance (if any) public health authorities in your community take on this issue. Is the disease more common in some regions or settings than in others? Can cat owners have their pets tested for the disease?

Virus, Virus Everywhere

In 336 B.C., 20-year-old Alexander the Great began carving out an empire that stretched across the Middle East and into India. He died twelve years later after he entered Babylon, the site of modern-day Baghdad.

By one account, Alexander became ill shortly after some ravens fell dead at his feet. When death came he was delirious with fever and paralyzed. To some researchers these and other symptoms suggest that Alexander was struck down by the West Nile virus. It causes encephalitis, a disease in which the brain becomes severely inflamed.

In 1999, researchers probing a disease outbreak in cows, horses, and people in and around New York City found West Nile virus in tissue samples. It was the first time anyone knew that the virus had entered the Western Hemisphere. Seven of the sixty-two people who got sick died. By 2003 nearly 9,000 human cases had been reported in North and Central America, with nearly 200 fatalities.

The West Nile virus usually is transmitted by mosquitoes, which pick it up while they are sucking the blood of a host—a bird, a farm animal, or a human. In North America, the virus has been found in more than forty different kinds of mosquitoes. As far

as we know, the only way to get the virus is to be bitten by a mosquito that carries it.

Birds also get the H5N1 virus, which causes avian (bird) flu. Migrating geese have carried it from Asia into Europe and North America. People in close contact with an infected bird can catch the virus, which causes a potentially deadly respiratory illness. The genetic makeup of viruses can change rapidly. So, although millions of chickens and other birds have been killed in an effort to keep the virus from spreading, the worry is that a version of H5N1 will arise that can spread directly from human to human, leading to a massive wave of illness around the world. At present there is no vaccine for H5N1.

Infectious diseases have always been with us. With modern modes of travel and other societal changes, however, pathogens that once were restricted to limited areas now can travel to distant places and have a global impact on human health.

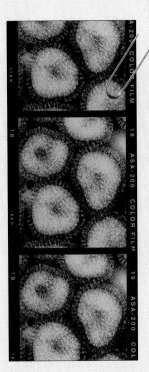

 How Would You Vote? Killing mosquitoes is the best defense against West Nile virus. Many local agencies now spray pesticides wherever mosquitoes are likely to breed. Some people object to spraying, fearing harmful effects on health or wildlife. Would you support a spraying program in your area? Cast your vote online at www.thomsonedu.com/biology/starr/humanbio.

 ## Key Concepts

THE NATURE OF INFECTIOUS DISEASE
A disease disrupts homeostasis so that one or more body functions do not occur normally. An infectious disease is one that can spread from person to person.

INFECTIOUS ORGANISMS
A wide variety of viruses, bacteria, protozoa, fungi, worms, and other pathogens can infect humans and cause mild to serious disease.

THE DISEASE PROCESS
Pathogens that cause infectious diseases all share some basic features, including the ability to evade the body's defenses. Such diseases are spread by direct or indirect contact, by inhalation, or by contact with a disease vector.

Links to Earlier Concepts

This chapter draws on your understanding of how innate and acquired immunity help protect the body from pathogens (10.1, 10.3). You will also see examples of how some pathogens use receptors on a cell's plasma membrane to gain entry to a cell (3.3) and how others use proteins or enzymes to attach to and invade cells or tissues (2.8).

18.1 Some General Principles of Infectious Disease

LINKS TO
SECTIONS
4.1, 6.4, 10.1,
AND 10.3

The body provides a hospitable environment for many normally harmless microorganisms. Pathogens must have features that allow them not only to overcome defenses, but to multiply and then move on to infect another host.

THE BODY IS HOME TO A GREAT MANY "FRIENDLY" MICROORGANISMS

You may recall from Chapter 10 that up to a thousand species of bacteria normally live on body surfaces. From birth onward, these and other "friendly" microorganisms colonize epithelial tissues of the skin, mouth, nasal cavity, the outer surface of the eye (the conjunctiva), the GI tract, the urethra, and in the case of females, the vagina (Figure 18.1). They all have some way of staying attached to us. For example, bacteria have projections coated with "sticky" proteins (aptly called *adhesins*) that "glue" them to epithelial cells. Even if you take a long, soapy shower, you can't remove many of the bacteria that normally live on your skin. By contrast, many pathogens and other microbial visitors don't have the necessary chemical or physical features to attach long-term to human cells, especially when the available skin surface is already packed with billions of resident microorganisms. Soap and water can easily carry pathogens away, which is why washing your hands helps protect against colds and flu.

DIFFERENT TYPES OF PATHOGENS CAUSE DISEASE IN DIFFERENT WAYS

"Disease" and "infection" are familiar words, even though you might not be able to explain exactly what they mean in biological terms. An **infection** occurs when a pathogen enters cells or tissues and multiplies. **Disease** develops when the body's defenses cannot be mobilized quickly enough to prevent a pathogen's activities from interfering with normal body functions. With infectious (contagious) diseases, the pathogen can move to another person, often in mucus, blood, or some other body fluid.

Harmful bacteria, viruses, and pathogens such as parasitic worms and fungi all cause disease in different ways (Table 18.1). Bacteria produce **toxins**, chemicals that poison or otherwise harm human cells. Many of the most powerful and dangerous bacterial toxins act directly on body cells. For example, the bacterial species *Clostridium botulinum* produces one of the most deadly biological toxins known—a neurotoxin that blocks transmission of nerve impulses by preventing neuron axons from releasing the neurotransmitter ACh (Section 6.4). Improperly canned, low-acid foods are a common source of the toxin, and eating only a tiny amount produces the potentially lethal disease called **botulism**.

Other bacterial toxins activate chemicals of immune responses, including cytokines and complement proteins (Section 10.1). They can trigger fever, diarrhea, and **septic shock**, which is a sudden and dangerous decline in blood pressure.

Viruses can cause disease by invading and destroying body cells. As an infected cell dies, new viral particles are released, spreading the infection. Some kinds of viruses also can become *latent* in cells—that is, they stay in an infected cell without multiplying, sometimes for a very long time, until stress or some other factor reactivates them. Other viruses alter cells in ways that lead to cancer.

Some pathogenic fungi release enzymes that break down tissues, as you will read shortly. Worms, protozoa, and other parasites also can enter the body and either damage cells or tissues directly, trigger harmful immune responses, or both. You will learn about some of these pathogens in Section 18.4.

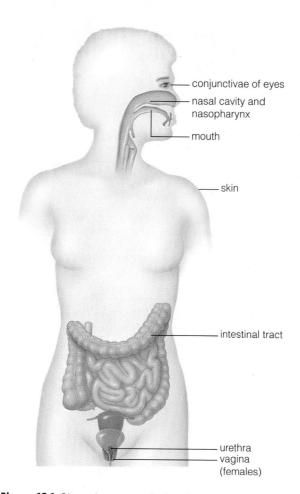

conjunctivae of eyes

nasal cavity and nasopharynx

mouth

skin

intestinal tract

urethra
vagina
(females)

Figure 18.1 Sites where normally harmless microorganisms colonize the body.

THE NATURE OF INFECTIOUS DISEASE

Table 18.1 Some Ways Pathogens Cause Disease

Type of Pathogen	Disease Mechanism
Bacteria	Produce toxins that poison cells and/or trigger damaging immune responses
Viruses	Kill body cells or cause cells to become cancerous
Fungi	Release enzymes that break down tissues
Parasites	Invade cells/tissues, trigger harmful immune responses

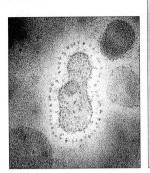 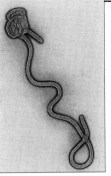

a SARS virus   **b** Ebola virus

Figure 18.2 Two examples of emerging pathogens.

Paradoxically, our immune system's responses to viral infection also cause symptoms we associate with disease, such as fever and fatigue. As you read in Chapter 10, a controlled inflammatory response helps rid the body of pathogens, but if the response goes awry it can lead to serious damage, especially if a major organ or organ system is involved. For instance, if the testicles of a male who contracts the sexually transmitted disease gonorrhea become inflamed, the inflammation can progress to an immune response that leaves the man sterile.

TO CAUSE AN INFECTION, PATHOGENS MUST MEET SEVERAL REQUIREMENTS

A pathogen's **host** is an organism that the pathogen can infect. In addition to having a suitable host, a pathogen usually also must have the following:

1. Access to a *reservoir*, a place where the pathogen can survive and remain infectious. Reservoirs include other infected organisms, soil, bodies of water, and **carriers**—organisms (including other people) in which the pathogen is living without causing disease symptoms.

2. A means of leaving the reservoir and entering a host—for example, contaminating substances that the host uses as food.

3. A way of attaching to a host's body.

4. A mechanism for entering a host's tissues.

5. The ability to avoid the host's defenses.

6. The ability to reproduce inside the host.

7. A way to return to the reservoir or move to a new host, as when virus particles are expelled by a sneeze.

Scientists use attributes of pathogens to help make sense of the bewildering array of infectious diseases. One approach is to group diseases by their reservoirs, such as the bodies of certain animals, water, or soil. For example, humans are the only reservoirs for pathogens that cause the common cold, measles, and gonorrhea.

For an infectious disease categorized as a **zoonosis** (zoe-uh-NOH-sis), animals other than humans are the main reservoirs, but the pathogens can infect humans. Rabies, West Nile virus, and avian flu are examples. HIV may have begun as a zoonosis in other primates. As we have seen with HIV, one danger with a zoonosis is that genetic changes in the pathogen may permit it to begin using the human body as a reservoir, allowing it to be passed from person to person.

EMERGING DISEASES PRESENT NEW CHALLENGES

Today health officials worry especially about so-called **emerging diseases**. Emerging diseases are caused by pathogens that until recently did not infect humans or were present only in limited areas. Many are caused by viruses. This group includes the encephalitis caused by West Nile virus and the severe respiratory disease caused by the SARS virus (Figure 18.2*a*). Other examples are hemorrhagic fevers that cause massive bleeding. In this latter group are dengue fever and the illness caused by the Ebola virus (Figure 18.2*b*). Lyme disease is a major emerging bacterial disease in the United States that we consider later in the chapter.

Why is all this happening? Broadly speaking, a few factors stand out. For one, there are simply many more humans on the planet, interacting with their surroundings and with each other. Each person is a potential target for pathogens. Also, more people are traveling, and carrying diseases along with them. Another important factor is the misuse and overuse of antibiotics, a topic we will take up in Section 18.3.

> *Many microorganisms live on body surfaces and normally cause no harm.*
>
> *In an infection, a pathogen enters cells or tissues. Disease develops when bacterial toxins, physical damage to cells and tissues, or other factors related to infection interfere with normal body functions.*

18.2 Viruses and Infectious Proteins

LINKS TO
SECTIONS
2.12 AND 10.7

Viruses aren't high on anyone's "most popular" list. The Nobel Prize–winning biologist Sir Peter Medawar once called them "bad news wrapped in protein."

A **virus** is a piece of genetic material inside a protein coat. It is not a cell, but can infect the cells of other organisms. To reproduce itself a virus must take over the host cell's metabolic machinery.

We classify viruses by several criteria. One is which type of nucleic acid—DNA or RNA—makes up the genetic material. Another is the general type of host organism the virus infects. Some viruses infect animals, others plants, and still others, called bacteriophages, infect bacteria.

In general, a virus particle consists of a nucleic acid core inside a protein coat called a capsid. Sometimes the capsid also is enclosed, in a lipid envelope.

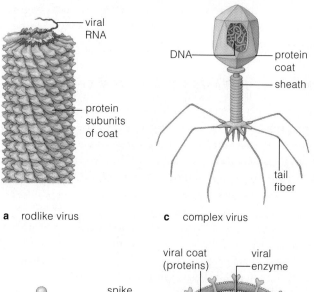

a   rodlike virus

c   complex virus

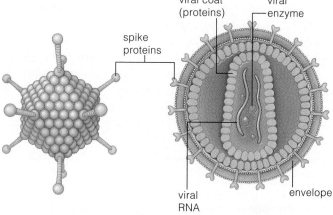

b   polyhedral virus

d   enveloped polyhedral virus

Figure 18.3 Virus structures. (**a**) A helical virus has a rod-shaped protein coat around its nucleic acid. (**b**) A polyhedral virus has a many-sided coat. (**c**) A complex virus, like this bacteriophage, has other structures attached to the coat. (**d**) An enveloped virus is surrounded by a membrane. HIV is shown here.

The coat around the particle consists of one or more protein subunits, often organized into the shapes shown in Figure 18.3. The coat protects the viral genetic material during the journey from one host cell to another. It also contains proteins that can bind with receptors on host cells. Some viruses also are surrounded by a membrane and are called *enveloped viruses*.

VIRUSES MULTIPLY INSIDE A HOST CELL

Different viruses multiply in different ways. In all cases, however, the process begins when a virus particle attaches to the host cell and either the whole virus or its genetic material enters the cell's cytoplasm. Next, the viral DNA or RNA is replicated (copied) and provides instructions for making enzymes and other proteins, using the host cell's metabolic machinery. These materials are used to assemble new virus particles, which then are released from the infected cell. Figure 18.4 shows these steps for an enveloped DNA virus infecting an animal cell.

A cell is a potential host if a virus can chemically recognize and lock on to receptors at the cell's surface. Differences in the receptors on different cell types are why a given virus often can attack only one type of cell. For instance, a flu virus can attack cells of respiratory epithelium but not liver cells, and a virus that causes hepatitis attacks liver cells but not cells of the airways.

When the multiplication cycle of a virus ends, the host cell usually dies and new virus particles are released. The virus may not kill the host cell outright, however. Instead, sometimes the virus becomes latent, as noted in Section 18.1. A common example is type 1 herpes simplex, which causes cold sores. It remains latent inside a ganglion (a cluster of neuron cell bodies) in facial tissue. Stressors such as a sunburn, falling ill, or emotional upsets can reactivate the virus. Then, virus particles move down the neurons to their tips near the skin. From there the virus particles infect epithelial cells, causing painful eruptions on the skin. Repeat outbreaks of genital herpes, caused by type 2 herpes simplex, happen the same way.

Another herpes virus, the Epstein-Barr virus (EBV), is commonly transmitted by kissing. An EBV infection can lead to **infectious mononucleosis**. Symptoms are a sore throat, overwhelming fatigue, and a mild fever. It can take a month or more to recover.

A **retrovirus** is an RNA virus that infects animal cells. Retroviruses establish yet another relationship with the host cell. This group's name comes from its "reverse" mode of pirating the host's genetic material. After the viral RNA chromosome enters the host cell cytoplasm, a viral enzyme called reverse transcriptase uses the RNA as a template for synthesizing a DNA molecule. The

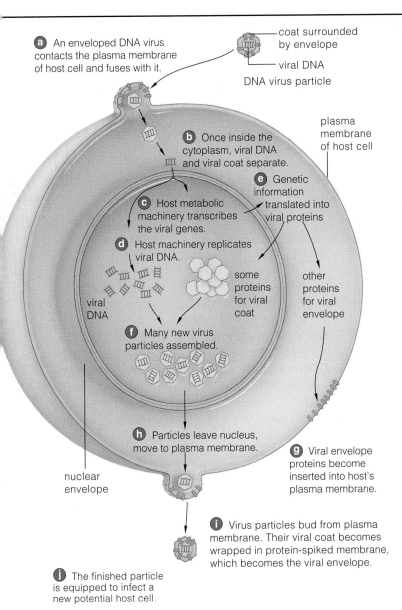

a An enveloped DNA virus contacts the plasma membrane of host cell and fuses with it.

coat surrounded by envelope

viral DNA

DNA virus particle

plasma membrane of host cell

b Once inside the cytoplasm, viral DNA and viral coat separate.

e Genetic information translated into viral proteins

c Host metabolic machinery transcribes the viral genes.

d Host machinery replicates viral DNA.

viral DNA

some proteins for viral coat

other proteins for viral envelope

f Many new virus particles assembled.

h Particles leave nucleus, move to plasma membrane.

nuclear envelope

g Viral envelope proteins become inserted into host's plasma membrane.

i Virus particles bud from plasma membrane. Their viral coat becomes wrapped in protein-spiked membrane, which becomes the viral envelope.

j The finished particle is equipped to infect a new potential host cell.

Figure 18.4 Multiplication cycle of an enveloped DNA virus infecting an animal cell.

new DNA molecule, called a *provirus*, then integrates into the host's DNA. This is what happens when a person becomes infected with HIV.

PRIONS ARE INFECTIOUS PROTEINS

A few rare, fatal degenerative diseases of the central nervous system are linked to small infectious proteins called **prions** (PREE-ons).

Prions are versions of otherwise normal proteins on brain neurons and some other types of cells. You may recall from Chapter 2 that proteins become folded into three-dimensional shapes. For some reason, prions are *mis*folded. Once in the body, they can cause normal versions of the protein to misfold and become prions

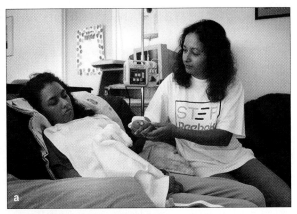

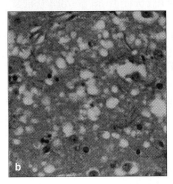

Figure 18.5
(**a**) Charlene Singh, the first known case of variant Creutzfeldt–Jakob disease, which is caused by prions. (**b**) Brain tissue damaged by BSE. The light-colored "holes" are areas where tissue was destroyed.

also. In the brain, massive clumps of them form and over time holes develop in the brain tissue, giving it a spongy appearance. As the disease progresses, muscle coordination and brain function are lost.

Prion diseases have been discovered in various species of animals. One example is BSE (for *bovine spongiform encephalitis*), which has come to be called "mad cow disease." Brain and spinal cord tissue from affected cattle can contain the prions. Humans who eat BSE-contaminated meat (such as brains, ground beef, sausage, and oxtails) are at risk for **variant Creutzfeldt–Jakob disease** (vCJD), which is caused by a prion that closely resembles the one that causes BSE. An outbreak of BSE in Britain in the late 1990s was linked to about 150 cases of vCJD in people who apparently consumed meat from the infected animals. One of them was an American woman who grew up in Britain (Figure 18.5).

Prions are also associated with a long-known form of CJD, which occurs worldwide. Most people with CJD are middle-aged or older, and about 15 percent of cases are thought to have a genetic origin.

A virus consists of nucleic acid enclosed in a protein coat and sometimes an outer envelope. Viruses multiply by pirating the metabolic machinery of a host cell.

The infectious proteins called prions cause various degenerative diseases of the nervous system.

18.3 Bacteria—The Unseen Multitudes

LINKS TO
SECTIONS
3.1, 7.3, 7.7,
AND 10.1

Bacteria, which are prokaryotes, were the first organisms on Earth. An estimated 500 to 1,000 species live harmlessly in each of us. Others are pathogens that cause disease.

Because bacteria are prokaryotes, they have no nucleus or other membrane-bound organelles. Most have a *cell wall* outside their plasma membrane (Figure 18.6). The wall is strong and fairly rigid, and helps maintain the shape of the bacterium. Although bacteria come in a range of shapes, three basic ones are common. A ball shape is a coccus (plural: cocci; from a word that means "berry"). A rod shape is a bacillus (plural: bacilli, which means "small staffs"). A "corkscrew"-shaped bacterial cell with one or more twists to it is a spirillum ("spiral"; plural: spirilla), sometimes called a *spirochete*.

coccus   bacillus   spirillum

Some kinds of bacteria have threadlike structures on the cell wall and plasma membrane. One type is a stiff protein filament called a *bacterial flagellum*. It rotates like a propeller to move the cell. Some kinds of bacteria have filaments called *pili* (singular: pilus) that help the cell stick to surfaces. The gonorrhea bacterium uses pili to attach to mucous membranes of the reproductive tract; the *Pseudomonas* bacterium in Figure 18.7 is using its pili to stick to the surface of a kitchen knife.

Bacteria reproduce by fission. The cell's DNA is copied, then the cell divides into two genetically identical daughter cells. Bacterial cells have just a single circular

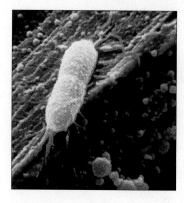

Figure 18.7
A *Pseudomonas* bacterium attached to the blade of a kitchen knife by way of its pili.

DNA molecule (a chromosome) to copy and parcel out to daughter cells. Under ideal conditions, some species can divide in about 20 minutes.

When some kinds of bacteria divide, the daughter cells can inherit one or more small circles of extra DNA. These DNA circles, which include a few genes, are called *plasmids*. One type, a "fertility" plasmid, carries genes that enable the bacterium to transfer plasmid DNA to another bacterial cell—genetic "instructions" that can include drug resistance, among other traits. Bacteria with this ability to transfer genes include some pathogens, such as *Salmonella* (a source of food-borne illness), and species of *Streptococcus* that cause various respiratory infections and **strep throat**.

BACTERIA PLAY BOTH POSITIVE AND NEGATIVE ROLES IN SOCIETY

We humans have many relationships with bacteria. We employ the metabolic machinery of some species to make cheeses, therapeutic drugs, and other useful items. As described in Chapter 22, bacteria (and plasmids) have been our partners in genetic engineering. Harmless *E. coli* bacteria in our colon make vitamin K, while other, pathogenic strains cause intestinal disease. In general, we tend to associate bacteria with disease.

After antibiotics were discovered in the 1940s, they were quickly harnessed to fight disease. An **antibiotic** is a substance that can destroy or inhibit the growth of bacteria and some other microorganisms. Bacteria and fungi produce most antibiotics. Some, such as penicillins, tetracyclines, and streptomycin, kill microorganisms by interfering with different life processes. For example, the penicillins break bonds that hold molecules together in the cell walls of susceptible bacteria.

Antibiotics don't work against viruses, which are not cells and so do not have a metabolism to disrupt. Body defenses such as the antiviral proteins called **interferons** (Section 10.1) may block replication of the virus inside

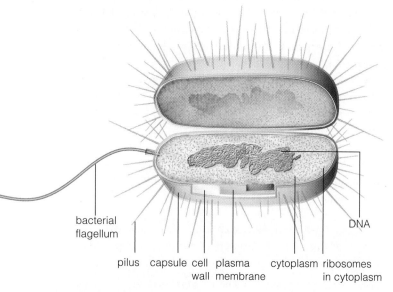

bacterial
flagellum

DNA

pilus   capsule cell   plasma   cytoplasm ribosomes
                 wall membrane   in cytoplasm

Figure 18.6 *Animated!*   The structure of one kind of bacterium.

INFECTIOUS ORGANISMS

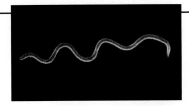

actual size: ●

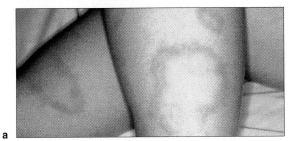

a

b

Figure 18.8 Examples of bacteria linked with emerging diseases. (**a**) *Borrelia burgdorferi*, which causes Lyme disease. The lower photograph shows the bull's-eye rash that is a symptom of Lyme disease, now the most common tick-borne disease in the United States. (**b**) *Mycobacterium tuberculosis*, which causes TB.

cells. Antiviral drugs prevent virus particles from exiting host cells or interfere with the viral "life cycle" in some other way.

Today there are scores of prescribed antibiotics. All are potent drugs. Different ones can have side effects such as triggering an allergic response or reducing the effectiveness of birth control pills. Even more serious, however, is the emergence of antibiotic-resistant microbes.

A BIOLOGICAL BACKLASH TO ANTIBIOTICS IS UNDER WAY

Over the years, some doctors prescribed antibiotics for patients who had viral—not bacterial—illnesses. In some countries, antibiotics are not prescription drugs, so people buy and take the drugs whenever they don't feel well. Some patients stop taking an antibiotic when they start to feel better, without finishing the full recommended course of treatment. Antibiotics also have been added to soaps, kitchen wipes, and many other consumer products. All these practices have contributed to the proliferation of bacteria that are genetically resistant to antibiotics that might otherwise kill them. A key reason is that ill-advised and improper use of antibiotics wipes out susceptible bacteria, leaving behind resistant "super bugs." As that happens over and over, soon the only disease-causing bacteria left are those equipped to fend off one or more antibiotics. At present, the list of drug-resistant bacteria includes strains that cause some cases of tuberculosis, urinary tract infections, strep throat, STDs such as gonorrhea and syphilis, dysentery, childhood middle-ear infections, and surgical-wound infections.

Staphylococcus aureus, or staph A, is a bacterium that can cause pneumonia and wound infections that rapidly destroy tissue. It is one of several microbes that may soon be resistant to *all* available antibiotics. Researchers are racing to develop new antibiotics that combat disease-causing organisms in new, effective ways.

BACTERIA CAUSE SOME IMPORTANT EMERGING AND REEMERGING DISEASES

A major emerging disease in the United States is **Lyme disease**, which apparently arose in the mid-1900s. It is caused by the spirochete *Borrelia burgdorferi* (Figure 18.8*a*), which is transmitted by ticks when they suck blood. Early symptoms resemble the flu, but if the disease progresses it can cause crippling arthritis, as well as heart and neurological problems. Fortunately, Lyme disease can be cured with antibiotics.

The situation is less bright with **tuberculosis** (TB), a "reemerging" disease that attacks the respiratory system and produces pneumonialike symptoms, among other ills. It is caused by a bacillus, *Mycobacterium tuberculosis* (Figure 18.8*b*). The bacteria are transmitted in droplets produced by coughing or sneezing. Normally the immune system kills the bacilli. If small lesions (called tubercles) do form in the lungs, their healing will leave a scar that can be seen on a chest X ray.

By the 1970s the advent of antibiotics helped make TB relatively rare in developed countries. Today, however, TB is reemerging as a major global health threat, in part due to increased travel and immigration, new strains that are resistant to antibiotics, and other factors. People who live in crowded, unsanitary conditions (such as refugee camps and urban slums) are at much higher risk. Many recent cases have developed in people infected with HIV, whose immune systems are weakened.

According to the National Institute of Allergy and Infectious Diseases, in parts of Eastern Europe *all* new TB patients have a drug-resistant strain. Screening programs try to ferret out early-stage cases for treatment. In the meantime, 3 to 5 million new cases are being diagnosed each year.

Bacteria are microscopic prokaryotic cells. They and some other pathogens often can be killed by antibiotics, although resistant strains are becoming more common. Antibiotics are not effective against viruses.

LINKS TO
SECTIONS
7.9, 10.4,
AND 16.11

Fungi and protozoa ("first animals") that infect humans are single eukaryotic cells. Various worms are parasites—they live on or in a host organism for at least part of its life cycle.

PARASITIC FUNGI AND PROTOZOA ARE SMALL BUT POTENTIALLY DANGEROUS

Common fungal pathogens that affect humans include those that cause yeast infections, the confusingly named "ringworm," and athlete's foot (Figure 18.9*a*). The fungi behind athlete's foot and ringworm infections release enzymes that break down the keratin in skin. They also can trigger painful inflammation. A yeast infection arises when a change in the chemical conditions in the vagina or elsewhere causes overgrowth of *C. albicans* cells.

Entamoeba histolytica, often simply called an "amoeba," is a protozoan that forms a cyst at one stage of its life cycle. The cyst is a tough body covering that helps the organism wait out "bad times," such as a lack of food. Contaminated food and water are common reservoirs for cysts. Inside human intestines, the amoeba completes its life cycle and in the process causes **amoebic dysentery**, a type of severe diarrhea (Figure 18.9*b*). In places where raw sewage carries cysts of *E. histolytica* into public water supplies, amoebic dysentery is a leading cause of death among small children.

Giardia intestinalis (Figure 18.9*c*) also forms cysts that enter water or food supplies in contaminated feces. It causes **giardiasis**, bringing awful "rotten egg" belches, explosive diarrhea, and other symptoms. Antibiotics can cure both *Giardia* and amoebic dysentery.

Trypanosoma brucei causes **African sleeping sickness** (Figure 18.9*d*). This severe form of encephalitis is passed from person to person by bites of the tsetse fly. Untreated, the disease is fatal. At any given time, at least a million people in Africa are infected.

Watery diarrhea is one symptom of **cryptosporidiosis**, caused by a protozoan called *Cryptosporidium parvum*. In the United States, cryptosporidiosis is an emerging infectious disease. *C. parvum* is a tough character. Its oocytes (immature eggs) can even survive a two-hour soak in full-strength bleach! Like cysts of *Giardia* and amoebas, the oocytes are spread in water or food contaminated with the feces of humans or other animal hosts.

WORMS ALSO CAN BE A SERIOUS THREAT

Poor sanitation isn't a factor only in diseases caused by protozoa; it also is how many worm infections spread. Another, less common route is eating contaminated meat or fish. For people in developed countries, such as the United States, pinworm infections are the most familiar. A pinworm is a small, white roundworm; it looks like a quarter-inch fleck of white thread. The infection route is oral–fecal; that is, first the person consumes worm eggs, then the eggs hatch and worms develop, then female pinworms lay *their* eggs just outside the anus, then those eggs find their way to a person's mouth, and so on.

Some other infectious worms, including tapeworms, hookworms, whipworms, and large *Ascaris* roundworms, can seriously damage body tissues and organs. Most common in the tropics, serious worm infections can be treated with therapeutic drugs. Even so, many millions of people in various parts of the world share their bodies with worms, often without knowing it.

Some species of protozoa are pathogens that cause serious diseases in humans. Often, they are spread by way of water or food contaminated with animal feces.

Infectious worms can do serious damage to many tissues and organs.

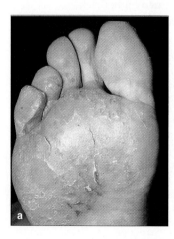
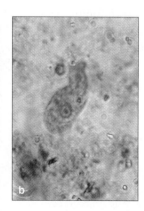

Figure 18.9 (**a**) Athlete's foot. (**b**) *Entamoeba histolytica*, which causes amoebic dysentery. This specimen has been stained, then photographed under a light microscope. (**c**) *Giardia intestinalis* causes intestinal disturbances. (**d**) *Trypanosoma brucei*, a protozoan that causes African sleeping sickness. It is shown next to a red blood cell.

18.5 Malaria: Efforts to Conquer a Killer

According to the World Health Organization, every year **malaria** kills nearly 3 million people, most of them children in Africa. And each year brings another 110 million new cases. The direct cause of malaria is a protozoan called *Plasmodium*. One of its life stages lives in the salivary glands of the female *Anopheles* mosquito. When the insect sucks blood from a person, *Plasmodium* travels from her salivary glands into the person's bloodstream. When it reaches the liver, severe illness and often a lifetime of frequent misery begin (Figure 18.10).

Shaking, chills, a high fever, and drenching sweats are classic symptoms of malaria. Symptoms abate for weeks or months, but relapses occur when dormant parasites become active. In time, a person with malaria may develop anemia (from the loss of red blood cells) and a greatly enlarged liver. As described in Section 20.5, some people, including those who are, or are descended from, West African blacks, may carry the gene for **sickle-cell anemia**—and this genetic heritage can help protect them from malaria. In people who inherit only one copy of the gene, red blood cells respond to infection in a way that cuts short the parasite life cycle.

Tropical Africa has always been malaria's stronghold. Today, many strains of *Plasmodium* have become drug-resistant, and the poorest people and countries in malaria-prone regions cannot easily afford the drugs that are available. These facts, and the awful toll malaria takes on the people of Africa, have spurred intensive efforts to develop a malaria vaccine. A major challenge is the complex life cycle of the pathogen, because it has turned out to be difficult to create a vaccine that will be effective against all its life stages.

A ray of hope has come from our increasing understanding of cell structure, and especially the different sorts of proteins that are present on the surfaces of cells. At this writing researchers are trying to develop a vaccine that recognizes a surface protein on the merozoite stage of *Plasmodium* (Figure 18.10d). This is the stage that invades red blood cells and causes the first real damage of malaria. If a vaccine can trigger a safe but effective immune response against this stage, it may be possible to end an initial infection before the pathogen becomes established in the liver. Clinical trials for this vaccine may have already begun by the time you are reading this book.

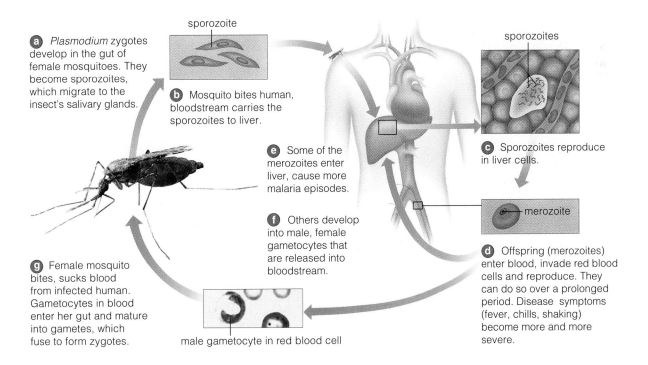

a *Plasmodium* zygotes develop in the gut of female mosquitoes. They become sporozoites, which migrate to the insect's salivary glands.

sporozoite

b Mosquito bites human, bloodstream carries the sporozoites to liver.

e Some of the merozoites enter liver, cause more malaria episodes.

f Others develop into male, female gametocytes that are released into bloodstream.

g Female mosquito bites, sucks blood from infected human. Gametocytes in blood enter her gut and mature into gametes, which fuse to form zygotes.

male gametocyte in red blood cell

sporozoites

c Sporozoites reproduce in liver cells.

merozoite

d Offspring (merozoites) enter blood, invade red blood cells and reproduce. They can do so over a prolonged period. Disease symptoms (fever, chills, shaking) become more and more severe.

Figure 18.10 *Animated!* Life cycle of one of the *Plasmodium* species that causes malaria.

18.6 Patterns of Infectious Diseases

LINKS TO SECTIONS 10.1, 10.9, AND 16.12

You can't avoid being exposed to infectious diseases, but you can limit your risk and be a better-informed health care consumer by understanding some basic characteristics of infectious microbes, including how they spread.

INFECTIOUS PATHOGENS SPREAD IN FOUR WAYS

By definition, an infectious disease can be transmitted from person to person. There are four common modes of transmission:

1. *Direct contact* with a pathogen, as by touching open sores or body fluids from an infected person. (This is where "contagious" comes from; the Latin *contagio* means "touch" or "contact.") Infected people can transfer pathogens from their hands, mouth, or genitals.

2. *Indirect contact*, as by touching doorknobs, tissues (or handkerchiefs), diapers, or other objects previously in contact with an infected person. As already noted, food and water can be contaminated by pathogens.

3. *Inhaling pathogens*, such as cold and influenza viruses, that have been spewed into the air by coughs and sneezes (Figure 18.11). This is the most common mode of transmission.

4. *Contact with a vector*, such as mosquitoes, flies, fleas, and ticks. A **disease vector** carries a pathogen from an infected person or contaminated material to new hosts. In some cases, part of the pathogen's life cycle must

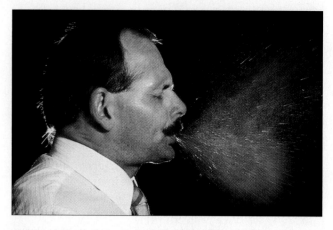

Figure 18.11 A full-blown sneeze. Inhaled pathogens spewed out during an unprotected sneeze can transmit colds, influenza, and other infectious diseases.

take place inside the vector, which is an intermediate host. Mosquitoes are the intermediate hosts for the West Nile virus and the *Plasmodium* parasites that cause malaria (Section 18.5).

Every year 5 to 10 percent of hospitalized people come down with a **nosocomial infection**—one that is acquired in a hospital, usually by direct contact. For example, despite the best precautions, bacteria may enter a person's urinary tract during a catheterization procedure. Why are nosocomial infections so common? Anyone who is sick enough to be hospitalized may have a compromised (and therefore less effective) immune system, and invasive medical procedures give bacteria easy access to tissues. Also, the intensive use of antibiotics in hospitals increases the chances that antibiotic-resistant pathogens will be present there. Hospitals usually are careful to monitor patients likely to be vulnerable to nosocomial infection.

DISEASES OCCUR IN FOUR PATTERNS

Infectious diseases sometimes are described in terms of the patterns in which they occur. In an **epidemic**, a disease rate increases to a level above what we would predict, based on experience. When cholera broke out all through Peru in 1991, that was an epidemic. The bubonic plague epidemic in fourteeth-century Europe killed 25 million people. When epidemics break out in several countries around the world in a given time span, they collectively are called a **pandemic**. AIDS is pandemic; it has spread worldwide since the first cases were identified in 1981.

A **sporadic disease**, such as whooping cough, breaks out irregularly and affects relatively few people. An **endemic disease**, such as the common cold, occurs more or less continuously. Many of the diseases noted in Table 18.2 are endemic in various parts of the world.

Table 18.2	Infectious Diseases: Global Health Threats*	
Disease	Type of Pathogen	Estimated Deaths per Year
Diarrheas (includes amoebic dysentery, cryptosporidiosis)	Protozoa, virus, and bacteria	31 million
Various respiratory infections (pneumonia, viral influenza, diphtheria, strep infections)	Virus, bacteria	7+ million
Malaria	Protozoan	2.7 million
Tuberculosis	Bacterium	2.4 million
Hepatitis (includes A, B, C, D, E)	Virus	1–2 million
Measles	Virus	220,000
Schistosomiasis	Worm	200,000
Whooping cough	Bacterium	100,000
Hookworm	Worm	50,000+

*Does not include AIDS-related deaths.

THE DISEASE PROCESS

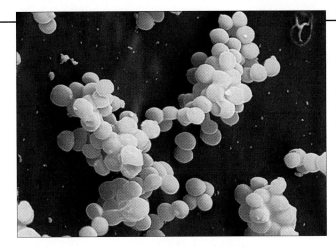

Figure 18.12 A virulent microbe: The bacterium *Staphylococcus aureus*, shown in a false-color micrograph.

VIRULENCE IS A MEASURE OF THE DAMAGE A PATHOGEN DOES

Pathogens are ranked according to their **virulence**—how likely it is that the pathogen will cause serious disease. Virulence depends on how fast the pathogen can invade tissues, how severe is the damage it causes, and which tissues it targets. For example, a virus that can cause pneumonia is more virulent than a virus that causes the sniffles. Rabies viruses are highly virulent because they target the brain. The SARS virus appears to be extremely virulent, because it makes people very sick very quickly.

Antibiotic resistance in certain bacteria, such as strains of *Staphylococcus aureus* (Figure 18.12), has made those microbes highly virulent. Infectious disease specialists have instituted a worldwide surveillance system to identify new resistant strains before they can become established.

THERE ARE MANY PUBLIC AND PERSONAL STRATEGIES FOR PREVENTING DISEASE

The best way to combat any disease is to prevent it in the first place. With infectious diseases, prevention depends on knowing how a disease is transmitted and what the pathogen reservoir is.

Figure 18.13 lists general strategies for preventing the transmission of pathogens that are present on the skin, in the respiratory tract, in the GI tract, and in blood. These strategies recognize that the human body, soil, water, and other animals all are reservoirs for a range of pathogens. Notice that regular hand washing tops the list for limiting your exposure to all but blood-borne pathogens. Public health measures include vaccination programs, standards for processing or treating supplies of food, drinking water, and blood products, and public dissemination of information on proper food-handling methods. Section 16.12 describes strategies for protecting yourself against sexually transmitted disease.

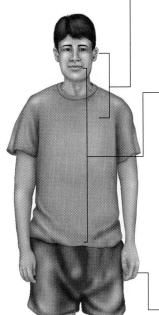

Respiratory tract
Preventative measures:
- Hand washing
- Cover mouth when coughing or sneezing
- Proper disposal of used tissues
- Vaccination programs

GI tract
Preventative measures:
- Hand washing
- Proper food storage, handling, and cooking
- Good public sanitation (sewage, drinking water)

Blood
Preventative measures:
- Avoid/prevent needle sharing/ IV drug abuse
- Maintain pure public blood supplies
- Vaccination programs against blood-borne pathogens (e.g., hepatitis B)

Skin
Preventative measures:
- Hand washing
- Limit contact with items used by an infected person

Figure 18.13 Disease prevention strategies are based on knowing how pathogens spread and what their reservoirs are.

Infectious diseases are transmitted by direct or indirect contact with pathogens, by being inhaled, or by vectors. A disease may turn up sporadically, it may become endemic to a particular region, or it may become epidemic or pandemic.

Some pathogens are extremely virulent—they can cause severe illness.

Preventing an infectious disease requires strategies for avoiding pathogen reservoirs and for limiting the chances that a given pathogen will be transmitted.

Summary

Section 18.1   An infection occurs when a pathogen enters the body and multiplies. Disease develops when the pathogen's activities interfere with normal body functioning. Pathogens cause disease in different ways: bacteria release toxins, viruses kill cells or trigger cancer, fungi produce enzymes that break down tissues, and parasites damage cells or tissues or trigger damaging immune responses such as inflammation.

Emerging diseases are ones that have never before affected large numbers of people around the world. They include the diseases caused by the West Nile virus, Ebola virus, and SARS virus. Some emerging diseases are zoonoses—they mainly infect other animals but also can infect humans. Reemerging diseases, such as tuberculosis, are diseases that have been well controlled in the past but are again being diagnosed in large numbers of people.

Section 18.2   A virus is a noncellular infectious particle. Viruses consist of nucleic acid (DNA or RNA) enclosed in a protein coat; some also have an outer envelope. A retrovirus, such as HIV, is an RNA virus that infects animal cells.

Viruses multiply by taking over the metabolic machinery of a host cell. They can't be killed by antibiotics, but may be susceptible to body defenses (interferons) and drugs that interfere with their ability to attach to host cells or reproduce inside them. Prion diseases are caused by infectious proteins smaller than viruses.

 Compare different forms of viruses and see how some of them can multiply.

Section 18.3   Bacteria are microscopic prokaryotic cells. Some species are used as living factories to produce therapeutic drugs and other products, but many other species cause disease. Overuse and misuse of antibiotics have contributed to the evolution of antibiotic-resistant bacteria.

Section 18.4   Pathogenic fungi include the species that cause athlete's foot, ringworm, and yeast infections. Some protozoa (single-celled eukaryotes) are harmful parasites. Often they are spread by way of water or food contaminated with feces from infected people or other animals. Some infections by worms also seriously harm tissues and organs.

 Observe the life cycle of the malaria-causing protozoan Plasmodium.

Section 18.6   An infectious disease may be transmitted by direct contact with a pathogen; by indirect contact (e.g., handling a contaminated object); by inhaling the pathogen; or through a disease vector that physically carries pathogens or contaminated material to new hosts. A nosocomial infection is one that is acquired in a hospital, usually by direct contact.

A disease epidemic makes more people ill than experience would have predicted. A pandemic occurs when epidemics break out more or less simultaneously in various places around the world. Virulence is a gauge of a pathogen's ability to cause serious disease.

Measures for preventing infectious diseases draw on our understanding of the reservoirs for pathogens and of the ways pathogens can be transmitted.

Review Questions

1. List at least four of the seven requirements a pathogen must meet in order to cause infectious disease.

2. What is a virus? What is a retrovirus? What is the difference between a virus and a prion?

3. What does it mean to say that a virus is latent?

4. Define the difference between a virus and a bacterium, and explain how bacteria reproduce.

5. What is an antibiotic? List the factors that have contributed to the emergence of antibiotic-resistant microbes.

6. What are the four main ways pathogens spread?

7. Explain what we mean by virulence, give an example of a highly virulent pathogen, and explain why it is classified this way.

Self-Quiz   *Answers in Appendix V*

1. A _____ is described as a noncellular infectious agent.
    a. virus   c. fungus
    b. bacterium   d. protozoan

2. _____ is a eukaryotic cell that causes disease when it is ingested in contaminated food or water.
    a. An adenovirus   c. *Entamoeba histolytica*
    b. *Chlamydia trachomatis*   d. A prion

3. Most pathogenic bacteria _____.
    a. cause disease by secreting toxins that damage body cells
    b. can enter the body in contaminated water or other substances
    c. may be able to multiply so rapidly they overwhelm the immune system
    d. a and b only
    e. a, b, and c are all correct

4. Antibiotic resistance is a result of _____.
    a. use of antibiotics against viruses
    b. patients' failure to take a full course of the drug
    c. self-prescribing of the drugs
    d. b and c only
    e. a, b, and c

5. An infectious disease may not be transmitted by
   a. a gene mutation.
   b. simply shaking hands with an infected person.
   c. being bitten by a flea.
   d. surgical instruments in a hospital.

6. Match the following terms and concepts:

   ____ endemic   a. causes serious disease
   ____ virulent   b. disease that is nearly always
         pathogen   present
   ____ prion   c. may become latent
   ____ epidemic   d. infectious protein
   ____ carrier   e. disease that occurs at a higher
   ____ virus   than expected rate
                   f. no apparent disease symptoms

Critical Thinking

1. A janitor in the cafe where you work has been diagnosed as HIV-positive, and a server has come down with type B hepatitis. Some other employees start a petition demanding that both people be required to wear a mask over the mouth and nose, not handle soiled dishes or food, and not use the employee restroom. Both infected employees strenuously object to the plan. You are asked to lead a discussion aimed at resolving the issue, and you decide to prepare a handout giving the scientific basis for making a decision in each case. What does the handout say?

2. A tourist returning from a tropical vacation comes down with monkeypox (Figure 18.14*a*), which is caused by a DNA virus. He rushes to the emergency room and pleads for an antibiotic, but the physician says she can't prescribe one. Why not?

3. The micrograph in Figure 18.14*b* shows bacteria on the tip of a pin. What general category of bacterium is shown? Thinking back to Chapter 10, what might occur in your body if you prick yourself with the pin and these bacteria are pathogenic?

Figure 18.15 Medieval attempt to deal with a bubonic plague epidemic—the Black Death—that may have killed half the people in Europe.

4. The mild pneumonia sometimes called "walking pneumonia" is caused by one species of an unusual group of bacteria called mycoplasmas. *Mycoplasma pneumoniae*, usually transmitted by respiratory droplets, causes pneumonia in about 10 percent of the people it infects. Antibiotics kill the pathogen. How would you classify (1) its mode of transmission and (2) its virulence?

5. In the Middle Ages bubonic plague struck Europe with a vengeance and killed an estimated one-third to one-half of the population. Not knowing that the disease is caused by a bacterium (*Yersinia pestis*) spread by fleas that live on rats, terrified people tried all sorts of methods to ward off the disease, including praying and dancing until they dropped (Figure 18.15). What is the disease vector in this instance? Which organism(s) is/are the reservoir?

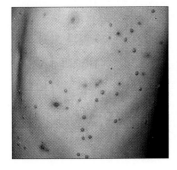

Figure 18.14 (a) Skin sores on a person infected with the virus that causes monkeypox. (b) Bacteria on the tip of a pin.

Explore on Your Own

Not long ago, researchers fighting rhinoviruses—the viruses that cause the common cold—reported they were nearing completion on the development of two new antiviral medicines. One prevents the cold virus from entering its target host cells (that is, it's an entry inhibitor). The other contains a protease inhibitor, which halts the actions of an enzyme many cold viruses need to replicate themselves. Both have been developed by "piggybacking" on research on HIV and AIDS.

A wealth of information on infectious diseases and their treatments is available on the World Wide Web. Starting, perhaps, with the website of the Centers for Disease Control and Prevention, go online and see what you can learn about these new cold remedies. What specific use is each designed for? What form does each take (e.g., nasal spray, pill, injection, etc.)? In the course of your search, can you uncover reports of any other new antiviral drugs?

Henrietta's Immortal Cells

Each human starts out as a fertilized egg. By the time of birth, cell division and other processes have given rise to a body of about a trillion cells. Even in an adult, many cells still divide, replacing damaged or worn-out ones.

In 1951, George and Margaret Gey of Johns Hopkins University were trying to develop a way to keep human cells dividing in the laboratory. An "immortal" cell line would help researchers study basic life processes as well as cancer and other diseases. It would be an alternative to experimenting directly on patients and risking lives.

The Geys' lab assistant, Mary Kubicek, tried again and again to establish a self-perpetuating line of human cancer cells. About to give up, she prepared one last culture and named them "HeLa" cells. The code name stood for the first two

Henrietta Lacks

letters of the patient's first and last names, Henrietta Lacks.

Those cells began to divide. Four days later, there were so many HeLa cells that the researchers subdivided them into more culture tubes.

Sadly, cancer cells in the patient were dividing just as rapidly. Six months after the cancer was diagnosed, the disease had spread throughout her body. Two months after that, Henrietta Lacks was dead. Yet some of her cells lived on in the Geys' laboratory as the first successful human cell culture.

In time, HeLa cells were shipped to research laboratories all over the world. Today, hundreds of important research projects draw on Henrietta's immortal cells.

Henrietta was only 31 when runaway cell division killed her. Now, decades later, her legacy helps humans everywhere, through descendants of her cells that are still dividing, day after day.

Understanding cell division—and ultimately how each of us is put together in the image of our parents—starts with the answers to three questions. First, what kind of information guides inheritance? Second, how is the information copied in a parent cell before being distributed into daughter cells? And third, what kinds of mechanisms actually parcel out the information to daughter cells?

 How Would You Vote? *It is illegal to sell your organs, but you can sell your cells, including eggs, sperm, and blood cells. Descendants of HeLa cells are sold all over the world by cell culture firms. Should the family of Henrietta Lacks share in the profits? Cast your vote online at www.thomsonedu.com/biology/starr/humanbio.*

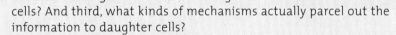

Key Concepts

BASIC PRINCIPLES OF CELL DIVISION
Cells reproduce by duplicating their chromosomes and when dividing the chromosomes and cytoplasm among daughter cells.

MITOSIS: GROWTH AND REPAIR
Body cells divide by mitosis, a process that divides the nucleus and maintains the parental chromosome number. The cytoplasm divides by a process of cytokinesis.

MEIOSIS AND SEXUAL REPRODUCTION
Gametes—sperm and oocytes—form by meiosis. A gamete receives half the number of parental chromosomes. Later, fertilization restores the parental chromosome number.

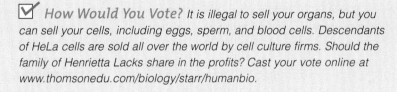

Links to Earlier Concepts

This chapter builds on what you have already learned about the cell nucleus and chromosomes (3.4) and explains how microtubules (3.9) play a major role in cell division. It also expands on the discussions in Chapter 16 of how sperm and eggs form—the processes of spermatogenesis and oogenesis (16.2 and 16.4).

19.1 Dividing Cells Bridge Generations

LINKS TO
SECTIONS
1.1, 3.3, 3.4,
16.2, AND 16.4

The continuity of life depends on the ability of cells—and whole organisms—to faithfully reproduce themselves. Dividing cells are the bridge between these generations.

DIVISION OF THE "PARENT" NUCLEUS SORTS DNA INTO NUCLEI FOR DAUGHTER CELLS

In biology, **reproduction** is when a parent cell produces a new generation of cells, or when parents produce a new individual. Reproduction is part of a **life cycle**, a recurring series of events in which individuals grow, develop, maintain themselves, and reproduce. The instructions for the human life cycle are encoded in our DNA, which we inherit from our parents. Reproduction begins with the division of single cells. It follows this basic rule: Each cell of a new generation must receive a copy of the parent cell's DNA. Otherwise the cell won't develop or function properly.

Each daughter cell also inherits some cytoplasm from the parent cell. This cytoplasm provides "start-up" machinery, such as enzymes and organelles, that will keep the new cell functioning until it has time to use its inherited DNA for growing and developing on its own.

A eukaryotic cell has only one nucleus, so the cell can't just split in two when it divides. First, the nucleus must be divided in a way that parcels out DNA and packages it in two nuclei, one for each new cell. Depending on the kind of cell, the nucleus will be divided by *mitosis* or by *meiosis* (Table 19.1). Both mechanisms sort out and package DNA molecules into new nuclei for daughter cells.

Mitosis is the division mechanism that divides the nucleus of a *somatic* (body) *cell*. The body grows, replaces worn-out or dead cells, and repairs tissues by way of mitosis.

In contrast to mitosis, **meiosis** is the mechanism for dividing the nucleus of **germ cells**—the oogonia in ovaries and spermatogonia in testes (Chapter 16). In these cells, meiosis must take place before gametes (sperm and eggs) form. Accordingly, meiosis is the first stage in sexual reproduction.

CHROMOSOMES ARE DNA "PACKAGES" IN THE CELL NUCLEUS

Both mitosis and meiosis divide a parent cell's DNA into new nuclei. To do this, mitosis and meiosis both must manipulate chromosomes—the DNA "packages" you first read about in Chapter 3. Each **chromosome** is one very long DNA molecule combined with protein. Each **gene** is a segment of DNA in a chromosome. Together, the DNA and protein are called *chromatin*. You will read more about chromosome structure in Section 19.2.

HAVING TWO SETS OF CHROMOSOMES MAKES A CELL DIPLOID

The sum of the chromosomes in cells of a given type is the **chromosome number**. Human DNA is carried on 23 chromosomes, but our somatic cells have two full sets of them—one set from each parent. Thus the chromosome number in body cells is 46 (Figure 19.1a). A cell that has two of each type of chromosome is called a **diploid** cell. The shorthand $2n$ indicates that a cell is diploid. The n stands for the number of chromosomes in one full set.

When the nucleus of a diploid parent cell divides by mitosis, the end result is two diploid daughter cells (Figure 19.2). One member of each chromosome pair is from the mother, and the other is from the father. Before a diploid cell divides, its chromosomes are duplicated, so it has four sets of chromosomes. Mitosis puts half of this doubled genetic material in each new cell, so each ends up with the diploid number of chromosomes.

In Figure 19.1a, there are 23 pairs of chromosomes. This is because each chromosome has been lined up with its partner from the other parent. Pairs numbered 1 through 22 are **autosomes**: the two members of each pair are the same length and shape and both carry hereditary instructions for the same traits. Pair number 23 consists of the **sex chromosomes**, which determine a person's biological sex. The two types of sex chromosomes are labeled X and Y. As you may know, a female has two X chromosomes, while a male is XY.

Table 19.1	Basics of Mitosis and Meiosis	
	Mitosis	**Meiosis**
Function	Growth, including repair and maintenance	Gamete production (sperm/eggs)
Occurs in	Somatic (body) cells	Germ cells in gonads (testes and ovaries)
Mechanism	One round of chromosome partitioning plus cytokinesis	Two rounds of chromosome partitioning and cytokinesis
Outcome	Maintains diploid chromosome number ($2n \rightarrow 2n$)	Reduces diploid chromosome number ($2n \rightarrow n$)
Effect	Two diploid daughter cells	Four haploid daughter cells

Paired corresponding chromosomes, one from each parent, are called **homologous chromosomes**, or simply *homologues* (from a Greek word meaning "to agree"). The X and Y sex chromosomes are considered homologues, even though they are of different size and form and for the most part they carry genes for different traits. You will read more about both types of chromosomes in Chapter 21.

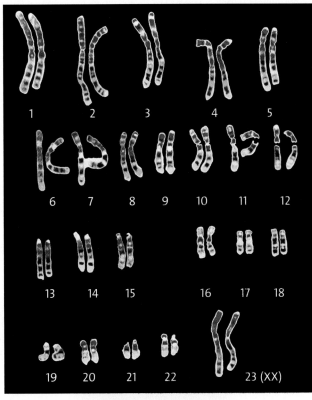

a

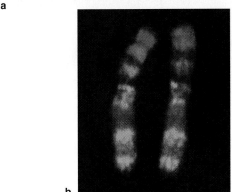

b

Figure 19.1 (**a**) Forty-six chromosomes from a human female. Each chromosome is duplicated. The presence of pairs of chromosomes (two of each type) tells you that they came from a diploid cell. One member of each pair contains genetic instructions inherited from the father. The other member contains instructions from the mother. (**b**) Close-up of a pair of homologous chromosomes from an animal cell.

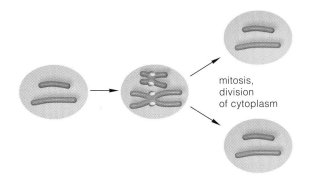

mitosis, division of cytoplasm

a Two of the chromosomes (unduplicated) in a parent cell at interphase

b The same two chromosomes, now duplicated, in that cell at interphase, prior to mitosis

c Two chromosomes (unduplicated) in the parent cell's daughter cells, which both start life in interphase

Figure 19.2 A simple way to think about how mitosis maintains the diploid chromosome number from one generation to the next.

HAVING JUST ONE SET OF CHROMOSOMES MAKES A CELL HAPLOID

Meiosis takes place in dividing germ cells in ovaries or testes. Remember from Chapter 16 that human germ cells are *spermatogonia* in males and *oogonia* in females. Spermatogonia and oogonia are diploid. However, unlike other body cells, they can give rise to haploid gametes—sperm or eggs—by way of meiosis.

A human sperm or oocyte typically contains just one set of 23 chromosomes, instead of 46 paired, homologous chromosomes. This is because meiosis is a **reductional division**. It halves the diploid number of chromosomes ($2n$) to a **haploid** number (n). And not just any half. Each haploid gamete ends up with one partner from each pair of homologous parent chromosomes.

With these basic concepts in mind, we now consider in more detail how cells divide. Chromosomes are central players in the story, so we begin in the next section by looking at how these structures are organized.

Mitosis and meiosis sort chromosomes, and therefore DNA, into new nuclei for daughter cells.

Mitosis takes place in somatic cells. The chromosome number remains the same from one cell generation to the next. The body grows and tissues are repaired by way of mitosis.

Meiosis takes place in germ cells in the ovaries and testes. It is the mechanism that produces gametes.

19.2 A Closer Look at Chromosomes

LINK TO SECTION 3.5

A chromosome is an awesomely organized structure that consists of coiled loops of DNA and proteins.

IN A CHROMOSOME, DNA INTERACTS WITH PROTEINS

Each chromosome in your body's cells consists of DNA and proteins that are attached to it (Figure 19.3). Some of these proteins are called *histones*. The DNA loops twice around certain histones, which then look a little like beads on a string when they are highly magnified. Each DNA-histone unit is called a *nucleosome*. When segments of DNA in part of a chromosome—that is, one or more genes—are to be "read out," the nucleosome packing there loosens up.

In mitosis and meiosis, before a cell begins dividing, it duplicates its chromosomes. For much of the division process, each chromosome and its copy stay together as **sister chromatids**:

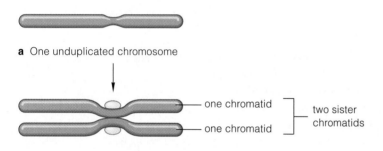

a One unduplicated chromosome

one chromatid
one chromatid
two sister chromatids

b One chromosome (duplicated)

As a cell nucleus is beginning to divide, the proteins and DNA interact and make the chromosome coil back on itself repeatedly. This condensed form may help keep chromosomes from getting tangled up when they are moved and sorted into parcels for daughter cells.

When the coiling is complete, a chromosome has its typical size and shape. Each of its sister chromatids also has at least one "pinched in" region called a **centromere**. Structures at the surface of centromeres are docking sites for microtubules that move chromosomes when a cell nucleus is dividing.

SPINDLES ATTACH TO CHROMOSOMES AND MOVE THEM

In both mitosis and meiosis, a cell's chromosomes move into new positions with the help of a spindle. The **spindle** is composed of two sets of microtubules extending from its two poles (Figure 19.4). These "poles" are centrioles, parts of a cell's cytoskeleton that were introduced in Section 3.8. One set of microtubules overlaps the other

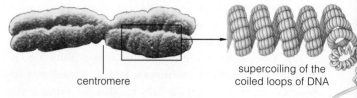

centromere

supercoiling of the coiled loops of DNA

Figure 19.3 *Animated!* A duplicated human chromosome in its most condensed form. Interacting proteins are holding loops of coiled DNA in a supercoil. At a deeper level of organization, proteins and DNA interact in a ropelike fiber.

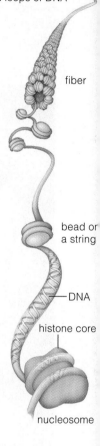

fiber

bead or a string

DNA

histone core

nucleosome

set at the spindle's equator, midway between its two poles. The spindle establishes where chromosomes will end up before a cell divides in two.

DNA and proteins interact in chromosomes. Some of these interactions condense chromosomes in cells that are ready to divide. When a cell's nucleus is about to divide, a microtubule apparatus called a spindle attaches to chromosomes and moves them into new positions.

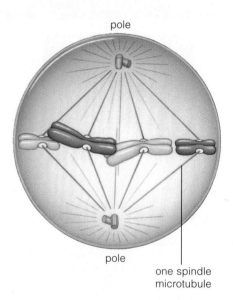

pole

pole

one spindle microtubule

Figure 19.4 Example of a spindle that moves chromosomes in a dividing cell. The barrel-shaped structures are centrioles.

19.3 The Cell Cycle

Every time a new cell comes into being, a multistep cell cycle begins anew.

The "lifetime" of a somatic cell is called the **cell cycle**. The cycle starts every time a new cell is produced, and it ends when the cell completes its own division. Mitosis is a part of the cycle, as sketched in Figure 19.5. Usually, the longest phase of the cell cycle is **interphase**. During this three-part phase, a cell increases its mass, more or less doubles the number of components in its cytoplasm, then duplicates its DNA. The parts of the cell cycle are often abbreviated this way:

G1 Of interphase, a gap (interval) before DNA synthesis begins

S Of interphase, the time when DNA is duplicated

G2 Of interphase, a second gap between the time DNA duplication ends and mitosis begins

M Mitosis; the phase in which chromosomes (duplicated DNA) are sorted into two sets, followed by division of the cytoplasm

The length of the cell cycle varies depending on the type of cell involved. For instance, the cycle lasts twenty-five hours in epithelial cells in your stomach lining and eighteen hours in bone marrow cells. New red blood cells form and replace your worn-out ones at an average rate of 2 to 3 million each second.

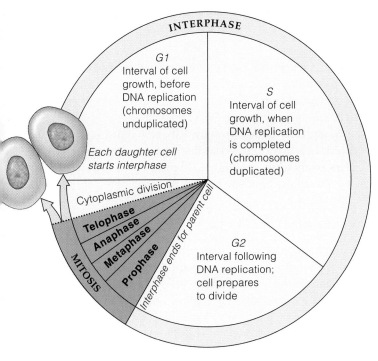

Figure 19.5 *Animated!* A general diagram of the cell cycle. The duration of each phase differs among different cell types.

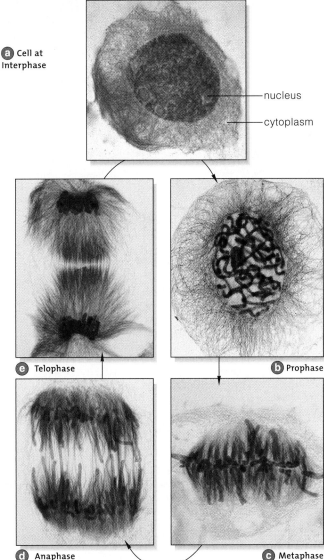

Figure 19.6 Light micrographs showing the progress of mitosis in a cell from the African blood lily (*Haemanthus*). The chromosomes are stained blue, and microtubules that are moving them about are stained red.

Figure 19.6 provides a glimpse of the changes in a cell as it leaves interphase and enters mitosis. The four stages of mitosis (prophase, metaphase, anaphase, and telophase) are the subject we turn to next.

> *A cell cycle begins at interphase, when a new cell (formed by mitosis and cytoplasmic division) increases its mass and the number of its cytoplasmic components, then duplicates its chromosomes. The cycle ends when the cell divides.*

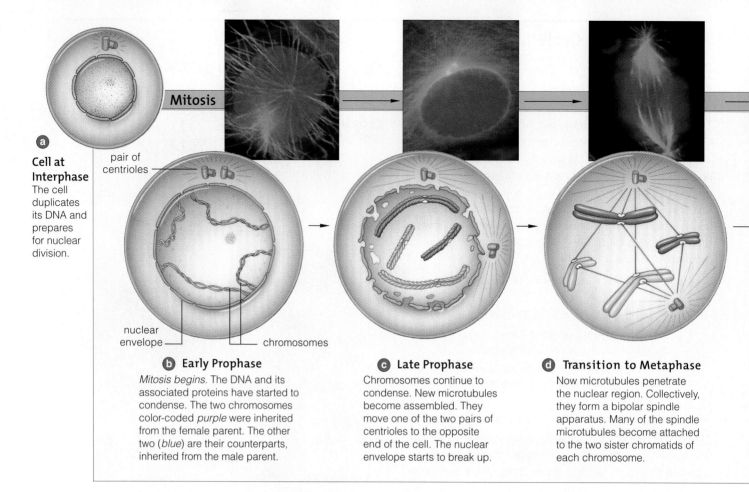

a Cell at Interphase
The cell duplicates its DNA and prepares for nuclear division.

pair of centrioles

Mitosis

nuclear envelope

chromosomes

b Early Prophase
Mitosis begins. The DNA and its associated proteins have started to condense. The two chromosomes color-coded *purple* were inherited from the female parent. The other two (*blue*) are their counterparts, inherited from the male parent.

c Late Prophase
Chromosomes continue to condense. New microtubules become assembled. They move one of the two pairs of centrioles to the opposite end of the cell. The nuclear envelope starts to break up.

d Transition to Metaphase
Now microtubules penetrate the nuclear region. Collectively, they form a bipolar spindle apparatus. Many of the spindle microtubules become attached to the two sister chromatids of each chromosome.

Figure 19.7 *Animated!* Mitosis. This mechanism ensures that daughter cells will have the same chromosome number as the parent cell. For clarity, the diagram shows only two pairs of chromosomes from a diploid (2*n*) animal cell.

19.4 The Four Stages of Mitosis

LINKS TO SECTIONS 3.5 AND 3.8

When interphase ends, a cell has stopped making new parts and its DNA has been replicated. Mitosis can now begin.

The four stages of mitosis are **prophase**, **metaphase**, **anaphase**, and **telophase**. Figure 19.7 shows mitosis in an animal cell.

MITOSIS BEGINS WITH PROPHASE

When prophase begins, a cell's chromosomes are threadlike. ("Mitosis" comes from the Greek *mitos*, for "thread.") During interphase the chromosomes were duplicated, so each is now two sister chromatids joined at the centromere. Early in prophase, the sister chromatids of each chromosome twist and fold into a more compact form. By the end of prophase, the chromosomes will be condensed into thick, rod shapes.

Meanwhile, in the cytoplasm, most microtubules of the cytoskeleton are breaking apart into their subunits (Section 3.9) and new microtubules are forming near the nucleus. The nuclear envelope physically prevents them from making contact with the chromosomes inside the nucleus, but not for long: The nuclear envelope starts to break up as prophase ends.

Many cells have two barrel-shaped centrioles. Each centriole was duplicated in interphase, so there are two pairs of them in prophase. Microtubules start moving one pair to the opposite pole of the developing spindle.

NEXT COMES METAPHASE

A lot happens between prophase and metaphase—so much that this transitional period has its own name, "prometaphase." The nuclear envelope breaks up into tiny, flattened vesicles. This allows the chromosomes to interact with microtubules extending toward them from the poles of the forming spindle. Microtubules from both poles harness each chromosome and start pulling on it. The two-way pulling orients the chromosome's two sister chromatids toward opposite poles. Meanwhile,

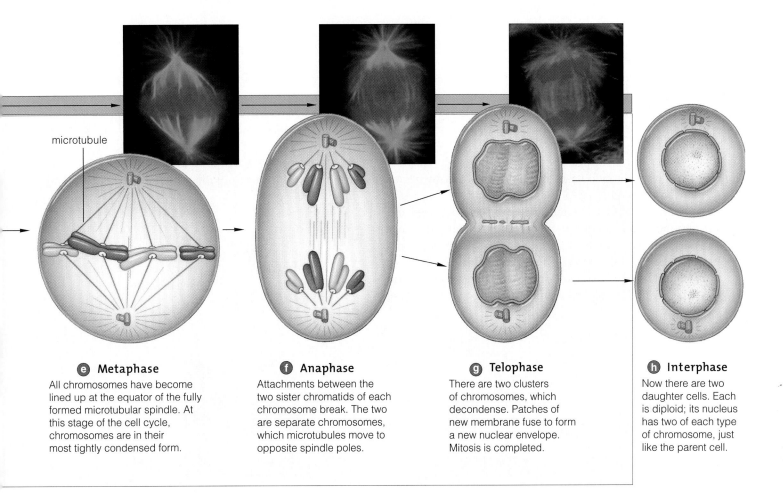

e Metaphase

All chromosomes have become lined up at the equator of the fully formed microtubular spindle. At this stage of the cell cycle, chromosomes are in their most tightly condensed form.

f Anaphase

Attachments between the two sister chromatids of each chromosome break. The two are separate chromosomes, which microtubules move to opposite spindle poles.

g Telophase

There are two clusters of chromosomes, which decondense. Patches of new membrane fuse to form a new nuclear envelope. Mitosis is completed.

h Interphase

Now there are two daughter cells. Each is diploid; its nucleus has two of each type of chromosome, just like the parent cell.

microtubule

---

overlapping spindle microtubules ratchet past each other and push the poles of the spindle apart. Soon the chromosomes reach the midpoint of the spindle.

When all the duplicated chromosomes are aligned midway between the poles of a completed spindle, we say the cell is in metaphase (*meta-* means midway between). This alignment of chromosomes is crucial for the next stage of mitosis, anaphase.

ANAPHASE, THEN TELOPHASE FOLLOW

At anaphase, sister chromatids of the chromosomes separate from each other and move to opposite spindle poles. Two mechanisms produce this movement. First, the microtubules attached to the centromeres shorten and *pull* the chromosomes to the poles. Second, the spindle elongates as overlapping microtubules continue to ratchet past each other and *push* the two spindle poles even farther apart. Once each chromatid is separated from its sister, it is an independent chromosome.

Telophase begins as soon as each of two clusters of chromosomes arrives at a spindle pole. The chromosomes are no longer harnessed to microtubules, and they return to threadlike form. Bit by bit, vesicles derived from the old nuclear envelope fuse around the chromosomes. Soon a new nuclear envelope separates each chromosome cluster from the cytoplasm. Each new nucleus has the same chromosome number as the parent nucleus. Once two nuclei form, telophase is over—and so is mitosis.

> *Before mitosis, each chromosome in a cell's nucleus is duplicated, so that it consists of two sister chromatids.*
>
> *Mitosis occurs in four consecutive stages, called prophase, metaphase, anaphase, and telophase.*
>
> *A spindle built of microtubules moves the sister chromatids of each chromosome apart, to opposite spindle poles. A new nuclear envelope forms around the two chromosome clusters. Both daughter nuclei have the same chromosome number as the parent cell's nucleus.*

19.5 How the Cytoplasm Divides

Dividing the nucleus into two daughter nuclei—each with the necessary "packet" of chromosomes—is a major step in cell division, but not the only one. The parent cell's cytoplasm also must be divided between the two new daughter cells.

Division of the cytoplasm, or **cytokinesis**, usually begins toward the end of anaphase. By this time, the two sister chromatids of each chromosome have been separated and are independent chromosomes. In an animal cell, about midway between the cell's two poles, a patch of plasma membrane sinks inward, forming a **cleavage furrow**. Microfilaments made of the contractile protein actin steadily pull the plasma membrane inward all around the cell, until the cell is pinched in two (Figure 19.8). Now there are two new cells, each with a nucleus, cytoplasm, and a plasma membrane.

This concludes our tour of mitotic cell division. Now we turn to meiosis, the nucleus-dividing mechanism that forms gametes. It is difficult not to be in awe of the astonishing precision with which both mitosis and meiosis take place.

> *Mitosis is a small part of the cell cycle. As it ends, the mechanism of cytokinesis cuts the cytoplasm into two daughter cells, each with a daughter nucleus.*

 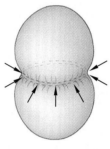 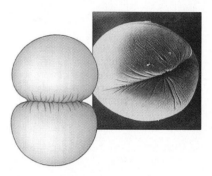 

1 Mitosis is over, and the spindle is disassembling.

2 At the former spindle equator, a ring of microfilaments attached to the plasma membrane contracts.

3 As the microfilament ring shrinks in diameter, it pulls the cell surface inward.

4 Contractions continue; the cell is pinched in two.

a

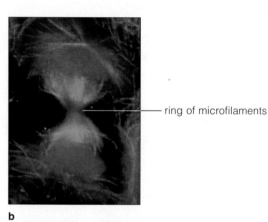

— ring of microfilaments

b

Figure 19.8 *Animated!* (**a**) Steps of cytokinesis in an animal cell. (**b**) The beltlike contractile ring of microfilaments inside a dividing animal cell that is undergoing cytokinesis.

19.6 Concerns and Controversies over Irradiation

What do a routine dental X ray and an irradiated side of beef have in common? Both are examples of ways we use ionizing radiation. Like some other technologies, this one can be a double-edged sword and can even fuel serious controversy.

IRRADIATION EFFECTS ON THE BODY Ionizing radiation includes various potentially harmful types of electromagnetic energy—for instance, radio waves, visible light, microwaves, cosmic rays from outer space, and radioactive radon gas in rocks and soil. Forms that can harm living cells, including radon and X rays, have enough energy to remove electrons from atoms and change them to positively charged ions (Section 2.4).

When ionizing radiation enters an organism, it may break apart chromosomes, alter genes, or both. If the chromosomes in an affected cell have been broken into fragments (Figure 19.9a), the spindle apparatus will not be able to harness and move the fragments when the cell divides. The cell or its descendants may then die. When ionizing radiation damage occurs in germ cells, the resulting gametes can carry damaged DNA. Therefore, an infant who inherits the DNA may have a genetic defect. If only somatic cells are affected, only the person exposed to the radiation will suffer damage.

When a person receives a sudden, large dose of ionizing radiation, it typically destroys cells of the immune system, epithelial cells of the skin and intestinal lining, and red blood cells, among other cell types. The results are raging infections, intestinal hemorrhages, anemia, and wounds that do not heal.

Small doses of ionizing radiation over a long period of time apparently cause less damage than the same total dosage given all at once. This may be due in part to the body's ability to repair damaged DNA. That said, ionizing radiation is associated with miscarriages, eye cataracts, and various cancers (Chapter 23).

On the other hand, medical X rays and diagnostic technologies such as magnetic resonance imaging (MRI) and PET scanning (Section 2.2) are valuable uses of ionizing radiation in health care. So is radiation therapy used in treating some cancers.

IRRADIATED FOOD Just as living body cells can be damaged or killed by radiation, so can harmful bacteria, fungi, and other microorganisms. As a result, foods ranging from grains and potatoes to fruits, spices, beef, pork, and other meats may be irradiated. Irradiated food sold in the United States carries an identifying logo (Figure 19.9).

Irradiated food is not radioactive, and some people are quite comfortable eating it because there is no scientific evidence that it presents a health hazard. In addition, irradiation limits spoilage, and proponents argue that it may reduce the incidence of food-borne illnesses. On the other hand, opponents worry that irradiation might promote the development of radiation-resistant microbes. Some also are concerned that irradiation may chemically change food in ways that could harm consumers. For the time being, however, there is no scientific evidence to support that fear.

LINKS TO SECTIONS 2.2 AND 2.4

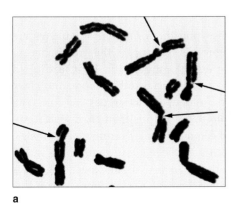

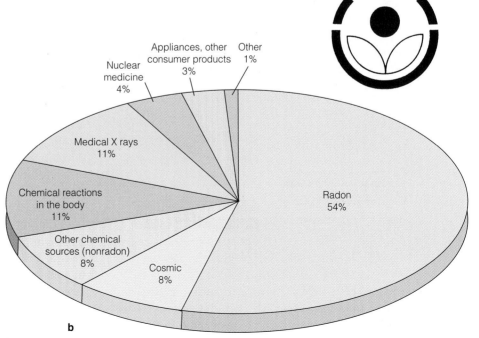

Figure 19.9 (a) Human chromosomes that have been exposed to ionizing radiation. Arrows point to broken pieces. (b) Sources of radiation exposure for people in the United States. (*far right*) Logo required to be placed on irradiated food products in the U.S.

b

Pie chart labels:
- Nuclear medicine 4%
- Appliances, other consumer products 3%
- Other 1%
- Medical X rays 11%
- Chemical reactions in the body 11%
- Other chemical sources (nonradon) 8%
- Cosmic 8%
- Radon 54%

19.7 Meiosis—The Beginnings of Eggs and Sperm

LINKS TO
SECTIONS
16.2 AND 16.4

Meiosis divides the nuclei of germ cells in a way that halves the number of chromosomes in daughter cells. It is the first step toward the gametes required for sexual reproduction.

IN MEIOSIS THERE ARE TWO DIVISIONS

Meiosis is like mitosis in some ways, even though the outcome is different. During interphase, a germ cell duplicates its DNA. Each duplicated chromosome consists of two sister chromatids attached to one another:

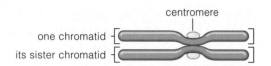

one chromosome in the duplicated state

As in mitosis, a spindle apparatus moves chromosomes into position for the formation of daughter nuclei. In meiosis, however, there are *two consecutive* divisions of the chromosomes. The result will be four haploid nuclei. There is no interphase between the two nuclear divisions, which we call meiosis I and meiosis II:

interphase (*DNA replication before meiosis I*)	MEIOSIS I	no interphase (*no DNA replication before meiosis II*)	MEIOSIS II
	PROPHASE I		PROPHASE II
	METAPHASE I		METAPHASE II
	ANAPHASE I		ANAPHASE II
	TELOPHASE I		TELOPHASE II

During meiosis I, each duplicated chromosome lines up with its partner, homologue to homologue. Then the two partners are moved apart:

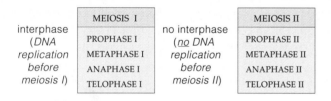

Each homologue in the cell pairs with its partner.

Then the partners separate.

The cytoplasm typically divides after each homologue is separated from its partner. The two daughter cells are haploid, with only one of each type of chromosome.

Later, during meiosis II, the two sister chromatids of each chromosome are separated from each other:

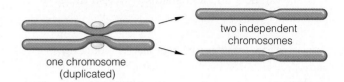

one chromosome (duplicated)

two independent chromosomes

Each sister chromatid is now a separate chromosome. After the four nuclei form, the cytoplasm divides again. The result is four haploid cells that eventually may function in reproduction.

MEIOSIS IS THE FIRST STEP IN THE FORMATION OF GAMETES

The human life cycle starts with meiosis. Next come the formation of gametes, fertilization, then growth of the new individual by way of mitosis:

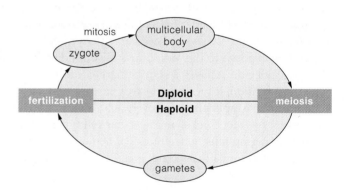

In a male, meiosis and the formation of gametes are called **spermatogenesis** because the forthcoming gametes will be sperm (Figure 19.10). First, a diploid germ cell increases in size. The resulting large, immature cell (a primary spermatocyte) undergoes meiosis. The resulting four haploid spermatids then change in form, develop tails, and become sperm—mature male gametes.

In females, meiosis and gamete formation are called **oogenesis** (Figure 19.11). As you might expect, oogenesis differs from spermatogenesis in some important ways. For example, compared to a primary spermatocyte, many more cytoplasmic components accumulate in a primary oocyte, the female germ cell that undergoes meiosis. Also, in females the cells formed after meiosis are of different sizes and have different functions.

The early stages of oogenesis unfold in a developing female embryo. Recall from Chapter 16, however, that

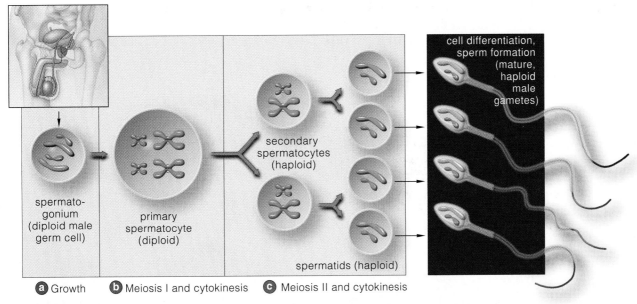

(a) Growth   (b) Meiosis I and cytokinesis   (c) Meiosis II and cytokinesis

Figure 19.10 *Animated!* General diagram of spermatogenesis.

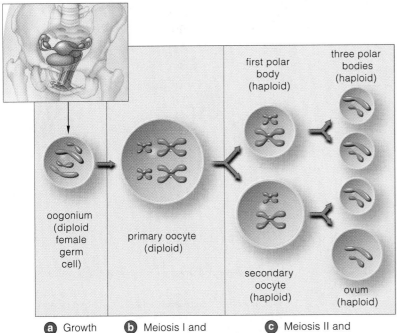

(a) Growth   (b) Meiosis I and cytokinesis   (c) Meiosis II and cytokinesis

Figure 19.11 *Animated!* Oogenesis. The diagram is not at the same scale as Figure 19.10. An egg becomes much larger than a sperm cell. It also is much larger than the polar bodies.

until a girl reaches puberty, her primary oocytes are arrested in prophase I. Then, each month, meiosis resumes in (usually) one oocyte that is ovulated. This cell, the secondary oocyte, receives nearly all the cytoplasm; the other, much smaller cell is a polar body. Both cells enter meiosis II, but the process is arrested again at metaphase II. If the secondary oocyte is fertilized, meiosis II continues. It results in one large cell and (often) three extremely small polar bodies. The polar bodies are "dumping grounds" for three sets of chromosomes so that the eventual egg ends up with only the necessary haploid number. The large cell develops into the mature egg (ovum). Its cytoplasm contains components that will help guide development of an embryo.

Oogenesis and spermatogenesis make gametes that are available for fertilization, the next life cycle stage. As you may remember from Chapter 17, fertilization restores the diploid number of chromosomes in a zygote.

> *Meiosis reduces the parental chromosome number by half—to the haploid number.*
>
> *Meiosis is the first step leading to the formation of gametes for sexual reproduction. In males, meiosis and gamete formation are called spermatogenesis. In females, these two processes are called oogenesis.*

A Visual Tour of the Stages of Meiosis

Meiosis I

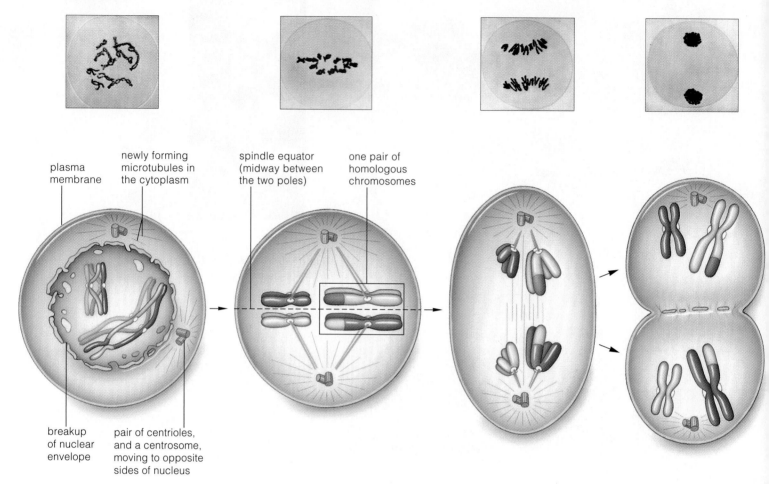

plasma membrane

newly forming microtubules in the cytoplasm

spindle equator (midway between the two poles)

one pair of homologous chromosomes

breakup of nuclear envelope

pair of centrioles, and a centrosome, moving to opposite sides of nucleus

ⓐ Prophase I

As prophase I begins, chromosomes become visible as threadlike forms. Each pairs with its homologue and usually swaps segments with it, as indicated by the breaks in color in the large chromosomes. Microtubules are forming a spindle. One pair is moved to the opposite side of the nuclear envelope, which is starting to break up.

ⓑ Metaphase I

Microtubules from one spindle pole have attached to one of each type of chromosome; microtubules from the other pole have attached to its homologue. By metaphase I, a tug-of-war between the two sets of microtubules has aligned all chromosomes midway between the poles.

ⓒ Anaphase I

Microtubules attached to each chromosome shorten and move it toward a spindle pole. Other microtubules, which extend from the poles and overlap at the spindle equator, move past each other and push the two poles farther apart.

ⓓ Telophase I

One of each type of chromosome has now arrived at the spindle poles. The cytoplasm divides, forming two haploid cells. All chromosomes are still duplicated.

Figure 19.12 *Animated!* Meiosis: the mechanism by which the parental number of chromosomes is reduced by half (to the haploid number) for forthcoming gametes. Only two pairs of homologous chromosomes are shown. Maternal chromosomes are shaded purple, and paternal ones blue.

Meiosis II

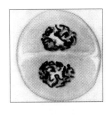
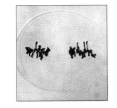

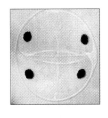

There is no DNA replication between the two nuclear divisions.

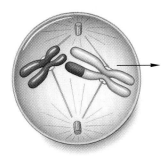
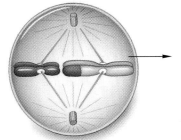
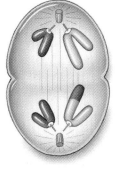
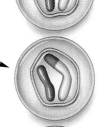

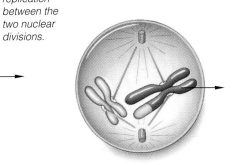
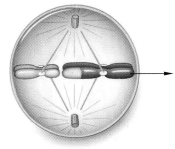
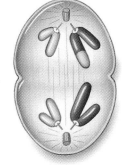
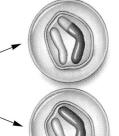

ⓔ Prophase II

A new spindle forms in each haploid cell. Microtubules have moved the centrioles to opposite ends of each cell. One chromatid of each chromosome becomes attached to one spindle pole, and its sister chromatid becomes attached to the opposite pole.

ⓕ Metaphase II

Microtubules from both spindle poles have assembled and disassembled in a tug-of-war that ended at metaphase II, when all chromosomes are positioned midway between the poles.

ⓖ Anaphase II

The attachment between sister chromatids of each chromosome breaks. Each is now a separate chromosome but is still attached to microtubules, which move it toward a spindle pole. Other microtubules push the poles apart. A cluster of unduplicated chromosomes ends up near each pole. One of each type of chromosome is present in each cluster.

ⓗ Telophase II

Four nuclei form as a new nuclear envelope encloses each cluster of chromosomes. After the cytoplasm divides, each of the resulting daughter cells has a haploid (*n*) number of chromosomes.

19.9 The Second Stage of Meiosis—New Combinations of Parents' Traits

The main function of meiosis is to reduce the diploid number of chromosomes by half, so gametes will have a haploid number of chromosomes. However, during prophase and metaphase of the second stage of meiosis, called meiosis II, two other events accomplish another function—variation in the genetic makeup of gametes.

IN PROPHASE I, GENES MAY BE REARRANGED

Genetic variation can be a major advantage for sexually reproducing organisms, like humans. This is because, as Chapter 24 describes more fully, variations in genetic traits can lead to modifications in body structures and functions. Passed from one generation to the next, such changes may eventually enable a population of organisms to adapt to changes in its environment.

Some genetic variations can come about during prophase I of meiosis. Why? This is a time when genes on the chromosomes in germ cells are rearranged. Notice Figure 19.13a, which shows two threadlike chromosomes. All the chromosomes in a germ cell condense this way. As they do, each is drawn close to its homologue. It is as if the homologues are stitched together along their length, without much space between them. (The X and Y chromosomes are "stitched" at one end only.) This side-by-side alignment favors **crossing over**. That is an interaction between two of the nonsister chromatids of a pair of homologues. In a crossover, *nonsister* chromatids break at the same places along their length. At these break points, they exchange corresponding segments— that is, genes. The X-shaped configuration shown in Figure 19.13e is evidence of a crossover.

a Both chromosomes shown here were duplicated during interphase, before meiosis. When prophase I is under way, sister chromatids of each chromosome are positioned so close together that they look like a single thread.

b Each chromosome becomes zippered to its homologue, so all four chromatids are tightly aligned. If the two sex chromosomes have different forms, such as X paired with Y, they still get zippered together, but only in a tiny region at their ends.

c We show the pair of chromosomes as if they already condensed only to give you an idea of what goes on. They really are in a tightly aligned, threadlike form during prophase I.

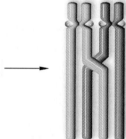

d The intimate contact encourages at least one crossover to happen at various intervals along the length of nonsister chromatids.

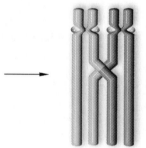

e Nonsister chromatids exchange segments at the crossover sites. They continue to condense into thicker, rodlike forms. By the start of metaphase I, they will be unzippered from each other.

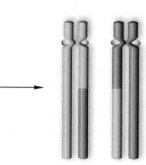

f Crossing over breaks up old combinations of genes and puts new ones together in the cell's pairs of homologous chromosomes.

Figure 19.13 *Animated!* Key events during prophase I, the first stage of meiosis. For clarity, this diagram of a cell shows only one pair of homologous chromosomes and one crossover. Here, the paternal chromosome is blue and its maternal homologue is purple.

MEIOSIS AND SEXUAL REPRODUCTION

The exchange of chromosome pieces is called **genetic recombination**. It leads to variation in inherited traits. How? Genes can come in alternative forms called **alleles**. For example, the gene for earlobe shape has two alleles— one calls for *attached* earlobes, and the other calls for *detached* earlobes. Often, many of the alleles on a chromosome differ a bit from the corresponding alleles on its homologue. Every crossover has the potential to create new combinations of alleles.

IN METAPHASE I, MATERNAL AND PATERNAL CHROMOSOMES ARE SHUFFLED

Chromosomes begin to be shuffled around during the transition from prophase I to metaphase I. Suppose this is happening right now in one of your germ cells. Crossovers have already made genetic mosaics of the chromosomes, but put this aside to simplify tracking. As in Figure 19.14, you can think of the 23 chromosomes from your mother as the *maternal* chromosomes and their 23 homologues from your father as the *paternal* chromosomes.

Kinetochore microtubules have already oriented one chromosome of each pair toward one spindle pole and its homologue toward the other (compare Section 19.8). Now they are moving all the chromosomes, which soon will be midway between the spindle poles.

Have all maternal chromosomes become tethered to one spindle pole and all paternal chromosomes to the other? Probably not, because the first contacts between microtubules and chromosomes are random. Accordingly, at metaphase I, maternal and paternal chromosomes aren't lined up at the spindle equator in any particular pattern, and after the homologues are separated at anaphase I, either member of each pair also can end up at either spindle pole.

Figure 19.14 shows the possibilities when there are only three pairs of homologues. By metaphase I, three pairs of homologues may be arranged in any one of four possible positions. In this case, eight combinations (2^3) of maternal and paternal chromosomes are possible for the forthcoming gametes.

Of course, a human germ cell has a full 23 pairs of homologous chromosomes, not just three. So a grand total of 2^{23}, or 8,388,608, combinations of maternal and paternal chromosomes are possible every time a germ cell gives rise to a gamete!

In each sperm or egg, the genetic instructions from the mother might differ slightly from those from the father. Chapter 20 explains more fully why this is so. But are you beginning to see why striking mixes of traits can show up even in the same family?

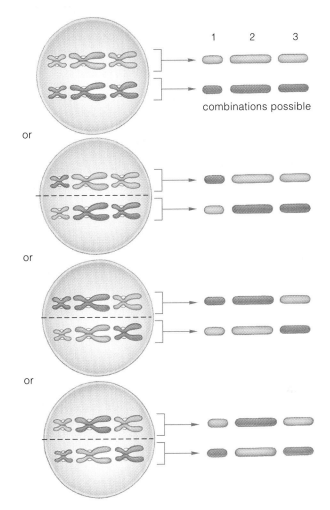

or

or

or

combinations possible

Figure 19.14 *Animated!* Possible outcomes of the random alignment of three pairs of homologous chromosomes at metaphase I of meiosis. Maternal chromosomes are purple; paternal ones blue.

During anaphase I, each homologue is separated from its partner. This separation is called **disjunction**. Each member of a set of sister chromatids moves to one pole of the spindle. As Chapter 21 describes, when homologues don't separate during meiosis, birth defects can result.

In prophase I, crossing over of segments of a pair of homologous chromosomes creates new gene combinations.

In metaphase I and anaphase I, maternal and paternal chromosomes are lined up at random at the spindle's equator, and then moved at random to one of the two spindle poles. As a result, gametes have new combinations of chromosomes, and children have varied combinations of parents' traits.

19.10 Meiosis and Mitosis Compared

We've considered the two ways that chromosomes—that is, the DNA—in cell nuclei can be divided. Mitosis occurs in somatic cells, while meiosis takes place in germ cells. The diagrams presented here summarize their similarities and key differences.

The end results of mitosis and meiosis differ in a crucial way (Figure 19.15). Mitosis produces genetically identical copies of a parent cell. (In this sense, they are the same as clones.) Meiosis, together with fertilization, promotes genetic variation in the traits offspring will have. First, crossing over during prophase I of meiosis puts new combinations of alleles in chromosomes. Next, during metaphase I, the members of each pair of homologous chromosomes are randomly moved to spindle poles, so that different mixes of maternal and paternal genes end up in gametes. Last but not least, simple chance brings together different combinations of genes when a sperm fertilizes an oocyte.

Figure 19.15 *Animated!* Comparison of mitosis and meiosis. This diagram is arranged to help you compare the similarities and differences between the two mechanisms. As in other diagrams in this chapter, maternal chromosomes are purple, and paternal chromosomes are blue.

Meiosis I

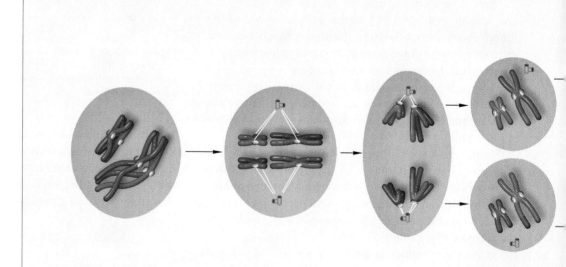

Prophase I
Chromosomes duplicated earlier in diploid (2*n*) germ cell during interphase. They condense. Spindle forms, attaches them to its poles. Crossovers occur between each pair of homologous chromosomes.

Metaphase I
Each maternal chromosome and its paternal homologue randomly aligned at the spindle equator; either one may get attached to either pole.

Anaphase I
Homologues separate from their partner, are moved to opposite poles.

Telophase I
Two haploid (*n*) clusters of chromosomes. New nuclear envelopes may form. Cytoplasm may divide before meiosis II gets under way.

MEIOSIS AND SEXUAL REPRODUCTION

Mitosis

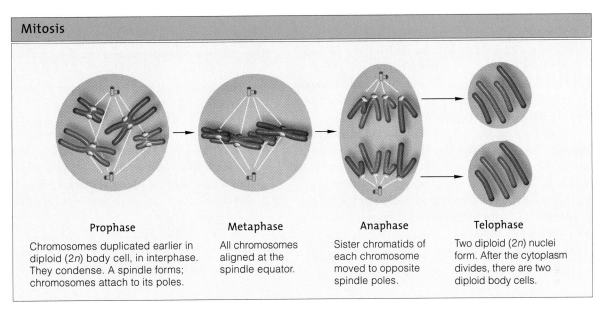

Prophase

Chromosomes duplicated earlier in diploid (2*n*) body cell, in interphase. They condense. A spindle forms; chromosomes attach to its poles.

Metaphase

All chromosomes aligned at the spindle equator.

Anaphase

Sister chromatids of each chromosome moved to opposite spindle poles.

Telophase

Two diploid (2*n*) nuclei form. After the cytoplasm divides, there are two diploid body cells.

Meiosis II

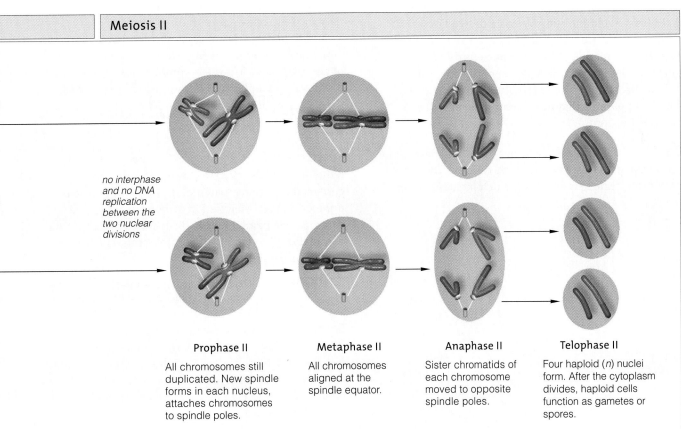

no interphase and no DNA replication between the two nuclear divisions

Prophase II

All chromosomes still duplicated. New spindle forms in each nucleus, attaches chromosomes to spindle poles.

Metaphase II

All chromosomes aligned at the spindle equator.

Anaphase II

Sister chromatids of each chromosome moved to opposite spindle poles.

Telophase II

Four haploid (*n*) nuclei form. After the cytoplasm divides, haploid cells function as gametes or spores.

Summary

Section 19.1 In reproduction, a parent cell produces a new generation of cells, or parents produce a new individual. Reproduction is part of a life cycle, a recurring sequence in which individuals grow, develop, and reproduce. Each cell of a new generation must receive a copy of the parental DNA and enough cytoplasm to start up its own operation.

Eukaryotic cells divide the cell nucleus, by either mitosis or meiosis. Mitosis is the division mechanism in somatic cells—body cells that are not specialized to make gametes. It functions in growth and tissue repair. Meiosis divides the nucleus in germ cells, cells in the gonads that give rise to gametes.

A chromosome is a DNA molecule and associated proteins. The sum of the chromosomes in a given cell type is the chromosome number. Human somatic cells have a diploid chromosome number of 46, or two copies of 23 types of chromosomes. Except for the sex chromosomes, pairs of homologous chromosomes are the same length and shape, and carry similar genes.

 Explore the structure of a chromosome.

Section 19.2 An unduplicated chromosome consists of one DNA molecule and proteins including histones. A duplicated chromosome consists of two DNA molecules that are temporarily attached to each other as sister chromatids. When the nucleus divides, microtubules attach to a centromere region on the paired chromatids. The microtubules are part of a spindle apparatus that moves chromosomes during nuclear division.

Section 19.3 The cell cycle begins when a new cell is produced and ends when that cell divides. The longest part of the cycle is interphase, which includes a growth stage (G1), when the cell roughly doubles the number of organelles and other components in its cytoplasm. In the S stage the cell's chromosomes are duplicated. There is also a final, brief growth stage (G2). The cell cycle proceeds at about the same rate in all cells of a given type.

 Investigate the stages of the cell cycle.

Section 19.4 Mitosis maintains the diploid number of chromosomes in the two daughter nuclei (Figure 19.16). Division of the cytoplasm, called cytokinesis, occurs toward the end of mitosis or at some point afterward.

Section 19.5 Mitosis has four stages:
  a. Prophase. Duplicated, threadlike chromosomes condense into rodlike structures; new microtubules start to assemble in organized arrays near the nucleus; they will form a spindle apparatus. The nuclear envelope disappears.
  b. Metaphase. Spindle microtubules orient the sister chromatids of each chromosome toward opposite spindle poles. The chromosomes align at the spindle equator.
  c. Anaphase. Sister chromatids of each chromosome separate. Both are now independent chromosomes, and they move to opposite poles.
  d. Telophase. Chromosomes decondense and a new nuclear envelope forms around the two clusters of chromosomes. Mitosis is completed. Cytokinesis divides the cytoplasm; the result is two diploid cells.

 Explore what happens during each stage of mitosis.

Sections 19.7, 19.8 Meiosis is a mechanism for reductional division. Two consecutive divisions of a germ cell reduce the parental diploid chromosome number by half. Meiosis I distributes homologous chromosome pairs. Meiosis II separates sister chromatids. The result, after cytokinesis, is four haploid cells (Figure 19.17).

 Explore what happens during each stage of meiosis, and learn how gametes form.

Section 19.9 Meiosis contributes to genetic variation. Crossing over during prophase I creates new combinations of genes (alleles) in chromosomes. Random alignment of pairs of homologues at metaphase I results in new combinations of maternal and paternal traits.

 Study how crossing over and other events during meiosis produce new combinations of genes on chromosomes.

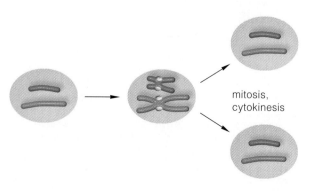

mitosis, cytokinesis

Figure 19.16 How mitosis maintains the chromosome number.

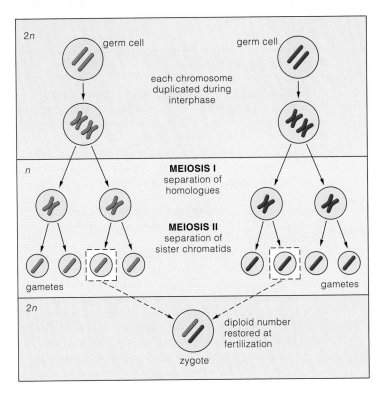

2n germ cell germ cell

each chromosome
duplicated during
interphase

n MEIOSIS I
separation of
homologues

MEIOSIS II
separation of
sister chromatids

gametes gametes

2n diploid number
restored at
fertilization

zygote

Figure 19.17 How the chromosome
number changes in meiosis.

Review Questions

1. Define the two mechanisms by which chromosomes are allotted in dividing human cells. What is cytokinesis?

2. Define somatic cell and germ cell. Which type of cell can undergo mitosis?

3. What is a chromosome? What is the difference between a diploid cell and a haploid cell?

4. What are homologous chromosomes?

5. Name the four main stages of mitosis, and describe the main features of each stage.

6. In a paragraph, summarize the similarities and differences between mitosis and meiosis.

Self-Quiz
Answers in Appendix V

1. DNA, packaged in chromosomes, is distributed to daughter cells by _____ or _____.

2. Each kind of organism contains a characteristic number of _____ in each cell; each of those structures is composed of a _____ molecule and proteins.

3. A pair of chromosomes that are similar in length, shape, and the traits they govern are called _____.
   a. diploid chromosomes
   b. mitotic chromosomes
   c. homologous chromosomes
   d. germ chromosomes

4. Somatic cells have a _____ number of chromosomes.

5. Interphase is the stage when _____.
   a. a cell ceases to function
   b. a germ cell forms its spindle apparatus
   c. a cell grows and duplicates its DNA
   d. mitosis takes place

6. After mitosis, each daughter cell contains genetic instructions that are _____ and _____ chromosome number of the parent cell.
   a. identical to the parent cell's; the same
   b. identical to the parent cell's; one-half the
   c. rearranged; the same
   d. rearranged; one-half the

7. All of the following are stages of mitosis *except*
   _____.
   a. prophase
   b. interphase
   c. metaphase
   d. anaphase

8. A duplicated chromosome has _____ chromatids.
   a. one   b. two   c. three   d. four

9. Crossing over in meiosis _____.
   a. occurs between sperm DNA and egg DNA at fertilization
   b. leads to genetic recombination
   c. occurs only rarely

10. Because of the _____ alignment of homologous chromosomes at metaphase I, gametes can end up with _____ mixes of maternal and paternal chromosomes.
   a. unvarying; different   c. random; duplicate
   b. unvarying; duplicate   d. random; different

11. Match stage of mitosis with the following key events.
   ___ metaphase
   ___ prophase
   ___ telophase
   ___ anaphase
   a. sister chromatids separate and move to opposite poles
   b. chromosomes condense and a microtubular spindle forms
   c. chromosomes decondense, daughter nuclei re-form
   d. chromosomes align at spindle equator

Critical Thinking

1. Under normal circumstances you can't inherit both copies of a homologous chromosome from the same parent. Why? Assuming that no crossing over has occurred, what is the chance that one of your non-sex chromosomes is an exact copy of the same chromosome possessed by your maternal grandmother?

2. Suppose you have a way of measuring the amount of DNA in a single cell during the cell cycle. You first measure the amount during the G1 phase. At what points during the remainder of the cycle would you predict changes in the amount of DNA per cell?

3. Adam has a pair of alleles (alternate forms) of a gene that influences whether a person is right-handed or left-handed. One allele says "right" and its partner says "left." Visualize one of his germ cells, in which chromosomes are being duplicated prior to meiosis. Visualize what happens to the chromosomes during anaphase I and II. (It might help to use toothpicks as models of the sister chromatids of each chromosome.) What fraction of Adam's sperm will carry the allele for right-handedness? For left-handedness?

4. Fresh out of college, Maria has her first job teaching school. When she goes for a pre-employment chest X ray required by the school district, the technician places a lead-lined apron over her abdomen but not over any other part of Maria's body. The apron prevents electromagnetic radiation from penetrating into the protected body area. What cells is the lead shield designed to protect, and why?

Explore on Your Own

Section 16.4 explains that oogenesis begins when a female is still a developing embryo. In the immature eggs (primary oocytes) of a female embryo, meiosis I begins, but is arrested in prophase I. Meiosis I won't resume until the female undergoes puberty. From then until menopause, just before an egg is ovulated, it will undergo the remaining stages of meiosis I. As the egg is traveling down an oviduct, meiosis II begins. This stage also is arrested, in metaphase II. Only if the egg is fertilized will all the stages of meiosis finally be completed (Figure 19.18).

Knowing this sequence, you can make a fairly accurate calculation of how long it took for the egg that helped give rise to you to pass through all the stages of meiosis. You only need to know the month and year your mother was born and the month and year you were conceived (or born). For your starting point, remember that a female's primary oocytes form during the third month of embryonic development, about 6 months before birth. If you have siblings, do the same calculation for them.

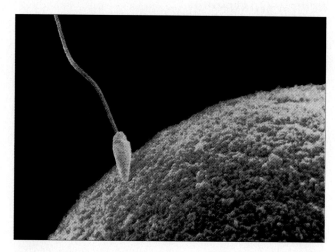

Figure 19.18 An egg with a sperm that will penetrate the egg and fertilize it—marking the end of meiosis in the egg and the beginning of bodily growth by mitosis.

Designer Genes?

Have you ever wished you could change one of your genetic traits? Although currently none of us can do much about the genes we have inherited, scientific advances have made it possible for parents to select certain genetic traits of their children.

You will read in Chapter 22 about genetic engineering and the mapping of the human genome, which is steadily pinpointing the locations of all the human genes (some 21,500 of them!) on our chromosomes. While many researchers who are working in this area focus on locating and fixing genes related to diseases, it's clear that another use for the technology could be something called *eugenic engineering*. This term refers to the manipulation of genes so that a pregnancy would produce a child having certain traits that the parents desire.

Deciding which forms of a trait are most "desirable" raises ethical and moral concerns. For instance, would it be okay to fiddle with the genes of parents to engineer taller or blue-eyed or fair-skinned boys or girls? Would it be okay to engineer "superhumans" with "genius" intelligence or amazing physical strength? Some

people might even want to commission a clone of themselves, with one or more genetic "enhancements."

Qualities such as intelligence and many other traits are controlled by complex gene interactions, and probably won't be prime candidates for eugenic engineering. Even so, in a recent survey of Americans, more than 40 percent of respondents said it would be fine to alter genes to make their children smarter or better looking. A poll of British parents found 18 percent willing to use genetic enhancement to prevent children from being aggressive and 10 percent to keep them from growing up to be homosexual.

With this chapter we begin to explore the topics of genes and the basic principles of inheritance. You will gain a much fuller understanding of how you came to have many of your biological characteristics. And you may also be better equipped to form opinions about the many issues of "genetic technology."

 How Would You Vote? Would you favor legislation that limits or prohibits engineering genes except for health reasons? Cast your vote online at www.thomsonedu.com/biology/starr/humanbio.

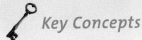

Key Concepts

GENES AND INHERITANCE
Inherited traits are specified by genes, which are segments of DNA. Genes have different forms called alleles. A gamete carries one copy of each gene.

PROBABILITY RULES
Chance determines which sperm will fertilize which egg. Thus, rules of probability apply to the inheritance of traits coded by a single gene.

INDEPENDENT ASSORTMENT
The paired copies of a gene on one chromosome are sorted into gametes independently of genes on other chromosomes.

GENE EFFECTS
Genetic traits are not always predictable. Examples are traits determined by more than one gene, and cases of genes that affect more than one trait.

Links to Earlier Concepts

Your reading in this chapter will flesh out the concept of inheritance introduced at the beginning of this textbook (1.1). It will also draw on what you have learned about diploid and haploid sets of chromosomes (19.1) and the formation of gametes during meiosis (19.7).

20.1 Basic Concepts of Heredity

You already know that meiosis halves the number of parental chromosomes and fertilization restores it, giving a new individual chromosomes from each parent. Here we pick up the story with the consequences of these events.

Having read Chapter 19, you already know a bit about chromosomes and genes. That's more than the monk Gregor Mendel knew in the 1850s. Mendel was curious about how sexually reproducing organisms inherited their traits, and to pursue this question he experimented with garden pea plants. Based on his research, Mendel hypothesized that "factors" from each parent were the units of heredity. Today we call such factors genes.

Figure 20.1 and the following list express some of Mendel's ideas in modern terms:

1. **Genes** are units of information about specific traits and are passed from parents to offspring. Today we know that humans have about 21,500 genes, and that genes are chemical instructions for building proteins. Each gene has a specific location, or locus, on a given chromosome.

2. Diploid cells have two copies of each gene, on pairs of homologous chromosomes.

3. The copies of a gene deal with the same trait, but their information about it may vary a little due to chemical differences between them. Each version of the gene is called an **allele**. Contrasting alleles produce much of the variation we see in traits, as when one person has attached earlobes and someone else's earlobes are detached (Figure 20.2).

4. If the two copies of a gene are identical alleles, this is a **homozygous condition** (*homo*: same; *zygo*: joined together). If the two allele copies are different, this is a **heterozygous condition** (*hetero*: different).

5. An allele is **dominant** when its effect on a trait masks that of any **recessive** allele paired with it. An uppercase letter represents a dominant allele, and a lowercase letter represents a recessive allele (for instance, *A* and *a* or *C* and *c*).

6. A *homozygous dominant* individual has a pair of dominant alleles (*AA*) for the trait being studied. A *homozygous recessive* individual has a pair of recessive alleles (*aa*). A *heterozygous* individual has a pair of nonidentical alleles (*Aa*).

7. Two terms help keep the distinction clear between genes and the traits they specify. The alleles an individual inherits are the **genotype**. Observable functional or physical traits, such as attached earlobes, are the **phenotype** (Table 20.1).

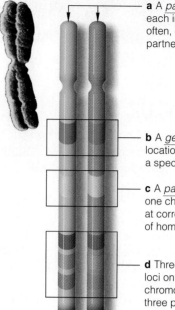

a A *pair of homologous chromosomes*, each in the unduplicated state (most often, one from a male parent and its partner from a female parent)

b A *gene locus* (plural, loci), the location for a specific gene on a specific type of chromosome

c A *pair of alleles* (each being one chemical form of a gene) at corresponding loci on a pair of homologous chromosomes

d Three *pairs of genes* (at three loci on this pair of homologous chromosomes); same thing as three pairs of alleles

Figure 20.1 *Animated!*   A few genetic terms illustrated.

Figure 20.2 Earlobe traits. (**a**) Actor Tom Cruise has attached earlobes. (**b**) Actress Joan Chen's are detached. The detached version is dominant. So are the alleles for long eyelashes and flat feet, among other human traits.

Genes are units of information about traits.

Different versions of a gene are called alleles. A person inherits two copies of each gene, one on each member of a pair of homologous chromosomes.

If the two inherited alleles of a gene are identical, the person is homozygous for the trait. If the two are different, the person is heterozygous for the trait.

a

b

Table 20.1	Genotype and Phenotype Compared	
Genotype	**Described as**	**Phenotype**
CC	homozygous dominant	chin fissure
Cc	heterozygous (one of each allele; dominant form of trait observed)	chin fissure
cc	homozygous recessive	smooth chin

20.2 One Chromosome, One Copy of a Gene

We all inherit pairs of genes on pairs of chromosomes. Some of Mendel's experiments revealed that each gamete—each egg or sperm—receives just one copy from each gene pair.

Mendel hypothesized that a diploid organism inherits two units (genes) for a trait, one from each parent. To test his idea he used a **monohybrid cross** (*mono-* means one). In this kind of cross, the two parents have different alleles for the gene in question. Although for ethical reasons we do not do experimental crosses between individual human beings, monohybrid matings do occur naturally, and they show patterns of inheritance at work.

The following example illustrates a monohybrid cross for the configuration of a person's chin. This trait is governed by a gene that has two allele forms. One allele, which calls for a chin fissure (actually a fissure in the skin), is dominant when it is present (Figure 20.3); we can use *C* to represent this allele. The recessive allele, which codes for a smooth chin, is *c*. In this example, one parent is homozygous for the *C* form of the gene and has a chin fissure, and the other is homozygous for the *c* form and has a smooth chin. The *CC* parent produces only *C* gametes; the *cc* parent produces only *c* gametes:

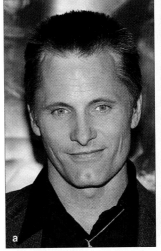

Figure 20.3 (**a**) The chin fissure, a trait that arises from one allele of a gene. Actor Viggo Mortensen received a gene that influences this trait from each of his parents. At least one of those genes was dominant. (**b**) What Mr. Mortensen's chin might have looked like if he had inherited identical alleles for "no chin fissure" instead.

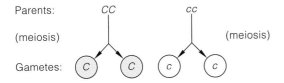

Each gamete receives one allele for the trait because each gamete has only one copy of each chromosome. We know already that the genes on homologues match up into equivalent pairs. So, when homologues separate into different gametes after meiosis II (Section 19.7), the two genes of each pair separate as well (Figure 20.4). This is called the principle of **segregation**. Because meiosis also halves the diploid chromosome number to the haploid number, each gamete has only one member of each chromosome pair—and also has the genes it carries.

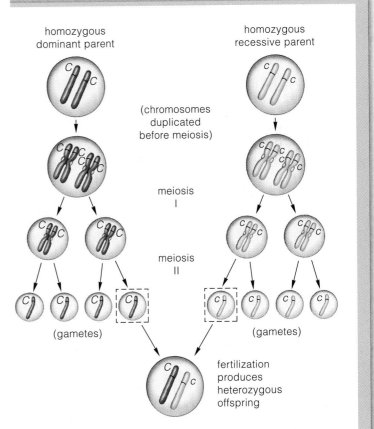

Figure 20.4 *Animated!* A monohybrid cross, showing how one allele of a pair segregates from the other. Two parents that are each homozygous for a different version of a trait give rise only to heterozygous offspring.

> The two copies of each gene in a diploid organism separate (segregate) from each other during meiosis in germ cells. As a result, each gamete contains only one copy of each gene.

20.3 Figuring Genetic Probabilities

When potential parents are concerned about passing a harmful trait to a child, genetic counselors must try to predict the likely outcome of the mating.

The parent generation of a cross is designated by a **P**. Offspring of a monohybrid cross are called the F_1 or *first filial* generation. In the present example, each child will inherit a pair of differing alleles for the trait, one from each homozygous parent. The children will thus each be heterozygous for the chin genotype, or *Cc*. Because *C* is dominant, each child will have a chin fissure.

Suppose now that one of the *Cc* children grows up and has a family with another *Cc* person. From this second cross we derive the F_2 (*second filial*) generation. Because half of each parent's gametes (sperm or eggs) are *C* and half are *c* (due to segregation at meiosis), four outcomes are possible every time a sperm fertilizes an egg:

Possible event:	Probable outcome:	
sperm *C* unites with egg *C*	1/4 *CC* offspring	
sperm *C* unites with egg *c*	1/4 *Cc* ⎤	chin fissure
sperm *C* unites with egg *c*	1/4 *Cc* ⎦ or 1/2 *Cc*	
sperm *c* unites with egg *c*	1/4 *cc*	smooth chin

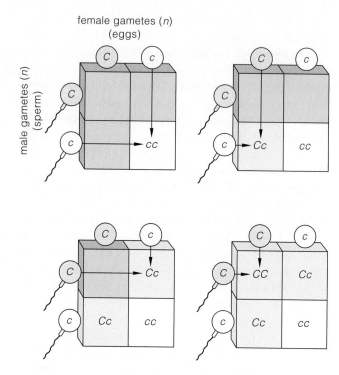

female gametes (*n*) (eggs)

male gametes (*n*) (sperm)

Figure 20.5 *Animated!* Punnett square method of determining the probable outcome of a genetic cross. Here the cross is between two heterozygous individuals. Circles represent gametes. Letters on gametes represent dominant or recessive alleles. The different squares depict the different genotypes possible among offspring.

A PUNNETT SQUARE CAN BE USED TO PREDICT THE RESULT OF A GENETIC CROSS

The diagrams in Figure 20.5 show how to construct a **Punnett square**—a convenient tool for determining the probable outcome of genetic crosses. In this case, there is a 75 percent chance that a child from a cross between two *Cc* parents will have at least one dominant *C* allele and a chin fissure. When a large number of offspring are involved, a ratio of 3:1 is likely (Figure 20.6).

The rules of probability apply to crosses because fertilization is a chance event. **Probability** is a number between zero and one that expresses the likelihood of a particular event. For example, an event with a probability of one will always occur and an event with a probability of zero will never occur. As you might guess, an event that has a probability of one-half (or 50 percent) is likely to occur in about half of all opportunities (Figure 20.7).

Having a chin fissure doesn't affect a person's health. However, a fair number of human genetic disorders, including cystic fibrosis and sickle-cell anemia, result from single-gene defects and so follow a Mendelian inheritance pattern. Chapter 21 looks more closely at genetic disorders. For the moment, it is important to realize two things:

1. The outcomes predicted by probability don't *have* to turn up in a given family. For instance, it's common to see families in which the parents have produced several children of the same sex. We consistently see predicted ratios only with a large number of events. You can test this for yourself by flipping a coin. Probability predicts that heads and tails should each come up about half the time, but you may have to flip the coin a hundred times to end up with a ratio close to 1:1. Likewise, probability predicts that parents with two or more children will have equal numbers of girls and boys, but as we all know, this doesn't always happen.

2. In a given genetic situation, *probability doesn't change.* The likelihood that a certain genotype will occur—say, a baby with the genotype for cystic fibrosis—is the same for every child no matter how many children a couple has. Based on the parents' genotypes, if the probability that a child will inherit a certain genotype is one in four, then each child of those parents has a one-in-four (25 percent) chance of inheriting the genotype. If the parents have three children without the trait, the fourth child still has only a one-in-four chance of inheriting it.

PROBABILITY RULES

A TESTCROSS ALSO CAN REVEAL GENOTYPES

Until the advent of high-tech genetic analysis, if you wanted to learn the genotype of a (nonhuman) organism you could do a **testcross**. In this method, an individual with an unknown genotype is crossed with an individual that is homozygous recessive for the trait being studied (say, it is *aa*). Then you observe the phenotype of offspring. If all offspring have a dominant form of the trait, you know that the "mystery" parent's genotype includes at least one dominant allele (the parent must be *Aa* or *AA*). If some offspring have the dominant phenotype and some have the recessive one, then the parent with the unknown genotype must be a heterozygote, or *Aa*.

Similar situations can shed light on the genotype of a human parent. Suppose that a woman has smooth cheeks and her husband has dimpled cheeks. "No dimples" is a recessive trait, so the woman is *dd*. If a child is born with no dimples, then the father must be a heterozygote for this trait, with a genotype of *Dd*. That is the only way he could himself have dimples and also father a *dd* child. If, on the other hand, the child has dimples, the father can be either *DD* or *Dd*. If he is *Dd*, a heterozygote, the probability that he will have a dimpled child is 1/2, a 50–50 chance every time. If he were *DD*, every child would have dimples.

> *Probability applies to the inheritance of single-gene traits. If the genotypes of parents are known, it is possible to establish a potential child's chances for inheriting a particular genotype and thus for having a particular phenotype (trait). Observing traits in offspring also can help reveal the genotypes of parents.*

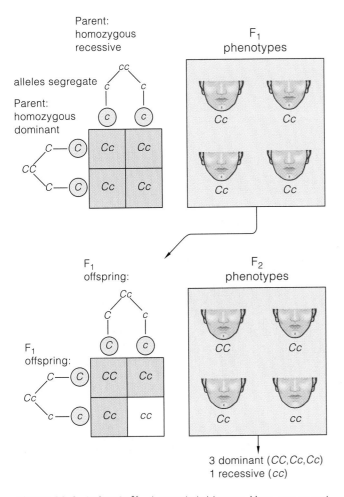

Figure 20.6 *Animated!* A monohybrid cross. Here, one parent is homozygous dominant for a trait. The other is homozygous recessive for the trait. Notice that the dominant-to-recessive ratio is 3:1 for the second generation (F₂) offspring.

How to Calculate Probability

Step 1. Actual genotypes of parental gametes

In the cross $Cc \times Cc$, gametes have a 50–50 chance of receiving either allele (*C* or *c*) from each parent. Said another way, the probability that a particular sperm or egg will be *C* is 1/2, and the probability that it will be *c* is also 1/2:

probability of *C*:   1/2

probability of *c*:   1/2

Step 2. Probable genotypes of offspring

Offspring receive one allele from each parent. Three different combinations of alleles are possible in this cross. To figure the probability that a child will receive a particular allele combination, simply multiply the probabilities of the individual alleles:

probability of *CC*: $1/2 \times 1/2 = 1/4$

probability of *Cc*: $1/2 \times 1/2 = 1/4$

probability of *cC*: $1/2 \times 1/2 = 1/4$ } 1/2

probability of *cc*: $1/2 \times 1/2 = 1/4$

Step 3. Probable phenotypes

Chin fissure: $1/4 + 1/4 + 1/4 = 3/4$
(*CC,Cc,cC*)

Smooth chin: 1/4
(*cc*)

Figure 20.7 Calculating probabilities. Simple multiplication lets you figure the probability that a child will inherit alleles for a particular phenotype.

PROBABILITY RULES

20.4 How Genes for Different Traits Are Sorted into Gametes

LINK TO SECTION 19.8

How are genes for different traits—say, cheek dimples and a chin fissure—allotted to gametes? Mendel also found an answer to this question.

Mendel was curious about how genes for two *different* traits are passed to offspring. Based on the results of experiments, he established another fundamental principle of inheritance, called **independent assortment**. This rule states that alleles for different traits are sorted into gametes independently of one another (Figure 20.8). In other words, either copy of a gene may end up in a particular gamete. (Genes that are very close to one another on the same chromosome are exceptions to this rule because they can be linked—that is, they tend to remain together.)

Evidence for independent assortment can come from a method called the **dihybrid cross**. As its name suggests, in this type of cross two traits are studied. And as with a monohybrid cross, "simple dominance" exists: that is, there are two contrasting alleles of each gene, one dominant and one recessive. The parents have different combinations of alleles. We don't use human subjects for genetic crosses, but the following example will help you grasp how a dihybrid cross shows independent assortment at work.

A dihybrid cross follows two generations of matings. Let's consider an example using dominant alleles C for chin fissure and D for cheek dimples and c and d as their recessive counterparts. In this case, let's say that the P generation consists of a man who is $ccdd$ and a woman who is $CCDD$. Each parent can produce just one type of gamete:

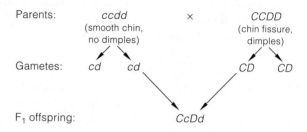

All of this couple's children (the F_1 generation) are $CcDd$ and have a chin fissure and cheek dimples.

What happens in a second generation if two $CcDd$ individuals mate? The man and woman can each produce four types of gametes in equal proportions:

$$1/4\ CD \quad 1/4\ Cd \quad 1/4\ cD \quad 1/4\ cd$$

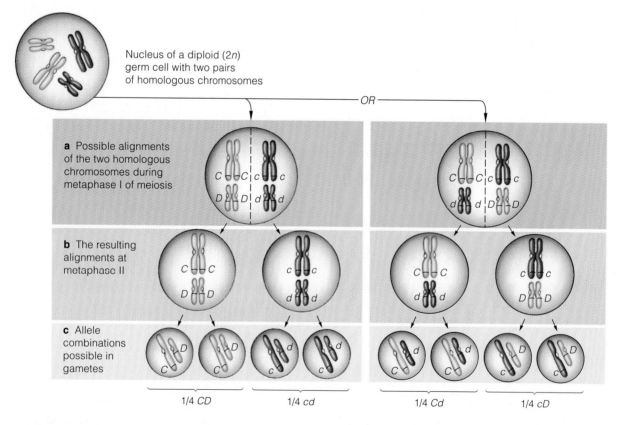

Figure 20.8 *Animated!* A case of independent assortment. Either chromosome of a pair may become attached to either pole of the spindle during meiosis. When two pairs are tracked, two different metaphase I lineups are possible.

INDEPENDENT ASSORTMENT

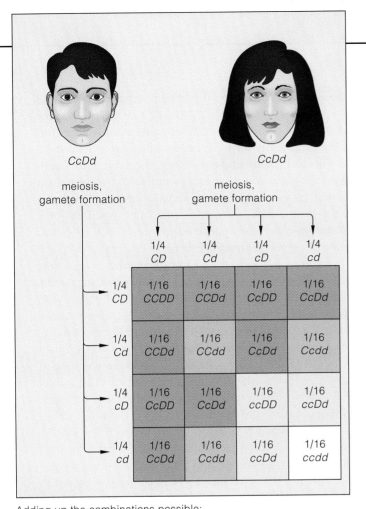

CcDd *CcDd*

meiosis, gamete formation meiosis, gamete formation

	1/4 CD	1/4 Cd	1/4 cD	1/4 cd
1/4 CD	1/16 CCDD	1/16 CCDd	1/16 CcDD	1/16 CcDd
1/4 Cd	1/16 CCDd	1/16 CCdd	1/16 CcDd	1/16 Ccdd
1/4 cD	1/16 CcDD	1/16 CcDd	1/16 ccDD	1/16 ccDd
1/4 cd	1/16 CcDd	1/16 Ccdd	1/16 ccDd	1/16 ccdd

Adding up the combinations possible:

▣ 9/16 or 9 chin fissure, dimples

▣ 3/16 or 3 chin fissure, no dimples

▢ 3/16 or 3 smooth chin, dimples

▢ 1/16 or 1 smooth chin, no dimples

Figure 20.9 *Animated!* Results from a mating in which both parents are heterozygous at both loci. *C* and *c* represent dominant and recessive alleles for a chin fissure. *D* and *d* represent dominant and recessive alleles for dimples. Rules of probability predict that certain combinations of phenotypes among offspring of this type of cross occur in a 9:3:3:1 ratio, on average.

This outcome is possible because the copies of genes on different chromosomes assort independently of one another during meiosis.

Punnet square analysis also can show the possibilities that a child will inherit a particular combination of single-gene traits. Simple multiplication (four kinds of sperm times four kinds of eggs) tells us that sixteen different gamete unions are possible when each parent in a dihybrid cross is heterozygous for the two genes in question. The Punnett square in Figure 20.9 shows the possibilities, using the chin fissure and dimple traits. It assumes that each parent is heterozygous at both gene loci. That is, both parents have the genotype *CcDd*. Notice that when

Probability in a Mating Where Both Parents Are Heterozygous at Two Loci

A dihybrid cross considers two traits. If both parents are heterozygous for both traits, a dihybrid cross produces the following 9:3:3:1 phenotype ratio (Figure 20.9):

9/16 or 9 chin fissure, dimples

3/16 or 3 chin fissure, no dimples

3/16 or 3 smooth chin, dimples

1/16 or 1 smooth chin, no dimples

Individually, these phenotypes have the following probabilities:

probability of chin fissure (12 of 16) = 3/4
probability of dimples (12 of 16) = 3/4
probability of smooth chin (4 of 16) = 1/4
probability of no dimples (4 of 16) = 1/4

To figure the probability that a child will show a particular *combination* of phenotypes, multiply the probabilities of the individual phenotypes in each possible combination:

Trait combination		Probability
Chin fissure, dimples	3/4 × 3/4	9/16
Chin fissure, no dimples	3/4 × 1/4	3/16
Smooth chin, dimples	1/4 × 3/4	3/16
Smooth chin, no dimples	1/4 × 1/4	1/16

Figure 20.10 Probability applied to independent assortment.

such individuals mate, there are nine possible ways for gametes to unite that produce a chin fissure and dimples, three for a chin fissure and no dimples, three for a smooth chin and dimples, and one for a smooth chin and no dimples. The probability of any one child having a chin fissure and dimples is 9/16; a chin fissure and no dimples, 3/16; a smooth chin and dimples, 3/16; and a smooth chin and no dimples, 1/16.

Figure 20.10 shows how to calculate the probability that a child will inherit genes for a particular set of two traits on different chromosomes.

Either copy of a gene may end up in a particular gamete. This is the inheritance principle of independent assortment.

A dihybrid cross looks at how two traits are inherited, and it can reveal evidence of independent assortment. That is, the cross can show that the paired copies of a given gene on one chromosome are assorted into gametes independently of other gene pairs on other chromosomes.

If both parents are heterozygous for two particular genes, sixteen genotypes and four phenotypes are possible.

INDEPENDENT ASSORTMENT

20.5 Single Genes, Varying Effects

LINKS TO SECTIONS 5.2, 8.4, 8.7, 10.2, 10.7 AND 18.5

Some traits have clearly dominant and recessive forms. For most traits, however, the story is not so simple.

Section 20.1 noted that genes are chemical instructions for building proteins. We say that a gene is "expressed" when its instructions are carried out and the cell makes the protein. In some cases, the expression of a gene leads to a single phenotype. Usually, however, the genetic underpinnings of traits is more complicated.

ONE GENE MAY AFFECT SEVERAL TRAITS

Expression of the gene at just a single locus (location) on a chromosome may affect two or more traits in positive or negative ways. This wide-ranging effect of a single gene is called **pleiotropy** (ply-AH-trow-pee, after the Greek *pleio-*, meaning "more," and *-tropic*, meaning "to change"). Much of what researchers have learned about how genes function has come from studies of genetic diseases. One example of pleiotropy is the recessive condition **CHH** (for *cartilage-hair hypoplasia*), which is caused by a mutant allele on chromosome 9. The mutation occurs in a gene called *RMRP*, and it affects many organ systems, including

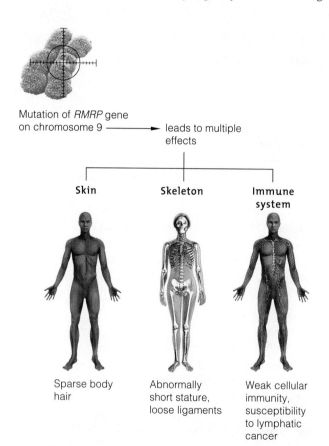

Mutation of *RMRP* gene on chromosome 9 ———▶ leads to multiple effects

Skin   Skeleton   Immune system

Sparse body hair

Abnormally short stature, loose ligaments

Weak cellular immunity, susceptibility to lymphatic cancer

Figure 20.11 Overview of the pleiotropic effects of CHH.

the skeletal system, the skin (integumentary system), and the immune system (Figure 20.11). We don't yet know what goes wrong when RMRP mutates, but people with CHH commonly have little body hair, abnormally short limbs, and loose ligaments. They also have faulty cell-mediated immunity and may be unusually susceptible to certain cancers, such as lymphoma, a cancer of lymphatic organs. It is not unusual for the various effects of a pleiotropic gene to appear over time as one effect has repercussions that lead to others.

Another example is **sickle-cell anemia**, which was introduced in the discussion of malaria in Chapter 18. This disabling and painful disease arises when a person is homozygous for a recessive allele. The normal allele, Hb^A, has instructions for building normal hemoglobin, the oxygen-transporting protein in red blood cells. When a person inherits two copies of the recessive mutant allele, Hb^S, he or she develops sickle-cell anemia. Section 8.7 described how red blood cells, which normally are biconcave disks, take on a sickle shape when the oxygen content of blood falls below a certain level. The sickled cells clump in blood capillaries and can rupture. The flow of blood can be so disrupted that the person's oxygen-deprived tissues are severely damaged (Figure 20.12). Homozygotes for the mutated hemoglobin gene (Hb^S/Hb^S) often die relatively young. Heterozygotes (Hb^A/Hb^S), on the other hand, have **sickle-cell trait**. They generally have few symptoms because the one Hb^A allele provides enough normal hemoglobin to prevent red blood cells from sickling.

During a crisis, sickle-cell anemia patients may receive blood transfusions, oxygen, antibiotics, and painkilling drugs. There is evidence that the food additive butyrate can reactivate "dormant" genes responsible for fetal hemoglobin, an efficient oxygen carrier that normally is produced only before birth. For this reason, some states require hospitals to screen newborn infants for sickle-cell anemia so that appropriate action can begin right away.

IN CODOMINANCE, MORE THAN ONE ALLELE OF A GENE IS EXPRESSED

As you have read, people who are heterozygous for a trait have two contrasting alleles for that trait. Sometimes, *both* alleles are expressed. We see a classic example of this **codominance** in people who are heterozygotes for alleles that confer A and B blood types. If you have type AB blood, for instance, you have a pair of codominant alleles that are both expressed in the stem cells that give rise to your red blood cells. You may remember from Chapter 8 that a polysaccharide (a sugar) on the surface of red

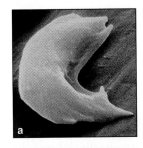

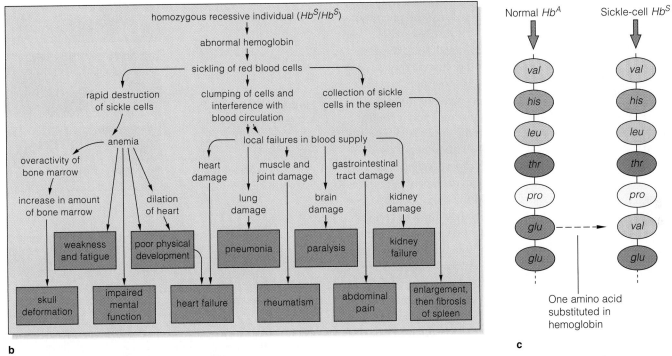

Figure 20.12 *Animated!* (**a**) Scanning electron micrograph of a sickled red blood cell. (**b**) The range of symptoms characteristic of sickle-cell anemia. (**c**) The genetic change that leads to sickle-cell anemia. The mutant allele codes for hemoglobin in which an incorrect amino acid has been substituted in the chain of amino acids making up the protein.

blood cells has different molecular forms that determine blood type. ABO blood typing (Section 8.4) can reveal which form(s) of the polysaccharide a person has.

The gene specifying an enzyme that synthesizes this polysaccharide has three alleles, which influence the sugar molecule's form in different ways. Two alleles, *IA* and *IB*, are codominant when paired with each other. A third allele, *i*, is recessive. When paired with either *IA* or *IB*, the effect of *i* is masked. A gene that has three or more alleles is called a **multiple allele system**. Many human genes fall into this category.

A single gene may affect two or more traits, a phenomenon called pleiotropy. The effects may not be simultaneous, but may have repercussions over time as one altered trait changes another trait and so on, as in CHH and sickle-cell anemia.

In some cases contrasting alleles for a trait are codominant—that is, both are expressed.

Some genes have more than two alleles. These multiple allele systems include the alleles for the ABO blood group.

20.6 Other Gene Impacts and Interactions

Many phenotypes, such as eye color, can't be predicted with certainty. Biologists have uncovered a variety of underlying causes for these variations.

A gene can have an all-or-nothing effect on a trait. Either you have dimples or you don't. But in some other cases, the expression of a gene varies due to gene interactions or nongenetic factors in the environment.

The term **penetrance** refers to the probability that someone who inherits an allele will have the phenotype associated with it. For example, the recessive allele that causes cystic fibrosis is completely penetrant; 100 percent of people who are homozygous for it develop CF. The dominant allele for having extra fingers or toes (called *polydactyly*) is incompletely penetrant. Some people who inherit it have the usual ten digits, while others have more (Figure 20.13). *Campodactyly* is caused by the abnormal attachment of muscles to bones of the little finger. Some people who inherit the allele for it have a stiff, bent little finger on both hands. Others have a bent pinkie on one hand only. When an allele can produce a range of phenotypes, its expression is said to be "variable." The campodactyly allele also is incompletely penetrant. In some people who inherit it, the trait doesn't show up at all.

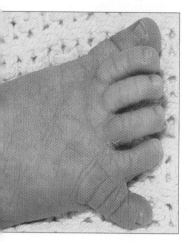

Figure 20.13
Six toes, an example of polydactyly.

Figure 20.14 Eye color, a polygenic trait. Alleles of more than one gene interact to produce and deposit melanin, a pigment that helps color the eye's iris (and skin, too).

Figure 20.15 *Animated!* Examples of continuous variation: Biology students (males, *left*; females, *right*) organized according to height.

4'11" 5'0" 5'1" 5'2" 5'3" 5'4" 5'5" 5'6" 5'7" 5'8" 5'9" 5'10" 5'11"
Height (feet/inches)

5'3" 5'4" 5'5" 5'6" 5'7" 5'8" 5'9" 5'10" 5'11" 6'0" 6'1" 6'2" 6'3" 6'4" 6'5"
Height (feet/inches)

POLYGENIC TRAITS: SEVERAL GENES COMBINED

Polygenic traits result from the combined expression of several genes. For example, skin color and eye color are the cumulative result of many genes involved in the stepwise production and distribution of melanin. Black eyes have abundant melanin in the iris. Dark-brown eyes have less melanin, and light-brown or hazel eyes have still less (Figure 20.14). Green, gray, and blue eyes don't have green, gray, or blue pigments. Instead, they have so little melanin that we readily see blue wavelengths of light being reflected from the iris. A person's hair color probably results from the interactions of several genes. This explains our real-world observation that there is lots of natural variation in human hair color.

For many polygenic traits a population may show **continuous variation**. That is, its members show a range of continuous, rather than incremental, differences in some trait. Continuous variation is especially evident in traits that are easily measurable, such as height (Figure 20.15).

DO GENES "PROGRAM" BEHAVIOR?

Identical twins have identical genes and look alike. In addition, they have many behaviors in common. Are these parallels coincidence, clever hoaxes, or evidence that aspects of behavior are inherited characteristics, like hair and eye color? Although the question is intriguing, clear answers have proven difficult to come by.

There is strong evidence that some basic behaviors, such as smiling to indicate pleasure and the crying of an infant when it is hungry, are genetically programmed. Scientists have also begun to look for links between genes and alcoholism, some types of mental illness, violent behavior, and even sexual orientation. Such studies raise controversial social issues. So far their greatest impact has been to point out how little we know about the biological basis of human behavior. In general, most human behavior is so complex that it is quite difficult to scientifically test hypotheses about genetic links.

Gene expression can sometimes vary, so that the resulting trait (phenotype) is unpredictable. Examples include alleles that are incompletely penetrant and polygenic traits that result from the combined expression of two or more genes.

SCIENCE COMES TO LIFE

20.7 Searching for Custom Cures

Everybody's different. And thanks to our genes, so is every *body*. Because each of us has our own personal mix of alleles—the varying chemical forms of genes—we also may respond differently to therapeutic drugs. A new field of study called *pharmacogenetics* aims to pinpoint genetic variations that influence how individuals respond to medications. The idea is to allow physicians to custom-prescribe the drugs that will be safest and most effective for each patient.

All medicines work for some people, but none is a perfect fit for all. Blood pressure drugs are an example. There are more than 100 different ones, partly because there are many individual differences in how well each one controls the condition. The medication and dose that work well for one patient may be only modestly effective for another, or may cause dangerous or unpleasant side effects. Figuring out the best course often is a matter of trial and error.

How can we take the guesswork out of prescribing drugs? A first step is identifying the genes that control common reactions to various drugs. That research is happening now (Figure 20.16). The technology for rapid, low-cost genetic screening also is becoming widely available. With these tools, a physician could order a genetic profile for a patient and use it to select the best medicine to deal with that person's illness.

Pharmacogenetics promises to improve patient care by allowing doctors to tailor treatments to their patients' genes. As medical treatment moves in this direction, it will be important to have safeguards in place to protect the privacy of patients' genetic records.

Figure 20.16 Dr. Stephen Liggett, a researcher in the field of pharmacogenetics at the University of Cincinnati. Dr. Liggett is studying genes that control patient responses to asthma drugs.

Summary

Section 20.1 Gregor Mendel's studies showed that genes are specific units passed to offspring. Each gene has a specific location on a chromosome. Chromosomes come in pairs, so a person has two copies of each gene. The copies are called alleles, and they may or may not be identical.

Someone who is homozygous for a trait (such as *AA*) has two identical alleles for the trait. Having two different alleles (*Aa*) is a heterozygous condition. Alleles (and traits) may be dominant (*A*) or recessive (*a*).

The term genotype refers to the particular alleles an individual has. Phenotype is the term used to refer to an individual's actual observable traits.

 Learn how Mendel crossed garden pea plants and the definitions of important genetic terms.

Section 20.2 A monohybrid cross (between parents with different versions of a single trait) provides evidence for the principle of gene segregation: (1) diploid organisms have two copies (alleles) of each gene, one on each of the two chromosomes of a homologous pair, and (2) the two copies of each gene segregate from each other during meiosis, so each gamete formed ends up with one gene or the other.

 Try your hand at carrying out monohybrid crosses.

Section 20.3 In a genetic cross, the parent generation is called the P generation and the first offspring are called the F_1 generation. A second generation of offspring is called the F_2 generation.

A Punnett square is a tool for figuring the probable outcome of a genetic cross. Probability is a number between zero and one that expresses the likelihood of a particular event. Matings between two heterozygous individuals (*Cc* × *Cc*) produce the following combinations of alleles in F_2 offspring:

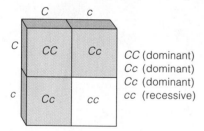

CC (dominant)
Cc (dominant)
Cc (dominant)
cc (recessive)

This results in a probability of 3/4 that any one child will have the dominant phenotype and 1/4 that the child will have the recessive phenotype.

A testcross is a tool in which the phenotypes of offspring are interpreted to identify the genotype of a parent. For ethical reasons this method is used only with organisms other than humans.

Section 20.4 A dihybrid cross occurs between two homozygous parents who show different versions of two traits.

The result in the F_1 generation is:

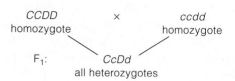

A mating between two heterozygous parents results in the following probable phenotypes:

9 dominant for both traits
3 dominant for *C*, recessive for *d*
3 dominant for *D*, recessive for *c*
1 recessive for both traits

Dihybrid crosses provide evidence of independent assortment of genes during meiosis. The members of each pair of homologous chromosomes sort into gametes independently of how the members of other chromosome pairs sort. Therefore, the genes on the chromosomes also sort independently.

If both parents are heterozygous for two genes (as in *CcDd*), sixteen genotypes and four phenotypes are possible.

 Observe the results of a dihybrid cross.

Section 20.5 In cases of pleiotropy, a single gene can influence many seemingly unrelated traits (as in sickle-cell anemia). In codominance, two contrasting alleles of a gene are both expressed. An example is seen in blood types: people who are type AB have codominant alleles for type A and type B blood. A gene that has three or more alleles is called a multiple allele system.

 Explore some patterns of inheritance that do not follow Mendel's rules.

Section 20.6 Various factors can influence the expression of genes. Penetrance refers to the probability that someone who inherits an allele will have the phenotype associated with it. Polygenic traits, such as eye color and height, are due to the expression of several genes. For polygenic traits we may see continuous variation in populations.

 See how environmental factors can affect genetic traits, and plot the continuous distribution for height for a group of students.

Review Questions

1. Define the difference between: (a) gene and allele, (b) dominant allele and recessive allele, (c) homozygote and heterozygote, and (d) genotype and phenotype.

2. State the theory of segregation. Does segregation occur during mitosis or during meiosis?

3. Distinguish between monohybrid and dihybrid crosses. What is a testcross, and why is it useful in genetic analysis?

4. What is independent assortment? Does independent assortment occur during mitosis or during meiosis?

Self-Quiz

Answers in Appendix V

1. Alleles are _____.
   a. alternate forms of a gene
   b. different molecular forms of a chromosome
   c. always homozygous
   d. always heterozygous

2. A heterozygote has _____.
   a. only one of the various forms of a gene
   b. a pair of identical alleles
   c. a pair of contrasting alleles
   d. a haploid condition, in genetic terms

3. The observable traits of an organism are its _____.
   a. phenotype   c. genotype
   b. pedigree   d. multiple allele system

4. Offspring of a monohybrid cross $AA \times aa$ are _____.
   a. all AA d. 1/2 AA and 1/2 aa
   b. all aa e. none of the above
   c. all Aa

5. Second-generation offspring from a cross between two homozygotes are the _____.
   a. F_1 generation   c. hybrid generation
   b. F_2 generation   d. none of the above

6. Assuming complete dominance, offspring of the cross $Aa \times Aa$ will show a phenotypic ratio of _____.
   a. 1:2:1   c. 9:1
   b. 1:1:1   d. 3:1

7. Which statement best fits the principle of segregation?
   a. Units of heredity are transmitted to offspring.
   b. Two genes of a pair separate from each other during meiosis.
   c. Members of a population become segregated.
   d. A segregating pair of genes is sorted out into gametes independently of how gene pairs located on other chromosomes are sorted out.

8. Dihybrid crosses of heterozygotes ($AaBb \times AaBb$) lead to F_2 offspring with phenotypic ratios close to _____.
   a. 1:2:1   c. 3:1
   b. 1:1:1:1   d. 9:3:3:1

9. Match each genetic term appropriately.
   ____dihybrid cross   a. $AA \times aa$
   ____monohybrid cross   b. Aa
   ____homozygous condition   c. $AABB \times aabb$
   ____heterozygous condition d. aa

Figure 20.17 A demonstration of the "tongue-roller" trait.

Critical Thinking

1. One gene has alleles A and a. Another has alleles B and b. For each genotype listed, what type(s) of gametes can be produced? (Assume independent assortment occurs.)
   a. $AABB$ c. $Aabb$
   b. $AaBB$ d. $AaBb$

2. Still referring to Problem 1, what will be the possible genotypes of offspring from the following matings? With what frequency will each genotype show up?
   a. $AABB \times aaBB$ c. $AaBb \times aabb$
   b. $AaBB \times AABb$ d. $AaBb \times AaBb$

3. A gene on one chromosome governs a trait involving tongue movement. If you have a dominant allele of that gene, you can curl the sides of your tongue upward (Figure 20.17). If you are homozygous for recessive alleles of the gene, you cannot roll your tongue. A gene on a different chromosome controls whether your earlobes are attached or detached. People with detached earlobes have at least one dominant allele of the gene. Because these two genes are on different chromosomes, they assort independently. Suppose a tongue-rolling woman with detached earlobes marries a man who has attached earlobes and can't roll his tongue. Their first child has attached earlobes and can't roll its tongue.
   a. What are the genotypes of the mother, father, and child?
   b. What is the probability that a second child will have detached earlobes and won't be a tongue-roller?

4. Go back to Problem 1, and assume you now study a third gene having alleles C and c. For each genotype listed, what type(s) of gametes can be produced?
   a. $AABBCC$ c. $AaBBCc$
   b. $AaBBcc$ d. $AaBbCc$

5. When you decide to breed your Labrador retriever Molly and sell the puppies, you discover that two of Molly's four siblings have developed a hip disorder that is traceable to the action of a single recessive allele. Molly herself shows no sign of the disorder. If you breed Molly to a male Labrador that does not carry the recessive allele, can you assure a purchaser that the puppies will also be free of the condition? Explain your answer.

6. The ABO blood system has been used to settle cases of disputed paternity. Suppose, as a geneticist, you must testify during a case in which the mother has type A blood, the child has type O blood, and the alleged father has type B blood. How would you respond to the following statements?

   a. *Man's attorney*: "The mother has type A blood, so the child's type O blood must have come from the father. Because my client has type B blood, he could not be the father."

   b. *Mother's attorney*: "Further tests prove this man is heterozygous, so he must be the father."

7. Soon after a couple marries, tests show that both the man and the woman are heterozygotes for the recessive allele that causes sickling of red blood cells; they are both Hb^A/Hb^S. What is the probability that any of their children will have sickle-cell trait? Sickle-cell anemia?

8. A man is homozygous dominant for 10 different genes that assort independently. How many genotypically different types of sperm could he produce? A woman is homozygous recessive for 8 of these genes and is heterozygous for the other 2. How many genotypically different types of eggs could she produce? What can you conclude about the relationship between the number of different gametes possible and the number of heterozygous and homozygous gene pairs that are present?

9. As is the case with the mutated hemoglobin gene that causes sickle-cell anemia, certain dominant alleles are crucial to normal functioning (or development). Some are so vital that when the mutant recessive alleles are homozygous, the combination is lethal and death results

before birth or early in life. However, such recessive alleles can be passed on by heterozygotes (*Ll*). In many cases, these are not phenotypically different from homozygous normals (*LL*). If two heterozygote parents mate (*Ll* × *Ll*), what is the probability that any of the children will be heterozygous?

10. Bill and Marie each have flat feet, long eyelashes, and "achoo syndrome" (chronic sneezing). All are dominant traits. The genes for these traits each have two alleles, which we can designate as follows:

	Dominant	Recessive
Foot arch	*A*	*a*
Sneezing	*S*	*s*
Eyelash length	*E*	*e*

Bill is heterozygous for each trait. Marie is homozygous for all of them. What is Bill's genotype? What is Marie's genotype? If they have four children, what is the probability that each child will have the same phenotype as the parents? What is the probability that a child will have short lashes, high arches, and no achoo syndrome?

11. You decide to breed a pair of guinea pigs, one black and one white. In guinea pigs, black fur is caused by a dominant allele (*B*) and white is due to homozygosity for a recessive allele (*b*) at the same locus. Your guinea pigs have 7 offspring, 4 black and 3 white. What are the genotypes of the parents? Why is there a 1:1 ratio in this cross?

Explore on Your Own

Many researchers are studying the possible genetic basis for personality and behavior traits, including violent behavior, the importance of "nature" versus "nurture," and other topics. To learn more about these efforts and find links to some other fascinating and reputable human genetics websites, visit the Personality Research website at www.personalityresearch.org.

Menacing Mucus

Cystic fibrosis (CF) is an eventually fatal genetic disorder. Its cause has been traced to a gene on chromosome 7 that codes for a membrane transport protein called CFTR. The protein helps chloride and water move into and out of exocrine cells, which secrete mucus or sweat. Several different mutations of the gene can result in a malfunctioning CFTR protein. Cystic fibrosis results when a person inherits a mutated CFTR gene from both parents.

More than 10 million people in the United States have a dominant, normal copy of the CFTR gene and one recessive, abnormal copy. Most don't know they are carriers. This knowledge gap, and the severe consequences of CF, eventually led to the first mass screening in the United States for carriers of a genetic disorder.

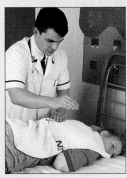
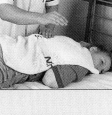
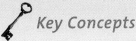

In cystic fibrosis, dry, thickened mucus clogs the airways, as in the filmstrip, and makes it hard to breathe. Chapter 11 described how ciliated cells lining the airways sweep out pathogens and debris that become trapped in mucus there. In people with CF, bacteria colonize the mucus and infections develop.

Although antibiotics can help control the lung infections, each day patients must undergo physiotherapy that includes vigorous thumping on the chest and back to loosen the mucus so it can be expelled. Most patients will suffer lung failure. CF patients rarely live beyond age 30.

Today hundreds of thousands of expectant parents have been screened for CF. In some early cases the results were misinterpreted, and a few couples may have unwittingly opted to abort normal fetuses.

Thinking about the power and challenges of genetic testing is an apt introduction to our topic in this chapter—the connections between chromosomes and inheritance, including the causes of many common inherited diseases. The more we learn about the genetic basis of health and disease, the more we all will be grappling with issues related to the proper use of that knowledge.

 How Would You Vote? Do we as a society want to encourage women to give birth only to offspring who will not develop serious gene-based medical problems? Cast your vote online at www.thomsonedu .com/biology/starr/humanbio.

Key Concepts

CHROMOSOMES AND GENES
We inherit two copies (alleles) of every gene, one from each parent. Each gene is located in a particular place on a specific chromosome. There are two kinds of chromosomes: autosomes and sex chromosomes.

PATTERNS OF INHERITANCE
Studies of genetic disorders can reveal patterns of inheritance. We now know that some traits arise from dominant or recessive alleles on an autosome or sex chromosome.

CHANGES IN CHROMOSOMES
Many genetic disorders arise from rare changes in the chromosome number or changes in chromosome structure. The effects of such changes can be harmful or lethal.

Links to Earlier Concepts

In this chapter you will learn more about the concept of homologous chromosomes (19.1). You will also gain a more complete understanding of how crossing over during meiosis creates new combinations of parent genes in sperm and eggs (19.9). You will expand your knowledge of how human chromosomes carry genetic information (20.2), and you will see some additional examples of how Mendelian patterns of inheritance operate in humans (20.3).

21.1 Genes and Chromosomes

LINKS TO
SECTIONS
19.1, 19.9,
AND 20.1

The two previous chapters gave a general idea of the structure of chromosomes and what happens to them during meiosis. We can now begin to correlate this information with some of the patterns of heredity.

UNDERSTANDING INHERITANCE STARTS WITH GENE–CHROMOSOME CONNECTIONS

Chapters 19 and 20 discussed how chromosomes carry genes, and basic "rules" for how genes are passed from one generation to another. To recap some of what you have learned thus far:

1. Each gene has a particular location (its **locus**) on a specific chromosome.

2. A diploid cell (2*n*) has pairs of homologous chromosomes. Except for the sex chromosomes (X and Y), the chromosomes of each pair are alike in length, shape, and the genes they include.

3. During meiosis in germ cells, homologous chromosomes interact, then separate from each other.

4. A given gene may have two or more alleles, but a diploid cell can carry only two of them, one on each member of a pair of homologous chromosomes.

5. In general, the process of independent assortment dictates that the different genes on a chromosome will move into gametes independently of one another.

6. At least one crossover takes place during meiosis I in germ cells. In a crossover, homologous chromosomes exchange corresponding segments.

CLOSELY LINKED GENES TEND TO STAY TOGETHER WHEN GAMETES FORM

The inheritance patterns we observe in humans (and other organisms) don't always jibe with the concept of independent assortment. Some of these exceptions are due to the fact that the genes on a chromosome are physically connected. When the distance between two genes is short, we say there is close **linkage** between them. Closely linked genes nearly always end up in the same gamete (Figure 21.1*a*). On the other hand, when two genes on a chromosome are far apart, it is more likely that crossing over will break up the linkage. Those genes are much more likely to assort into gametes independently of each other (Figure 21.1*b*). The patterns in which genes are distributed into gametes are so regular they can be used to map the positions of the genes on a chromosome.

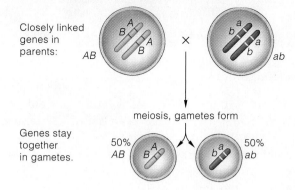

a Strong linkage between two genes. Crossing over does not separate them. Half of the gametes have one parent's genotype and half have the other.

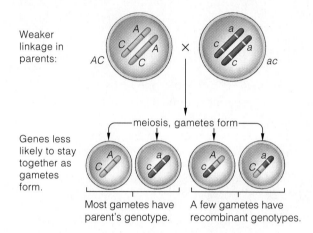

b Weak linkage; crossing over puts new gene combinations in some gametes.

Figure 21.1 Examples of linkage.

THE X AND Y CHROMOSOMES ARE QUITE DIFFERENT GENETICALLY

It has been said that "men are from Mars, women from Venus." While that topic goes far beyond biology class, there *are* biological differences between the sexes. A male's diploid cells each have one X chromosome and one Y chromosome (XY). A female's diploid cells each have two X chromosomes (XX). Each X chromosome carries 2,062 genes, but a Y chromosome is much smaller and carries only 330 genes.

Despite their differences, the X and Y chromosomes can synapse (be joined together briefly) in a small region along their length. This allows the X and Y to function as homologues during meiosis.

Major patterns of inheritance in humans reflect factors such as gene linkage, in which the physical distance of genes on a chromosome influences whether the genes move into the same or different gametes.

The two human sex chromosomes, X and Y, differ in size, shape, and the number and kinds of genes they carry.

21.2 Picturing Chromosomes with Karyotypes

A diagram called a *karyotype* can help answer questions about a person's chromosomes. Chromosomes are the most condensed and easiest to identify at the phase of meiosis called metaphase (Section 19.10).

A technician who wants to prepare a karyotype doesn't assume that it will be possible to find a body cell that is dividing. Instead, cells are cultured in the laboratory along with chemicals that stimulate cells to grow and to divide by mitosis. Blood cells are often used for this purpose.

Once the cell culture is established, a chemical called colchicine is added. It arrests mitosis at metaphase. After the colchicine treatment, the "soup" of cultured cells is placed into glass tubes that are whirled in a centrifuge. The spinning force moves the cells to the bottom of the test tubes.

Next the cells are transferred to a saline (salt–water) solution. They swell (by osmosis) and separate, as do the metaphase chromosomes. At this point the cells are ready to be placed on a microscope slide, "fixed" (stabilized by air-drying or some other method), and stained so that they are easy to see.

The chromosomes are photographed through the microscope, and the image is enlarged. Then the photograph is cut apart, one chromosome at a time. The cutouts are arranged in order, essentially from largest to smallest (Figure 21.2). The sex chromosomes are placed last. All pairs of homologous chromosomes are aligned horizontally, by their centromeres. Figure 21.2*f* shows a karyotype diagram prepared this way.

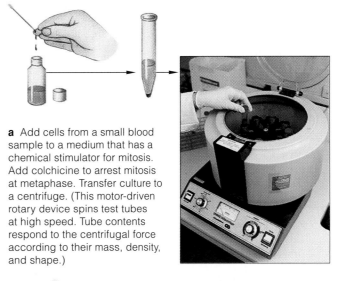

a Add cells from a small blood sample to a medium that has a chemical stimulator for mitosis. Add colchicine to arrest mitosis at metaphase. Transfer culture to a centrifuge. (This motor-driven rotary device spins test tubes at high speed. Tube contents respond to the centrifugal force according to their mass, density, and shape.)

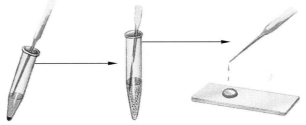

b Centrifugation forces cells to bottom of tube. Draw off culture medium. Add a dilute saline solution to tube. Add a fixative.

c Prepare and stain cells for microscopy.

d Put cells on a microscope slide. Observe.

e Photograph one cell through microscope. Enlarge image of its chromosomes. Cut the image apart. Arrange chromosomes as a set.

Figure 21.2 *Animated!* How to prepare a karyotype (**a–e**). (**f**) A human karyotype. Human somatic cells have twenty-two pairs of autosomes and one pair of sex chromosomes (XX or XY). These are metaphase chromosomes from a female; and each is in the duplicated state. In the orange box at the far right are the two sex chromosomes (XY) of a male.

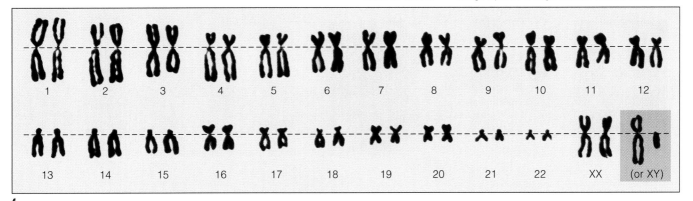

f

21.3 How Sex Is Determined

LINKS TO
SECTIONS
16.2 AND 16.4

Whether an embryo develops into a female or a male depends on which sex chromosomes are carried by the egg and sperm that unite at fertilization.

SEX IS A QUESTION OF X OR Y

Every normal egg that is produced by a female has an **X chromosome**. Half the sperm cells produced by a male carry an X chromosome and half carry a **Y chromosome**.

The father's sperm determines a baby's sex. If an X-bearing sperm fertilizes an X-bearing egg, the embryo will develop into a female. If the sperm has a Y chromosome, the embryo will develop into a male (Figure 21.3*a*).

The genes on a Y chromosome include the master gene for male sex determination, which has been dubbed *SRY*.

When the gene is expressed, a sequence of steps leads to the formation of testes, the primary male reproductive organs (Figure 21.3*b*). When that gene is absent—that is, when no Y chromosome is present—ovaries form automatically, and the developing embryo is female.

As you've read, an X chromosome carries more than 2,000 genes. Like other chromosomes, it has some genes associated with sexual traits, such as the distribution of body fat and hair. However, most of its genes deal with nonsexual traits, such as blood-clotting functions. These genes can be expressed in males as well as in females. (Males, recall, also carry one X chromosome.) The genes on X and Y chromosomes are sometimes called **X-linked genes** and **Y-linked genes**, respectively.

Figure 21.3 *Animated!* (**a**) How the sex of a human is determined. Males transmit their Y chromosome to sons but not daughters. Males get their X chromosome from their mother. (**b**) Duct system in the early embryo that develops into a male or a female reproductive system.

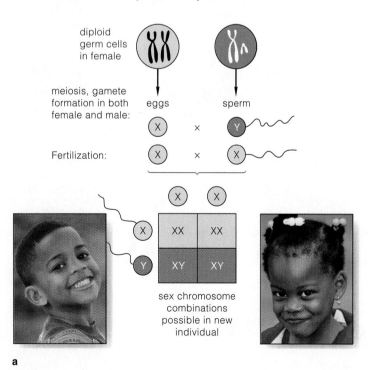

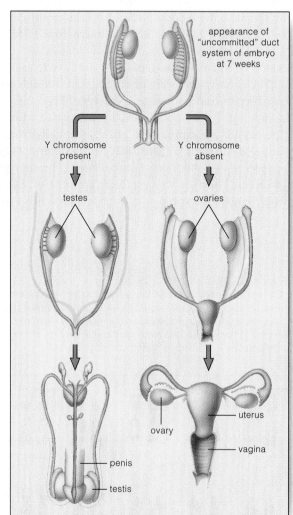

CHROMOSOMES AND GENES

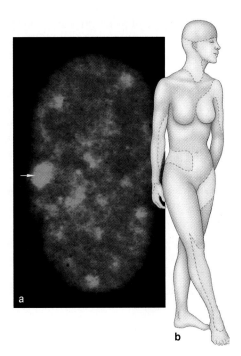

Figure 21.4 (a) Light micrograph showing a Barr body (a condensed X chromosome) in a cell's nucleus. The X chromosome in cells of males is not condensed this way. (b) A patchy, "mosaic" tissue effect that shows up in females who have anhidrotic ectodermal dysplasia. Some patches of skin have sweat glands, but other patches (colored yellow here) have none.

IN FEMALES, ONE X IS INACTIVATED

Since females have two X chromosomes and males have only one, do females have twice as many X-linked genes as males do and therefore have a double dose of their gene products? Generally no, because a compensating phenomenon called **X inactivation** occurs in females. Although the mechanism is not completely understood, it appears that most or all of the genes on one of a female's X chromosomes are "switched off" soon after the first cleavages of the zygote. In a cell, inactivation can occur in either one of the two X chromosomes. The inactivated X is condensed into a **Barr body** (Figure 21.4a), which can be seen under the microscope. From then on, the same X chromosome will be inactivated in all the descendants of the cell. Once X inactivation takes place, the embryo continues to develop. Typically, a female's body has patches of tissue where the genes of the maternal X chromosome are expressed, and other patches where the genes of the paternal X chromosome are expressed.

Some females have **anhidrotic ectodermal dysplasia** (Figure 21.4b). This condition occurs in females in whom the active X chromosome in certain tissues carries a mutated gene that blocks the formation of sweat glands. A similar phenomenon gives a female calico cat her varied coat color, which is determined by an allele on the cat's active X chromosome.

Human sex is determined by the father's sperm, which can carry either an X chromosome or a Y chromosome. XY embryos develop as males, and XX embryos as females.

In females, one of the two X chromosomes is inactivated soon after embryonic development begins.

21.4 Human Genetic Analysis

LINKS TO
SECTIONS
20.1 AND 20.3

How can prospective parents assess their risk of conceiving a child with an inherited disorder? If a relative has the disorder, a first step can be to construct a genetic family history.

A PEDIGREE SHOWS GENETIC CONNECTIONS

In nonhuman organisms, geneticists use experimental crosses to do genetic analysis. Since we can't experiment with humans, however, a basic tool is a "genetic family history" called a **pedigree chart**. The chart tracks several generations of a family, showing who exhibited the trait being investigated. The example shown in Figure 21.5 includes definitions of some of the symbols used.

When analyzing pedigrees, geneticists rely on their knowledge of probability and of Mendelian inheritance patterns, which may yield clues to a trait's genetic basis.

For instance, they might figure out that the allele that causes a disorder is dominant or recessive or that it is located on an autosome or a sex chromosome. Pedigrees often are used to identify those at risk of transmitting or developing the trait in question—including any children that a couple may have.

Gathering numerous family pedigrees increases the numerical base for analysis. Figure 21.6 shows a series of pedigrees for Huntington disease, in which the nervous system progressively degenerates. Genetic researcher Nancy Wexler constructed the pedigrees for an extended family in Venezuela that includes some 10,000 people.

A person who is heterozygous for a recessive trait can be designated as a *carrier*. A carrier shows the dominant phenotype (no disease symptoms) but still can produce gametes with the recessive allele and potentially pass on that allele to a child. If both parents are carriers for a

Polydactyly is one of several genetic disorders that are common among the Old Order Amish of Lancaster County, Pennsylvania. A different gene mutation causes the fatal disorder known as Amish microencephaly, which has been traced back nine generations to one couple. The defective gene's effects include a severely underdeveloped brain. The story isn't all negative, however. Research on the condition holds promise for better understanding of some other brain defects in newborns.

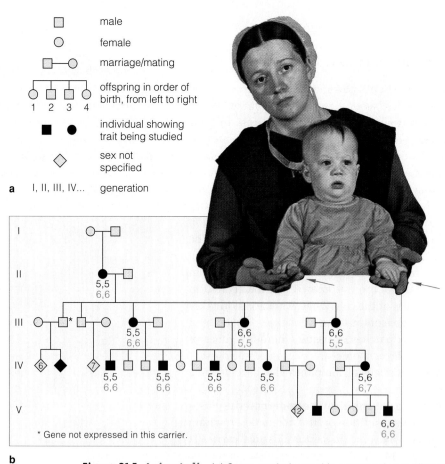

Figure 21.5 *Animated!* (**a**) Some symbols used in constructing pedigree diagrams. (**b**) A pedigree for polydactyly, which confers extra fingers, extra toes, or both. As described in Section 20.6, expression of the gene governing polydactyly can vary. Here, black numerals designate the number of fingers on each hand. Blue ones designate the number of toes on each foot.

PATTERNS OF INHERITANCE

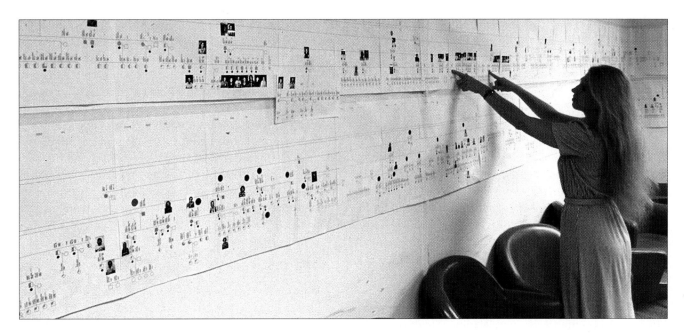

Figure 21.6 From the human genetic researcher Nancy Wexler, a pedigree for Huntington disease, or HD. Wexler has a special interest in HD because it runs in her family.

genetic disorder, a child has a 25 percent chance of being homozygous for the harmful recessive allele.

In thinking about genetics, it's good to keep in mind the difference between an abnormality and a disorder. A **genetic abnormality** is simply deviation from the average, such as having six toes on each foot instead of five. A **genetic disorder** causes mild to severe medical problems. A **syndrome** is a set of symptoms that usually occur together and characterize a disorder.

Each of us carries an average of three to eight harmful recessive alleles. Why don't alleles that cause severe disorders simply disappear from human populations? There are several reasons. For example, new gene alleles can come about by way of mutation, as described in Section 21.8. Also, in heterozygotes, a recessive allele is paired with a normal dominant one that prevents the recessive phenotype from showing up. However, the recessive allele still can be transmitted to offspring.

It is not uncommon for some genetic disorders to be described as diseases (for example, Huntington disease and cystic fibrosis), but in other situations the terms are not interchangeable. For instance, a person's genes may increase her or his susceptibility or weaken the response to infection by a virus, bacterium, or some other pathogen. Strictly speaking, however, the resulting illness isn't a genetic disease. A common example is a genetic predisposition to develop allergies or asthma that runs in some families.

GENETIC ANALYSIS MAY PREDICT DISORDERS

Some prospective parents suspect that they are likely to produce a severely afflicted child. Their first child, a close relative, or they themselves may have a genetic disorder, and they wonder how likely it is that future children also will be affected. Psychologists, geneticists, and other specialists may be called in to provide answers.

A common first step is determining the genotype of each parent. Family pedigrees can aid the diagnosis. For disorders that follow a simple Mendelian inheritance pattern, it's possible to predict the chances a given child will be affected—but not all follow Mendelian patterns. And those that do can be influenced by other factors, some identifiable, others not. Even when the extent of risk has been determined with confidence, prospective parents must understand that the risk is the same for *each* pregnancy. If a pregnancy has one chance in four of producing a child with a genetic disorder, the same odds apply to every subsequent pregnancy.

For many genes, pedigree analysis may reveal Mendelian inheritance patterns that provide information about the probability the genes may be transmitted to children.

A genetic abnormality is an uncommon version of an inherited trait. A genetic disorder is an inherited condition that produces mild to severe medical problems.

21.5 Inheritance of Genes on Autosomes

LINKS TO SECTIONS 9.10, 13.10, 20.1, AND 20.3

Research on the underlying causes of genetic disorders and abnormalities has revealed patterns in the way dominant and recessive genes are inherited. In this section we look at genes—that is, alleles—on autosomes.

INHERITED RECESSIVE TRAITS CAUSE A VARIETY OF DISORDERS

For some traits, inheritance patterns reveal two clues that point to a recessive allele on an autosome. *First*, if both parents are heterozygous, any child of theirs will have a 50 percent chance of being heterozygous and a 25 percent chance of being homozygous recessive (Figure 21.7). *Second*, if both parents are homozygous recessive, any child of theirs will be too.

Cystic fibrosis, which you read about in the chapter introduction, is an *autosomal recessive* condition. So is **phenylketonuria** (PKU), which results from abnormal buildup of the amino acid phenylalanine. Affected people are homozygous for a recessive allele that fails to provide instructions for an enzyme that converts phenylalanine to tyrosine. Excess phenylalanine builds up, and if it is diverted into other metabolic pathways, compounds including phenylpyruvic acid may be produced. At high levels, phenylpyruvic acid can cause mental retardation. Fortunately, a diet low in phenylalanine will prevent PKU symptoms. A wide range of diet soft drinks and other products are sweetened with aspartame, which contains phenylalanine. Such products must carry a warning label so people with PKU can avoid using them.

In some autosomal recessive disorders, the defective gene product is an enzyme needed to metabolize lipids. Infants born with **Tay-Sachs disease** lack hexosaminidase A, which is an enzyme required for the metabolism of sphingolipids, a type of lipid that is especially abundant in the plasma membrane of cells in nerves and the brain. Affected babies seem normal at birth, but over time they lose motor functions and also become deaf, blind, and mentally retarded. Most die in early childhood.

Tay-Sachs disease is most common among children of eastern European Jewish descent. Biochemical tests before conception can determine whether either member of a couple carries the recessive allele.

SOME DISORDERS ARE DUE TO DOMINANT GENES

Other kinds of clues indicate that an *autosomal dominant* allele is responsible for a trait. First, the trait usually appears in each generation because the allele is usually expressed even in heterozygotes. Second, if one parent is heterozygous and the other is homozygous for the normal, recessive allele, there is a 50 percent chance that any child of theirs will be heterozygous (Figure 21.8). A few dominant alleles that cause severe genetic disorders persist in populations. Some result from spontaneous mutations. In other cases, expression of a dominant allele may not prevent reproduction, or affected people have children before the disorder's symptoms become severe.

An example is **Huntington disease**, in which the basal nuclei of the brain (Section 13.8) degenerate. In about

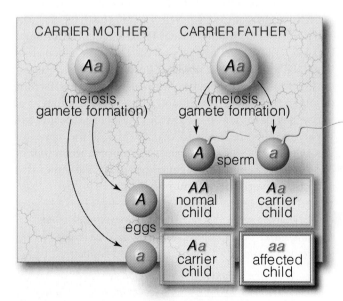

Figure 21.7 *Animated!* A pattern for autosomal recessive inheritance. In this case both parents are heterozygous carriers of the recessive allele (shaded red here).

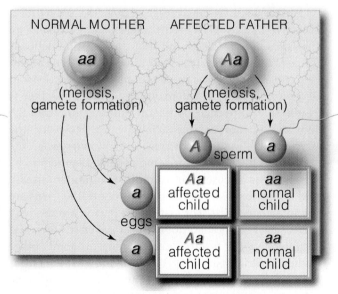

Figure 21.8 *Animated!* A pattern for autosomal dominant inheritance. In this example, the dominant allele (red) is fully expressed in the carriers.

PATTERNS OF INHERITANCE

Figure 21.9 Animated! People with Marfan syndrome are tall and thin. Flo Hyman (left) was captain of the United States volleyball team that won an Olympic silver medal in 1984. Two years later, during a game in Japan, she slid silently to the floor and died. A dime-sized weak spot in the wall of her aorta had burst. Besides Hyman, at least two other affected college basketball stars have died abruptly as a result of the syndrome.

Figure 21.10 Three males affected by achondroplasia. One (center) has a big audience: Actor Verne Troyer, also known as Mini Me in the *Austin Powers* series of spy movies. Troyer stands two feet, eight inches tall.

half the cases, symptoms emerge after age 30—when the person may already have had children (so the allele may be passed on). Homozygotes for the Huntington allele die as embryos, so affected adults are always heterozygous. A person who has one parent with the disorder thus has a 50 percent chance of having the dominant allele (Figure 21.9). Testing can reveal the disease-causing allele, which is on chromosome 4. Unfortunately, because of the nature of the Huntington defect, there is no cure. Some at-risk people opt not to have the diagnostic test, and many elect to remain childless to avoid passing on the disorder.

Marfan syndrome is another autosomal dominant condition. The responsible allele codes for a defective form of the protein fibrillin, which is found in connective tissue. The abnormal fibrillin has a variety of effects, including disrupting both the structure and the function of smooth muscle cells in the wall of the aorta, the large vessel that carries blood away from the heart. Over time the wall thins and weakens, and it can rupture suddenly during strenuous exercise. Marfan syndrome affects 1 in 10,000 people throughout the world. Until recent medical advances, it killed most affected people before they reached age 50 (Figure 21.9).

Another example, **achondroplasia**, also affects about 1 in 10,000 people. Homozygous dominant infants usually are stillborn. Heterozygotes can reproduce, but while they are young and their limb bones are forming, the cartilage elements of those bones cannot form properly. For that reason, at maturity affected people have abnormally short arms and legs (Figure 21.10). Adults who have

achondroplasia don't grow taller than about 4 feet, 4 inches. In many cases, the dominant allele has no other phenotypic effects.

About 1 person in 500 is heterozygous for a dominant autosomal allele that causes a condition called **familial hypercholesterolemia**. This allele leads to dangerously elevated blood cholesterol because it fails to encode the normal number of cell receptors for LDLs (low-density lipoproteins). You may remember from Chapter 9 that LDLs bind cholesterol in the blood, the first step in removing it from the body. A person who is homozygous for the allele may develop severe cholesterol-related heart disease as a child. Affected people usually die in early adulthood.

If both parents are heterozygous carriers of a recessive allele, there is a 25 percent chance that a child of theirs will be homozygous for the trait and exhibit the recessive phenotype.

A dominant trait may appear in each generation, because the dominant allele is expressed even in heterozygotes.

IMPACTS, ISSUES

The mutant gene that causes CF has pleiotropic effects. Lung infections are the greatest threat, but CF also increases the risk of gallstones. Too much chloride is lost in sweat, shifting the body's salt–water balance and making the heart beat irregularly. Also, mucus thickens. This blocks secretion of digestive enzymes from the pancreas, so affected people may be malnourished. It also blocks ducts in the male reproductive tract. Because sperm cannot be ejaculated, males with CF are sterile.

21.6 Inheritance of Genes on the X Chromosome

LINKS TO
SECTIONS
6.6, 8.6,
AND 14.10

Geneticists have also discovered that, as with genes on autosomes, genes on the X chromosome are inherited according to predictable patterns. Here, too, clues have come from studies of abnormalities and disorders.

X-LINKED RECESSIVE INHERITANCE

When a recessive allele on an X chromosome causes a trait, two clues often point to this source. First, many more males than females are affected. This is because a recessive allele can be masked in females, who may inherit a dominant allele on their other X chromosome. It cannot be masked in males, who have only one X chromosome (Figure 21.11). Second, only a daughter can inherit the recessive allele from an affected father, because his sons will receive a copy of his Y chromosome, not the X.

Two forms of the bleeding disorder **hemophilia** are inherited as X-linked recessive traits. The most common one, hemophilia A, is caused by a mutation in the gene for the blood-clotting protein called factor VIII (see Section 8.6). A male with a recessive allele on his X chromosome is always affected, and risks death from anything that causes bleeding, even a bruise. The blood of a heterozygous female clots normally, because the nonmutated gene on her normal X chromosome makes enough factor VIII. (Hemophilia B has similar symptoms, but the clotting factor is factor IX.)

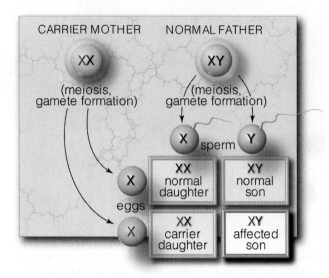

Figure 21.11 *Animated!* One pattern for X-linked inheritance. This example shows the outcomes possible when the mother carries a recessive allele on one of her X chromosomes (*red*).

Hemophilia A affects about 1 in 7,000 males. Among nineteenth-century European royal families, however, the frequency was unusually high because close relatives often married. Queen Victoria of England and two of her daughters were carriers. In a pedigree developed some years ago, more than 15 of her 69 descendants at that time were affected males or female carriers (Figure 21.12).

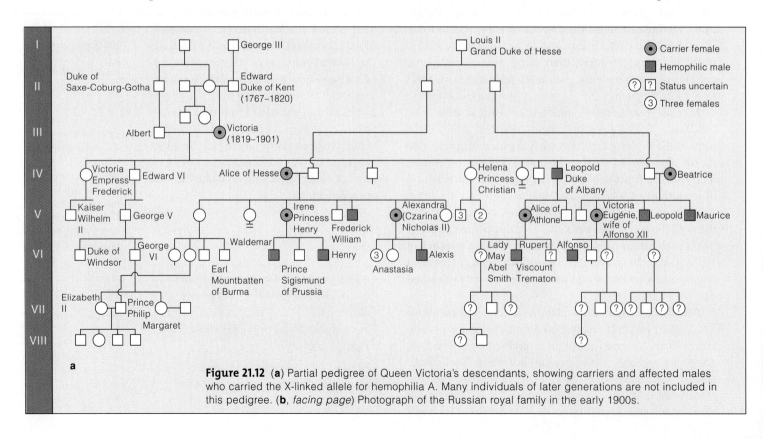

Figure 21.12 (a) Partial pedigree of Queen Victoria's descendants, showing carriers and affected males who carried the X-linked allele for hemophilia A. Many individuals of later generations are not included in this pedigree. (**b**, *facing page*) Photograph of the Russian royal family in the early 1900s.

PATTERNS OF INHERITANCE

You may recall from Chapter 6 that diseases lumped under "muscular dystrophy" involve progressive wasting of muscle tissue. **Duchenne muscular dystrophy** (DMD) is X-linked. It affects 1 in 3,500 males, usually in childhood. As muscles degenerate, affected boys become weak and unable to walk. They usually die by age 20 from cardiac or respiratory failure. The gene that is mutated in DMD normally encodes the protein *dystrophin*, which gives structural support to the muscle cell plasma membrane. In DMD, muscle cells lack dystrophin, so they can't withstand the physical stress of contraction and break down. In time the whole muscle is destroyed.

Red/green color blindness is an X-linked recessive trait. About 8 percent of males in the United States have this condition. It arises from mutation of an allele that codes for the protein opsin, which binds visual pigments in cone cells of the retina. Females also can have red/green color blindness, but this occurs rarely because a girl must inherit the recessive allele from both parents.

SOME TYPES OF X-LINKED ABNORMALITIES ARE QUITE RARE

The **faulty enamel trait** is one of a few known examples of a trait caused by a dominant mutant allele that is X-linked. With this disorder, the hard, thick enamel coating that normally protects teeth fails to develop properly (Figure 21.13). The allele is inherited in a way that resembles the pattern for X-linked recessive alleles, except that the trait is expressed in heterozygous females. An affected father will transmit the allele to all of his daughters but none of his sons. A heterozygous mother will transmit the allele to half her offspring, regardless of their gender.

Figure 21.13 The discolored, abnormal tooth enamel of a person with the faulty enamel trait.

On rare occasions, someone whose sex chromosomes are XY develops as a female. The result is **testicular feminizing syndrome**, or "androgen insensitivity." A gene mutation on the X chromosome has produced defective receptors for androgens (male sex hormones), including testosterone. Normally, cells in the testes and other male reproductive organs bind one or more of the hormones and then proceed to develop. With defective receptors, however, they can't bind the hormones. As a result, the embryo develops externally as a female but has no uterus or ovaries.

MANY FACTORS COMPLICATE GENETIC ANALYSIS

This section's examples of autosomal and X-linked traits give you a general idea of the kinds of clues that geneticists look for. Genetic analysis is usually a difficult task, however. To start with, these days few people have large families, so it may be necessary to pool several pedigrees. Typically the geneticist will make detailed analyses of clinical data and keep abreast of current research, in part because more than one gene may be responsible for a given phenotype. For example, we know of dozens of conditions that can arise from a mutated gene on an autosome *or* from a mutated gene on the X chromosome. Also, some genes on autosomes are dominant in males but recessive in females—so they may initially appear to be due to X-linked recessive inheritance, even though they are not.

Czarina Alexandra (a carrier; descendant of Queen Victoria)

Czar Nicholas II (free of allele for hemophilia A)

Alexis (hemophilic son)

b The Russian royal family members. All are believed to have been executed near the end of the Russian Revolution. They were recently exhumed from their hidden graves, but DNA fingerprinting indicated that the remains of Alexis and one daughter, Anastasia, were not among them. At this writing, the search is on for other graves where the two missing children might have been buried.

A trait that shows up most often in males and that a son can inherit only from his mother is most likely passed on through X-linked recessive inheritance.

A few rare mutant traits are passed to offspring via X-linked dominant inheritance. A heterozygous mother will pass the allele to half her offspring. An affected father will pass the allele only to his daughters.

21.7 Sex-Influenced Inheritance

LINKS TO SECTIONS 2.13, 8.1, 19.1, AND 19.7

A person's sex can influence certain traits, even when the gene involved is not on a sex chromosome.

Sex-influenced traits appear more often in one sex than in the other, or the phenotype differs depending on whether the person is male or female. The difference may reflect the varying influences of male and female sex hormones on how genes are expressed. Genes for such traits are on autosomes, not on sex chromosomes. Pattern baldness (Figure 21.14) is an example. We can designate the "no baldness" form of the responsible gene b^+ and the "baldness" form b. A man will develop pattern baldness if he is homozygous (bb) and *also* if he is a heterozygote (b^+b). A woman will develop pattern baldness, usually later in life, only if she is bb.

> Sex-influenced traits appear more frequently in one sex, or the phenotype differs among males and females.

Genotype	Male Phenotype	Female Phenotype
b^+b^+	Hair	Hair
b^+b	Bald	Hair
bb	Bald	Bald

Figure 21.14 A typical case of pattern baldness. Geneticists usually denote the alleles for a sex-influenced trait with lowercase letters. Here, b^+ is dominant and b is recessive. Because of the influence of sex hormones, female heterozygotes have a full head of hair. Male heterozygotes, and all homozygous recessives, show pattern baldness.

21.8 Changes in a Chromosome or Its Genes

The structure of a chromosome isn't "written in stone." It can change in a variety of ways.

As you know, DNA packaged in chromosomes consists of various types of nucleotides linked by chemical bonds (Section 2.13). A **gene mutation** is a change in one or more of the nucleotides that make up a particular gene. Mutations can arise in various ways that we will discuss in detail in Chapter 22. In this section we are concerned with changes in the structure of whole chromosomes. During meiosis, pieces of chromosomes can be deleted, duplicated, or moved around in other ways. The result often is harmful.

VARIOUS CHANGES IN A CHROMOSOME'S STRUCTURE MAY CAUSE A GENETIC DISORDER

A chromosome region may be deleted spontaneously, or by a virus, by irradiation, chemical assaults, or some other environmental factor:

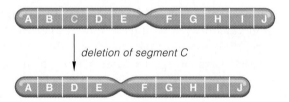

deletion of segment C

Any part of a chromosome can be lost. Wherever such a *deletion* happens, it permanently removes one or more of the chromosome's genes. The loss of a gene can lead to serious problems. For example, one deletion from human chromosome 5 leads to abnormal mental development and an abnormally shaped larynx. When an affected infant cries, the sounds produced resemble meowing—hence the name of the disorder, **cri-du-chat** (French, meaning "cat cry"). Figure 21.15 shows an affected child.

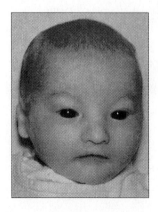

Figure 21.15 Male infant who developed cri-du-chat syndrome. Ears are low on the head relative to the eyes.

CHANGES IN CHROMOSOMES

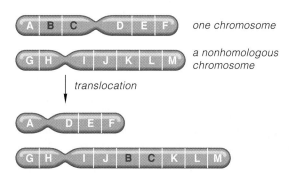

one chromosome

a nonhomologous chromosome

translocation

Figure 21.16 *Animated!* Simple diagram of a chromosomal translocation.

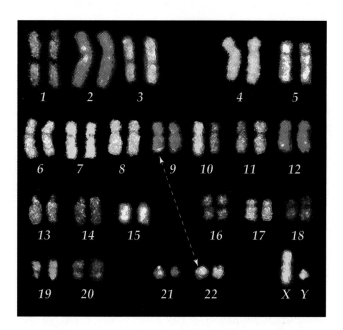

Figure 21.17 A spectral karyotype of the forty-six chromosomes in a human diploid cell. The arrow indicates the bit of chromosome 22 fused to the Philadelphia chromosome—chromosome number 9—before it was altered.

Given that humans have diploid cells, do genes on the affected chromosome's homologue ever compensate for the loss? In fact, this is often the case if a segment deleted from one chromosome is present—and normal—on the homologous chromosome. However, if the remaining, homologous segment is abnormal or carries a harmful recessive allele, nothing will mask *its* effects.

Even normal chromosomes contain the changes called *duplications*, which are gene sequences that are repeated. Often the same gene sequence is repeated thousands of times. You might guess that so much duplicate DNA would be harmful, but no genetic disorder has yet been linked to duplication.

Another kind of chromosome change is called a *translocation*. Here, part of one chromosome exchanges places with a corresponding part of another chromosome that is *not* its homologous partner (Figure 21.16). This sort of change to a chromosome's structure is virtually sure to be harmful. For instance, in some people a region of chromosome 8 has been translocated to chromosome 14—and the result can be several rare types of cancer. The disease develops because genes in that region are no longer properly regulated, a topic that is discussed further in Chapter 22.

Patients who have a chronic type of leukemia (a blood cell cancer) have an abnormally long chromosome 9—called the Philadelphia chromosome after the city where it was discovered. The extra length is actually a piece of chromosome 22. By chance, both chromosomes break in a stem cell in bone marrow. Then, each broken piece reattaches to the *wrong* chromosome—and a gene located at the end of chromosome 9 becomes fused with a gene in chromosome 22 (Figure 21.17). Instructions from this altered gene lead to the synthesis of an abnormal protein. In some way that researchers do not yet understand, that protein promotes the runaway multiplication of white blood cells.

In the next section we turn to some genetic effects that can arise when a gamete or new embryo receives an abnormal number of chromosomes—either too few or too many.

Genetic disorders can result from mutations on genes on chromosomes or from changes in the structure of one or more chromosomes.

Factors such as viral infection and irradiation can delete part of a chromosome. Sometimes the genes of a normal segment on the affected chromosome's homologue can compensate for the loss.

In a duplication, a gene sequence is repeated in a chromosome.

In a translocation, part of a chromosome is transferred to another, nonhomologous, chromosome.

CHANGES IN CHROMOSOMES

Chromosomes and Human Genetics   399

21.9 Changes in Chromosome Number

Several kinds of events can increase or decrease the number of chromosomes in gametes.

Sometimes gametes—and subsequent embryos—end up with the wrong chromosome number. The effects range from minor physical ones to deadly disruption of body function. More often, an affected fetus is *miscarried*, or spontaneously aborted before birth.

About half of fertilized eggs have a lethal condition called *aneuploidy* (AN-yoo-ploy-dee). In this situation, the embryo doesn't have an exact multiple of the normal haploid set of 23 chromosomes. A *polyploid* embryo has three, four, or more sets of the normal haploid set of 23 chromosomes. All but 1 percent of human polyploids die before birth, and the rare newborns die soon afterward.

Chromosome numbers can change during mitosis or meiosis or even at fertilization. For instance, a cell cycle might advance through DNA duplication and mitosis; then for some reason it stops before the dividing cell's cytoplasm divides. The cell then is polyploid—it has four of each type of chromosome.

NONDISJUNCTION IS A COMMON CAUSE OF ABNORMAL NUMBERS OF AUTOSOMES

In **nondisjunction**, one or more pairs of chromosomes fail to separate during mitosis or meiosis. Here again, some or all of the resulting cells end up with too many or too few chromosomes (Figure 21.18).

If fertilization involves a gamete that has an extra chromosome ($n + 1$), the result will be **trisomy**: the new individual will have three of one type of chromosome ($2n + 1$). If the gamete is missing a chromosome, then the result is **monosomy** ($2n - 1$). Most changes in the number

of autosomes arise through nondisjunction when meiosis is forming gametes. About 1 in every 1,000 children is born with trisomy 21—three copies of chromosome 21.

A person with trisomy 21 will have **Down syndrome** (Figure 21.19). Symptoms vary, but most affected people are mentally retarded. About 40 percent develop heart defects. Because of abnormal skeletal development, older children have shortened body parts, loose joints, and poorly aligned hip, finger, and toe bones. Their muscles and muscle reflexes are weak, and their motor functions develop slowly. With special training, though, people with Down syndrome often engage in normal activities.

For women, the incidence of nondisjunction increases with age. The probability that a woman will conceive an embryo with Down syndrome rises steeply after age 35. Yet 80 percent of trisomic 21 infants are born to younger mothers. This statistic reflects the fact that between the ages of 18 and 35 women are the most fertile, so more babies are born to mothers in this age range.

NONDISJUNCTION ALSO CAN CHANGE THE NUMBER OF SEX CHROMOSOMES

Most sex chromosome abnormalities come about as a result of nondisjunction as gametes are forming. Let's look at a few of the resulting phenotypes.

TURNER SYNDROME AND XXX FEMALES About 1 in every 5,000 newborns has **Turner syndrome**, in which a nondisjunction has reduced the chromosome number to 45 (Figure 21.20). Most people with Turner syndrome are missing an X chromosome (in most cases, the one that would have come from the father), and the condition is symbolized as XO. Turner syndrome occurs less often than

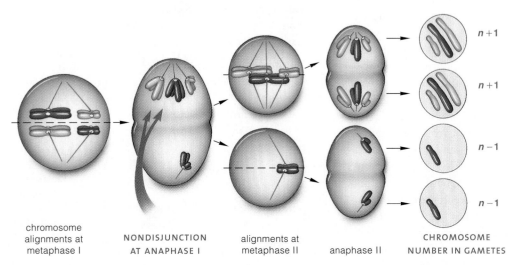

chromosome
alignments at
metaphase I

NONDISJUNCTION
AT ANAPHASE I

alignments at
metaphase II

anaphase II

CHROMOSOME
NUMBER IN GAMETES

$n + 1$

$n + 1$

$n - 1$

$n - 1$

Figure 21.18 *Animated!*
Nondisjunction. In this example, chromosomes fail to separate during anaphase I of meiosis and so there is a change in the chromosome number in resulting gametes.

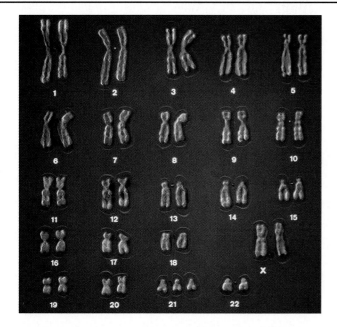

Figure 21.19 (**a**) Karyotype of a girl with Down syndrome; note the three copies of chromosome 21. (**b**) A boy with Down syndrome.

other sex chromosome abnormalities, probably because most XO embryos are miscarried early in the pregnancy. Affected people are female and have a webbed neck and other phenotypic abnormalities. Their ovaries don't function, they are sterile, and secondary sexual traits don't develop at puberty. People who have Turner syndrome often age prematurely and have shortened life expectancies.

Roughly 1 in 1,000 females has three X chromosomes. Two of these X chromosomes are condensed to Barr bodies, and most XXX females develop normally.

KLINEFELTER SYNDROME In **Klinefelter syndrome**, nondisjunction produces the genotype XXY. This sex chromosome abnormality occurs in about 1 in 500 males. XXY males have low fertility and usually some mental

retardation. They have abnormally small testes, sparse body hair, and may develop enlarged breasts. Testosterone injections can reverse some aspects of the phenotype.

XYY CONDITION About 1 in every 1,000 males has one X and two Y chromosomes, a condition due to nondisjunction of duplicated Y chromosomes during meiosis. XYY males tend to be taller than average, but otherwise they have a normal male phenotype.

Most changes in the number of chromosomes arise due to nondisjunction during meiosis and the formation of gametes in parents.

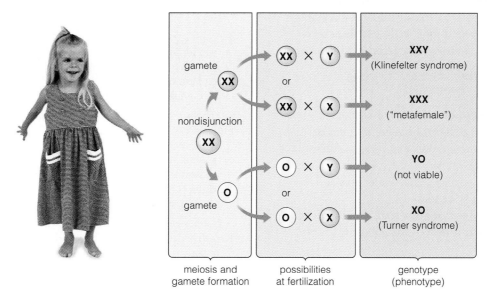

Figure 21.20 Genetic disorders that result from nondisjunction of X chromosomes followed by fertilization by normal sperm. The photograph shows a girl with Turner syndrome.

Summary

Section 21.1 Each gene has a specific location on a specific chromosome. Diploid cells have two homologous chromosomes of each type, one from each parent. The homologues pair up at meiosis and crossing over exchanges segments between them. The genes on a chromosome are physically linked. Those that are closest together usually end up in the same gamete.

Chromosomes that are the same in males and females are called autosomes. They are roughly the same in size and shape and carry genes for the same traits. Sex chromosomes (X and Y) differ from each other in size, shape, and the genes they carry.

Section 21.2 In karyotyping, a diagnostic tool, a person's metaphase chromosomes are prepared for microscopy, photographed, and arranged in sequence in a chart based on their characteristics.

 Learn how to create a karyotype.

Section 21.3 A person's sex is determined by the father's sperm, which can carry an X chromosome or a Y chromosome. Males have an XY genotype, females an XX. Genes on the X and Y chromosomes are called X-linked genes and Y-linked genes, respectively. In a female, a mechanism of X inactivation shuts down the expression of genes carried on one of her X chromosomes.

 See how gender is determined in a developing human embryo.

Section 21.4 A pedigree chart diagrams the genetic relationships among relatives. It can help establish inheritance patterns and track genetic abnormalities through several generations. Table 21.1 lists some common genetic disorders discussed in this textbook.

 Learn more about making a human pedigree.

Section 21.5 Studies of genetic disorders provide information about patterns of gene inheritance. In disorders that involve autosomal recessive inheritance, a person who is homozygous for a recessive allele shows the recessive phenotype. Heterozygotes generally have no symptoms.

In autosomal dominant inheritance, a dominant allele usually is expressed to some extent.

 Investigate autosomal inheritance.

Sections 21.6–21.7 Many genetic disorders are X-linked—the mutated gene occurs on the X chromosome. Males, who inherit only one X chromosome, typically are affected. Sex-influenced traits, such as pattern baldness, appear more frequently in one sex. They may reflect the varying influences of sex hormones.

 Investigate X-linked inheritance.

Section 21.8 A chromosome's structure can be changed by deletions, duplications, or translocations. Such alterations usually lead to harmful changes in traits.

Section 21.9 Chromosome number can be altered by nondisjunction, in which one or more pairs of chromosomes do not separate during meiosis or mitosis. In trisomy, a new individual inherits an extra copy of one type of chromosome. If one copy of a given chromosome is missing, the condition is called monosomy.

Review Questions

1. How do X and Y chromosomes differ?

2. What do we mean when we say someone is a carrier of a genetic trait?

3. What evidence indicates that a trait is coded by a dominant allele on an autosome?

4. Explain the difference between an X-linked trait and a sex-influenced trait.

5. Explain what nondisjunction is, and give two examples of phenotypes that can result from it.

Self-Quiz   *Answers in Appendix V*

1. _____ segregate during _____.
   a. Homologues; mitosis
   b. Genes on one chromosome; meiosis
   c. Homologues; meiosis
   d. Genes on one chromosome; mitosis

2. The alleles of a gene on homologous chromosomes end up in separate _____.
   a. body cells
   b. gametes
   c. nonhomologous chromosomes
   d. offspring
   e. both b and d are possible

3. Genes on the same chromosome tend to remain together during _____ and end up in the same _____.
   a. mitosis; body cell   d. meiosis; gamete
   b. mitosis; gamete   e. both a and d
   c. meiosis; body cell

Table 21.1 Examples of Human Genetic Disorders and Genetic Abnormalities

Disorder or Abnormality*	Main Consequences	Disorder or Abnormality*	Main Consequences
AUTOSOMAL RECESSIVE INHERITANCE		**X-LINKED RECESSIVE INHERITANCE**	
Albinism *21 CT*	Absence of pigmentation	Red-green color blindness *14.10, 21.6*	Inability to distinguish all or some colors of visible light
Cystic fibrosis *21 CI, 21.5, 22.9*	Oversecretion of mucus leading to organ damage	Duchenne muscular dystrophy *21.6*	Progressive wasting of muscles
Phenylketonuria (PKU) *21.5*	Mental retardation	Hemophilia *21.6*	Impaired blood clotting
Sickle-cell anemia *20.5*	Harmful pleiotropic effects on organs throughout body	Testicular feminizing syndrome *21.6, 21 CT*	XY individual has some female traits and is sterile
		Fragile X syndrome *22.2*	Mental retardation
AUTOSOMAL DOMINANT INHERITANCE		**SEX-INFLUENCED INHERITANCE**	
Achondroplasia *21.5*	One form of dwarfism	Pattern baldness *21.7*	Loss of hair on the top and upper sides of the head
Achoo syndrome *20 CT*	Chronic sneezing		
Amyotrophic lateral sclerosis (ALS) *22.8*	Loss of all muscle function	**CHANGES IN CHROMOSOME NUMBER**	
Familial hypercholesterolemia *21.5*	High cholesterol levels in blood; eventually clogged arteries	Down syndrome *21.9, 21 CT*	Mental retardation, heart defects
Huntington disease *21.5, 22.2*	Nervous system degenerates progressively, irreversibly	Klinefelter syndrome *21.9*	Sterility, retardation in many cases
Polydactyly *21.4*	Extra fingers, toes, or both	Turner syndrome *21.9*	Sterility; abnormal ovaries, abnormal sexual traits
Tay-Sachs disease *21.5*	Progressive deterioration of the nervous system	XYY condition *21.9*	Mild retardation or no symptoms
Marfan syndrome *21.5*	Absence or abnormal formation of connective tissue	**CHANGES IN CHROMOSOME STRUCTURE**	
		Cri-du-chat syndrome *21.8*	Mental retardation, abnormal larynx
X-LINKED DOMINANT INHERITANCE		Philadelphia chromosome (chronic myelogenous leukemia) *21.8*	Overproduction of white blood cells followed by organ malfunction
Faulty enamel trait *21.6*	Abnormal tooth enamel		

*Italic numbers indicate sections in which a disorder is described. *CI* signifies Chapter Introduction. *CT* signifies an end-of-chapter Critical Thinking question.

4. The probability of a crossover occurring between two genes on the same chromosome is _____.
   a. unrelated to the distance between them
   b. increased if they are closer together on the chromosome
   c. increased if they are farther apart on the chromosome
   d. zero

5. A chromosome's structure can be altered by _____.
   a. deletions   c. translocations
   b. duplications   d. all of the above

6. Nondisjunction can be caused by _____.
   a. crossing over in meiosis
   b. segregation in meiosis
   c. failure of chromosomes to separate during meiosis
   d. multiple independent assortments

7. A gamete affected by nondisjunction could have _____.
   a. a change from the normal chromosome number
   b. one extra or one missing chromosome
   c. the potential for a genetic disorder
   d. all of the above

8. Genetic disorders can be caused by _____.
   a. gene mutations
   b. changes in chromosome structure
   c. changes in chromosome number
   d. all of the above

9. A person who is a carrier for a genetic trait _____.
   a. is heterozygous for a dominant trait
   b. is heterozygous for a recessive trait
   c. is homozygous for a recessive trait
   d. could be either a or b but not c

10. Match the following chromosome terms appropriately.
    ____ crossing over
    ____ deletion
    ____ nondisjunction
    ____ translocation
    ____ gene mutation

    a. a chemical change in DNA that may affect genotype and phenotype
    b. movement of a chromosome segment to a nonhomologous chromosome
    c. disrupts gene linkages during meiosis
    d. causes gametes to have abnormal chromosome numbers
    e. loss of a chromosome segment

Critical Thinking

1. A female runner is disqualified from a race because testing shows an XY phenotype. Later, a medical exam reveals "androgen insensitivity," a condition in which an embryo's cells lack receptors for the male hormone testosterone. As a result, the person's phenotype is clearly female. Discuss whether you agree or disagree with the disqualification, and why.

2. If a couple has six boys, what is the probability that a seventh child will be a girl?

3. Human sex chromosomes are XX for females and XY for males.
   a. From which parent does a male inherit his X chromosome?
   b. With respect to an X-linked gene, how many different types of gametes can a male produce?
   c. If a female is homozygous for an X-linked allele, how many different types of gametes can she produce with respect to this allele?
   d. If a female is heterozygous for an X-linked allele, how many different types of gametes can she produce with respect to this allele?

4. People with Down syndrome have an extra copy of chromosome 21, for a total of 47 chromosomes in body cells. However, in a few cases of Down syndrome, 46 chromosomes are present. This total includes two normal-looking chromosomes 21, one normal chromosome 14, and a longer-than-normal chromosome 14. Interpret this observation. How can these individuals have 46 chromosomes?

5. If a trait appears only in males, is this good evidence that the trait is due to a Y-linked allele? Explain why you answered as you did.

6. A woman unaffected by hemophilia A whose father had hemophilia A marries a man who also has hemophilia A. If their first child is a boy, what is the probability he will have the disorder?

7. Among people of European descent, about 4 percent have the allele for cystic fibrosis. Yet only about 1 in 2,500 people actually has the disorder. What is the most likely reason for this finding?

8. The young woman shown at left has albinism—very pale skin, white hair, and pale blue eyes. This phenotype, typically caused by a recessive allele, is due to the absence of melanin, which imparts color to the skin, hair, and eyes. Suppose a person with albinism marries a person with typical pigmentation and they have one child with albinism and three with typical pigmentation. What is the genotype of the parent with typical pigmentation? Why is the ratio of this couple's offspring 3:1?

Explore on Your Own

Several mutant genes are known to be associated with neurobiological disorders (NBDs) such as schizophrenia, which affects one of every hundred people worldwide. Schizophrenia is characterized by delusions, hallucinations, disorganized speech, and abnormal social behavior. Another facet of schizophrenia and other NBDs (such as bipolar disorder) is that affected people often are exceptionally creative. One example is John Nash (Figure 21.21a), the brilliant mathematician and Nobel Prize winner whose battle with schizophrenia was portrayed in the film *A Beautiful Mind*. Another was the writer Virginia Woolf (Figure 21.21b), who committed suicide after a long mental breakdown.

Evidence suggests to some researchers that a number of other highly creative, distinguished historical figures, possibly including Abraham Lincoln, suffered from some type of NBD. To explore this topic further, do a Web search and make a list of ten well-known people from the past who may have had an NBD. What behaviors or other characteristics have been cited to support the hypothesis that each individual was affected by an NBD?

Figure 21.21
(a) Mathematician John Nash. (b) Writer Virginia Woolf. Both of these highly creative people suffered from a genetic neural disorder.

Ricin and Your Ribosomes

In 2003, police acting on a tip stormed a London apartment, where they collected castor beans and laboratory equipment. They arrested several young men and reawakened the world to the potential threat of ricin as a biochemical weapon.

Ricin is a protein produced by the castor plant. Its lethal effects on humans have been known since the 1880s. When Germany unleashed mustard gas against Allied troops during World War I, the United States and England feverishly investigated whether ricin, too, could be used as a weapon. Both countries shelved the research when the war ended.

In 1969, Georgi Markov, a Bulgarian writer, defected to the West at the height of the Cold War. As he strolled down a London street, an assassin used a modified umbrella to fire a tiny ball laced with ricin into Markov's thigh. Markov died in agony three days later. His attacker escaped.

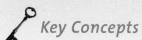

Seeds of the castor plant are used to make castor oil, which is an ingredient in a variety of consumer and industrial products. Ricin is simply a dangerous by-product of the process. How does ricin exert its deadly effects? It inactivates ribosomes, the cell's machinery for making proteins. Ricin itself is a protein. Like all proteins, it is synthesized in cells based on information contained in DNA. One of its polypeptide chains, shown in the filmstrip, helps ricin insert itself into cells. The other is an enzyme that damages part of the ribosome where amino acids are joined together. Protein synthesis stops and cells begin to die. So does the affected person, because there is no antidote.

This chapter discusses DNA and how cells use it to make the proteins all of us need to survive. We will also consider how that knowledge is being applied in biotechnology. As you'll see, the ability to alter and manipulate genes, and understand the biological roles of particular ones, is changing our world in dramatic ways.

 How Would You Vote? *There is evidence that in recent years terrorists have explored developing ricin-based weapons. Scientists are working to develop a vaccine to protect against it. Given the threat of biochemical warfare, would you be willing to be vaccinated—or does the threat seem too remote? Cast your vote online at www.thomsonedu .com/biology/starr/humanbio.*

Key Concepts

GENETIC INSTRUCTIONS IN DNA
Genes are sequences of nucleotides in DNA, which consists of two strands of nucleotides twisted into a double helix. When DNA replicates, a "parent" strand is joined to a new, complementary strand.

MAKING PROTEINS
Genes are the genetic code for proteins. Cells build proteins in two steps. First, an mRNA molecule is transcribed from DNA. Then mRNA is translated into a string of amino acids, the primary structure of proteins.

ENGINEERING AND EXPLORING GENES
Biotechnology is a tool for altering genes and studying their effects. Practical applications include gene therapy, DNA fingerprinting, and studying the human genome.

Links to Earlier Concepts

This chapter expands on some basic concepts introduced in the early chapters of this textbook. It draws on what you know about DNA and RNA (2.13) and revisits how the primary structure of a protein develops (2.11). Here you will learn how RNA participates in the process, and you will also gain a much fuller understanding of the key role played by a cell's ribosomes (3.6). The chapter's discussion of DNA replication will deepen your understanding of events that take place during the cell cycle, before chromosomes are replicated and assorted into new cells (19.3, 19.4, 19.7).

22.1 DNA: A Double Helix

LINKS TO
SECTIONS
2.11, 2.13, 3.5,
19.2, AND 20.1

DNA is built of nucleotides—the components of the biological molecules called nucleic acids. These nucleotides are arranged to form a double helix.

DNA IS BUILT OF FOUR KINDS OF NUCLEOTIDES

Earlier chapters noted that a chromosome consists of a DNA molecule and proteins. A molecule of DNA, in turn, is built from four kinds of **nucleotides**, the building blocks of nucleic acids (Figure 22.1).

A DNA nucleotide is built of a five-carbon sugar (deoxyribose), a phosphate group, and one of these four nitrogen-containing bases:

adenine	guanine	thymine	cytosine
A	G	T	C

Many researchers were in the race to discover DNA's structure, but James Watson and Francis Crick were the first to realize that DNA consists of two strands of nucleotides twisted into a double helix. Nucleotides in a strand are linked together, like boxcars in a train, by strong covalent bonds. Weaker hydrogen bonds link the bases of one strand with bases of the other. The two strands run in opposite directions, as sketched in Figure 22.2 on the facing page.

CHEMICAL "RULES" DETERMINE WHICH NUCLEOTIDE BASES IN DNA CAN PAIR UP

The bases in the four DNA nucleotides have different shapes, and different sites where hydrogen bonds can form. These factors determine which bases can pair up. Adenine pairs with thymine, and guanine pairs with cytosine. Therefore, two kinds of **base pairs** occur in DNA: A—T and G—C. In a double-stranded DNA molecule, the amount of adenine equals the amount of thymine, and the amount of guanine equals the amount of cytosine.

While base pairs must form as we've just described—A with T and G with C—the nucleotides can line up in *any* order. For example, these are just three possibilities of the pattern you might find in a stretch of DNA:

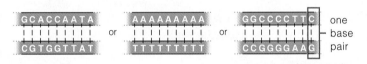

Figure 22.1

Figure 22.1 *Animated!* The four kinds of nucleotides in DNA. The five-carbon sugars are shaded orange. Each one has a phosphate group attached to its ring structure (below, on the left). Small numerals on the structural formulas identify the carbon atoms to which various parts of the molecule are attached. The photograph shows James Watson (left) and Francis Crick, who figured out the structure of DNA.

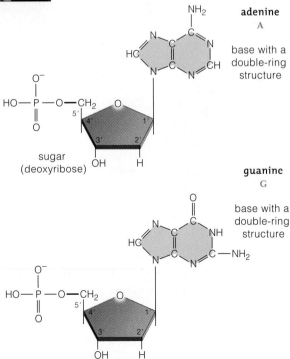

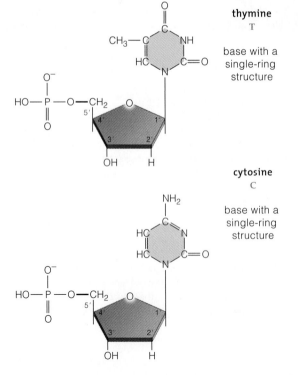

GENETIC INSTRUCTIONS IN DNA

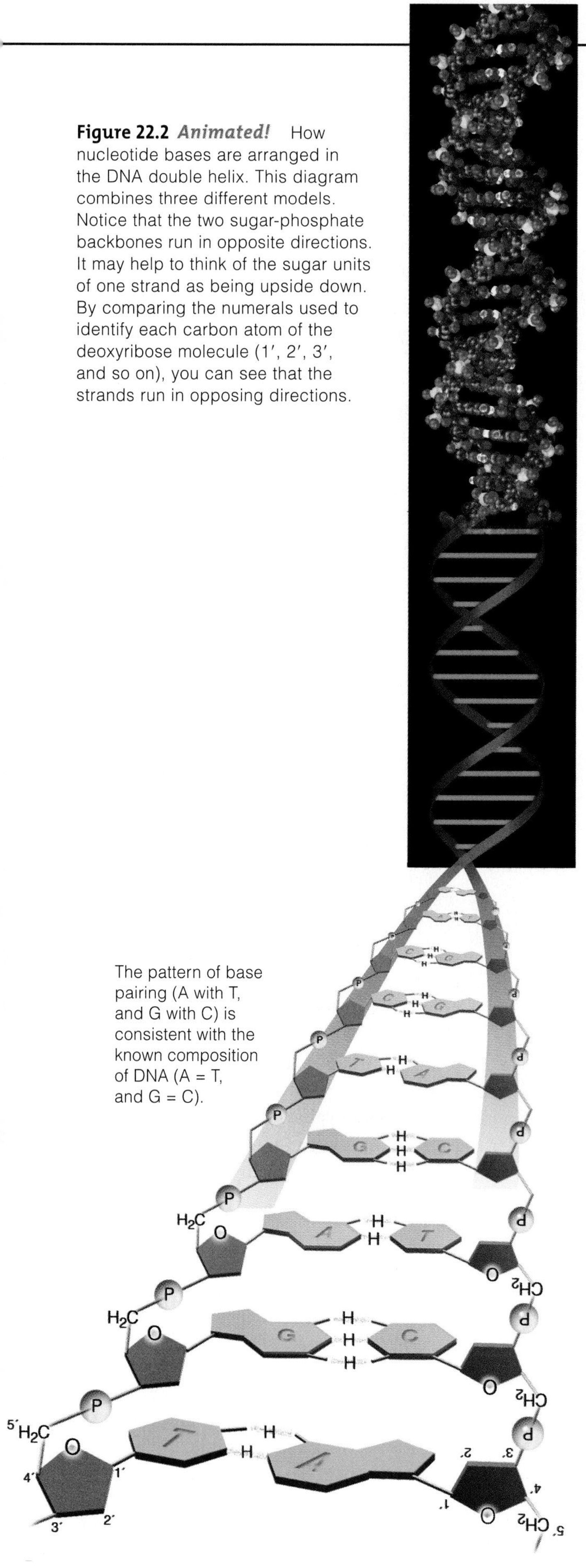

Figure 22.2 *Animated!* How nucleotide bases are arranged in the DNA double helix. This diagram combines three different models. Notice that the two sugar-phosphate backbones run in opposite directions. It may help to think of the sugar units of one strand as being upside down. By comparing the numerals used to identify each carbon atom of the deoxyribose molecule (1′, 2′, 3′, and so on), you can see that the strands run in opposing directions.

The pattern of base pairing (A with T, and G with C) is consistent with the known composition of DNA (A = T, and G = C).

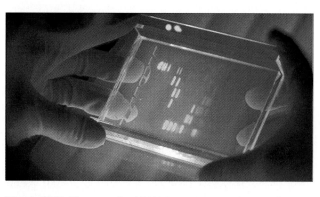

Figure 22.3 Photograph of DNA fragments as visualized through a process called gel electrophoresis.

A GENE IS A SEQUENCE OF NUCLEOTIDES

Previously, you have read that genes are the units of heredity. Chemically, a **gene** is a sequence of nucleotides in a DNA molecule. The **nucleotide sequence** of each gene codes for a specific polypeptide chain. Polypeptide chains, remember, are the basic structural units of proteins. We'll see how they form later on. Figure 22.3 shows one way researchers can visualize small fragments of DNA.

Table 22.1 summarizes the main concepts of DNA structure we have discussed thus far. With these in mind, we now turn to a key feature of each DNA molecule—how it can replicate so that the information it contains is faithfully passed on to new generations.

A DNA molecule consists of two strands of nucleotides held together at their bases by hydrogen bonds. The two strands run in opposite directions and twist to form a double helix.

DNA's nucleotides are built of the sugar deoxyribose, a phosphate group, and one of the nitrogen-containing bases adenine (A), guanine (G), thymine (T), and cytosine (C).

In a DNA molecule, two kinds of base pairing occur: A—T and G—C.

Chemically, a gene is a sequence of nucleotides in DNA.

Table 22.1	A DNA Summary
Nucleotide	Building blocks of nucleic acids; composed of phosphate and a nitrogen-containing base (A, G, T, or C)
Base pair	Two bases (A + T or G + C) held together by hydrogen bonds
Gene	A sequence of nucleotides in a DNA molecule

22.2 Passing on Genetic Instructions

LINKS TO
SECTIONS
20.6 AND 21.5

Knowing a little about DNA's structure lets you take the next step—understanding how a DNA molecule can be faithfully duplicated so parents may pass on their traits to offspring.

HOW IS A DNA MOLECULE DUPLICATED?

Chapter 19 described how DNA must be duplicated before a cell divides. This process of duplication is called **DNA replication**.

Enzymes easily break the hydrogen bonds between the two nucleotide strands of a DNA molecule. When enzymes and other proteins act on the molecule, one strand unwinds from the other and exposes stretches of its nucleotide bases. Cells contain stockpiles of free nucleotides that can pair with the exposed bases (A with T and G with C), and they are linked by hydrogen bonds. Each parent strand remains intact while a new strand is assembled on it, nucleotide by nucleotide.

As replication occurs, the newly formed double-stranded molecule twists back into a double helix. One strand is from the starting molecule, so *that* strand is said to be conserved. Only the second strand has been freshly synthesized—so each DNA molecule is really half new and half "old." The DNA replication mechanism is thus called **semiconservative replication** (Figure 22.4).

During replication, enzymes called **DNA polymerases** and other proteins unwind the DNA molecule, keep the two unwound strands separated, and assemble and seal a new strand on each one. DNA polymerases also link the individual nucleotides on a parent strand.

MISTAKES AND DAMAGE IN DNA CAN BE REPAIRED

DNA polymerases and other enzymes also act in **DNA repair**. There is plenty of opportunity for errors and damage to occur. For example, DNA replication takes place at tremendous speed—between ten and twenty nucleotides per second are added at a replication site. Every time a cell replicates its store of DNA before dividing, at least 3 billion nucleotides must be assembled properly. It's no wonder, then, that mistakes are made. It has been estimated that a human cell must repair breaks in a strand of its DNA up to 2 million times every hour! Luckily, if an error takes place during replication, enzymes may detect and correct the problem, restoring the proper DNA sequence. When an error is not fixed, the result is a mutation.

Previous chapters have noted that DNA is vulnerable to damage from certain chemicals, ionizing radiation, and ultraviolet light (as in sunlight or the rays of tanning lamps). One type of damage is the formation of *thymine dimers*. UV light causes two neighboring thymine bases to become linked (forming a dimer; Figure 22.5*a*). The new structure distorts the affected DNA molecule in a way that prevents effective DNA repair. A thymine dimer can lead to the genetic disorder **xeroderma pigmentosum** (Figure 22.5*b*), in which radiation damage to DNA cannot be fixed.

Unrepaired mutations in some genes may be responsible for a high percentage of cancers. For example, people who have xeroderma pigmentosum are at high risk for lethal skin cancer.

A MUTATION IS A CHANGE IN THE SEQUENCE OF A GENE'S NUCLEOTIDES

Every so often, genes do change. Sometimes one base gets substituted for another in the nucleotide sequence. At other times, an extra base is inserted into the sequence or a base is deleted from it. These kinds of small-scale changes in the nucleotide sequence of genes are **gene mutations**.

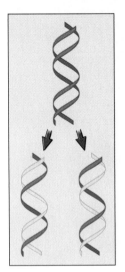

Figure 22.4 *Animated!*
Semiconservative nature of DNA replication. The original two-stranded DNA molecule is shown in blue. A new strand (yellow) is assembled on each parent strand.

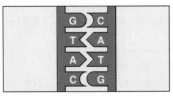

a Parent DNA molecule; two complementary strands of base-paired nucleotides.

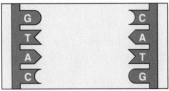

b Replication begins; the two strands unwind and separate from each other at specific sites along the length of the DNA molecule.

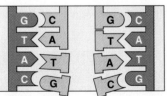

c Each "old" strand serves as a structural pattern (a template) for the addition of bases according to the base-pairing rule.

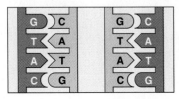

d Bases positioned on each old strand are joined together into a "new" strand. Each half-old, half-new DNA molecule is just like the parent molecule.

Figure 22.6 shows two common kinds of gene mutations. Figure 22.6*b* diagrams a **base-pair substitution**, in which the wrong nucleotide is paired with an exposed base while DNA is being replicated. Proofreading enzymes may fix the error. But if they don't, a mutation will be established in the DNA in the next round of replication. As a result of this mutation, one amino acid might replace another during protein synthesis. This is what has happened in people who have sickle-cell anemia (Section 20.6). Figure 22.6*b* shows a **deletion**, in which a base has been lost.

In an **expansion mutation**, a nucleotide sequence is repeated over and over, sometimes many hundreds of times. Expansion mutations cause several genetic disorders, including Huntington disease and **fragile X syndrome** (Figure 22.7*a*), in which the brain does not develop properly.

Mutations also can result when segments of DNA called **transposable elements** move around. These bits of DNA can move from one location to another in the same DNA molecule or in a different one. The DNA of a human diploid cell typically contains hundreds of thousands of copies of one transposable element (called Alu). When it is inserted in a particular location, a genetic disorder called **neurofibromatosis** results (Figure 22.7*b*).

While mutations can occur in the DNA in any cell, they are inherited only when they take place in germ cells that give rise to gametes. Whether a mutation turns out to be harmful, neutral, or beneficial depends on a variety of factors, including how the resulting protein affects body functions.

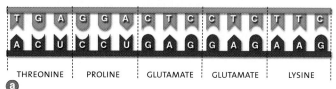
part of DNA template

mRNA transcribed from DNA

THREONINE PROLINE GLUTAMATE GLUTAMATE LYSINE

a resulting amino acid sequence

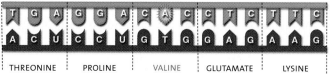
base substitution in DNA

altered mRNA

THREONINE PROLINE VALINE GLUTAMATE LYSINE

b altered amino acid sequence

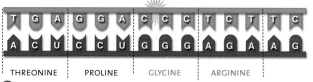
deletion in DNA

altered mRNA

THREONINE PROLINE GLYCINE ARGININE

c altered amino acid sequence

Figure 22.6 *Animated!* Examples of gene mutation. (**a**) Part of a gene (blue), the mRNA (brown), and the specified amino acid sequence. (**b**) A base-pair substitution. (**c**) A deletion in DNA.

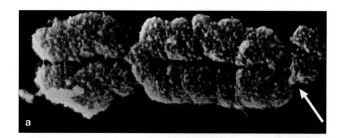

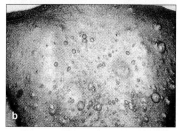

Figure 22.7 (**a**) A human chromosome showing the constriction that occurs with fragile X syndrome (arrow). (**b**) Soft skin tumors of a person with neurofibromatosis.

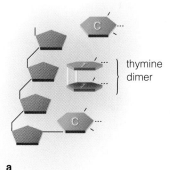
thymine dimer

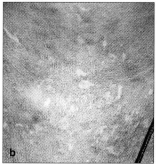

Figure 22.5 (**a**) Sketch of a thymine dimer. Covalent bonds have formed between two thymines, and the two nucleotides to which the thymines belong form an abnormal bulky structure that may interfere with replication of the DNA molecule. (**b**) Skin lesions that are typical of xeroderma pigmentosum.

DNA replication is semiconservative. After the double helix unwinds, each parent strand stays intact and enzymes assemble a new, complementary strand on it.

Enzymes involved in replication also may repair damage in a DNA molecule.

A gene mutation is an alteration in one or more bases in the nucleotide sequence of DNA.

A mutation may have harmful, neutral, or beneficial effects on body structures and functions.

22.3 DNA into RNA—The First Step in Making Proteins

LINKS TO SECTIONS 2.11, 2.13, AND 15.2

How do you get from DNA's bases to a protein? The answer starts with the sequence of bases.

The path from genes to proteins involves two processes, called transcription and translation. In both, molecules of ribonucleic acid, or **RNA**, have crucial roles. Most often, RNA consists of a single strand. Structurally, it is much like a strand of DNA. Its nucleotides each consist of a sugar (ribose), a phosphate group, and a nitrogen-containing base. However, its bases are adenine, cytosine, guanine, and **uracil**, not thymine. The differences between DNA and RNA are summarized here:

	DNA	RNA
Sugar:	deoxyribose	ribose
Bases:	adenine, cytosine, guanine, thymine	adenine, cytosine, guanine, uracil

Like the thymine in DNA, the uracil in RNA base-pairs with adenine.

In **transcription**, molecules of RNA are assembled on DNA templates in the nucleus. In **translation**, which you'll read about in Section 22.5, RNA molecules move from the nucleus into the cytoplasm, where they in turn become templates for assembling polypeptide chains. When translation is complete, one or more polypeptide chains are folded into protein molecules. Said another way, DNA guides the synthesis of RNA, then RNA guides the synthesis of proteins.

Genes are transcribed into three types of RNA:

ribosomal RNA (rRNA) — a nucleic acid chain that combines with certain proteins to form a ribosome, a structure on which a polypeptide chain is assembled

messenger RNA (mRNA) — a linear sequence of nucleotides that carries protein-building instructions; this "code" is delivered to the ribosome for translation into a polypeptide chain

transfer RNA (tRNA) — another nucleic acid chain that can pick up a specific amino acid and pair with an mRNA code word for that amino acid

An important point to remember is that only mRNA eventually is translated into a protein. The other two types of RNA operate in the process of translation.

IN TRANSCRIPTION, DNA IS DECODED INTO RNA

In transcription, a strand of RNA is assembled on a DNA template according to the base-pairing rules:

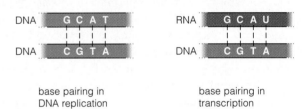

base pairing in DNA replication

base pairing in transcription

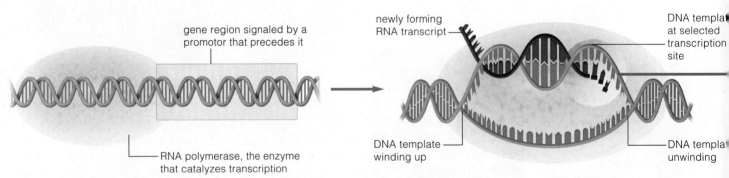

gene region signaled by a promotor that precedes it

RNA polymerase, the enzyme that catalyzes transcription

newly forming RNA transcript

DNA template at selected transcription site

DNA template winding up

DNA template unwinding

a RNA polymerase initiates transcription at a promoter in DNA. After binding to a promoter, RNA polymerases recognize a base sequence in DNA as a template for making a strand of RNA from free ribonucleotides, which have the bases adenine, cytosine, guanine, and uracil.

b During transcription, the DNA double helix becomes unwound in front of the RNA polymerase. Short lengths of the newly forming RNA strand briefly wind up with its DNA template strand. New stretches of RNA unwind from the template (and the two DNA strands wind up again).

Figure 22.8 *Animated!* Gene transcription. In this process, an mRNA molecule is assembled on a DNA template. The sketch in (**a**) shows a gene in part of a DNA double helix. The base sequence of one of the two nucleotide strands is used as the template. (**b–d**) Transcribing that gene results in a molecule of mRNA.

Transcription takes place in the cell nucleus, but it is *not* the same as DNA replication. One key difference is that only the gene serves as the template—not the whole DNA strand. A second difference is that enzymes called **RNA polymerases** are involved. Also, transcription produces a single-stranded molecule, not one with two strands.

Transcription starts at a **promoter**—a sequence of bases that signals the start of a gene. Proteins help position an RNA polymerase on the DNA so that it binds with the promoter. As transcription starts, a nucleotide "cap" is added to the beginning of the mRNA for protection. This capped end, which is designated 5', is also where the mRNA will bind to a ribosome when the time comes for translation. The trailing end of the forming mRNA molecule is designated 3'.

As RNA polymerase moves along the DNA, it joins nucleotides together (Figure 22.8). When it reaches a termination sequence of bases, the RNA transcript is released. A protective "tail" is added to its 3' end.

The new RNA transcript is an unfinished molecule. It must be modified before its protein-building instructions can be used. For example, when many human genes are transcribed, this "pre-mRNA" contains sections called **introns**—in some cases including as many as 100,000 nucleotides! Introns may be a sort of genetic gibberish; researchers have not discovered any that code for proteins.

All new mRNA transcripts also contain regions called exons. Unlike introns, **exons** are the nucleotide sequences that carry DNA's vital protein-building instructions. Before an mRNA leaves the nucleus, its introns are snipped out and its exons are spliced together. The result is a mature mRNA that is ready to enter the cell cytoplasm and be translated into a protein.

GENE TRANSCRIPTION CAN BE TURNED ON OR OFF

Most of the cells of your body carry the same genes. Many of those genes carry instructions for synthesizing proteins that are essential to any cell's structure and functioning. Yet each type of cell also uses a small subset of genes in specialized ways. For example, every cell carries the genes for hemoglobin, but only the precursors of red blood cells activate those genes. Each cell determines which genes are active and which gene products appear, when, and in what amounts. Some genes might be switched on and off throughout a person's life. Other genes might be turned on only in certain cells and only at certain times.

Genes are regulated by molecules that interact with DNA, RNA, or other substances. For example, **regulatory proteins** speed up or halt transcription. Some regulatory proteins also may bind with noncoding DNA sequences, triggering the transcription of a neighboring gene (or shutting it down). This is how steroid hormones act. As described in Section 15.2, once molecules of steroid hormones such as estrogen and testosterone are inside a cell, they enter the nucleus. There they bind to receptor proteins. Each hormone-receptor complex then can bind to a DNA sequence that acts as a regulator. It activates the promoter of a gene (or genes), which then is transcribed.

Protein synthesis has two steps, transcription and translation.

In transcription, a sequence of bases in one strand of a DNA molecule is the template for assembling a strand of RNA. The RNA strand is built according to base-pairing rules (adenine pairs with uracil, cytosine with guanine).

Before leaving the nucleus, new RNA transcripts are modified into their final form.

Genes are activated or suppressed in different ways to produce specialized cell structures or functions.

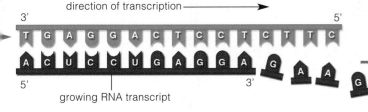

direction of transcription ⟶

3'   5'

growing RNA transcript

c In the gene region, RNA polymerase catalyzed the covalent bonding of ribonucleotides to one another to form an RNA strand. The base sequence in the new strand is complementary to the exposed bases on the DNA as a template. Many other proteins assist in transcription.

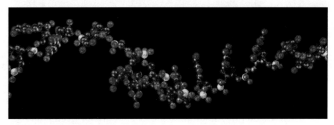

d At the end of the gene region, the last stretch of the new transcript is unwound and released from the DNA template. Shown below it is a model for a transcribed strand of RNA.

22.4 Reading the Genetic Code

LINK TO SECTION 2.11

Like a DNA strand, an mRNA molecule is a linear sequence of nucleotides. What are the protein-building "words" encoded in that sequence?

CODONS ARE mRNA "WORDS" FOR BUILDING PROTEINS

Each "word" in the mRNA instructions for building a protein is a set of three nucleotide bases that are "read" by enzymes. These base triplets are called **codons**. In Figure 22.9 you can see how the order of different codons in an mRNA molecule determines the order of amino acids added to the sequence in a growing polypeptide chain. There are sixty-four kinds of codons (Figure 22.10). Together they are the **genetic code**. The code is each cell's basic instructions for making proteins.

Most of the twenty kinds of amino acids can be "ordered up" by more than one **start codon**. (For example, glutamate corresponds to the code words GAA *or* GAG.) The codon AUG also establishes the reading frame for translation. That is, ribosomes start their "three-bases-at-a-time" selections at an AUG that is the start signal in an mRNA strand. Three **stop codons** (UAA, UAG, and UGA) signal ribosomes to stop adding amino acids to the polypeptide chain.

FIRST BASE	Amino acids that correspond to base triplets:				THIRD BASE
	SECOND BASE OF A CODON				
	U	C	A	G	
U	phenylalanine	serine	tyrosine	cysteine	U
	phenylalanine	serine	tyrosine	cysteine	C
	leucine	serine	STOP	STOP	A
	leucine	serine	STOP	tryptophan	G
C	leucine	proline	histidine	arginine	U
	leucine	proline	histidine	arginine	C
	leucine	proline	glutamine	arginine	A
	leucine	proline	glutamine	arginine	G
A	isoleucine	threonine	asparagine	serine	U
	isoleucine	threonine	asparagine	serine	C
	isoleucine	threonine	lysine	arginine	A
	methionine (or START)	threonine	lysine	arginine	G
G	valine	alanine	aspartate	glycine	U
	valine	alanine	aspartate	glycine	C
	valine	alanine	glutamate	glycine	A
	valine	alanine	glutamate	glycine	G

Figure 22.10 *Animated!* The genetic code. The codons in mRNA are nucleotide bases, "read" in blocks of three. Sixty-one of these base triplets correspond to specific amino acids. Three others serve as signals that stop translation. The left column of the diagram shows the first of the three nucleotides in each codon in mRNA. The middle columns show the second nucleotide. The right column shows the third. Reading from left to right, for instance, the triplet U G G corresponds to tryptophan. Both U U U and U U C correspond to phenylalanine.

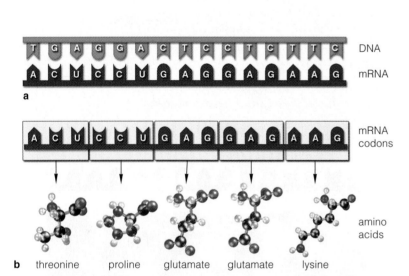

Figure 22.9 The steps from genes to proteins. (**a**) This region of a DNA double helix was unwound during transcription. (**b**) Bases on one strand were a template for assembling an mRNA strand. In the mRNA, blocks of three nucleotide bases equaled one codon. Each codon called for one amino acid in a polypeptide chain.

tRNA TRANSLATES THE GENETIC CODE

There are unattached amino acids and tRNA molecules in the cell cytoplasm. Each tRNA has a "hook" site where it can attach to a specific amino acid. As shown in Figure 22.11, a tRNA also has an **anticodon**, a nucleotide triplet that can base-pair with codons. When a series of tRNAs bind to a series of codons, the matching up of codons and anticodons automatically lines up the amino acids attached to tRNAs in the order specified by mRNA.

GENETIC INSTRUCTIONS IN DNA

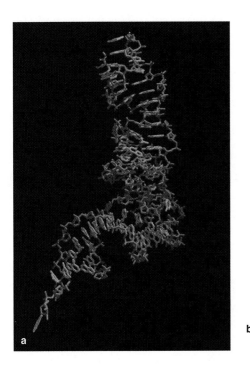

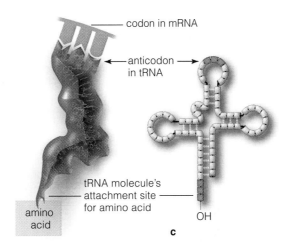

codon in mRNA

anticodon in tRNA

tRNA molecule's attachment site for amino acid

amino acid

OH

b

c

Figure 22.11 *Animated!* (**a**) A three-dimensional model of one type of tRNA molecule. (**b**) Simplified model of tRNA that is used in later illustrations. The "hook" at the lower end is the site where a specific amino acid can be attached. (**c**) Structural features that are found in all tRNAs.

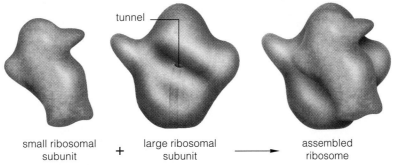

tunnel

Figure 22.12 *Animated!* Model for the small and large subunits of a ribosome. Polypeptide chains are assembled on the small subunit's platform. Newly forming chains may move through the large subunit's tunnel.

small ribosomal subunit + large ribosomal subunit ⟶ assembled ribosome

A cell has more than sixty kinds of codons but fewer kinds of tRNAs. How do the needed match-ups take place? Remember that by the base-pairing rules, adenine pairs with uracil, and cytosine with guanine. However, for codon–anticodon interactions, the rules loosen up for the third base. For example, CCU, CCC, CCA, and CCG all specify the amino acid proline but require only two tRNAs. This kind of flexibility in the codon–anticodon pairing at the third base is called the *wobble effect*.

tRNAS ARE RIBOSOME BUILDING BLOCKS

A cell's tRNAs interact with mRNA at binding sites on the surface of ribosomes. Each ribosome has two subunits (Figure 22.12). The subunits are built in the nucleus from rRNA and proteins; then they are shipped to the cytoplasm. There they will combine into ribosomes during translation, as described in the next section.

Instructions for building proteins are encoded in the nucleotide sequence of DNA and mRNA. The genetic code is a set of sixty-four base triplets (nucleotide bases, read in blocks of three). A codon is a base triplet in mRNA.

Different combinations of codons specify the amino acid sequences of polypeptide chains.

Specific tRNA anticodons bind specific codons in mRNA. In this way amino acids line up in the order specified by mRNA. Thus tRNAs translate mRNA into a corresponding sequence of amino acids.

mRNAs carry protein-building instructions from a cell's DNA to its cytoplasm. rRNAs are components of ribosomes, where amino acids are assembled into polypeptide chains.

22.5 Translating the Genetic Code into Protein

The protein-building instructions built into mRNA transcripts are translated into proteins at ribosomes in the cytoplasm.

TRANSLATION HAS THREE STAGES

The translation phase of protein synthesis has three stages, called initiation, elongation, and termination. Here we describe the basics of this complex process.

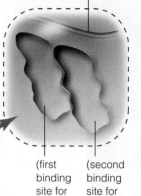

binding site for mRNA

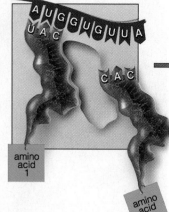

(first binding site for tRNA)　(second binding site for tRNA)

amino acid 1

amino acid 2

Elongation

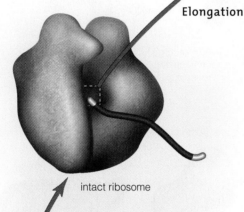

c As the final step of initiation a large ribosome subunit joins with the small one. Once this structure has formed, chain *elongation*—the second stage of translation—can get under way.

intact ribosome

Initiation

b *Initiation*, the first stage of translating the mRNA transcript, is about to begin. An initiator tRNA (one that can start this stage) is loaded onto a platform of a small ribosome subunit. The small subunit / tRNA complex attaches to the 5' end of the mRNA. It moves along the mRNA and "scans" it for an AUG start codon.

a A mature mRNA transcript leaves the nucleus by passing through pores across the nuclear envelope. It enters the cytoplasm, which contains pools of many free amino acids, tRNAs, and ribosome subunits.

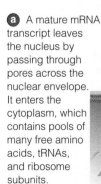

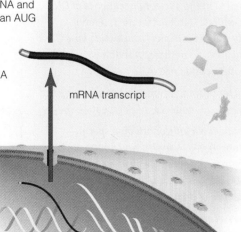

mRNA transcript

d This close-up of the small ribosome subunit's platform shows the relative positions of binding sites for an mRNA transcript and for tRNAs that deliver amino acids to the intact ribosome.

e The initiator tRNA has become positioned in the first tRNA binding site on the ribosome platform. Its anticodon matches up with the start codon (AUG) of the mRNA, which also has become positioned in *its* binding site. Another tRNA is about to move into the platform's second tRNA binding site. It is one that can bind with the codon following the start codon.

During *initiation*, a tRNA (which can start transcription) and an mRNA are both loaded onto a ribosome. First, the initiator tRNA binds with the small ribosome subunit. AUG, the start codon for the mRNA transcript, matches up with this tRNA's anticodon. The AUG also binds with the small subunit. Next, a large ribosome subunit binds with the small subunit. When joined together in this way, the three elements form an initiation complex (Figure 22.13*b*). Now the next stage can begin.

In the *elongation* stage of translation, a polypeptide chain forms as the mRNA strand passes between the ribosome subunits, like a thread moving through the eye of a needle. Some proteins in the ribosome are enzymes. They join amino acids together in the sequence dictated by the codon sequence in the mRNA molecule. Figure 22.13*f–i* shows that a peptide bond forms between the most recently attached amino acid and the next amino acid being delivered to the ribosome. (Section 2.11 explains how a peptide bond forms.)

During the last stage of translation, *termination*, a stop codon in the mRNA moves onto the ribosome platform, and no tRNA has a corresponding anticodon. Now proteins called release factors bind to the ribosome. They trigger enzyme action that detaches the mRNA *and* the new chain from the ribosome (Figure 22.13*j–l*).

Figure 22.13 *Animated!* Translation, the second step of protein synthesis.

MAKING PROTEINS

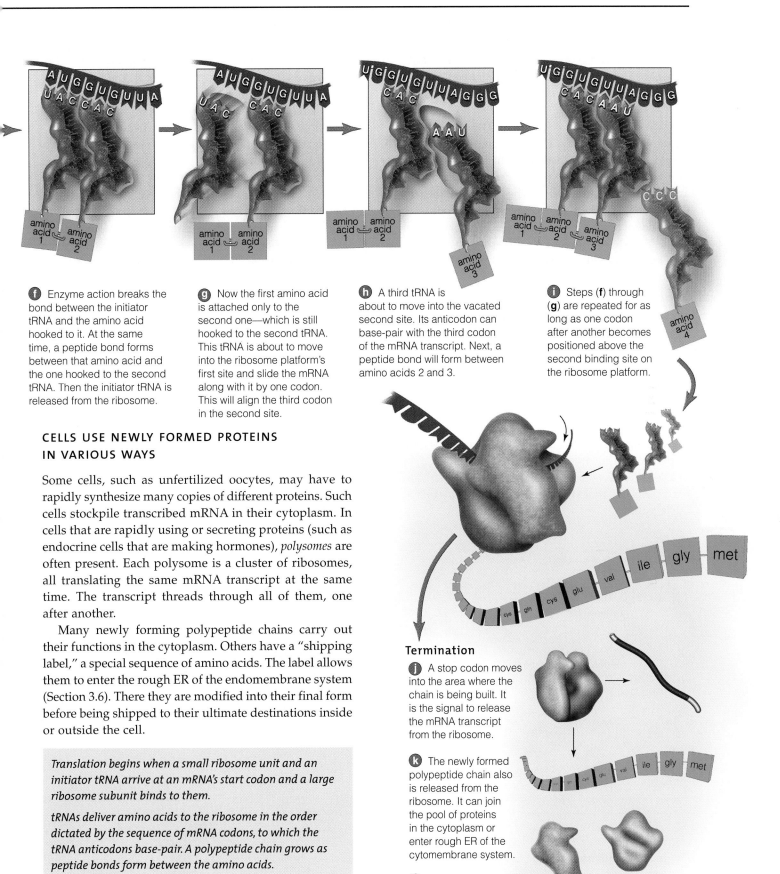

(f) Enzyme action breaks the bond between the initiator tRNA and the amino acid hooked to it. At the same time, a peptide bond forms between that amino acid and the one hooked to the second tRNA. Then the initiator tRNA is released from the ribosome.

(g) Now the first amino acid is attached only to the second one—which is still hooked to the second tRNA. This tRNA is about to move into the ribosome platform's first site and slide the mRNA along with it by one codon. This will align the third codon in the second site.

(h) A third tRNA is about to move into the vacated second site. Its anticodon can base-pair with the third codon of the mRNA transcript. Next, a peptide bond will form between amino acids 2 and 3.

(i) Steps (**f**) through (**g**) are repeated for as long as one codon after another becomes positioned above the second binding site on the ribosome platform.

CELLS USE NEWLY FORMED PROTEINS IN VARIOUS WAYS

Some cells, such as unfertilized oocytes, may have to rapidly synthesize many copies of different proteins. Such cells stockpile transcribed mRNA in their cytoplasm. In cells that are rapidly using or secreting proteins (such as endocrine cells that are making hormones), *polysomes* are often present. Each polysome is a cluster of ribosomes, all translating the same mRNA transcript at the same time. The transcript threads through all of them, one after another.

Many newly forming polypeptide chains carry out their functions in the cytoplasm. Others have a "shipping label," a special sequence of amino acids. The label allows them to enter the rough ER of the endomembrane system (Section 3.6). There they are modified into their final form before being shipped to their ultimate destinations inside or outside the cell.

> *Translation begins when a small ribosome unit and an initiator tRNA arrive at an mRNA's start codon and a large ribosome subunit binds to them.*
>
> *tRNAs deliver amino acids to the ribosome in the order dictated by the sequence of mRNA codons, to which the tRNA anticodons base-pair. A polypeptide chain grows as peptide bonds form between the amino acids.*
>
> *Translation ends when a stop codon triggers events that cause the chain and the mRNA to detach from the ribosome.*

Termination

(j) A stop codon moves into the area where the chain is being built. It is the signal to release the mRNA transcript from the ribosome.

(k) The newly formed polypeptide chain also is released from the ribosome. It can join the pool of proteins in the cytoplasm or enter rough ER of the cytomembrane system.

(l) The two ribosome subunits separate.

22.6 Tools for "Engineering" Genes

LINKS TO
SECTIONS
18.3 AND 19.1

Nature has conducted countless genetic experiments through mutation and other events. We humans now can use advanced technology to change gene-based traits.

Today researchers use **recombinant DNA technology** to create genetic changes. They can cut and splice DNA from different species, then insert the modified molecules into bacteria or other types of cells that can replicate genetic material and divide. The cells copy the foreign DNA along with their own. Copying soon produces large quantities of *recombinant DNA* molecules. This technology also is the basis of **genetic engineering**, in which genes are isolated, modified, and inserted back into the same organism or into a different one.

ENZYMES AND PLASMIDS FROM BACTERIA ARE BASIC TOOLS

Recombinant DNA technology depends on the genetic workings of bacteria. You may recall from Chapter 18 that many bacteria have plasmids, which are small circular molecules of extra DNA that contain a few genes. The bacterium's replication enzymes can copy plasmid DNA. Bacteria also have *restriction enzymes*—enzymes that can recognize and cut apart specific short sequences of bases in DNA. Today, plasmids and restriction enzymes are basic parts of a tool kit for doing genetic *recombination* in the laboratory.

Many restriction enzymes make staggered cuts. The cuts leave single-stranded "tails" on the end of a DNA fragment. Depending on the particular molecule being cut, the resulting fragments may be tens of thousands of bases long. That is long enough to be useful for studying the organization of a **genome**, which is all the DNA in a haploid set of a given species' chromosomes. The human genome is about 2.9 billion base pairs long.

DNA fragments with staggered cuts have so-called sticky ends. "Sticky" means that a restriction fragment's single-stranded tail can base-pair with a complementary tail of any other DNA fragment or molecule cut by the same restriction enzyme. If you mix together some DNA fragments cut by the same restriction enzyme, the sticky ends of any two fragments that have complementary base sequences will base-pair and form a recombinant DNA molecule. Then another enzyme seals the nicks.

Using the necessary enzymes, it's possible to insert foreign DNA into bacterial plasmids. The result is called a **DNA clone**, because when a bacterium replicates a plasmid, it makes many identical, "cloned" copies of it.

A DNA clone is sometimes called a *cloning vector*. It taxies foreign DNA into a host cell that can divide rapidly (such as a bacterium or a yeast cell). This can be the start of a cloning factory—a population of rapidly dividing descendants, all with identical copies of the foreign DNA (Figure 22.14). As they divide they amplify—make much more of—the foreign DNA.

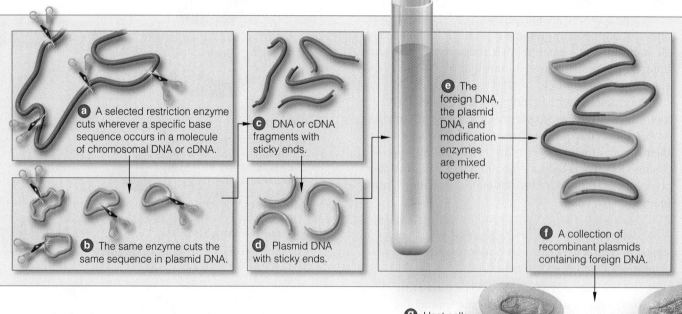

Figure 22.14 Animated! (**a–f**) Formation of recombinant DNA—in this case, a collection of chromosomal DNA fragments or cDNA that are sealed into bacterial plasmids. (**g**) Recombinant plasmids are inserted into host cells that can rapidly amplify the foreign DNA of interest.

(a) A selected restriction enzyme cuts wherever a specific base sequence occurs in a molecule of chromosomal DNA or cDNA.

(b) The same enzyme cuts the same sequence in plasmid DNA.

(c) DNA or cDNA fragments with sticky ends.

(d) Plasmid DNA with sticky ends.

(e) The foreign DNA, the plasmid DNA, and modification enzymes are mixed together.

(f) A collection of recombinant plasmids containing foreign DNA.

(g) Host cells able to divide rapidly take up recombinant plasmids.

a PCR starts with a fragment of double-stranded DNA.

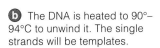

b The DNA is heated to 90°–94°C to unwind it. The single strands will be templates.

c Primers designed to base-pair with ends of the DNA strands will be mixed with the DNA.

d The mixture is cooled. The lower temperature promotes base pairing between the primers and the ends of the DNA strands.

e DNA polymerases recognize the primers as start tags. They assemble complementary sequences on the strands. This doubles the number of identical DNA fragments.

f The mixture is heated again. The higher temperature makes all of the double-stranded DNA fragments unwind.

g The mixture is cooled. The lower temperature promotes base pairing between more primers added to the mixture and the single strands.

h DNA polymerase action again doubles the number of identical DNA fragments.

The sequence of reactions just described continues. Each time, the number of DNA fragments in the mixture doubles. Billions of fragments are synthesized very quickly.

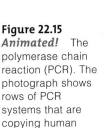

Figure 22.15
Animated! The polymerase chain reaction (PCR). The photograph shows rows of PCR systems that are copying human DNA.

THE POLYMERASE CHAIN REACTION IS A FAST WAY TO COPY DNA

The **polymerase chain reaction**, or **PCR**, is an even faster way to copy DNA. These reactions occur in test tubes, and primers get them started.

A **primer** is a manmade, short nucleotide sequence that base-pairs with any complementary sequences in DNA. The workhorses of DNA replication—the DNA polymerases—chemically recognize primers as start tags. Following a computer program, machines synthesize a primer one step at a time.

PCR also uses a DNA polymerase that is not destroyed at the elevated temperatures required to unwind a DNA double helix. (Such high temperatures will denature and destroy the activity of most DNA polymerases.)

In the lab, primers, the polymerase, DNA from an organism, and nucleotides are all mixed together. Next, the mixture is exposed to precise temperature cycles. During each temperature cycle, the two strands of all the DNA molecules in the mixture unwind from each other.

Primers line up on exposed nucleotides at the targeted site according to base-pairing rules (Figure 22.15). Each round of reactions doubles the number of DNA molecules amplified from the target site. For example, if there are 10 such molecules in the test tube, there soon will be 20, then 40, 80, 160, 320, and so on. In short order, there will be billions of copies of a target piece of DNA.

PCR can amplify samples that contain tiny amounts of DNA, and it is used in laboratories all over the world. As you'll read shortly, it can copy DNA from even a single hair follicle or a drop of blood left at a crime scene.

Restriction enzymes and modification enzymes cut apart DNA and splice it into plasmids and other cloning vectors. The recombinant plasmids can be inserted into bacteria or other rapidly dividing cells that can make multiple, identical copies of the foreign DNA.

PCR is a rapid method of amplifying DNA in test tubes.

IMPACTS, ISSUES

In spite of its dangers, ricin shows promise as a cancer-fighting toxin. The protein that inactivates ribosomes is attached to cancer-specific antibodies or other weapons. Once inside cancer cells, ricin kills them by blocking protein synthesis. One concern: the highly toxic ricin seeps out of dying cancer cells.

22.7 "Sequencing" DNA

Genes are sequences of nucleotides. One way recombinant DNA technology has helped us gain a fuller understanding of genes is through automated DNA sequencing.

Once you have cloned DNA or amplified it using PCR, you can find out the sequence of its nucleotides in just a few hours using **automated DNA sequencing**. The method uses standard and modified versions of the four nucleotides (A, T, G, and C). Each modified version has been attached to another molecule that fluoresces (lights up) a preselected color during the sequencing process. Before the reactions, all the nucleotides are mixed with millions of copies of the DNA to be sequenced, a primer, and DNA polymerase. After the DNA is separated into single strands, a series of chemical steps produce a DNA "soup" containing millions of copies of DNA fragments that are tagged with the fluorescing molecules. These can be separated by length into *sets* of fragments. Each set corresponds to one nucleotide in the entire base sequence. After several more steps—each taking place at lightning speed—the DNA sequencing machine assembles information from all the nucleotides in the sample and reveals the original DNA's sequence. Figure 22.16 shows what a printout from a DNA sequencer might look like.

For genetic engineering—that is, changing the chemical makeup of a gene—researchers fashion short stretches of DNA that have been radioactively labeled. This *probe* can be added to a mixture of DNA fragments called a **gene library**. The "library" is inserted into bacteria. Probes are designed to match up with DNA that contains the gene of interest. Once the probe has located the matching nucleotide sequence, the bacteria containing that gene can be cultured and the gene isolated for further research.

Automated DNA sequencing can determine the order of nucleotides in a DNA fragment.

A probe can be used to identify a particular gene among the many in a gene library.

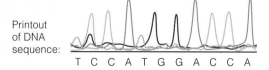

Printout of DNA sequence:

T C C A T G G A C C A

Figure 22.16 *Animated!* What part of a printout from a DNA sequencing machine might look like.

22.8 Mapping the Human Genome

In 2002, after an awesome duel of supercomputers and advanced DNA sequencing machines, competing teams of researchers in Europe and the U.S. jointly announced that they had determined the sequence of nearly all the nucleotide bases in the human genome. This rough draft of the human Book of Life already is yielding valuable insights into the genetic basis of many disorders and diseases.

Thanks to the **Human Genome Project**, we now know that the genome consists of about 2.9 billion nucleotide bases— As, Ts, Gs, and Cs. The bases are subdivided into roughly 21,500 genes, which provide all the instructions needed to build and operate the body.

GENOME MAPPING PROVIDES BASIC BIOLOGICAL INFORMATION

Why be interested in the order of the nucleotides in our DNA? To begin with, we now have key basic knowledge about our genome, such as the remarkable fact that the protein-coding portions of our genes, the exons, make up only about 1.5 percent of our DNA! The rest is so-called noncoding DNA, and over half of it appears to be in the form of repeated sequences of various types (such as the Alu repeat mentioned in Section 22.2). We cannot assume that all noncoding DNA is "junk DNA," so one of the challenges biologists now face is figuring out the possible roles of noncoding regions (Figure 22.17).

It turns out that our DNA also is sprinkled with SNPs ("snips"). Each SNP (for *single nucleotide polymorphism*)

Figure 22.17 Gene-sequencing supercomputers being used in human genome research at Celera Genomics in Maryland. The map of the human genome will have many applications, including more efficiently designed drugs and better understanding of genetic disorders.

ENGINEERING AND EXPLORING GENES

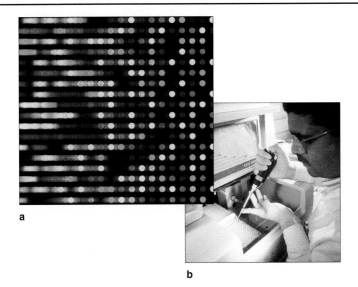

a

b

Figure 22.18 (**a**) Representation of part of a DNA chip, used to analyze thousands of genes at once. Sequences from a gene library are fixed onto a slide or microchip. Each gene's position in the array is known. Labeled probes will bind to complementary base sequences on the gene chip. Active genes glow in fluorescent light. (**b**) Cancer researcher pipetting copies of a gene into an automated sequence analyzer.

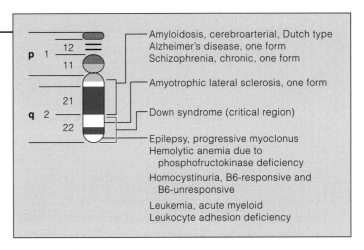

Figure 22.19 Map of human chromosome 21 showing genes correlated with specific diseases. In all, researchers have mapped 225 genes to this chromosome. The upper arm of the chromosome is marked p and the lower arm q. In this case, each arm consists of a single region, labeled 1 and 2, respectively. In other chromosomes, each arm may have two or more regions. Stains used in chromosome analysis produce a series of bands within each region. The small numbers just to the left of each arm indicate specific bands. A combination of letters and numbers indicates the chromosome region where a given gene is found; for instance, the gene for one form of amyotrophic lateral sclerosis (ALS) is 21q2.22.

is a change in one nucleotide in a sequence. It appears there are around 1.4 million SNPs in the human genome. Many result in different gene alleles—the different versions of a gene that encode slightly different traits.

DNA CHIPS HELP IDENTIFY MUTATIONS AND DIAGNOSE DISEASES

Unraveling the human genome has vast implications for human biology and medicine. One new tool is the **DNA chip**, a microarray of thousands of DNA sequences that are stamped onto a glass plate about the size of a business card (Figure 22.18). Within days the chips can pinpoint which genes are silent and which are being expressed in a body tissue, including a cancerous tumor. Such chips are already being used to home in on mutations, diagnose genetic diseases, and test how drugs or other therapies affect the expression of genes.

As new genes are identified, biologists are exploring the roles of proteins they encode. We already know how having a particular allele can set the course of diseases such as sickle-cell anemia and forms of breast cancer. Soon it will be possible to apply the deeper understanding of human genes to many more health concerns. For instance, a simple blood test might be able to provide a complete genetic profile of a person's inborn predisposition to various cancers, heart disease, asthma, and diabetes. Armed with your profile, your doctor might be able to diagnose and treat problems earlier and more effectively. Drugs could be customized for various genetic situations.

CHROMOSOME MAPPING SHOWS WHERE GENES ARE LOCATED

Genome sequencing also can identify where specific genes are located on chromosomes. For instance, we know that genes on chromosome 21 are responsible for early-onset Alzheimer's disease, some forms of epilepsy, and one type of *amyotrophic lateral sclerosis*, or ALS, a disease that destroys motor neurons (Figure 22.19). More than 60 disorders have been mapped to chromosome 14. In 2002, a consortium of public and private laboratories began a $100 million effort to correlate disorders with individual genetic differences (SNPs).

Along with the promise of genome sequencing come serious cautions. As discussed elsewhere in this book, there is the possibility that genetic profiling could lead to discrimination against people seeking employment or insurance, based solely on their genetic makeup. New issues could arise in the genetic screening of embryos, already a major ethical concern to some. Clearly, a challenging genetic future awaits us.

Researchers have determined the sequence of nucleotides in the entire human genome.

The Human Genome Project is mapping the locations of specific genes on human chromosomes. The project holds great promise for scientific and medical research.

22.9 Some Applications of Biotechnology

LINKS TO
SECTIONS
10.1 AND 21.5

Practical applications of biotechnology are coming along almost as fast as the leaps in our understanding of human genetics. They include genetic fixes for diseases and using DNA as a personal identifier.

RESEARCHERS ARE EXPLORING GENE THERAPY

There are 15,500 known human genetic disorders. Some, such as cystic fibrosis and hemophilia, develop when just one single gene mutates, producing a nonfunctional protein. Many cancers arise the same way. **Gene therapy** aims to replace mutated genes with normal ones that will encode functional proteins, or to insert genes that restore normal controls over gene activity. Scientists around the world are exploring these possibilities.

GENES CAN BE INSERTED TWO WAYS

The size of a gene (how many base pairs it has) helps determine how it might be inserted into a host cell. Smaller genes can be carried into animal cells by viruses; larger ones must enter a host cell some other way.

In *transformation*, cells cultured in the laboratory are exposed to DNA that contains a gene of interest, and some of the foreign DNA may become integrated into the host cell's genome. Exposing the host cells to a weak electric current seems to help. Even so, sometimes only one cell in 10 million takes up a new gene.

In *transfection*, a gene is inserted into a virus (often, a retrovirus). The first step is to remove from the virus its genetic instructions that would allow it to replicate and cause disease. In their place goes the gene to be transferred. Next, the virus is allowed to infect target cells (Figure 22.20). Once the virus is inside the host cell, the desired foreign DNA usually becomes integrated into the host cell's DNA.

Transfection can be used only with genes that are expressed in the tissues into which the new DNA will be inserted. In addition, many introduced genes turn off within a few days or weeks. And a few patients have died from complications of the procedure, which raises serious safety questions.

RESULTS OF GENE THERAPY HAVE BEEN MIXED

The first federally approved gene therapy test on humans began in the early 1990s. It aimed to replace a defective gene that causes one type of **severe combined immune deficiency**, called SCID-X1. In this disorder, stem cells in the affected person's bone marrow fail to make the immune system's infection-fighting lymphocytes (Section 10.1). Affected children (called "bubble babies") must

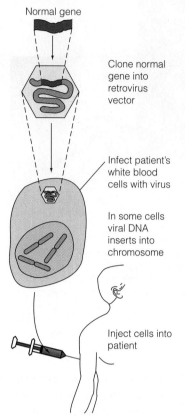

Figure 22.20 One method of gene therapy. Here, white blood cells are removed from a patient with a genetic disorder. Next, a retrovirus is used as a vector to insert a normal gene into the DNA. After the normal gene begins to produce its protein, the cells are placed back into the patient. If enough of the normal protein is present in the patient, the disorder's symptoms diminish. Because the genetically altered blood cells live but a few months, however, the procedure must be repeated regularly.

live in germ-free isolation tents. Several years ago, a retrovirus was used to insert copies of a normal allele into stem cells taken from the bone marrow of eleven children with the disorder. Then the genetically modified stem cells were infused back into the marrow, which began to make lymphocytes. Some months later, seven of the treated children, including the boy shown in Figure 22.21, left their isolation tents. Their immune system had kicked in. Since that early success, however, several of the children have developed leukemia.

Cystic fibrosis has been another target. The idea is to introduce normal copies of the defective CFTR gene into cells of the respiratory system, using a viral vector in a nasal spray. In the trials thus far, only about 5 percent of affected cells have taken up the normal gene. One major problem has been that the retrovirus vectors used are not effective enough at delivering the new genes to cells that need them.

At this writing, gene therapy has been most successful in treating cancer. Early trials have targeted the deadly skin cancer malignant melanoma, leukemia, a fast-growing form of lung cancer, and cancers of the brain, ovaries, and other organs. In some approaches, tumor cells are first removed from a patient and grown in the laboratory. Genes for an interleukin (which helps activate white blood cells of the immune system) are then introduced

Figure 22.21 A boy who was born with SCID-X1. His immune system was not properly developed, so he couldn't fight infections. A gene transfer freed him from life in a germ-free isolation tent. However, it also may have increased his risk of developing leukemia.

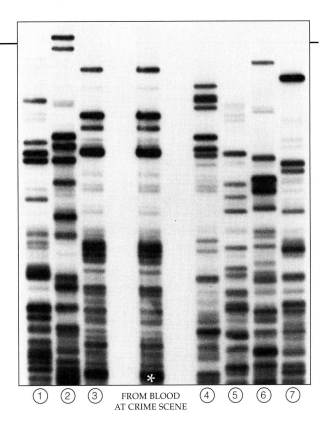

① ② ③   FROM BLOOD AT CRIME SCENE   ④ ⑤ ⑥ ⑦

Figure 22.22 *Animated!* DNA fingerprint from a crime-scene bloodstain, with DNA fingerprints from blood of seven suspects (circled numbers). Which one of the seven was a match?

into the cells, and the cells are returned to the body. In theory, interleukins produced by the tumor cells may act as "suicide tags" that stimulate T cells of the immune system to recognize cancerous cells and attack them. An exciting variation on this theme involves structures called "lipoplexes," laboratory-made packets in which a plasmid is encased in a lipid coat that helps it gain entry into a cell. The plasmid carries a gene encoding a protein marker that can trigger an immune system attack on cancer cells. Several dozen melanoma patients have been treated with lipoplex therapy, with encouraging results.

Other trials are adding the gene for a powerful anti-tumor agent to lymphocytes that home in on tumors. At present, however, gene therapy is still experimental, extremely costly, and available to only a few. Even as the methods are refined, it is unlikely that gene therapy will soon become a widely used approach for treating and curing disease.

GENETIC ANALYSIS ALSO IS USED TO READ DNA FINGERPRINTS

As you know, each of us has a unique set of fingerprints. Likewise, no two people (other than identical twins) have exactly the same sequence of bases in their DNA. Thus each of us also has a **DNA fingerprint**—a unique set of certain DNA fragments that we have inherited from our parents in a Mendelian pattern. DNA finger-prints are extremely accurate—so accurate, in fact, that they can easily distinguish between tissues taken from full siblings.

How does DNA fingerprinting work? More than 99 percent of human DNA is exactly the same in all people, regardless of race and gender. Thus DNA fingerprinting focuses only on the part that tends to differ from one

person to the next. Throughout the human genome are tandem repeats—short regions of repeated DNA—that are very different from person to person. ("Tandem" means "two together" and refers to the fact that the repeated DNA region is double-stranded.) Each person has a unique combination of repeats.

Figure 22.22 shows some tandem-repeat DNA fragments that were separated by a procedure called gel electrophoresis (see Figure 22.3). They were taken from blood collected at a crime scene. See how different the DNA fingerprints are—and how only one exactly matches the pattern from the crime scene? Forensic scientists use DNA fingerprinting to identify criminals and crime victims from the DNA in blood, semen, or bits of tissue left at a crime scene.

The variation in tandem repeats in DNA also can be detected as *restriction fragment length polymorphisms*, or **RFLPs** ("riff-lips"). These are DNA fragments of different sizes that have been cut out of DNA by restriction enzymes. In this chapter's *Explore on Your Own*, you'll read about a controversy in which RFLPs have been used in an effort to trace a famous family history.

In gene therapy, one or more normal genes are inserted into body cells to correct a genetic defect or enhance the activity of specific genes.

DNA fingerprinting is another application of biotechnology.

22.10 Issues for a Biotechnological Society

Some people contend that no matter what the species of organism, DNA should never be altered. But nature alters DNA all the time (by mutations). We humans have been doing genetic experiments for centuries by manipulating matings to create modern crop plants and new breeds of dogs, cats, birds, cattle, and other animals. The real issue is how to bring about beneficial changes without harming ourselves or the environment.

In recent years, a host of concerns have been brought to the table for discussion. Some hotly debated topics include the following:

• Most recombinant bacteria or viruses are altered in ways that will prevent them from reproducing outside a laboratory. In theory, though, transgenic (Section 22.11) bacteria or viruses could mutate, possibly becoming new pathogens in the process.

• Bioengineered plants (Section 22.11) could escape from test plots and pose an ecological risk by becoming "superweeds" resistant to herbicides and other controls.

• Crop plants with added insect resistance could set in motion an evolutionary seesaw in which new, even more formidable, insect pests arise.

• Transgenic fish that feed voraciously and grow rapidly could displace natural species, wrecking the delicate ecological balance in streams and lakes.

Experts point out that some of these possibilities—such as transgenic bacteria or viruses becoming pathogenic strains or bioengineered plants becoming "superweeds"—are not very likely. However, there are documented cases of engineered plant genes turning up in wild plants.

Genetically modified ("GM") food is a particularly hot topic these days. In the United States, it is probably

Figure 22.24 Bioremediation, the removal of oil by natural biodegradation, is accelerated by the application of fertilizers that provide oil-eating bacteria with nitrogen and phosphorus nutrients to augment the oil's natural carbon content. The rocks pictured here were treated with bioremediation to repair damage from the 1989 oil spill from the Exxon Valdez tanker in Alaska.

impossible to avoid such foods—a whopping 38 percent of soybeans and 25 percent of corn crops are engineered to withstand weedkillers or make their own pesticides. For years, the corn and soybeans have found their way into breakfast cereals, soy sauce, vegetable oils, beer, soft drinks, and lots of other food products. They also are fed to farm animals.

In Europe, public resistance to genetically modified food is strong. Many people speak out against what the tabloids call "Frankenfoods." Protesters routinely vandalize crops (Figure 22.23). Worries abound that such foods may be more toxic, less nutritious, and promote antibiotic resistance.

On the other hand, biotechnologists envision a new **Green Revolution** in which genetically modified food plants can help feed the hungry. They argue that growing such plants can hold down food production costs, reduce the use of polluting agricultural chemicals, improve crop yields, and offer enhanced flavor, nutrition, salt tolerance, and drought tolerance. Already, some modified plants can safely make useful proteins, including hemoglobin. Genetically enhanced "oil-eating" bacteria have been used to help clean up oil spills, an example of what is called *bioremediation* (Figure 22.24).

Meanwhile, there have been threats of boycotts of genetically modified food exports from the United States—a $50 billion enterprise annually. Even in the U.S., some upscale restaurants and "gourmet" food producers have banned such foods from their kitchens. On the other hand, restrictions on genetic engineering will have a major impact on U.S. agriculture, an impact that will trickle down to what we eat and how much we pay for it.

You can easily look up scientific research on this issue and form your own opinions. That's preferable to being swayed by catchy phrases (such as Frankenfood) or by biased reports from chemical manufacturers. You might start with "Politics, Misinformation, and Biotechnology," a *Science* article (February 18, 2000). For other views on biotechnology's usefulness and safety, you can check out www.actionbioscience.org.

Figure 22.23 (Left) A malnourished child in southeast Ethiopia. (Right) British protesters uprooting genetically modified crop plants.

ENGINEERING AND EXPLORING GENES

22.11 Engineering Bacteria, Animals, and Plants

Any genetically engineered organism that carries one or more foreign genes is transgenic. Here we look briefly at just a few examples.

Bacteria were the first organisms to be bioengineered. Plasmids in modified bacteria can carry a range of human genes, which are expressed to produce large quantities of useful human proteins. Many of these proteins, such as human growth hormone, once were available only in tiny amounts and were costly because they had to be chemically extracted from endocrine tissues. Other human proteins produced today by bacteria include insulin and interferons.

Many types of animal cells can be "micro-injected" with foreign DNA. For instance, when researchers introduced the gene for human growth hormone into mice, the result was the "super mouse" shown in Figure 22.25a. Recombinant DNA technology also has been used to transfer human and cow growth hormones into pigs, which then grow much faster. Transgenic goats are being used to produce CFTR protein (used to treat cystic fibrosis), as well as tissue plasminogen activator, or tPA, which dissolves blood clots and is given to some heart attack and stroke patients. Both of these medically useful proteins show up in the goats' milk.

People with hemophilia A now can obtain the needed blood-clotting factor VIII from a drug that is produced by hamster ovary cells in which genes for human factor VIII have been inserted. Factor VIII produced in this way eliminates the need to obtain it from human blood, and thus the risk of transmitting blood-borne diseases.

Plants have been intriguing genetic engineers for a long time. Researchers now routinely grow crop plants and many other plant species from cells cultured in the laboratory. They use a variety of methods to pinpoint genes that confer desirable traits, such as resistance to salt, a pathogen, or a herbicide (Figure 22.25b). Later, whole plants with the trait can be regenerated from the cultured cells.

> *Applications of genetic engineering include efforts to develop transgenic animals or animal cells capable of producing medically useful substances.*
>
> *Genetic engineering of plants may produce crop plants with new, beneficial traits.*

Figure 22.25 (**a**) Mouse littermates. The larger mouse grew from a fertilized egg into which the gene for human growth hormone had been inserted. (**b**) Aspen seedlings genetically altered for a higher ratio of cellulose, compared to the control plant at left. Wood from such trees might make it easier to manufacture paper and some clean-burning fuels, such as ethanol. The altered trees also grew 25–30 percent faster than unaltered ones. (**c**) Some of the USDA-approved crop plants.

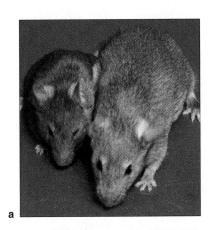

a

b

USDA-Approved Crop Plant	Modified Trait
Tomato, potato, corn, rice, sugar beet; canola bean; cotton, flax	Resistance to the 218 kilograms of weed-killing herbicides applied annually on United States croplands
Potato, squash, papaya	More resistance to viral, bacterial, and fungal attacks
Tomato	Delayed ripening; easier to ship, with less bruising
Corn, chicory	Sterility; unable to cross-pollinate wild stocks

c

Summary

Section 22.1 Chemically, a gene is a sequence of nucleotide bases in DNA. The bases are adenine, thymine, guanine, and cytosine (A, T, G, and C). The nucleotide sequence of most genes codes for the sequence of amino acids in a polypeptide chain.

Section 22.2 A DNA molecule consists of two strands of nucleotides twisted together in a double helix. DNA duplicates by semiconservative replication. One strand in each new DNA molecule is from the parent molecule, and one strand is new. DNA polymerases and other enzymes unwind the existing DNA molecule, keep the strands separate, and assemble a new strand on each one.

A gene mutation is a change in the nucleotide sequence of DNA. Mutations may be harmful, neutral, or beneficial. Mutations in germ cells (cells that produce gametes) can be passed to the next generation.

 Investigate the effects of a gene mutation.

Section 22.3 The path from genes to proteins requires steps called transcription and translation (Figure 22.26). In transcription, DNA provides the instructions for synthesizing RNA (using nucleotides in the cell). The double-stranded DNA is unwound at a gene region, then RNA polymerases use the exposed bases as a template to build a corresponding strand of RNA. Transcription starts at a promoter, and base-pairing rules govern which bases pair up. In RNA, guanine pairs with cytosine, and uracil (not thymine) pairs with adenine.

DNA encodes three kinds of RNA molecules. Messenger RNA (mRNA) carries instructions for building proteins. Ribosomal RNA (rRNA) forms the subunits of ribosomes (along with proteins), the structures on which amino acids are assembled into polypeptide chains. Different kinds of transfer RNA (tRNA) pick up amino acids and deliver them to ribosomes in the order specified by mRNAs.

A new mRNA transcript consists of introns (nucleotide sequences that do not code for proteins) and exons. Exons are the mRNA sequences that carry protein-building instructions. Regulatory proteins stimulate or suppress gene transcription and so control gene activity.

 Learn how genes are transcribed and how these transcripts are processed.

Section 22.4 In translation, RNAs interact in the synthesis of polypeptide chains. Amino acids are linked one after another, in the sequence required to produce a specific polypeptide chain.

Translation follows the genetic code, a set of sixty-four base triplets—nucleotide bases that ribosome proteins "read" three at a time.

A base triplet in an mRNA molecule is a codon. A given combination of codons specifies the amino acid sequence of a polypeptide chain.

Thomson NOW! *Explore the genetic code.*

Section 22.5 Translation has three stages. In chain initiation, a small ribosome subunit and an initiator tRNA bind with an mRNA transcript and move along it until they reach an AUG start codon. The small subunit binds with a large ribosome subunit.

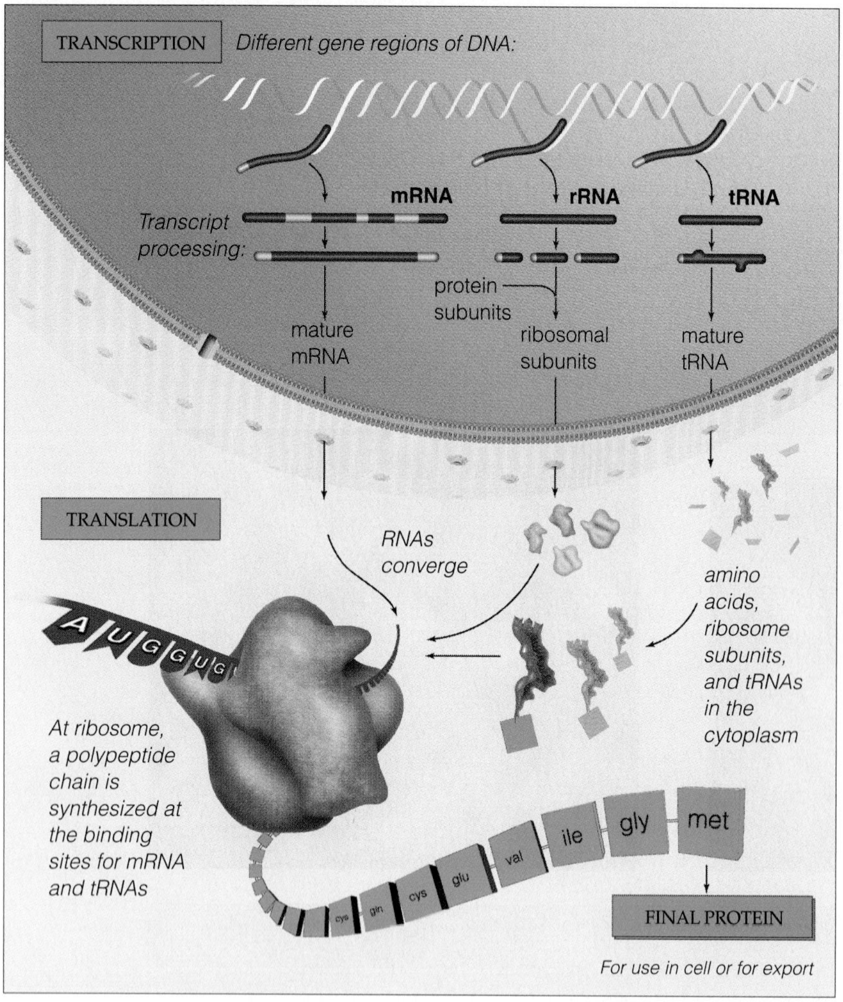

TRANSCRIPTION *Different gene regions of DNA:*

mRNA   rRNA   tRNA

Transcript processing:

protein subunits

mature mRNA   ribosomal subunits   mature tRNA

TRANSLATION

RNAs converge

amino acids, ribosome subunits, and tRNAs in the cytoplasm

AUGGUG

At ribosome, a polypeptide chain is synthesized at the binding sites for mRNA and tRNAs

cys   gln   cys   glu   val   ile   gly   met

FINAL PROTEIN

For use in cell or for export

Figure 22.26 Animated! Summary of transcription and translation, the two steps leading to protein synthesis in cells. DNA is transcribed into RNA in the nucleus. RNA is translated into proteins in the cytoplasm.

In chain elongation, tRNAs deliver amino acids to the ribosome. Their anticodons base-pair with mRNA codons. The amino acids are joined (by peptide bonds) to form a new polypeptide chain.

In chain termination, an mRNA stop codon moves onto the ribosome; then the polypeptide chain and mRNA detach from the ribosome.

 Observe the translation of mRNA.

Section 22.6 Recombinant DNA technology is the foundation for genetic engineering. Restriction enzymes are used to cut DNA molecules into fragments, which are inserted into a cloning vector (such as a plasmid) and then multiplied in rapidly dividing cells.

A DNA clone is a foreign DNA sequence that has been introduced and amplified in dividing cells. DNA sequences also can be amplified in test tubes by the polymerase chain reaction.

The genome of a species is all the DNA in a haploid set of chromosomes. The human genome, including our species' 21,500 genes, consists of about 2.9 billion nucleotides. Some tools of recombinant DNA technology can be used to identify genes in the genome.

 Explore the tools used to make recombinant DNA, and learn how researchers isolate and copy genes.

Sections 22.7–22.8 Automated DNA sequencing can quickly determine the exact sequence of nucleotides in segments of DNA—and, accordingly, in genes. The human genome has been sequenced, and researchers have begun identifying genes responsible for a variety of traits and genetic disorders.

 Investigate DNA sequencing, and observe the process of DNA fingerprinting.

Sections 22.9–22.11 Recombinant DNA technology and genetic engineering have enormous potential for research and applications in medicine, agriculture, and industry. Both also may pose ecological and social risks.

Review Questions

1. Why is DNA replication called "semiconservative"?

2. Name one kind of mutation that produces an altered protein. What determines whether the altered protein will have beneficial, neutral, or harmful effects?

3. How are the polypeptide chains of proteins that are specified by DNA assembled?

4. How does RNA differ from DNA?

5. Name the three classes of RNA and describe their functions.

6. Distinguish between a codon and an anticodon.

7. Describe the three steps of translation.

8. What is a restriction enzyme?

9. What is a "gene sequence"?

Self-Quiz   *Answers in Appendix V*

1. Nucleotide bases, read _____ at a time, serve as the "code words" of genes.

2. DNA contains genes that are transcribed into _____.
   a. proteins   d. tRNAs
   b. mRNAs   e. b, c, and d
   c. rRNAs

3. mRNA is produced by _____.
   a. replication   c. transcription
   b. duplication   d. translation

4. _____ carries coded instructions for an amino acid sequence to the ribosome.
   a. DNA   c. mRNA
   b. rRNA   d. tRNA

5. tRNA _____.
   a. delivers amino acids to ribosomes
   b. picks up genetic messages from rRNA
   c. synthesizes mRNA
   d. all of the above

6. An anticodon pairs with the bases of _____.
   a. mRNA codons   c. tRNA anticodons
   b. DNA codons   d. amino acids

7. The loading of mRNA onto the small ribosomal subunit occurs during _____.
   a. initiation of transcription c. translation
   b. transcript processing   d. chain elongation

8. Use the genetic code (Figure 22.10) to translate the mRNA sequence AUGCGCACCUCAGGAUGAGAU. (Human reading frames start with AUG.) Which amino acid sequence is being specified?
   a. meth-arg-thr-ser-gly-stop-asp . . .
   b. meth-arg-thr-ser-gly . . .
   c. meth-arg-tyr-ser-gly-stop-asp . . .
   d. none of the above

9. Match the terms related to protein building.
   _____ alters genetic   a. initiation, elongation,
         instructions   termination
   _____ codon   b. conversion of genetic
   _____ transcription   messages into
   _____ translation   polypeptide chains
   _____ stages of   c. base triplet for an amino
         transcription,   acid
         translation   d. RNA synthesis
                              e. mutation

10. Rejoined cut DNA fragments from different organisms are best known as _____.
    a. cloned genes
    b. mapped genes
    c. recombinant DNA
    d. conjugated DNA

11. The polymerase chain reaction _____.
    a. is a natural reaction in bacterial DNA
    b. cuts DNA into fragments
    c. amplifies DNA sequences in test tubes
    d. inserts foreign DNA into bacterial DNA

Critical Thinking

1. Which mutation would be more harmful: a mutation in DNA or one in mRNA? Explain your answer.

2. Jimmie's Produce put out a bin of tomatoes having beautiful red color and looking "just right" in terms of ripeness. A sign above the bin identified them as genetically modified produce. Most shoppers selected unmodified tomatoes in the neighboring bin, even though those tomatoes were pale pink and hard as rocks. Which ones would you pick? Why?

3. Previous chapters have discussed various types of cloning, including the cloning of stem cells and of embryos. Dolly, the sheep shown in Figure 22.27a, grew from a cloned cell. Is cloning the same as genetic engineering? Explain your answer.

4. Scientists at Oregon Health Sciences University produced the first transgenic primate by inserting a jellyfish gene into a fertilized egg of a rhesus monkey. The gene encodes a bioluminescent protein that fluoresces green. The egg was implanted in a surrogate monkey's uterus, where it developed into a male that was named ANDi (Figure 22.27b).

The long-term goal of this gene transfer project is not to make glowing-green monkeys. It is the transfer of human genes into primates whose genomes are most like ours. Transgenic primates could then be studied to gain insight into genetic disorders, which might lead to the development of cures.

However, something more controversial is at stake. Will the time come when foreign genes can be inserted into human embryos? Would it be ethical to transfer a chimpanzee or monkey gene into a human embryo to cure a genetic defect? Or to bestow immunity against a potentially fatal disease such as AIDS?

Explore on Your Own

Thomas Jefferson (Figure 22.28) was the third president of the United States and the main author of the Declaration of Independence. The proprietor of a Virginia plantation, Mr. Jefferson owned African-American slaves, including a woman named Sally Hemings who eventually bore five children, including several sons. In 1998, writing in the prestigious scientific journal *Nature*, British researchers cited genetic analysis purportedly showing that a modern male descendant of Hemings had the same Y chromosome as males descended from the Jefferson family. Some in the media reported it as solid genetic evidence supporting an old rumor that Thomas Jefferson had fathered at least one of his slave's sons. Other researchers and historians soon pointed out that there was no way the study could present conclusive findings, and eventually the original researchers agreed. Even so, tales of the "scientifically proven" intimate relationship between Jefferson and Hemings are still part of popular lore.

Numerous websites explore this story and the larger issue of ethical guidelines for such "biohistory." To learn more, go to www.tjheritage.org, which provides links to discussions of problems and issues this incident raised.

Figure 22.27 *Animated!* (a) The cloned sheep Dolly with her first lamb. The lamb was conceived the old-fashioned way. (b) ANDi, the first transgenic primate. His cells have a jellyfish gene for bioluminescence.

Figure 22.28 Thomas Jefferson.

Between You and Eternity

Cancer strikes one in three people in the United States and kills one in four. According to the American Cancer Society, there are about 1,500 cancer deaths every day, over half a million in a year. On average, cancer strikes more males than females, but the pattern varies depending on the type of cancer involved. Take breast cancer, for example. Each year more than 200,000 women in the United States are diagnosed with the disease. About 5 to 10 percent of them carry a gene mutation that greatly increases their risk of developing breast cancer. There are two of these "breast cancer susceptibility genes," called BRCA1 and BRCA2.

Rather than worry constantly, some healthy women who carry BRCA1 or BRCA2 choose to have their breasts removed before cancer can strike. The surgery, called preventive (or prophylactic) mastectomy, is also an option for women who have already had cancer in one breast, or have a strong family history of the disease.

Although preventive mastectomy may seem like a drastic step, most women who opt for the procedure report no major negative aftereffects.

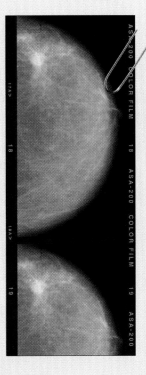

Yet there are concerns associated with removing normal breasts as insurance against future disease. For one, just because a woman has an elevated risk doesn't mean that she will actually develop breast cancer. Most such women won't. And because it is impossible for a surgeon to remove every last bit of breast tissue, a woman who goes through with the surgery still has as much as a 10 percent chance of developing breast cancer. When a breast is removed (a procedure called radical mastectomy), usually other tissue or a prosthetic breast is used to reconstruct it—a procedure that also brings risks and complications.

Fortunately, intensive research is rapidly increasing our understanding of and ability to treat many kinds of cancer, including those of the breast, ovary, colon, and skin. Most cancers are treatable, and many are curable if the disease is discovered early.

 How Would You Vote? Some young women with an elevated genetic risk of developing breast cancer have chosen to have their breasts removed even before cancer develops. Should health insurers help pay for the cost of this surgery? Cast your vote online at www.thomsonedu.com/ biology/starr/humanbio.

 ## Key Concepts

CANCER: UNCONTROLLED CELL DIVISION
Cancer cells are abnormal in both structure and function. Cancer develops when gene changes remove the normal controls over cell division.

DIAGNOSIS, TREATMENT, AND PREVENTION
Cancer is diagnosed by biopsy and other tools. Early detection increases the chances of successful treatment. Lifestyle decisions that promote health can also limit a person's risk of developing cancer.

MAJOR TYPES OF CANCER
Cancer is categorized according to the type of tissue in which it first develops. Every organ system in the body is susceptible to cancer.

Links to Earlier Concepts

In cancer, normal controls over cell division, a part of the cell cycle, are lost. Often, one or more genetic changes are part of the process. Hence this chapter draws on what you have learned about normal cell structure (3.2) and the cell cycle (19.3), as well as about gene mutations and mechanisms of DNA repair (22.2). You will also read about links between cancer and operations of the immune system (10.1, 10.7, 10.9) and effects of ionizing radiation (19.6). You will learn more about how monoclonal antibodies are used to combat certain forms of cancer, and about uses of immunotherapy in cancer treatment (10.8).

23.1 Cancer: Cell Controls Go Awry

LINKS TO
SECTIONS
3.2, 17.1, 19.3,
AND 22.2

As genes switch on and off, they determine when and how fast the cell will grow and divide, when it will stop dividing, and even when it will die. Cancer can result when controls over cell division are lost.

SOME TUMORS ARE CANCER, OTHERS ARE NOT

If cells in a tissue overgrow—an abnormal enlargement called **hyperplasia**—the result is a defined mass of tissue called a **tumor**. Technically, a tumor is a *neoplasm*, which means "new growth."

A tumor may not be "cancer." As Figure 23.1*a* shows, the cells of a *benign* tumor are often enclosed by a capsule of connective tissue, and inside the capsule they have an orderly organization. They also tend to grow slowly and to be well differentiated (structurally specialized), much like normal cells of the same tissue (Section 17.1). Benign

Table 23.1	Comparison of Benign and Malignant Tumors	
	Malignant Tumor	**Benign Tumor**
Rate of growth	Rapid	Slow
Nature of growth	Invades surrounding tissue	Expands in the same tissue
Spread	Metastasizes via the bloodstream and the lymphatic system	Does not spread
Cell differentiation	Usually poor	Nearly normal

tumors usually stay put in the body, push aside but don't invade surrounding tissue, and generally can be easily removed by surgery. Benign tumors *can* threaten health, as when they occur in the brain. Nearly everyone has at least several of the benign tumors we call moles. Most of us also have or have had some other type of benign neoplasm, such as a cyst. Often, the immune system destroys benign growths. Table 23.1 compares the main features of malignant and benign tumors.

Dysplasia ("bad form") is an *abnormal* change in the sizes, shapes, and organization of cells in a tissue. Such change is often an early step toward **cancer**. Under the microscope, the edges of a cancerous tumor usually look ragged (Figure 23.1*b*), and its cells form a disorganized clump. Most cancer cells also have characteristics that enable them to behave differently from normal cells.

CANCER CELLS HAVE ABNORMAL STRUCTURE

Cancer is basically a genetic disorder, as you will read in Section 23.2. It is the outcome of a series of mutations in a cell's genes. One result of these changes is that the structure of a cancer cell is not normal. Often, the nucleus is abnormally large, and there is less cytoplasm than usual. Cancer cells also are poorly differentiated. That is, they often do not have the structural specializations of healthy cells in mature body tissues. The extent to which cancer cells are specialized can be medically important. In general, the less differentiated cancer cells are, the more likely they are to break away from the primary tumor and spread the disease.

When a normal cell is transformed into a cancerous one, more changes take place. The cytoskeleton shrinks, becomes disorganized, or both. Proteins that are part of the plasma membrane are lost or altered, and new, different ones appear. These changes are passed on to the cell's descendants: When a transformed cell divides, its daughter cells are cancerous cells too.

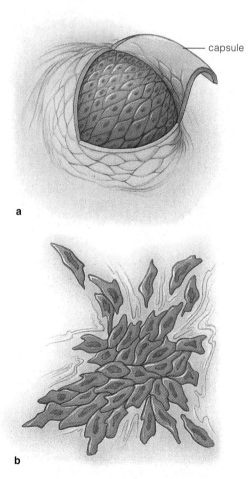

capsule

a

b

Figure 23.1 (**a**) Sketch of a benign tumor. Cells appear nearly normal, and connective tissue encapsulates the tumor mass. (**b**) A cancerous neoplasm. Due to the abnormal growth of cancer cells, the tumor is a disorganized heap of cells. Some of the cells may break off and invade surrounding tissues, a process called metastasis.

Figure 23.2 Scanning electron micrograph of a cancer cell surrounded by some of the body's white blood cells that may or may not be able to destroy it.

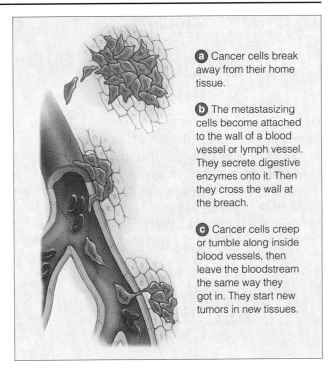

a Cancer cells break away from their home tissue.

b The metastasizing cells become attached to the wall of a blood vessel or lymph vessel. They secrete digestive enzymes onto it. Then they cross the wall at the breach.

c Cancer cells creep or tumble along inside blood vessels, then leave the bloodstream the same way they got in. They start new tumors in new tissues.

Figure 23.3 *Animated!* Stages in metastasis.

CANCER CELLS LACK NORMAL CONTROLS OVER CELL DIVISION

Contrary to popular belief, cancer cells don't necessarily divide more rapidly than normal cells do, but they do increase in number faster. Why? Normally, the death of cells closely balances the production of new ones by mitosis. In a cancerous tumor, however, at any given moment more cells are dividing than are dying. As this runaway cell division continues, the cancer cells do not respond to crowding, as normal cells do. A normal cell stops dividing once it comes into contact with another cell, so the arrangement of cells in a tissue remains orderly. A cancer cell, however, keeps on dividing. Therefore, cancer cells pile up in a disorganized heap. This is why cancerous tumors are often lumpy.

Cancer cells also do not adhere well to the cells next to them in a tissue. As Figure 23.2 shows, they may form extensions (pseudopodia, "false feet") that enable them to move about. These extensions allow cancer cells to break away from the parent tumor and invade other tissues, including the lymphatic system and circulatory system (Figure 23.3). This invasive activity is called **metastasis**. The ability to metastasize comes about by way of additional mutations in cancer cells. And metastasis is what makes a cancer malignant—that is, capable of causing harm.

Some kinds of cancer cells produce the hormone HCG, human chorionic gonadotropin. (Recall from Chapter 17 that HCG maintains the uterine lining when a pregnancy begins. It also helps prevent the mother's immune system from attacking the fetus as "foreign.") The presence of HCG in the blood can serve as a red flag that a cancer exists somewhere in a person's body.

Some cancer cells produce a chemical that stimulates cell division, *and* they have receptors for that chemical. Cancer cells also secrete a growth factor, *angiogenin*, that encourages new blood vessels to grow around the tumor. The blood vessels can "feed" the tumor with the large supply of nutrients and oxygen it needs to continue growing. A key area of cancer research is work on developing drugs that "starve" tumors to death by blocking the effects of angiogenin.

Cancer cells have genetic mutations that have removed normal controls over cell division and organization into tissues. The cells are abnormal in both their structure and their function.

Cancer cells are often poorly differentiated. They can leave a primary tumor, invade surrounding tissue, and metastasize to other areas of the body.

23.2 The Genetic Triggers for Cancer

LINKS TO
SECTIONS
10.7, 10.9, 19.3,
18.2, AND 19.6

We know that cancer is basically a genetic disorder—and we know that the changes occur in a sequence of steps.

CANCER DEVELOPS IN A MULTISTEP PROCESS

The transformation of a normal cell into a cancer cell is a multistep process called **carcinogenesis** (Figure 23.4). Cancer develops in many kinds of organisms, even some plants. There is fossil evidence that some dinosaurs developed cancer. Wherever it turns up, cancer almost always develops through two or more steps in which genetic changes upset normal controls over cell division.

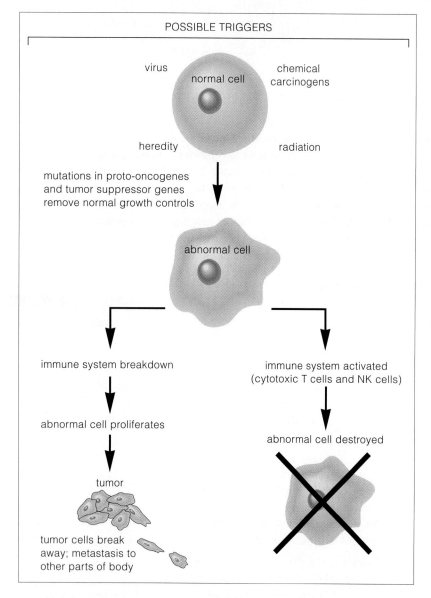

Figure 23.4 The steps in carcinogenesis.

ONCOGENES INDUCE CANCER WHEN TUMOR SUPPRESSOR GENES ARE MUTATED

Certain genes are **proto-oncogenes** (*proto* "before"). An **oncogene** is a gene that can induce cancer. What's the difference? Proto-oncogenes are normal genes that regulate cell growth and development. They encode proteins that include growth factors (signals sent by one cell to trigger growth in other cells), proteins that regulate cell adhesion, and the protein signals for cell division. If something alters the structure of a proto-oncogene or the way its protein-making instructions are read out, it may be converted into an oncogene that does not respond to controls over cell division.

An oncogene acting alone does not cause malignant cancer. That usually requires mutations in several genes, including at least one **tumor suppressor gene**. These are genes that can halt cell growth and division, preventing cancers from developing.

The roles of some tumor suppressor genes are beginning to come into focus. For example, we know that the childhood eye cancer **retinoblastoma** is likely to develop when a child has only one functional copy of a tumor suppressor gene on chromosome 13. The two genes known to be associated with a predisposition to breast cancer, BRCA1 and BRCA2 (also on chromosome 13), also are tumor suppressor genes. As noted in the chapter introduction, women who inherit mutant forms of these genes are at high risk of developing breast cancer (Figure 23.5).

Research has revealed a lot about a tumor suppressor gene called p53. This gene codes for a regulatory protein that stops cell division when cells are stressed or damaged. When p53 mutates, the controls don't operate. Then an affected cell may begin runaway division. To make matters worse, when mutated, the p53 gene's faulty protein may activate an oncogene. More than half of cancers involve a mutated or absent p53 gene. This discovery has led to the development of tests that screen for p53 mutations. Gene therapy in which normal copies of p53 are inserted into cancerous tumors has had promising early results.

Oncogenes don't always mean trouble. For instance, a tumor suppressor gene might prevent an oncogene from ever being expressed. On the other hand, an oncogene or a suppressor gene also may mutate in a way that has the opposite effect. A change in a chromosome's structure may move an oncogene into a position that allows it to be expressed. Or genetic material that interferes with controls may enter a cell—for example, by way of a viral infection.

CANCER: UNCONTROLLED CELL DIVISION

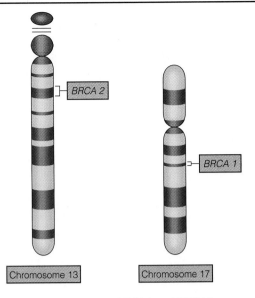

Figure 23.5 The locations for BRCA1 and BRCA2 on chromosomes 13 and 17, respectively. Most cases of inherited breast cancer are due to mutation of one or the other of these tumor suppressor genes.

THERE ALSO ARE OTHER ROUTES TO CANCER

INHERITED SUSCEPTIBILITY TO CANCER Heredity plays a major role in about 5 percent of cancers, including cases of familial breast cancer, colorectal cancer, and lung cancer. If a mutation in a germ cell or a gamete (sperm or egg) alters a proto-oncogene or tumor suppressor gene, the defect can be passed on from parent to child. An affected person may be more likely to develop cancer if later mutations occur in other proto-oncogenes, in tumor suppressor genes, or in genes that control aspects of cell metabolism and responses to hormones.

VIRUSES Viruses cause some cancers. Sometimes a viral infection can alter a proto-oncogene when the viral DNA is inserted at a certain position in the host cell's DNA. For example, the viral gene could take the place of a regulatory DNA sequence that normally prevents a proto-oncogene from switching on (or off) at the wrong time. Other viruses carry oncogenes as part of their genetic material and insert them into the host's DNA. Most viruses linked to human cancer are DNA viruses.

CHEMICAL CARCINOGENS There are thousands of known **carcinogens**, cancer-causing substances that can cause damage leading to a subsequent mutation in DNA. The list includes many chemicals that are by-products of industrial activities, such as asbestos, vinyl chloride, and benzene. The list also includes hydrocarbons in cigarette smoke and on the charred surfaces of barbecued meats, and substances in dyes and pesticides. Some of the first carcinogens to be identified were substances in fireplace soot, which frequently caused cancer of the scrotum in chimney sweeps. Some plants and fungi also produce carcinogens. Aflatoxin, which is produced by a fungus that attacks stored grain and other seeds, causes liver cancer. For this reason, some authorities advise against eating unprocessed peanut butter.

Certain chemicals are "pre-carcinogens" that cause gene changes only *after* being altered by metabolic events in the body or cell. Others are cancer "promoters"—alone they are not carcinogenic, but they can make a carcinogen more potent if the carcinogen *and* the promoter act on a cell at the same time.

RADIATION Section 19.6 noted that radiation can cause cancer-related mutations in DNA. Common sources include ultraviolet radiation from sunlight and tanning lamps, medical and dental X rays, and some radioactive materials used to diagnose diseases. Other sources are radon gas in soil and water, background radiation from cosmic rays, and the gamma rays emitted from nuclear reactors and radioactive wastes. Sun exposure is probably the greatest radiation risk factor for most people.

BREAKDOWNS IN IMMUNITY When a normal cell turns cancerous, altered proteins at its surface function like foreign antigens—the "nonself" tags that mark a cell for destruction by cytotoxic T cells and natural killer cells. A healthy immune system can detect and destroy some types of cancer cells, but this protective function can break down as a person ages. This natural deterioration is why the risk of cancer increases as we grow older.

Some researchers hypothesize that a person's cancer risk rises whenever the immune system is suppressed for an extended period. In addition to factors such as infection by HIV, some therapeutic drugs can suppress immunity, as can anxiety and severe depression.

Finally, for various reasons, the cells of a growing cancer may not display the antigens (or other molecules) that must bind with receptors on lymphocytes in order to trigger an immune response. For all intents and purposes, the immune system is "blind" to the cancer threat.

Cancer develops through a multistage process in which gene changes remove normal controls over cell division.

Oncogenes have a major role in changing a normal cell to a cancerous one. Proto-oncogenes may become oncogenes if there is a change in their structure or how they are expressed.

The development of cancer typically also requires the absence or mutation of at least one tumor suppressor gene.

23.3 Assessing the Cancer Risk from Environmental Chemicals

According to the American Cancer Society, factors in our environment lead to about half of all cancers. This statistic includes exposure to UV light and radiation, and it also includes agricultural and industrial chemicals. How are people exposed to these chemicals? And how dangerous are they? Let's begin with the first question.

Government statistics indicate that about 40 percent of the food in American supermarkets contains detectable residues of one or more of the active ingredients in commonly used pesticides. The residues are especially likely to be found in tomatoes, grapes, apples, lettuce, oranges, potatoes, beef, and dairy products. Imported crops, such as fruits, vegetables, coffee beans, and cocoa, can also carry significant pesticide residues—sometimes including pesticides, such as DDT, that are banned in the United States. The pesticide category includes chemicals used as fungicides, insecticides, and herbicides. There are roughly 600 different chemicals in this category, which are used alone or in combination.

Avoiding exposure to pesticides is difficult. Although residues of some pesticides can (and should) be removed from the surfaces of fruits and vegetables by washing before eating, it can be difficult to avoid coming into contact with pesticides used in community spraying programs to control mosquitoes and other pests, or used to eradicate animal and plant pests on golf courses and along roadsides. We have more control over the chemicals we use in gardens and on lawns (Figure 23.6).

Table 23.2	Some Industrial Chemicals Linked to Cancer
Chemical/Substance	**Type of Cancer**
Benzene	Leukemias
Vinyl chloride	Liver, various connective tissues
Various solvents	Bladder, nasal epithelium
Ether	Lung
Asbestos	Lung, epithelial linings of body cavities
Arsenic	Lung, skin
Radioisotopes	Leukemias
Nickel	Lung, nasal epithelium
Chromium	Lung
Hydrocarbons in soot, tar smoke	Skin, lung

Agricultural chemicals are not the only potential threats to human health. A variety of industrial chemicals also have been linked to cancer. In one way or another, the industrial chemicals in Table 23.2 all can cause carcinogenic mutations in DNA.

Some years ago biochemist Bruce Ames developed a test that could be used to assess the ability of chemicals to cause mutations. This Ames test uses *Salmonella* bacteria as the "guinea pigs," because chemicals that cause mutations in bacterial DNA may also have the same effect on human DNA. After extensive experimentation, Ames arrived at some interesting conclusions. First, he found that more than 80 percent of known cancer-causing chemicals do produce mutations. However, Ames testing at many different laboratories has *not* revealed a "cancer epidemic" caused by synthetic chemicals.

Ames's findings do not mean we should carelessly expose our crops, our gardens, or ourselves to environmental chemicals. For instance, the National Academy of Sciences has warned that the active ingredients in 90 percent of all fungicides, 60 percent of all herbicides, and 30 percent of all insecticides in use in the United States *have the potential* to cause cancer in humans. At the same time responsible scientists recognize that it is virtually impossible to determine that a certain level of a specific chemical caused a particular cancer or some other harmful effect. Given these facts, it is probably prudent to err on the side of caution and, as much as possible, limit our exposure to the potential carcinogens in our increasingly chemical world.

Figure 23.6 A common source of pesticide exposure. Home garden chemicals are just one route by which we can come into contact with mutagenic or carcinogenic substances.

23.4 Diagnosing Cancer

Table 23.3 lists seven common cancer warning signs. These can help you spot cancer in its early stages, when treatment is most effective. Routine cancer screening becomes important as a person ages; Table 23.4 lists some recommended cancer screening tests.

To maximize the chances that a cancer can be cured, early and accurate diagnosis of cancer is extremely important. To confirm—or rule out—cancer, various types of tests can refine the diagnosis. Blood tests can detect **tumor markers**, substances produced by specific types of cancer cells or by normal cells in response to the cancer. For example, as we noted earlier, the hormone HCG is a highly specific marker for certain cancers. Prostate-specific antigen, or PSA, is a useful marker for detecting prostate cancer, and a marker has been identified for ovarian cancer as well. Radioactively labeled monoclonal antibodies, which home in on tumor antigens, are useful for pinpointing the location and sizes of tumors of the colon, brain, bone, and some other tissues.

Medical imaging of tumors includes methods such as magnetic resonance imaging (MRI), X rays, ultrasound, and computerized tomography (CT). Unlike a standard X ray, an MRI scan can reveal tumors that are obscured by bone, such as in the brain (Figure 23.7).

The definitive cancer detection tool is **biopsy**. A small piece of suspect tissue is removed from the body through a hollow needle or exploratory surgery. A pathologist then microscopically examines cells of the tissue sample for the characteristic features of cancer cells.

LINK TO SECTION 2.2

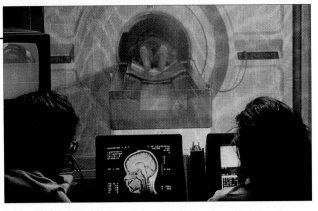

Figure 23.7 MRI scanning. MRI does not use X rays. The patient is placed inside a chamber that is surrounded by a magnet. The machine generates a magnetic field in which nuclei of hydrogen and some other atoms in the body align and absorb energy. A computer analyzes the information and uses it to generate an image of soft tissues.

A *DNA probe* is a segment of radioactively labeled DNA. Such probes can be used to locate gene mutations or alleles associated with some types of inherited cancers. The procedure is expensive, and most people don't have health insurance that will cover it. On the other hand, genetic screening can allow people with increased genetic susceptibility to make medical and lifestyle choices that could significantly reduce their cancer risk.

Biopsy is the definitive tool for diagnosing cancer. Other diagnostic procedures include blood testing for substances produced by cancer cells and medical imaging.

Table 23.3	The Seven Warning Signs of Cancer*
Change in bowel or bladder habits and function	
A sore that does not heal	
Unusual bleeding or bloody discharge	
Thickening or lump	
Indigestion or difficulty swallowing	
Obvious change in a wart or mole	
Nagging cough or hoarseness	

* Notice that the first letters of the signs spell the advice CAUTION. *Source:* American Cancer Society.

Table 23.4	Recommended Cancer Screening Tests			
Test or Procedure	**Cancer**	**Sex**	**Age**	**Frequency**
Breast self-examination	Breast	Female	20+	Monthly
Mammogram	Breast	Female	40–49 50+	Every 1–2 years Yearly
Testicle self-examination	Testicle	Male	18+	Monthly
Sigmoidoscopy	Colon	Male, Female	50+	Every 3–5 years
Fecal occult blood test	Colon	Male, Female	50+	Yearly
Digital rectal examination	Prostate, colorectal	Male, Female	40+	Yearly
Pap test	Uterus, cervix	Female	18+ and all sexually active women	Every other year until age 35; yearly thereafter
Pelvic examination	Uterus, ovaries, cervix	Female	18–39 40+	Every 1–3 years w/Pap, yearly
General checkup		Male, Female	20–39 40+	Every 3 years Yearly

23.5 Some Major Types of Cancer

In general, a cancer is named according to the type of tissue in which it first forms. For instance, cancers of connective tissues such as muscle and bone are **sarcomas**. Various types of **carcinomas** arise from epithelium, including cells of the skin and epithelial linings of internal organs. When a cancer begins in a gland or its ducts, it is called an **adenocarcinoma**. **Lymphomas** are cancers of lymphoid tissues in organs such as lymph nodes, and cancers arising in blood-forming regions—mainly stem cells in bone marrow—are **leukemias**.

The remainder of this chapter surveys some common cancers in the United States. Figure 23.8 summarizes the most recent data for males and females.

> A cancer is categorized according to the tissue in which it arises. Examples include sarcomas (connective tissue), carcinomas (epithelium), adenocarcinomas (glandular tissue), lymphomas (lymphoid tissues), and leukemias (blood-forming regions such as bone marrow).

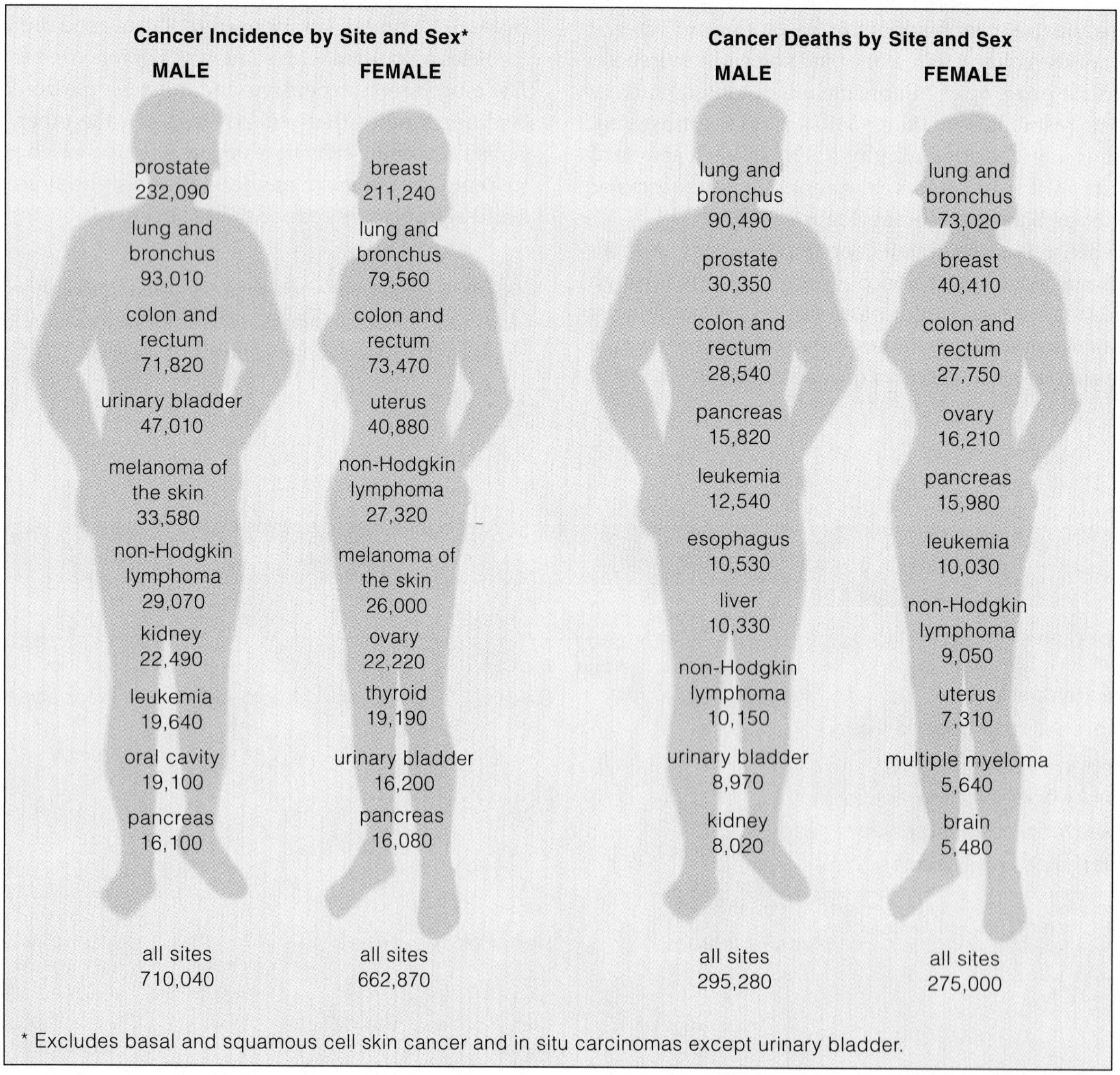

Cancer Incidence by Site and Sex*		Cancer Deaths by Site and Sex	
MALE	**FEMALE**	**MALE**	**FEMALE**
prostate 232,090	breast 211,240	lung and bronchus 90,490	lung and bronchus 73,020
lung and bronchus 93,010	lung and bronchus 79,560	prostate 30,350	breast 40,410
colon and rectum 71,820	colon and rectum 73,470	colon and rectum 28,540	colon and rectum 27,750
urinary bladder 47,010	uterus 40,880	pancreas 15,820	ovary 16,210
melanoma of the skin 33,580	non-Hodgkin lymphoma 27,320	leukemia 12,540	pancreas 15,980
non-Hodgkin lymphoma 29,070	melanoma of the skin 26,000	esophagus 10,530	leukemia 10,030
kidney 22,490	ovary 22,220	liver 10,330	non-Hodgkin lymphoma 9,050
leukemia 19,640	thyroid 19,190	non-Hodgkin lymphoma 10,150	uterus 7,310
oral cavity 19,100	urinary bladder 16,200	urinary bladder 8,970	multiple myeloma 5,640
pancreas 16,100	pancreas 16,080	kidney 8,020	brain 5,480
all sites 710,040	all sites 662,870	all sites 295,280	all sites 275,000

* Excludes basal and squamous cell skin cancer and in situ carcinomas except urinary bladder.

Figure 23.8 Summary of annual incidence of and deaths from common cancers, by site and sex. Data are estimates for the United States, 2005. Reprinted by permission of the American Cancer Society, Inc.

23.6 Treating and Preventing Cancer

When a person is diagnosed with cancer, a variety of weapons are available to combat it. And anyone can adopt an "anticancer lifestyle."

LINKS TO SECTIONS 10.8 AND 19.3

Surgery, drugs, and irradiation of tumors are the major weapons against cancer. Surgery may even be a complete cure when a tumor is fully accessible and has not spread.

CHEMOTHERAPY AND RADIATION KILL CANCER CELLS DIRECTLY

Chemotherapy uses drugs to kill cancer cells. Most anticancer drugs are designed to kill dividing cells. They disrupt DNA replication during the S phase of the cell cycle or prevent mitosis by blocking formation of the mitotic spindle. Unfortunately, such drugs are also toxic to rapidly dividing healthy cells such as hair cells, stem cells in bone marrow, immune system lymphocytes, and epithelial cells of the gut lining. As a result, chemotherapy patients can suffer side effects such as hair loss, nausea, vomiting, and reduced immune responses. Radiation also kills both cancer cells and healthy cells in the irradiated area. A treatment option called *adjuvant therapy* (adjuvant means "helping") combines surgery and a less toxic dose of chemotherapy. A cancer patient might receive enough chemotherapy to shrink a tumor, for instance, then have surgery to remove what's left.

Section 10.8 described how monoclonal antibodies can be used to deliver lethal doses of anticancer drugs to tumor cells while sparing healthy cells. The antibodies target antigens (cell surface markers) on various types of tumors. The idea is to link tumor-specific monoclonal antibodies with lethal doses of cytotoxic (cell-killing) drugs. Recent experiments have shown promising results in some patients with one form of leukemia, a type of breast cancer, and certain gliomas (cancers that arise in glia in the central nervous system). Interferons also can activate cytotoxic T cells and natural killer cells, which then may recognize and kill some types of cancer cells. So far, interferon therapy has been useful only against some rare forms of cancer.

ANTICANCER DRUGS MAY BE MATCHED TO GENETIC CHARACTERISTICS OF CANCER CELLS

Traditionally, drugs used in chemotherapy have been matched to the organ in which a cancer occurs—this drug for breast cancer, that one for lung cancer, and so on. A promising new strategy aims instead to match chemotherapy with the particular genetic characteristics of a patient's cancerous cells. This approach recognizes that there are hundreds of genetically different subgroups

Figure 23.9 Young women taking steps to limit their risk of lung cancer, the chief cancer killer

of cancer, and that some subgroups share the same gene mutations—and chemical features—regardless of where the cancer develops. For example, the drug Gleevec works well against some types of leukemia and also against some sarcomas.

YOU CAN LIMIT YOUR CANCER RISK

None of us can control factors in our heredity or biology that might lead one day to cancer, but we can all make lifestyle decisions that promote health. The American Cancer Society recommends the following strategies for limiting cancer risk:

1. Avoid tobacco in any form, including secondary smoke from others (Figure 23.9).

2. Maintain a desirable weight. Being more than 40 percent overweight increases the risk of several cancers.

3. Eat a low-fat diet that includes plenty of vegetables and fruits. As noted in Chapter 7, antioxidants such as vitamin E may help prevent some kinds of cancer.

4. Drink alcohol in moderation. Heavy alcohol use, especially in combination with smoking, increases risk for cancers of the mouth, larynx, esophagus, and liver.

5. Learn whether your job or residence exposes you to such industrial agents as nickel, chromate, vinyl chloride, benzene, asbestos, and agricultural pesticides, which are associated with various cancers.

6. Protect your skin from excessive sunlight.

Surgery, chemotherapy, and other treatment strategies are used to fight cancer. However, the best defense is a good offense—making lifestyle choices that promote health. Not smoking and eating a healthful diet are high on the list.

23.7 Cancers of the Breast and Reproductive System

LINK TO SECTION 16.11

In the United States, breast cancer is a major killer of women. In both females and males, reproductive system cancers also are major causes for concern.

BREAST CANCER IS A MAJOR CAUSE OF DEATH

In the United States, about one woman in eight develops breast cancer. Breast cancer is much rarer in males, but of all cancers in women, it currently ranks second only to lung cancer as a cause of death. In women, obesity, late childbearing, early puberty, late menopause, and excessive levels of estrogen (and perhaps some other hormones) are risk factors. As you've already read, a family history of breast cancer may indicate that a tumor suppressor gene such as BRCA1 or BRCA2 is mutated. Although 80 percent of breast lumps are *not* cancer, a woman should seek medical advice about any breast lump, thickening, dimpling, breast pain, or discharge.

Chances for cure are excellent if breast cancer is detected early and treated promptly. Hence a woman should examine her breasts every month (about a week after her menstrual period, during her reproductive years). Figure 23.10 shows the steps of a self-exam. Low-dose mammography (breast X ray) is the most effective method for detecting small breast cancers. It is 80 percent reliable. The American Cancer Society recommends an annual mammogram for women over 50 and for younger women at high risk.

Treatment depends mainly on the extent of the disease. In *modified radical mastectomy*, the affected breast tissue, overlying skin, and nearby lymph nodes are removed, but muscles of the chest wall are left intact. For a small tumor, the treatment may be *lumpectomy*, which is less disfiguring than a mastectomy because it leaves some breast tissue in place. In both cases, removed lymph nodes are examined to see if the cancer has begun to spread.

Drugs are also used in the fight against breast cancer. A few (such as tamoxifen) can sometimes shrink tumors.

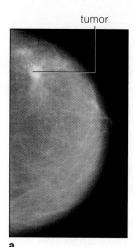

tumor

Figure 23.10 (**a**, *right*) Mammogram showing a breast cancer tumor. (**b**, *below*) How to perform a breast self-examination.

a

1. Lie down and put a folded towel under your left shoulder, then put your left hand behind your head. With the right hand (fingers flat), begin the examination of your left breast by following the outer circle of arrows shown. Gently press the fingers in small, circular motions to check for any lump, hard knot, or thickening. Next, follow the inner circle of arrows. Continue doing this for at least three more circles, one of which should include the nipple. Then repeat the procedure for the right breast. For a complete examination, repeat the procedure while standing in a shower. Hands glide more easily over wet skin.

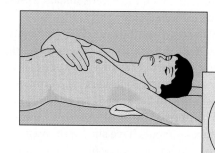

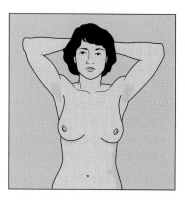

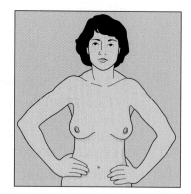

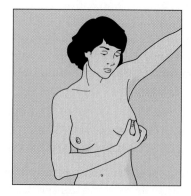

2. Stand before a mirror, lift your arms over your head, and look for any unusual changes in the contour of your breasts, such as a swelling, dimpling, or retraction (inward sinking) of the nipple. Also check for any unusual discharge from the nipple.

If you discover a lump or any other change during a breast self-examination, it's important to see a physician at once. Most changes are not cancerous, but let the doctor make the diagnosis.

b

MAJOR TYPES OF CANCER

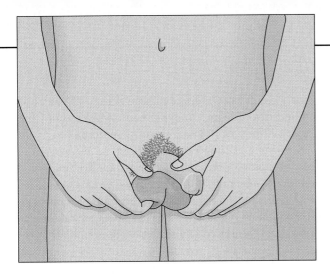

Figure 23.11 The method for testicular self-examination recommended by the American Cancer Society.

Do the exam when the scrotum is relaxed, as it is after a warm bath or shower. Simply roll each testicle between the thumb and forefinger, feeling for any lumps or thickening. By performing the exam regularly, you can detect changes early on and discuss them with your physician. As with breast lumps, most such changes are *not* cancer, but only a doctor can make the diagnosis. (*right*) International cycling champion Lance Armstrong, a survivor of testicular cancer.

UTERINE AND OVARIAN CANCER AFFECT WOMEN

Cancers of the uterus most often affect the endometrium (uterine lining) and the cervix. Various types are treated by surgery, radiation, or both. The incidence of uterine cancers is falling, in part because precancerous phases of cervical cancer can be easily detected by the *Pap smear* that is part of a routine gynecological examination. The risk factors for cervical cancer include having many sex partners, early age of first intercourse, cigarette smoking, and genital warts (Section 16.11). Endometrial cancer is more common during and after menopause.

Ovarian cancer is often lethal because its chief symptom, an enlarged abdomen, doesn't show up until the cancer is advanced and has already metastasized. Sometimes the first sign is abnormal vaginal bleeding or vague abdominal discomfort. Risk factors include family history of the disease, childlessness, and a history of breast cancer. Taxol, a compound originally derived from the Pacific yew tree, is now used to treat ovarian cancer, with moderate success. It can even shrink tumors significantly in 20–30 percent of advanced cases, helping to prolong a patient's life.

TESTICULAR AND PROSTATE CANCER AFFECT MEN

Several thousand cases of testicular cancer are diagnosed annually in the United States. In its early stages, this cancer is painless. However, it can spread to lymph nodes in a male's abdomen, chest, neck, and, eventually, his lungs. Once a month from high school onward, men should examine each testicle separately after a warm bath or shower (when the scrotum is relaxed). The testis should be rolled gently between the thumb and forefinger

to check for any unusual lump, enlargement, or hardening (Figure 23.11). Because the epididymis may be confused with a lump, the important thing is to *compare the two testes*. A lump may or may not be painful, but only a physician can rule out the possibility of disease. Surgery is the usual treatment, and the success rate is high when the cancer is caught before it can spread.

Prostate cancer is second only to lung cancer in causing cancer deaths in men. There are no confirmed risk factors. Symptoms include various urinary problems, although these can also signal simply a noncancerous enlarged prostate. For men over 40, an annual digital rectal examination, which enables a physician to feel the prostate, is the first step in detecting unusual lumps. The PSA blood test can screen for suspiciously large amounts of that tumor marker. If a physician suspects cancer after these two tests have been performed, the next step is a tissue biopsy. The cure rate for prostate tumors detected early is over 90 percent.

Many prostate cancers grow slowly and seem to cause few problems. In such cases a physician may recommend simply monitoring the tumor, until there is a clear threat to health.

Breast cancer affects about one woman in eight. Rarely, it also occurs in males.

The most common reproductive cancers in women develop in the ovaries and uterus. In males, cancers of the testis and prostate gland are the most common reproductive cancers.

Chances for a cure are best when cancer is detected early. Monthly self-examination is a crucial tool for early detection of breast cancer and testicular cancer.

23.8 A Survey of Other Common Cancers

LINKS TO
SECTIONS
4.9, 11.7,
AND 16.11

It seems that no tissue or organ system is immune from cancer's attack. This section surveys five major groups.

ORAL AND LUNG CANCERS

Cancers of the lips, mouth, tongue, salivary glands, and throat are most common among smokers and people who use smokeless tobacco (snuff), especially if they are also heavy alcohol drinkers. In 2005, the American Cancer Society reported 29,300 new cases of oral cancer, which can be especially deadly. Of people who are diagnosed with oral cancer, only about half survive for 5 years, even with treatment.

Lung cancer kills more people than any other cancer. Long-term tobacco smoking is the overwhelming risk factor. (Recent evidence suggests that one chemical in tobacco smoke, benzoypyrene, specifically mutates the p53 tumor suppressor gene described in Section 23.2.) Other risk factors are exposure to asbestos, to industrial chemicals such as arsenic, and to radiation. For a smoker, a combination of these factors greatly boosts the odds of developing cancer. For nonsmokers, especially spouses and children of smokers, inhaling secondhand smoke also poses a significant cancer risk. The Environmental Protection Agency estimates that lung cancer resulting from regularly breathing secondhand smoke kills 3,000 people in the United States each year.

In recent years, the incidence of lung cancer has decreased in men. However, a rise in the relative number of female smokers in the past several decades is now reflected in the fact that lung cancer is far and away the leading cancer killer of women. Warning signs include a nagging cough, shortness of breath, chest pain, blood in coughed-up phlegm, unexplained weight loss, and frequent respiratory infections or pneumonia.

Four types of lung cancer account for 90 percent of cases. About one-third of lung cancers are **squamous cell carcinomas**, affecting squamous epithelium in the bronchi. Another 48 percent are either **adenocarcinomas** or **large-cell carcinomas**. A fourth type, called **small-cell carcinoma**, spreads rapidly and kills most of its victims within 5 years.

CANCERS OF THE STOMACH, PANCREAS, COLON, AND RECTUM

Cancers that develop in the stomach and pancreas are usually adenocarcinomas of duct cells. They're often not detected until they have spread. Cigarette smoking is a risk factor for pancreatic cancer, while stomach cancer may be associated with heavy alcohol consumption and a diet rich in smoked, pickled, and salted foods. Liver cancer is uncommon in the United States, but we now know that hepatitis B infection can trigger it.

Most cancers of the colon and rectum—or colorectal cancers—are adenocarcinomas, and many get their start in growths called polyps (Figure 23.12). Warning signs include a change in bowel habits, rectal bleeding, and blood in the feces. A family history of colorectal cancer or inflammatory bowel disease is a major risk factor.

URINARY SYSTEM CANCERS

Carcinomas of the bladder and kidney account for about 100,000 new cancer cases each year. The incidence is higher in males, and smoking and exposure to certain industrial chemicals are major risk factors. Kidney cancer easily metastasizes via the bloodstream to the lungs, bone, and liver. An inherited type, called Wilms tumor, is one of the most common of all childhood cancers.

CANCERS OF THE BLOOD AND LYMPHATIC SYSTEM

Lymphomas develop in organs such as the lymph nodes, spleen, and thymus. They include the diseases known as **non-Hodgkin lymphoma**, **Hodgkin's disease**, and **Burkitt lymphoma**. Risk seems to increase along with viral infections—such as HIV—that impair functioning of the immune system. Lymphoma symptoms include enlarged lymph nodes, rashes, weight loss, and fever. Intense itching and night sweats also are typical of Hodgkin's disease. Chemotherapy and radiation are the standard treatments, and new treatments using targeted monoclonal antibodies are being developed.

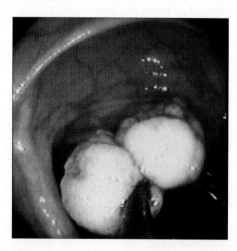

Figure 23.12 A polyp on the colon wall. Polyps can be a precursor to colorectal cancer. This picture was taken during a colonoscopy, a key screening test in which a slender, flexible, lighted tube is inserted into the colon. The tube is connected to a video camera and display monitor. Suspicious polyps can be removed and examined for signs of cancerous cells.

MAJOR TYPES OF CANCER

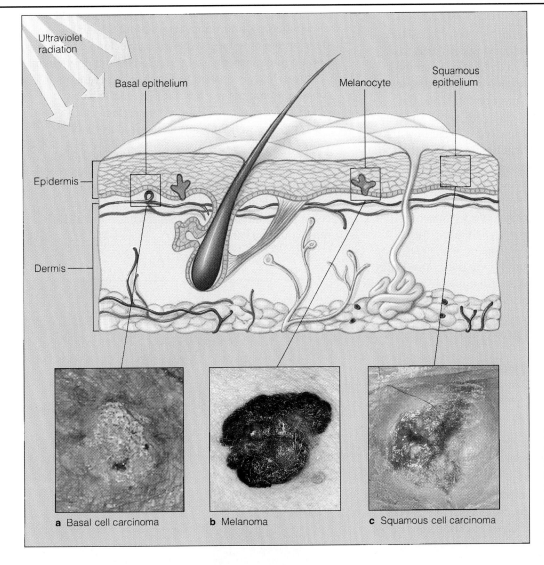

Figure 23.13 (**a**) Basal cell carcinoma, (**b**) malignant melanoma, and (**c**) squamous cell carcinoma. Each type of skin cancer has a distinctive appearance. Malignant melanoma can be deadly because it metastasizes aggressively.

a Basal cell carcinoma

b Melanoma

c Squamous cell carcinoma

Leukemias are cancers in which stem cells in bone marrow overproduce white blood cells. Some types are the most common childhood cancers, but other types are diagnosed more often in adults. Risk factors include Down syndrome, exposure to chemicals such as benzene, and radiation exposure. Many leukemias can be treated effectively with chemotherapy using two compounds—vincristine and vinblastine—derived from species of periwinkle plants that grow only on the African island of Madagascar. Environmentalists often cite these drugs and other plant-derived cancer-fighting substances as reasons why we should try to preserve wilderness areas where medicinal plants may be discovered.

SKIN CANCER

Skin cancers are the most common of all cancers. For all types, fair skin and exposure to ultraviolet (UV) radiation in sunlight (or tanning salons) are risk factors. UV damage to the DNA in melanocytes, the melanin-producing cells in the skin's epidermis, is the most dangerous kind,

called malignant melanoma (see Section 4.9). Malignant melanoma is a grave threat because in its later stages it metastasizes quickly.

Squamous cell (epidermal) carcinomas start out as scaly, reddened bumps (Figure 23.13c). They grow rapidly and can spread to nearby lymph nodes unless they are surgically removed. Basal cell carcinomas begin as small, shiny bumps and slowly grow into ulcers with beaded margins. Basal cell carcinoma and squamous cell carcinoma are much more common than malignant melanoma, and usually they are easily treated by minor surgery in a doctor's office.

Common cancers include oral and lung cancers and those of the digestive system (especially colorectal cancer), the urinary system, the blood and lymphatic systems (leukemias and lymphoma), and the skin. Each year, these collectively account for more than 750,000 cases of cancer in the United States alone.

Summary

Sections 23.1–23.2 Overgrowing cells lead to a tissue mass called a tumor. In dysplasia, a common precursor to cancer, cells develop abnormalities in size, shape, and organization. Cancer results when the genetic controls over cell division are lost completely. Cancer cells differ from normal cells. To begin with, they have structural abnormalities, including (typically) poor differentiation and altered surface proteins. They also grow uncontrolled and can invade surrounding tissues, a process called metastasis.

Cancer develops through a process of carcinogenesis, which involves a series of genetic changes. Initially, mutation may alter a proto-oncogene into a cancer-causing oncogene. Infection by a virus can also insert an oncogene into a cell's DNA or disrupt normal controls over a proto-oncogene. One or more tumor suppressor genes probably must be missing or become mutated before a normal cell can be transformed into a cancerous one (Table 23.5).

A predisposition to a certain type of cancer can be inherited. Other causes of carcinogenesis are viral infection, chemical carcinogens, radiation, faulty immune system functioning, and possibly a breakdown in DNA repair.

Thomson NOW! *Learn more about cancer and metastasis.*

Section 23.4 Common methods for cancer diagnosis include blood testing for the presence of substances produced either by specific types of cancer cells or by normal cells in response to the cancer. Medical imaging (as by magnetic resonance imaging) also can aid diagnosis. Biopsy provides a definitive diagnosis.

Section 23.5 In general, a cancer is named according to the type of tissue in which it arises. Common ones include sarcomas (connective tissues such as muscle and bone), carcinomas (epithelium), adenocarcinomas (glands or their ducts), lymphomas (lymphoid tissues), and leukemias (blood-forming regions).

Table 23.5	Cancer Causes and Contributing Factors
Cause/Factor	**Impact**
Oncogene	May alter control of cell division
Faulty tumor suppressor gene	Fails to halt runaway cell division
Viral infection	Switches proto-oncogene to oncogene or inserts an oncogene into the host cell DNA
Carcinogen	Damages DNA
Radiation	Damages DNA
Faulty immunity	Fails to tag cancer cells for destruction

Section 23.6 Cancer treatments include surgery, chemotherapy, and tumor irradiation. Under development are target-specific monoclonal antibodies and immune therapy using interferons and interleukins.

Lifestyle choices such as the decision not to use tobacco, to maintain a low-fat diet, and to avoid overexposure to direct sunlight and chemical carcinogens can help limit personal cancer risk.

Sections 23.7–23.8 Among adult females, the most prevalent cancer sites are the breasts, colon and rectum, lungs, and uterus. Among adult males the most prevalent sites are the prostate gland, lung, colon and rectum, and bladder. Skin cancer is the most common cancer overall.

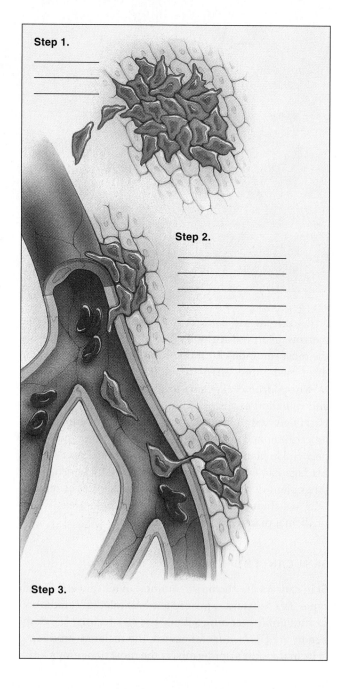

Step 1.

_____
_____
_____

Step 2.

_____
_____
_____
_____
_____
_____

Step 3.

_____
_____

Review Questions

1. How are cancer cells structurally different from normal cells of the same tissue? What is the relevance of altered surface proteins to uncontrolled growth?

2. What are the differences between a benign tumor and a cancerous one?

3. Write a short paragraph that summarizes the roles of proto-oncogenes, oncogenes, and tumor suppressor genes in carcinogenesis.

4. List the four main categories of cancer tumors.

5. What are the seven warning signs of cancer? (Remember the American Cancer Society's clue word, CAUTION.)

6. Using the diagram on the facing page as a guide, indicate the major steps in cancer metastasis.

Self-Quiz

Answers in Appendix V

1. A tumor is _____.
   a. malignant by definition
   b. always enclosed by connective tissue
   c. a mass of tissue that may be benign or malignant
   d. always slow-growing

2. Cancer cells _____.
   a. lack normal controls over cell division
   b. secrete the growth factor angiogenin
   c. display altered surface proteins
   d. are not inhibited by contact with other cells
   e. all of the above

3. The onset of cancer seems to require the activity of an oncogene plus the absence or mutation of at least one _____.

4. Chemical carcinogens _____.
   a. include viral oncogenes
   b. can damage DNA and cause a mutation
   c. must be ingested in food
   d. are not found in foods

5. So far as we know, carcinogenesis is *not* triggered by _____.
   a. breakdowns in DNA repair
   b. a breakdown in immunity
   c. radiation
   d. protein deficiency
   e. inherited gene defects

6. Tumor suppressor genes _____.
   a. occur normally in cells
   b. promote metastasis
   c. are brought into cells by viruses
   d. only rarely affect the development of cancer

7. _____ is the definitive method for detecting cancer.
   a. Blood testing
   b. Physician examination
   c. Biopsy
   d. Medical imaging

8. The most common therapeutic approaches to treating cancer include all of the following except _____.
   a. chemotherapy
   b. irradiation of tumors
   c. surgery to remove cancerous tissue
   d. administering doses of vitamins

9. The goal of immune therapy is to _____.
   a. cause defective T cells in the thymus to disintegrate
   b. activate cytotoxic T cells
   c. dramatically increase the numbers of circulating macrophages
   d. promote the secretion of monoclonal antibodies

10. Currently, _____ cancer is the leading cause of death among adult females; _____ cancer is the leading cause of cancer death among adult males.
   a. lung; prostate
   b. breast; colon
   c. lung; lung
   d. breast; lung

Critical Thinking

1. Propose an explanation of the observation that higher rates of cancer are associated with increasing age.

2. A textbook on cancer contains the following statement: "Fundamentally, cancer is a failure of the immune system." Why does this comment make sense?

3. Ultimately, cancer kills because it spreads and disturbs homeostasis. Consider, for example, a kidney cancer that metastasizes to the lungs and liver. What are some specific homeostatic mechanisms that the spreading disease could disrupt?

4. Over the last few months, your best friend Mark has noticed a small, black-brown, raised growth developing on his arm. When you suggest that he have it examined by his doctor, he says he's going to wait and see if it gets any larger. You know that's not a very smart answer. Give at least three arguments that you can use to try to convince Mark to seek medical advice as soon as possible.

5. Some desperate cancer patients consume pills or other preparations containing shark cartilage, which the manufacturers tout as an anti-angiogenesis compound. The basis for these claims is the fact that blood vessels do not grow into cartilage. Responsible researchers point out that, regardless of the properties of cartilage, there is no way that *eating* it could provide any anticancer benefit. Why is this correct?

Explore on Your Own

Most families have been touched by cancer in one way or another. The American Cancer Society website is a portal to a huge amount of reliable information on the risks for, causes of, and treatments for virtually any cancer. Choose a cancer to investigate and see how much you can learn about it in just half an hour. Does your research give you any new insights into your own risk for the cancer? What is your reaction to the stories of cancer survivors that are posted on the website?

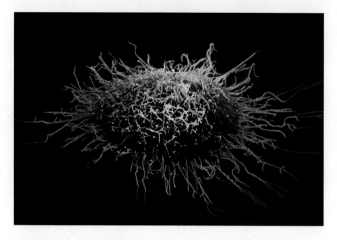

Scanning electron micrograph of a cervical cancer cell.

Measuring Time

How do you measure time? Probably you can relate to a few centuries of human events. But geologic time? Thinking about the distant past takes us deep into unknown territory. Yet there are clues about past events if we know where to look for them.

Consider asteroids—rocky, metallic objects hurtling through space. They range in size from a few meters to 1,000 kilometers (about 750 miles) across. At least 6,000 of them orbit the sun and dozens cross Earth's orbit.

In the past, many asteroid impacts altered the course of evolution. A layer of iridium all around the world tells us that one struck the Earth about 65 million years ago. The impact led to the extinction of the dinosaurs and many other life forms. Iridium is rare on Earth, but not in asteroids. Fossils of dinosaurs have never been found above that layer.

It has been only about 100,000 years since modern humans evolved. In fact, dozens of humanlike species evolved in Africa during the 5 million years before our species—*Homo sapiens*—even showed up. So why are we the only ones left?

Early species of humans lived in small bands. What if most were casualties of the estimated twenty asteroids that struck when they were alive? What if *our* ancestors were just plain lucky? About 2.3 million years ago, one huge object from space hit the ocean, west of what is now Chile. If it had collided with the rotating Earth just a few hours earlier, our ancestors in southern Africa would have been wiped out, and we wouldn't be around today. Looking toward the future, astrophysicists predict that in 2028 a large asteroid will sweep close to the Earth or the Moon. Probably, some say, a little too close for comfort.

Thinking about time and the changes it can bring is our task in this chapter. We will start by considering the basic principles of evolution and the kinds of evidence evolutionary biologists use in their work. Later we'll survey key trends in human evolution, and close the chapter with a look at what is currently known about the chemical origins of life on our planet.

 How Would You Vote? A large asteroid impact could obliterate civilization and much of Earth's biodiversity. Should we spend millions, even billions, of dollars to search for and track asteroids? Cast your vote online at www.thomsonedu.com/biology/starr/humanbio.

 ## Key Concepts

BASICS OF EVOLUTION
The theory of evolution by natural selection reflects observations of the living world and ideas about how populations of organisms interact with their environment.

EVIDENCE FOR EVOLUTION
Clues from biogeography, fossils, and comparisons of body form, development, and biochemistry provide evidence of evolution—inherited changes in lines of descent.

HUMAN EVOLUTION
Trends in the evolution of humans include upright walking, refined vision and hand movements, and development of a complex brain and behaviors.

LIFE'S ORIGINS
Molecules of life and the first living cells are thought to have emerged on Earth about 3.8 billion years ago.

Links to Earlier Concepts

This chapter returns to a basic concept in biology, the evolution of life on Earth (1.2). The discussion in Chapter 1 of the use of the term "theory" in science also applies to this chapter's discussion of one of the cornerstone ideas in biology, the theory of evolution by natural selection (1.7).

Your reading here will draw on your knowledge of genes (20.1, 22.1) and mutations and changes in the structure and functioning of chromosomes (21.9), and of how environmental factors can modify the expression of genes (20.8). The section on life's origins builds on the earlier discussion of enzymes and how organic compounds are assembled (2.8) and of amino acids (2.11) and cell membranes (3.1).

24.1 A Little Evolutionary History

In Latin, the word evolutio *means "the act of unrolling." That image is an apt way to begin thinking of biological evolution.*

Biologists define **evolution** as genetic change in a line of descent through successive generations. All told, the evolution of life on Earth has resulted in many hundreds of thousands of species. Of those living today, some are closely related, others more distantly. As you'll read later in this chapter, we still can find biochemical evidence that we humans share a common ancestor with life forms as different from us as bacteria and corn plants. Shared ancestry is one of the core ideas in evolution.

In 1831, the source of the Earth's amazing diversity of life forms was a matter of hot dispute. Many people believed that all species had come into existence at the same time in the distant past, at the same center of creation, and had not changed since. Yet, by the mid-nineteenth century, several hundred years of exploration and advances in the sciences of geology and comparative anatomy had raised questions. For example, why were some species found in particular isolated regions, and nowhere else? Why, on the other hand, were similar (but not identical) species found in widely separated parts of the world? Why did species as different as humans, whales, and bats have certain strikingly similar anatomical structures? What was the significance of geologists' discoveries of similar fossil organisms in similar layers of the Earth's sedimentary rocks—regardless of where in the world the layers occurred?

Enter Charles Darwin, who in 1831 was 22 years old, had a theology degree from Cambridge, and wasn't sure what he wanted to do with his life. He wasn't interested in becoming a clergyman. All he had ever wanted to do was hunt, fish, collect shells, or simply watch insects and birds. John Henslow, a botanist who befriended Darwin during his university years, saw the young man's real interests. He arranged for Darwin to work as a naturalist aboard the HMS *Beagle*, which was about to embark on a five-year voyage around the world (Figure 24.1). The *Beagle* sailed first to South America, to finish up work on mapping the coastline. During the long Atlantic crossing, Darwin studied geology and collected and examined marine life. During stops along the coast and at various islands, he observed other species of organisms in environments ranging from sandy shores to high mountains. After returning to England in 1836, Darwin began talking with other naturalists about a topic that was on many scholars' minds—the growing evidence that life forms evolve, changing over time.

But *how* could organisms evolve? One clue came from Thomas Malthus, a British clergyman and economist, who proposed that any population will tend to outgrow its

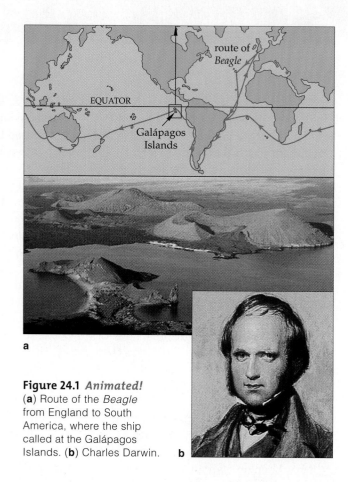

Figure 24.1 *Animated!*
(**a**) Route of the *Beagle* from England to South America, where the ship called at the Galápagos Islands. (**b**) Charles Darwin.

resources, and so in time its members must compete for what is available. Darwin's observations also suggested that any population can produce more individuals than the environment can support, yet populations tend to be stable over time. For instance, a starfish can release 2,500,000 eggs a year, but the oceans are not filled with starfish. What determines who lives and who dies as predators, starvation, and environmental events take their toll on a population? Chance could be a factor, but a second major clue to the evolution puzzle came from observations by Darwin and others that even members of the same species vary slightly in their traits.

Darwin's melding of his observations of the natural world with the ideas of other thinkers led him to propose that evolution could occur by way of a process called **natural selection**. Widespread acceptance of this theory would not come until nearly 70 years later, when a new field, genetics, provided insights into how traits could vary. Now let's consider some current views of the mechanisms of evolution.

Combining observations of the natural world with ideas about interactions of populations with their environment, Darwin forged his theory of evolution by natural selection.

24.2 A Key Evolutionary Idea: Individuals Vary

Evolution occurs in populations of organisms. It begins when the genetic makeup of a population changes.

The history of life on Earth spans nearly 4 billion years. It is a story of how species originated, survived or went extinct, and stayed put or spread into new environments. The overall "plot" of the story is evolution, genetic change in lines of descent over time. **Microevolution** is the name for cumulative genetic changes that may give rise to new species. **Macroevolution** is the name for the large-scale patterns, trends, and rates of change among *groups* of species. Later on you'll get a fuller picture of these two patterns. Here, we begin with a fundamental principle of evolution: the traits of individuals in a population vary.

INDIVIDUALS DON'T EVOLVE—POPULATIONS DO

An individual fish, flower, bacterium, or person does not evolve. Evolution occurs only when there is change in the genetic makeup of *populations of organisms*. In biology, the definition of a population is very specific: A **population** is a group of individuals of the same species occupying a given area. As you know from your own experience with other human beings, there is a great deal of genetic variation within and among populations of the same species (Figures 24.2 and 24.3).

Overall, the members of a population have similar traits—that is, phenotypes. They have the same general form and appearance (*morphological* traits), their body

LINKS TO SECTIONS 20.1 AND 21.1

Figure 24.3 Fabulous variation among shell patterns and colors of individuals of the snail populations on Caribbean islands.

structures function in the same way (*physiological* traits), and they respond the same way to certain basic stimuli (*behavioral* traits). However, the details of traits vary quite a bit. For instance, individual humans vary in the color of their body hair, as well as in its texture, amount, and distribution over the body. As we know from previous chapters, this example only hints at the immense genetic variation in human populations. Populations of most other species show the same kinds of variation.

VARIATION COMES FROM GENETIC DIFFERENCES

In theory, the members of a population have inherited the same number and kinds of genes. These genes make up the population's **gene pool**. Remember, though, that each kind of gene in the pool may have slightly different forms, called alleles. Variations in traits in a population—hair color, say—result when individuals inherit different combinations of alleles. Whether your hair is black, brown, red, or blond depends on *which* alleles of certain genes you inherited from your mother and father.

If you go through a bag of chocolate candies that have different-colored sugar coatings, you'll see that some colors turn up more or less often than others do. The same holds true for the gene alleles in a population. Some are much more common than others. The manufacturer can adjust the number of "reds" or "blues" in the overall mix of candy pieces, but where genes are concerned, changes come about by mutation, natural selection, and other processes. Those changes can lead to microevolution, which we consider next.

Figure 24.2 A small sample of the outward variation in *Homo sapiens*. Variation in traits arises from different combinations of alleles carried by different members of populations.

 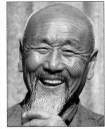 

A population is a group of individuals of the same species occupying a given area.

Individuals in a population share many general phenotypes, but in most populations there also is a great deal of underlying genetic variation.

Evolution is genetic change in lines of descent over time. It results from changes in the genetic makeup of populations of organisms.

24.3 Microevolution: How New Species Arise

LINKS TO
SECTIONS
1.7, 21.7, AND 21.9

DNA mutations in the germ cells that form gametes can be passed from generation to generation. Most people carry several mutations. From the perspective of evolution, mutation is significant because mutations are the source of alleles—the alternative forms of genes.

MUTATION PRODUCES NEW FORMS OF GENES

Because mutations are accidental changes in DNA, most are probably harmful, altering traits such that an affected individual can't survive or reproduce as well as other individuals. For example, for us humans, small cuts or other minor injuries are common. But before effective medical treatments existed, people with hemophilia, whose blood does not clot properly, could die young from such minor injuries. As a result, the various hemophilias were rare, because the faulty genes were rarely passed on. By contrast, a *beneficial trait* improves some aspect of an individual's functioning in the environment and so improves the chances of surviving and reproducing. A *neutral trait*, such as attached earlobes in humans, neither helps nor hinders survival under prevailing conditions.

NATURAL SELECTION CAN RESHAPE THE GENETIC MAKEUP OF A POPULATION

Microevolution leads to new species. Actually, several processes are included in this category, but the one that probably accounts for most changes in the mix of alleles in a gene pool is natural selection. As Section 24.1 noted, Darwin formulated his **theory of evolution by natural selection** by correlating his understanding of inheritance with certain features of populations. In 1859 he published his ideas in a classic book, *On the Origin of Species*. In modern terms, the main points of Darwin's insight might be expressed as follows:

1. **The individuals of a population vary in their body form, functioning, and behavior**.

2. **Many variations can be passed from generation to generation**. This simply means that different versions of genes—alleles—can pass from parents to offspring.

3. **In every set of circumstances, some versions of a trait are more advantageous than others**. That is, some traits impart a better chance of surviving and reproducing. The expression "survival of the fittest" is verbal shorthand for this advantage.

4. **Natural selection is the difference in survival and reproduction** that we observe in individuals who have different versions of a trait. The "fittest" traits—and the gene alleles that govern them—are more likely to be "selected" for survival.

5. **A population is evolving when some forms of a trait are becoming more or less common** relative to the other forms. The shifts are evidence that the corresponding versions of genes are becoming more or less common.

6. **Over time, shifts in the makeup of gene pools** have been responsible for the amazing diversity of life forms on Earth.

As natural selection occurs over time, organisms come to have characteristics that suit them to the conditions in a particular environment. We call this trend **adaptation**.

You have probably heard references to the "theory" of evolution. Yet decades of careful scientific observations have yielded so much evidence of evolution and natural selection that both ideas are more properly considered fundamental principles of the living world. Later we will consider a few examples of this evidence.

CHANCE CAN ALSO CHANGE A GENE POOL

Natural selection is not the only process that can adjust the relative numbers of different alleles in a gene pool. Chance can also play a major role. This kind of gene pool "tweaking" is called **genetic drift**. Often the change is most rapid in small populations. In one type of genetic drift, called the *founder effect*, a few individuals leave a population and establish a new one. Purely by chance, the

Finnish boy

relative numbers of various alleles in the new population are likely to differ from those in the old group. For example, geneticists have strong evidence that ethnic Finns are descended from a small band of people who settled in what is now Finland about 4,000 years ago. Today, blond hair and blue eyes are distinctive Finnish features. In addition, at least 30 genetic disorders that are rare or absent elsewhere are common in Finland.

The makeup of a gene pool also can change as new individuals enter a population or other ones leave it. This physical movement of alleles, or **gene flow**, helps keep neighboring populations genetically similar. Over time it tends to counter the differences between populations that are brought about through mutation, genetic drift, and natural selection. In our modern age of international travel, the pace of gene flow among human populations has increased dramatically.

In Finland, there has historically been little gene flow. Climate and geography have isolated Finns from the rest of Europe for centuries. Even now, Finns have much less genetic variation than other Europeans.

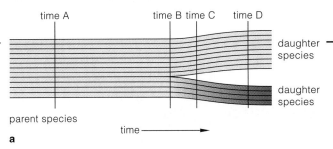

Figure 24.4 (**a**) Divergence, the first step on the path to speciation. Each horizontal line represents a different population. Because evolution is gradual, we cannot say at any one point in time that there are now two species rather than one. At time A there is only one species. At D there are two. At B and C the split has begun but isn't complete. (**b**) Artist's view of *Homo habilis* males in an East African woodland.

THE ABILITY TO INTERBREED DEFINES A SPECIES

For all sexually reproducing organisms, including us humans, a **species** is a genetic unit consisting of one or more populations of organisms that usually closely resemble each other physically and physiologically. Members of a species can interbreed and produce fertile offspring under natural conditions. No matter how diverse their traits are, those individuals belong to the same species as long as they can interbreed successfully and so share a common gene pool. From this perspective, although a female lawyer in India may never meet up with an Icelandic fisherman, there is no biological reason why the two could not mate and produce children. But neither person could mate successfully with a chimpanzee, even though chimps and humans are closely related species and have more than 90 percent of their genes in common. In evolutionary terms, humans and chimpanzees are "reproductively isolated." Their genetic differences ensure that they can't mate and produce fertile offspring.

Reproductive isolation develops when gene flow between two populations stops. This often occurs when two populations become separated geographically. For example, the fossil record indicates that early members of the genus *Homo* diverged genetically, possibly when one or more populations emigrated out of Africa toward Asia and Europe. When populations are in different environments, mutation, natural selection, and genetic drift begin to operate independently in each one. These processes can change the gene pools of each in different ways, and eventually the differences can result in aspects of structure, function, or behavior that reduce the chances of successful interbreeding. The two populations may breed in different seasons, for example, or they may have different mating rituals. There may be changes in body structure that physically interfere with mating. Other types of isolating mechanisms prevent zygotes or hybrid offspring from developing properly.

The buildup of genetic differences between isolated populations is called *divergence* (Figure 24.4a). When the genetic differences are so great that members of the two populations can't interbreed successfully, **speciation** has occurred: the populations are now separate species. Figure 24.4b is an artist's rendition of an extinct human species, *Homo habilis*. Clearly, there are many differences between *H. habilis* and our own species, *Homo sapiens*.

SPECIATION CAN OCCUR GRADUALLY OR IN "BURSTS"

Although we know that in the final analysis factors such as gene mutation and natural selection trigger speciation, biologists disagree about the pace and timing of microevolution. According to the traditional model, called *gradualism*, new species emerge through many small changes in form over long spans of time. In other words, microevolution is constantly going on in tiny increments; new species result. By contrast, according to the model called *punctuated equilibrium*, most evolutionary change occurs in bursts. That is, each species undergoes a spurt of changes in form when it first branches from the parental lineage, then changes very little for the rest of its time on Earth.

There is evidence that the driving force behind rapid changes in species may be dramatic changes in climate or some other aspect of the physical environment. This type of sudden change (in evolutionary terms) alters the physical conditions to which populations of organisms have become adapted. For example, the onset of an ice age changes the living conditions for species in affected land areas. A major shift in ocean currents (which help determine water temperature) might have the same effect on marine species. Punctuated equilibrium could help explain why the fossil record doesn't give much evidence of a continuum of microevolution—the record often lacks forms between closely related species. It's likely that both models have a place in explaining the history of life.

Natural selection and chance can alter the kinds and relative numbers of various gene alleles in a population. Building genetic changes can result in the evolution of new species. A species is a unit of one or more populations of individuals that can successfully interbreed under natural conditions.

24.4 Looking at Fossils and Biogeography

The fossil record and biogeography are important tools in reconstructing the intertwined journey of life and Earth.

A **fossil** is recognizable, physical evidence of ancient life (Figure 24.5). The more Darwin and others learned about the similarities and differences among fossils and living organisms, the stronger the evidence became that such diversity is the result of evolution—and specifically of evolution by natural selection as populations adapted to their surroundings.

FOSSILS ARE FOUND IN SEDIMENTARY ROCK

The most common fossils are bones, teeth, shells, seeds, and other hard parts of organisms. (When an organism dies, soft parts usually decompose first.) **Fossilization** begins with burial in sediments or volcanic ash. With time, water seeps into the organic remains, infusing them with dissolved metal ions and other inorganic compounds. As more and more sediments accumulate above the burial site, the remains are subjected to increasing pressure. Over great spans of time, the chemical changes and pressure transform them to stony hardness.

Organisms are more likely to be preserved when they are buried rapidly in the absence of oxygen. Entombment by volcanic ash or anaerobic mud satisfies this condition very well. Preservation also is favored when a burial site is not disturbed. Usually, though, fossils are broken, crushed, deformed, or scattered by erosion, rock slides, and other geologic events.

Fossil-containing layers of sedimentary rock formed long ago, when silt, volcanic ash, and other materials were gradually deposited, one above the other (Figure 24.6). This layering of sedimentary deposits is called *stratification*. Although most sedimentary layers form horizontally, geologic disturbances can tilt or break them.

Figure 24.6 The Grand Canyon of the American Southwest, once part of an ocean basin. Its sedimentary rock layers gradually formed over hundreds of millions of years.

COMPLETENESS OF THE FOSSIL RECORD VARIES

We currently have fossils of about 250,000 species, far more than in Darwin's day. However, judging from present diversity, there must have been millions of ancient, now-extinct species, and we will not be able to recover fossils for most of them. Why? There are several reasons. Most of the important, large-scale movements in the Earth's crust have wiped out evidence from crucial periods in the history of life. In addition, most members of ancient communities simply have not been preserved. For example, plenty of hard-shelled mollusks and bony fishes are represented in the fossil record. Jellyfishes and soft-bodied worms are not, even though they may have been common. Population density and body size skew the record more. A population of ancient plants may have produced millions of spores in one growing season, while the earliest members of the human lineage lived in small groups and produced few young. Therefore, the chance of finding a fossilized skeleton of an early human is small compared to the chance of finding spores of plant species that lived at the same time.

The fossil record is also heavily biased toward certain environments. Most species for which we have fossils lived on land or in shallow seas that, through geologic uplifting, became part of continents. We have only a few fossils from sediments beneath the ocean, which covers three-fourths of the Earth's surface. It's worth noting that most fossils have been discovered in the Northern Hemisphere. The reason is probably that most geologists have lived and worked there.

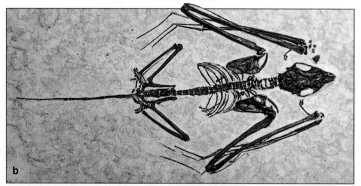

Figure 24.5 Examples of fossils. (**a**) Fossilized leaf of an ancient fern. (**b**) The skeleton of a bat that lived 50 million years ago. Fossils of this quality are rare.

EVIDENCE FOR EVOLUTION

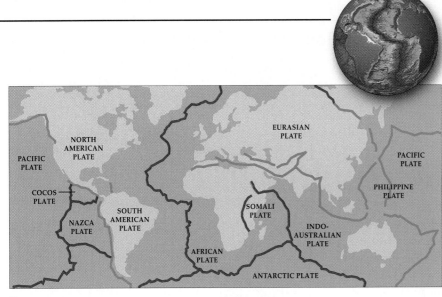

a

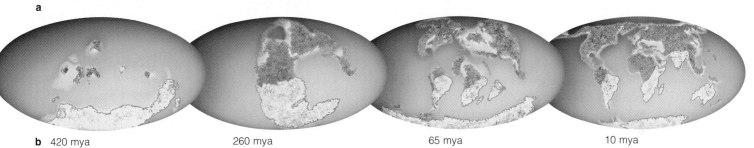

b   420 mya   260 mya   65 mya   10 mya

Figure 24.7 *Animated!*
(**a**) Plate tectonics, the arrangement of the Earth's crust in rigid plates that split apart, move about, and collide with one another. (**b**) About 240 million years ago (mya) Earth's land masses were joined in a massive supercontinent, Pangea. By about 40 million years ago plate movements had split Pangea into isolated land masses, including Africa, South America, Australia, and Eurasia.

How do we know how old a fossil is? Sedimentary rocks that contain fossils are dated by determining their position relative to nearby volcanic rocks. The age of the volcanic rocks is determined by **radiometric dating**. This method tracks the radioactive decay of an isotope of some element that had been trapped inside the rock when the rock formed. Like the ticking of a perfect clock, the decay rate is constant. Radiometric dating is about 90 percent accurate.

BIOGEOGRAPHY PROVIDES OTHER CLUES

Darwin also believed that the concept of evolution by natural selection could help shed light on the subject we know today as **biogeography**—the study of the world distribution of plants and animals. Biogeography asks the basic question of why certain species (and higher groupings) occur where they do. For example, why do Australia, Tasmania, and New Guinea have species of monotremes (egg-laying mammals such as the duckbilled platypus), while such animals are absent from other regions of the world where the living conditions are similar? And why do we find the greatest diversity of life forms in tropical regions? The simplest explanation for such biogeographical patterns is that species occur where they do either because they evolved there from ancestral species or because they dispersed there from somewhere else.

Charles Darwin probably would have been fascinated to learn of modern *plate tectonics*, the movement of plates of the Earth's crust (Figure 24.7). From studying evidence of such movements, we know that early in our planet's history all present-day continents, including Africa and South America, were parts of a massive "supercontinent" called Pangea (Figure 24.7*b*). By determining the locations of plates at different times in Earth's history, researchers can shed light on possible dispersal routes for some groups of organisms and when (in geological history) the movements took place.

Fossils are present in layers of sedimentary rocks. The deeper the layers, the older the fossils. The completeness of the fossil record varies, depending on the kinds of organisms represented, where they lived, and how stable their burial sites have been.

Biogeographical patterns can provide clues to where a species arose. Along with evidence from plate tectonics, the patterns also may shed light on the routes by which some groups of organisms dispersed.

IMPACTS, ISSUES

Iridium is rare on Earth but not on asteroids. A thin, dark layer of it dates to the Cretaceous-Tertiary (K-T) boundary, a sign that the Earth took a huge hit at that time, about 65 million years ago. Iridium-rich dust blew skyward and then drifted all over the globe. It marks the K-T mass extinction.

24.5 Comparing the Form and Development of Body Parts

Beyond considering fossils and how different life forms are distributed, early evolutionary thinkers also noted patterns in the form of body parts. Modern biologists have shed light on evolutionary history by comparing stages of development in major groups of organisms.

COMPARING BODY FORMS MAY REVEAL EVOLUTIONARY CONNECTIONS

Comparative morphology uses information contained in patterns of body form to reconstruct evolutionary history. When populations of a species branch out in different evolutionary directions, they diverge in their appearance, the functions of certain body parts, or both. Yet the related species also remain alike in many ways, because their evolution modifies a shared body plan. In such species we typically see **homologous structures**. These are the same body parts that have been modified in different ways in different lines of descent from a common ancestor (recall that *homo-* means "same").

For example, the same ancestral organism probably gave rise to most land-dwelling vertebrates, which have homologous structures. Apparently their common ancestor had four five-toed limbs. The limbs diverged in form and became wings in pterosaurs, birds, and bats (Figure 24.8). All these wings are homologous—they have the same parts. The five-toed limb also evolved into the flippers of porpoises and the anatomy of your own forearms and fingers.

Can body parts in organisms that *don't* have a recent common ancestor come to resemble one another in form and function? Yes. These **analogous structures** (from *analagos*, meaning "similar") arise when different lineages evolve in the same or similar environments. Different body parts, which were put to similar uses, were modified through natural selection and ended up resembling one another. This pattern of change is called **morphological convergence**. For example, a dolphin, a fast-swimming marine mammal, has a sleek, torpedo-shaped torso—and so does a tuna, a fast-swimming fish.

DEVELOPMENT PATTERNS ALSO PROVIDE CLUES

Vertebrates include fishes, amphibians, reptiles, birds, and mammals. Yet despite how different these groups are, comparing the ways in which their embryos develop provides strong evidence of their evolutionary links.

Early in development, the embryos of all the different vertebrate lineages go through strikingly similar stages (Figure 24.9). During vertebrate evolution, mutations that disrupted an early stage of development would have had devastating effects on the organized interactions required for later stages. Evidently, embryos of different

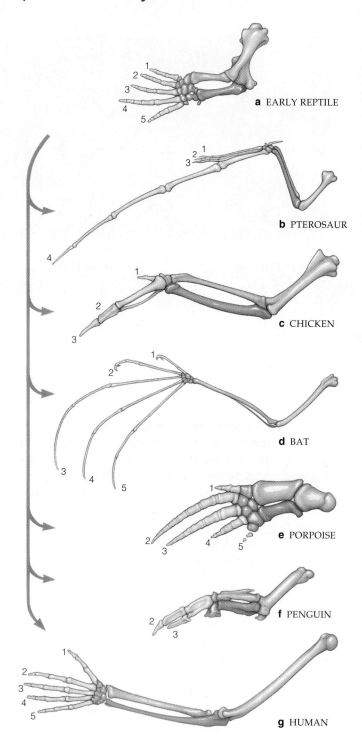

Figure 24.8 *Animated!* Morphological divergence in the vertebrate forelimb, starting with a generalized form of ancestral early reptiles. Diverse forms evolved even as similarities in the number and position of bones were preserved. The drawings are not to the same scale.

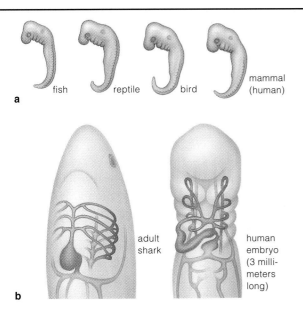

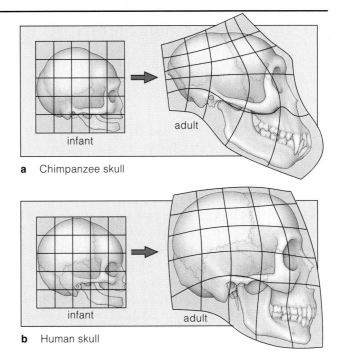

Figure 24.9 *Animated!* From comparative embryology, some evidence of evolutionary relationships among vertebrates. (**a**) Adult vertebrates show great diversity, yet the very early embryos retain striking similarities. This is evidence of change in a common program of development. (**b**) Fishlike structures still form in early embryos of reptiles, birds, and mammals. For example, a two-chambered heart (orange), certain veins (blue), and portions of arteries called aortic arches (red) develop in fish embryos and persist in adult fishes. The same structures form in an early human embryo.

Figure 24.10 *Animated!* Proportional changes in a chimpanzee skull (**a**) and a human skull (**b**). The skulls are quite similar in infants. Imagine that those represented here are paintings on a blue rubber sheet divided into a grid. Stretching the sheet deforms the grid's squares. For the adult skulls, differences in size and shape in corresponding grid sections reflect differences in growth patterns.

groups remained similar because mutations that altered early steps in development were selected against.

So how did the *adults* of different vertebrate groups come to be so different? At least some differences resulted from mutations that altered the onset, rate, or time of completion of certain developmental steps. Such mutations would bring about changes in shape through increases or decreases in the size of body parts. They also could lead to adult body forms that still have some juvenile features. For instance, Figure 24.10 illustrates how change in the growth rate at a key point in development may have produced differences in the proportions of chimpanzee and human skull bones, which are alike at birth. Later on, they change dramatically for chimps but only slightly for humans. Chimpanzees and humans arose from the same ancestral stock. Their genes are nearly identical. Even so, at some point on the separate evolutionary road leading to humans, some regulatory genes probably mutated in ways that proved adaptive. From then on, instead of promoting the rapid growth required for dramatic changes in skull bones, the mutated genes have blocked it.

As Figure 24.11 indicates, the bodies of humans, pythons, and other organisms can have what seem like useless *vestigial* structures. For example, consider your own ear-wiggling muscles—which four-footed mammals (such as dogs) use to orient their ears. In humans, such body parts are left over from a time when more functional versions were important for an ancestor.

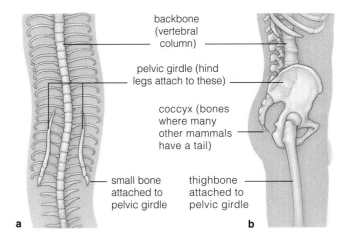

Figure 24.11 *Animated!* Python bones (**a**) corresponding to the pelvic girdle of other vertebrates, including humans. A snake has small vestigial hind limbs on its underside, remnants from a limbed ancestor. (**b**) The human coccyx is a similar vestige from an ancestral species that had a bony tail, although muscles still attach to it.

Comparative morphology gleans evolutionary information from observed patterns of body form. Similarities in patterns of development may be clues to evolutionary relationships among organisms.

24.6 Comparing Biochemistry

LINKS TO
SECTIONS
22.3 AND 22.4

Automated DNA sequencing now makes it relatively easy to identify biochemical similarities and differences that can shed light on evolutionary connections.

The kinds and numbers of traits species do (or don't) share are clues to how closely they are related. The same holds true for biochemical traits. Remember, the DNA of each species contains instructions for making RNAs and then proteins. This means that comparisons of DNA, RNA, and proteins from different species are additional ways of evaluating evolutionary relationships.

For example, by comparing body form you might suppose that monkeys, humans, chimpanzees, and other **primates** are related. You can then test your idea by looking for differences in the amino acid sequence of a protein, such as hemoglobin, that occurs in all primates. You also could decide to see whether the nucleotide sequences in their DNA match closely or not much at all. Logically, the species that are most similar in their biochemistry are the most closely related.

Suppose two species have the same gene. The amino acid sequences of the resulting protein are the same or nearly so. The absence of mutation implies that the species are closely related. What if the sequences are markedly different? Then many neutral mutations must have accumulated in them. A very long time must have passed since the species shared a common ancestor.

For instance, a wide range of organisms, from certain bacteria to corn plants to humans, produce cytochrome *c*, a protein of electron transport chains. Studies show that the gene coding for the protein has changed very little over vast spans of time. In humans, cytochrome *c* has a primary structure of 104 amino acids. Chimps have the identical amino acid sequence. It differs by 1 amino acid in rhesus monkeys, 18 in chickens, 19 in turtles, and 56 in yeasts. On the basis of this biochemical information, would you assume humans are more closely related to a chimpanzee or to a rhesus monkey? A chicken or a turtle?

Nucleotide sequences from various species also can be compared to estimate when a divergence took place. Neutral mutations are the key. Because they have little or no effect on survival or reproduction, neutral mutations become part of the DNA. Some researchers estimate that neutral mutations have accumulated in highly conserved genes (such as that for cytochrome *c*) at a constant rate. They are like a series of predictable ticks of a "molecular clock." The total number of ticks will "unwind" down through the past, stopping at the approximate time when lineages diverged onto separate evolutionary paths.

Species that are more closely related will be more similar biochemically than are distantly related species.

24.7 How Species Come and Go

The history of life on Earth is marked by extinction, and by the evolution of new species.

IN EXTINCTION, SPECIES ARE LOST

Extinction is the irrevocable loss of a species. Overall, species disappear at a fairly steady rate of "background extinction." A **mass extinction** is an abrupt, widespread rise in extinction rates above the background level. It is a catastrophic event in which major groups are wiped out simultaneously. Dinosaurs and many marine groups died out during a mass extinction that occurred around 65 million years ago—possibly the result of environmental changes that occurred after one or more large meteorites struck the Earth during a short period of time.

You may be aware that in the past few hundred years human activities have caused the extinction of thousands of species. The extinction rate is accelerating as we cut down forests, fill in wetlands, and otherwise destroy the habitats of other animals and plants with which we share the Earth. Global climate change is another factor that has contributed to patterns of extinctions.

IN ADAPTIVE RADIATION, NEW SPECIES ARISE

In **adaptive radiation**, new species of a lineage move into a wide range of habitats during bursts of microevolution. Many adaptive radiations have occurred during the first few million years after a mass extinction. Fossil evidence suggests this happened after dinosaurs went extinct. Many new species of mammals arose and radiated into habitats where dinosaurs had once lived.

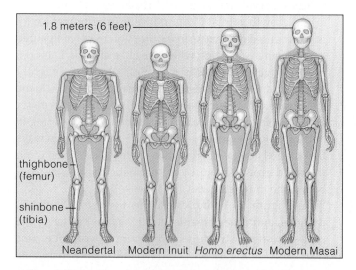

Figure 24.12 Climate and body build. Humans adapted to cold climates have a heat-conserving body—stockier, with shorter legs—compared to humans adapted to hot climates.

EVIDENCE FOR EVOLUTION

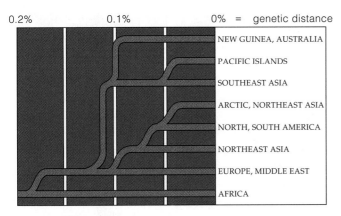

0.2%   0.1%   0% = genetic distance

NEW GUINEA, AUSTRALIA

PACIFIC ISLANDS

SOUTHEAST ASIA

ARCTIC, NORTHEAST ASIA

NORTH, SOUTH AMERICA

NORTHEAST ASIA

EUROPE, MIDDLE EAST

AFRICA

Figure 24.13 *Animated!* One proposed family tree for populations of modern humans (*Homo sapiens*) that are native to different regions of the world. The diagram's branch points (suggested by biochemical studies) show presumed, small genetic divergences between populations.

Adaptive radiations also have occurred in the human lineage. The ancestors of modern humans, including the tool-using species *Homo habilis* ("handy man," pictured in Figure 24.4*b*), apparently remained in Africa until about 2 million years ago. Around that time, genetic divergence led to new species, including *Homo erectus*, a human species that the fossil record places on the evolutionary road to modern humans. *H. erectus* coexisted for a time with *H. habilis*. But while members of its sister species also were upright walkers, *H. erectus* did the name justice; its populations walked out of Africa, going left into Europe or right into Asia. Judging from Middle Eastern fossils, *Homo sapiens* had evolved by 100,000 years ago.

What kinds of selection pressures triggered this adaptive radiation? We don't know, although this *was* a time of physical changes. One group of early humans, the Neandertals, had large brains and were massively built. Some Neandertal populations were the first humans to adapt to the coldest regions (Figure 24.12). Subsequent genetic modifications gave rise to anatomically modern humans. Figure 24.13 shows a possible family tree for those groups. The tree's branches represent separate lines of descent from a common ancestor.

As lineages evolve, they may undergo adaptive radiation, in which new species rapidly (on a geological time scale) fill a range of habitats. Lineages also lose species to extinction.

24.8 Endangered Species

No giant asteroids have hit the Earth for 65 million years. Yet a major extinction event is under way. Throughout the world, human activities are swiftly driving many species to extinction. An **endangered species** is an endemic species extremely vulnerable to extinction. *Endemic* means it originated in only one geographic region and lives nowhere else.

We are just one species among millions. Yet because of our rapid population growth, we are threatening others by way of removing natural habitats, overharvesting (such as of whales and many fishes), illegal wildlife trading, and a variety of other activities.

Habitat loss—the physical reduction in suitable places to live, or serious chemical pollution of them—is one of the major threats to more than 90 percent of endangered species. At the same time, many of these species are under pressure from exotic species—those that have been introduced from other areas, and which may outcompete the native creatures for food or living space.

Another problem is that human nature is not always nature's friend. For example, it seems that the rarer a wild animal becomes, the more its value soars in the black market (Figure 24.14). A live Amazon macaw can bring $30,000, the hide of a Bengal tiger $100,000. Other contraband, such as grizzly bear gallbladders, shark fins, and rhinoceros horns, can be equally pricey.

In time, nature may recover from some of these assaults. But the evidence suggests that if recoveries from mass extinction occur, they can take millions of years. In the meantime, will future generations of humans ask why their forebears saw fit to exterminate so many other life forms?

Figure 24.14 Confiscated products made from endangered species. It's estimated that 90 percent of the illegal wildlife trade, which threatens hundreds of species of animals and plants, goes undetected.

24.9 Evolution from a Human Perspective

Shared ancestry is one of the core ideas in evolution. Like other life forms, we humans have a well-defined place in the evolutionary scheme of things.

All told, the evolution of life on Earth has given rise to many hundreds of thousands of species—some closely related, others more distantly. An organism's scientific name, always shown in italic type, has been a kind of shorthand for its place in the living world. In a binomial system devised centuries ago, the name has two parts—as in *Homo sapiens*, the binomial that is used for humans. The first part is the genus name. A **genus** (plural: genera) encompasses all the species that are similar to one another and distinct from others in certain traits. The second part of the name indicates the particular species within the genus.

Species are organized into a hierarchy of groupings. Table 24.1 lists them, using humans as the example. Each group above the genus level includes a larger array of organisms that share more general features.

Each lineage of life forms has its defining traits. As primates, humans share certain characteristics with all other primates (such as being land-dwellers). At the same time, we differ in major ways from other primate lineages, such as the New World monkeys (Figure 24.15a). We are genetically closer to the great apes (Figure 24.15b), and closest to the bonobos (Figure 24.15c). Like us, bonobos walk upright—one of five key evolutionary trends in the lineage that led to humans.

FIVE TRENDS MARK HUMAN EVOLUTION

Primates evolved from ancestral mammals more than 60 million years ago. Fossils suggest that the first primates resembled small rodents. They may have foraged in the forest for insects, seeds, buds, and eggs. Between 54 and 38 million years ago, some primates were living in the trees—a habitat where natural selection would strongly favor some traits over others.

Figure 24.15 Some nonhuman primates. (**a**) A spider monkey, one of the New World monkeys. (**b**) A male gorilla. (**c**) Bonobos, our closest primate relative. Like us, bonobos walk upright.

PRECISION GRIP AND POWER GRIP The first mammals spread their toes apart to help support the body as they walked or ran on four legs. Primates still spread their toes or fingers. Many also make cupping motions, as when a monkey lifts food to its mouth. Other hand movements also developed in our ancient tree-dwelling relatives. Changes in hand bones allowed fingers to be wrapped around objects (that is, *prehensile* hand movements were possible), and the thumb and tip of each finger could touch (opposable movements).

In time, hands also began to be freed from load-bearing functions—they were not needed to support the body. Much later, refinements in hand movements led to the precision grip and the power grip:

These hand positions would enable early humans to make and use tools. They helped form the foundation for the development of early technologies and culture.

IMPROVED DAYTIME VISION Early primates had an eye on each side of the head. Later ones had forward-directed eyes, an arrangement that is better for detecting shapes

Table 24.1	Classification of Humans
Domain	Eukarya
Kingdom	Animalia
Phylum	Chordata
Class	Mammalia
Order	Primates
Family	Hominidae
Genus	*Homo*
Species	*sapiens* (only living species of this genus)

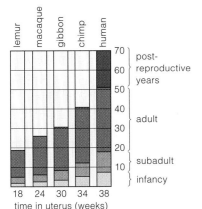

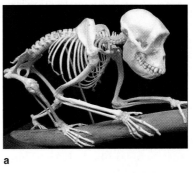

a

Figure 24.16 Trend toward longer life spans and longer dependency among primates.

and movements in three dimensions. Later modifications allowed the eyes to respond to variations in color and light intensity (dim to bright)—another advantage for life in the trees.

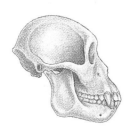

Jaw shape and teeth of an early primate

CHANGES IN DENTITION Changes in the teeth and jaws of early primates accompanied a shift from eating insects, to fruits and leaves, and on to a mixed diet. Later on, rectangular jaws and long canine teeth came to be further defining features of monkeys and apes. Along the road leading to humans, a bow-shaped jaw and teeth that were smaller and all about the same length evolved.

CHANGES IN THE BRAIN AND BEHAVIOR Living on tree branches also favored shifts in reproductive and social behavior. Imagine the advantages of single births over litters, for example, or of clinging longer to the mother. In many primate lineages, parents started to invest more effort in fewer offspring. They formed strong bonds with their young, maternal care became more intense, and the learning period grew longer (Figure 24.16).

Brain regions, such as the cerebral cortex, that are involved in information processing began to expand, and the brain case enlarged. New behavior promoted more brain development—which in turn stimulated more new behavior. In other words, brain modifications and behavioral complexity became closely linked. We see the links clearly in the parallel evolution of the human brain and culture. **Culture** is the sum total of behavior patterns of a social group, passed between generations by learning and symbolic behavior—especially language. The capacity for language arose among ancestral humans through changes in the skull bones and expansion of parts of the brain.

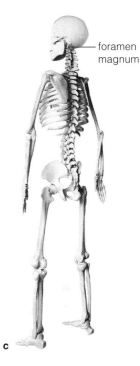

foramen magnum

foramen magnum

b   c

Figure 24.17 *Animated!* The skeletal organization of (**a**) monkeys, (**b**) apes (the gorilla is shown here), and (**c**) humans.

UPRIGHT WALKING Of all primates, only humans can stride freely on two legs for long periods of time. This habitual two-legged gait, called **bipedalism**, emerged as elements of the ancestral primate skeleton were reorganized. As shown in Figure 24.17, compared with apes and monkeys, humans have a shorter, S-shaped backbone. In addition, in monkeys and apes the foramen magnum, the opening at the base of the skull where the spinal cord can connect with the brain, is at the back of the skull. In humans, the foramen magnum is close to the center of the base of the skull. These and other features, such as the position and shape of the knee and ankle joints and pelvic girdle, make bipedalism possible. By current thinking, the evolution of bipedalism was the key modification in the origin of human and humanlike species, both past and present.

Five trends emerged along the evolutionary road leading to humans: refined hand movements, refined vision, generalized dentition, the interconnected development of a more complex brain and cultural behavior, and upright walking.

24.10 Emergence of Early Humans

By 36 million years ago tree-dwelling primates called anthropoids had evolved in tropical forests. One or more types were on or very close to the evolutionary road that would lead to monkeys, apes, and humans.

Between 23 and 5 million years ago, apelike forms—the first *hominoids*—evolved and spread through Africa, Asia, and Europe. At that time, shifts in Earth's land masses and ocean circulation caused a long-term change in climate. Lush African forests began to give way to open woodlands and later to grasslands. Food became harder to find. In these new circumstances, most of the hominoids went extinct. A survivor was the common ancestor of two enduring lineages that arose by 7 million years ago. One gave rise to the great apes, and the other to the first *hominids*.

EARLY HOMINIDS LIVED IN CENTRAL AFRICA

Sahelanthropus tchadensis was an ape or a hominid that lived in Central Africa about 6 or 7 million years ago, during the time when the ancestors of humans were becoming distinct from the apes (Figure 24.18*a*). The remains of another form, *Australopithecus afarensis*, are shown in Figure 24.18*b*. This individual, dubbed "Lucy,"

lived about 3.2 million years ago and had a slight build, unlike some other African hominids. Half a million years later, at what is now Laetoli, Tanzania, *A. afarensis* walked across fresh volcanic ash during a light rain, which turned the ash to quick-drying cement. We do not know much about how different hominids were related or whether they used tools, but the footprints at Laetoli (Figure 24.18*c*), as well as fossil hip and limb bones, confirm that they walked upright.

IS *HOMO SAPIENS* "OUT OF AFRICA"?

By a little over 2 million years ago, species of **humans**—members of the genus *Homo*—were living in woodlands of eastern Africa. One was *Homo habilis* (Figure 24.18*d*). Another was *Homo rudolfensis* (Figure 24.18*e*).

Compared to hominids, these early humans had a larger brain, smaller face, and thickly enameled teeth. They consumed a mixed diet of plant and animal foods, and used tools. Fossil hunters have found many stone tools dating to the time of *H. habilis*.

A divergence produced *Homo erectus*, a species related to modern humans. Its name means "upright man." *H. erectus* coexisted for a time with *H. habilis*. The fossil record indicates that between 2 million and 500,000 years ago, *H. erectus* began leaving Africa in waves. Some still lived in Southeast Asia between 53,000 and 37,000 years ago. As recently as 30,000 years ago, the massively built and large-brained Neandertals lived in Europe and the Near East (Figure 24.19). Their extinction coincided with the origin of modern humans between 40,000 and 30,000

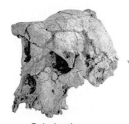
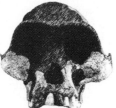

Sahelanthropus tchadensis
7–6 million years

Australopithecus afarensis
3.6–2.9 million years

a

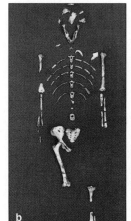

H. habilis
1.9–1.6 million years

Homo rudolfensis
2.4–1.8 million years

b **c** **d** **e**

Figure 24.18 Evidence of African hominids and early humans. (**a**) Representative fossils of African hominids. (**b**) Remains of "Lucy" (*Australopithecus afarensis*), who lived 3.2 million years ago. (**c**) At Laetoli in Tanzania, Mary Leakey found these footprints made 3.7 million years ago. (**d**) Fossil skull of *H. habilis*. (**e**) Fossil skull of *H. rudolfensis*.

H. erectus
2 million–53,000? years

H. neanderthalensis
200,000–30,000 years

Figure 24.19 Fossils of early modern human forms.

years ago. Did the two groups interbreed or make war on one another? We don't know. Neandertal DNA has gene sequences that are not present in modern-day gene pools.

But *where* did our species, *Homo sapiens*, originate? Researchers are divided on this point, although all base their hypotheses on measurements of the small genetic differences among modern human populations. We know that by 1 million years ago, *H. erectus* was living in many regions. The *multiregional model* holds that *H. erectus* evolved along different paths in different regions, in response to local selection pressures. Subpopulations of *H. sapiens* may have evolved from these groups, with gene flow preventing speciation. In 2003, fossils of early humans that date to 18,000 years ago turned up on the Indonesian island of Flores. The species was named *H. floresiensis*. Its features suggest that it may be descended from *H. erectus*, which vanished 200,000 years ago.

In the *African emergence model*, humans arose in sub-Saharan Africa between 200,000 and 100,000 years ago, *then* moved out of Africa (Figure 24.20). In each region where

they settled, they replaced *H. erectus* populations that had preceded them. Regional differences in phenotype that we associate with races evolved later.

Various lines of evidence support this model. A fossil from Ethiopia, in North Africa, indicates that *Homo sapiens* had evolved by 160,000 years ago. Also, gene patterns from forty-three ethnic groups in Asia suggest that modern humans moved from Central Asia, along India's coast, then on into Southeast Asia and southern China. Later on they dispersed north and west into China and Siberia, then down into the Americas.

Cave painting made by early modern humans at Lascaux, France.

Whatever the case, for the past 40,000 years, cultural evolution has been outpacing biological evolution of the human species—and so we leave our story. If you find yourself pondering this subject, here's a point to keep in mind: Humans spread rapidly through the world by devising cultural means to deal with a diverse range of environments. Hunters and gatherers still persist in parts of the world even as other groups have moved from "stone-age" technology to the age of high-tech. Such differences attest to the remarkable behavioral plasticity and depth of human adaptations.

Various species of hominids evolved in Africa about 7 million years ago. The genus Homo *arose about 2 million years ago.* Homo sapiens, *modern humans, is the only remaining species in this genus.*

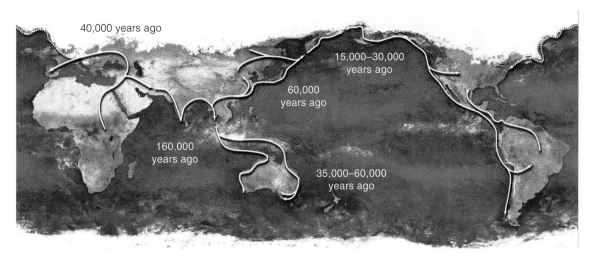

40,000 years ago

15,000–30,000 years ago

60,000 years ago

160,000 years ago

35,000–60,000 years ago

H. sapiens
fossil from Ethiopia,
160,000 years old

Figure 24.20 Estimated times when populations of early *Homo sapiens* were colonizing different regions of the world, based on radiometric dating of fossils. The presumed dispersal routes (white lines) seem to support the African emergence model.

24.11 Earth's History and the Origin of Life

LINK TO SECTION 1.1

Four billion years ago, the Earth was a thin-crusted inferno (Figure 24.21). Yet within 200 million years, life had originated on its surface! We have no record of the event. As far as we know, movements in the young Earth's crust and mantle (a rocky region just below the crust), volcanic activity, and erosion obliterated all traces of it. Still, researchers from various "walks of science" have been able to put together a plausible explanation of how life originated.

CONDITIONS ON EARLY EARTH WERE INTENSE

What were the prevailing physical and chemical conditions on Earth at the time of life's origin? To answer this question, we need to know a little bit about what the young Earth was like. When patches of its crust were forming, heat and gases blanketed the Earth. This first atmosphere probably consisted of gaseous hydrogen (H_2), nitrogen (N_2), carbon monoxide (CO), and carbon dioxide (CO_2). Were gaseous oxygen (O_2) and water also present? Probably not. Rocks don't release much oxygen during volcanic eruptions. Even if oxygen *were* released, those small amounts would have reacted at once with other elements, and any water would have evaporated because of the intense heat.

When the crust finally cooled and solidified, water condensed into clouds and the rains began. For millions of years, runoff from rains stripped mineral salts and other compounds from the Earth's parched rocks. Salt-laden waters collected in depressions in the crust and formed the early seas.

The foregoing events were crucial to the beginning of life. Without an oxygen-free atmosphere, the organic compounds that started the story of life never would have formed on their own. Why? Oxygen would have attacked them and disrupted their functioning (as free radicals of oxygen do in cells). Without liquid water, cell membranes would not have formed, because cell membranes take on their bilayer organization only in water.

As you know, cells are the basic units of life. Each has a capacity for independent existence. Cells could never simply have appeared one day on the early Earth. Their emergence required the existence of biological molecules built from organic compounds. It also required metabolic pathways, which could be organized and controlled inside the confines of a cell membrane. These essential events are our next topic.

BIOLOGICAL MOLECULES PAVED THE WAY FOR CELLS TO EVOLVE

Some scientists believe that the structure shown in Figure 24.22 is a fossilized string of cells that is 3.5 billion years old. The first living cells probably emerged around 3.8 billion years ago. They resembled modern anaerobic bacteria, which do not require or use oxygen. Before

Figure 24.21 Representation of the primordial Earth, about 4 billion years ago. Within another 500 million years, diverse types of living cells would be present on the surface.

a

↑

self-replicating system enclosed in a selectively permeable, protective lipid sphere

DNA ——→ RNA ——→ enzymes and other proteins

formation of protein–RNA systems, evolution of DNA

formation of lipid spheres

spontaneous formation of lipids, carbohydrates, amino acids, proteins, nucleotides under abiotic conditions

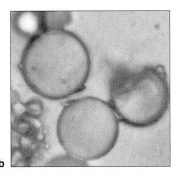

b

Figure 24.23 (**a**) One possible sequence of events that led to the first self-replicating systems, then to the first living cells. (**b**) Lab-grown proto-cells. When amino acids were heated, they formed protein chains. When moistened, the chains assembled into a membrane.

something as complex as a cell was possible, however, biological molecules must have come about through **chemical evolution**. Researchers have been able to put together several reasonable scenarios by which life on Earth could have emerged.

Rocks collected from Mars, meteorites, and the Earth's moon—which all formed at the same time as the Earth, from the same cosmic cloud—contain precursors of biological molecules. Possibly, sunlight, lightning, or heat escaping from the Earth's crust supplied enough energy to drive chemical reactions that yielded more complex organic molecules. In various experiments that recreated conditions on the early Earth, molecules such as amino acids, glucose, ribose, deoxyribose, and other sugars were produced from formaldehyde. Adenine was produced from hydrogen cyanide. Adenine plus ribose occur in ATP, NAD, and other nucleotides vital to cells.

How did complex compounds such as proteins form? By one scenario, these kinds of molecules could have assembled on clay in the muck of tidal flats. Clay is formed of thin, stacked layers of aluminosilicates with metal ions at its surface. Clay and metal ions attract amino acids. From experiments we know that when clay is first warmed by sunlight, then

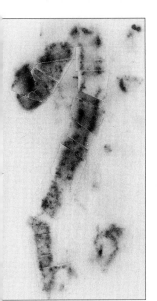

Figure 24.22 A 3.5 billion-year-old fossil? Not everyone agrees, but some researchers believe that this is a string of walled cells. It was unearthed in the Warrawoona rocks of Western Australia.

alternately dried out and moistened, it actually promotes reactions that produce complex organic compounds.

In another hypothesis, a variety of complex organic compounds formed near deep-sea hydrothermal vents. Today, species of primitive bacteria thrive in and around the vents. In laboratory tests, when amino acids were heated and immersed in water, they ordered themselves into small proteinlike molecules.

You may recall from Chapter 1 that metabolism and reproduction are two basic characteristics of life. During the first 600 million years of Earth history, enzymes, ATP, and other molecules that are important in metabolism could have assembled spontaneously in places where they were near one another. If so, their proximity would have promoted chemical interactions—and perhaps the beginning of metabolic pathways.

In the area of reproduction, it's possible that the first "molecule of life" was not DNA but RNA. Simple self-replicating systems of RNA, enzymes, and coenzymes have been created in some laboratories. If so, how DNA entered the picture is still a mystery.

Figure 24.23*a* summarizes some key events that may have led to the first cells. Before cells could exist, there must have been membrane-bound sacs that sheltered molecules such as DNA (or RNA), key amino acids, and the like. Here again, this kind of structure has formed in the laboratory. In one case, heated amino acids formed protein chains, which self-assembled into small spheres (Figure 24.23*b*). Later, the spheres picked up lipids in their surroundings and a lipid–protein film formed at their surface. In other experiments, lipids self-assembled into sacs that had many properties of cell membranes. While there still are major gaps in our knowledge of life's origins, experiments continue to provide clues as to how chemical evolution could have occurred.

Many experiments provide indirect evidence that the complex organic molecules characteristic of life could have formed under conditions that existed on the early Earth.

Summary

Section 24.1 Evolution modifies existing species. Therefore, broadly speaking, all species, past and present, share a common ancestry. The naturalist Charles Darwin proposed that evolution could occur by way of a process he named natural selection.

Section 24.2 In biology, a population is a group of individuals of the same species occupying a given area. The totality of genes in a population make up its gene pool. Each kind of gene may have different alleles, and these genetic variations produce variations in traits.

The relative numbers of different alleles—that is, the different versions of genes—can change as a result of four processes of microevolution: mutation, genetic drift, gene flow, and natural selection (Table 24.2). The large-scale patterns, trends, and rates of change among groups of species over time are called macroevolution.

Section 24.3 According to the theory of evolution by natural selection, there may be a difference in survival and reproduction among members of a population that vary in one or more traits. That is, under prevailing conditions, one form of a trait may be favored because individuals that have it tend to survive and therefore to reproduce more often, so it becomes more common than other forms of the trait.

Organisms tend to come to have characteristics that suit them to the conditions in a particular environment. This trend is called adaptation.

A species is a genetic unit consisting of populations of organisms that closely resemble each other and that can interbreed and produce fertile offspring under natural conditions. New species come into being when the differences between isolated populations become so great that their members are not able to interbreed successfully in nature.

 Learn more about genetic drift, directional selection, and how species become reproductively isolated.

Section 24.4 Evidence of evolutionary relationships comes in part from fossils and studies of biogeography. Fossils may be dated using radiometric dating.

 Learn more about fossil formation and the geologic time scale.

Sections 24.5, 24.6 Comparative morphology often reveals similarities in embryonic development that indicate an evolutionary relationship. Similarities may reveal homologous structures, shared as a result of descent from a common ancestor. Alternatively, analogous structures arise when different lineages evolve in the same or similar environments. In comparative biochemistry, gene mutations that have accumulated in different species provide evolutionary clues.

Section 24.7 In a mass extinction, major lineages perish abruptly. Adaptive radiation is a burst of evolutionary activity; a lineage rapidly produces many new species. Both kinds of events have changed the course of biological evolution many times.

Section 24.9 Five major evolutionary trends in the primate lineage leading to *H. sapiens* included (1) a transition to bipedalism, with related changes in the skeleton; (2) increased motor skills related to structural modification of the hands; (3) more reliance on daytime vision, including color vision and depth perception; (4) transition away from specialized eating habits, with corresponding modification of dentition; and (5) the enlargement and reorganization of the brain.

Section 24.10 Humans (*Homo sapiens*) are classified in the hominid family of the primate order, and are members of the only existing species of the genus *Homo*. In human evolution, the development of a larger, more complex brain correlated with increasingly sophisticated technology and with the development of complex behaviors and culture. For the last 40,000 years, human cultural evolution has outpaced our species' biological evolution.

Section 24.11 Life originated on Earth about 3.8 billion years ago. Various experiments provide indirect evidence that life originated under conditions that presumably existed on the early Earth.

Comparative analyses of the composition of cosmic clouds and of rocks from other planets and the Earth's moon suggest that precursors of the complex molecules associated with life were available.

In simulations of primordial conditions, chemical precursors assembled into sugars, amino acids, and other organic compounds.

Metabolic pathways could have evolved as a result of chemical competition for the limited supplies of organic molecules that had accumulated in the seas.

Table 24.2	Major Processes of Microevolution
Mutation	A heritable change in DNA
Genetic drift	Random fluctuation in allele frequencies over time, due to chance occurrences alone
Gene flow	Movement of alleles among populations through migration of individuals
Natural selection	Change or stabilization of allele frequencies due to differences in survival and reproduction among variant members of a population

Self-replicating systems of RNA, enzymes, and coenzymes have been synthesized in the laboratory. How DNA entered the picture is not yet understood.

Lipid and lipid–protein membranes that show properties of cell membranes form spontaneously under the prescribed conditions.

 See experiments on how organic compounds can form spontaneously, and investigate the scientific history of life on Earth.

Review Questions

1. Distinguish between microevolution and macroevolution.

2. Explain how natural selection differs from adaptation.

3. Explain the difference between:
   a. divergence and gene flow
   b. homologous and analogous structures

4. Explain the difference between a primate and a hominid.

5. Describe the chemical and physical environment in which the first living cells may have evolved.

Self-Quiz   *Answers in Appendix V*

1. A _____ is a genetic unit consisting of one or more populations of organisms that usually closely resemble one another physically and physiologically.

2. The relative numbers of different genes (alleles) in a gene pool change as a result of four processes of microevolution: _____, _____, _____, and _____.

3. A difference in survival and reproduction among members of a population that vary in one or more traits is called _____.

4. The fossil record of evolution correlates with evidence from _____.
   a. the geologic record
   b. radiometric dating
   c. comparing development patterns and morphology
   d. comparative biochemistry
   e. all of the above

5. Comparative biochemistry _____.
   a. is based mainly on the fossil record
   b. often reveals similarities in embryonic development stages that indicate evolutionary relationships
   c. is based on mutations that have accumulated in the DNA of different species
   d. compares the proteins and the DNA from different species to reveal relationships
   e. both c and d are correct

Figure 24.24 A few of the world's remaining cheetahs.

6. Comparative morphology _____.
   a. is based mainly on the fossil record
   b. shows evidence of divergences and convergences in body parts among certain major groups
   c. compares the proteins and the DNA from different species to reveal relationships
   d. both b and c are correct

7. In _____, new species of a lineage move into a wide range of habitats by way of bursts of microevolutionary events.
   a. an adaptive radiation
   b. natural selection
   c. genetic drift
   d. punctuated equilibrium

8. The pivotal modification in hominid evolution was _____.
   a. the transition to bipedalism
   b. hand modification that increased manipulative skills
   c. a shift from omnivorous to specialized eating habits
   d. less reliance on smell, more on vision
   e. expansion and reorganization of the brain

Critical Thinking

1. The cheetahs shown in Figure 24.24 are among the approximately 20,000 of these sleek, swift cats left in the world. One reason cheetahs have become endangered is that about 10,000 years ago cheetahs experienced a severe loss in their numbers, and since then the survivors have been inbreeding. As a result, the modern cheetah gene pool has very little variation and includes alleles that reduce the sperm counts of males and impair the animals' resistance to certain diseases. Based on what you've read in this chapter about genetic variation, does it seem likely that a "gene therapy" program might be able to correct the genetic problems and help cheetahs make a comeback? Explain your answer.

2. Humans can inherit various alleles for the liver enzyme ADH (alcohol dehydrogenase), which breaks down ingested alcohol. People of Italian and Jewish descent

Figure 24.25 Rama the cama displaying his unexpectedly short temper.

commonly have a form of ADH that detoxifies alcohol very rapidly. People of northern European descent have forms of ADH that are moderately effective in alcohol breakdown, while people of Asian descent typically have ADH that is less efficient at processing alcohol. Explain why researchers have been able to use this information to help trace the origin of human use of alcoholic beverages.

3. In 1992 the frozen body of a Stone Age man was discovered in the Austrian Alps. Although the "Iceman" died about 5,300 years ago, his body is amazingly intact and researchers have analyzed DNA extracted from bits of his tissue. Can these studies tell us something about early human evolution? Explain your reasoning.

4. *Rama the cama*, a llama–camel hybrid, was born in 1997 (Figure 24.25). Camels and llamas have a shared ancestor but have been separated for 30 million years. Veterinarians collected semen from a male camel weighing close to 1,000 pounds, then used it to artificially inseminate a female llama one-sixth his weight. Rama resembles both parents, with a camel's long tail and short ears but no hump, and llama-like hooves rather than camel footpads. The idea was to breed an animal having a camel's strength and endurance and a llama's gentle disposition.

Rama is smaller than expected and has a camel's short temper. Now old enough to mate, he is too short to get together with a female camel and too heavy to mount a female llama. He will have to wait several years to try to mate with a female cama born in 2002, and only time will tell if any offspring from such a match will be fertile.

Does Rama's story tell you anything about the genetic changes required for irreversible reproductive isolation in nature? Explain why a biologist might not view Rama as evidence that llamas and camels are the same species.

Explore on Your Own

About 50,000 years ago humans began domesticating wild dogs. By about 14,000 years ago, people started to favor new breeds of dogs using artificial selection. Dogs having desired forms of traits were selected from each litter and later encouraged to breed. Those with undesired forms of traits were passed over.

This process has produced scores of domestic dog breeds, including sheep-herding border collies, sled-pulling huskies, and dogs as strikingly different as Great Danes and chihuahuas (Figure 24.26).

With a little bit of sleuthing on the Web or in a library, you should be able to discover numerous other examples of how humans have used artificial selection to develop desired animal breeds or plant varieties. How has artificial selection affected aspects of your own life, such as pets, foods you eat, and ornamental garden plants?

Figure 24.26 Two designer dogs, the Great Dane (legs, left) and the chihuahua.

The Human Touch

In 1722, on Easter morning, a European explorer landed on a small volcanic island off the west coast of South America and found a few hundred hungry Polynesians living in caves. He saw dry grasses and scorched shrubs, but no trees. He noticed about 200 massive stone statues near the coast and 700 unfinished, abandoned ones in inland quarries. Some weighed 50 tons.

Two years later Captain James Cook visited and found the island empty of people. Nearly all the statues had been tipped over, often onto face-shattering spikes.

Easter Island, as it came to be called, is only 165 square kilometers (64 square miles) in size. From archeological and historical records, we know that it was populated by voyagers from the Marquesas around A.D. 350. Their new home was a paradise with fertile soil, dense forests, and lush grasses. As the population grew rapidly, the inhabitants cut down trees for fuel wood and canoes.

By 1400, possibly as many as 15,000 people were living on the island. Crop yields had declined as erosion and harvesting depleted the soil's nutrients. In time, fish vanished from nearshore waters around the island, all the native birds had been eaten, and people were raising rats for food.

With survival at stake, those in power appealed to the gods. Stoneworkers carved images of divine beings from stones of massive size. A system of greased logs moved the stones over miles of rough terrain to the coast.

By about 1550, Easter Island's trees and fishes were gone. Having devastated the natural resources that supported them, the islanders turned to their last source of protein: they began to hunt and eat each other. In short order, the Easter Island society fell apart.

Today, what are 6.5 *billion* people doing to themselves and to the global environment? The Earth is our island in space. Are we making a crisis of our own?

For us humans, as for all living things, ecological principles govern how populations grow and can be sustained over time. In this chapter we'll survey these principles, see how they apply to our species, and then consider some actual and potential impacts of human activity on Earth's ecosystems and resources.

 How Would You Vote? Goods can be manufactured in ways that protect the environment, but often cost more than comparable goods produced without regard for their ecological impact. Are you willing to pay extra for "green" products? Cast your vote online at www.thomsonedu.com/biology/starr/humanbio.

Key Concepts

PRINCIPLES OF ECOLOGY
Ecology is the study of interactions of organisms with one another and the environment. Energy flows through ecosystems, passing from organism to organism by way of food webs.

CHEMICAL CYCLES
Nutrients cycle in ecosystems. In biogeochemical cycles, water and nutrients move from the physical environment to organisms, then back to the environment.

HUMAN IMPACTS ON THE ENVIRONMENT
Humans have sidestepped certain natural limits on population growth. Human population size has soared as a result, leading to many impacts on the natural world.

Links to Earlier Concepts

This chapter aims to place your study of human biology in the broader context of the whole living world (1.3).

25.1 Some Basic Principles of Ecology

LINK TO SECTION 1.3

The biosphere consists of those regions of the Earth's crust, waters, and atmosphere where organisms live.

The Earth's surface is remarkably diverse. In climate, soils, vegetation, and animal life, its deserts differ from hardwood forests, which differ from tropical rain forests, prairies, and arctic tundra. Oceans, lakes, and rivers differ physically and in their arrays of organisms. Each of these realms is called a **biome** (Figure 25.1).

Ecology is the study of the interactions of organisms with one another and with the physical environment. The general type of place in which a species normally lives is its **habitat**. For example, muskrats live in a stream habitat, damselfish in a coral reef habitat. The habitat of any organism has certain characteristic physical and chemical features. Every species also interacts with others that occupy the same habitat. Humans live in "disturbed" habitats, which we have deliberately altered for purposes such as agriculture and urban development. In any given habitat the populations of all species directly or indirectly associate with one another as a **community**.

Have you ever heard someone speak of an "ecological niche"? The **niche** (pronounced "nitch") of a species consists of the various physical, chemical, and biological conditions the species needs to live and reproduce in an ecosystem. Examples of those conditions include the amount of water, oxygen, and other nutrients a species needs, the temperature ranges it can tolerate, the places it finds food, and the type of food it consumes. *Specialist* species have narrow niches. They may be able to use only one or a few types of food or live only in one type of habitat. For example, the red-cockaded woodpecker builds its nest mainly in longleaf pines that are at least 75 years old. Humans and houseflies are examples of *generalist* species with broad niches. Both can live in a range of habitats and eat many types of food.

An **ecosystem** consists of one or more communities of organisms interacting with one another *and* with the physical environment through a flow of energy and a cycling of materials. Figure 25.2 shows some of the organisms interacting in an arctic tundra ecosystem.

Communities of organisms make up the *biotic*, or living, portions of an ecosystem. New communities may develop in habitats that were once empty of life, such as land exposed by a retreating glacier, or in a previously disturbed inhabited area such as an abandoned pasture. Through a process called **succession**, the first species to thrive in the habitat are then replaced by others, which are replaced by others in orderly progression until the composition of species becomes steady as long as other conditions remain the same. This more or less stable array of species is called a **climax community**.

Figure 25.1 Major types of ecosystems on Earth are called *biomes*. They include land biomes such as the warm desert near Tucson, Arizona, shown in (**a**), and aquatic realms such as the mountain lake in (**b**), which lies in the Canadian Rockies.

In *primary succession*, changes begin when pioneer species colonize a newly available habitat, such as a recently deglaciated region (Figure 25.3). In *secondary succession*, a community develops toward the climax state after parts of a habitat have been disturbed. For example, this pattern occurs in abandoned fields, where wild grasses and other plants spring up when cultivation stops. Changing climate, natural disasters, such as forest fires caused by lightning, and other factors often interfere with succession toward climax conditions. Hence, we rarely see truly stable climax communities.

> *The general type of place where a species normally lives is its habitat. A community includes the populations of all species in a given habitat.*
>
> *An ecosystem consists of one or more communities of organisms interacting with one another and with the physical environment through a flow of energy and a cycling of materials.*

Figure 25.2 Some of the producers, consumers, and decomposers of the arctic tundra: sedges, mosses, and other plants, along with the lemming (eater of plant parts); the snowy owl (eater of lemmings); and a fungal decomposer. As in all ecosystems, these organisms interact with one another and with the physical environment through a one-way flow of energy and a cycling of materials.

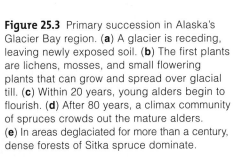

Figure 25.3 Primary succession in Alaska's Glacier Bay region. (**a**) A glacier is receding, leaving newly exposed soil. (**b**) The first plants are lichens, mosses, and small flowering plants that can grow and spread over glacial till. (**c**) Within 20 years, young alders begin to flourish. (**d**) After 80 years, a climax community of spruces crowds out the mature alders. (**e**) In areas deglaciated for more than a century, dense forests of Sitka spruce dominate.

25.2 Feeding Levels and Food Webs

LINK TO
SECTION
1.3

Although there are many different types of ecosystems, they are all alike in many aspects of their structure and function.

Nearly every ecosystem runs on energy from the sun. Plants and other photosynthetic organisms are **producers** (or *autotrophs*, which means "self-feeders"). They capture sunlight energy and use it to build organic compounds from inorganic raw materials (Figure 25.4).

All other organisms in an ecosystem are **consumers** (or *heterotrophs*, "other-feeders"). One way or another, consumers take in energy that has been stored in the tissues of producers. For the most part, consumers fall into four categories:

Herbivores such as grazers and insects eat plants; they are *primary* consumers.

Carnivores such as lions eat animals. Carnivores are *secondary* or *tertiary* (third-level) consumers.

Omnivores such as humans, dogs, and grizzly bears feed on a variety of foods, either plant or animal.

Decomposers such as fungi, bacteria, and worms get energy from the remains or products of organisms.

Producers obtain an ecosystem's nutrients *and* its initial pool of energy. As they grow, they take up water and carbon dioxide (which provide oxygen, carbon, and hydrogen), as well as minerals such as nitrogen. These materials, recall, are the building blocks for biological molecules. When decomposers get their turn at this organic matter, they can break it down to inorganic bits. If those bits are not washed away or otherwise removed from the system, producers can reuse them as nutrients.

It is important to remember that ecosystems must have an ongoing input of energy, as from the sun. Often they depend on outside sources of nutrients as well (as when erosion carries minerals into a lake). Ecosystems also *lose* energy and nutrients. Most of the energy that producers capture eventually is lost to the environment in the form of metabolic heat. Though generally recycled, some nutrients also are lost, as when minerals are leached out of soil by water seeping down through it.

ENERGY MOVES THROUGH A SERIES OF ECOSYSTEM FEEDING LEVELS

Each species in an ecosystem has its own position in a hierarchy of feeding levels (also called *trophic levels*; *troph* means "nourishment"). A key factor in how any ecosystem functions is the transfer of energy from one of its feeding levels to another.

Primary producers, which gain energy directly from sunlight, make up the first feeding level. Corn plants in a field or waterlilies in a pond are examples. Snails and other herbivores that feed on the producers are at the next feeding level. Birds and other primary carnivores that prey on the herbivores form a third level. A hawk that eats a snake is a secondary carnivore. Decomposers, humans, and many other organisms can obtain energy from more than one source. They can't be assigned to a single feeding level.

FOOD CHAINS AND WEBS SHOW WHO EATS WHOM

A linear sequence of who eats whom in an ecosystem is sometimes called a **food chain**. However, you won't often find such a simple, isolated chain as this one. Most species belong to more than one food chain, especially when they are at a low feeding level. It's more accurate to view food chains as cross-connecting with one another in **food webs**. Figure 25.5 shows a typical food web in a prairie ecosystem.

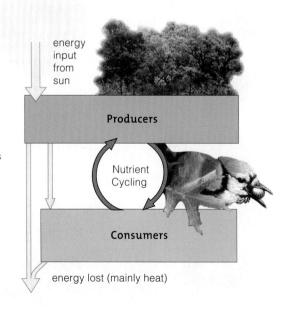

Figure 25.4
Animated! Simple model of ecosystems. Energy flows in one direction: into the ecosystem, through its living organisms, and then out from it. Its nutrients are cycled among autotrophs and heterotrophs. For this model, energy flow starts with autotrophs that can capture energy from the sun.

energy input from sun

Producers

Nutrient Cycling

Consumers

energy lost (mainly heat)

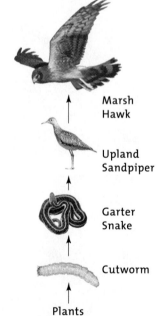

Marsh Hawk

Upland Sandpiper

Garter Snake

Cutworm

Plants

A simple food chain.

PRINCIPLES OF ECOLOGY

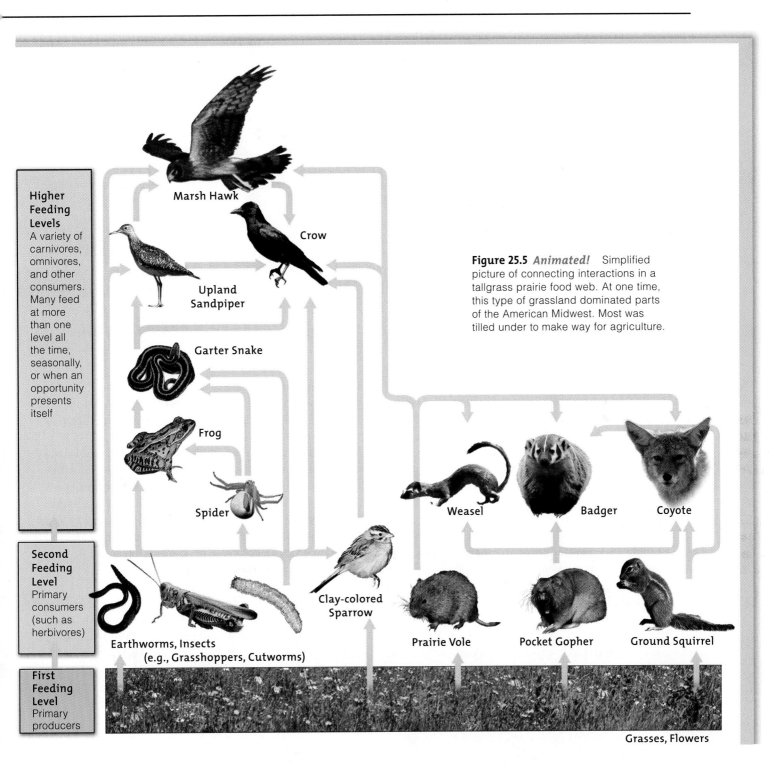

Higher
Feeding
Levels
A variety of
carnivores,
omnivores,
and other
consumers.
Many feed
at more
than one
level all
the time,
seasonally,
or when an
opportunity
presents
itself

Marsh Hawk

Crow

Upland
Sandpiper

Garter Snake

Frog

Spider

Weasel

Badger

Coyote

Figure 25.5 *Animated!* Simplified picture of connecting interactions in a tallgrass prairie food web. At one time, this type of grassland dominated parts of the American Midwest. Most was tilled under to make way for agriculture.

**Second
Feeding
Level**
Primary
consumers
(such as
herbivores)

**First
Feeding
Level**
Primary
producers

Earthworms, Insects
(e.g., Grasshoppers, Cutworms)

Clay-colored
Sparrow

Prairie Vole

Pocket Gopher

Ground Squirrel

Grasses, Flowers

A food web is a network of crossing, interlinked food chains. It is made up of primary producers and a variety of consumers and decomposers.

By way of food webs, different species in an ecosystem are interconnected.

25.3 Energy Flow through Ecosystems

Energy flows into food webs from an outside source, most often the sun. Energy leaves mainly by losses of metabolic heat, which each organism generates.

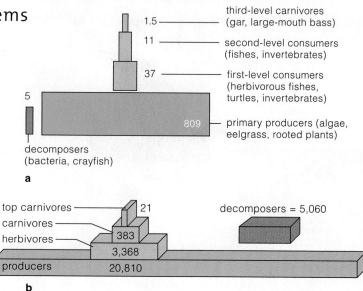

Figure 25.7 (a) A biomass pyramid. (b) An energy pyramid.

PRODUCERS CAPTURE AND STORE ENERGY

In land ecosystems, the usual primary producers are plants. The rate at which they secure and store energy in their own tissues during a given period of time is the **primary productivity** of the ecosystem. How much energy actually gets stored in the tissues of plants depends on how many individual plants live there, and on the balance between energy trapped (by photosynthesis) and energy used in the plants' life processes.

Other factors also affect the final amount of stored energy in an ecosystem at any given time. For example, how much energy ecosystems trap and store depends partly on the physical size of producers, the availability of mineral nutrients, the amounts of sunlight and rainfall during each growing season, and the temperature range. The harsher the conditions, the less new plant growth per season—so the productivity will be lower. As shown in Figure 25.6, there are big differences in the primary productivity in different ecosystems.

CONSUMERS SUBTRACT ENERGY FROM ECOSYSTEMS

An **ecological pyramid** is a way to represent the energy relationships of an ecosystem. In these pyramids, the primary producers form a base for successive tiers of consumers above them.

Biomass is the combined weight of all of an ecosystem's organisms at each tier. Figure 25.7a shows a biomass pyramid (measured in grams per square meter) for one

small aquatic ecosystem. This kind of biomass pyramid is common in nature. There are lots of primary producers—plants—and few top carnivores such as bears, lions, or killer whales.

Some biomass pyramids are "upside-down," with the smallest tier on the bottom. A pond or the sea is like this. Its primary producers have less biomass than consumers feeding on them. The producers consist of phytoplankton (tiny floating photosynthetic organisms), which grow and reproduce fast enough to provide a steady supply of food for a much greater biomass of zooplankton (small floating animals). Zooplankton in turn become food for larger animals.

An *energy pyramid*, too, shows how usable energy diminishes as it flows through an ecosystem. It has a large energy base at the bottom (Figure 25.7b). An energy pyramid is always "right-side up." It gives a more accurate picture of ever-diminishing amounts of energy flowing through successive feeding levels of the ecosystem.

Energy enters ecosystems from outside—usually from the sun. Energy leaves ecosystems mainly by loss of metabolic heat from organisms.

Primary productivity is the total energy stored by an ecosystem's photosynthesizers in a given amount of time, after the plants' own energy needs are met.

Ecological pyramids depict the tiered feeding (trophic) structure of ecosystems. A biomass pyramid shows the combined weight of organisms in each tier. An energy pyramid shows the loss of usable energy from the base tier to higher levels.

Figure 25.6 A summary of satellite data on the Earth's primary productivity during 2002. Productivity is coded as red (highest) down through orange, yellow, green, blue, and purple (lowest).

PRINCIPLES OF ECOLOGY

25.4 Biogeochemical Cycles—An Overview

Ecosystems depend on primary productivity. For this reason, the availability of water, carbon dioxide, and the mineral ions that serve as nutrients for producers has a profound impact on ecosystems.

In a **biogeochemical cycle**, ions or molecules of a nutrient are moved from the environment to organisms, then back to the environment—part of which serves as a reservoir for them. They generally move slowly through the reservoir, compared to their rapid movement between organisms and the environment. Figure 25.8 provides an overview of the relationship between most ecosystems and the biogeochemical part of the cycles.

Each year, the amount of a nutrient that cycles through a major ecosystem is more than enters or leaves. Together with nutrient recycling, fresh inputs help maintain an ecosystem's nutrient reserves. For instance, rainfall, snowfall, and the slow weathering of rocks help replenish the reserves. At the other end of the balance sheet, ecosystems also can lose some of their nutrient reserves. For example, various minerals may be washed away by runoff from irrigated cropland.

There are three categories of biogeochemical cycles, based on the part of the environment that has the largest supply of the ion or molecule being considered. As you will see later, in the *global water cycle*, oxygen and hydrogen move in the form of water molecules. In *atmospheric cycles*, much of the nutrient—carbon and nitrogen, for example—is in the form of a gas. *Sedimentary cycles* move phosphorus and other chemicals that do not occur as gases. Such solid nutrients move from land to the seafloor and return to dry land only through geological uplifting, which may take millions of years. The Earth's crust is the major storehouse for these substances.

> *In a biogeochemical cycle, nutrients move from the environment to organisms, then back to their reservoir in the environment.*
>
> *Nutrients generally move slowly through the environment but rapidly between organisms and the environment.*

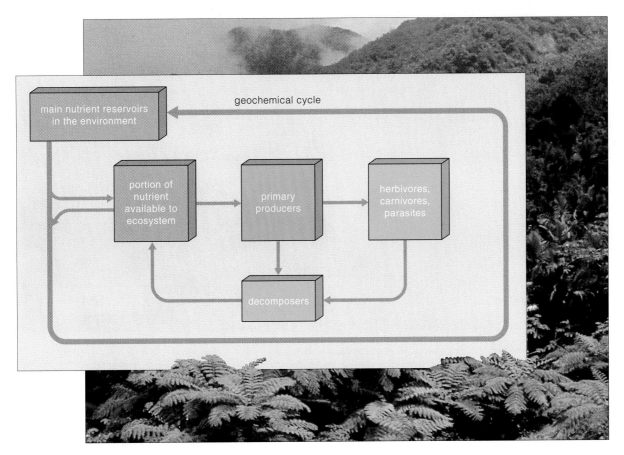

Figure 25.8 Overview of nutrient flow through a land ecosystem. The movement of nutrients from the physical environment, through organisms, and back to the environment is a biogeochemical cycle.

25.5 The Water Cycle

You already know that water is vital for all organisms. It also carries nutrients into and out of ecosystems.

Driven by solar energy, the Earth's waters move slowly, on a vast scale, from the ocean into the atmosphere, to land, and back to the ocean—the main reservoir. Water evaporating into the lower atmosphere initially stays aloft in the form of vapor, clouds, and ice crystals. It returns to Earth as precipitation, mostly rain and snow. Ocean currents and prevailing wind patterns influence this global *hydrologic cycle*, or **water cycle** (Figure 25.9).

Have you ever heard a news report about concerns over changes in a watershed? A *watershed* is any area in which the precipitation is funneled into a single body of water, such as a stream, river, or bay. Watersheds can be vast. The Mississippi River watershed extends across roughly one-third of the United States.

Most water entering a watershed seeps into soil or becomes surface runoff that moves into streams. Plants take up water and dissolved minerals from soil, then lose water through their leaves. Research shows that the plant life in a watershed can be vital to preventing the loss of soil nutrients in runoff.

Measurements of the inputs and losses of water at watersheds have practical applications. For example, cities and towns that depend on surface water supplies in watersheds (such as reservoirs) can adjust their water use based on seasonal variations.

In the water cycle, water slowly moves from the oceans through the atmosphere, onto land, then back to the oceans.

A watershed is an area where the precipitation is funneled into a single body of water, such as a stream, river, or bay.

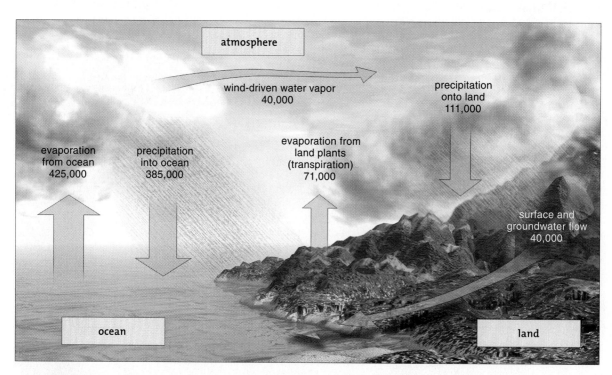

Figure 25.9 *Animated!* The water cycle. Water moves from oceans to the atmosphere, the land, and back to the ocean. Yellow boxes signify the main reservoirs. Arrow labels identify the processes involved in water movement between reservoirs, measured in cubic kilometers per year.

CHEMICAL CYCLES

25.6 Cycling Chemicals from the Earth's Crust

We continue our look at biogeochemical cycling with one of the sedimentary cycles—one that involves chemicals locked in the Earth's crust.

Living organisms can't survive without small amounts of phosphorus. This mineral is a key component of ATP, NADPH, phospholipids, nucleic acids, and many other organic compounds.

In the **phosphorus cycle**, the mineral phosphorus moves from land to sediments in the seas and then back to land (Figure 25.10). The Earth's crust is the main storehouse for phosphorus and for other minerals such as calcium and potassium.

Phosphorus is typically present in rock formations on land, in the form of phosphates. Through the natural processes of weathering and erosion, phosphates enter rivers and streams, which eventually transport them to the ocean. There, mainly on the continental shelves, phosphorus accumulates with other minerals. Millions of years pass. Where crustal plates collide, part of the seafloor may be uplifted and drained. Over geologic time, weathering releases phosphates from the newly exposed rocks—and the geochemical phase of the phosphorus cycle begins again.

The ecosystem phase of the phosphorus cycle is much more rapid. Plants can take up dissolved phosphorus so quickly and efficiently that they may deplete the supply in soil. Herbivores obtain phosphorus by eating plants, carnivores get it by consuming herbivores, and all animals excrete phosphorus as a waste product in urine and feces. It is also released to the soil as organic matter decomposes. Plants then take up phosphorus, rapidly recycling it in the ecosystem.

Phosphorus is linked to some ecological problems, including the *eutrophication* of lakes. This is the name for nutrient enrichment of a body of water. Although eutrophication is a natural process, human activities can speed it up and upset the natural balance. An example is Lake Washington in Seattle, where runoff containing large amounts of phosphorus from fertilizer or detergents (in sewage) triggered the growth of thick, slimy mats of algae. The solution was to severely limit discharges into the lake. With time, the algal "blooms" retreated.

> *The Earth's crust is the main storehouse for phosphorus and other minerals that move through ecosystems as part of sedimentary cycles. The geochemical phase of these cycles advances extremely slowly.*

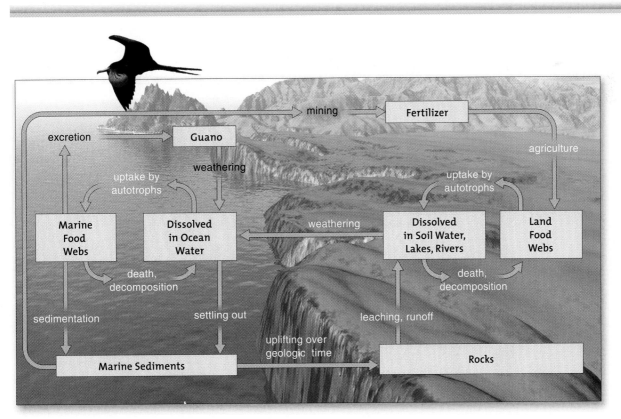

Figure 25.10 *Animated!* The phosphorus cycle. This is an example of a sedimentary cycle.

25.7 The Carbon Cycle

Carbon moves through the atmosphere and food webs on its way to and from the oceans, sediments, and rocks.

Figure 25.11 sketches the global **carbon cycle**. Sediments and rocks hold most of the carbon, followed by the ocean, soil, atmosphere, and land biomass. Carbon enters the atmosphere as cells engage in aerobic respiration, as fossil fuels burn, and when volcanoes erupt and release it from rocks in the Earth's crust. Most atmospheric carbon occurs as carbon dioxide (CO_2). Most carbon dissolved in the ocean is in the forms of bicarbonate and carbonate.

You've probably seen bubbles of CO_2 escaping from a glass of carbonated soda. Why doesn't the CO_2 dissolved in warm ocean surface waters escape to the atmosphere? Driven by winds and regional differences in water density, water makes a gigantic loop from the surface of the Pacific and Atlantic Oceans to the Atlantic and Antarctic seafloors. There its CO_2 moves into deep storage reservoirs before water loops up. Figure 25.12 diagrams this looping movement. It is a factor in carbon's distribution and global budget.

Photosynthesizers capture billions of metric tons of carbon atoms in organic compounds every year. However, the average length of time that a carbon atom is held in any given ecosystem varies quite a bit. For example, organic wastes and remains decompose rapidly in tropical rain forests, so not much carbon accumulates at the surface of soils. In marshes, bogs, and other anaerobic areas, decomposers cannot break down organic compounds completely, so carbon gradually builds up in peat and other forms of compressed organic matter. Also, in food webs of ancient aquatic ecosystems, carbon was incorporated in shells and other hard parts. The shelled organisms died and sank, then became buried in sediments. The same things are happening today. Carbon remains buried for many millions of years in deep sediments until part of

Figure 25.11 *Animated!* Global carbon cycle. Part (**a**) shows the cycle through typical marine ecosystems. Part (**b**) shows how carbon cycles through land ecosystems. Yellow boxes indicate the main carbon reservoirs. The vast majority of carbon atoms are in sediments and rocks, followed by ever lesser amounts in ocean water, soil, the atmosphere, and biomass. Here are typical annual fluxes in the global distribution of carbon, in gigatons:

From atmosphere to plants by carbon fixation	120
From atmosphere to ocean	107
To atmosphere from ocean	105
To atmosphere from plants	60
To atmosphere from soil	60
To atmosphere from fossil-fuel burning	5
To atmosphere from net destruction of plants	2
To ocean from runoff	0.4
Burial in ocean sediments	0.1

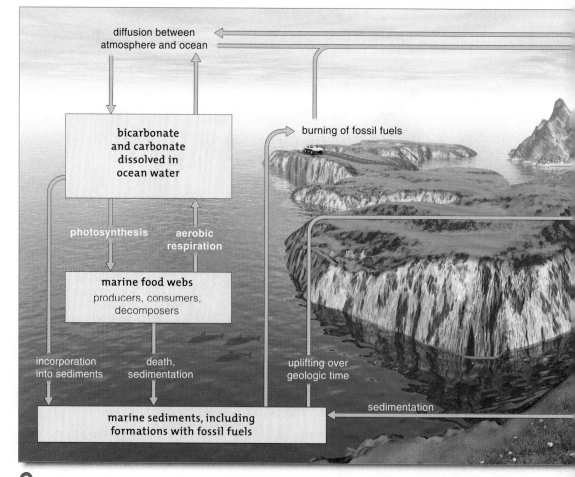

the seafloor is uplifted above the ocean surface through geologic forces. Other buried carbon is slowly converted to long-standing reserves of gas, petroleum, and coal, which we tap for use as fossil fuels.

Human activities, including the burning of fossil fuels, are putting more carbon into the atmosphere than can be cycled to the ocean reservoir. This is amplifying the greenhouse effect, which contributes to global warming. *Science Comes to Life* in Section 25.8 describes the greenhouse effect and some possible outcomes of increases—or decreases—in it.

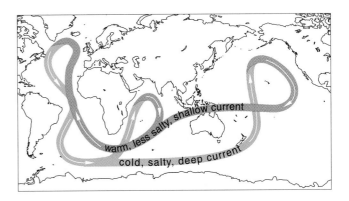

Figure 25.12 Loop of ocean water that delivers carbon dioxide to its deep ocean reservoir. It sinks in the cold, salty North Atlantic and rises in the warmer Pacific.

The ocean and atmosphere interact in the global cycling of carbon. Fossil-fuel burning and other human activities may be contributing to imbalances in the global carbon budget.

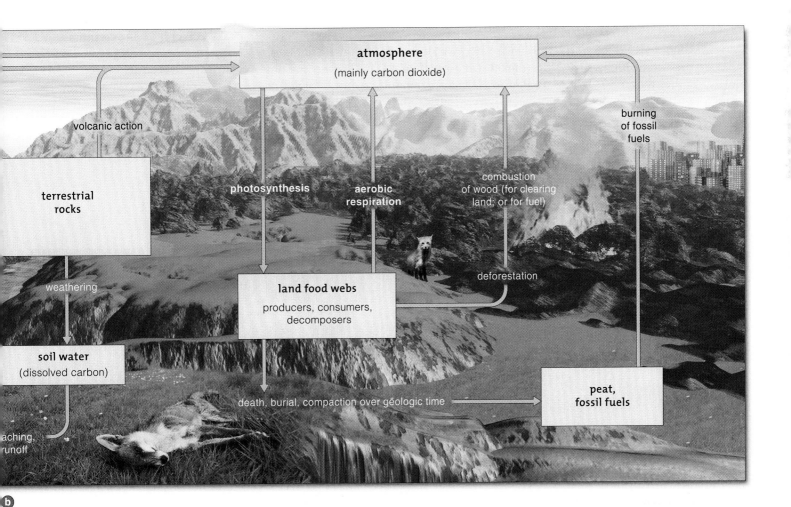

25.8 Global Warming

Concentrations of a variety of gaseous molecules in Earth's atmosphere play a central role in shaping the average temperature near its surface. Temperature, in turn, has huge effects on global and regional climates.

Atmospheric molecules of carbon dioxide, water, ozone, methane, nitrous oxide, and chlorofluorocarbons are key players in interactions that affect global temperature. Collectively, the gases act like the panes of glass in a greenhouse—hence their name, "greenhouse gases." Wavelengths of visible light pass through these gases to Earth's surface, which absorbs them and then emits them as heat. Greenhouse gases impede the escape of this heat back into space, radiating much of it back toward the Earth's surface (Figure 25.13).

With time, heat builds up in the lower atmosphere and the air temperature near the surface rises. This warming action is known as the **greenhouse effect**. Without it, Earth's surface would be so cold that it could not support life.

In the 1950s, researchers in a laboratory on Hawaii's highest volcano began a long-term program of measuring the concentrations of different greenhouse gases in the atmosphere. That remote site is almost free of local airborne contamination. It is also representative of conditions for the Northern Hemisphere. What did they find? Briefly, that carbon dioxide concentrations follow the annual cycle of plant growth (that is, primary production) in the Northern Hemisphere. They drop in summer, when photosynthesis rates are highest (and plants use lots of CO_2). They rise in winter, when photosynthesis by plants slows.

Figure 25.14 *Animated!* *Facing page*, graphs of recent increases in four categories of atmospheric greenhouse gases. A key factor is the sheer number of gasoline-burning vehicles in large cities. *Above*, Mexico City on a smoggy morning. With more than 10 million residents, it's the world's largest city.

The troughs and peaks in Figure 25.14a are annual lows and highs of global carbon dioxide concentrations. For the first time, scientists have been able to see the integrated effects of the carbon balances for an entire hemisphere. Notice the midline of the troughs and peaks in the cycle. It shows that the concentration of carbon dioxide is steadily increasing, as are the concentrations of other greenhouse gases.

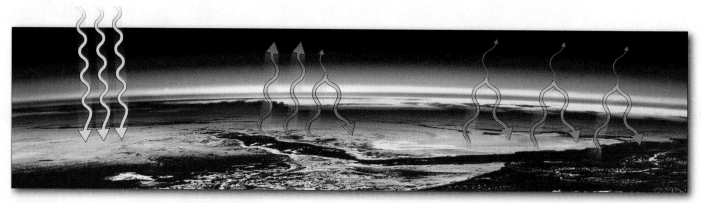

a Rays of sunlight penetrate the lower atmosphere and warm the Earth's surface.

b The surface radiates heat (infrared wavelengths) to the lower atmosphere. Some heat escapes into space. But greenhouse gases and water vapor absorb some infrared energy and radiate a portion of it back toward Earth.

c Increased concentrations of greenhouse gases trap more heat near Earth's surface. Sea surface temperature rises, more water evaporates into the atmosphere, and Earth's surface temperature rises.

Figure 25.13 *Animated!* The greenhouse effect.

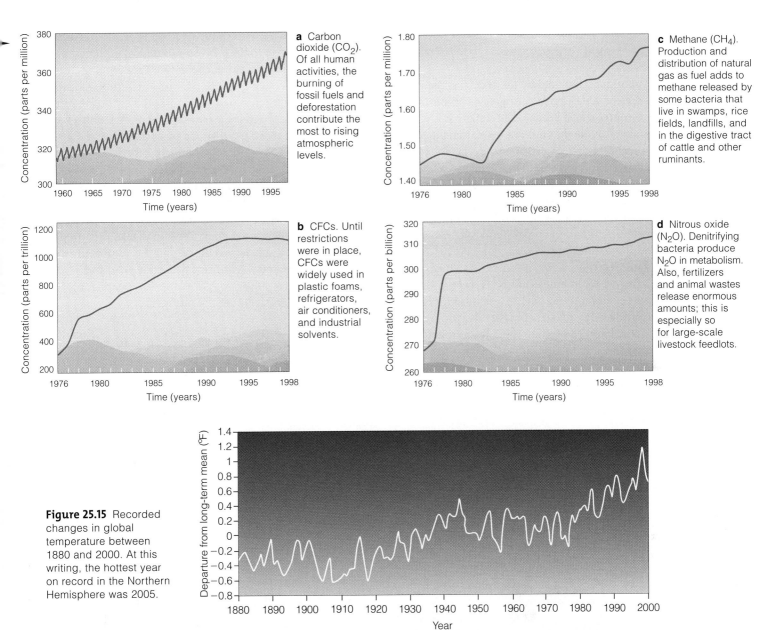

a Carbon dioxide (CO_2). Of all human activities, the burning of fossil fuels and deforestation contribute the most to rising atmospheric levels.

b CFCs. Until restrictions were in place, CFCs were widely used in plastic foams, refrigerators, air conditioners, and industrial solvents.

c Methane (CH_4). Production and distribution of natural gas as fuel adds to methane released by some bacteria that live in swamps, rice fields, landfills, and in the digestive tract of cattle and other ruminants.

d Nitrous oxide (N_2O). Denitrifying bacteria produce N_2O in metabolism. Also, fertilizers and animal wastes release enormous amounts; this is especially so for large-scale livestock feedlots.

Figure 25.15 Recorded changes in global temperature between 1880 and 2000. At this writing, the hottest year on record in the Northern Hemisphere was 2005.

Atmospheric levels of greenhouse gases are far higher than they were in most of the past. Carbon dioxide may be at its highest level since 420,000 years ago, and possibly since 20 million years ago.

There is a growing consensus that the increase in greenhouse gases is a factor in **global warming**, a long-term rise in temperature near Earth's surface. Since direct atmospheric readings started in 1861, the lower atmosphere's temperature has risen by more than 1°F, mostly since 1946 (Figure 25.15). Also since then, nine of the ten hottest years occurred between 1990 and the present. Data from satellites, weather stations and balloons, research ships, and supercomputer programs suggest that irreversible climate changes are under way. Polar ice is melting and glaciers are retreating. This past century, the sea level may have risen as much as 20 centimeters (eight inches).

Continued temperature increases will have drastic effects on climate. As evaporation increases, so will global precipitation. Intense rains and flooding are expected to become more frequent in some regions. As investigations continue, a key research goal is to identify all the variables in play. The crucial factor may be the one we do not yet know.

25.9 The Nitrogen Cycle

Nitrogen, a component of our proteins and nucleic acids, moves in an atmospheric cycle called the nitrogen cycle.

Gaseous nitrogen (N_2) makes up about 80 percent of the atmosphere, the largest nitrogen reservoir. Among organisms, only a few kinds of bacteria can break the triple covalent bonds that hold its two atoms together.

Figure 25.16 shows the **nitrogen cycle**. As you can see, bacteria play key roles. They convert nitrogen to forms plants can use, and they also release nitrogen to complete the cycle. Land ecosystems lose more nitrogen through leaching of soils, although leaching provides nitrogen inputs to aquatic ecosystems such as streams, lakes, and the oceans.

Nitrogen fixation is the process in which certain bacteria convert N_2 to ammonia (NH_3), which then dissolves to form ammonium (NH_4^+). Plants assimilate and use this nitrogen to make amino acids, proteins, and nucleic acids. Plant tissues are the only nitrogen source for humans and other animals.

Bacteria and some fungi also break down nitrogen-containing wastes and remains of organisms. The decomposers use part of the released proteins and amino acids for their own life processes. But most of the nitrogen is still in the decay products, in the form of ammonia or ammonium, which plants take up. In a process called **nitrification**, bacteria convert these compounds to nitrite (NO_2^-). Other bacteria use the nitrite in metabolism and produce nitrate (NO_3^-), which plants also use.

Certain plants are better than others at securing nitrogen. The best are legumes such as peas and beans, which have mutually beneficial associations with nitrogen-fixing bacteria. In addition, most land plants have similar associations with fungi, forming specialized roots that enhance the plant's ability to take up nitrogen.

Some nitrogen is lost to the air by **denitrification**, when bacteria convert nitrate or nitrite to N_2 and a bit of nitrous oxide (N_2O). Much of the N_2 escapes into the atmosphere, completing the nitrogen cycle.

> *Nitrogen cycles from the atmosphere, through nitrogen-fixing organisms in soil and water, into plants and then to consumers, and ultimately back into the atmosphere.*

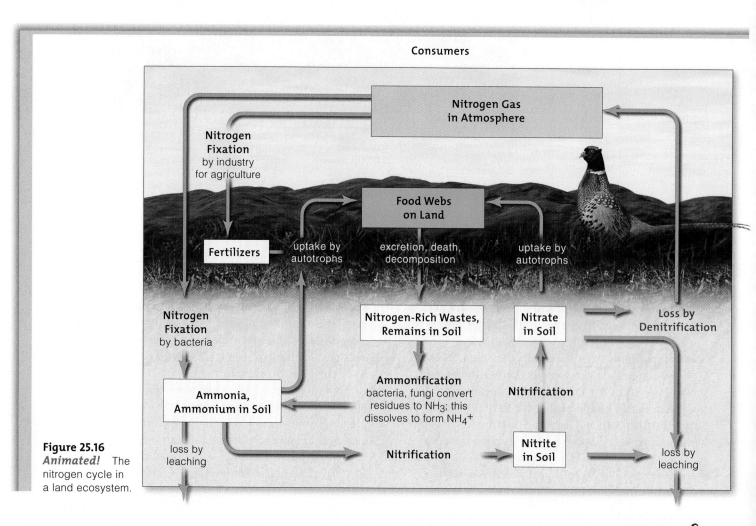

Figure 25.16 *Animated!* The nitrogen cycle in a land ecosystem.

25.10 Biological Magnification

DDT, a synthetic organic pesticide, was first used during World War II. In mosquito-infested regions of the tropical Pacific, people were vulnerable to malaria. DDT helped control the mosquitoes that were transmitting the disease agents (*Plasmodium japonicum*). In war-ravaged cities of Europe, people were suffering from the crushing headaches, fevers, and rashes associated with typhus. DDT helped control the body lice that were transmitting *Rickettsia rickettsii*, the bacterium that causes this terrible disease. After the war, it seemed like a good idea to use DDT against insects that were agricultural or forest pests, transmitters of pathogens, or merely nuisances in homes and gardens.

DDT is nearly insoluble in water, so you might think that it would act only where applied. But winds can carry DDT in vapor form, and water can transport fine particles of it. Being highly soluble in fat, DDT can collect in tissues, where it undergoes **biological magnification**. This term refers to an increase in concentration of a nondegradable (or slowly degradable) substance in organisms as it is passed upward through food chains (Figure 25.17). Most of the DDT from all the organisms that a consumer eats during its lifetime will become concentrated in its tissues. Many organisms also can partially metabolize DDT to DDE and other modified compounds with different but still disruptive effects. Both DDT and the modified compounds are toxic or physiologically disruptive to *many* aquatic and terrestrial animals.

After the war, DDT began to move through the global environment, infiltrate food webs, and affect organisms in ways that no one had predicted. In cities where DDT was sprayed to control Dutch elm disease, songbirds started dying. In streams flowing through forests where DDT was sprayed to control spruce budworms, salmon started dying. In croplands sprayed to control one kind of pest, new kinds of pests moved in. DDT was indiscriminately killing off the natural predators that had been keeping pest populations in

Figure 25.18 A peregrine falcon, a top carnivore in some food webs. This raptor almost became extinct as a result of biological magnification of DDT. A wildlife management program successfully brought back its population sizes.

check. It took no great leap of the imagination to make the connection. All of those organisms were dying at the same time and the same places as the DDT applications.

Much later, side effects of biological magnification started showing up far from the areas of DDT application. Most devastated were species at the end of food chains, including bald eagles, peregrine falcons, ospreys, and brown pelicans (Figure 25.18). A product of DDT breakdown interferes with physiological processes. As one consequence, birds produced eggs with brittle shells—and many of the chick embryos didn't make it to hatching time. Some species were facing extinction.

For decades now, DDT has been banned in the United States, except for limited applications where public health is endangered. Even today, however, some birds lay thin-shelled eggs. They pick up DDT at their winter ranges in Latin America, where DDT is still widely used. As recently as 1990, the California State Department of Health called for a fishery off the coast of Los Angeles to be closed. DDT from industrial waste discharges that ended 25 years ago is still contaminating that ecosystem.

DDT Residues (ppm wet weight of whole live organism)	
Ring-billed gull fledgling (*Larus delawarensis*)	75.5
Herring gull (*Larus argentatus*)	18.5
Osprey (*Pandion haliaetus*)	13.8
Green heron (*Butorides virescens*)	3.57
Atlantic needlefish (*Strongylura marina*)	2.07
Summer flounder (*Paralichthys dentatus*)	1.28
Sheepshead minnow (*Cyprinodon variegatus*)	0.94
Hard clam (*Mercenaria mercenaria*)	0.42
Marsh grass shoots (*Spartina patens*)	0.33
Flying insects (mostly flies)	0.30
Mud snail (*Nassarius obsoletus*)	0.26
Shrimps (composite of several samples)	0.16
Green alga (*Cladophora gracilis*)	0.083
Plankton (mostly zooplankton)	0.040
Water	0.00005

Figure 25.17 Biological magnification in a Long Island, New York, estuary. Some birds in this ecosystem died of DDT poisoning, and the biologists who studied the system noted various other harmful effects. Summer flounder, one of the fish species in which DDT accumulated, is a popular human food.

25.11 Human Population Growth

Advances in agriculture, industrialization, sanitation, and health care have fueled ever faster growth of the human population—with major implications for Earth's ecological future.

THE HUMAN POPULATION IS GROWING RAPIDLY

In 2004, there were 6.3 billion people on Earth (Figure 25.19). It took a long time, 2.5 million years, for the human population to reach the 1 billion milestone. It took less than 200 years more to reach 6 billion!

In any population, the growth rate is determined by the balance between births and deaths, plus gains and losses from immigration and emigration. Populations in different parts of the world grow at different rates, but overall, birth rates have been coming down worldwide. Death rates are falling, too, mainly because improved nutrition and health care are lowering infant mortality rates (the number of infants per 1,000 who die within their first year). However, the HIV/AIDS pandemic has sent death rates soaring in some African countries.

The six countries expected to show the most growth are India, China, Pakistan, Nigeria, Bangladesh, and Indonesia, in that order. China (with 1.3 billion people) and India (with 1.1 billion) combined dwarf all other countries in population size. They make up 38 percent of the world population. The United States is next in line. But with 292 million people, it represents only 4.6 percent of the world population.

The **total fertility rate** (TFR) is the average number of children born to women of a population during their reproductive years. In 1950, the worldwide TFR averaged 6.5. Currently it is 2.8, which is still far above replacement level of 2.1—the number of children a couple must have to replace themselves.

These numbers are averages. TFRs are at or below replacement levels in many developed countries; the developing countries in western Asia and Africa have the highest rates. Figure 25.20 has some examples of the disparities in the population distribution.

Even if every couple decides to bear no more than two children, the world population will keep growing for 60 years. It is projected to reach nearly 9 billion by 2050. Do you wonder how we can achieve corresponding increases in food production, drinkable water, energy reserves, and all the wood, steel, and other materials we use to meet everyone's basic needs—something we are not doing even now?

Figure 25.19 Growth curve (red) of the human population. Its vertical axis represents world population, in billions. (The dip between the years 1347 and 1351 is the time when 60 million people died from bubonic plague in Asia and Europe.) Agricultural revolutions, industrialization, and improvements in health care have sustained the accelerated growth of the last two centuries. The blue box lists how long it took for the human population to increase from 5 million to 6 billion.

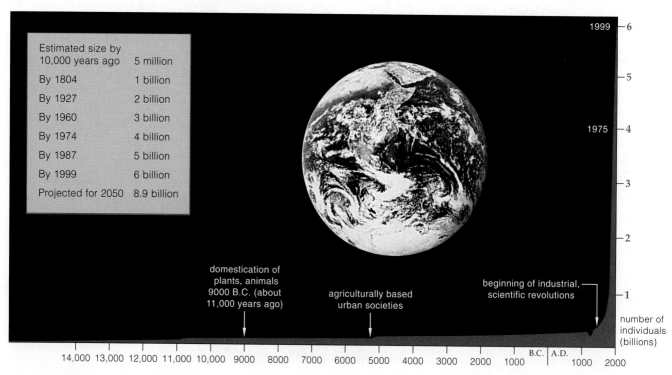

Estimated size by	
10,000 years ago	5 million
By 1804	1 billion
By 1927	2 billion
By 1960	3 billion
By 1974	4 billion
By 1987	5 billion
By 1999	6 billion
Projected for 2050	8.9 billion

domestication of plants, animals 9000 B.C. (about 11,000 years ago)

agriculturally based urban societies

beginning of industrial, scientific revolutions

number of individuals (billions)

14,000 13,000 12,000 11,000 10,000 9000 8000 7000 6000 5000 4000 3000 2000 1000 | B.C. A.D. | 1000 2000

HUMAN IMPACTS ON THE ENVIRONMENT

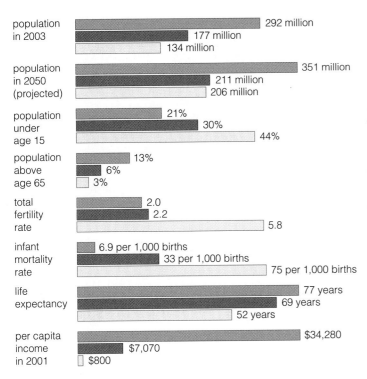

population in 2003	292 million	
	177 million	
	134 million	
population in 2050 (projected)	351 million	
	211 million	
	206 million	
population under age 15	21%	
	30%	
	44%	
population above age 65	13%	
	6%	
	3%	
total fertility rate	2.0	
	2.2	
	5.8	
infant mortality rate	6.9 per 1,000 births	
	33 per 1,000 births	
	75 per 1,000 births	
life expectancy	77 years	
	69 years	
	52 years	
per capita income in 2001	$34,280	
	$7,070	
	$800	

Figure 25.20 Key demographic indicators for three countries, mainly in 2003. The United States (gold bar) is highly developed, Brazil (brown bar) is moderately developed, and Nigeria (ivory bar) is less developed. The photograph shows just a few of India's 1.1 billion people.

POPULATION STATISTICS HELP PREDICT GROWTH

A population's **demographics**—its vital statistics, such as its size, age structure, and density—strongly influence its growth and its impact on ecosystems. Ecologists define population size as the number of individuals in the population's gene pool. **Population density** is the total number of individuals in a given area of habitat, such as the number of people who live within a hectare of land. Another demographic is the general pattern in which the population's members are distributed in their habitat. We humans are social animals. We tend to cluster in villages, towns, and cities, where we interact with one another and have access to jobs and other resources.

A population's **age structure** tracks the relative numbers of individuals of each age. These are often divided into prereproductive, reproductive, and postreproductive age categories. In theory, people in the first category will be able to produce offspring when they are sexually mature. Along with the people in the second category, they help make up the population's **reproductive base**. The population of the United States has a narrow base and is an example of slow growth. As you can see in Figure 25.21, this state of affairs contrasts with the age structure in rapidly growing populations, which have a broad reproductive base.

> *Advances in agriculture, industrialization, sanitation, and health care have allowed human population growth to surge in the last several centuries.*
>
> *Differences in population growth among countries correlate with economic development. The human population will soon reach a level that will severely strain resources.*

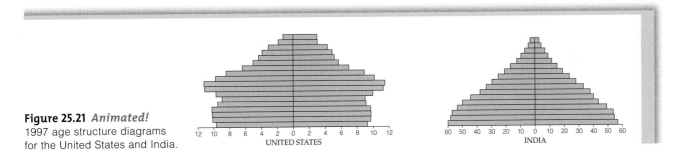

Figure 25.21 *Animated!*
1997 age structure diagrams for the United States and India.

UNITED STATES

INDIA

25.12 Nature's Controls on Population Growth

Can the human population grow faster and faster and faster? Simply put, no. Even when conditions are ideal, there is a maximum rate at which any population can grow.

Human populations can grow at a maximum rate of 2 to 5 percent per year. The rate is determined by how soon individuals begin to reproduce, how often they reproduce, and how many offspring are born each time.

Even when a population is not at its full reproductive potential, it can grow exponentially (in doubling increments from 2 to 4, then 8, 16, 32, 64, and so on). For instance, it is biologically possible for human females to bear twenty children or more, but few do so. Yet since the mid-1700s, our population has been growing *exponentially*. It can't do so forever, however, because "Mother Nature" doesn't work that way.

THERE IS A LIMIT ON HOW MANY PEOPLE THE EARTH CAN SUSTAIN

Environmental factors prevent most populations of organisms from reaching their full biotic potential. For instance, when a basic resource such as food or water is in short supply, it becomes a limiting factor on growth. Other kinds of limiting factors include predation (as by pathogens) and competition for living space.

The concept of limiting factors is important because it defines the **carrying capacity**—the number of individuals of a species that can be sustained indefinitely by the resources in a given area. Some experts believe that Earth has the resources to support from 7 to 12 billion humans, with a reasonable standard of living for many. Others believe that the current human population of 6.3 billion is already exceeding its carrying capacity. These viewpoints share the basic premise that overpopulation is the root of many, if not most, of the environmental problems the world now faces.

A low-density population starts to grow slowly, then goes through a rapid growth phase, and then growth levels off once the carrying capacity is reached. This pattern is called **logistic growth**. A plot of logistic growth gives us an S-shaped curve (Figure 25.22). This curve is only a simple approximation of what goes on in the natural world.

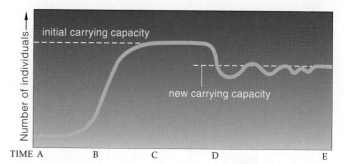

Figure 25.22 *Animated!* An S-shaped curve of logistic growth. The curve flattens out as the carrying capacity is reached. Changed environmental conditions can reduce the environment's carrying capacity. This happened to the human population in Ireland in the late 19th century, when a fungus wiped out the potatoes that were the mainstay of the diet.

SOME NATURAL POPULATION CONTROLS ARE RELATED TO POPULATION DENSITY

When a growing population's density increases, high density and overcrowding result in competition for resources. They also put individuals at increased risk of being killed by infectious diseases and parasites, which are more easily spread in crowded living conditions. These are **density-dependent controls** on population growth. Once such factors take their toll on a population and its density decreases, the pressures ease and the population may grow once more.

A classic example is the bubonic plague that killed 60 million Asians and Europeans—about one-third of the population—during the fourteenth century. *Yersinia pestis*, the bacterium responsible, normally lives in wild rodents; fleas transmit it to new hosts. It spread like wildfire through the cities of medieval Europe because dwellings were crowded together, sanitation was poor, and rats were everywhere. In 1994, bubonic plague and a related disease, pneumonic plague, raced through rat-infested cities in India where garbage and animal carcasses had piled up for months in the streets. Only crash efforts by public health officials averted a public health crisis.

Density-independent controls can also operate. These are events such as floods, earthquakes, or other natural disasters that cause deaths regardless of whether the members of a population are crowded or not.

Because resources are not unlimited, over the long term a given area can support only a finite number of individuals of a species. Other factors that limit population growth include disease organisms and effects of pollution.

25.13 Assaults on Our Air

Let's start this survey of the impact of human activities by defining pollution, which is central to our problems.

Pollutants are substances that adversely affect the health, activities, or survival of a population. Air pollutants include carbon dioxide, oxides of nitrogen and sulfur, and chlorofluorocarbons (CFCs). Others are photochemical oxidants formed as the sun's rays interact with certain chemicals. The United States releases more than 700,000 metric tons of air pollutants every day. Some of the worst air pollution disasters have been associated with the atmospheric condition called *smog*.

Where winters are cold and wet, **industrial smog** forms as a gray haze over industrialized cities that burn coal and other fossil fuels for manufacturing, heating, and generating electric power. The burning releases airborne dust, smoke, ashes, soot, asbestos, oil, bits of lead and other heavy metals, and sulfur oxides. Most industrial smog forms in cities of developing countries, including China and India, as well as in eastern Europe.

In warm climates, **photochemical smog** forms as a brown, smelly haze over large cities. The key culprit is nitric oxide. After it is released from vehicles, nitric oxide reacts with oxygen in the air to form nitrogen dioxide. When exposed to sunlight, nitrogen dioxide can react with hydrocarbons (such as partly burned gasoline) to form photochemical oxidants. Some of those in smog resemble tear gas; even traces can sting the eyes, irritate lungs, and damage crops.

Oxides of sulfur and nitrogen are among the worst air pollutants. These substances come mainly from power plants and factories fueled by coal, oil, and gas, as well as from motor vehicles. Dissolved in atmospheric water, they form weak sulfuric and nitric acids that winds may disperse over great distances. If they fall to Earth in rain and snow, they form **acid rain**. Acid rain can be much more acidic than normal rainwater, sometimes as acidic as lemon juice (pH 2.3). The acids eat away at marble, metals, even nylon. They also seriously damage the chemistry of ecosystems.

Canadian researchers have reported that inhaled soot particles from a steel mill caused DNA mutations that showed up in the sperm of male mice, and were passed on to offspring. More study is needed to learn whether such pollution-spurred mutations can cause disease in mice—or in humans who also breathe the polluted air.

THE OZONE LAYER HAS BEEN DAMAGED

Ozone is a molecule of three oxygen atoms (O_3). It occurs in two regions of Earth's atmosphere. In the troposphere, the region closest to the Earth's surface, ozone is part of

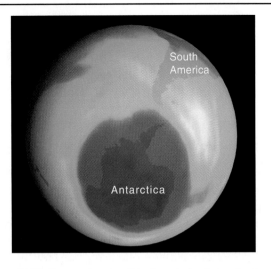

Figure 25.23 Seasonal ozone thinning above Antarctica in 2001. Darkest blue indicates the area with the lowest ozone level, at that time the largest recorded.

smog and can damage the respiratory system (as well as other organisms). On the other hand, ozone in the next atmospheric layer—the stratosphere (17–48 kilometers, or 11–30 miles above Earth)—intercepts harmful ultraviolet radiation that can cause skin cancer and eye cataracts. **Ozone thinning** has damaged this protective screen. September through mid-October, an ozone "hole" appears over the Antarctic, extending over an area about the size of the continental United States. Since 1987 the ozone layer over the Antarctic has been thinning by about half every year; a new hole, over the Arctic, appeared in 2001. That year the Antarctic hole was the biggest ever, covering an area greater than North America (Figure 25.23).

Chlorofluorocarbons (CFCs) are the main ozone depleters. These gases (compounds of chlorine, fluorine, and carbon) are used as coolants in refrigerators and air conditioners and in solvents and plastic foams. They slowly escape into the air and resist breakdown. Through a series of chemical steps, each of their chlorine molecules can destroy over 10,000 molecules of ozone! A widely used fungicide called methyl bromide is even worse. It will account for 15 percent of ozone thinning in future years unless production stops.

CFC production in developed countries has been phased out. Developing countries may phase it out by 2010. Methyl bromide production will also end at that time. Even assuming this happens, 100 to 200 years will pass before the ozone layer fully recovers.

Pollutants adversely affect the health, activities, or survival of a population. Air pollution can have global repercussions, as when CFCs and other compounds contribute to a thinning of the ozone layer that shields life on Earth from the sun's ultraviolet radiation.

25.14 Water, Wastes, and Other Problems

Three of every four humans do not have enough clean water to meet basic needs. Most of Earth's water is salty (in oceans). Of every million liters of water on our planet, only 6 liters are readily usable for human activities.

PROBLEMS WITH WATER ARE SERIOUS

As the human population grows exponentially, so do the demands and impacts on the Earth's limited supply of fresh water.

About a third of the world's food grows on land that is irrigated with water piped in from groundwater, lakes, or rivers. Irrigation water often contains large amounts of mineral salts. Where soil drainage is poor, evaporation may cause salt buildup, or *salinization*. Globally, salinization is estimated to have reduced yields on 25 percent of all irrigated cropland. Large-scale irrigation (Figure 25.24) is depleting groundwater stored in the Ogallala aquifer, which extends from South Dakota to Texas and has been providing about 30 percent of the groundwater used for irrigation in the United States. In some areas the farmers are working to reduce their water use by switching to more efficient irrigation systems.

Communities located in deserts have historically been notorious water wasters; in a 1999 estimate, Las Vegas, Nevada—desert home of golf courses, swimming pools, and lush lawns—was said to use more water per resident than any other city in the world. Since then, dwindling supplies have forced the city to launch a program of water conservation. In coastal areas, overuse of groundwater can cause saltwater intrusion into human water supplies.

In many regions, agricultural runoff pollutes public water sources with sediments, pesticides, and fertilizers.

Figure 25.24 A center-pivot sprinkler system, which is about 70 to 80 percent efficient.

Power plants and factories pollute water with chemicals (including carcinogens), radioactive materials, and excess heat. Such pollutants may accumulate in lakes, rivers, and bays before reaching their ultimate destination, the oceans. Contaminants from human activities have begun to turn up even in supposedly "pure" water in underground aquifers.

Many people view the oceans as convenient refuse dumps. Cities throughout the world dump untreated sewage, garbage, and other noxious debris into coastal waters. Cities along rivers and harbors maintain shipping channels by dredging the polluted muck and barging it out to sea. We don't yet know the full impact such practices may have on fisheries that provide human food.

WHERE WILL WE PUT SOLID WASTES AND PRODUCE FOOD?

Billions of metric tons of solid wastes are dumped, burned, and buried annually in the United States alone. This includes 50 billion nonreturnable cans and bottles. Paper products make up one-half of the total volume of solid wastes.

In natural ecosystems, solid wastes are recycled, but we humans bury them in landfills or incinerate them. Incinerators can add heavy metals and other pollutants to the air and leave a highly toxic ash that must be disposed of safely. Land that is both available and acceptable for landfills is scarce and becoming scarcer. All landfills eventually "leak," posing a threat to groundwater supplies. That is one reason why communities increasingly take the "not in my back yard" (NIMBY) approach to landfills. On the plus side, more and more people now participate in recycling programs. In addition to recycling, you can play a role by refusing to buy goods that are lavishly boxed and wrapped, packaged in indestructible containers, or designed for one-time use.

Today, almost a quarter of the Earth's land is being used for agriculture. Scientists have made valiant efforts to improve crop production on existing land. Under the banner of the "green revolution," research has been geared to improving the varieties of crop plants for higher yields and exporting modern agricultural practices and equipment to developing countries. Unfortunately, the green revolution is based on massive inputs of fertilizers and pesticides and ample irrigation to sustain high-yield crops. It is based also on fossil fuel energy to drive farm machines. Crop yields *are* four times as high as from traditional methods. But the modern practices use up to 100 times more energy and minerals. Also there are signs that limiting factors are coming into play to slow down further increases in crop yields.

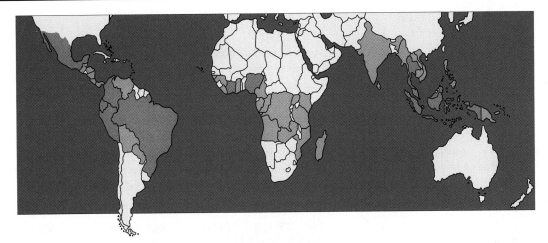

Figure 25.25 Countries that permit the greatest destruction of tropical forests. Red shading denotes where 2,000 to 14,800 square kilometers are deforested annually. Orange denotes "moderate" deforestation (100 to 1,900 square kilometers).

Overgrazing of livestock on marginal lands is a prime cause of *desertification*—the conversion of grasslands or cropland to a less productive, desertlike state. Worldwide, this has happened to about 9 million square kilometers over the past 50 years, and the trend is continuing.

DEFORESTATION HAS GLOBAL REPERCUSSIONS

The world's great forests influence the biosphere in many ways. Like giant sponges, forested watersheds absorb, hold, and gradually release water. Forests also help control soil erosion, flooding, and sediment buildup in rivers, lakes, and reservoirs.

Deforestation is the name for removal of all trees from large tracts of land for logging, agricultural, or grazing operations. The loss of vegetation exposes the soil, and this promotes leaching of nutrients and erosion, especially on steep slopes. Cleared plots soon become infertile and are abandoned. The photograph in Figure 25.25 shows forest destruction in the Amazon basin of South America.

Deforestation is linked with a variety of ecological problems; it can even lead to a major decline in rainfall over a wide area. But one of the most troubling effects relates to the global carbon cycle. Tropical forests absorb much of the sunlight reaching equatorial regions of the Earth's surface. When the forests are cleared, the land

becomes "shinier," and reflects more incoming energy back into space. The many millions of photosynthesizing trees in these vast forests help sustain the global cycling of carbon and oxygen. When trees are harvested or burned, carbon stored in their biomass is released to the atmosphere in the form of carbon dioxide—and this may be amplifying the greenhouse effect.

About half the world's tropical forests have been cut down for cropland, fuel wood, grazing land, and timber. Deforestation is greatest in Brazil, Indonesia, Colombia, and Mexico. If clearing continues at present rates, only Brazil and Zaire will have large tropical forests in 2010. By 2035, most of their forests will be gone. In other parts of the world, such as British Columbia (in western Canada) and Siberia, vast tracts of temperate forests are also disappearing fast.

Conservation biologists are attempting to reverse the trend. For example, in Brazil, a coalition of 500 groups is working to preserve the country's remaining tropical forests. In Kenya, women have planted millions of trees. Their success has inspired similar programs in more than a dozen countries in Africa.

Like other ecological problems, those related to land use are linked directly to the rapid growth of the human population, and the fact that people must have places to live and grow food, as well as materials to build homes and use as fuel. In the coming years one of the major challenges for humanity will be finding ways to balance human needs with the Earth's finite ability to provide the required resources.

Worldwide, human water supplies are threatened by overuse and by pollution with agricultural and industrial wastes.

Large-scale deforestation can have serious repercussions on the biosphere. Both tropical forests and temperate forests are rapidly being destroyed as trees are cut for timber and fuel, and to clear land for agriculture and livestock grazing.

25.15 Concerns about Energy

Paralleling the growth of the human population is a steep rise in total and per capita energy consumption.

The Earth has abundant energy supplies. However, there is a huge difference between the *total* and the *net* amounts available. Net energy is what is left over after subtracting the energy used to locate, extract, transport, store, and deliver energy to consumers. Some sources, such as direct solar energy, are renewable (Figure 25.26*a*). Others, such as coal and petroleum, are not. Figure 25.26*b* shows the percentages of different energy sources used globally. Overall, people in developed countries use far more energy per person than those in developing countries.

FOSSIL FUELS ARE GOING FAST

Oil, coal, and natural gas are **fossil fuels**, the fossilized remains of ancient forests. Fossil fuels are nonrenewable resources. Known oil and gas reserves may be used up in this century, and as the reserves run out in accessible areas, there is mounting pressure to explore wilderness

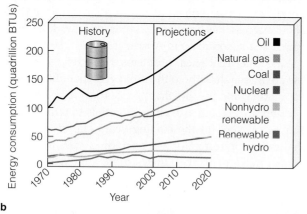

b

Figure 25.26 (**a**) Electricity-producing photovoltaic cells in panels that collect sunlight, a renewable source of energy. (**b**) Global energy consumption by fuel type, 1970–2003, projected to 2020. The values are in BTUs (British thermal units), a commonly used measure of heat energy. The data graphed here are from the U.S. Department of Energy, *Annual Energy Review, 2003 and 2004.*

areas in Alaska and other fragile environments. Due in part to the associated environmental costs, many people oppose the idea.

In theory, world reserves of coal can meet human energy needs for at least several centuries. But coal burning has been the main source of air pollution, because most coal reserves contain low-quality, high-sulfur material. Fossil-fuel burning also releases carbon dioxide and adds to the greenhouse effect.

CAN OTHER ENERGY SOURCES MEET THE NEED?

Today, nuclear power generates electricity at a relatively low cost. So why hasn't there been a "nuclear power revolution"? Safety concerns are one reason. Compared to coal-burning plants of the same capacity, a nuclear plant emits less radioactivity and pollutants. However, there is always the potential for a meltdown if the reactor core becomes overheated. This happened at Chernobyl in Ukraine in 1986, when the reactor core melted through its concrete containment slab, contaminating groundwater and releasing lethal amounts of radiation. Millions of people throughout Europe were exposed to dangerous levels of radioactive fallout and increased risk of certain cancers. Radioactive wastes also are highly dangerous. Some must be isolated for 10,000 years or longer, and there is little agreement on the best way to store them. The decision to store nuclear wastes generated in the United States in a remote area of Nevada was hotly debated.

There are other alternatives, although none has yet proven itself for widespread use. In windy places, solar energy is converted into the mechanical energy of winds, which generate electricity as they turn giant turbines. Entrepreneurs are developing solar cells that use sunlight energy to generate the electricity required to produce hydrogen gas. There has been considerable research aimed at developing hydrogen fuel cells to power cars. So far, though, the cost is high and the process generates high levels of greenhouse gases. Using different technology, many automakers now sell hybrid vehicles that run on a combination of gasoline and electric power generated by batteries.

A futuristic alternative is fusion power. The idea there is to fuse hydrogen atoms to form helium atoms, a reaction that releases considerable energy. It's not clear yet what the most likely practical applications for fusion power will be.

The need for energy supplies is growing along with the human population. As fossil fuels are depleted, the demand is increasing for safe, renewable, cost-efficient alternatives.

HUMAN IMPACTS ON THE ENVIRONMENT

25.16 Loss of Biodiversity

On March 24, 1900, in Ohio, a young hunter shot the last known wild passenger pigeon. Today humans are a major cause of the rapid loss of other species.

Recently, biologist John Musick of the Virginia Institute of Marine Science reported his latest findings on the decline in populations of sharks (Figure 25.27). Over the past 40 years, Musick has tracked the numbers of shark species that spend much of their lives off the mid-Atlantic coast of the United States and have become popular as human food. Musick's numbers show that, like codfish in parts of the North Atlantic, today several species of sharks have all but disappeared. Based on similar evidence from many scientists, the National Marine Fisheries Service has imposed catch restrictions on commercial shark fishing operations, over the strong protests of fleet operators. Since then, additional fish species have been recognized as threatened, including numerous kinds of salmon, the bluefin tuna, and others.

Sooner or later all species become extinct, but humans have become a major factor in the premature disappearance of more and more of them. By one current estimate, our actions are leading to the premature extinction of at least six species per hour.

Many biologists consider this epidemic of extinction to be an even more serious problem than depletion of stratospheric ozone or global warming because it is occurring faster and can't be reversed. Neither can rapid extinction be balanced by speciation, because it takes thousands of generations for new species to evolve.

Globally, tropical deforestation is the greatest killer of species, followed by destruction of coral reefs. However, plant extinctions can be more important ecologically than animal extinctions since most animal species depend directly or indirectly on plants for food, and often for shelter as well. Humans also have traditionally

Figure 25.28 The rosy periwinkle found in the threatened tropical forests of Madagascar. The plant is a source of two anticancer drugs.

depended on plants as sources of medicines. Indeed, 40 percent of prescription drug sales in the United States involve natural plant products. One example is the rosy periwinkle of Madagascar (Figure 25.28). This charming plant is the source of the anticancer drugs vincristine and vinblastine. Unfortunately, humans have destroyed 90 percent of the vegetation on Madagascar, so we may never know if it was home to plant sources of other life-saving drugs as well. Meanwhile, in the United States we are going about the business of habitat destruction at a dizzying pace. For instance, we have logged 92 percent of old-growth forests and drained half the wetlands, which filter human water supplies and provide homes for waterfowl and juvenile fishes. Hundreds of native species in these areas have been driven to extinction, and dozens more are endangered.

The underlying causes of wildlife extinction are human population growth and economic policies that don't value the environment but instead promote unsustainable exploitation. As our population grows, we clear, occupy, and damage more land to supply food, fuel, timber, and other resources. In wealthy countries, affluence leads to greater than average resource use per person. Elsewhere, the combination of rapid population growth and poverty pushes the poor to cut forests, grow crops on marginal land, overgraze grasslands, and poach endangered animals.

How can we help stem the tide of extinctions? Among other strategies, individuals can support efforts to reduce deforestation, global warming, ozone depletion, and poverty—the greatest threats to Earth's wildlife and to the human species.

Figure 25.27 A shark in its natural habitat. Worldwide, sharks are among the animals increasingly at risk of overexploitation by human enterprises.

Human population growth and other factors have led to widespread destruction of the habitats where many other species live. As a result, the Earth is experiencing an epidemic of extinctions.

Summary

Section 25.1 Ecology is the study of interactions of organisms with one another and with their physical environment. The regions of the Earth's crust, waters, and atmosphere where organisms live make up the biosphere. Every kind of organism has a habitat where it normally lives; the species within a given habitat associate with each other as a community.

An ecosystem encompasses producers, consumers, and decomposers and their physical environment. All interact with their environment and with one another through a flow of energy and a cycling of materials. A species' niche consists of the combined physical, chemical, and biological conditions it needs to live and reproduce in an ecosystem.

In ecological succession, the first species that take hold in a habitat are later replaced by others which themselves are replaced until conditions support a more or less stable array of species in the habitat.

Section 25.2 Ecosystems gain and lose energy and nutrients. Sunlight is the main energy source. Primary producers (photosynthesizing plants or algae) convert solar energy to forms that consumers can use. Primary producers also assimilate many of the nutrients that are transferred to other members of the system.

Consumers include herbivores that feed on plants, carnivores that feed on animals, and omnivores that have combination diets. Consumers also include decomposers, such as fungi and bacteria that feed on particles of dead or decomposing material.

An ecosystem's energy supply is transferred through feeding (trophic) levels. Primary producers make up the first feeding level, herbivores make up the next, carnivores the next, and so on.

Organisms that get energy from more than one source cannot be assigned to a single feeding level.

A food chain is a straight-line sequence of who eats whom in an ecosystem. Food chains usually cross-connect in intricate food webs.

Thomson NOW! *Learn about energy flow, the cycling of materials, and food webs.*

Section 25.3 Primary productivity is the rate at which producers capture and store a given amount of energy in a given time period in an ecosystem. The amount of energy flowing through consumer levels drops at each energy transfer through the loss of metabolic heat and as food energy is shunted into organic wastes. Energy relationships in an ecosystem can be represented as an ecological pyramid in which producers form a base for successive levels of consumers.

Sections 25.4–25.9 In biogeochemical cycles, substances move from the physical environment, to organisms, then back to the environment. Water enters and leaves ecosystems through the water cycle. In the phosphorus cycle (a sedimentary cycle), phosphorus in the Earth's crust moves to marine sediments, then back to the land. In the carbon cycle and the nitrogen cycle, where the elements exist mainly as atmospheric gases, the elements move through food webs and then ultimately return to the atmosphere.

Fossil fuel burning and the conversion of natural ecosystems to agriculture or grazing land are boosting atmospheric concentrations of carbon dioxide. The increase is a major factor in global warming.

Thomson NOW! *Learn more about the carbon cycle and other biogeochemical cycles.*

Section 25.11 Recent, rapid growth of the human population has been due mainly to advances in agriculture, industrialization, sanitation, and health care. Differences in growth among countries correlate partly with levels of economic development and partly with demographics such as population density and age structure. Populations of countries with a large reproductive base generally grow the fastest.

Thomson NOW! *Learn how to estimate population size, and observe patterns of logistic growth and exponential growth.*

Section 25.12 Carrying capacity is the number of individuals of a species that can be sustained indefinitely by the available resources in an area.

Populations initially show logistic growth—rapid growth that levels off when carrying capacity is reached. Carrying capacity, competition, and other factors limit population growth. Density-dependent controls on growth include competition for resources, disease, and predation.

Sections 25.13–25.16 The exponential growth of the human population has brought increased pollution and demands for energy, water, food, and waste disposal sites.

Air pollutants are present in industrial and photochemical smog and acid rain. Widespread deforestation is associated with leaching of soil nutrients, erosion, and possible disruption of the global carbon cycle.

Fossil fuels are nonrenewable and are major sources of air pollution. To meet the increasing demand for energy, conservation and the development of alternative energy sources will be crucial.

Human population growth and related activities also are contributing to the rapid loss of other species.

Review Questions

1. Explain what an ecosystem is, and name the central roles that producers play in all ecosystems.

2. Define and give examples of the different feeding levels in ecosystems.

3. Explain the difference between a food chain and a food web in an ecosystem.

4. If you were growing a vegetable garden, what variables might affect its net primary production?

5. Describe the greenhouse effect. Make a list of twenty agricultural and other products that you depend on. Are any implicated in global warming?

6. Describe the reservoirs and organisms involved in the carbon cycle and the nitrogen cycle.

Self-Quiz   *Answers in Appendix V*

1. _____ can be thought of as an ecosystem.
   a. A freshwater spring   c. A city
   b. A rain forest   d. All of the above

2. Ecosystems have _____.
   a. energy gains and losses   c. one feeding level
   b. nutrient cycling but not losses   d. a and b

3. _____ is the study of how organisms interact with one another as well as with their physical and chemical environment.

4. Feeding levels can be described as _____.
   a. structured feeding relationships
   b. who eats whom in an ecosystem
   c. a hierarchy of energy transfers
   d. all of the above

5. A feeding relationship that proceeds from algae to a fish, then to a fisherman, and then to a shark is _____.
   a. a food chain   c. a and b
   b. a food web

6. Primary productivity is affected by _____.
   a. photosynthesis and energy use by plants
   b. how many plants are neither eaten nor decomposed
   c. rainfall
   d. temperatures
   e. all of the above

7. The number of individuals that can be sustained indefinitely by the resources in a given region is the

   _____.
   a. biotic potential
   b. carrying capacity
   c. environmental resistance
   d. density-dependent control

8. Which of the following does *not* affect sustainable population size?
   a. predation
   b. competition
   c. pollution
   d. each of the above can affect population size

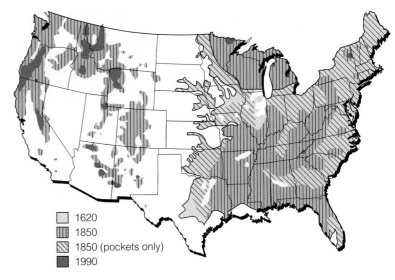

1620
1850
1850 (pockets only)
1990

Figure 25.29 Map showing the extent of deforestation in the United States from 1620 through 1990.

9. Match the following terms with the suitable description.
   ____ pollutant
   ____ biogeochemical cycle
   ____ ecosystem parts
   ____ density-dependent control
   ____ carrying capacity
   ____ ecological pyramid

   a. water or nutrients moving from the environment, to organisms, then back
   b. producers, consumers, decomposers
   c. substances that adversely affect health or survival
   d. maximum sustainable population size
   e. infectious disease
   f. energy relationships in an ecosystem

Critical Thinking

1. Imagine and describe a situation whereby you would be a participant (not the top predator) in a food chain.

2. Vast forests once covered much of the United States (Figure 25.29). Some researchers and policymakers are proposing that we should protect entire ecosystems, such as forests, rather than just endangered species. Opponents say that doing so might endanger property values. Would you favor setting aside large tracts of land for restoration? Or would you consider such efforts too intrusive on individual rights of property ownership?

3. How have humans increased the carrying capacity of their environments? Have they avoided some of the limiting factors on population growth?

Explore on Your Own

Benjamin Franklin once remarked, "It's not until the well runs dry that we know the worth of water." Everybody needs water, and as noted in this chapter, that precious fluid is in increasingly short supply. Have you ever wondered about your own "water impact"? You can get an idea using the following average statistics:

In a typical U.S. home, flushing toilets, washing hands, and bathing account for about 78% of the water used.

Nearly all toilets installed in the United States since 1994 use 1.6 gallons for each flush. Older toilets use about 4 gallons.

A shower uses about 5 gallons per minute (less if you have a low-flow shower head). Brushing teeth with the water running uses about 2 gallons. Shaving with the water running full blast can use up to 20 gallons.

Washing dishes with an automatic dishwasher uses about 15 gallons; handwashing dishes with the water running doubles the water use.

Using these numbers as a guide, keep track of your water use for a typical day. Are there ways you could conserve and still meet your basic needs?

Appendix I Concepts in Cell Metabolism

The Nature and Uses of Energy

Any time an object is not moving, it has a store of **potential energy**—a capacity to do work, simply owing to its position in space and the arrangement of its parts. If a stationary runner springs into action, some of the runner's potential energy is transformed into **kinetic energy**, the energy of motion.

Energy on the move does work when it imparts motion to other things—for example, when you throw a ball. In skeletal muscle cells in your arm, the energy currency ATP (adenosine triphosphate, Section 3.13) gave up some of its potential energy to molecules of contractile units and set them in motion. The combined motions in many muscle cells resulted in the movement of whole muscles. The transfer of energy from ATP also resulted in the release of another form of kinetic energy called **heat**, or *thermal energy*.

The potential energy of molecules is called **chemical energy** and is measured as kilocalories. A **kilocalorie** is the amount of energy it takes to heat 1,000 grams of water from 14.5°C to 15.5°C at standard pressure.

As noted in Chapter 3, cells use energy for chemical work, to stockpile, build, rearrange, and break apart substances. They channel it into *mechanical work*—to move cell structures and the whole body or parts of it. They also channel it into *electrochemical work*—to move charged substances into or out of the cytoplasm or an organelle compartment.

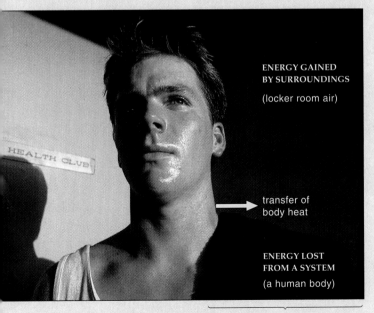

ENERGY GAINED
BY SURROUNDINGS

(locker room air)

transfer of
body heat

ENERGY LOST
FROM A SYSTEM

(a human body)

NET ENERGY CHANGE = 0

Figure A-1

Laws of Thermodynamics

We cannot create energy from scratch; we must first get it from someplace else. Why? According to the **first law of thermodynamics**, the total amount of energy in the universe remains constant. More energy cannot be created; existing energy cannot vanish or be destroyed. It can only be converted from one form to some other form. For instance, when you eat, your cells extract energy from food and convert it to other forms, such as kinetic energy for moving about.

With each metabolic conversion, some of the energy escapes to your surroundings, as heat. Even when you "do nothing," your body gives off about as much heat as a 100-watt lightbulb because of conversions in your cells. The energy being released is transferred to atoms and molecules that make up the air, and in this way it heats up the surroundings, as shown in Figure A-1. In general, the body cannot recapture energy lost as heat, but the energy still exists in the environment outside the body. Overall, there is a one-way flow of energy in the universe.

The human body obtains its energy mainly from the covalent bonds in organic compounds, such as glucose and glycogen. When the compounds enter metabolic reactions, specific bonds break or are rearranged. For example, your cells release usable energy from glucose by breaking all of its covalent bonds. After many steps, six molecules of carbon dioxide and six of water remain. Compared with glucose, these leftovers have more stable arrangements of atoms, but chemical energy in their bonds is much less. Why? Some energy was lost at each breakdown step leading to their formation. This is why glucose is a much better source of usable energy than, for example, water is.

As the molecular events just described take place, some heat is lost to the surroundings and cannot be recaptured. Said another way, no energy conversion can ever be 100 percent efficient. Therefore, the total amount of energy in the universe is spontaneously flowing from forms rich in energy (such as glucose) to forms having less and less of it. This is the main point of the **second law of thermodynamics**.

Examples of Energy Changes

When cells convert one form of energy to another, there is a change in the amount of potential energy that is available to them. Cells of photosynthetic organisms, notably green plants, convert energy in sunlight into chemical energy, which is stored in the bonds of organic compounds. The outcome is a net increase in energy in

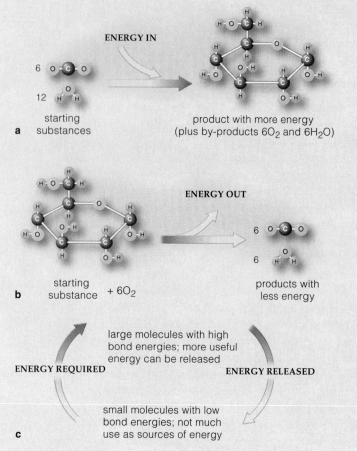

a starting substances

ENERGY IN

product with more energy
(plus by-products $6O_2$ and $6H_2O$)

b starting substance + $6O_2$

ENERGY OUT

products with less energy

6 CO_2

6 H_2O

c

ENERGY REQUIRED

large molecules with high bond energies; more useful energy can be released

ENERGY RELEASED

small molecules with low bond energies; not much use as sources of energy

Figure A-2

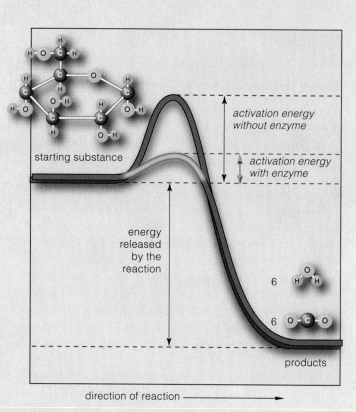

starting substance

activation energy without enzyme

activation energy with enzyme

energy released by the reaction

6 H_2O

6 CO_2

products

direction of reaction ⟶

Figure A-3

the product molecule (such as glucose), as diagrammed in Figure A-2*a*. A reaction in which there is a net increase in energy in the product compound is an **endergonic reaction** (meaning energy in). By contrast, reactions in cells that break down glucose (or another energy-rich compound) release energy. They are called **exergonic reactions** (meaning energy out).

The Role of Enzymes in Metabolic Reactions

The catalytic molecules called **enzymes** are crucial actors in metabolism. To better understand why, it helps to begin with the idea that in cells, molecules or ions of substances are always moving at random. As a result of this random motion, they are constantly colliding. Metabolic reactions may take place when participating molecules collide— but only *if* the energy associated with the collisions is great enough. This minimum amount of energy required for a chemical reaction is called **activation energy**. Activation energy is a barrier that must be surmounted one way or another before a reaction can proceed.

Nearly all metabolic reactions are reversible. That is, they can run "forward," from starting substances to products, or in "reverse," from a product back to starting substances. Which way such a reaction runs depends partly on the ratio of reactant to product. When there is a high concentration of reactant molecules, the reaction is likely to run strongly in the forward direction. On the other hand, when the product concentration is high enough, more molecules or ions of the product are available to revert spontaneously to reactants. Any reversible reaction tends to run spontaneously toward **chemical equilibrium**—the point at which it will be running at about the same pace in both directions.

As just described, before reactants enter a metabolic reaction they must be activated by an energy input; only then will the steps leading to products proceed. And while random collisions *might* provide the energy for reactions, our survival depends on thousands of reactions taking place with amazing speed and precision. This is the key function of enzymes, for *enzymes lower the activation energy barrier* (Figure A-3). As Section 3.13 described, substrates and enzymes interact at the enzyme's active site. According to the **induced fit model**, a surface region of each substrate has chemical groups that are almost but not quite complementary to chemical groups in an active site. However, as substrates settle into the site, the contact strains some of their bonds, making them easier to break. There also are interactions among charged or polar groups that prime substrates for conversion to an activated state. With these changes, substrates fit precisely in the enzyme's active site. They now are in an activated state, in which they will react spontaneously.

Glycolysis: The First Stage of the Energy-Releasing Pathway

Energy that is converted into the chemical bond energy of adenosine triphosphate—ATP—fuels cell activities. Cells make ATP by breaking down carbohydrates (mainly glucose), fats, and proteins. During the breakdown reactions, electrons are stripped from intermediates, then energy associated with the liberated electrons drives the formation of ATP.

Recall that cells rely mainly on **aerobic respiration**, an oxygen-dependent pathway of ATP formation. The main energy-releasing pathways of aerobic respiration all start with the same reactions in the cytoplasm. During this initial stage of reactions, called **glycolysis**, enzymes break apart and rearrange a glucose molecule into two molecules of pyruvate, which has a backbone of three carbon atoms. Following up on the discussion in Section 3.14, here you can track in a bit more detail on what happens to a glucose molecule in the first stage of aerobic respiration.

Glucose is one of the simple sugars. Each molecule of glucose contains six carbon, twelve hydrogen, and six oxygen atoms, all joined by covalent bonds (Figure A-4). The carbons make up the backbone. With glycolysis, glucose or some other carbohydrate in the cytoplasm is partially broken down, the result being two molecules of the three-carbon compound pyruvate:

$$glucose \longrightarrow glucose\text{-}6\text{-}phosphate \longrightarrow 2\ pyruvate$$

The first steps of glycolysis require energy. As diagrammed in Figure A-5 on page A-4, they advance only when two ATP molecules each transfer a phosphate group to glucose and so donate energy to it. Such transfers, recall, are phosphorylations. In this case, they raise the energy content of glucose to a level that is high enough to allow the *energy-releasing* steps of glycolysis to begin.

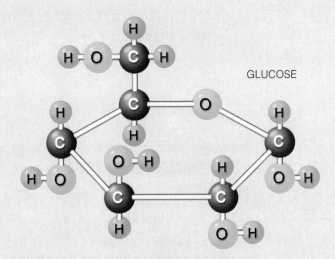

GLUCOSE

Figure A-4

The first energy-releasing step breaks the activated glucose into two molecules. Each of these molecules is called PGAL (phosphoglyceraldehyde). Next, each PGAL is converted to an unstable intermediate that allows ATP to form by giving up a phosphate group to ADP. The next intermediate in the sequence does the same thing. Thus, a total of four ATP form by **substrate-level phosphorylation**. This metabolic event is the direct transfer of a phosphate group from a substrate of a reaction to some other molecule—in this case, ADP. Remember, though, two ATP were invested to jump-start the reactions. So the *net* energy yield is only two ATP.

Meanwhile, the coenzyme NAD^+ picks up electrons and hydrogen atoms liberated from each PGAL, thus becoming NADH. When the NADH gives up its cargo at a different reaction site, it reverts to NAD^+. Said another way, like other coenzymes NAD^+ is reusable.

In sum, glycolysis converts energy stored in glucose to a transportable form of energy, in ATP. NAD^+ picks up electrons and hydrogen that are removed from each glucose molecule. The electrons and hydrogen have roles in the next stage of reactions. So do the end products of glycolysis—the two molecules of pyruvate.

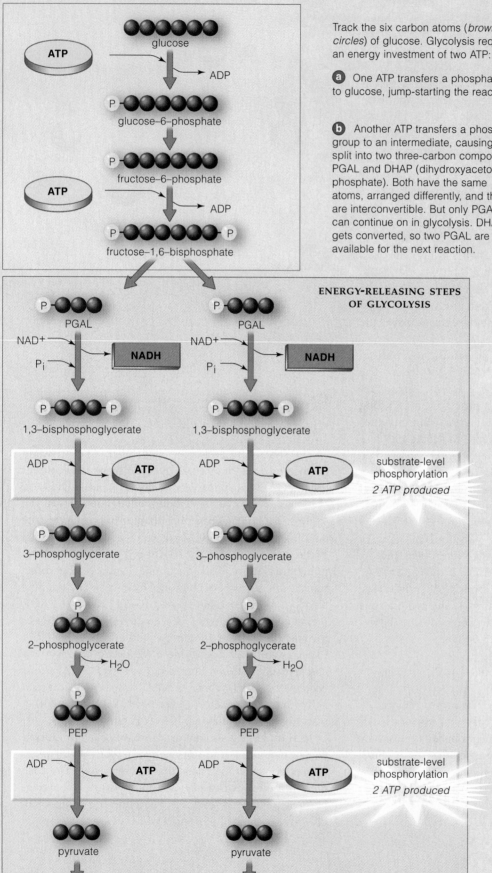

ENERGY-REQUIRING STEPS OF GLYCOLYSIS

2 ATP invested

glucose

ATP

ADP

P–glucose–6–phosphate

P–fructose–6–phosphate

ATP

ADP

P–fructose–1,6–bisphosphate–P

ENERGY-RELEASING STEPS OF GLYCOLYSIS

P–PGAL

NAD+

Pᵢ

NADH

P–1,3–bisphosphoglycerate–P

ADP

ATP

substrate-level phosphorylation
2 ATP produced

P–3–phosphoglycerate

P–2–phosphoglycerate

H₂O

P–PEP

ADP

ATP

substrate-level phosphorylation
2 ATP produced

pyruvate

P–PGAL

NAD+

Pᵢ

NADH

P–1,3–bisphosphoglycerate–P

ADP

ATP

P–3–phosphoglycerate

P–2–phosphoglycerate

H₂O

P–PEP

ADP

ATP

pyruvate

to second set of reactions

Track the six carbon atoms (*brown circles*) of glucose. Glycolysis requires an energy investment of two ATP:

a One ATP transfers a phosphate group to glucose, jump-starting the reactions.

b Another ATP transfers a phosphate group to an intermediate, causing it to split into two three-carbon compounds: PGAL and DHAP (dihydroxyacetone phosphate). Both have the same atoms, arranged differently, and they are interconvertible. But only PGAL can continue on in glycolysis. DHAP gets converted, so two PGAL are available for the next reaction.

c Two NADH form when each PGAL gives up two electrons and a hydrogen atom to NAD+.

d Two intermediates each transfer a phosphate group to ADP. *Thus, two ATP have formed by direct phosphate group transfers.* The original energy investment of two ATP is now paid off.

e Two more intermediates form. Each gives up one hydrogen atom and an —OH group. These combine as water. Two molecules called PEP form by these reactions.

f Each PEP transfers a phosphate group to ADP. *Once again, two ATP have formed by substrate-level phosphorylation.*

In sum, glycolysis has a net energy yield of two ATP for each glucose molecule. Two NADH also form during the reactions, and two molecules of pyruvate are the end products.

Figure A-5 Glycolysis, first stage of the main energy-releasing pathways. The reaction steps proceed inside the cytoplasm of every living prokaryotic and eukaryotic cell. In this example, glucose is the starting material. By the time the reactions end, two pyruvate, two NADH, and four ATP have been produced. Cells invest two ATP to start glycolysis, however, so the *net* energy yield of glycolysis is two ATP.

Depending on the type of cell and on environmental conditions, the pyruvate may enter the second set of reactions of the aerobic pathway, which includes the Krebs cycle. Or it may be used in other reactions, such as a fermentation pathway.

Second Stage of the Aerobic Pathway

When pyruvate molecules formed by glycolysis leave the cytoplasm and enter a mitochondrion, the scene is set for both the second and the third stages of the aerobic pathway. Figure A-6 diagrams these steps in detail.

a Pyruvate from glucose enters a mitochondrion. It undergoes initial conversions before entering cyclic reactions (the Krebs cycle).

b Pyruvate is stripped of a carboxyl group (COO^-), which departs as CO_2. It also gives up hydrogen and electrons to NAD^+, forming NADH. Then a coenzyme joins with the remaining two-carbon fragment, forming acetyl–CoA.

c The acetyl–CoA transfers its two-carbon group to oxaloacetate, a four-carbon compound that is the entry point into the Krebs cycle. The result is citrate, with a six-carbon backbone. Addition and then removal of H_2O changes citrate to isocitrate.

d Isocitrate enters conversion reactions, and a COO^- group departs (as CO_2). Hydrogen and electrons are transferred to NAD^+, forming NADH.

e Another COO^- group leaves (as CO_2) and another NADH forms. The resulting intermediate attaches to a coenzyme A molecule, forming succinyl–CoA. *At this point, three carbon atoms have been released, balancing out the three that entered the mitochondrion (in pyruvate).*

f Now the attached coenzyme is replaced by a phosphate group (donated by the substrate GTP). That phosphate group becomes attached to ADP. Thus, for each turn of the cycle, one ATP forms by substrate-level phosphorylation.

g Electrons and hydrogen are transferred to FAD, forming $FADH_2$.

h In the final conversion reactions, the oxaloacetate is regenerated, and hydrogen and electrons are transferred to NAD^+, forming NADH.

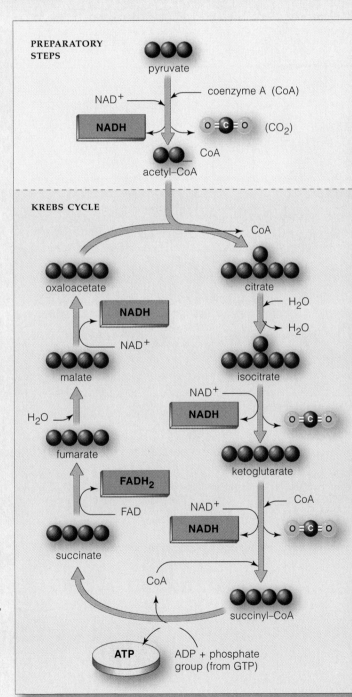

Figure A-6 Second stage of aerobic respiration: the Krebs cycle and reaction steps that precede it. For each three-carbon pyruvate molecule entering the cycle, three CO_2, one ATP, four NADH, and one $FADH_2$ molecules form. The steps shown proceed *twice*, because each glucose molecule was broken down earlier to *two* pyruvate molecules.

Third Stage of Aerobic Cellular Respiration

Most ATP is produced in the third stage of the aerobic pathway. Electron transport systems and neighboring proteins called ATP synthases serve as the production machinery. They are embedded in the inner membrane that divides the mitochondrion into two compartments (Figure A-7). They interact with electrons and H^+ ions, which coenzymes deliver from reaction sites of the first two stages of the aerobic pathway.

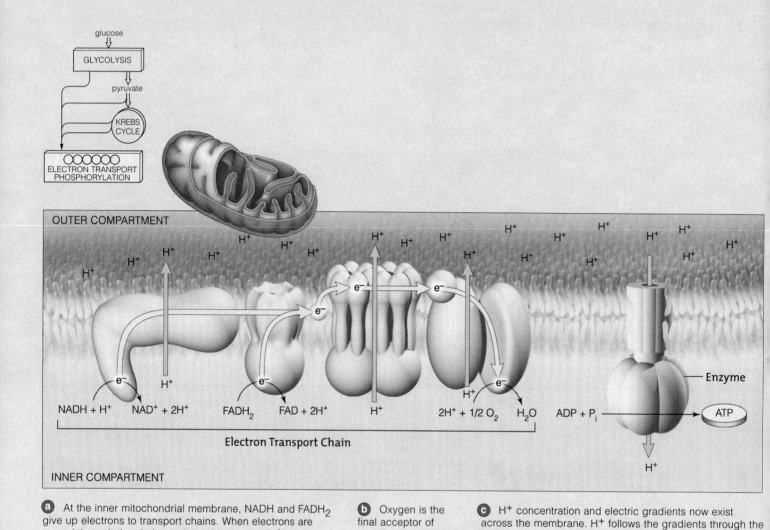

Electron Transport Chain

a At the inner mitochondrial membrane, NADH and $FADH_2$ give up electrons to transport chains. When electrons are moved through the chains, unbound hydrogen (H^+) is shuttled across the membrane to the outer compartment.

b Oxygen is the final acceptor of electrons at the end of the transfer chain.

c H^+ concentration and electric gradients now exist across the membrane. H^+ follows the gradients through the interior of enzymes, to the inner compartment. The flow drives the formation of ATP from ADP and phosphate (P_i).

Figure A-7 Electron transfer phosphorylation, the third and final stage of aerobic respiration.

Summary of Glycolysis and Aerobic Cellular Respiration

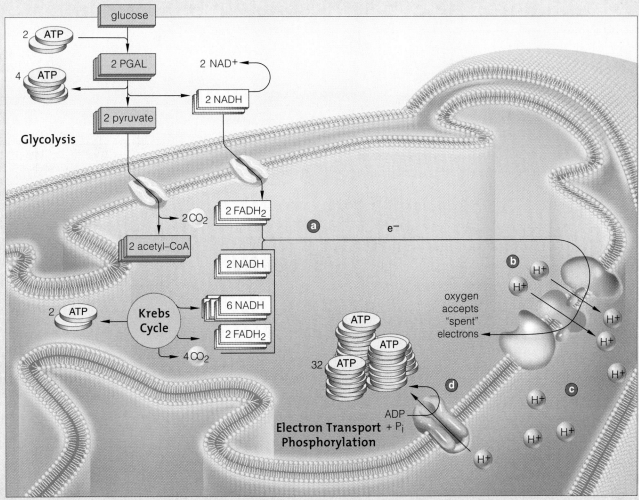

a Electrons and hydrogen from NADH and FADH$_2$ that formed during the first and second stages enter electron transport chains.

b As electrons are being transferred through these chains, H$^+$ ions are shuttled across the inner membrane, into the outer compartment.

c More H$^+$ accumulates in the outer compartment than in the inner one. Chemical and electric gradients have been established across the inner membrane.

d Hydrogen ions follow the gradients through the interior of enzymes, driving ATP formation from ADP and phosphate (P$_i$).

Figure A-8 Summary of the harvest from the energy-releasing pathway of aerobic respiration. Commonly, thirty-six ATP form for each glucose molecule that enters the pathway. But the net yield varies according to shifting concentrations of reactants, intermediates, and end products of the reactions. It also varies among different types of cells.

Cells differ in how they use the NADH from glycolysis, which cannot enter mitochondria. At the outer mitochondrial membrane, these NADH give up electrons and hydrogen to transport proteins, which shuttle the electrons and hydrogen across the membrane. NAD$^+$ or FAD already inside the mitochondrion accept them, thus forming NADH or FADH$_2$.

Any NADH inside the mitochondrion delivers electrons to the highest possible entry point into a transport system. When it does, enough H$^+$ is pumped across the inner membrane to make *three* ATP. By contrast, any FADH$_2$ delivers them to a lower entry point. Fewer hydrogen ions can be pumped, so only *two* ATP can form.

In liver, heart, and kidney cells, for example, electrons and hydrogen from glycolysis enter the highest entry point of transport systems, so the energy harvest is thirty-eight ATP. More commonly, as in skeletal muscle and brain cells, they are transferred to FAD—so the harvest is thirty-six ATP.

Appendix II Periodic Table of the Elements

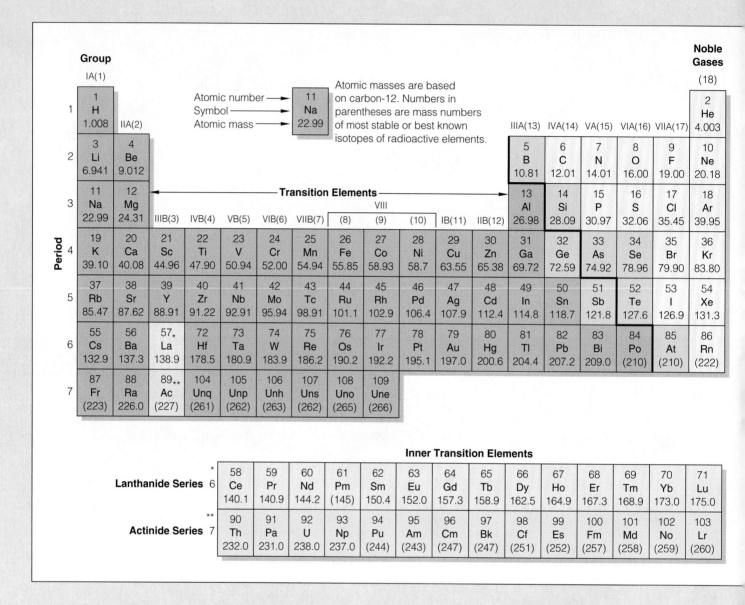

Group																	Noble Gases
IA(1)																	(18)

Atomic number → 11
Symbol → Na
Atomic mass → 22.99

Atomic masses are based on carbon-12. Numbers in parentheses are mass numbers of most stable or best known isotopes of radioactive elements.

Transition Elements

Inner Transition Elements

Lanthanide Series 6

*	58 Ce 140.1	59 Pr 140.9	60 Nd 144.2	61 Pm (145)	62 Sm 150.4	63 Eu 152.0	64 Gd 157.3	65 Tb 158.9	66 Dy 162.5	67 Ho 164.9	68 Er 167.3	69 Tm 168.9	70 Yb 173.0	71 Lu 175.0

Actinide Series 7

**	90 Th 232.0	91 Pa 231.0	92 U 238.0	93 Np 237.0	94 Pu (244)	95 Am (243)	96 Cm (247)	97 Bk (247)	98 Cf (251)	99 Es (252)	100 Fm (257)	101 Md (258)	102 No (259)	103 Lr (260)

Metric-English Conversions

Length

English		Metric
inch	=	2.54 centimeters
foot	=	0.30 meter
yard	=	0.91 meter
mile (5,280 feet)	=	1.61 kilometer

To convert	multiply by	to obtain
inches	2.54	centimeters
foot	30.00	centimeters
centimeters	0.39	inches
millimeters	0.039	inches

Weight

English		Metric
grain	=	64.80 milligrams
ounce	=	28.35 grams
pound	=	453.60 grams
ton (short) (2,000 pounds)	=	0.91 metric ton

To convert	multiply by	to obtain
ounces	28.3	grams
pounds	453.6	grams
pounds	0.45	kilograms
grams	0.035	ounces
kilograms	2.2	pounds

Volume

English		Metric
cubic inch	=	16.39 cubic centimeters
cubic foot	=	0.03 cubic meter
cubic yard	=	0.765 cubic meters
ounce	=	0.03 liter
pint	=	0.47 liter
quart	=	0.95 liter
gallon	=	3.79 liters

To convert	multiply by	to obtain
fluid ounces	30.00	milliliters
quart	0.95	liters
milliliters	0.03	fluid ounces
liters	1.06	quarts

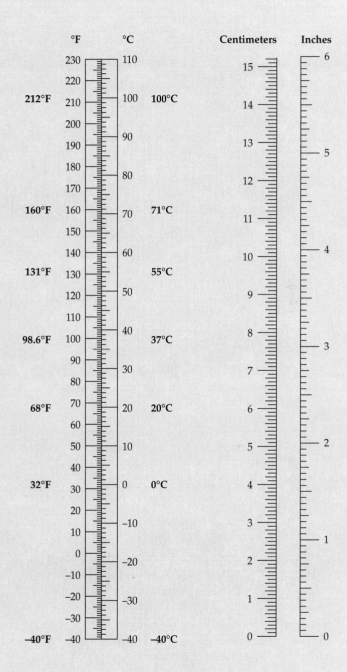

To convert temperature scales:

Fahrenheit to Celsius: $°C = 5/9(°F - 32)$

Celsius to Fahrenheit: $°F = 9/5(°C + 32)$

Appendix IV Answers to Genetics Problems

CHAPTER 20

1. a: *AB*

   b: *AB* and *aB*

   c: *Ab* and *ab*

   d: *AB, aB, Ab,* and *ab*

2. a: *AaBB* will occur in all the offspring.

   b: 25% *AABB*; 25% *AaBB*; 25% *AABb*; 25% *AaBb*

   c: 25% *AaBb*; 25% *Aabb*; 25% *aaBb*; 25% *aabb*

   d: 1/16 *AABB* (6.25%)
   1/8 *AaBB* (12.5%)
   1/16 *aaBB* (6.25%)
   1/8 *AABb* (12.5%)
   1/4 *AaBb* (25%)
   1/8 *aaBb* (12.5%)
   1/16 *AAbb* (6.25%)
   1/8 *Aabb* (12.5%)
   1/16 *aabb* (6.25%)

3. a: Mother must be heterozygous for both genes; father is homozygous recessive for both genes. The first child is also homozygous recessive for both genes.

   b: The probability that the second child will not be able to roll the tongue and will have detached earlobes is 1/4 (25%).

4. a: ABC

   b: ABc, aBc

   c: ABC, ABc, aBC, aBc

   d: ABC, ABc, AbC, Abc, aBC, aBc, abC, abc

5. Because Molly does not exhibit the recessive hip disorder, she must be either homozygous dominant (HH) for this trait, or heterozygous (Hh). If the father is homozygous dominant (HH), then he and Molly cannot produce offspring that are homozygous recessive (hh), and so none of their offspring will have the undesirable phenotype. However, if Molly is a heterozygote for the trait, notice that the probability is 1/2 (50%) that a puppy will be heterozygous (Hh) and so carry the trait.

6. a: The mother must be heterozygous ($I^A i$). The man having type B blood could have fathered the child if he were also heterozygous ($I^B i$).

   b: If the man is heterozygous, then he *could be* the father. However, because any other type B heterozygous male could also be the father, one cannot say that this particular man absolutely must be. Actually, any male who could contribute an O allele (*i*) could have fathered the child. This would include males with type O blood (*ii*) or type A blood who are heterozygous.

7. The probability is 1/2 (50%) that a child of this couple will be a heterozygote and have sickle cell trait. The probability is 1/4 (25%) that a child will be homozygous for the sickling allele and so will have sickle cell anemia.

8. For these ten traits, all the man's sperm will carry identical genes. He cannot produce genotypically different sperm. The woman can produce eggs with four genotypes. This example underscores the fact that the more heterozygous gene pairs that are present, the more genetically different gametes are possible.

9. The mating between two carriers of a lethal trait is *Ll* × *Ll*.

   Progeny genotypes: 1/4 *LL* + 1/2 *Ll* + 1/4 *ll*.
   Phenotypes: 1/4 homozygous survivors (*LL*)

   1/2 heterozygous survivors (*Ll*)

   1/4 lethal (*ll*) nonsurvivors

10. Bill's genotype: *Aa Ss Ee*
    Marie's genotype: *AA SS EE*

No matter how many children Bill and Marie have, the probability is 100% that each child will have the parents' phenotype. Because Marie can produce only dominant alleles, there is no way that a child could inherit a *pair* of recessive alleles for any of these three traits—and that is what would be required in order for the child to show the recessive phenotype. Thus the probability is zero that a child will have short lashes, high arches, and no achoo syndrome.

11. The white-furred parent's genotype is *bb*; the black-furred parent must be *Bb*; because if it were *BB* all offspring would be heterozygotes (*Bb*) and would have black fur. A monohybrid cross between a heterozygote and a homozygote typically yields a 1:1 phenotype ratio. Four black and three white guinea pigs is close to a 1:1 ratio.

CHAPTER 21

1. Key: Among other complicated issues, here you must consider whether (in your view) a person's gender is determined by "genetic sex" or by reproductive structures. Also, review the discussion of sex determination at the beginning of Chapter 14. In addition, note that the secondary sex characteristics of normal males, which are due to the sensitivity of cells to testosterone, include greater muscle mass—which is why males typically are physically stronger than females.

2. The probability is the same for each child: 1/2, or 50%.

3. a: From his mother.

   b: A male can produce two types of gametes with respect to an X-linked gene. One type will possess only a Y chromosome and so lack this gene; the other type will have an X chromosome and will have the X-linked gene.

   c: A female homozygous for an X-linked gene will produce just one type of gamete containing an X chromosome with the gene.

   d: A female heterozygous for an X-linked gene will produce two types of gametes. One will contain an X chromosome with the dominant allele, and the other type will contain an X chromosome with the recessive allele.

4. Most of chromosome 21 has been translocated to chromosome 14. While this individual has 46 chromosomes, there are in fact three copies of chromosome 21. The third copy of chromosome 21 is attached to chromosome 14.

5. No. Many traits are sex-influenced and controlled by genes on autosomes.

6. 50%. His mother is heterozygous for the allele, so there is a 50% chance that any male offspring will inherit the allele. Since males do not inherit an X chromosome from their fathers, the genotype of the father is irrelevant to this question.

7. The allele for cystic fibrosis is recessive. Most of the carriers are heterozygous for the allele and do not have the cystic fibrosis phenotype.

8. The parent with typical pigmentation is heterozygous for the albinism allele. The probability that any one child will have the albinism phenotype is 50%. However, with a small sample of only four offspring, there is a high probability of a deviation from a 1:1 ratio due to the random mixes of alleles that occur during meiosis and fertilization (discussed in Chapter 18).

Appendix V Answers to Self-Quizzes

CHAPTER 1: 1. DNA, 2. cell, 3. Homeostasis, 4. vertebrates, mammals, 5. nine, 6. c, 7. d, 8. c, 9. b

CHAPTER 2: 1. carbon, 2. a, 3. c, 4. c, 5. d, 6. b, 7. b, 8. c, 9. c, e, b, d, a, 10. a (plus R group interactions)

CHAPTER 3: 1. d, 2. cytoskeleton, 3. c, 4. a, 5. d, 6. b, g, f, d, c, a, e, 7. d, 8. d, 9. c, e, d, a, b, 10. b, 11. b, a, c, 12. The electron transport systems and carrier proteins required for ATP formation are embedded in the membrane between the inner and outer compartment of the mitochondrion.

CHAPTER 4: 1. d, 2. c, 3. b, 4. b, 5. a, 6. c, 7. c, 8. Receptors, integrator, effectors, 9. d, e, c, b, a

CHAPTER 5: 1. Skeletal and muscular, 2. d, 3. d, 4. c, 5. b, 6. a, 7. skull, rib cage, and vertebral column; pectoral girdles, pelvic girdle, and bones of extremities, 8. g, d, f, e, c, b, h, a

CHAPTER 6: 1. Skeletal and muscular, 2. skeletal, cardiac, smooth, 3. d, 4. b, 5. d, 6. f, d, e, a, b, g, c

CHAPTER 7: 1. digesting, absorbing, eliminating, 2. caloric, energy, 3. carbohydrates, 4. essential amino acids, essential fatty acids, 5. b, 6. c, 7. d, 8. c, 9. e, d, a, b, c

CHAPTER 8: 1. d, 2. d, 3. c, 4. b, 5. d, a, c, e, b

CHAPTER 9: 1. a, 2. systole, diastole, 3. b, 4. c, 5. b, 6. a, 7. a, 8. d, 9. d, b, a, c, 10. c, a, b

CHAPTER 10: 1. d, 2. f, 3. d, 4. d, 5. b, 6. d, 7. a, 8. Tears are surface barriers that may wash away pathogens, 9. c, b, a, e, d, f

CHAPTER 11: 1. d, 2. b, 3. d, 4. b, 5. c, 6. e, 7. c, 8. d, 9. d, 10. d, 11. f, a, e, d, b, c

CHAPTER 12: 1. d, 2. f, 3. a, 4. b, 5. c, 6. d, 7. a, 8. d, 9. b, d, e, c, a

CHAPTER 13: 1. stimuli, 2. action potentials, 3. neurotransmitter, 4. c, 5. b, 6. c, 7. e, d, b, c, a

CHAPTER 14: 1. stimulus, 2. sensation, 3. Perception, 4. d, 5. b, 6. d, 7. c, 8. e, 9. b, 10. c, 11. b, 12. e, c, d, a, b

CHAPTER 15: 1. f, 2. d, 3. b, 4. d, 5. e, 6. b, 7. b, 8. b, 9. b, e, g, d, f, a, c, 10. d, a, b, c, e

CHAPTER 16: 1. hypothalamus, 2. b, 3. d, 4. c, 5. b, 6. d

CHAPTER 17: 1. a, 2. a, 3. c, 4. morphogenesis, 5. c, 6. c, 7. d, a, f, e, c, b, 8. e, 9. e

CHAPTER 18: 1. a, 2. c, 3. e, 4. e, 5. a, 6. d, b, a, c, f, e

CHAPTER 19: 1. mitosis; meiosis, 2. chromosomes; DNA, 3. c, 4. diploid, 5. c, 6. a, 7. b, 8. b, 9. b, 10. d, 11. d, b, c, a

CHAPTER 20: 1. a, 2. c, 3. a, 4. c, 5. b, 6. d, 7. b, 8. d, 9. c, a, d, b

CHAPTER 21: 1. c, 2. e, 3. e, 4. c, 5. d, 6. c, 7. d, 8. d, 9. d, 10. c, e, d, b, a

CHAPTER 22: 1. three, 2. b, 3. c, 4. c, 5. a, 6. a, 7. c, 8. a, 9. e, c, d, a, b, 10. c, 11. c

CHAPTER 23: 1. c, 2. e, 3. tumor suppressor gene, 4. b, 5. d, 6. a, 7. c, 8. d, 9. b, 10. c

CHAPTER 24: 1. species, 2. mutation, genetic drift, gene flow, and natural selection, 3. natural selection, 4. e, 5. e, 6. b, 7. a, 8. a

CHAPTER 25: 1. d, 2. d, 3. Ecology, 4. d, 5. a, 6. e, 7. b, 8. d, 9. c, a, b, e, d, f

Glossary

abdominal cavity Body cavity that holds the stomach, liver, pancreas, most of the intestine, and several other organs.

ABO blood typing Method of characterizing an individual's blood according to whether one or both of two protein markers, A and B, are present at the surface of red blood cells. The O signifies that neither marker is present.

abortion Spontaneous or induced expulsion of the embryo or fetus from the uterus.

absorption The movement of nutrients, fluid, and ions across the gastrointestinal tract lining and into the internal environment.

accommodation In the eye, adjustments of the lens position that move the focal point forward or back so that incoming light rays are properly focused on the retina.

acetylcholine (ACh) A neurotransmitter that can excite or inhibit various target cells in the brain, spinal cord, glands, and muscles.

acetyl–CoA (uh-SEED-ul) Coenzyme A with a two-carbon fragment from pyruvate attached. In the second stage of aerobic respiration, it transfers the fragment to oxaloacetate for the Krebs cycle.

achondroplasia Genetic disorder that results in abnormally short arms and legs in affected adults.

acid A substance that releases hydrogen ions in water.

acid–base balance State in which extracellular fluid is neither too acidic nor too basic, an outcome of controls over its concentrations of dissolved ions.

acidity Of a solution, an excess of hydrogen ions relative to hydroxyl ions.

acid rain Wet acid deposition; falling of rain (or snow) rich in sulfur and nitrogen oxides.

acromegaly Oversecretion of growth hormone during adulthood; bones and facial features become abnormally enlarged.

acrosome An enzyme-containing cap that covers most of the head of a sperm and helps the sperm penetrate an egg at fertilization.

actin (AK-tin) A globular contractile protein. In muscle cells, actin interacts with another protein, myosin, to bring about contraction.

action potential An abrupt, brief reversal in the steady voltage difference (resting membrane potential) across the plasma membrane of a neuron.

active site A crevice on the surface of an enzyme molecule where a specific reaction is catalyzed.

active transport The pumping of one or more specific solutes through a transport protein that spans the lipid bilayer of a cell membrane. Most often, the solute is transported against its concentration gradient. The protein is activated by an energy boost, as from ATP.

adaptation [L. *adaptare*, to fit] In evolutionary biology, the process of becoming adapted (or more adapted) to a given set of environmental conditions. Of sensory neurons, a decrease in the frequency of action potentials (or their cessation) even when a stimulus is maintained at constant strength.

adaptive immunity Immune responses that the body develops in response to antigens of specific pathogens, toxins, or abnormal body cells.

adaptive radiation A burst of speciation events, with lineages branching away from one another as they partition the existing environment or invade new ones.

adaptive trait A trait from gene mutation that helps an organism survive and reproduce under a given set of environmental conditions.

adenine (AH-de-neen) A purine; a nitrogen-containing base in certain nucleotides.

adenocarcinoma Cancer of a gland or its ducts.

ADH Antidiuretic hormone. Produced by the hypothalamus and released by the posterior pituitary, it stimulates reabsorption in the kidneys and so reduces urine volume.

adhering junctions Cell junctions that cement cells together.

adipose tissue A type of connective tissue having an abundance of fat-storing cells and blood vessels for transporting fats.

adrenal cortex (ah-DREE-nul) Outer portion of the adrenal gland; its hormones have roles in metabolism, inflammation, maintaining extracellular fluid volume, and other functions.

adrenal medulla Inner region of the adrenal gland; its hormones help control blood circulation and carbohydrate metabolism.

aerobic respiration (air-OH-bik) [Gk. *aer*, air, and *bios*, life] The main energy-releasing metabolic pathway of ATP formation, in which oxygen is the final acceptor of electrons stripped from glucose or some other organic compound. The pathway proceeds from glycolysis through the Krebs cycle and electron transport phosphorylation. A typical net yield is 36 ATP for each glucose molecule.

afferent arteriole In the urinary system, an arteriole that delivers blood to each nephron.

African sleeping sickness A severe form of encephalitis passed from person to person by bites of the tsetse fly.

age structure Of a population, the number of individuals in each of several or many age categories.

agglutination (ah-glue-tin-AY-shun) The clumping together of foreign cells that have invaded the body (as pathogens or in tissue grafts or transplants). Clumping is induced by cross-linking between antibody molecules that have already bound antigen at the surface of the foreign cells.

aging A range of processes, including the breakdown of cell structure and function, by which the body gradually deteriorates.

agranulocyte Class of white blood cells that lack granular material in the cytoplasm; includes the precursors of macrophages (monocytes) and lymphocytes.

AIDS Acquired immunodeficiency syndrome. A set of chronic disorders following infection by the human immunodeficiency virus (HIV), which destroys key cells of the immune system.

alcohol Organic compound that includes one or more hydroxyl groups (—OH); it dissolves readily in water. Sugars are examples.

aldosterone (al-DOSS-tuh-roan) Hormone secreted by the adrenal cortex that helps regulate sodium reabsorption from the kidney.

allantois (ah-LAN-twahz) [Gk. *allas*, sausage] One of four extraembryonic membranes that form during embryonic development. In humans, it functions in early blood formation and development of the urinary bladder.

allele (uh-LEEL) For a given location on a chromosome, one of two or more slightly different molecular forms of a gene that code for different versions of the same trait.

allele frequency The relative abundances of each kind of allele carried by the individuals of a population.

allergen Any normally harmless substance that provokes inflammation, excessive mucus secretion, and other immune responses.

allergy An immune response made against a normally harmless substance.

all-or-none principle Principle that states that individual cells in a muscle's motor units always contract fully in response to proper stimulation. If the stimulus is below a certain threshold, the cells do not respond at all.

alveolus (al-VEE-uh-lus), plural alveoli [L. *alveus*, small cavity] Any of the many cup-shaped, thin-walled out-pouchings of respiratory bronchioles. A site where oxygen diffuses from air in the lungs to the blood, and carbon dioxide diffuses from blood to the lungs.

Alzheimer's disease Degenerative brain disorder that leads to memory loss.

amine hormone A hormone derived from the amino acid tyrosine.

amino acid (uh-MEE-no) A small organic molecule having a hydrogen atom, an amino group, an acid group, and an R group covalently bonded to a central carbon atom. The subunit of polypeptide chains, which represent the primary structure of proteins.

ammonification (uh-MOAN-ih-fih-KAY-shun) A process by which certain microorganisms break down nitrogen-containing wastes and remains of other organisms.

amnesia A loss of fact memory.

amnion (AM-nee-on) One of four extraembryonic membranes. It becomes a fluid-filled sac in which the embryo (and fetus) can grow, move freely, and be protected from sudden temperature shifts and impacts.

amoebic dysentery Type of severe diarrhea caused by a protozoan.

anabolism A metabolic activity that assembles small molecules into more complex molecules that store energy.

anaerobic pathway (AN-uh-ROW-bik) [Gk. *an*, without, and *aer*, air] Metabolic pathway in which a substance other than oxygen serves as the final acceptor of electrons that have been stripped from substrates.

anal canal The canal from the rectum to the anus through which feces pass.

analogous structures Body parts, once different in separate lineages, that were put to comparable uses in similar environments and that came to resemble one another in form and function. They are evidence of morphological convergence.

anaphase (AN-uh-faze) The stage at which microtubules of a spindle apparatus separate sister chromatids of each chromosome and move them to opposite spindle poles. During anaphase I of meiosis, the two members of each pair of homologous chromosomes separate. During anaphase II, sister chromatids of each chromosome separate.

anaphylactic shock A whole-body allergic response in which a person's blood pressure plummets, among other symptoms.

anemias Disorders that indicate that red blood cells, which contain hemoglobin, are not delivering enough oxygen to meet the body's needs.

aneuploidy (AN-yoo-ploy-dee) A change in the chromosome number following inheritance of one extra or one fewer chromosome.

aneurysm A pouchlike weak spot in an artery.

anhidrotic ectodermal dysplasia Condition that occurs in females whose active X chromosome has a mutated gene that blocks the formation of sweat glands.

antibiotic [Gk. *anti*, against] A normal metabolic product of certain microorganisms that kills or inhibits the growth of other microorganisms.

antibody Any of a variety of Y-shaped receptor molecules with binding sites for specific antigens. Only B cells produce antibodies, then position them at their surface or secrete them.

antibody-mediated immunity The B cell defensive response to pathogens in the body wherein antibodies are produced.

anticodon In a tRNA molecule, a sequence of three nucleotide bases that can pair with an mRNA codon.

antigen (AN-tih-jen) [Gk. *anti*, against, and *genos*, race, kind] Substance that is recognized as foreign to the body and that triggers an immune response. Most antigens are protein molecules at the surface of infectious agents or tumor cells.

antigen–MHC complex Unit consisting of fragments of an antigen molecule bound to MHC proteins. MHC complexes displayed at the surface of an antigen-presenting cell such as a macrophage promote an immune response by lymphocytes.

antigen-presenting cell A macrophage or other cell that displays antigen–MHC complexes at its surface and so promotes an immune response by lymphocytes.

antioxidant A chemical that can give up an electron to a free radical before the free radical damages DNA or some other cell constituent.

anus Terminal opening of the gastrointestinal tract.

aorta (ay-OR-tah) [Gk. *airein*, to lift, heave] Main artery of systemic circulation; carries oxygenated blood away from the heart to all body regions except the lungs.

aortic body Any of several receptors in artery walls near the heart that respond to changes in levels of carbon dioxide and oxygen in arterial blood.

apoptosis (APP-oh-TOE-sis) Genetically programmed cell death. Molecular signals lead to self-destruction in body cells that have finished their prescribed functions or have become altered, as by infection or transformation into a cancerous cell.

appendicular skeleton (ap-en-DIK-yoo-lahr) Bones of the limbs, hips, and shoulders.

appendix A slender projection from the cup-shaped pouch (cecum) at the start of the colon.

arrhythmia Irregular or abnormal heart rhythm, sometimes caused by stress, drug effects, or coronary artery disease.

arteriole (ar-TEER-ee-ole) Any of the blood vessels between arteries and capillaries. They are control points where the volume of blood delivered to different body regions can be adjusted.

artery Any of the large-diameter blood vessels that conduct deoxygenated blood to the lungs and oxygenated blood to all body tissues. The thick, muscular artery wall allows arteries to smooth out pulsations in blood pressure caused by heart contractions.

asexual reproduction Mode of reproduction by which offspring arise from a single parent and inherit the genes of that parent only.

asthma Lung disorder in which the bronchioles narrow when smooth muscle in their walls spasms.

astigmatism Vision problem where one or both corneas have an uneven curvature and cannot bend incoming light rays to the same focal point.

atherosclerosis Condition in which an artery's wall thickens and loses its elasticity, and the vessel becomes clogged with lipid deposits. In the artery wall, abnormal smooth muscle cells accumulate, and there is an increase in connective tissue. Plaques consisting of lipids, calcium salts, and fibrous material extend into the artery lumen, disrupting blood flow.

atherosclerotic plaque Cholesterol and other lipids that build up in the arterial wall, leaving less room for flowing blood.

atmosphere A region of gases, airborne particles, and water vapor enveloping the Earth; 80 percent of its mass is distributed within 17 miles of the Earth's surface.

atmospheric cycle A biogeochemical cycle in which the atmosphere is the largest reservoir of an element. The carbon and nitrogen cycles are examples.

atom The smallest unit of matter that is unique to a particular element.

atomic number The number of protons in the nucleus of each atom of an element; it differs for each element.

ATP Adenosine triphosphate (ah-DEN-uh-seen try-FOSS-fate) A nucleotide composed of adenine, ribose, and three phosphate groups. As the main energy carrier in cells, it directly or indirectly delivers energy to or picks up energy from nearly all metabolic pathways.

ATP/ADP cycle In cells, a mechanism of ATP renewal. When ATP donates a phosphate group to other molecules (and so energizes them), it reverts to ADP, then forms again by phosphorylation of ADP.

atrioventricular (AV) node In the septum dividing the heart atria, a site that contains bundles of conducting fibers. Stimuli arriving at the AV node from the cardiac pacemaker (sinoatrial node) pass along the bundles and continue on via Purkinje fibers to contractile muscle cells in the ventricles.

atrioventricular valve One-way flow valve between the atrium and ventricle in each half of the heart.

atrium (AY-tree-um) Upper chamber in each half of the heart; the right atrium receives deoxygenated blood (from tissues) entering the pulmonary circuit of blood flow, and the left atrium receives oxygenated blood from pulmonary veins.

australopith (oss-TRAH-low-pith) [L. *australis*, southern, and Gk. *pithekos*, ape] Any of the earliest known species of hominids; that is, the first species of the evolutionary branch leading to humans.

autoimmune response Misdirected immune response in which lymphocytes mount an attack against normal body cells.

automated DNA sequencing Machine method of determining the sequence of nucleotides in DNA using standard and modified versions of the four DNA nucleotides.

autonomic nerves (ah-toe-NOM-ik) Those nerves leading from the central nervous system to the smooth muscle, cardiac muscle, and glands of internal organs and structures—that is, to the visceral portion of the body.

autosomal dominant inheritance Condition arising from the presence of a dominant allele on an autosome (not a sex chromosome). The allele is always expressed to some extent, even in heterozygotes.

autosomal recessive inheritance Condition arising from a recessive allele on an autosome (not a sex chromosome). Only recessive homozygotes show the resulting phenotype.

autosome Any chromosome that is not a sex (gender-determining) chromosome.

autotroph (AH-toe-trofe) [Gk. *autos*, self, and *trophos*, feeder] An organism able to build its own large organic molecules by using carbon dioxide and energy from the physical environment. Compare heterotroph.

axial skeleton (AX-ee-uhl) The skull, backbone, ribs, and breastbone (sternum).

axon Of a neuron, a long, cylindrical extension from the cell body, with finely branched endings. Action potentials move rapidly, without alteration, along an axon; their arrival at axon endings may trigger the release of neurotransmitter molecules that influence an adjacent cell.

B lymphocyte, or **B cell** The only white blood cell that produces antibodies, then positions them at the cell surface or secretes them as weapons in immune responses.

bacterial conjugation The transfer of plasmid DNA from one bacterial cell to another.

baroreceptor reflex The short-term control over arterial pressure. It keeps blood pressure within normal limits in the face of sudden, but not long-term, changes in blood pressure.

Barr body In the cells of females, a condensed X chromosome that was inactivated during early embryonic development.

basal body A centriole that, after having given rise to the microtubules of a flagellum or cilium, remains attached to its base in the cytoplasm.

basal metabolic rate Amount of energy required to sustain body functions when a person is resting, awake, and has not eaten for 12–18 hours.

base A substance that accepts H^+ in water.

base pair A pair of hydrogen-bonded nucleotide bases in two strands of nucleic acids. In a DNA double helix, adenine pairs with thymine, and guanine with cytosine. When an mRNA strand forms on a DNA strand during transcription, uracil (U) pairs with the DNA's adenine.

base-pair substitution A mutation established in a replicating DNA molecule (a chromosome) when one base is wrongly substituted for another in a base pair.

basement membrane Noncellular layer of mostly proteins and polysaccharides that is sandwiched in between an epithelium and underlying connective tissue.

basophil Fast-acting white blood cell that secretes histamine and other substances during inflammation.

B cell *See* B lymphocyte.

bicarbonate–carbon dioxide buffer system A system used to restore the body's normal pH level by neutralizing excess H^+ and allowing for the exhalation of carbon dioxide formed during the reaction. It does not eliminate the excess H^+ and therefore has only a temporary effect.

biogeochemical cycle The movement of an element such as carbon or nitrogen from the environment to organisms, then back to the environment.

biogeography [Gk. *bios*, life, and *geographein*, to describe the surface of the Earth] The study of major land regions, each having distinguishing types of plants and animals and generally retaining its identity because of climate and geographic barriers to gene flow.

biological clock Internal time-measuring mechanism that has a role in adjusting an organism's daily activities, seasonal activities, or both in response to environmental cues.

biological magnification The increasing concentration of a nondegradable or slowly degradable substance in body tissues as it is passed along food chains.

biology The scientific study of life.

biomass The combined weight of all the organisms at a particular feeding (trophic) level in an ecosystem.

biome A broad, vegetational subdivision of a biogeographic realm shaped by climate, topography, and composition of regional soils.

biopsy Diagnostic procedure in which a small piece of tissue is removed from the body through a hollow needle or exploratory surgery, and then examined for signs of a particular disease (often cancer).

biosphere [Gk. *bios*, life, and *sphaira*, globe] All regions of the Earth's waters, crust, and atmosphere in which organisms live.

biosynthetic pathway A metabolic pathway in which small molecules are assembled into large organic molecules.

biotic potential Of a population, the maximum rate of increase per individual under ideal conditions.

bipedalism A habitual standing and walking on two feet, as by humans.

blastocyst (BLASS-tuh-sist) [Gk. *blastos*, sprout, and *kystis*, pouch] In embryonic development, a blastula stage consisting of a hollow ball of surface cells and an inner cell mass.

blastomere One of the small, nucleated cells that form during cleavage of a zygote.

blastula (BLASS-chew-lah) An embryonic stage consisting of a ball of cells produced by cleavage.

blood A fluid connective tissue composed of water, solutes, and formed elements (blood cells and platelets); it carries substances to and from cells and helps maintain an internal environment that is favorable for cell activities.

blood pressure Fluid pressure, generated by heart contractions, that keeps blood circulating.

blood–brain barrier Term applied to modified structure of brain capillaries that helps control which bloodborne substances reach neurons in the brain.

bolus Softened, lubricated ball of food, created by chewing and mixing of food with saliva.

bones Organs that function in movement and locomotion, protection of other organs, mineral storage, and (in some bones) blood cell production.

bone marrow A connective tissue where blood cells are formed.

bone remodeling Process of ongoing calcium deposits and withdrawals from bone that adjusts bone strength and maintains levels of calcium and phosphorus in blood.

bone tissue Mineral-hardened connective tissue; the main tissue in bone.

botulism Disease caused by exposure to a biological toxin produced by the bacterium *Clostridium botulinum.*

Bowman's capsule Cup-shaped portion of a nephron that receives water and solutes being filtered from blood.

brain Organ that receives, integrates, stores, and retrieves information, and coordinates appropriate responses by stimulating and inhibiting the activities of different body parts.

brain case The eight bones that together surround and protect the brain.

brain stem The midbrain, pons, and medulla oblongata, the core of which contains the reticular formation that helps govern activity of the nervous system as a whole.

bronchiole A component of the finely branched bronchial tree inside each lung.

bronchitis Inflammation of the bronchial walls that destroys wall tissue and can eventually block parts of the respiratory tract.

bronchus, plural bronchi (BRONG-cuss, BRONG-kee) [Gk. *bronchos*, windpipe] Tubelike branchings of the trachea that lead into the lungs.

buffer system A weak acid and the base that forms when it dissolves in water. The two work as a pair to counter slight shifts in pH.

bulbourethral glands Two glands of the male reproductive system that secrete mucus-rich fluid into the urethra when the male is sexually aroused.

bulk A volume of fiber and other undigested material that absorption processes in the colon cannot decrease.

Burkitt lymphoma Type of lymphoma that involves sites other than the lymph nodes.

cancer A malignant tumor, the cells of which show profound abnormalities in the plasma membrane and cytoplasm, abnormal growth and division, and weakened capacity for adhesion within the parent tissue (leading to metastasis). Unless eradicated, cancer is lethal.

candidiasis Vaginal yeast infection; symptoms include vaginal discharge, itching, and irritation.

capillary [L. *capillus*, hair] A thin-walled blood vessel that functions in the exchange of gases and other substances between blood and interstitial fluid.

capillary bed Dense capillary networks containing true capillaries where exchanges occur between blood and tissues, and also thoroughfare channels that link arterioles and venules.

carbaminohemoglobin A hemoglobin molecule that has carbon dioxide bound to it; $HbCO_2$.

carbohydrate [L. *carbo*, charcoal, and *hydro*, water] A simple sugar or large molecule composed of sugar units. All cells use carbohydrates as structural materials, energy stores, and transportable forms of energy. The three classes of carbohydrates include monosaccharides, oligosaccharides, and polysaccharides.

carbon cycle A biogeochemical cycle in which carbon moves from its reservoir in the atmosphere, through oceans and organisms, then back to the atmosphere.

carbonic anhydrase Enzyme in red blood cells that catalyzes the conversion of unbound carbon dioxide to carbonic acid and its dissociation products, thereby helping maintain the gradient that keeps carbon dioxide diffusing from interstitial fluid into the blood.

carcinogen (kar-SIN-uh-jen) An environmental agent or substance, such as ultraviolet radiation, that can trigger cancer.

carcinogenesis The transformation of a normal cell into a cancerous one.

carcinoma Cancer of the epithelium, including skin cells and epithelial linings of internal organs.

cardiac conduction system (KAR-dee-ak) Set of noncontractile cells in heart muscle that spontaneously produce and conduct the electrical events that stimulate heart muscle contractions.

cardiac cycle The sequence of muscle contraction and relaxation constituting one heartbeat.

cardiac muscle Type of muscle found only in the heart wall; cardiac muscle cells contract as a single unit.

cardiac output The amount of blood each ventricle of the heart pumps in one minute.

cardiac pacemaker Sinoatrial (SA) node; the basis of the normal rate of heartbeat. The self-excitatory cardiac muscle cells that spontaneously generate rhythmic waves of excitation over the heart chambers.

cardiovascular system Organ system that is composed of the heart and blood vessels and that functions in the rapid transport of blood to and from tissues.

carnivore [L. *caro, carnis*, flesh, + *vovare*, to devour] Animal that eats other animals.

carotid artery Artery in the neck that contains baroreceptors, which monitor arterial pressure.

carotid body Any of several sensory receptors that monitor carbon dioxide and oxygen levels in blood; located at the point where carotid arteries branch to the brain.

carrier An organism in which a pathogen is living without causing disease symptoms.

carrier protein A protein that binds specific substances and changes shape in ways that shunt the substances across a plasma membrane. Some carrier proteins function passively; others require an energy input.

carrying capacity The maximum number of individuals in a population (or species) that can be sustained indefinitely by a given environment.

cartilage A type of connective tissue with solid yet pliable intercellular material that resists compression.

cartilaginous joint Type of joint in which cartilage fills the space between adjoining bones; only slight movement is possible.

catabolism Metabolic activity that breaks down large molecules into simpler ones, releasing the components for use by cells.

cataracts Clouding of the eye's lens associated with aging.

cell [L. *cella*, small room] The smallest living unit; an organized unit that can survive and reproduce on its own, given DNA instructions and suitable environmental conditions, including appropriate sources of energy and raw materials.

cell cortex An array of cross-linked, bundled, gel-like microfilaments that reinforces the cell's plasma membrane.

cell count The number of cells of a given type in a microliter of blood.

cell cycle Events during which a cell increases in mass, roughly doubles its number of cytoplasmic components, duplicates its DNA, then undergoes nuclear and cytoplasmic division. It extends from the time a new cell is produced until it completes its own division.

cell determination Process that determines what an embryonic cell will become; nervous tissue or epithelium, for example.

cell differentiation The gene-guided process by which cells in different locations in the embryo become specialized.

cell-mediated immunity The T cell defensive response to pathogens in the body, wherein cytotoxic T cells attack the invaders directly.

cell theory A theory in biology, the key points of which are that (1) all organisms are composed of one or more cells, (2) the cell is the smallest unit that still retains a capacity for independent life, and (3) all cells arise from pre-existing cells.

cell-to-cell junction A point of contact that physically links two cells or that provides functional links between their cytoplasm.

cellular respiration The process by which cells break apart carbohydrates, lipids, or proteins to form ATP.

central nervous system The brain and spinal cord.

centriole (SEN-tree-ohl) A cylinder of triplet microtubules that gives rise to the microtubules of cilia and flagella.

centromere (SEN-troh-meer) [Gk. *kentron*, center, and *meros*, a part] A small, constricted region of a chromosome having attachment sites for microtubules that help move the chromosome during nuclear division.

cerebellum (ser-ah-BELL-um) [L. diminutive of *cerebrum*, brain] Hindbrain region with reflex centers for maintaining posture and refining limb movements.

cerebral cortex Thin surface layer of the cerebral hemispheres. Some regions of the cortex receive sensory input, others integrate information and coordinate appropriate motor responses.

cerebral hemispheres The left and right sides of the cerebrum, which are separated by a deep fissure.

cerebrospinal fluid Clear extracellular fluid that surrounds and cushions the brain and spinal cord.

cerebrum (suh-REE-bruhm) Part of the forebrain; the most complex integrating center.

channel protein Type of transport protein that serves as a pore through which ions or other water-soluble substances move across the plasma membrane. Some channels remain open, while others are gated and open and close in controlled ways.

chemical bond A union between the electron structures of two or more atoms or ions.

chemical evolution Process by which biological molecules evolved.

chemical synapse (SIN-aps) [Gk. *synapsis*, union] A small gap, the synaptic cleft, that separates two neurons (or a neuron and a muscle cell or gland cell) and that is bridged by neurotransmitter molecules released from the presynaptic neuron.

chemoreceptor (KEE-moe-ree-sep-tur) Sensory receptor that detects chemical energy (ions or molecules) dissolved in the surrounding fluid.

chemotherapy The use of therapeutic drugs to kill cancer cells.

CHH Cartilage-hair hypoplasia; a disease, caused by mutation of a single gene, that affects multiple organ systems, including the skeletal, integumentary, and immune systems.

chlamydia Sexually transmitted disease caused by infection by the bacteria *Chlamydia trachomatis.* The bacterium infects cells of the genital organs and urinary tract.

chlorofluorocarbon (klore-oh-FLOOR-oh-car-bun), or **CFC** One of a variety of odorless, invisible compounds of chlorine, fluorine, and carbon, widely used in commercial products, that are contributing to the destruction of the ozone layer above the Earth's surface.

chorion (CORE-ee-on) One of four extraembryonic membranes; it encloses the embryo and the three other membranes. Absorptive structures (villi) that develop at its surface are crucial for the transfer of substances between the embryo and mother.

chromatid Of a duplicated eukaryotic chromosome, one of two DNA molecules and its associated proteins. One chromatid remains attached to its "sister" chromatid at the centromere until they are separated from each other during a nuclear division; then each is a separate chromosome.

chromatin A cell's collection of DNA and all of the proteins associated with it.

chromosome (CROW-moe-soam) [Gk. *chroma,* color, and *soma,* body] A DNA molecule with its associated proteins.

chromosome number The number of each type of chromosome in all cells except dividing germ cells or gametes.

chyme The mixture of swallowed food boluses and acidic gastric fluid in the stomach that enters the small intestine during digestion.

cilium (SILL-ee-um), plural cilia [L. *cilium,* eyelid] Of eukaryotic cells, a short, hairlike projection that contains a regular array of microtubules. Cilia serve as motile structures, help create currents of fluids, or are part of sensory structures.

circadian rhythm (ser-KAYD-ee-un) [L. *circa,* about, and *dies,* day] A cycle of physiological events that is completed every 24 hours or so, even when environmental conditions remain constant.

clavicle Long, slender collarbone that connects the pectoral girdle with the sternum (breastbone).

cleavage Stage of development when mitotic cell divisions convert a zygote to a ball of cells, the blastula.

cleavage furrow Of a cell undergoing cytoplasmic division, a shallow, ringlike depression that forms at the cell surface as contractile microfilaments pull the plasma membrane inward. It defines where the cytoplasm will be cut in two.

cleavage reaction Enzyme action that splits a molecule into two or more parts; hydrolysis is an example.

climate Prevailing weather conditions for an ecosystem, including temperature, humidity, wind speed, cloud cover, and rainfall.

climax community Following primary and secondary succession, the array of species that remains more or less steady under prevailing conditions.

clonal selection hypothesis Hypothesis that lymphocytes activated by a specific antigen will rapidly multiply and differentiate into huge subpopulations of cells, all having the parent cell's specificity against that antigen.

cloning vector Plasmid that has been modified in the laboratory to accept foreign DNA.

cochlea Coiled, fluid-filled chamber of the inner ear. Sound waves striking the eardrum become converted to pressure waves in the cochlear fluid, and the pressure waves ultimately cause a membrane to vibrate and bend sensory hair cells. Signals from bent hair cells travel to the brain, where they may be interpreted as sound.

codominance Condition in which a pair of nonidentical alleles are both expressed, even though they specify two different phenotypes.

codon One of a series of base triplets in an mRNA molecule, most of which code for a sequence of amino acids of a specific polypeptide chain. (Of 64 codons, 61 specify different amino acids and three of these also serve as start signals for translation; one other serves only as a stop signal for translation.)

coenzyme A type of nucleotide that transfers hydrogen atoms and electrons from one reaction site to another. NAD^+ is an example.

cofactor A metal ion or coenzyme; it helps catalyze a reaction or serves briefly as an agent that transfers electrons, atoms, or functional groups from one substrate to another.

coitus Sexual intercourse.

colon (CO-lun) The large intestine.

community The populations of all species occupying a habitat; also applied to groups of organisms with similar lifestyles in a habitat.

compact bone Type of dense bone tissue that makes up the shafts of long bones and outer regions of all bones. Narrow channels in compact bone contain blood vessels and nerves.

comparative morphology [Gk. *morph,* form] Anatomical comparisons of major lineages.

complement system A set of about 20 proteins circulating in blood plasma with roles in nonspecific defenses and in immune responses. Some induce lysis of pathogens, others promote inflammation, and others stimulate phagocytes to engulf pathogens.

compound A substance in which the relative proportions of two or more elements never vary. Organic compounds have a backbone of carbon atoms arranged as a chain or ring structure. The simpler, inorganic compounds do not have comparable backbones.

concentration gradient A difference in the number of molecules (or ions) of a substance between two adjacent regions, as in a volume of fluid.

conclusion In scientific reasoning, a statement that evaluates a hypothesis based on test results.

concussion Temporary upset of the electrical activity of brain neurons resulting from a blow to the head.

condensation reaction Chemical step in which two molecules become covalently bonded into a larger molecule, and water often forms as a byproduct.

cone cell In the retina, a type of photoreceptor that responds to intense light and contributes to sharp daytime vision and color perception.

conjunctivitis Inflammation of the conjunctiva, the transparent membrane that lines the inside of the eyelids and the white of the eye.

connective tissue A category of animal tissues, all having mostly the same components but in different proportions. These tissues contain fibroblasts and other cells, the secretions of which form fibers (of collagen and elastin) and a ground substance (of modified polysaccharides).

consumer [L. *consumere,* to take completely] Of ecosystems, a heterotrophic organism that obtains energy and raw materials by feeding on the tissues of other organisms. Herbivores, carnivores, omnivores, and parasites are examples.

continuous variation A more or less continuous range of small differences in a given trait among all the individuals of a population.

control group In a scientific experiment, a group used to evaluate possible side effects of a test involving an experimental group. Ideally, the control group should differ from the experimental group only with respect to the variable being studied.

controlled experiment An experiment that tests only one prediction of a hypothesis at a time.

core temperature The body's internal temperature, as opposed to temperatures of the tissues near its surface. Normal human core temperature is about 37°C (98.6°F).

cornea Transparent tissue in the outer layer of the eye, which causes incoming light rays to bend.

coronary artery Either of two arteries leading to capillaries that service cardiac muscle.

corpus callosum (CORE-pus ka-LOW-sum) A band of 200 million axons that functionally link the two cerebral hemispheres.

corpus luteum (CORE-pus LOO-tee-um) A glandular structure; it develops from cells of a ruptured ovarian follicle and secretes progesterone and some estrogen, both of which maintain the lining of the uterus (endometrium).

cortex [L. *cortex,* bark] In general, a rindlike layer; the kidney cortex is an example.

covalent bond (koe-VAY-lunt) [L. *con,* together, and *valere,* to be strong] A sharing of one or more electrons between atoms or groups of atoms. When electrons are shared equally, the bond is nonpolar. When electrons are shared unequally, the bond is polar—slightly positive at one end and slightly negative at the other.

cranial cavity Body cavity that houses the brain.

creatine phosphate Organic compound that transfers phosphate to ADP in a rapid, short-term, ATP-generating pathway.

cri-du-chat Genetic disorder that leads to abnormal mental development and an abnormally shaped larynx.

critical thinking Objective evaluation of information.

cross-bridge The interaction between actin and myosin filaments that is the basis of muscle cell contraction.

crossing over During prophase I of meiosis, an interaction between a pair of homologous chromosomes. Their nonsister chromatids break at the same place along their length and exchange corresponding segments at the break points. Crossing over breaks up old combinations of alleles and puts new ones together in chromosomes.

culture The sum total of behavior patterns of a social group, passed between generations by learning and by symbolic behavior, especially language.

cyclic AMP Cyclic adenosine monophosphate. A nucleotide that has roles in intercellular communication, as when it serves as a second messenger (a cytoplasmic mediator of a cell's response to signaling molecules).

cystic fibrosis Genetic disorder in which dry, thickened mucus clogs the airways, making breathing difficult and leading to bacterial infections.

cystitis Inflammation of the bladder usually resulting from an infection of the urinary tract.

cytokine Any of the chemicals released by white blood cells that help muster or strengthen defense responses.

cytokinesis (sigh-toe-kih-NEE-sis) [Gk. *kinesis,* motion] Cytoplasmic division; the splitting of a parent cell into two daughter cells.

cytomembrane system [Gk. *kytos,* hollow vessel] Organelles that function as a system to modify, package, and distribute newly formed proteins and lipids. Endoplasmic reticulum, Golgi bodies, lysosomes, and a variety of vesicles are its components.

cytoplasm (SIGH-toe-plaz-um) [Gk. *plassein*, to mold] All cellular parts, particles, and semifluid substances enclosed by the plasma membrane except for the nucleus.

cytosine (SIGH-toe-seen) A pyrimidine; one of the nitrogen-containing bases in nucleotides.

cytoskeleton A cell's internal "skeleton." Its microtubules and other components structurally support the cell and organize and move its internal components.

cytosol The jellylike fluid portion of the cytoplasm.

cytotoxic T cell Type of T lymphocyte that directly kills infected body cells and tumor cells by lysis.

decomposer [L. *de-*, down, away, and *companere*, to put together] A heterotroph that obtains energy by chemically breaking down the remains, products, or wastes of other organisms. Decomposers help cycle nutrients to producers in ecosystems. Certain fungi and bacteria are examples.

deductive logic Pattern of thinking by which a person makes inferences about specific consequences or specific predictions that must follow from a hypothesis.

deforestation The removal of all trees from a large tract of land, such as the Amazon Basin or the Pacific Northwest.

deletion At the cellular level, loss of a segment from a chromosome. At the molecular level (in a DNA molecule), loss of one to several base pairs from a DNA molecule.

demographics A population's vital statistics.

denaturation (deh-nay-chur-AY-shun) Of a protein, the loss of three-dimensional shape following disruption of hydrogen bonds and other weak bonds.

dendrite (DEN-drite) [Gk. *dendron*, tree] A short, slender extension from the cell body of a neuron.

dendritic cell A type of white blood cell that alerts the adaptive immune system when an antigen is present in tissue fluid of the skin or body linings.

denitrification (dee-nite-rih-fih-KAY-shun) The conversion of nitrate or nitrite by certain bacteria to gaseous nitrogen (N_2) and a small amount of nitrous oxide (N_2O).

dense connective tissue A type of fibrous connective tissue with more collagen fibers than loose connective tissue; it is strong but not very flexible.

density-dependent controls Factors, such as predation, parasitism, disease, and competition for resources, that limit population growth by reducing the birth rate, increasing the rates of death and dispersal, or all of these.

density-independent controls Factors such as storms or floods that increase a population's death rate more or less independently of its density.

dentition The type, size, and number of an animal's teeth.

dermis The layer of skin underlying the epidermis, consisting mostly of dense connective tissue.

desertification (dez-urt-ih-fih-KAY-shun) The conversion of grasslands, rain-fed cropland, or irrigated cropland to desertlike conditions, with a drop in agricultural productivity of 10 percent or more.

development The genetically guided emergence of specialized, morphologically distinct body parts.

diabetes insipidus Excessive urination which can lead to dehydration; sometimes a result of damage to the posterior lobe of the pituitary.

diabetes mellitus Lack of insulin causes glucose to build up in the blood, leading to excess water loss and disruption of the body's water–solute balance.

diaphragm (DIE-uh-fram) [Gk. *diaphragma*, to partition] Muscular partition between the thoracic and abdominal cavities, the contraction and relaxation of which contributes to breathing. Also, a contraceptive device used temporarily to prevent sperm from entering the uterus during sexual intercourse.

diastole Relaxation phase of the cardiac cycle.

diffusion Net movement of like molecules (or ions) down their concentration gradient. In the absence of other forces, molecular motion and random collisions cause their net outward movement from one region into a neighboring region where they are less concentrated (because collisions are more frequent where the molecules are most crowded together).

digestion The breakdown of food particles into nutrient molecules small enough to be absorbed.

digestive system An internal tube from which ingested food is absorbed into the internal environment; divided into regions specialized for food transport, processing, and storage.

dihybrid cross In genetics, an experimental cross in which offspring inherit two gene pairs, each consisting of two nonidentical alleles.

diploid (DIP-loyd) **number** The chromosome number of somatic cells and of germ cells prior to meiosis. Such cells have two chromosomes of each type (that is, pairs of homologous chromosomes). Compare haploid number.

disaccharide (die-SAK-uh-ride) [Gk. *di*, two, and *sakcharon*, sugar] A type of simple carbohydrate, of the class called oligosaccharides; two monosaccharides covalently bonded.

disease Condition that develops when the body's defenses cannot prevent a pathogen's activities from interfering with normal body functions.

disease vector Something (such as an insect) that carries a pathogen from an infected person or contaminated material to new hosts.

disjunction The separation of each homologue from its partner during anaphase I of meiosis.

distal tubule The tubular section of a nephron most distant from the glomerulus; a major site of water and sodium reabsorption.

divergence Accumulation of differences in allele frequencies between populations that have become reproductively isolated from one another.

DNA Deoxyribonucleic acid (dee-OX-ee-rye-bow-new-CLAY-ik). For all cells (and many viruses), the molecule of inheritance. A category of nucleic acids, each usually consisting of two nucleotide strands twisted together helically and held together by hydrogen bonds. The nucleotide sequence encodes the instructions for assembling proteins, and, ultimately, a new individual.

DNA chip A microarray of thousands of DNA sequences that are stamped onto a glass plate; can help identify mutations and diagnose diseases by pinpointing which genes are silent and which are being expressed in a body tissue.

DNA clone An identical copy of foreign DNA that was inserted into plasmids (typically, bacteria).

DNA fingerprint Of each individual, a unique array of RFLPs, resulting from the DNA sequences inherited (in a Mendelian pattern) from each parent.

DNA polymerase (poe-LIM-uh-rase) Enzyme that assembles a new strand on a parent DNA strand during replication; also takes part in DNA repair.

DNA repair Following an alteration in the base sequence of a DNA strand, a process that restores the original sequence, as carried out by DNA polymerases, DNA ligases, and other enzymes.

DNA replication Of cells, the process by which the hereditary material is duplicated for distribution to daughter nuclei. An example is the duplication of eukaryotic chromosomes during interphase, prior to mitosis.

dominant allele In a diploid cell, an allele that masks the expression of its partner on the homologous chromosome.

Down syndrome Genetic disorder in which affected people are mentally retarded and have limited motor function.

drug addiction Chemical dependence on a drug, which assumes an "essential" biochemical role in the body following habituation and tolerance.

dry acid deposition The falling to Earth of sulfur and nitrogen oxides; called *wet acid deposition,* or acid rain, when it occurs in rain or snow.

Duchenne muscular dystrophy Genetic disorder linked to the X chromosome that leads to wasting of muscle tissue.

duplication A change in a chromosome's structure resulting in the repeated appearance of the same gene sequence.

dysplasia An abnormal change in the sizes, shapes, and organization of cells in a tissue.

ecological pyramid A way to represent the energy relationships of an ecosystem.

ecology [Gk. *oikos,* home, and *logos,* reason] Study of the interactions of organisms with one another and with their physical and chemical environment.

ecosystem [Gk. *oikos,* home] An array of organisms and their physical environment, all of which interact through a flow of energy and a cycling of materials.

ectoderm [Gk. *ecto,* outside, and *derma,* skin] The outermost primary tissue layer (germ layer) of an embryo, which gives rise to the outer layer of the integument and to tissues of the nervous system.

effector A muscle (or gland) that responds to signals from an integrator (such as the brain) by producing movement (or chemical change) that helps adjust the body to changing conditions.

effector cell Of the differentiated subpopulations of lymphocytes that form during an immune response, the type of cell that engages and destroys the antigen-bearing agent that triggered the response.

efferent arteriole In the urinary system, the arteriole that carries filtered blood from the nephron.

egg A mature female gamete; also called an ovum.

elastic connective tissue A form of dense connective tissue found in organs that must stretch.

El Niño Massive eastward movement of warm surface waters of the western equatorial Pacific that displaces cooler water off the coast of South America. A recurring event that disrupts the climate and ecosystems throughout the world.

electrocardiogram (ECG) A recording of the electrical activity of the heart's cardiac cycle.

electrolyte Any chemical substance, such as a salt, that ionizes and dissociates in water and is capable of conducting an electrical current.

electron Negatively charged unit of matter, with both particulate and wavelike properties, that occupies one of the orbitals around the atomic nucleus. Atoms can gain, lose, or share electrons with other atoms.

electron transfer A molecule donates one or more electrons to another molecule.

electron transport system An organized array of enzymes and cofactors, bound in a cell membrane, that accept and donate electrons in sequence. When such systems operate, hydrogen ions (H^+) flow across the membrane, and the flow drives ATP formation and other reactions.

element Any substance that cannot be decomposed into substances with different properties.

elimination The excretion of undigested and unabsorbed residues from the end of the gastrointestinal tract.

embryo (EM-bree-oh) [Gk. *en*, in, and probably *bryein*, to swell] Of animals generally, the stage formed by way of cleavage, gastrulation, and other early developmental events.

embryonic disk In early development, the oval, flattened cell mass that gives rise to the embryo shortly after implantation.

emerging disease Disease caused by a newly mutated strain of an existing pathogen or one that is now exploiting an increased availability of human hosts.

emphysema Lung disorder in which the lungs become so distended and inelastic that gas exchange is inefficient.

emulsification In digestion, the breaking of large fat globules, a suspension of fat droplets coated with bile salts.

encapsulated receptor Receptor surrounded by a capsule of epithelial or connective tissue; common near the body surface.

encephalitis Inflammation of the brain, usually caused by a virus.

end product Substance present at the end of a metabolic pathway.

endangered species Endemic (native) species highly vulnerable to extinction.

endemic disease A disease that occurs more or less continuously in a region.

endergonic reaction Chemical reaction resulting in a net gain in energy.

endocrine gland A ductless gland that secretes hormones, which usually enter interstitial fluid and then the bloodstream.

endocrine system System of cells, tissues, and organs that is functionally linked to the nervous system and that exerts control by way of hormones and other chemical secretions.

endocytosis (en-doe-sigh-TOE-sis) Movement of a substance into cells; the substance becomes enclosed by a patch of plasma membrane that sinks into the cytoplasm, then forms a vesicle around it. Phagocytic cells also engulf pathogens in this manner.

endoderm [Gk. *endon*, within, and *derma*, skin] The inner primary tissue layer, or germ layer, of an embryo, which gives rise to the inner lining of the gut and organs derived from it.

endomembrane system System in cells that includes the endoplasmic reticulum, Golgi bodies, and various kinds of vesicles, and in which new proteins are modified into final form and lipids are assembled.

endometrium (en-doh-MEET-ree-um) [Gk. *metrios*, of the womb] Inner lining of the uterus consisting of connective tissues, glands, and blood vessels.

endoplasmic reticulum (ER) (en-doe-PLAZ-mik reh-TIK-yoo-lum) An organelle that begins at the nucleus and curves through the cytoplasm. In rough ER (which has many ribosomes on its cytoplasmic side), new polypeptide chains acquire specialized side chains. In many cells, smooth ER (with no attached ribosomes) is the main site of lipid synthesis.

endotherm Organism such as a human that maintains body temperature from within, generally by metabolic activity and controls over heat conservation and dissipation.

energy The capacity to do work.

energy carrier A molecule that delivers energy from one metabolic reaction site to another. ATP is the premier energy carrier; it readily donates energy to nearly all metabolic reactions.

energy pyramid A pyramid-shaped representation of an ecosystem's trophic structure (feeding levels), illustrating the energy losses at each transfer to a different feeding level.

enzyme (EN-zime) One of a class of proteins that greatly speed up (catalyze) reactions between specific substances. The substances that each type of enzyme acts upon are called its substrates.

eosinophil Fast-acting, phagocytic white blood cell that targets worms, fungi, and other large pathogens.

epidemic A disease rate in an area or population that increases above predicted or normal levels.

epidermis The outermost tissue layer of skin.

epiglottis A flaplike structure at the start of the larynx, the position of which directs the movement of air into the trachea or food into the esophagus.

epinephrine (ep-ih-NEF-rin) Adrenal hormone that raises blood levels of glucose and fatty acids; also increases the heart's rate and force of contraction.

epiphyseal plate Region of cartilage that covers either end of a growing long bone, permitting the bone to lengthen. The epiphyseal plate is replaced by bone when growth stops in late adolescence.

epithelium (ep-ih-THEE-lee-um) A tissue consisting of one or more layers of adhering cells that covers the body's external surfaces and lines its internal cavities and tubes. Epithelium has one free surface; the opposite surface rests on a basement membrane between it and an underlying connective tissue. Epidermis is an example.

erythrocyte (eh-RITH-row-site) [Gk. *erythros*, red, and *kytos*, vessel] Red blood cell.

esophagus (ee-SOF-uh-gus) Tubular portion of the digestive system that receives swallowed food and leads to the stomach.

essential amino acid Any of eight amino acids that the body cannot synthesize and must be obtained from food.

essential fatty acid Any of the fatty acids that the body cannot synthesize and must be obtained from food.

estrogen (ESS-tro-jen) A sex hormone that helps oocytes mature, induces changes in the uterine lining during the menstrual cycle and pregnancy, and maintains secondary sexual traits; also influences body growth and development.

eukaryotic cell (yoo-carry-AH-tic) [Gk. *eu*, good, and *karyon*, kernel] A cell that has a "true nucleus" and other membrane-bound organelles. Compare prokaryotic cell.

evolution, biological [L. *evolutio*, act of unrolling] Change within a line of descent over time. A population is evolving when some forms of a trait are becoming more or less common relative to the other kinds of traits. The shifts are evidence of changes in the relative abundances of alleles for that trait, as brought about by mutation, natural selection, genetic drift, and gene flow.

excitatory postsynaptic potential or **EPSP** One of two competing signals at an input zone of a neuron; a graded potential that brings the neuron's plasma membrane closer to threshold.

excretion Any of several processes by which excess water, excess or harmful solutes, or waste materials leave the body by way of the urinary system.

exercise Activity that increases the level of contractile activity in muscles.

exergonic reaction Chemical reaction that shows a net loss in energy.

exocrine gland (EK-suh-krin) [Gk. *es*, out of, and *krinein*, to separate] Glandular structure that secretes products, usually through ducts or tubes, to a free epithelial surface.

exocytosis (ek-so-sigh-TOE-sis) Movement of a substance out of a cell by means of a transport vesicle, the membrane of which fuses with the plasma membrane, so that the vesicle's contents are released outside.

exon Of eukaryotic cells, any of the nucleotide sequences of a pre-mRNA molecule that are spliced together to form the mature mRNA transcript and are ultimately translated into protein.

expansion mutation A gene mutation in which a nucleotide sequence is repeated over and over.

experiment A test in which some phenomenon in the natural world is manipulated in controlled ways to gain insight into its function, structure, operation, or behavior.

expiration Expelling air from the lungs; exhaling.

exponential growth (ex-po-NEN-shul) Pattern of population growth in which greater and greater numbers of individuals are produced during the successive doubling times; the pattern that emerges when the per capita birth rate remains even slightly above the per capita death rate, putting aside the effects of immigration and emigration.

external respiration Movement of oxygen from alveoli into the blood, and of carbon dioxide from the blood into alveoli.

extinction Irrevocable loss of a species.

extracellular fluid All the fluid not inside cells; includes plasma (the liquid portion of blood) and tissue fluid (which occupies the spaces between cells and tissues).

extracellular matrix A material, largely secreted, that helps hold many animal tissues together in certain shapes; it consists of fibrous proteins and other components in a ground substance.

extraembryonic membranes Membranes that form along with a developing embryo, including the yolk sac, amnion, allantois, and chorion.

eyes Sensory organs that allow vision; they contain tissue with a dense array of photoreceptors.

F_1 (first filial generation) The offspring of an initial genetic cross.

F_2 (second filial generation) The offspring of parents who are the first filial generation from a genetic cross.

facilitated diffusion A form of passive transport where transport proteins provide a channel through which solutes cross a cell membrane.

FAD Flavin adenine dinucleotide, a nucleotide coenzyme. When delivering electrons and unbound protons (H^+) from one reaction to another, it is abbreviated $FADH_2$.

familial hypercholesterolemia Genetic disorder that leads to dangerously high blood cholesterol.

fat A lipid with a glycerol head and one, two, or three fatty acid tails. The tails of saturated fats have only single bonds between carbon atoms and hydrogen atoms attached to all other bonding sites. Tails of unsaturated fats additionally have one or more double bonds between certain carbon atoms.

fatty acid A long, flexible hydrocarbon chain with a COOH group at one end.

faulty enamel trait Genetic disorder linked to the X chromosome in which tooth enamel fails to develop properly.

femur Thighbone; longest and strongest bone of the body.

fermentation [L. *fermentum*, yeast] A type of anaerobic pathway of ATP formation; it starts with glycolysis, ends when electrons are transferred back to one of the breakdown products or intermediates, and regenerates the NAD^+ required for the reaction. Its net yield is two ATP per glucose molecule broken down.

fertilization [L. *fertilis*, to carry, to bear] Fusion of a sperm nucleus with the nucleus of an egg, which thereupon becomes a zygote.

fetus Term applied to an embryo after it reaches the age of eight weeks.

fever Body temperature that has climbed above the normal set point, usually in response to infection. Mild fever promotes an increase in body defense activities.

fibrous connective tissue A specialized form of connective tissue that is strong and stretchy; the three types are loose, dense, and elastic.

fibrous joint Type of joint in which fibrous connective tissue unites the adjoining bones and no cavity is present.

fight–flight response The combination of sympathetic and parasympathetic nerve responses that prompt the body to react quickly to intense arousal.

filtration In urine formation, the process by which blood pressure forces water and solutes out of glomerular capillaries and into the cupped portion of a nephron wall (glomerular capsule).

flagellum (fluh-JELL-um), plural flagella [L., *whip*] Tail-like motile structure of many free-living eukaryotic cells; it has a distinctive 9 + 2 array of microtubules.

fluid mosaic model Model of membrane structure in which proteins are embedded in a lipid bilayer or attached to one of its surfaces. The lipid molecules give the membrane its basic structure, impermeability to water-soluble molecules, and (through packing variations and movements) fluidity. Proteins carry out most membrane functions, such as transport, enzyme action, and reception of signals or substances.

follicle (FOLL-ih-kul) In an ovary, a primary oocyte (immature egg) together with the surrounding layer of cells.

food chain A straight-line sequence of who eats whom in an ecosystem.

food pyramid Chart of a purportedly well-balanced diet; continually being refined.

food web A network of cross-connecting, interlinked food chains encompassing primary producers and an array of consumers, detritivores, and decomposers.

forebrain Brain region that includes the cerebrum and cerebral cortex, the olfactory lobes, and the hypothalamus.

fossil Physical remains or other evidence of an organism that lived in the distant past. Most fossils are skeletons, shells, leaves, seeds, and tracks that were buried in rock layers before they decomposed.

fossil fuels The fossilized remains of ancient forests. Examples include oil, coal, and natural gas. Fossil fuels are nonrenewable resources.

fossilization How fossils form. An organism or traces of it become buried in sediments or volcanic ash. Water and dissolved inorganic compounds infiltrate the remains. Accumulating sediments exert pressure above the burial site. Over time, the pressure and chemical changes transform the remains to stony hardness.

fovea Funnel-shaped depression in the center of the retina where photoreceptors are densely arrayed and visual acuity is the greatest.

free nerve endings Thinly myelinated or unmyelinated branched endings of sensory neurons in skin and internal tissues. They serve as mechanoreceptors, thermoreceptors, or pain receptors.

free radical Any highly reactive molecule or molecule fragment having an unpaired electron.

FSH Follicle-stimulating hormone. The name comes from its function in females, in whom FSH helps stimulate follicle development in ovaries. In males, it acts in the testes as part of a sequence of events that trigger sperm production.

functional group An atom or group of atoms that is covalently bonded to the carbon backbone of an organic compound and that influences its behavior.

gallbladder Organ of the digestive system that stores bile secreted from the liver.

gamete (GAM-eet) A haploid cell that functions in sexual reproduction. Sperm and eggs are examples.

ganglion (GANG-lee-un), plural ganglia [Gk. *ganglion*, a swelling] A distinct clustering of cell bodies of neurons in regions other than the brain or spinal cord.

gap junctions Channels that connect the cytoplasm of adjacent cells and help cells communicate by promoting the rapid transfer of ions and small molecules between them.

gastric juice Highly acidic mix of water and secretions from the stomach's glandular epithelium (HCl, mucus, pepsinogens, etc.) that kills ingested microbes and begins food breakdown.

gastrointestinal (GI) tract The digestive tube, extending from the mouth to the anus and including the stomach, small and large intestines, and other specialized regions with roles in food transport and digestion.

gastrulation (gas-tru-LAY-shun) The stage of embryonic development in which cells become arranged into two or three primary tissue layers (germ layers); in humans, the layers are an inner endoderm, an intermediate mesoderm, and a surface ectoderm.

gene A unit of information about a heritable trait that is passed on from parents to offspring. Each gene has a specific location on a chromosome.

gene flow A microevolutionary process; a physical movement of alleles out of a population as individuals leave (emigrate) or enter (immigrate); allele frequencies change as a result.

gene frequency More precisely, allele frequency: the relative abundances of all the different alleles for a trait that are carried by the individual members of a population.

gene library A mixed collection of bacteria that contain many different cloned DNA fragments.

gene mutation Small-scale change in the nucleotide sequence of a gene.

gene pair In diploid cells, the two alleles at a given locus on a pair of homologous chromosomes.

gene pool Sum total of all genotypes in a population. More accurately, allele pool.

gene therapy Generally, the transfer of one or more normal genes into body cells in order to correct a genetic defect.

genetic abnormality An uncommon version of an inherited trait.

genetic code [After L. *genesis*, to be born] The correspondence between nucleotide triplets in DNA (then in mRNA) and specific sequences of amino acids in the resulting polypeptide chains; the basic language of protein synthesis.

genetic disorder An inherited condition that results in mild to severe medical problems.

genetic drift A microevolutionary process; a change in allele frequencies over the generations due to chance events alone.

genetic engineering Altering the information content of DNA through use of recombinant DNA technology.

genetic recombination Presence of a new combination of alleles in a DNA molecule compared to the parental genotype; the result of processes such as crossing over at meiosis, chromosome rearrangements, gene mutation, and recombinant DNA technology.

genital herpes Infection of tissues in the genital area by a herpes simplex virus; an extremely contagious sexually transmitted disease.

genital warts Painless growths caused by infection of epithelium by the human papillomavirus.

genome All the DNA in a haploid number of chromosomes of a species.

genotype (JEEN-oh-type) Genetic constitution of an individual. Can mean a single gene pair or the sum total of the individual's genes. Compare phenotype.

genus, plural genera (JEEN-us, JEN-er-ah) [L. *genus*, race, origin] A taxonomic grouping of species exhibiting certain phenotypic similarities and evolutionary relationships.

germ cell The cell of sexual reproduction; germ cells give rise to gametes. Compare somatic cell.

germ layer One of three primary tissue layers that forms during gastrulation and that gives rise to certain tissues of the adult body. Compare ectoderm; endoderm; mesoderm.

gigantism Excessive growth caused by the overproduction of growth hormone during childhood.

gland A secretory cell or multicellular structure derived from epithelium and often connected to it.

glaucoma Disorder in which too much aqueous humor builds up inside the eyeball and collapses the blood vessels serving the retina.

glial cells *See* neuroglia.

global warming A long-term increase in the temperature of the Earth's lower atmosphere.

glomerular capillaries The set of blood capillaries inside the Bowman's capsule of a nephron.

glomerulonephritis A general term for a large number of kidney disorders.

glomerulus (glow-MARE-you-luss), plural glomeruli [L. *glomus*, ball] The first portion of the nephron, where water and solutes are filtered from blood.

glucagon (GLUE-kuh-gone) Hormone that stimulates conversion of glycogen and amino acids to glucose; secreted by alpha cells of the pancreas when the flow of glucose decreases.

glucocorticoid Hormone secreted by the adrenal cortex that influences metabolic reactions that help maintain the blood glucose level.

gluconeogenesis The process by which liver cells synthesize glucose.

glyceride (GLISS-er-eyed) One of the molecules, commonly called fats and oils, that has one, two, or three fatty acid tails attached to a glycerol backbone. They are the body's most abundant lipids and its richest source of energy.

glycerol (GLISS-er-all) [Gk. *glykys*, sweet, and L. *oleum*, oil] A three-carbon molecule with three hydroxyl groups attached; together with fatty acids, a component of fats and oils.

glycogen (GLY-kuh-jen) A storage polysaccharide that can be readily broken down into glucose subunits.

glycolysis (gly-CALL-ih-sis) [Gk. *glykys*, sweet, and *lysis*, loosening or breaking apart] Initial reactions of both aerobic and anaerobic pathways by which glucose (or some other organic compound) is partially broken down to pyruvate with a net yield of two ATP. Glycolysis proceeds in the cytoplasm of all cells, and oxygen has no role in it.

glycoprotein A protein having linear or branched oligosaccharides covalently bonded to it. Most human cell surface proteins and many proteins circulating in blood are glycoproteins.

Golgi body (GOHL-gee) Organelle in which newly synthesized polypeptide chains as well as lipids are modified and packaged in vesicles for export or for transport to specific locations within the cytoplasm.

gonad (GO-nad) Primary reproductive organ in which gametes are produced. Ovaries and testes are gonads.

gonorrhea The sexually transmitted disease caused by the bacterium *Neisseria gonorrhoeae*. This bacterium can infect epithelial cells of the genital tract, rectum, eye membranes, and the throat.

graded potential Of neurons, a local signal that slightly changes the voltage difference across a small patch of the plasma membrane. Such signals vary in magnitude, depending on the stimulus. With prolonged or intense stimulation, they may spread to a trigger zone of the membrane and initiate an action potential.

granulocyte Class of white blood cells that have a lobed nucleus and various types of granules in the cytoplasm; includes neutrophils, eosinophils, and basophils.

granulosa cell An estrogen-secreting cell of the epithelial lining of a follicle.

Graves disease A toxic goiter caused by an excess of thyroid hormones in the blood.

gray matter The dendrites, neuron cell bodies, and neuroglial cells of the spinal cord and cerebral cortex.

Green Revolution In developing countries, the use of improved (sometimes genetically modified) crop varieties, modern agricultural practices (including massive inputs of fertilizers and pesticides), and equipment to increase crop yields.

greenhouse effect Warming of the lower atmosphere due to the presence of greenhouse gases: carbon dioxide, methane, nitrous oxide, ozone, water vapor, and chlorofluorocarbons.

ground substance The intercellular material made up of cell secretions and other noncellular components.

growth factor A type of signaling molecule that can influence growth by regulating the rate at which target cells divide.

guanine A nitrogen-containing base; one of those present in nucleotide building blocks of DNA and RNA.

habitat [L. *habitare,* to live in] The type of place where an organism normally lives, characterized by physical features, chemical features, and the presence of certain other species.

hair A flexible structure of mostly keratinized cells, rooted in skin with a shaft above its surface.

hair cell Type of mechanoreceptor that may give rise to action potentials when bent or tilted.

half-life The time it takes for half of a quantity of radioisotope to decay into a different, more stable isotope.

haploid (HAP-loyd) **number** The chromosome number of a gamete that, as an outcome of meiosis, is only half that of the parent germ cell (it has only one of each pair of homologous chromosomes). Compare diploid number.

HCG Human chorionic gonadotropin. A hormone that helps maintain the lining of the uterus during the menstrual cycle and during the first trimester of pregnancy.

HDL A high-density lipoprotein in blood; it transports cholesterol to the liver for further processing.

headache Pain in the head caused by tension in muscles or blood vessels in the face, neck, and scalp.

heart Muscular pump that keeps blood circulating through the animal body.

heart attack Damage to or death of heart muscle.

heart failure Disorder in which the heart is weakened and does not pump blood with normal efficiency.

helper T cell Type of T lymphocyte that produces and secretes chemicals that promote formation of large effector and memory cell populations.

hemoglobin (HEEM-oh-glow-bin) [Gk. *haima,* blood, and L. *globus,* ball] Iron-containing, oxygen-transporting protein that gives red blood cells their color.

hemophilia Genetic disorder in which blood does not clot normally; the affected person risks dying from any injury that causes bleeding.

hemostasis (hee-mow-STAY-sis) [Gk. *haima,* blood, and *stasis,* standing] Stopping of blood loss from a damaged blood vessel through coagulation, blood vessel spasm, platelet plug formation, and other mechanisms.

hepatic portal vein Vessel that receives nutrient-laden blood from villi of the small intestine and transports it to the liver, where excess glucose is removed. The blood then returns to the general circulation via a hepatic vein.

hepatitis B An extremely contagious, sexually transmitted disease caused by infection by the hepatitis B virus. Chronic hepatitis can lead to liver cirrhosis or cancer.

hepatitis C Bloodborne virus that causes severe liver cirrhosis and sometimes cancer.

herbivore [L. *herba,* grass, and *vovare,* to devour] Plant-eating animal.

herpes infection Caused by various forms of herpes simplex; can result in blindness.

heterotroph (HET-er-oh-trofe) [Gk. *heteros,* other, and *trophos,* feeder] Organism that cannot synthesize its own organic compounds and must obtain nourishment by feeding on autotrophs, each other, or organic wastes. Animals, fungi, many protists, and most bacteria are heterotrophs. Compare autotroph.

heterozygous condition (het-er-oh-ZYE-guss) [Gk. *zygoun,* join together] For a given trait, having nonidentical alleles at a particular locus on a pair of homologous chromosomes.

hindbrain One of the three divisions of the brain; the medulla oblongata, cerebellum, and pons; includes reflex centers for respiration, blood circulation, and other basic functions; also coordinates motor responses and many complex reflexes.

histamine Local signaling molecule that fans inflammation; makes arterioles dilate and capillaries more permeable (leaky).

histone Any of a class of proteins that are intimately associated with DNA and that are largely responsible for its structural (and possibly functional) organization in eukaryotic chromosomes.

histoplasmosis Lung disease caused by a fungus.

Hodgkin's disease Malignant lymphoma whose symptoms include intense itching and night sweats.

homeostasis (hoe-me-oh-STAY-sis) [Gk. *homo,* same, and *stasis,* standing] A physiological state in which the physical and chemical conditions of the internal environment are being maintained within tolerable ranges.

homeostatic feedback loop An interaction in which an organ (or structure) stimulates or inhibits the output of another organ, then shuts down or increases this activity when it detects that the output has exceeded or fallen below a set point.

hominid [L. *homo*, man] All species on the evolutionary branch leading to modern humans. *Homo sapiens* is the only living representative.

hominoids Apes, humans, and their recent ancestors.

Homo erectus A hominid lineage that emerged between 1.5 million and 300,000 years ago and that may include the direct ancestors of modern humans.

Homo habilis A type of early hominid that may have been the maker of stone tools that date from about 2.5 million years ago.

Homo sapiens The species of modern humans that emerged between 300,000 and 200,000 years ago.

homologous chromosome (huh-MOLL-uh-gus) [Gk. *homologia*, correspondence] (also called *homologue*) One of a pair of chromosomes that resemble each other in size, shape, and the genes they carry, and that line up with each other at meiosis I. The X and Y chromosomes differ in these respects but still function as homologues.

homologous structure The same body part, modified in different ways, in different lines of descent from a common ancestor.

homozygous condition (hoe-moe-ZYE-guss) Having two identical alleles at a given locus (on a pair of homologous chromosomes).

homozygous dominant condition Having two dominant alleles at a given locus (on a pair of homologous chromosomes).

homozygous recessive condition Having two recessive alleles at a given gene locus (on a pair of homologous chromosomes).

hormone [Gk. *hormon*, to stir up, set in motion] Any of the signaling molecules secreted from endocrine glands, endocrine cells, and some neurons that the bloodstream distributes to nonadjacent target cells (any cell having receptors for that hormone).

host An organism that can be infected by a pathogen.

Human Genome Project A research project in which the estimated 3 billion nucleotides present in the DNA of human chromosomes were sequenced.

human immunodeficiency virus (HIV) A retrovirus; the pathogen that causes AIDS (acquired immune deficiency syndrome).

human papillomavirus (HPV) Virus that causes genital warts; HPV infection is suspected of having a role in the development of some cases of cervical cancer.

humans Members of the genus *Homo*, evolved from hominids.

humerus The long bone of the upper arm.

Huntington disease Genetic disorder in which the basal nuclei of the brain degenerate.

hybrid offspring Of a genetic cross, offspring with a pair of nonidentical alleles for a trait.

hydrocarbon A molecule having only hydrogen atoms attached to a carbon backbone.

hydrogen bond A weak attraction between an electronegative atom and a hydrogen atom that is already taking part in a polar covalent bond.

hydrogen ion A free (unbound) proton; a hydrogen atom that has lost its electron and so bears a positive charge (H^+).

hydrologic cycle A biogeochemical cycle, driven by solar energy, in which water moves slowly through the atmosphere, on or through surface layers of land masses, to the ocean and back again.

hydrolysis (high-DRAWL-ih-sis) [L. *hydro*, water, and Gk. *lysis*, loosening or breaking apart] Enzyme-mediated reaction in which covalent bonds break, splitting a molecule into two or more parts, and H^+ and OH^- (derived from a water molecule) become attached to the exposed bonding sites.

hydrophilic substance [Gk. *philos*, loving] A polar substance that is attracted to the polar water molecule and so dissolves easily in water. Sugars are examples.

hydrophobic substance [Gk. *phobos*, dreading] A nonpolar substance that is repelled by the polar water molecule and so does not readily dissolve in water. Oil is an example.

hydroxide ion Ionized compound of one oxygen and one hydrogen atom (OH^-).

hyperopia Also known as farsightedness; nearby objects appear blurry because their images are focused behind the retina, not on it.

hyperplasia An abnormal enlargement of tissue that leads to a tumor.

hypertonic solution A fluid having a greater concentration of solutes relative to another fluid.

hypodermis A subcutaneous layer having stored fat that helps insulate the body; although not part of skin, it anchors skin while allowing it some freedom of movement.

hypothalamus [Gk. *hypo*, under, and *thalamos*, inner chamber] A brain center that monitors visceral activities (such as salt–water balance and temperature control) and that influences related forms of behavior (as in hunger, thirst, and sex).

hypothesis A possible explanation of a specific phenomenon.

hypotonic solution A fluid that has a lower concentration of solutes relative to another fluid.

immune response A series of events by which B and T lymphocytes recognize a specific antigen, undergo repeated cell divisions that form huge lymphocyte populations, and differentiate into subpopulations of effector and memory cells. Effector cells engage and destroy antigen-bearing agents. Memory cells enter a resting phase and are activated during subsequent encounters with the same antigen.

immune system Interacting white blood cells that defend the body through self/nonself recognition, specificity, and memory. T and B cell antigen receptors ignore the body's own cells yet collectively recognize a billion specific threats. Some B and T cells formed in a primary response are set aside as memory cells for future battles with the same antigen.

immunity The body's overall ability to resist and combat any substance foreign to itself.

immunization Various processes, including vaccination, that promote increased immunity against specific diseases.

immunodeficiency Disorder in which a person's immune system is weakened or absent.

immunoglobulin Any of the five classes of antibodies that participate in specific ways in defense and immune responses. Examples are IgM antibodies (first to be secreted during immune responses) and IgG antibodies (which activate complement proteins and neutralize many toxins).

immunotherapy Procedures that enhance a person's immunological defenses against tumors or certain pathogens.

implantation Series of events in which a blastocyst (pre-embryo) invades the endometrium (lining of the uterus) and becomes embedded there.

in vitro fertilization Conception outside the body ("in glass" petri dishes or test tubes).

independent assortment Mendelian principle that each gene pair tends to assort into gametes independently of other gene pairs located on nonhomologous chromosomes.

induced-fit model Model of enzyme action whereby a bound substrate induces changes in the shape of the enzyme's active site, resulting in a more precise molecular fit between the enzyme and its substrate.

inductive logic Pattern of thinking by which a person derives a general statement from specific observations.

industrial smog A type of gray-air smog that develops in industrialized regions when winters are cold and wet.

infection Invasion and multiplication of a pathogen in a host. *Disease* follows if defenses are not mobilized fast enough; the pathogen's activities interfere with normal body functions.

infectious mononucleosis A disease caused by the Epstein-Barr virus, which affects white blood cells, causing them to overproduce agranulocytes.

inflammation Process in which, in response to tissue damage or irritation, phagocytes and plasma proteins, including complement proteins, leave the bloodstream, then defend and help repair the tissue. Proceeds during both nonspecific and specific (immune) defense responses.

influenza Disease in which an infection begins in the nose or throat and spreads to the lungs.

inheritance The transmission, from parents to offspring, of structural and functional patterns that have a genetic basis.

inhibiting hormone A signaling molecule produced and secreted by the hypothalamus that controls secretions by the anterior lobe of the pituitary gland.

inhibitory postsynaptic potential, or **IPSP** Of neurons, one of two competing types of graded potentials at an input zone; tends to drive the resting membrane potential away from threshold.

innate immunity The body's inborn, preset immune responses, which act quickly when tissue is damaged or microbes have invaded.

inner cell mass In early development, a clump of cells in the blastocyst that will give rise to the embryonic disk.

insertion The end of a muscle that is attached to the bone that moves most when the muscle contracts.

inspiration The drawing of air into the lungs; inhaling.

insulin Pancreatic hormone that lowers the level of glucose in blood by causing cells to take up glucose; also promotes the synthesis of fat and protein and inhibits the conversion of protein to glucose.

integration, neural [L. *integrare*, to coordinate] Moment-by-moment summation of all excitatory and inhibitory synapses acting on a neuron; occurs at each level of synapsing in a nervous system.

integrator Of homeostatic systems, a control point where different bits of information are pulled together in the selection of a response. The brain is an example.

integument An organ system that provides a protective body covering; in humans, the skin, oil and sweat glands, hair, and nails.

interferon Protein produced by T cells that interferes with viral replication. Some interferons also stimulate the tumor-killing activity of macrophages.

interleukin One of a variety of communication signals, secreted by macrophages and by helper T cells, that drive immune responses.

intermediate Substance that forms between the start and end of a metabolic pathway.

intermediate filament A ropelike cytoskeletal element that mechanically strengthens cells.

internal environment The fluid bathing body cells and tissues; it consists of blood plus interstitial fluid.

internal respiration Movement of oxygen into tissues from the blood, and of carbon dioxide from tissues into the blood.

interneuron Any of the neurons in the brain and spinal cord that integrate information arriving from sensory neurons and that influence other neurons in turn.

interphase Of cell cycles, the time interval between nuclear divisions in which a cell increases its mass, roughly doubles the number of its cytoplasmic components, and finally duplicates its chromosomes (replicates its DNA).

interstitial fluid (in-ter-STISH-ul) [L. *interstitus,* to stand in the middle of something] That portion of the extracellular fluid occupying spaces between cells and tissues.

intervertebral disk One of a number of disk-shaped structures containing cartilage that serve as shock absorbers and flex points between bony segments of the vertebral column.

intracellular fluid The fluid inside cells.

intron A noncoding portion of a newly formed mRNA molecule.

inversion A change in a chromosome's structure after a segment separated from it was then inserted at the same place, but in reverse. The reversal alters the position and order of the chromosome's genes.

ion (EYE-on) An atom or a compound that has gained or lost one or more electrons and hence has acquired an overall negative or positive charge.

ionic bond An association between ions of opposite charge.

iris Of the eye, a circular pigmented region behind the cornea with a "hole" in its center (the pupil) through which incoming light enters.

isotonic solution A fluid having the same solute concentration as a fluid against which it is being compared.

isotope (EYE-so-tope) For a given element, an atom with the same number of protons as the other atoms but with a different number of neutrons.

J-shaped curve A curve, obtained when population size is plotted against time, that is characteristic of unrestricted, exponential growth.

joint An area of contact or near-contact between bones.

juxtaglomerular apparatus In kidney nephrons, a region of contact between the arterioles of the glomerulus and the distal tubule. Cells in this region secrete renin, which triggers hormonal events that stimulate increased reabsorption of sodium.

karyotype (CARRY-oh-type) The number of metaphase chromosomes in somatic cells and their defining characteristics.

keratin A tough, water-insoluble protein manufactured by most epidermal cells.

keratinization (care-at-in-iz-AY-shun) Process by which keratin-producing epidermal cells of skin die and collect at the skin surface as keratinized "bags" that form a barrier against dehydration, bacteria, and many toxic substances.

keratinocytes Cells of the epidermis that make keratin.

kidney One of a pair of organs that filter mineral ions, organic wastes, and other substances from the blood and help regulate the volume and solute concentrations of extracellular fluid.

kidney stones Deposits of uric acid, calcium salts, and other substances that have settled out of urine and collected in the renal pelvis.

kilocalorie 1,000 calories of heat energy, or the amount of energy needed to raise the temperature of 1 kilogram of water by 1°C; the unit of measure for the caloric value of foods.

kinetochore A specialized group of proteins and DNA at the centromere of a chromosome that serves as an attachment point for several spindle microtubules during mitosis or meiosis. Each chromatid of a duplicated chromosome has its own kinetochore.

Klinefelter syndrome Disorder that produces the genotype XXY; affected males have low fertility and usually some mental retardation.

Krebs cycle Together with a few conversion steps that precede it, the stage of aerobic respiration in which pyruvate is completely broken down to carbon dioxide and water. Coenzymes accept the unbound protons (H^+) and electrons stripped from intermediates during the reactions and deliver them to the next stage.

lactate fermentation Anaerobic pathway of ATP formation in which pyruvate from glycolysis is converted to the three-carbon compound lactate, and NAD^+ (a coenzyme used in the reactions) is regenerated. Its net yield is two ATP.

lactation The production of milk by hormone-primed mammary glands.

lacteal Small lymph vessel in villi of the small intestine that receives absorbed triglycerides. Triglycerides move from the lymphatic system to the general circulation.

large-cell carcinoma Type of lung cancer; along with adenocarcinomas, it accounts for 48 percent of all lung cancers.

large intestine The colon; a region of the GI tract that receives unabsorbed food residues from the small intestine and concentrates and stores feces until they are expelled from the body.

larynx (LARE-inks) A tubular airway that leads to the lungs. It contains vocal cords, where sound waves used in speech are produced.

latency Of viruses, a period of time during which viral genes remain inactive inside the host cell.

LDL Low-density lipoprotein that transports cholesterol; cells take it up, but excess amounts contribute to atherosclerosis.

lens Of the eye, a saucer-shaped region behind the iris containing multiple layers of transparent proteins. Ligaments can move the lens, which functions to focus incoming light onto photoreceptors in the retina.

leukemia Cancers of the white blood cells that cause the runaway multiplication of abnormal cells and the destruction of bone marrow.

Leydig cell In testes, cells in connective tissue around the seminiferous tubules that secrete testosterone and other signaling molecules.

LH Luteinizing hormone, secreted by the anterior lobe of the pituitary gland. In males it acts on Leydig cells of the testes and prompts them to secrete testosterone. In females, LH stimulates follicle development in the ovaries.

life cycle Recurring series of genetically programmed events from the time individuals are produced until they themselves reproduce.

ligament A strap of dense, regular connective tissue that connects two bones at a joint.

limbic system Brain regions that, along with the cerebral cortex, collectively govern emotions.

limiting factor Any essential resource that is in short supply and so limits population growth.

lineage (LIN-ee-age) A line of descent.

linkage The tendency of genes located on the same chromosome to end up in the same gamete. For any two of those genes, the probability that crossing over will disrupt the linkage is proportional to the distance separating them.

lipid A greasy or oily compound of mostly carbon and hydrogen that shows little tendency to dissolve in water, but that dissolves in nonpolar solvents (such as ether). Cells use lipids as energy stores and structural materials, especially in membranes.

lipid bilayer The structural basis of cell membranes, consisting of two layers of mostly phospholipid molecules. Hydrophilic heads force all fatty acid tails of the lipids to become sandwiched between the hydrophilic heads.

lipoprotein Molecule that forms when proteins circulating in blood combine with cholesterol, triglycerides, and phospholipids absorbed from the small intestine.

liver Organ with roles in storing and interconverting carbohydrates, lipids, and proteins absorbed from the gut; disposing of nitrogen-containing wastes; and other tasks.

local signaling molecule Cellular secretion that alters chemical conditions in the immediate vicinity where it is secreted, then is swiftly broken down.

locus (LOW-cuss) The location of a particular gene on a chromosome.

logic Thought patterns by which a person draws a conclusion that does not contradict evidence used to support that conclusion.

logistic growth (low-JIS-tik) Pattern of population growth in which a low-density population slowly increases in size, goes through a rapid growth phase, then levels off once the carrying capacity is reached.

loop of Henle The hairpin-shaped, tubular region of a nephron that functions in reabsorption of water and solutes.

loose connective tissue Flexible fibrous connective tissue with few fibers and cells.

lung Saclike organ that serves as an internal respiratory surface.

lung cancer Growth of a tumor in the lungs, most often caused by cigarette smoke.

lymph (limf) [L. *lympha*, water] Tissue fluid that has moved into the vessels of the lymphatic system.

lymph capillary A small-diameter vessel of the lymph vascular system that has no obvious entrance; tissue fluid moves inward by passing between overlapping endothelial cells at the vessel's tip.

lymph node A lymphoid organ that serves as a battleground of the immune system; each lymph node is packed with organized arrays of macrophages and lymphocytes that cleanse lymph of pathogens before it reaches the blood.

lymph vascular system [*vasculum*, a small vessel] The vessels of the lymphatic system, which take up and transport excess tissue fluid and reclaimable solutes as well as fats absorbed from the digestive tract.

lymphatic system An organ system that supplements the circulatory system. Its vessels take up fluid and solutes from interstitial fluid and deliver them to the bloodstream; its lymphoid organs have roles in immunity.

lymphocyte A T cell or B cell.

lymphoid organ The lymph nodes, spleen, thymus, tonsils, adenoids, and other organs with roles in immunity.

lymphoma Cancer of lymphoid tissues in organs such as lymph nodes.

lysis [Gk. *lysis*, a loosening] Gross structural disruption of a plasma membrane that leads to cell death.

lysosome (LYE-so-sohm) In cells, the main organelle of digestion, with enzymes that can break down polysaccharides, proteins, nucleic acids, and some lipids.

lysozyme Present in mucous membranes that line body surfaces, an infection-fighting enzyme that attacks and destroys various types of bacteria by digesting the bacterial cell wall.

macroevolution The large-scale patterns, trends, and rates of change among groups of species.

macrophage One of the phagocytic white blood cells. It engulfs anything detected as foreign. Some also become antigen-presenting cells that serve as the trigger for immune responses by T and B lymphocytes. Compare antigen-presenting cell.

macular degeneration Disorder in which part of the retina breaks down and causes a blind spot.

malabsorption disorder Disease caused by anything that interferes with the uptake of nutrients across the lining of the small intestine.

malaria Disease due to infection by a plasmodium; symptoms include shaking, chills, and a high fever.

malignant melanoma Common eye cancer that typically develops in the middle layer of the eye (the choroid).

malnutrition A state in which body functions or development suffer due to inadequate or unbalanced food intake.

mammal A type of vertebrate; the only animal having offspring that are nourished by milk produced by mammary glands of females.

mandible The lower jaw; the largest single facial bone.

Marfan syndrome Genetic disorder that disrupts the structure and function of smooth muscle cells in the wall of the aorta.

mass extinction An abrupt rise in extinction rates above the background level; a catastrophic, global event in which major taxa are wiped out simultaneously.

mass number The total number of protons and neutrons in an atom's nucleus. The relative masses of atoms are also called atomic weights.

mast cell A type of white blood cell that releases enzymes and histamine during tissue inflammation.

matrix In connective tissue, fiberlike structural proteins together with a "ground substance" of polysaccharides that give each kind of tissue its particular properties.

mechanoreceptor Sensory cell or cell part that detects mechanical energy associated with changes in pressure, position, or acceleration.

medical imaging Any of several diagnostic methods including MRI, X ray, ultrasound, and CT.

medulla oblongata Part of the brain stem with reflex centers for respiration, blood circulation, and other vital functions.

meiosis (my-OH-sis) [Gk. *meioun*, to diminish] Two-stage nuclear division process in which the chromosome number of a germ cell is reduced by half, to the haploid number. (Each daughter nucleus ends up with one of each type of chromosome.) Meiosis is the basis of gamete formation.

melanocytes Cells in the deepest layer of epidermis; melanocytes produce the brown-black pigment melanin found in keratinocytes.

membrane attack complexes Structures that form pores in the plasma membrane of a pathogen, causing lysis (disintegration).

memory The storage and retrieval of information about previous experiences; underlies the capacity for learning.

memory cell Any of the various B or T lymphocytes of the immune system that are formed in response to invasion by a foreign agent and that circulate for some period, available to mount a rapid attack if the same type of invader reappears.

meninges Membranes of connective tissue that are layered between the skull bones and the brain and cover and protect the neurons and blood vessels that service the brain tissue.

meningitis Inflammation of the meninges covering the brain and/or spinal cord.

menopause (MEN-uh-pozz) [L. *mensis*, month, and *pausa*, stop] End of the reproductive portion of a human female's life cycle.

menstrual cycle The cyclic release of oocytes and priming of the endometrium (lining of the uterus) to receive a fertilized egg; the complete cycle averages about 28 days in female humans.

menstruation Periodic sloughing of the blood-enriched lining of the uterus when pregnancy does not occur.

mesoderm (MEH-so-derm) [Gk. *mesos*, middle, and *derm*, skin] In an embryo, a primary tissue layer (germ layer) between ectoderm and endoderm. Gives rise to muscle; organs of circulation, reproduction, and excretion; most of the internal skeleton (when present); and connective tissue layers of the gastrointestinal tract and integument.

messenger RNA (mRNA) A linear sequence of ribonucleotides transcribed from DNA and translated into a polypeptide chain; the only type of RNA that carries protein-building instructions.

metabolic acidosis Lower than optimal blood pH caused by diabetes mellitus.

metabolic pathway An orderly sequence of enzyme-mediated reactions by which cells maintain, increase, or decrease the concentrations of particular substances.

metabolic syndrome Slightly elevated blood sugar that increases the risk of developing type 2 diabetes.

metabolism (meh-TAB-oh-lizm) [Gk. *meta*, change] All controlled, enzyme-mediated chemical reactions by which cells acquire and use energy. Through these reactions, cells synthesize, store, break apart, and eliminate substances in ways that contribute to growth, survival, and reproduction.

metaphase Of mitosis or meiosis II, the stage when each duplicated chromosome has become positioned at the midpoint of the microtubular spindle, with its two sister chromatids attached to microtubules from opposite spindle poles. Of meiosis I, the stage when all pairs of homologous chromosomes are positioned at the spindle's midpoint, with the two members of each pair attached to opposite spindle poles.

metastasis The process in which cancer cells break away from a primary tumor and migrate (via blood or lymphatic tissues) to other locations, where they establish new cancer sites.

MHC marker Any of a variety of proteins that are self markers. Some occur on all body cells of an individual; others occur only on macrophages and lymphocytes.

micelle (my-CELL) Tiny droplet of bile salts, fatty acids, and monoglycerides; plays a role in fat absorption from the small intestine.

microevolution Changes in allele frequencies brought about by mutation, genetic drift, gene flow, and natural selection.

microfilament [Gk. *mikros*, small, and L. *filum*, thread] One of a variety of cytoskeletal components. Actin and myosin filaments are examples.

micrograph Photograph of an image brought into view with the aid of a microscope.

microorganism Organism, usually single-celled, too small to be observed without a microscope.

microtubule Hollow cylinder built mainly of tubulin subunits; a cytoskeletal element with roles in cell shape, motion, and growth and in the structure of cilia and flagella.

microtubule organizing center MTOC; mass of substances in the cell cytoplasm that dictate the orientation and organization of the cell's microtubules.

microvillus (my-crow-VILL-us) [L. *villus*, shaggy hair] A slender, cylindrical extension of the cell surface that functions in absorption or secretion.

midbrain A brain region that evolved as a coordination center for reflex responses to visual and auditory input; together with the pons and medulla oblongata, part of the brain stem, which includes the reticular formation.

mineral An inorganic substance required for the normal functioning of body cells.

mineralocorticoid Hormone secreted by the adrenal cortex that mainly regulates the concentrations of mineral salts in extracellular fluid.

mitochondrion (my-toe-KON-dree-on), plural mitochondria. Organelle that specializes in ATP formation; it is the site of the second and third stages of aerobic respiration.

mitosis (my-TOE-sis) [Gk. *mitos*, thread] Type of nuclear division that maintains the parental chromosome number for daughter cells. It is the basis of bodily growth and, in many eukaryotic species, asexual reproduction.

mixture Atoms of two or more elements intermingled in proportions that can and usually do vary.

molecule A unit of matter in which chemical bonding holds together two or more atoms of the same or different elements.

monoclonal antibody [Gk. *monos*, alone] Antibody produced in the laboratory by a population of genetically identical cells that are clones of a single "parent" antibody-producing cell.

monohybrid cross In genetics, an experimental cross in which offspring inherit a pair of nonidentical alleles for a single trait being studied, so that they are heterozygous.

monomer A small molecule that is commonly a subunit of polymers, such as the sugar monomers of starch.

monosaccharide (mon-oh-SAK-ah-ride) [Gk. *sakharon*, sugar] The simplest carbohydrate, with only one sugar unit. Glucose is an example.

monosomic Abnormal genetic condition in which one chromosome of diploid cells has no homologue.

monosomy Condition in which one of the chromosomes in a gamete has no homologue.

morphogenesis (more-foe-JEN-ih-sis) [Gk. *morphe*, form, and *genesis*, origin] Processes by which differentiated cells in an embryo become organized into tissues and organs, under genetic controls and environmental influences.

morphological convergence Process in which lineages that are only distantly related evolve in response to similar environmental pressures, becoming similar in appearance, functions, or both. Analogous structures are evidence of this evolutionary pattern.

morula A compact ball of sixteen embryonic cells formed after the third round of cleavage.

motility In digestion, the movement of ingested material through the GI tract.

motor neuron A neuron that delivers signals from the brain and spinal cord that can stimulate or inhibit the body's effectors (muscles, glands, or both).

motor protein A type of protein that can move cell parts in a sustained, directional way.

motor unit A motor neuron and the muscle fibers under its control.

multiple allele system A gene that has three or more different molecular forms (alleles).

multiple sclerosis Autoimmune disease that can be triggered by a viral infection and destroys myelin sheaths of neurons in the central nervous system.

muscle fatigue A decline in tension of a muscle that has been kept in a state of tetanic contraction as a result of continuous, high-frequency stimulation.

muscle tension A mechanical force, exerted by a contracting muscle, that resists opposing forces such as gravity and the weight of objects being lifted.

muscle tissue Tissue having cells able to contract in response to stimulation, then passively lengthen and so return to their resting state.

muscle tone In muscles, a steady low-level contracted state that helps stabilize joints and maintain general muscle health.

muscle twitch Muscle response in which the muscle contracts briefly, then relaxes, when a brief stimulus activates a motor unit.

muscular system Skeletal muscles, which attach to bones and pull on them to move the body and its parts.

mutagen (MEW-tuh-jen) An environmental agent that can permanently modify the structure of a DNA molecule. Certain viruses and ultraviolet radiation are examples.

mutation, gene [L. *mutatus*, a change] A heritable change in DNA due to the deletion, addition, or substitution of one or several bases in the nucleotide sequence.

myelin sheath Of many sensory and motor neurons, an axonal sheath that affects how fast action potentials travel; formed from the plasma membranes of Schwann cells that wrap repeatedly around the axon and are separated from each other by a small node.

myocardium The cardiac muscle tissue.

myofibril (MY-oh-fy-brill) One of many threadlike structures inside a muscle cell; each is functionally divided into sarcomeres, the basic units of contraction.

myopia Also known as nearsightedness; distant objects appear blurry because their images are focused in front of the retina, not on it.

myosin (MY-uh-sin) A contractile protein. In muscle cells, it interacts with the protein actin to bring about contraction.

NAD⁺ Nicotinamide adenine dinucleotide; a nucleotide coenzyme. When carrying electrons and unbound protons (H^+) between reaction sites, it is abbreviated NADH.

NADP Nicotinamide adenine dinucleotide phosphate; a phosphorylated nucleotide coenzyme. When carrying electrons and unbound protons (H^+) between reaction sites, it is abbreviated $NADPH_2$.

nasal cavity The region of the respiratory system where inhaled air is warmed, moistened, and filtered of airborne particles and dust.

natural killer cell Cell of the immune system, a type of lymphocyte, that kills tumor cells (by lysis) or cells infected by a virus.

natural selection A microevolutionary process; a difference in survival and reproduction among members of a population that vary in one or more traits.

negative feedback mechanism A homeostatic feedback mechanism in which an activity changes some condition in the internal environment and so triggers a response that reverses the change.

nephritis Inflammation of the kidneys.

nephron (NEFF-ron) [Gk. *nephros*, kidney] Of the kidney, a slender tubule in which water and solutes filtered from blood are selectively reabsorbed and in which urine forms.

nerve Cordlike communication line of the peripheral nervous system, composed of axons of sensory neurons, motor neurons, or both, encased in connective tissue. In the brain and spinal cord, similar cordlike bundles are called nerve tracts.

nerve impulse *See* action potential.

nerve tract A bundle of myelinated axons of interneurons inside the spinal cord and brain.

nervous system System of neurons oriented relative to one another in precise message-conducting and information-processing pathways.

nervous tissue Tissue composed of neurons and (in the central nervous system) neuroglia.

neural tube Embryonic forerunner of the brain and spinal cord.

neuroendocrine control center The parts of the hypothalamus and pituitary gland that interact to control many body functions.

neurofibromatosis Genetic disorder that occurs when a transposable element is inserted in a particular location in a gene.

neuroglia (nur-oh-GLEE-uh) Cells that structurally and metabolically support neurons. They make up about half the volume of nervous tissue in the human body.

neuromodulator A signaling molecule that influences the effects of transmitter substances by enhancing or reducing membrane responses in target neurons.

neuromuscular junction Chemical synapse between axon terminals of a motor neuron and a muscle cell.

neuron A nerve cell; the basic unit of communication in the nervous system. Neurons collectively sense environmental change, integrate sensory inputs, then activate muscles or glands that initiate or carry out responses.

neurotransmitter Any of the class of signaling molecules that are secreted from neurons, act on adjacent cells, and are then rapidly degraded or recycled.

neutral trait A trait from gene mutation that neither harms nor helps the individual's ability to survive and reproduce.

neutron Unit of matter, one or more of which occupies the atomic nucleus. Neutrons have mass but no electric charge.

neutrophil Phagocytic white blood cell that takes part in inflammatory responses against bacteria.

niche (nitch) [L. *nidas*, nest] The full range of physical and biological conditions under which members of a species can live and reproduce.

nitrification (nye-trih-fih-KAY-shun) A process in which certain bacteria strip electrons from ammonia or ammonium present in soil. The end product, nitrite (NO_2^2), is broken down to nitrate (NO_3^2) by different bacteria.

nitrogen cycle Biogeochemical cycle in which gaseous nitrogen is captured by nitrogen-fixing microorganisms and then moves through organisms and ecosystems before being returned to the atmosphere. The atmosphere is the largest reservoir of nitrogen.

nitrogen fixation Process by which a few kinds of bacteria convert gaseous nitrogen (N_2) to ammonia.

nociceptor A receptor, such as a free nerve ending, that detects stimuli causing tissue damage.

nondisjunction Failure of one or more chromosomes to separate properly during mitosis or meiosis.

nongonococcal urethritis (NGU) An inflammation of the urethra; often caused by infection by the bacterium that causes chlamydia and considered a sexually transmitted disease.

non-Hodgkin lymphoma Lymphoma consisting of a group of malignant tumors of B or T lymphocytes.

nonsteroid hormone A type of water-soluble hormone, such as a protein hormone, that cannot cross the lipid bilayer of a target cell. These hormones enter the cell by receptor-mediated endocytosis, or they bind to receptors that activate membrane proteins or second messengers within the cell.

nosocomial infection An infection that is acquired in a hospital, usually by direct contact.

nuclear envelope A double membrane (two lipid bilayers and associated proteins) that is the outermost portion of a cell nucleus.

nucleic acid (noo-CLAY-ik) A long, single- or double-stranded chain of four different nucleotides joined at their phosphate groups. Nucleic acids differ in which nucleotide base follows the next in the sequence. DNA and RNA are examples.

nucleolus (noo-KLEE-oh-lus) [L. *nucleolus*, a little kernel] Within the nucleus of a nondividing cell, a site where the protein and RNA subunits of ribosomes are assembled.

nucleosome (NOO-KLEE-oh-sohm) Of chromosomes, one of many organizational units, each consisting of a small stretch of DNA looped twice around a "spool" of histone molecules, which another histone molecule stabilizes.

nucleotide (NOO-klee-oh-tide) A small organic compound having a five-carbon sugar (deoxyribose), nitrogen-containing base, and phosphate group. Nucleotides are the structural units of adenosine phosphates, nucleotide coenzymes, and nucleic acids.

nucleotide coenzyme A protein that transports hydrogen atoms (free protons) and electrons from one reaction site to another in cells.

nucleotide sequence The order of nucleotides in a gene; it codes for a specific polypeptide chain.

nucleus (NOO-klee-us) Of atoms, the central core consisting of one or more positively charged protons and (in all but hydrogen) electrically neutral neutrons. In cells, a membranous organelle that physically isolates and organizes the DNA, out of the way of cytoplasmic machinery.

nutrient Element with a direct or indirect role in metabolism that no other element fulfills.

nutrition All those processes by which food is ingested, digested, absorbed, and later converted to the body's own organic compounds.

obesity An excess of fat in the body's adipose tissues, often caused by imbalances between caloric intake and energy output.

olfactory receptors Receptors in the nasal epithelium that detect water-soluble or volatile substances.

oligosaccharide A carbohydrate consisting of a short chain of two or more covalently bonded sugar units. One subclass, disaccharides, has two sugar units. Compare monosaccharide; polysaccharide.

omnivore [L. *omnis*, all, and *vovare*, to devour] An organism that feeds on a variety of food types, such as plant and animal tissues.

oncogene (ON-coe-jeen) A gene that has the potential to induce cancerous transformations in a cell.

oocyte An immature egg.

oogenesis (oo-oh-JEN-uh-sis) Formation of a female gamete, from a germ cell to a mature haploid ovum (egg).

oral cavity The mouth.

orbital Volume of space around the nucleus of an atom in which electrons are likely to be at any instant.

organ A body structure of definite form and function that is composed of more than one tissue.

organ of Corti Membrane region of the inner ear that contains the sensory hair cells involved in hearing.

organ system Two or more organs that interact chemically, physically, or both in performing a common task.

organelle Of cells, an internal, membrane-bounded sac or compartment that has a specific, specialized metabolic function.

organic compound A compound having a carbon backbone, often with carbon atoms arranged as a chain or ring structure, and at least one hydrogen atom.

organogenesis Stage of development in which primary tissue layers (germ layers) split into subpopulations of cells, and different lines of cells become unique in structure and function; foundation for growth and tissue specialization, when organs acquire specialized chemical and physical properties.

orgasm The culmination of the sex act which involves muscle contractions and sensations of warmth, release, and relaxation.

origin The end of a muscle that is attached to the bone that remains relatively stationary when the muscle contracts.

osmoreceptor Sensory receptor that detects changes in water volume (solute concentration) in the fluid bathing it.

osmosis (oss-MOE-sis) [Gk. *osmos*, act of pushing] The tendency of water to move across a cell membrane in response to a concentration gradient.

osteoblast A bone cell that forms bone.

osteoclast A bone cell that breaks down the matrix of bone tissue.

osteocyte A living bone cell.

osteon A set of thin, concentric layers of compact bone tissue surrounding a narrow canal carrying blood vessels and nerves; arrays of osteons make up compact bone.

ovarian cycle Cycle during which a primary oocyte matures and is ovulated.

ovary (OH-vuh-ree) The primary female reproductive organ, where eggs form.

oviduct (OH-vih-dukt) Duct through which eggs travel from the ovary to the uterus. Also called Fallopian tube.

ovulation (ahv-you-LAY-shun) During each turn of the menstrual cycle, the release of a secondary oocyte (immature egg) from an ovary.

ovum (OH-vum) A mature female gamete (egg).

oxaloacetate A four-carbon compound with roles in metabolism (e.g., the point of entry into the Krebs cycle).

oxidation-reduction reaction An electron transfer from one atom or molecule to another. Often hydrogen is transferred along with the electron or electrons.

oxidative phosphorylation (foss-for-ih-LAY-shun) Final stage of aerobic respiration, in which ATP forms after hydrogen ions and electrons (from the Krebs cycle) are sent through a transport system that gives up the electrons to oxygen.

oxygen debt Lowered O_2 level in blood when muscle cells have used up more ATP than they have formed by aerobic respiration.

oxyhemoglobin A hemoglobin molecule that has oxygen bound to it; HbO_2.

ozone thinning Pronounced seasonal thinning of the Earth's ozone layer, as in the lower stratosphere above Antarctica.

P (parent) generation The designation for the parent generation in a genetic cross.

palate Structure that separates the nasal cavity from the oral cavity. The bone-reinforced hard palate serves as a hard surface against which the tongue can press food as it mixes it with saliva.

pancreas (PAN-cree-us) Gland that secretes enzymes and bicarbonate into the small intestine during digestion, and that also secretes the hormones insulin and glucagon.

pancreatic islets Any of the two million clusters of endocrine cells in the pancreas, including alpha cells, beta cells, and delta cells.

pandemic A situation in which epidemics of a disease break out in several countries around the world within a given time span.

paralysis Loss of sensation and/or motor function, often due to spinal cord injuries.

parasite [Gk. *para*, alongside, and *sitos*, food] An organism that obtains nutrients directly from the tissues of a living host, which it lives on or in and may or may not kill.

parasympathetic nerve Of the autonomic nervous system, any of the nerves carrying signals that tend to slow the body down overall and divert energy to basic tasks; parasympathetic nerves also work continually in opposition with sympathetic nerves to bring about minor adjustments in internal organs.

parathyroid glands (pare-uh-THY-royd) Endocrine glands embedded in the thyroid gland that secrete parathyroid hormone, which helps restore blood calcium levels.

Parkinson's disease Degenerative brain disorder that affects normal muscle function.

parturition Birth.

passive immunity Temporary immunity conferred by deliberately introducing antibodies into the body.

passive transport Diffusion of a solute through a channel or carrier protein that spans the lipid bilayer of a cell membrane. Its passage does not require an energy input; the protein passively allows the solute to follow its concentration gradient.

pathogen (PATH-oh-jen) [Gk. *pathos*, suffering] An infectious, disease-causing agent, such as a virus or bacterium.

PCR *See* polymerase chain reaction.

pectoral girdle Set of bones, including the scapula (shoulder blade) and clavicle (collarbone), to which the long bone of each arm attaches. The pectoral girdles form the upper part of the appendicular skeleton and are only loosely attached to the rest of the body by muscles.

pedigree chart A chart of genetic connections among individuals, as constructed according to standardized methods.

pelvic cavity Body cavity in which the reproductive organs, bladder, and rectum are located.

pelvic girdle Set of bones including coxal bones that form an open basin, the pelvis; the lower part of the appendicular skeleton. The upper portions of the two coxal bones are the hipbones; the thighbones (femurs) join the coxal bones at hip joints. The pelvic girdle bears the body's weight when a person stands and is much more massive than the pectoral girdle.

pelvic inflammatory disease (PID) Generally, a bacterially caused inflammation of the uterus, oviducts, and ovaries. Often a complication of gonorrhea, chlamydia, or some other sexually transmitted disease.

penetrance In a given population, the percentage of individuals in which a particular genotype is expressed (that is, the percentage of individuals who have the genotype and also exhibit the corresponding phenotype).

penis Male organ that deposits sperm into the female reproductive tract; also houses the urethra.

pepsin Any of several digestive enzymes that are part of gastric fluid in the stomach.

peptide hormone A hormone that consists of a short chain of amino acids.

perception The conscious interpretation of some aspect of the external world created by the brain from nerve impulses generated by sensory receptors.

perforin A type of protein secreted by a natural killer cell of the immune system, and which creates holes (pores) in the plasma membrane of a target cell.

peripheral nervous system (per-IF-ur-uhl) [Gk. *peripherein*, to carry around] The nerves leading into and out from the spinal cord and brain and the ganglia along those communication lines.

peripheral vasoconstriction The reduction in blood flow to capillaries near the body's surface to retain body heat.

peripheral vasodilation The dilation of blood vessels in the skin that allows excess heat in the blood to dissipate.

peristalsis (pare-ih-STAL-sis) Rhythmic contraction of muscles that moves food forward through the gastrointestinal tract.

peritoneum Lining of the coelom that also covers and helps maintain the position of internal organs.

peritubular capillaries The set of blood capillaries that threads around the tubular parts of a nephron; they function in reabsorption of water and solutes and in secretion of hydrogen ions and some other substances in the forming of urine.

peroxisome Enzyme-filled vesicle in which fatty acids and amino acids are digested first into hydrogen peroxide (which is toxic), then to harmless products.

PGA Phosphoglycerate (foss-foe-GLISS-er-ate). A key intermediate in glycolysis.

PGAL Phosphoglyceraldehyde. A key intermediate in glycolysis.

pH scale A scale used to measure the concentration of free hydrogen ions in blood, water, and other solutions; pH 0 is the most acidic, 14 the most basic, and 7, neutral.

phagocyte (FAYG-uh-sight) [Gk. *phagein*, to eat, and *-kytos*, hollow vessel] A macrophage or other white blood cell that engulfs and destroys foreign agents.

phagocytosis (fayg-uh-sigh-TOE-sis) [Gk. *phagein*, to eat, and *-kytos*, hollow vessel] Engulfment of foreign cells or substances by specialized white blood cells by means of endocytosis.

pharynx (FARE-inks) A muscular tube by which food enters the gastrointestinal tract; the dual entrance for the tubular part of the digestive tract and windpipe (trachea).

phenotype (FEE-no-type) [Gk. *phainein*, to show, and *-typos*, image] Observable trait or traits of an individual; arises from interactions between genes, and between genes and the environment.

phenylketonuria (PKU) Genetic disorder in which the amino acid phenylalanine builds up abnormally in the affected person.

pheromone (FARE-oh-moan) [Gk. *phero*, to carry, and *-mone*, as in hormone] A type of signaling molecule secreted by exocrine glands that serves as a communication signal between individuals of the same species.

phospholipid A type of lipid that is the main structural component of cell membranes. Each has a hydrophobic tail (of two fatty acids) and a hydrophilic head that incorporates glycerol and a phosphate group.

phosphorus cycle Movement of phosphorus from rock or soil through organisms, then back to soil.

phosphorylation (foss-for-ih-LAY-shun) The attachment of unbound (inorganic) phosphate to a molecule; also the transfer of a phosphate group from one molecule to another, as when ATP phosphorylates glucose.

photochemical smog A brown-air smog that develops over large cities when the surrounding land forms a natural basin.

photoreceptor A light-sensitive sensory cell.

pigment A light-absorbing molecule.

pilomotor response Contraction of smooth muscle controlling the erection of body hair when outside temperature drops. This creates a layer of still air that reduces heat losses from the body. (It is most effective in mammals that have more body hair than humans do.)

pineal gland (PY-neel) A light-sensitive endocrine gland that secretes melatonin, a hormone that influences reproductive cycles and the development of reproductive organs.

pituitary dwarfism Underproduction of growth hormone, causing an affected adult to be abnormally small.

pituitary gland An endocrine gland that interacts with the hypothalamus to coordinate and control many physiological functions, including the activity of many other endocrine glands. Its posterior lobe stores and secretes hypothalamic hormones; the anterior lobe produces and secretes its own hormones.

placenta (pluh-SEN-tuh) Of the uterus, an organ composed of maternal tissues and extraembryonic membranes (the chorion especially); it delivers nutrients to the fetus and accepts wastes from it, yet allows the fetal circulatory system to develop separately from the mother's.

plasma (PLAZ-muh) Liquid portion of blood; consists of water, various proteins, ions, sugars, dissolved gases, and other substances.

plasma cell In adaptive immunity, an effector B cell that quickly floods the bloodstream with antibodies.

plasma membrane The outermost cell membrane. Its lipid bilayer structure and proteins carry out most functions, including transport across the membrane and reception of extracellular signals.

plasmid Of many bacteria, a small, circular molecule of extra DNA that carries only a few genes and replicates independently of the bacterial chromosome.

platelet (PLAYT-let) A cell fragment in blood that releases substances necessary for clot formation.

pleiotropy (pleye-AH-troe-pee) [Gk. *pleon,* more, and *trope,* direction] A type of gene interaction in which a single gene exerts multiple effects on seemingly unrelated aspects of an individual's phenotype.

pleura Thin, double membrane surrounding each lung.

pneumonia Infection that causes inflammation in lung tissue, followed by buildup of fluid in the lungs and difficulty in breathing.

polar body Any of up to three cells that form during the meiotic cell division of an oocyte; the division also forms the mature egg, or ovum.

pollutant Any substance with which an ecosystem has had no prior evolutionary experience in terms of kinds or amounts, and that can accumulate to disruptive or harmful levels. Can be naturally occurring or synthetic.

polycystic kidney disease An inherited disorder in which cysts form in both kidneys.

polygenic trait Trait that results from the combined expression of several genes.

polymer (PAH-lih-mur) [Gk. *polus,* many, and *meris,* part] A molecule composed of three to millions of small subunits that may or may not be identical.

polymerase chain reaction (PCR) DNA amplification method; DNA containing a gene of interest is split into single strands, which enzymes (polymerases) copy; the enzymes also act on the accumulating copies, multiplying the gene sequence by the millions.

polymorphism (poly-MORE-fizz-um) [Gk. *polus,* many, and *morphe,* form] Of a population, the persistence through the generations of two or more forms of a trait, at a frequency greater than can be maintained by new mutations alone.

polypeptide chain Three or more amino acids joined by peptide bonds.

polyploidy (PAHL-ee-ployd-ee) A change in the chromosome number following inheritance of three or more of each type of chromosome.

polysaccharide [Gk. *polus,* many, and *sakharon,* sugar] A straight or branched chain of hundreds of thousands of covalently linked sugar units of the same or different kinds. The most common polysaccharides are cellulose, starch, and glycogen.

polysome Of protein synthesis, several ribosomes all translating the same messenger RNA molecule, one after the other.

pons Hindbrain region; traffic center for signals between centers of the cerebellum and forebrain.

population A group of individuals of the same species occupying a given area.

population density The number of individuals of a population that are living in a specified area or volume.

population size The number of individuals that make up the gene pool of a population.

positive feedback mechanism Homeostatic mechanism by which a chain of events is set in motion that intensifies a change from an original condition.

precapillary sphincter A wispy ring of smooth muscle at the branch of a capillary that regulates the flow of blood into the capillary.

prediction A claim about what you can expect to observe in nature if a theory or hypothesis is correct.

primary immune response Activity of white blood cells and their products elicited by a first-time encounter with an antigen; includes both antibody-mediated and cell-mediated responses.

primary productivity Of ecosystems, *gross* primary productivity is the rate at which the producer organisms capture and store a given amount of energy during a specified interval. *Net* primary productivity is the rate of energy storage in the tissues of producers in excess of their rate of aerobic respiration.

primate The mammalian lineage that includes prosimians, tarsioids, and anthropoids (monkeys, apes, and humans).

primer A man-made short nucleotide sequence designed to base-pair with any complementary DNA sequence; later, DNA polymerases recognize it as a start tag for replication.

prion (PREE-on) Small infectious protein that causes rare, fatal degenerative diseases of the nervous system.

probability With respect to any chance event, the most likely number of times it will turn out a certain way, divided by the total number of all possible outcomes.

probe Very short stretch of DNA designed to base-pair with part of a gene being studied and labeled with an isotope to distinguish it from DNA in the sample being investigated.

producer, primary. Of ecosystems, any of the organisms that secure energy from the physical environment, as by photosynthesis or chemosynthesis. Green plants are the Earth's main primary producers.

progesterone (pro-JESS-tuh-rown) Female sex hormone secreted by the ovaries.

prokaryotic cell (pro-carry-OH-tic) [L. *pro,* before, and Gk. *karyon,* kernel] A bacterium; a single-celled organism that has no nucleus or any of the other membrane-bound organelles characteristic of eukaryotic cells.

promoter Of transcription, a base sequence that signals the start of a gene; the site where RNA polymerase initially binds.

prophase Of mitosis, the stage when each duplicated chromosome starts to condense, microtubules form a spindle apparatus, and the nuclear envelope starts to break up.

prophase I Of meiosis, the stage at which the microtubular spindle starts to form, the nuclear envelope starts to break up, and each duplicated chromosome also condenses and pairs with its homologous partner. At this time, their sister chromatids typically undergo crossing over and genetic recombination.

prophase II Of meiosis, a brief stage after interkinesis during which each chromosome still consists of two chromatids.

prostaglandin Any of various lipids present in tissues throughout the body that can act as local signaling molecules. Prostaglandins typically cause smooth muscle to contract or relax, as in blood vessels, the uterus, and airways.

prostate gland Gland in males that wraps around the urethra and ejaculatory ducts; its secretions become part of semen.

protein A large organic compound composed of one or more chains of amino acids held together by peptide bonds. Proteins have unique sequences of different kinds of amino acids in their polypeptide chains; such sequences are the basis of a protein's three-dimensional structure and chemical behavior.

protein hormone A hormone that consists of a long amino acid chain.

proto-oncogene A gene sequence similar to an oncogene but that codes for a protein required in normal cell function; may trigger cancer, generally when specific mutations alter its structure or function.

proton Positively charged particle, one or more of which is present in the atomic nucleus.

proximal tubule The tubular region of a nephron that receives water and solutes filtered from the blood.

psychoactive drug A chemical that acts on the central nervous system, altering the activity of brain neurons and associated mental and physical states.

puberty Period of human development that marks the arrival of sexual maturity as the reproductive organs begin to function.

pubic lice A parasite transmitted by close body contact; causes itching and irritation.

pulmonary circuit Blood circulation route leading to and from the lungs.

pulse Rhythmic pressure surge of blood flowing in an artery, created during each cardiac cycle when a ventricle contracts.

Punnett square method A method to predict the probable outcome of a mating or an experimental cross in simple diagrammatic form.

purine Nucleotide base having a double ring structure. Adenine and guanine are examples.

pyelonephritis Inflammation of the kidney usually resulting from an infection of the bladder.

pyrimidine (pie-RIM-ih-deen) Nucleotide base having a single ring structure. Cytosine and thymine are examples.

pyruvate (pie-ROO-vate) A compound with a backbone of three carbon atoms. Two pyruvate molecules are the end products of glycolysis.

r Designates net population growth rate; the birth and death rates are assumed to remain constant and so are combined into this one variable for population growth equations.

radiation therapy Cancer treatment that relies on radiation from radioisotopes to destroy or impair the activity of cancerous cells.

radioisotope An unstable atom that has dissimilar numbers of protons and neutrons and that spontaneously decays (emits electrons and energy) to a new, stable atom that is not radioactive.

radiometric dating A method of dating fossils that tracks the radioactive decay of material in the specimen.

radius One of two long bones of the forearm that extend from the humerus (at the elbow joint) to the wrist. The radius runs along the thumb side of the forearm, parallel to the ulna.

reabsorption In the kidney, the diffusion or active transport of water and usable solutes out of a nephron and into capillaries leading back to the general circulation; regulated by ADH and aldosterone.

receptor, sensory. A sensory cell or cell part that may be activated by a specific stimulus.

recessive (allele or trait) [L. *recedere*, to recede] In heterozygotes, an allele whose expression is fully or partially masked by expression of its partner; fully expressed only in the homozygous recessive condition.

recognition protein Protein at cell surface recognized by cells of like type; helps guide the ordering of cells into tissues during development and functions in cell-to-cell interactions.

recombinant DNA Any molecule of DNA that incorporates one or more nonparental nucleotide sequences. It is the outcome of microbial gene transfer in nature or resulting from recombinant DNA technology.

recombinant DNA technology Procedures by which DNA (genes) from different species may be isolated, cut, spliced together, and the new recombinant molecules multiplied in quantity in a population of rapidly dividing cells such as bacteria.

recombination Any enzyme-mediated reaction that inserts one DNA sequence into another.

rectum Final region of the gastrointestinal tract, which receives and temporarily stores undigested food residues (feces).

red blood cell Erythrocyte; an oxygen-transporting cell in blood.

red-green color blindness Vision problem in which the retina lacks cone cells that normally respond to light of red or green wavelengths.

red marrow A substance in the spongy tissue of many bones that serves as a major site of blood cell formation.

reductional division In meiosis, mode of cell division in which daughter cells end up with one-half the normal diploid number of chromosomes, one partner from each pair of homologous parent chromosomes.

reflex [L. *reflectere*, to bend back] A simple, stereotyped movement elicited directly by sensory stimulation.

reflex arc [L. *reflectere*, to bend back] Type of neural pathway in which signals from sensory neurons directly stimulate or inhibit motor neurons without intervention by interneurons.

refractory period Of neurons, the period following an action potential at a given patch of membrane when sodium gates are shut and potassium gates are open, so that the patch is insensitive to stimulation.

regulatory protein A protein that enhances or suppresses the rate at which a gene is transcribed.

releasing hormone A hypothalamic signaling molecule that stimulates or slows down secretion by target cells in the anterior lobe of the pituitary gland.

renal corpuscle Bowman's capsule plus the glomerular capillaries it cups around.

reproduction In biology, processes by which a new generation of cells or multicellular individuals is produced. Sexual reproduction requires meiosis, formation of gametes, and fertilization. Asexual reproduction refers to the production of new individuals by any mode that does not involve gametes.

reproductive base The number of actually and potentially reproducing individuals in a population.

reproductive isolating mechanism Any aspect of structure, functioning, or behavior that restricts gene flow between two populations.

reproductive isolation An absence of gene flow between populations.

reproductive success Production of viable offspring by an individual.

reproductive system An organ system consisting of a pair of gonads (testes in males, ovaries in females). Its sole function is the continuation of the species.

respiration [L. *respirare*, to breathe] The overall exchange of oxygen from the environment for carbon dioxide wastes from cells by way of circulating blood. Compare aerobic respiration.

respiratory bronchiole Smallest airway in the respiratory system; opens onto alveoli.

respiratory cycle One inhalation, one exhalation of air into and out of the lungs.

respiratory membrane The multilayer structure of an alveolus wall and associated capillaries.

respiratory surface In alveoli of the lungs, the thin, moist membrane across which gases diffuse.

respiratory system An organ system specialized for respiration, human lungs, and airways.

resting membrane potential Of neurons and other excitable cells that are not being stimulated, the steady voltage difference across the plasma membrane.

restriction enzymes Class of bacterial enzymes that cut apart foreign DNA injected into them, as by viruses; also used in recombinant DNA technology.

reticular formation Of the brain stem, a major network of interneurons that helps govern activity of the whole nervous system.

retina A thin layer of neural tissue in the eye that contains densely packed photoreceptors.

retinal detachment Separation of the retina from the underlying choroid.

retinoblastoma A cancer of the retina; usually hereditary.

retrovirus An RNA virus that infects animal cells and with reverse transcriptase creates an RNA template to synthesize a DNA molecule that integrates itself into the host's DNA.

reverse transcription Assembly of DNA on a single-stranded mRNA molecule by viral enzymes.

RFLPs Restriction fragment length polymorphisms. Of DNA samples from different individuals, slight but unique differences in the banding pattern of fragments of the DNA that have been cut with restriction enzymes.

Rh blood typing A method of characterizing red blood cells on the basis of a protein that serves as a self marker at their surface; Rh^+ signifies its presence and Rh^-, its absence.

rhodopsin Substance in rod cells of the eye consisting of the protein opsin and a side group, cis-retinal. When the side group absorbs incoming light energy, a series of chemical events follows that result in action potentials in associated neurons.

ribosome In all cells, the structure at which amino acids are strung together in specified sequence to form the polypeptide chains of proteins. An intact ribosome consists of two subunits, each composed of ribosomal RNA and protein molecules.

ribosomal RNA (rRNA) Type of RNA molecule that combines with proteins to form ribosomes, on which the polypeptide chains of proteins are assembled.

ribs Twelve pairs of bones that provide a scaffolding for the upper torso.

rigor mortis A stiffening of skeletal muscles caused when a person dies and the body cells stop making ATP.

RNA Ribonucleic acid. A category of single-stranded nucleic acids that function in processes by which genetic instructions are used to build proteins.

RNA polymerase Enzyme that catalyzes the assembly of RNA strands on DNA templates.

rod cell Of the retina, a photoreceptor sensitive to very dim light that contributes to coarse perception of movement.

S-shaped curve A curve that is characteristic of logistic growth; it is obtained when population size is plotted against time.

salinization A salt buildup in soil as a result of evaporation, poor drainage, and often the importation of mineral salts in irrigation water.

salivary amylase Starch-degrading enzyme in saliva.

salivary gland Any of the glands that secrete saliva, a fluid that initially mixes with food in the mouth and starts the breakdown of starch.

salt Compound that releases ions other than H^+ and OH^- in solution.

saltatory conduction In myelinated neurons, rapid, node-to-node hopping of action potentials.

sampling error Error that develops when an experimenter uses a sample (or subset) of a population, an event, or some other aspect of nature for an experimental group that is not large enough to be representative of the whole.

sarcoma Cancer of connective tissues such as muscle and bone.

sarcomere (SAR-koe-meer) The basic unit of muscle contraction; a region of myosin and actin filaments organized in parallel between two Z bands of a myofibril inside a muscle cell.

sarcoplasmic reticulum (sar-koe-PLAZ-mik reh-TIK-you-lum) In muscle cells, a membrane system that takes up, stores, and releases the calcium ions required for cross-bridge formation in sarcomeres, hence for contraction.

scapula Flat, triangular bone on either side of the pectoral girdle; the scapulae form the shoulder blades.

Schwann cell A specialized neuroglial cell that grows around a neuron axon, forming a myelin sheath.

scientific method A systematic way of gathering knowledge about the natural world.

second messenger A molecule inside a cell that mediates and generally triggers an amplified response to a hormone.

secondary immune response Rapid, prolonged response by white blood cells, memory cells especially, to a previously encountered antigen.

secondary oocyte An oocyte (unfertilized egg cell) that has completed meiosis I; it is this haploid cell that is released at ovulation.

secondary sexual trait Trait associated with maleness or femaleness, but not directly involved with reproduction. Beard growth in males and breast development in females are examples.

secretion A cell acting on its own or as part of glandular tissue releases a substance across its plasma membrane, into the surroundings.

sedimentary cycle Biogeochemical cycle. An element having no gaseous phase moves from land, through food webs, to the seafloor, then returns to land through long-term uplifting.

segmentation In tubular organs such as the large intestine, an oscillating movement produced by rings of muscle in the tube wall.

segregation, Mendelian principle of [L. *se-*, apart, and *grex*, herd] The principle that diploid organisms inherit a pair of genes for each trait (on a pair of homologous chromosomes) and that the two genes segregate during meiosis and end up in separate gametes.

seizure disorder Also known as epilepsy; disease characterized by recurring seizures.

selective permeability Of a cell membrane, a capacity to let some substances but not others cross it at certain times. The capacity arises as an outcome of the membrane's lipid bilayer structure and its transport proteins.

semen [L. *serere*, to sow] Sperm-bearing fluid expelled from the penis during male orgasm.

semicircular canals Fluid-filled canals positioned at different angles within the vestibular apparatus of the inner ear and that contain sensory receptors that detect head movements, deceleration, and acceleration.

semiconservative replication [Gk. *hēmi*, half, and L. *conservare*, to keep] Reproduction of a DNA molecule when a complementary strand forms on each of the unzipping strands of an existing DNA double helix, the outcome being two "half-old, half-new" molecules.

semilunar valve In each half of the heart; during each heartbeat, it opens and closes in ways that keep blood flowing in one direction from the ventricle to the arteries leading away from it.

seminal vesicle Part of the male reproductive system; secretes fructose that nourishes sperm.

seminiferous tubules Coiled tubes inside the testes where sperm develop.

senescence (sen-ESS-cents) [L. *senescere*, to grow old] Sum total of processes leading to the natural death of an organism or some of its parts.

sensation The conscious awareness of a stimulus.

sensory adaptation In a sensory system, a state in which the frequency of action potentials eventually slows or stops even when the strength of a stimulus is constant.

sensory neuron Any of the nerve cells that act as sensory receptors, detecting specific stimuli (such as light energy) and relaying signals to the brain and spinal cord.

sensory receptor A sensory cell or specialized cell adjacent to it that can detect a particular stimulus.

sensory system Element of the nervous system consisting of sensory receptors (such as photoreceptors), nerve pathways from the receptors to the brain, and brain regions that process sensory information.

septic shock Sudden and dangerous decline in blood pressure.

septum Of the heart, a thick wall that divides the heart into right and left halves.

Sertoli cell Any of the cells in seminiferous tubules that nourish and otherwise aid the development of sperm.

severe combined immune deficiency (SCID-X1) Genetic disorder in which stem cells in the affected person's bone marrow fail to make lymphocytes; affected children must live in germ-free isolation tents.

sex chromosome A chromosome that determines a new individual's gender. Compare autosomes.

sexually transmitted disease (STD) Infection passed from person to person through sexual contact.

shell model Model of electron distribution in atoms in which orbitals available to electrons occupy a nested series of shells.

shifting cultivation The cutting and burning of trees, followed by tilling of ashes into the soil; once called "slash-and-burn agriculture."

sickle-cell anemia An inherited disorder in which red blood cells are shaped like sickles, impeding their ability to carry oxygen.

simple goiter Enlargement of the thyroid gland due to an iodine-deficient diet.

sinoatrial (SA) node Region of conducting cells in the upper wall of the right atrium; the cells generate periodic waves of excitation that stimulate the atria to contract.

sinus In the skull, an air-filled space lined with mucous membrane that functions to lighten the skull.

sister chromatid Of a duplicated chromosome, one of two DNA molecules (and associated proteins) that remain attached at their centromere only during nuclear division. Each ends up in a separate daughter nucleus.

skeletal muscle Type of muscle that interacts with the skeleton to bring about body movements. A skeletal muscle typically consists of bundles of many long cylindrical cells encapsulated by connective tissue.

skeletal system The bones of the skeleton along with cartilages, joints, and ligaments.

sliding filament model Model of muscle contraction in which myosin filaments physically slide along and pull actin filaments toward the center of the sarcomere, which shortens. The sliding requires ATP energy and cross-bridge formation between the actin and myosin.

small-cell carcinoma Type of lung cancer that spreads rapidly and kills most victims within five years.

small intestine The portion of the digestive system where digestion is completed and most nutrients are absorbed.

smooth muscle One of the three main muscle types; occurs in the walls of internal organs and generally is not under voluntary control.

sodium–potassium pump A transport protein spanning the lipid bilayer of the plasma membrane. When activated by ATP, its shape changes and it selectively transports sodium ions out of the cell and potassium ions in.

solute (SOL-yoot) [L. *solvere*, to loosen] Any substance dissolved in a solution. In water, this means spheres of hydration surround the charged parts of individual ions or molecules and keep them dispersed.

solvent Fluid in which one or more substances is dissolved.

somatic cell (so-MAT-ik) [Gk. *somā*, body] Any body cell that is not a germ cell (which gives rise to gametes).

somatic nerve Nerves leading from the central nervous system to skeletal muscles.

somatic sensation Awareness of touch, pressure, heat, cold, pain, and limb movement.

somatosensory cortex Part of the gray matter of the cerebral hemispheres that controls somatic sensations.

somites In a developing embryo, paired blocks of mesoderm that will give rise to most bones and to the skeletal muscles of the neck and trunk.

special senses Vision, hearing, olfaction, or another sensation that arises from a particular location, such as the eyes, ears, or nose.

speciation (spee-cee-AY-shun) The evolutionary process by which species originate. One speciation route starts with divergence of two reproductively isolated populations of a species. They become separate species when accumulated differences in allele frequencies prevent them from interbreeding successfully under natural conditions.

species (SPEE-ceez) [L. *species*, a kind] A unit consisting of one or more populations of individuals that can interbreed under natural conditions to produce fertile offspring that are reproductively isolated from other such units.

sperm [Gk. *sperma*, seed] A mature male gamete.

spermatogenesis (sperm-at-oh-JEN-ih-sis) Formation of a male gamete, from a germ cell to a mature sperm.

sphere of hydration A clustering of water molecules around the individual molecules of a substance placed in water. Compare solute.

sphincter (SFINK-tur) Ring of muscle between regions of a tubelike system (as between the stomach and small intestine).

spina bifida Birth defect in which the neural tube doesn't close and separate from ectoderm.

spinal cavity Body cavity that houses the spinal cord.

spinal cord The portion of the central nervous system threading through a canal inside the vertebral column and providing direct reflex connections between sensory and motor neurons, as well as communication lines to and from the brain.

spindle apparatus A bipolar structure that forms during mitosis or meiosis and that moves the chromosomes. It consists of two sets of microtubules that extend from the opposite poles and that overlap at the spindle's equator.

spleen The largest lymphoid organ; it is a filtering station for blood and a reservoir of lymphocytes, red blood cells, and macrophages.

spongy bone Type of bone tissue in which hard, needlelike struts separate large spaces filled with marrow. Spongy bone occurs at the ends of long bones and within the breastbone, pelvis, and bones of the skull.

sporadic disease A disease that breaks out irregularly and affects relatively few people.

squamous cell carcinoma Type of lung cancer that affects squamous epithelium in the bronchi.

start codon Of protein synthesis, a base triplet in a strand of mRNA that serves as the start signal for mRNA translation.

stem cell Unspecialized cell that can give rise to descendants that differentiate into specialized cells.

sternum Elongated flat bone (also called the breastbone) to which the upper ribs attach and so form the rib cage.

steroid (STAIR-oid) A lipid with a backbone of four carbon rings and with no fatty acid tails. Steroids differ in their functional groups. Different types have roles in metabolism, intercellular communication, and cell membranes.

steroid hormone A type of lipid-soluble hormone synthesized from cholesterol. Many steroid hormones move into the nucleus and bind to receptors there; others bind to receptors in the cytoplasm, and the entire complex moves into the nucleus.

sterol A type of lipid with a rigid backbone of four fused carbon rings. Sterols occur in cell membranes; cholesterol is the main type in human tissues.

stimulant A psychoactive drug that initially increases alertness and physical activity, and then depresses the central nervous system. Caffeine, nicotine, cocaine, and amphetamines are examples.

stimulus [L. *stimulus*, goad] A specific change in the environment, such as a variation in light, heat, or mechanical pressure, that the body can detect through sensory receptors.

stomach A muscular, stretchable sac that receives ingested food; the organ between the esophagus and intestine in which considerable protein digestion occurs.

stop codon Of protein synthesis, a base triplet in a strand of mRNA that serves as the stop signal for translation, so that no more amino acids are added to the polypeptide chain.

strep throat Common bacterial infection of the throat and tonsils.

substrate A reactant or precursor molecule for a metabolic reaction; a specific molecule or molecules that an enzyme can chemically recognize, briefly bind to, and modify in a specific way.

substrate-level phosphorylation The direct, enzyme-mediated transfer of a phosphate group from the substrate of a reaction to another molecule. An example is the transfer of phosphate from an intermediate of glycolysis to ADP, forming ATP.

succession, primary and secondary (suk-SESH-un) [L. *succedere*, to follow after] Orderly changes from the time pioneer species colonize a barren habitat through replacements by various species until the climax community, when the composition of species remains steady under prevailing conditions.

suppressor T cell Any of the cells that produce chemical signals that help shut down an immune response.

surface-to-volume ratio A mathematical relationship in which volume increases with the cube of the diameter, but surface area increases only with the square. In growing cells, the volume of cytoplasm increases more rapidly than the surface area of the plasma membrane that must service the cytoplasm. Because of this constraint, cells generally remain small or elongated, or have elaborately folded membranes.

sympathetic nerve Of the autonomic nervous system, any of the nerves generally concerned with increasing overall body activities during times of heightened awareness, excitement, or danger; sympathetic nerves also work in opposition with parasympathetic nerves to bring about minor adjustments in internal organs.

synaptic integration (sin-AP-tik) The moment-by-moment combining of excitatory and inhibitory signals arriving at a trigger zone of a neuron.

syndrome A set of symptoms that may not individually be telling, but collectively characterize a disorder or disease.

synovial joint Freely movable joint in which adjoining bones are separated by a fluid-filled cavity and stabilized by straplike ligaments. An example is the ball-and-socket joint at the hip.

syphilis The sexually transmitted disease caused by infection by the spirochete bacterium *Treponema pallidum*. Untreated syphilis can lead to lesions in mucous membranes, the eyes, bones, skin, liver, and central nervous system.

systemic circuit (sis-TEM-ik) Circulation route in which oxygenated blood flows from the lungs to the left half of the heart, through the rest of the body (where it gives up oxygen and takes on carbon dioxide), then back to the right side of the heart.

systole Contraction phase of the cardiac cycle.

target cell Any cell that has receptors for a specific signaling molecule and that may alter its behavior in response to the molecule.

taste receptors Chemoreceptors in the taste buds.

Tay-Sachs disease Genetic disorder in which an enzyme for lipid metabolism is missing; it results in neural degeneration.

TCR Antigen-binding receptor of T cells.

tectorial membrane Inner ear structure against which sensory hair cells are bent, producing action potentials that travel to the brain via the auditory nerve.

telophase (TEE-low-faze) Of mitosis, the final stage when chromosomes decondense into threadlike structures and two daughter nuclei form. Of meiosis I, the stage when one of each pair of homologous chromosomes has arrived at one or the other end of the spindle pole. At telophase II, chromosomes decondense and four daughter nuclei form.

telophase II Of meiosis, final stage when four daughter nuclei form.

temporal summation The adding together (summing) of several muscle contractions, resulting in a single, stronger contraction, when stimulatory signals arrive in rapid succession.

tendon A cord or strap of dense, regular connective tissue that attaches a muscle to bone or to another muscle.

teratogens Agents that can cause birth defects.

test An attempt to produce actual observations that match predicted or expected observations.

testcross In genetics, an experimental cross to reveal whether an organism is homozygous dominant or heterozygous for a trait. The organism showing dominance is crossed to an individual known to be homozygous recessive for the same trait.

testicular feminizing syndrome Genetic disorder in which androgen receptors are defective; an affected XY embryo will develop as a female, but without uterus or ovaries.

testis, plural testes. Male gonad; primary reproductive organ in which male gametes and sex hormones are produced.

testosterone (tess-TOSS-tuh-rown) In males, a major sex hormone that helps control reproductive functions.

tetanus Condition in which a muscle motor unit is maintained in a state of contraction for an extended period.

thalamus Of the forebrain, a coordinating center for sensory input and a relay station for signals to the cerebrum.

theory A testable explanation of a broad range of related phenomena. In modern science, only explanations that have been extensively tested and can be relied on with a very high degree of confidence are accorded the status of theory.

thermal inversion Situation in which a layer of dense, cool air becomes trapped beneath a layer of warm air; can cause air pollutants to accumulate to dangerous levels close to the ground.

thermoreceptor Sensory cell that can detect radiant energy associated with temperature.

thoracic cavity The chest cavity; holds the heart and lungs.

thirst center Cluster of nerve cells in the hypothalamus that can inhibit saliva production, resulting in mouth dryness that the brain interprets as thirst and leading a person to seek out drinking fluids.

threshold Of neurons and other excitable cells, a certain minimum amount by which the voltage difference across the plasma membrane must change to produce an action potential.

thymine Nitrogen-containing base in some nucleotides.

thymus A lymphoid organ with endocrine functions; lymphocytes of the immune system multiply, differentiate, and mature in its tissues, and its hormone secretions affect their functions.

thyroid gland An endocrine gland that produces hormones that affect overall metabolic rates, growth, and development.

tidal volume Volume of air, about 500 milliliters, that enters or leaves the lungs in a normal breath.

tight junction Cell junction where strands of fibrous proteins oriented in parallel with a tissue's free surface collectively block leaks between the adjoining cells.

tissue A group of cells and intercellular substances that function together in one or more specialized tasks.

T lymphocyte, or **T cell** One of a class of white blood cells that carry out immune responses. The helper T and cytotoxic T cells are examples.

tonicity The relative concentrations of solutes in two fluids, such as inside and outside a cell. When solute concentrations are isotonic (equal in both fluids), water shows no net osmotic movement in either direction. When one fluid is hypotonic (has less solutes than the other), the other is hypertonic (has more solutes) and is the direction in which water tends to move.

total fertility rate (TFR) The average number of children born to the women in a given population during their reproductive years.

touch-killing Mechanism by which cytotoxic T cells directly release perforins and toxins into a target cell and cause its destruction.

toxin A normal metabolic product of one species with chemical effects that can hurt or kill individuals of another species.

trace element Any element that represents less than 0.01 percent of body weight.

tracer A substance with a radioisotope attached to it so that its pathway or destination in a cell, organism, ecosystem, or some other system can be tracked, as by scintillation counters that detect its emissions.

trachea (TRAY-kee-uh) The windpipe, which carries air between the larynx and bronchi.

trachoma A bacterial infection that damages the eyeball and the conjunctiva, sometimes resulting in blindness.

transcription [L. *trans,* across, and *scribere,* to write] Of protein synthesis, the assembly of an RNA strand on one of the two strands of a DNA double helix; the base sequence of the resulting transcript is complementary to the DNA region on which it was assembled.

transfer RNA (tRNA) Of protein synthesis, any of the type of RNA molecules that bind and deliver specific amino acids to ribosomes and pair with mRNA code words for those amino acids.

translation In protein synthesis, the conversion of the coded sequence of information in mRNA into a particular sequence of amino acids to form a polypeptide chain; depends on interactions of rRNA, tRNA, and mRNA.

translocation A change in a chromosome's structure following the insertion of part of a nonhomologous chromosome into it.

transport protein One of many kinds of membrane proteins involved in active or passive transport of water-soluble substances across the lipid bilayer of a plasma membrane. Solutes on one side of the membrane pass through the protein's interior to the other side.

transposable element DNA element that can spontaneously "jump" to new locations in the same DNA molecule or a different one. Such elements often inactivate the genes into which they become inserted.

trichomoniasis Vaginal inflammation due to a parasite; symptoms include vaginal discharge, burning, and itching.

trigger zone The region of a motor neuron where proper stimulation can trigger an action potential (nerve impulse).

triglyceride (neutral fat) A lipid having three fatty acid tails attached to a glycerol backbone. Triglycerides are the body's most abundant lipids and richest energy source.

trisomy (TRY-so-mee) Of diploid cells, the abnormal presence of three of one type of chromosome.

trophoblast Surface layer of cells of the blastocyst that secrete enzymes that break down the uterine lining where the embryo will implant.

T tubules Tubelike extensions of a muscle cell's plasma membrane.

tuberculosis (TB) Serious bacterial lung infection that eventually destroys patches of lung tissue.

tumor A tissue mass composed of cells that are dividing at an abnormally high rate.

tumor marker A substance that is produced by a specific type of cancer cell or by normal cells in response to cancer.

tumor suppressor gene A gene whose protein product operates to keep cell growth and division within normal bounds, or whose product has a role in keeping cells anchored in place within a tissue.

Turner syndrome Genetic disorder in which one of two X chromosomes is missing; affected people are female, but are sterile and have no secondary sexual traits.

tympanic membrane The eardrum, which vibrates when struck by sound waves.

type 1 diabetes Also called juvenile-onset diabetes; an autoimmune response destroys beta cells in the pancreas.

type 2 diabetes A kind of diabetes where insulin levels are normal, but target cells don't respond properly; capillaries become damaged, resulting in less blood flow and eventual tissue death.

ulna One of two long bones of the forearm; the ulna extends along the little finger side of the forearm, parallel to the radius on the thumb side.

ultrafiltration Bulk flow of a small amount of protein-free plasma from a blood capillary when the outward-directed force of blood pressure is greater than the inward-directed osmotic force of interstitial fluid.

umbilical cord Structure containing blood vessels that connect a fetus to its mother's circulatory system by way of the placenta.

uracil (YUR-uh-sill) Nitrogen-containing base found in RNA molecules; can base-pair with adenine.

urea The main waste product when cells break down proteins.

ureter Tubular channel that carries urine from each kidney to the urinary bladder.

urethra Tube that carries urine from the bladder to the body surface.

urinary bladder Storage organ for urine.

urinary excretion A mechanism by which excess water and solutes are removed by way of the urinary system.

urinary incontinence Urine leakage due to age-related weakening of the bladder and urethra.

urinary system An organ system that adjusts the volume and composition of blood and so helps maintain extracellular fluid.

urine Fluid formed by filtration, reabsorption, and secretion in kidneys; consists of wastes, excess water, and solutes.

uterus (YOU-tur-us) [L. *uterus*, womb] Chamber in which the developing embryo is contained and nurtured during pregnancy.

vaccine Antigen-containing preparation injected into the body or taken orally; it elicits an immune response leading to the proliferation of memory cells that offer long-lasting protection against that particular antigen.

vagina Chamber of the female reproductive system that receives sperm, forms part of the birth canal, and channels menstrual flow to the exterior.

variable Of a scientific experiment, the only factor that is not the same in the experimental group as it is in the control group.

variant Creutzfeldt–Jakob disease Disease of the central nervous system caused by a prion similar to the one responsible for mad cow disease.

vas deferens Tube leading to the ejaculatory duct; one of several tubes through which sperm move after they leave the testes just prior to ejaculation.

vasoconstriction Decrease in the diameter of an arteriole, so that blood pressure rises; may be triggered by the hormones epinephrine and angiotensin.

vasodilation Enlargement of arteriole diameter, so that blood pressure falls; may be triggered by hormones including epinephrine and angiotensin.

veins Of the circulatory system, the large-diameter vessels that lead back to the heart.

ventricle (VEN-tri-kul) Of the heart, one of two chambers from which blood is pumped out. Compare atrium.

venule Small blood vessel that receives blood from tissue capillaries and merges into larger-diameter veins; a limited amount of diffusion occurs across venule walls.

vertebra, plural vertebrae. One of a series of hard bones arranged with intervertebral disks into a backbone.

vertebrate Animal having a backbone of bony segments, the vertebrae.

vesicle (VESS-ih-kul) [L. *vesicula*, little bladder] One of a variety of small membrane-bound sacs in the cell cytoplasm that function in the transport, storage, or digestion of substances or in some other activity.

vestibular apparatus A closed system of fluid-filled canals and sacs in the inner ear that functions in the sense of balance. Compare semicircular canals.

villus (VIL-us), plural villi. Any of several types of absorptive structures projecting from the free surface of an epithelium.

viroid An infectious particle consisting only of very short, tightly folded strands or circles of RNA. Viroids might have evolved from introns, which they resemble.

virulence The relative ability of a pathogen to cause serious disease.

virus A noncellular infectious agent consisting of DNA or RNA and a protein coat; can replicate only after its genetic material enters a host cell and subverts its metabolic machinery.

vision Precise light focusing onto a layer of photoreceptive cells that is dense enough to sample details concerning a given light stimulus, followed by image formation in the brain.

visual cortex Part of the brain that receives signals from the optic nerves.

vital capacity Maximum volume of air that can move out of the lungs after a person inhales as deeply as possible.

vitamin Any of more than a dozen organic substances that the body requires in small amounts for normal cell metabolism but generally cannot synthesize for itself.

vocal cords A pair of elastic ligaments on either side of the larynx wall. Air forced between them causes the cords to vibrate and produce sounds.

water (hydrologic) cycle The movement of water from oceans to the atmosphere, the land, and back to the ocean.

watershed Any region in which all precipitation drains into a single stream or river.

white blood cell Leukocyte; any of the macrophages, eosinophils, neutrophils, and other cells that are the central components of the immune system.

white matter Of the spinal cord, major nerve tracts so named because of the glistening myelin sheaths of their axons.

X chromosome A sex chromosome with genes that cause an embryo to develop into a female, provided that it inherits a pair of these.

xeroderma pigmentosum Genetic disorder in which radiation damage to DNA cannot be fixed.

X inactivation A compensating phenomenon in females that "switches off" one X chromosome soon after the first cleavages of the zygote.

X-linked gene Any gene on an X chromosome.

X-linked recessive inheritance Recessive condition in which the responsible, mutated gene occurs on the X chromosome.

Y chromosome A sex chromosome with genes that cause the embryo that inherited it to develop into a male.

Y-linked gene Any gene on a Y chromosome.

yellow marrow Bone marrow that consists mainly of fat and hence appears yellow. It can convert to red marrow and produce red blood cells if the need arises.

yolk sac One of four extraembryonic membranes. Part becomes a site of blood cell formation and some of its cells give rise to the forerunners of gametes.

zero population growth State in which the number of births in a population is balanced by the number of deaths over a specified period, assuming immigration and emigration also are balanced.

zona pellucida A noncellular coating around an oocyte.

zoonosis An infectious disease that mainly affects animals other than humans, but can also be passed on to humans.

zygote (ZYE-goat) The first cell of a new individual, formed by the fusion of a sperm nucleus with the nucleus of an egg (fertilization).

Credits and Acknowledgments

This page constitutes an extension of the copyright page. We have made every effort to trace the ownership of all copyrighted material and to secure permission from copyright holders. In the event of any question arising as to the use of any material, we will be pleased to make the necessary corrections in future printings. Thanks are due to the following authors, publishers, and agents for permission to use the material indicated.

CHAPTER 1 Page 1 (top right) © Pete Turnley/ Corbis; (bottom right) Antoinette Jongen/FPG/Getty Images; **Page 2** © Hubert Stadler/Corbis; **1.1** © Daniel McDonald/The Stock Shop; **1.3** (a) © Rich Buzzelli/ Tom Stack & Associates; (b) © Open Door Images/ Picture Quest; **1.4** (b) Lisa Starr, with PDB file from NYU Scientific Visualization Laboratory; (d) Ed Reschke; (g) Evan Cerasoli; (h) © Antoinette Jongen/ FPG/Getty Images; (i) © Gabe Palmer/Corbis; (j) Ed Reschke; (k) NASA; **1.6** (a) Gary Head; (b) © George Musil/Visuals Unlimited; **1.7** (b) © Superstock; **1.8** (right) © Michael Newman/PhotoEdit; (left) Dr. Douglas Coleman/The Jackson Laboratory; **1.9** (a) © Raghu Rai/Magnum Photos; (b) National Science Foundation; **Page 10** © Royalty-Free/CORBIS; **1.11** © Carolyn A. McKeone/Photo Researchers, Inc.; **Page 13** © Digital Vision/Picture Quest; **1.12** (a–d) Gary Head.

CHAPTER 2 Page 15 (top right) PBD file from NYU Scientific Visualization Laboratory; (bottom left) © Owaki-Kulla/CORBIS; (top right) Lisa Starr with PDB file from NYU Scientific Visualization Laboratory; **2.2** Gary Head; **2.3** (a) John Greim/Medichrome/The Stock Shop; (d) Dr. Harry T. Chugani, M.D., UCLA School of Medicine; **Page 18** © Michael S. Yamashita/ Corbis; **2.4** (a–c) Rendered with Atom in a Box, © Dauger Research, Inc.; **2.7** (a) (top) Gary Head; (bottom) © Bruce Iverson; (c) Lisa Starr with PDB ID:ID:1BNA; H.R. Drew, R.M. Wing, T. Takano, C. Broka, S. Tanaka, K. Itakura, R.E. Dickerson. Structure of a B-DNA Dodecamer, Conformation and Dynamics, PNAS V. 78 2179, 1981; **2.8** © Alan Craft/PhotoFile; **2.9** (top) PDB file from NYU Scientific Visualization Lab; (bottom) © Lester Lefkowitz/CORBIS; **2.11** Lisa Starr; **2.12** Michael Grecco/Picture Group; **2.14** (c) © 2000 Liz Von Hoenel/Stone/Getty Images; **Page 28** Lisa Starr; **2.15** © David M. Phillips/Visuals Unlimited; **2.16** © Nancy J. Pierce/Photo Researchers, Inc.; **2.17** (a) PBD file courtesy of Dr. Christina A. Bailey, Department of Chemistry & Biochemistry, California Polytechnic State University, San Luis Obispo, CA; **2.18** (a) PBD file courtesy of Dr. Christina A. Bailey, Department of Chemistry & Biochemistry, California Polytechnic State University, San Luis Obispo, CA; **2.19** PBD files from NYU Scientific Visualization Laboratory; (right) Lewis Lainey; **2.20** (all images) © PhotoDisc/Getty Images; **2.22** (a–d) PBD files from NYU Scientific Visualization Laboratory; **2.24** PDB ID: 1BBB; Silva, M.M., Rogers, P.H., Arnone, A.; A third quaternary structure of human hemoglobin A at 1.7-A resolution; J Biol Chem 267 pp. 17248 (1992); **2.25** (a) © Ron Davis/Shooting Star; (b) © Frank Trapper/Corbis

Sygma; **2.26** PDB files from Klotho Biochemical Compounds Declarative Database; **2.27** (b) Lisa Starr; **2.28** (right) Sue Hartzell; (left) © Inga Spence/Tom Stack & Associates.

CHAPTER 3 Page 41 © Prof. P. Motta & T. Naguro/ SPL/Photo Researchers, Inc.; **3.1** © Manfred Cage/ Peter Arnold; **3.3** (a) Ed Reschke; (b) © Carolina Biological Supply/Phototake; **3.6** G.L. Decker, Baylor College of Medicine; **3.8** (compound light microscope) Leica Microsystems, Inc., Deerfield, IL; (a) Lennart Nilsson from *Behold Man* © 1974 by Albert Bonniers. Forlag & Little Brown & Co., Boston; (transmission electron microscope) © George Musil/Visuals Unlimited; (b, top) © David M. Phillips/Visuals Unlimited; (b, bottom) © Science Photo Library/ Photo Researchers, Inc.; (c) © Driscoll, Youngquist, Baldeschwieler/Caltech/Science Source/Photo Researchers, Inc.; **3.9** Stephen Wolfe; **3.10** © Don W. Fawcett/Visuals Unlimited; (b) A.C. Faberge, Cell and Tissue Research, 151:403–415, 1974; **3.11** (b, c) © Don W. Fawcett/Visuals Unlimited; (d) Gary Grimes; **3.12** (b) Keith R. Porter; **3.13** (a) © J.W. Shuler/Photo Researchers, Inc.; (b) © Lennart Nilsson; **3.14** From *Tissue and Cell* vol. 27, pp. 421–427, courtesy of Bjorn Afzelius, Stockholm University; after Stephen L. Wolfe, *Molecular and Cellular Biology*, Wadsworth, 1993; **3.18** M. Sheetz, R. Painter, and S. Singer, *Journal of Cell Biology*, 70:193 (1976), by copyright permission of the Rockefeller University Press; **Page 57** © Dennis Kunkel/Phototake USA; (mitochondria) © Prof. P. Motta & T. Naguro/SPL/Photo Researchers, Inc.; **Page 59** Gary Head; (top right) © Prof. P. Motta & T. Naguro/SPL/Photo Researchers, Inc.; **3.27** Gary Head; **3.28** © Randy Faris/CORBIS; **Page 63** © Prof. P. Motta & T. Naguro/SPL/Photo Researchers, Inc.; **3.29** Gary Head.

CHAPTER 4 Page 67 (top left) © Jerry Ohlinger/ Corbis Sygma; (bottom left) © Ron Sachs/Corbis Sygma; (top right) © Science Photo Library/Photo Researchers, Inc.; **4.1** (a, left) © Manfred Kage/Peter Arnold; (a, right) © Focus on Sports; (b) © Ray Simons/Photo Researchers, Inc.; (c) © Ed Reschke/ Peter Arnold, Inc.; (d) Don D. Fawcett; **4.2** (a–c) Ed Reschke; (d) © Science Photo Library/Photo Researchers, Inc.; (e) Ed Reschke; (f) © University of Cincinnati, Raymond Walters College, Biology; **4.3** © Science Photo Library/Photo Researchers, Inc.; **Page 72** (top left) © Science Photo Library/Photo Researchers, Inc.; (bottom right) Courtesy of Muscular Dystrophy Association; **4.4** (a) Ed Reschke; (b) © Biophoto Associates/Photo Researchers, Inc.; (c) Ed Reschke; (bottom) ©Tony McConnell/SPL/ Photo Researchers, Inc.; **4.5** (a) © Triarch/Visuals Unlimited; (b) Nancy Kedersha/Harvard Medical School; **Page 73** (top right) © Science Photo Library/ Photo Researchers, Inc.; (bottom right) © Stanley Flegler/Visuals Unlimited; **4.7** (c) © Fabian/Corbis Sygma; (d) From *Perspectives in Human Biology* by L. Knapp, Wadsworth, 1998; **Page 78** (top left) © Science Photo Library/Photo Researchers, Inc.; (bottom left) © Mauro Fermariello/Photo Researchers, Inc.; **Page 79** © Biophoto Associates/Science Source/

Photo Researchers, Inc.; **4.10** (a) From Sherwood, *Human Physiology from Cells to Systems*, 4th ed.; (b) © John D. Cunningham/Visuals Unlimited; **4.11** © CNRI/SPL/Photo Researchers, Inc.; **Page 80** From Sherwood, *Human Physiology from Cells to Systems*, 5th ed., p. 11; **4.13** © VVG/SPL/Photo Researchers, Inc.; **4.14** Colin Monteath, Hedgefog House, New Zealand; **Page 85** (upper right) © Ed Reschke/Peter Arnold, Inc; (upper right, lower left) Ed Reschke; (lower right) © University of Cincinnati, Raymond Walters College, Biology; **4.15** © Sean Sprague/Stock Boston; **Page 86** © Royalty-Free/CORBIS.

CHAPTER 5 Page 87 © Prof. P. Motta, Dept. of Anatomy, Univ. of La Sapienza, Rome/SPL/Photo Researchers, Inc.; **5.3** (a, b) © Prof. P. Motta, Dept. of Anatomy, Univ. of La Sapienza, Rome/SPL/Photo Researchers, Inc.; **Page 89** Prof. P. Motta, Dept. of Anatomy, Univ. of La Sapienza, Rome/SPL/Photo Researchers, Inc.; **5.5** Yokochi and J. Rohen, *Photographic Anatomy of the Human Body*, 2/e. Igaku-Shoin, Ltd., 1979; **5.8** From *Perspectives in Human Biology* by L. Knapp, Wadsworth, 1998; **5.9** From *Perspectives in Human Biology* by L. Knapp, Wadsworth, 1998; **Page 95** Prof. P. Motta, Dept. of Anatomy, Univ. of La Sapienza, Rome/SPL/Photo Researchers, Inc.; **5.10** Ed Reschke; (a) From *Human Anatomy and Physiology*, fourth edition, by A. Spence and E. Mason, p. 763, Brooks/Cole, 1992; (b) From *Perspectives in Human Biology* by L. Knapp, Wadsworth, 1998; **5.12** © James Stevenson/Photo Researchers, Inc.; **5.13** From *Perspectives in Human Biology* by L. Knapp, Wadsworth, 1998; **5.14** (a) Paul Sponseller, MD/Johns Hopkins Medical Center; (b) Courtesy of the family of Tiffany Manning; **Page 99** © Prof. P. Motta, Dept. of Anatomy, Univ. of La Sapienza, Rome/SPL/Photo Researchers, Inc.; (bottom) © Mike Devlin/SPL/Photo Researchers, Inc.; **5.15** Biophoto Associates/SPL/Photo Researchers, Inc.; **Page 102** (top and bottom left) From *Human Anatomy and Physiology*, fourth edition, by A. Spence and E. Mason, p. 763, Brooks/Cole, 1992.

CHAPTER 6 Page 103 © Steve Cole/PhotoDisc Green/Getty Images; **6.1** From Knapp, *Perspectives in Human Biology*, Brooks/Cole, 1998; **6.6** (a) © Gladden Willis, MD/Visuals Unlimited; (b) © Tim Davis/ Photo Researchers, Inc.; **Page 108** Dance Theatre of Harlem, by Frank Capri; **6.7** (a, b) © Don Fawcett/ Visuals Unlimited; **Page 111** © Steve Cole/PhotoDisc Green/Getty Images; **6.12** From Knapp, *Perspectives in Human Biology*, Brooks/Cole, 1998; **6.15** © Don Fawcett, Bloom and Fawcett, 11/e, after J. Desaki and Y. Uehara/Photo Researchers, Inc.; **6.16** (d) Painting by Sir Charles Bell, 1809, Courtesy of Royal College of Surgeons, Edinburgh; **6.17** (c) Adam Pretty/ Allsport Concepts/Getty Images; **6.18** © Russ Schleipman/CORBIS; **6.19** (a) © Richard B. Levine; (b) © Michael Neveux; **Page 115** © Steve Cole/ PhotoDisc Green/Getty Images; **6.20** (b) From Knapp, *Perspectives in Human Biology*, Brooks/Cole, 1998.

CHAPTER 7 Pages 119, 127, 139 © Gusto/Photo Researchers, Inc.; **Page 124** (bottom left) Courtesy of

Lisa Brown; **7.7** After *Human Physiology: Mechanisms of Body Function,* fifth edition, by A. Vander et al. Copyright © 1990 McGraw-Hill. Used by permission of the McGraw-Hill Companies; (b, above) Microslide courtesy of Mark Nielsen, University of Utah; (b, below) © Don Fawcett/Visuals Unlimited; **7.8** From Knapp, *Perspectives in Human Biology,* Brooks/Cole, 1998; **Page 128** After *Human Physiology: Mechanisms of Body Function,* fifth edition, by A. Vander et al. Copyright © 1990 McGraw-Hill. Used by permission of The McGraw-Hill Companies; **7.11** From Knapp, *Perspectives in Human Biology,* Brooks/Cole, 1998; **7.13** From Knapp, *Perspectives in Human Biology,* Brooks/Cole, 1998; **7.14** (a) Gary Head; (b) Courtesy of Dr. Michael S. Donnenberg; (c) © Stanley Flegler/Visuals Unlimited; (d) © P. Hawtin, University of Southampton/SPL/Photo Researchers, Inc.; **7.15** (a) © Michael A. Keller/FPG/Getty Images; (b) © Scott Camazine/Photo Researchers, Inc.; **7.17** (a, b) © Ralph Pleasant/FPG/Getty Images; (c) © 2001 PhotoDisc, Inc.; (d) © Paul Poplis Photography, Inc. Stock Food America; (e) © 2001 PhotoDisc, Inc.; (f) © 2001 PhotoDisc, Inc.; (g) Gary Head; **7.18** Gary Head; **Page 142** © Nancy J. Pierce/Photo Researchers, Inc.; **7.20** © Reuters NewsMedia/Corbis.

CHAPTER 8   Pages 143, 147 © David Scharf/Peter Arnold, Inc.; **8.1** © National Cancer Institute/Photo Researchers, Inc.; **8.2** © 2001 Eyewire, modified by Lisa Starr; (art) After *Bloodline Image Atlas,* University of Nebraska-Omaha, and Sherri Wicks, Human Physiology and Anatomy, University of Wisconsin Web Education System, and others; **8.3** © David Scharf/Peter Arnold, Inc.; **8.4** From Sherwood, *Human Physiology from Cells to Systems,* 4th ed; **Page 148** (table) From Cummings, *Human Heredity* 6th ed., Table 15.5, p. 380; **8.5** (a) From Cummings, *Human Heredity* 7th ed., Brooks/Cole, 2006; (b) © Lester V. Bergman & Associates, Inc.; **8.6** After *Principles of Anatomy and Physiology,* sixth edition, by Gerard J. Tortora and Nicholas P. Anagnostakos. Copyright 1990 Biological Sciences Textbooks, Inc., A & P Textbooks, Inc., and Elia-Sparta, Inc. Reprinted by permission of John Wiley & Sons Inc.; © 2006 Alliance Pharmaceutical Corp. All Rights Reserved; **8.7** © A. Ramey/PhotoEdit; **8.8** From Solomon et al., *Biology* 7th ed., Brooks/Cole, 2006; © Prof. P. Motta, Dept. of Anatomy, Univ. of La Sapienza, Rome/SPL/Photo Researchers, Inc.; **8.9** From Knapp, *Perspectives in Human Biology,* Brooks/Cole, 1998; **8.10** (a, b) © Stanley Flegler/Visuals Unlimited; (c) © Moredum Animal Health, Ltd./Photo Researchers, Inc.; **8.11** From Maslak, P., Blast Crisis of Chronic Myelogenous Leukemia (posted online December 5, 2001). ASH Image Bank. Copyright American Society of Hematology, used with permission; **8.12** © Oliver Meckes/Ottawa/Photo Researchers, Inc.; **Page 156** © Dennis Kunkel/Phototake.

CHAPTER 9   9.4 © Lester V. Bergman/Corbis; **9.6** © Jose Palaez, Inc./Corbis; **Page 163** Lewis L. Lainey; **9.9** © Shiela Terry/SPL/Photo Researchers, Inc.; **9.11** Based on *Basic Human Anatomy* by A. Spence, Benjamin-Cummings, 1982; **9.13** (a) © Dr. John D. Cunningham/Visuals Unlimited; **9.14** (a) © Biophoto Associates/Photo Researchers, Inc.; (b) Lennart Nilsson from *Behold Man* © 1974; by Albert Bonniers. Forlag & Little Brown & Co., Boston; (c) From Sherwood, *Human Physiology* 5th ed., Brooks/Cole, 2004; **9.16** From Sherwood, *Human Physiology* 5th ed.,

Brooks/Cole, 2004; **9.17** (a) Ed Reschke; (b) © Biophoto Associates/Photo Researchers, Inc.; **Page 174** © Stone/Getty Images.

CHAPTER 10   Page 175 (top right) © K.G. Murti/Visuals Unlimited; (bottom left) © Lowell Tindell; **10.1** After *Bloodline Image Atlas,* University of Nebraska-Omaha, and Sherri Wicks, Human Physiology and Anatomy, University of Wisconsin Web Education System, and others; **10.4** (a) © Kwangshin Kim/Photo Researcher, Inc.; (b) www.zahnarzt-shuttgart.com; **10.5** © Biology Media/Photo Researchers, Inc.; **10.6** Robert R. Dourmashkin, Courtesy of Clinical Research Center, Harrow, England; **10.8** © NSIBC/SPL/Photo Researchers, Inc.; **10.14** © Dr. A. Liepins/SPL/Photo Researchers, Inc.; **Page 187** Courtesy of MU Extension and Agricultural Information; **10.16** © Matt Meadows/Peter Arnold, Inc.; **10.17** © Simon Fraser/Photo Researchers, Inc.; **10.18** (a) © David Scharf/Peter Arnold, Inc.; **10.19** © James Stevenson/Photo Researchers, Inc.; **Page 193** © Larry Williams/Corbis; **10.20** © Dr. P. Marazzi/Photo Researchers, Inc.; **Page 194** © The Granger Collection, New York.

CHAPTER 11   Page 195 (top right) © James Stevenson/Photo Researchers, Inc.; (bottom left) © Ariel Skelley/Corbis; **11.2** © CNRI/SPL/Photo Researchers, Inc.; **11.3** (a–e) Courtesy of Kay Elemetrics Corp.; modified from *Human Anatomy and Physiology,* fourth edition, by A. Spence and E. Mason, Brooks/Cole, 1992; **11.6** (a) © Galen Rowell/Peter Arnold, Inc.; (b) Ron Romanosky; **11.7** (a, b) © SIU/Visuals Unlimited; **11.10** (a) © R. Kessel/Visuals Unlimited; **11.14** Lennart Nilsson from *Behold Man* © 1974 by Albert Bonniers. Forlag & Little Brown & Co., Boston; **11.15** (a, b) © O. Auerbach/Visuals Unlimited; **11.16** © Larry Mulvehill/Photo Researchers, Inc.; **11.17** Courtesy of Dr. Joe Losos; **Page 210** © Don Hopey/Pittsburgh Post-Gazette, 2002, all rights reserved. Reprinted with permission.

CHAPTER 12   Page 211 © Ed Kashi/Corbis; **Page 213** U.S. Department of Agriculture; **12.1** (top) © Greg Mancuso/Jeroboam; (right) Gary Head; **12.8** From Sherwood, *Human Physiology* 5th ed., Brooks/Cole, 2004; **Page 220** Gary Head; **12.11** © 2000 David Joel/Stone/Getty Images; **Page 224** © David Jennings/The Image Works.

CHAPTER 13   Pages 225 (right), **227** © Manni Mason's Pictures; **Page 225** (left) © Empics; **13.1** © Manfred Kage/Peter Arnold, Inc.; **13.2** © Nancy Kedersha/UCLA/Photo Researchers, Inc.; **13.8** (a) © Don Fawcett, Bloom & Fawcett, 11/e, after J. Desaki & Y. Uehara/Photo Researchers, Inc.; **Page 231** (bottom right) © Cordelia Molloy/Photo Researchers, Inc.; **Page 235** © Wiley/Wales/ProFiles West/Index; **13.15** © Manfred Kage/Peter Arnold, Inc.; **13.17** Yokochi and J. Rohen, *Photographic Anatomy of the Human Body,* 2/e. Igaku-Shoin, Ltd., 1979; **13.18** © Colin Chumbley/Science Source/Photo Researchers, Inc.; **13.19** (a) From James W. Kalat, *Introduction to Psychology,* third edition, Brooks/Cole Publishing Company, 1993; (b) © Marcus Raichle, Washington University of Medicine; **13.20** After Wilder Penfield and Theodore Rasmussen, *The Cerebral Cortex of Man,* 1950, Macmillan Publishing Company; © renewed by Theodore Rasmussen; © Colin Chumbley/Science Source/Photo Researchers,

Inc.; **13.23** (b) © David Stocklein/Corbis; **13.24** (b) From Sherwood, *Human Physiology* 5th ed., Brooks/Cole, 2004; (c) © Lauren Greenfield/Corbis Sygma; **13.25** (a) © Bettmann/Corbis; (b) © S. Carmona/Corbis; **13.26** (a) Edy the D. London et al., *Archives of General Psychiatry,* 47:567–574 (1990); (b) Edy the D. London et al., *Archives of General Psychiatry,* 47:567–574 (1990); (c) © Ogden Gigli/Photo Researchers, Inc.; **13.27** From *Perspectives in Human Biology* by L. Knapp, Wadsworth, 1998.

CHAPTER 14   Pages 249, 254 © Ted Beaudin/FPG/Getty Images; **14.1** From Hensel and Bowman, *Journal of Physiology,* 23:564-568, 1960; (a) © David Turnley/Corbis; **14.3** © Colin Chumbley/Science Source/Photo Researchers, Inc.; after Wilder Penfield and Theodore Rasmussen, *The Cerebral Cortex of Man,* 1950, Macmillan Publishing Company; © renewed by Theodore Rasmussen; **14.4** From Knapp, *Perspectives in Human Biology,* Brooks/Cole, 1998; **14.6** © Omikron/SPL/Photo Researchers, Inc.; **Page 255** Sue Hartzell; **14.9** (a) © Fabien/Corbis Sygma; (e, top) Dr. Thomas R. Van DeWater, University of Miami Ear Institute; **14.10** From Solomon et al., *Biology* 7th ed., Brooks/Cole, 2006; **14.12** © AFP Photo/Timothy A. Clary/Corbis; **14.13** (a) Robert E. Preston, courtesy Joseph E. Hawkins, Kresge Hearing Research Institute, University of Michigan Medical School; (b) Robert E. Preston, courtesy Joseph E. Hawkins, Kresge Hearing Research Institute, University of Michigan Medical School; **14.14** © Ted Beaudin/FPG/Getty Images; **14.15** © Richard Megna/Fundamental Photographs; **14.17** Lennart Nilsson © Boehringer Ingelheim International GmbH; **14.18** © Ophthalmoscopic image from Webvision: http://webvision.med.utah.edu/; **14.21** (a, b) Gerry Ellis/The Wildlife Collection; **14.22** © Dr. P. Marazzi/Photo Researchers, Inc.

CHAPTER 15   Page 269, 270, 285 © Peg Skorpinski; **15.5** (a) Courtesy of Dr. Erica Eugster; (b) Courtesy of Dr. William H. Daughaday, Washington University School of Medicine, from A.I. Mendelhoff and D.E. Smith, eds., *American Journal of Medicine,* 20:133 (1956); **15.6** Gary Head; **15.7** © Scott Camazine/Photo Researchers, Inc.; **15.11** © Yoav Levy/Phototake; **Page 284** Evan Cerasoli; **15.12** Thomas Zimmerman/FPG/Getty Images; **15.13** (a, b) © James King-Holmes/Photo Researchers, Inc.; **Page 288** Courtesy of G. Baumann, M.D., Northwestern University.

CHAPTER 16   Page 289 © Don Fawcett/Photo Researchers, Inc.; **Page 291** © Laura Dwight/Corbis; **16.3** (b) Ed Reschke; **16.6** Lennart Nilsson from *A Child Is Born,* © 1966, 1977 Dell Publishing Co., Inc.; **Page 301** © David Frazier/Photo Researchers, Inc.; **16.9** Lennart Nilsson from *A Child Is Born,* © 1966, 1977 Dell Publishing Co., Inc.; **16.11** From Solomon et al., *Biology* 7th ed., Brooks/Cole, 2006; **Page 303** © Andy Walker, Midland Fertility Services/Photo Researchers, Inc.; **16.12** (right) © Lester Lefkowitz/Corbis; (left) © Sandy Roessler/FPG/Getty Images; **16.13** From Cummings, *Human Heredity* 7th ed., Brooks/Cole, 2006; **16.14** (a) © David M. Phillips/Visuals Unlimited; (b) © CNRI/SPL/Photo Researchers, Inc.; **16.15** (a) © Biophoto Associates/Photo Researchers, Inc.; (b) © CNRI/Photo Researchers, Inc.; **16.16** © Zeva Oelbaum/Peter

Arnold, Inc.; **16.17** (a) © NIBSC/Photo Researchers, Inc.; (a–f) After *Molecular and Cellular Biology* by Stephen L. Wolfe, Wadsworth, 1993; **16.18** (a) © George Musil/Visuals Unlimited; (b) © Kenneth Greer/Visuals Unlimited; **16.19** (a) © E. Gran/SPL/ Photo Researchers, Inc.; (b) © David M. Phillips/ Visuals Unlimited; **16.20** ©Todd Warshw/Getty Images; **16.21** Lennart Nilsson from *A Child Is Born*, © 1966, 1977 Dell Publishing Co., Inc.

CHAPTER 17 **Pages 313** (right), **315** (top right), **327, 332** Lennart Nilsson from *A Child Is Born*, © 1966, 1977 Dell Publishing Co., Inc.; **Page 313** (bottom left) © 1999 Dana Fineman/Corbis Sygma; **17.2** Lennart Nilsson from *A Child Is Born*, © 1966, 1977 Dell Publishing Co., Inc.; (art) Adapted from *Mechanisms of Development*, by R.G. Ham and M.J. Veomett, C.V. Mosby Co., 1980; **Page 315** (lower right, top) Courtesy of Amy Waddell/UCLA's Mattel Children's Hospital; (lower right, bottom) © Reuters NewMedia/ Corbis; **Page 317** © Dennis Degnan/Corbis; **17.6** After B. Burnside, *Developmental Biology*, 1970, 26:416–441. Copyright © 1971 by Academic Press, reproduced by permission of the publisher; **17.7** (a, b) Courtesy of Dr. Kathleen K. Sulik, Bowles Center for Alcohol Studies, the University of North Carolina at Chapel Hill; (c) © John DaSiai, MD/Custom Medical Stock Images; **17.10** (a–c) Lennart Nilsson from *A Child Is Born*, © 1966, 1977 Dell Publishing Co., Inc.; **17.11** After Patten, Carlson & others; **17.12** Lennart Nilsson from *A Child Is Born*, © 1966, 1977 Dell Publishing Co., Inc.; **Page 326** Evan Cerasoli; **17.16** © Sandy Roessler/FPG/Getty Images; **17.17** Modified after *The Developing Human: Clinically Oriented Embryology*, fourth edition, by Keith L. Moore, W.B. Saunders Co., 1988; (a) © Zeva Oelbaum/Corbis; (b) James Hanson, M.D.; **17.18** Fran Heyl Associates; **17.19** Lennart Nilsson from *A Child Is Born*, © 1966, 1977 Dell Publishing Co., Inc.; **17.20** Lisa Starr; **17.21** Gary Head; adapted from *Developmental Anatomy* by L.B. Arey, W.B. Saunders Co., 1965; **17.22** Jose Carillo/PhotoFile; **17.23** (a) © G. Musil/ Visuals Unlimited; (b) © Cecil Fox/Science Source/ Photo Researchers, Inc.; (c) © Stone/Getty Images; **17.24** © Dr. E. Walker/Photo Researchers, Inc.; **Page 338** Lauren and Homer. Photography by Gary Head.

CHAPTER 18 **Page 339** © K.G. Murti/Visuals Unlimited; **18.1** From Ingraham and Ingraham, *Microbiology* 2nd ed., Brooks/Cole, 2000; **18.2** (a) © Sercomi/Photo Researchers, Inc.; (b) © Camr/B. Dowsett/SPL/Photo Researchers, Inc.; **18.4** (c) E.A. Zottola, University of Minnesota; **18.5** (a) © Lily Echeverria/Miami Herald; (b) ©APHIS photo by Dr. Al Jenny; **18.8** (a) © Stem Jems/Photo Researchers, Inc.; (top left) Center for Disease Control; (top right) Edward S. Ross; (b) © CAMR, Barry Dowsett/ SPL/Photo Researchers, Inc.; **18.9** (a) © Dr. P. Marazzi/SPL/Photo Researchers, Inc.; (b) © Larry Jensen/Visuals Unlimited; (c) © Jerome Paulin/ Visuals Unlimited; (d) © Oliver Meckes/Photo Researchers, Inc.; **18.10** (f) Micrograph Steven L'Hernault; (g) © Sinclair Stammers/Photo Researchers, Inc.; **18.11** © Kent Wood/Photo Researchers, Inc.; **18.12** © Oliver Meckes/Ottawa/ Photo Researchers, Inc.; **18.14** (a) University of Erlangen; (b) © Tony Brain and David Parker/SPL/ Photo Researchers, Inc.; **18.15** © Bettmann/Corbis; **Page 352** © Mednet/Phototake.

CHAPTER 19 **Pages 353** (right), **360, 368** Dr. Pascal Madaule, Paris; **Page 353** (bottom left) Courtesy of the family of Henrietta Lacks; **19.1** L. Willatt, East Anglian Regional Genetics Service/SPL/Photo Researchers, Inc.; (b) Courtesy of Carl Zeiss MicroImaging, Thornwood, NY; **19.6** (a–e) Andrew S. Bajer, University of Oregon; **19.7** (b–g) Jennifer W. Shuler/Science Source/Photo Researchers, Inc.; **19.8** (a) © D.M. Phillips/Visuals Unlimited; (b) © Jennifer W. Shuler/Science Source/Photo Researchers, Inc.; **19.9** (a) © John W. Gofman and Arthur R. Tamplin. From *Poisoned Power: The Case Against Nuclear Power Plants Before and After Three Mile Island*, Rodale Press, 1979; **19.12** (a–h) With thanks to the John Innes Foundation Trustees; **19.15** (left) Dr. Stan Erlandsen, University of Minnesota; (right) © Cytographics; **19.18** © Don W. Fawcett/Photo Researchers, Inc.

CHAPTER 20 **Page 373** Lisa Starr; **20.1** (right) C.J. et al., *Cytogenetics and Cell Genetics*, 35:21-27, © 1983, S. Karger, A.G. Basel; **20.2** (a) © Zuma Press; (b) © Fabian/Corbis Sygma; **20.3** (a, b) © Lisa O'Connor/ ZUMA/CORBIS; **20.11** (a) © Andrew Syred/Photo Researchers, Inc.; **20.12** (a) © Stanley Flegler/Visuals Unlimited; **20.13** © Jim Stevenson/SPL/Photo Researchers, Inc.; **20.14** (a, b) © Frank Cezus/FPG/ Getty Images; (c) © PhotoDisc/Getty Images; (d) © Ted Beaudin/FPG/Getty Images; (e) © Stan Sholik/ FPG/Getty Images; **20.15** (a, b) Courtesy of Ray Carson, University of Florida News and Public Affairs; **Page 383** Gary Head; **20.16** Dr. Stephen Liggett, University of Cincinnati. Photo by Daniel Davenport, University of Cincinnati; **20.17** Evan Cerasoli.

CHAPTER 21 **Pages 387** (right), **392, 395, 399** © SPL/Photo Researchers, Inc.; **Page 387** (bottom left) © Simon Fraser/RVI, Newcastle-upon-Tyne/SPL/ Photo Researchers, Inc.; **21.2** (a) © Charles D. Winters/Photo Researchers, Inc.; (b) © Omikron/ Photo Researchers, Inc.; **21.3** (a) © Eyewire, Inc.; (b) © PhotoDisc/Getty Images; after *Human Heredity: Principles and Issues*, third edition, by M. Cummings, p. 126, Brooks/Cole, 1994; **21.4** © Dr. Karen Dyer Montgomery; **Page 391** © Stapleton Collection/ Corbis; **21.5** Dr. Viktor A. McKusick; **21.6** Steve Uzzell; **21.9** © Bettmann/Corbis; **21.10** © Frank Trapper/Corbis Sygma; **21.12** After *Human Genetics*, second edition, by Victor A. McKusick, Copyright © 1969 by Prentice-Hall, Inc., Englewood Cliffs, NJ; (b) © Bettmann/Corbis; **21.13** W.M. Carpenter; **21.14** © Rivera Collection/Superstock, Inc.; **21.15** Courtesy of G.H. Valentine; **21.17** From "Multicolor Spectral Karyotyping of Human Chromosomes" by E. Schrock, T. Ried, et al., *Science*, 26 July 1996, Volume 273, pp 495. Used by permission of E. Schrock and T. Ried and The American Association for the Advancement of Science; **21.19** (a) © CNRI/Photo Researchers, Inc.; (b) © Lauren Shear/Photo Researchers, Inc.; **21.20** © UNC Medical Illustration & Photographs; **Page 404** (bottom left) © Rick Guidotti, Positive Exposure; (top right) © Reuters/CORBIS; (bottom right) © Hulton-Deutsch Collection/CORBIS.

CHAPTER 22 **UN p. 405** (right), **413, 417:** PDB ID:2 AAI; E. Rutenber, B.J. Katzin, S. Ernst, E.J. Collins, D. Mlsna, M.P. Ready, J.D. Robertus: Crystallographic Refinement of Ricin to 2.5 A. Proteins, 10 p 240 (1991); **Page 405** (bottom left) © Vaughan Fleming/ SPL/Photo Researchers, Inc.; **22.1** © A. Barrington Brown/Photo Researchers, Inc.; **22.2** (top) PDB ID: 1BBB; Silva, M. M., Rogers, P. H., Arnone, A.: A third quaternary structure of human hemoglobin A at 1.7-A resolution. J Biol Chem 267 pp. 17248 (1992); **22.3** Damon Biotech, Inc.; **22.5** (b) Shabaz A. Janqua, MD/Dermatlas; http://www.dermatlas.org; **22.7** (a) C.J. Harrison; (b) Photo courtesy of National Neurofibromatosis Foundation; **22.8** Lisa Starr; **22.11** tRNA model by Dr. David B. Goodin, The Scripps Research Institute; **22.15** © David Parker/ SPL/Photo Researchers, Inc.; **22.16** © TEK Image/ Photo Researchers, Inc.; **22.17** © Volker Steger/ SPL/Photo Researchers, Inc.; **22.18** (a) © Alfred Pasieka/SPL/Photo Researchers, Inc.; (b) © Colin Cuthbert/SPL/Photo Researchers, Inc.; **22.20** After *Human Heredity*, fifth edition by M. Cummings, p. 332, Brooks/Cole, 2000; **22.21** © Empics; **22.22** Cellmark Diagnostics; **22.23** (a) © AP/Wide World Photos; (b) © Adrian Arbib/Still Pictures; **22.24** Patrick J. Endres/Visuals Unlimited **22.25** (a) R. Brinster & R.E. Hammer, School of Veterinary Medicine, University of Pennsylvania; (b) Dr. Vincent Chiang, School of Forestry & Wood Products, Michigan Technology University; **22.27** (a) © Empics; (b) © Mike Stewart/Corbis Sygma; **22.28** © Bettmann/Corbis.

CHAPTER 23 **Page 427, 429** Miles-Peninsula Hospitals; **23.2** Lennart Nilsson © Boehringer Ingelheim International GmbH; **23.6** © Tony Freeman/PhotoEdit; **23.7** © Paul Shambroom/Photo Researchers, Inc.; **23.8** From American Cancer Society's Cancer Facts and Figures—2000. Reprinted by permission of the American Cancer Society, Inc.; **23.9** © Billy E. Barnes/Jeroboam; **23.10** (a) Miles-Peninsula Hospitals; **23.11** © AFP/Pascal Parani/ Corbis; **23.12** © S. Benjamin/Custom Medical Stock Photo; **23.13** (a) © Biophoto Associates/Photo Researchers, Inc.; (b) © James Stevenson/SPL/Photo Researchers, Inc.; (c) © Biophoto Associates/Science Source/Photo Researchers, Inc.; **Page 442** © Science Photo Library/Photo Researchers, Inc.

CHAPTER 24 **Page 443** NASA Galileo Imaging Team. Art by Don Davis; **24.1** (a) © Dieter & Mary Plage/Survival Anglia/Oxford Scientific; (b) Portrait of Charles Darwin (1809–1882), 1840 (w/c on paper), Richmond, George (1809–1896)/Down House, Kent, UK/The Bridgeman Art Library; **24.2** (a) © Peter Bowater/Photo Researchers, Inc.; (b) © Sam Kleinman/Corbis; (c) © Owen Franken/Corbis; (d) © Christopher Briscoe/Photo Researchers, Inc.; **24.3** Alan Solem; **Page 446** © Layne Kennedy/Corbis; **24.4** (b) Art by Jean-Paul Tibbles. *The Book of Life*, © 1993 Ebury Press; **24.5** (a) © Martin Land/Photo Researchers, Inc.; (b) Department of Geosciences, Princeton University; **24.6** © Danny Lehman/Corbis; **24.7** NOAA/NGDC; (b) After A.M. Ziegler, C.R. Scotese, and S.F. Barrett, "Mesozoic and Cenozoic Paleogeographic Maps," and J. Krohn and J. Sundermann (Eds.), *Tidal Frictions and the Earth's Rotation II*, Springer-Verlag, 1983; **Page 449** David A. Kring/NASA/University of Arizona Space Imagery Center; **24.9** (b) Adapted from *General Zoology*, sixth edition, by T. Storer et al. Copyright © 1979 McGraw-Hill; **24.14** Steve Hillebrand, U.S. Fish & Wildlife Service; **Page 454** (table) Bruce Coleman Ltd.; **24.15** (b) © Dallas Zoo, Robert Cabello; (c) Courtesy of Dr. Takeshi Furuichi, biology, Meiji-Gakuin

University, Yokohama; **24.17** (a) Allen Gathman, Biology Dept., Southeast Missouri State University; (b) Bone Clones®, www.boneclones.com; (c) Gary Head; **24.18** (a, left) Sahelanthropus tchadensis, MPFT/Corbis Sygma; (a, right) Sahelanthropus tchadensis, MPFT/Corbis Sygma; (b) Dr. Donald Johanson, Institute of Human Origins; (c) Louise M. Robbins; (d) Lisa Starr; (e) National Museum of Ethiopia, Addis Ababa. © 1985 David L. Brill; **24.19** (a, b) Lisa Starr; **24.20** (left) NASA; (right) National Museum of Ethiopia, Addis Ababa. © 1985 David L. Brill; **Page 457** Douglas Mazonowicz/ Gallery of Prehistoric Art; **24.21** Painting by Chesley Bonestell; **24.22** Stanley W. Awramik; **24.23** Sidney W. Fox; **24.24** © Kjell B. Sandved/Visuals Unlimited; **24.25** © Gulf News, Dubai, UAE; **24.26** © Elliot Erwitt/Magnum Photos, Inc.

CHAPTER 25 **Pages 463, 480** © Tom Till/Stone/ Getty Images; **25.1** (a) © George H. Huey/Corbis; (b) Jack Carey; **25.2** (a, b, d) Roger K. Burnard; (c) © Darrell Gulin/Corbis; **25.3** (a) © Doug Peebles/ Corbis; (b) © Pat O'Hara/Corbis; (c, d) © Tom Bean/ Corbis; (e) © Duncan Merrell/Taxi/Getty Images; **25.6** NASA's Earth Observatory; **25.8** Gerry Ellis/ The Wildlife Collection; **25.14** © Kevin Schafer/Getty Images; **25.17** Gary Head; **25.18** © David T. Grewcock/ CORBIS; **25.19** NASA; **25.20** © Antoinette Jongen/ FPG/Getty Images; **25.23** NASA; **25.24** © George H. Huey/Corbis; **25.25** © R. Bieregaard/Photo Researchers, Inc.; after G.T. Miller, Jr., *Living in the Environment*, eighth edition, Brooks/Cole, 1993. All rights reserved; **25.26** (a) © Alex MacLean/ Landslides; **25.27** © Erwin Christian/ZEFA/CORBIS; **25.28** © Richard Parker/Photo Researchers, Inc.; **Page 488** E.F. Benfield, Virginia Tech.

Index

The letter i *designates illustrations;* t *designates tables.*

Erythropoietin, 147, 147i, 156, 214, 277t
Escherichia coli, 130, 132, 132i, 344
Esophagus, 68t, 120, 120i, 121, 122, 123i, 124, 124i, 197, 201, 206i, 434t, 435
Essential amino acids, 134
Essential fatty acids, 134
Estrogen, 26i, 31, 270–272, 277t, 280, 285, 294–298, 301, 317, 320, 326, 327, 333, 411, 436
Estrogen receptor, 89, 191
Ethmoid bone, 92, 92i
Eugenic engineering, 373
Eukarya, 3, 3i, 454t
Eukaryotic cells, 42, 42i, 43, 44–45, 44i, 48, 49, 52, 346, 354
Euphoria, 199
Eustachian tube, 257, 259
Eutrophication, 471
Evaporation, 212, 212t
Evaporative heat loss, 83
Evolution, 3, 9, 136, 443–462. *See also* Human evolution
Excitatory postsynaptic potentials (EPSPs), 231
Exhalation. *See* Expiration/exhalation
Exocrine cells, 126, 126i, 282, 387
Exocrine glands, 68, 185, 270
Exocytic vesicles, 51i, 57i
Exocytosis, 56–57, 57i, 129, 129i
Exons, 411, 418
Expansion mutation, 409
Experimental group, 7, 7i, 8
Experimentation, 8
Experiments, 6–7, 7i, 8, 14, 416
Expiration/exhalation, 200, 200i, 201i, 204, 204i
Expiratory reserve volume, 201, 201i
Extension of joints, 97i
External environment, 84, 249
External respiration, 202
External stimuli, 77i
External urethral sphincter, 217
Extinct species, 484
Extinction, 449, 452, 453, 485
Extracellular fluid (ECF), 80, 80i, 144, 144i, 159, 169, 212–213, 213i, 218, 220, 221, 239, 250t, 280
Extraembryonic membranes, 320, 320i, 336t
Extrinsic clotting mechanism, 153
Eye muscles, 235i, 261
Eye sockets, 92
Eye tissues, 187
Eyeballs, 260, 260t, 262, 263i, 264, 265
Eyes, 68t, 112, 134, 236i, 238, 240, 241i, 242i, 251, 252i, 258–261, 260t, 324, 340
  aging and, 335
  cataracts in, 265, 361, 481
  disorders in, 261, 264–265, 264i, 265i
  gene effects on, 382, 382i, 383

infections in, 264–265, 304
iris scanning technology for, 249
teratogen vulnerability in, 328i
vision sense, 260–261, 261i, 262–263

F

F_1. *See* First filial generation (F_1)
F_2. *See* Second filial generation (F_2)
Facial bones, 92
Facial muscles, 104, 114, 235i
Facial nerves, 235
Facial wrinkles, 111
Facilitated diffusion, 56, 56i
Fact memory, 242, 242i
Factor IX, 396
Factor VIII, 396, 423
Factor X, 152
FAD. *See* Flavin adenine dinucleotide (FAD)
Fainting, 268
Fallopian tube, 294, 303. *See also* Oviducts
"False feet," 319
Familial hypercholesterolemia, 395
Family (classification), 454t
Farsightedness, 261, 264, 264i, 335
FAS. *See* Fetal alcohol syndrome (FAS)
Fascicles of muscles, 72
Fast/white muscle cells, 107, 107i
Fat, 22, 28, 30–31, 30i, 44, 44i, 63, 78, 119, 124–128, 131, 131i, 134, 135, 144i, 179, 280, 282, 331
  adipose tissues store, 71, 71i
  aging and, 332, 333
  animal, 30, 134, 270
  brown, 83
  calories and, 138, 139t
  diet and, 134, 135
  glucose disorders and, 283
  malabsorption disorders and, 133
  neutral, 30
  vitamins and, 136t
Fat cells, 63, 285
Fat deposits, 127, 130i, 294
Fat digestion, 31, 127t, 128
Fat metabolism, 120i
Fat-soluble vitamins, 134, 136t, 137, 144
Fat-storing tissues. *See* Adipose tissues
Fatty acids, 26i, 30, 30i, 51, 63, 112, 128, 128t, 129, 129i, 130, 134, 136t, 277t, 280, 281i
Faulty enamel trait, 397, 397i
FDA. *See* United States Food and Drug Administration (FDA)
Fecal contamination, 130
Fecal occult blood test, 433t
Feces, 120i, 127, 130, 132, 134, 212, 212t, 338, 346

Feedback loop, 242, 278i, 293, 294
Feedback mechanisms, 80–81, 277
Feeding levels, ecosystem, 466, 467i
Female condom, 301
Femoral artery, 158i
Femoral vein, 158i
Femur, 88, 88i, 89, 90i, 91i, 95, 95i, 96i, 99, 276. *See also* Thighbone
Fermentation, 60, 63, 63i
Fertile period, 300, 316
Fertility, 300–301, 333, 400
"Fertility awareness," 300, 312
Fertility drugs, 302, 313
Fertility rate, 478, 479i
Fertilization, 77i, 290, 292, 294, 294t, 295i, 297, 299, 302, 303, 320, 322i, 326, 328, 331t, 372i, 390, 390i, 400, 401i
  early development and, 314, 315i, 316–317, 316i, 317i, 318
  in vitro, 302–303, 302i, 330, 332
Fertilized egg, 67
Fertilizers, 45
Fetal alcohol syndrome (FAS), 329, 329i
Fetal cells, 330, 330i
Fetal circulation, 321i, 327
Fetal period, 325i, 328i
Fetoscopy, 330, 330i
Fetus, 67, 77i, 81, 96, 143, 280, 326, 326i, 331t, 400
  birth control and, 301
  cystic fibrosis and, 387
  detecting birth defects and, 330, 330i
  developing, 185, 324–325
  development disorders and, 328, 329
  malformed, 136t
  prenatal development and, 321, 323, 325i
  Rh blood typing and, 150, 150i
  toxoplasmosis and, 338
Fever-reducing drugs, 83
Fever, 83, 154, 155, 173, 176, 177t, 180, 181, 185, 190, 207, 244, 305, 339, 340, 341, 342, 347, 347i, 438
Fiber, 22, 29, 133, 133i, 134
Fibrillin, 395
Fibrin, 152, 152i
Fibrinogen, 144i, 152, 152i
Fibroblasts, 70i, 333
Fibrocartilage, 71, 93
Fibrous connective tissues, 70, 71t, 96, 127
Fibrous joints, 96
Fibrous tunic, 260t
Fibula, 91i, 95, 95i, 96i
Fight-flight response, 237, 280, 281
Filtration, in nephrons, 216, 216i, 216t, 217i
Fimbriae, 297
Finger bones, 91i, 94. *See also* Phalanges

Finger joints, 106
First filial generation (F_1), 176, 377i
First polar body, 296i, 297
Fission, 344i
Flagella, 53, 53i, 292
Flat bones, 90
Flavin adenine dinucleotide (FAD), 59, 60, 61i
Flex points, 93
Flexibility, 94
Flexion, of joints, 97i
Flexor digitorum superficialis, 118i
Flu symptoms, 207, 306, 308, 338, 345
Flu virus, 182, 340, 342
Fluoride, 15
Fluorine, 15, 137t
Foam spermicide, 300i
Folate, 136t, 328. *See also* Folic acid
Folic acid, 154, 328. *See also* Folate
Follicle, hair, 78, 78i, 79, 252, 333, 417
Follicle, ovarian, 296, 296i, 297, 297i, 298i, 302
Follicle-stimulating hormone (FSH), 272t, 274t, 275, 292, 293, 293i, 295, 296, 297, 297i, 298i, 301, 302, 333
Fontanels, 96
Food, ecology and, 463, 464, 482, 483, 485
Food allergies, 190
Food and Drug Administration, 103
Food-borne illness, 344, 361
Food chain, example of, 466i
Food groups, servings in, 135
Food poisoning, 132
Food processing, 120, 122–123
Food production, 37
Food proteins, 59
Food pyramid, 135
Food supplements, 103
Food webs, 466, 467i, 468, 472, 473i, 476i, 477, 477t
Foot bones, 95
Foramen magnum, 92, 92i, 455, 455i
Foramen ovale, 325, 325i, 327
Forearm, 105i, 107i, 252i, 450
Forearm muscles, 113, 118i
Forebrain, 238, 239, 239i, 274, 322i
Formaldehyde, 459
Fossil fuels, 25, 473, 473i, 481, 484
Fossil record, 447, 448, 456
Fossilization, 448
Fossils, 443, 448–449, 448i, 450, 456, 456i, 457, 457i, 458
Founder effect, 446
Fovea, 260i, 260t, 262, 263i
Fractures, 94, 98i, 99
Fragile X syndrome, 409, 409i
Fraternal twins, 317
Free nerve endings, 252

Nondisjunction, 400–401, 400i, 401i
Nonhomologous chromosome, 399i
Nonpolar covalent bond, 20, 21, 21i
Nonpolar molecules, 54, 54i
Nonpolar substances, 23, 30
Nonshivering heat production, 83
Nonsister chromatids, 366, 366i
Nonsteroid hormones, 272
Norepinephrine, 237, 244, 245, 271i, 272t, 277t, 280, 281, 284
Norplant, 300i, 301
Nosocomial infection, 348
Notochord, 318
Noxious substances, respiratory disorders and, 206
Nuclear envelope, 44i, 45i, 48–51, 273i, 343i, 358, 358i, 359, 359i, 364i, 365i, 368i, 414i
Nuclear medicine, 17
Nuclear membrane, 48
Nuclear pores, 49, 49i
Nuclear power, 484
Nucleic acid metabolism, 136t
Nucleic acids, 26, 36, 51, 126, 128, 136t, 137t, 213, 342, 342i, 405, 410, 476
Nucleolus, 44i, 45i, 48i, 48t, 49
Nucleoplasm, 44i, 48, 48i, 48t, 49i
Nucleosome, 356, 356i
Nucleotide bases, 128t, 406–407, 407i, 408, 412i, 418
Nucleotide sequence, 407, 408, 409, 411, 417, 418, 452
Nucleotides, 26i, 36, 36i, 58, 128t, 398, 406–412, 417, 418, 459, 459i
Nucleus, 19, 42t, 44i, 45i, 48–49, 48i, 48t, 49i, 50i, 71i, 73, 104, 147, 226, 239, 272, 273i, 292, 354, 356, 357i, 360, 391i
  atomic, 16, 18
  cancer and, 428, 433i
  genetic code and, 413
  making proteins and, 410, 411, 414i
  meiosis and, 362, 364i, 365i, 368, 369i
  mitosis and, 358, 359, 368, 369i
  prokaryotic vs. eukaryotic, 42
Nutrient molecules, 121, 127, 129
Nutrients, 51i, 88, 115, 119–121, 125, 126, 128–131, 154, 159, 166, 198i, 211, 213, 213i, 217i, 255, 316
  cancer cells and, 429
  development disorders and, 328
  diet and, 134
  digestive system disorders and, 132
  gastric bypass surgery and, 124
  malabsorption disorders and, 133

managing, 131i
  prenatal development and, 320–321, 324, 325, 325i
Nutrition, 79, 119, 120, 133–139, 136t, 137t, 328, 335
NuvaRing, 301

O

Obesity, 87, 119, 124, 138, 165i, 167, 170, 170t, 283, 285, 436
Oblique muscle, 124i
Observations, 6, 7, 9
Occipital bone, 92, 92i
Occipital lobes, 240, 241i
Octuplets, 313
Odor molecules, 254
Odors, 249
OI. See Osteogenesis imperfecta (OI)
Oil glands, 68, 79, 333
Old age, 331t
Oleic acid, 30i
Olestra, 6, 7
Olfaction, 254–255
Olfactory bulbs, 239, 254, 255i
Olfactory epithelium, 254, 255
Olfactory nerve, 235i
Olfactory nerve tract, 255i
Olfactory neurons, 254
Olfactory receptors, 250t, 254, 255, 255i, 289
Oligodendrocytes, 232, 235
Oligosaccharides, 28, 35, 128t, 176
Omnivores, 466, 467i, 468i, 469i
On the Origin of Species, 446
Oncogenes, 430, 431
Oocytes, 294–297, 298i, 302, 303, 316, 332, 346, 355, 362, 363, 368, 372, 415
Oogenesis, 362, 363i, 372
Oogonia, 354, 355
Oogonium, 363i
Opiates, 253
Opposing interaction, 270
Opsin, 262, 397
Optic chiasma, 271i
Optic disk, 260i
Optic nerves, 235i, 236i, 239i, 251, 260, 260i, 260t, 262, 263, 263i, 265
Oral cancer, 438
Oral cavity, 120, 120i, 122, 196i. See also Mouth
Oral contraceptives, 300i, 301
Orbitals, 18, 19i
Order, 454t
Organ of Corti, 257, 257i
Organ systems, 4, 4i, 67, 75–81, 119, 121i, 198i, 211, 213i, 225, 318, 319, 323–325, 341, 380, 438
Organ transplants, 187
Organelle membranes, 50i
Organelles, 41, 42, 43, 44, 48, 50, 50i, 52, 52i, 115, 147, 187, 226, 344, 354
Organic compounds, 26, 27, 35, 36, 37, 52, 58, 458, 459
Organic molecules, 59, 129, 459

Organogenesis, 314, 318
Organs, 4, 4i, 42, 43, 52, 63, 70, 70i, 72, 74–76, 89i, 90, 104, 120, 121, 126–127, 146, 162i, 163, 169, 179, 236i, 237, 241, 250t, 252, 253i, 274, 284, 327. See also Specific organ systems
  abdominal, 82i
  aging and, 332, 332i, 334, 335
  defined, 67
  development disorders and, 328i, 329
  diet and, 134, 135
  early development of, 314–315, 314i, 315i
  gas exchange and, 196i
  homeostasis and, 80–81, 80i
  hormone-producing, 270
  hypothermia and, 83
  infectious diseases and, 341, 346
  internal (See Internal organs)
  pig, 187
  prenatal development and, 319, 320, 324, 325
  replacement, 78
  sensory, 91i, 232, 254, 260, 335
Orgasm, 299
Origin, of life, 106, 458–459, 459i
Orphan drugs, 41
Osmoreceptors, 250, 250t
Osmosis, 54, 55, 55i, 57, 129, 130, 144, 165, 216, 217i, 218, 280, 389
Osteoarthritis, 87, 98, 138, 333
Osteoblasts, 88, 89, 89i, 99
Osteoclasts, 89, 279
Osteocytes, 88, 88i
Osteogenesis imperfecta (OI), 99
Osteon, 88
Osteoporosis, 89, 89i, 124, 137, 137t, 333
Osteosarcoma, 99
OT. See Oxytocin (OT)
Otitis media, 259
Otolith organ, 258
Otoliths, 258, 258i
Outer ear, 256, 256i
Outer epidermal layer, 79i
Outer mitochondrial membrane, 52i
Output zones of neurons, 226, 226i, 230
Oval window, 256, 256i, 257i
Ovarian cancer, 420, 427, 433, 434t, 437
Ovarian cycle, 296–297, 296i, 297i, 298, 298i
Ovarian hormones, 294, 312
Ovaries, 68t, 69i, 253i, 271i, 272, 274t, 275i, 277t, 289, 354, 354t, 355, 390, 390i, 397, 401
  aging and, 333
  cancer of, 420, 427
  early development and, 316, 316i, 317
  menstrual cycle and, 298, 298i
  ovarian cycle and, 296, 296i, 297, 297i
  STDs and, 304
Overdose, drug, 225
Overexercise, 139

Overweight, 278. See also Obesity
Oviducts, 294–297, 299, 300, 303–305, 316, 316i, 317, 338, 372. See also Fallopian tube
Ovulation, 294, 295t, 296–297, 296i, 297i, 298i, 299–302, 316i, 372
Ovum, 297, 316, 363i. See also Eggs
Oxidation, 22
Oxygen, 15, 18, 19, 19i, 20, 21, 21i, 54, 54i, 77i, 121, 121i, 159, 162, 163, 195, 211, 213, 213i, 223
  aging and, 335
  antioxidants and, 22
  ATP production and, 60, 61, 61i, 62i, 63
  blood disorders and, 154
  blood transporting, 146, 146i
  blood vessels and, 166, 167, 168, 169
  in body, 15t, 26
  breathing and, 200, 205, 205i
  cancer cells and, 429
  early Earth and, 448, 458
  in Earth's crust, 15t
  in ecosystems, 469
  exercise and, 115
  gas exchange and, 196–197, 198, 198i, 199, 199i, 202–203, 203i
  mitochondria and, 52
  muscle contraction and, 112, 112i
  myoglobin and, 107
  red blood cell production and, 147, 147i
  sickle-cell anemia and, 380
  smoking and, 206i
  in water, 22, 22i
Oxygen debt, 112
Oxygen receptors, 205i
Oxygen-transporting protein, 34, 35i
Oxygent, 151
Oxyhemoglobin, 146, 202
Oxytocin (OT), 81, 270, 271i, 272t, 274–275, 274i, 274t, 284, 326, 327, 327i
Ozone, 9, 9i, 79, 474, 481, 481i, 485

P

P. See Parent generation (P)
Pacemaker cells, 164
Pacemakers, 17, 164, 204, 204i
Pacinian corpuscles, 252
Pain, 37, 79, 136t, 155, 171, 181, 181i, 206, 231, 244, 249, 250, 250i, 250t, 252–253, 253i, 283t, 295, 304, 312, 326, 381i, 438
Pain receptors, 250, 250i, 250t, 252. See also Nociceptors
Painkillers, 87, 93, 231, 244, 245
Palate, 92, 92i, 122, 254, 328i
Palatine bone, 92, 92i

muscle, 72, 72i, 103, 104–111, 115, 138, 169, 314, 397
nervous, 73, 90, 237, 274t, 314, 319
prenatal development and, 319i, 320i, 321i, 323i, 325, 325i
regeneration of, 127
scar, 114, 127, 207, 265, 295
Tobacco, 195, 206, 210, 435, 438
Toe bones, 91i, 400. *See also* Phalanges
Tolerance, to drugs, 245
Tongue, 121, 122, 123i, 136t, 197, 197i, 235i, 240, 241i, 250t, 252, 252i, 254, 254i, 276, 385, 438
Tonic water, 255
Tonicity, 55, 55i
Tonsils, 178i, 179, 254i
Tooth decay, 15, 132, 132i, 137t, 189
Tooth development, 31
Tooth sockets, 92
Total fertility rate (TFR), 478, 479i
Touch, 250t, 252
Toxemia, 155
Toxic chemicals, 255, 477, 477t
Toxic fumes, respiratory disorders and, 206
Toxic substances, 59
Toxic waste product, 127
Toxoplasmosis, 338
Trace elements, 15
Tracers, 17
Trachea, 68, 68t, 71, 122, 123, 123i, 196i, 197, 201, 278, 278i. *See also* Windpipe
Trachoma, 265
Traits
 bacterial, 344
 behavioral, 445
 beneficial, 446
 development/aging and, 317
 engineering, 416, 423, 423i
 evolution and, 444, 445, 445i, 446–447, 452, 454, 462
 human evolution and, 454
 inherited (*See* Inherited traits)
 morphological, 445
 neutral, 446
 nonsexual, 390
 physiological, 445
 polygenic, 382i, 383
Tranquilizers, 329
Trans fats. *See* Trans fatty acids
Trans fatty acids, 30, 134, 135, 170, 171
Transcription, 306, 307i, 410–411, 410i, 411i, 412i, 414, 424
Transfection, 420
Transfer RNA (tRNA), 410, 412–413, 413i, 414, 414i, 415i
Transfers, 26–27, 60, 61i
Transfusions, 148, 148t, 149i, 150, 151, 151i, 308, 380
Translation, 410, 411, 412, 412i, 413, 414–415, 414i–415i, 424
Translocation, 399, 399i
Transmission electron micrograph, 47i, 52i

Transmission electron microscope, 47, 47i
Transplantation, organ, 187
Transplanted tissues, 187
Transplants, 127, 187, 221, 265
Transport molecules, 60
Transport proteins, 32, 46, 46i, 48, 54, 56–57, 56i, 129, 217, 228, 231, 239
Transposable elements, 409
Transverse colon, 130, 130i
Transverse plane, 76i
Trapezius, 105i
Treponema pallidum, 305
Treponeme, 305
TRH. *See* Thyrotropin-releasing hormones (TRH)
Triceps, 106, 107i
Triceps brachii, 105i
Trichomonas vaginalis, 309, 309i
Trichomoniasis, 309, 309i, 312t
Tricuspid valve, 160, 160i
Trigger zone, 226, 226i, 227, 228, 228i, 231
Triglycerides, 30, 30i, 35, 63, 128, 128t, 129, 129i, 134, 142, 163, 283
Triiodothyronine (T$_3$), 271i, 272t, 277t, 278
Triple covalent bond, 20
Triplets, 313, 332
Trisomy 21, 400
tRNA. *See* Transfer RNA (tRNA)
Trophoblast, 316, 317i, 320
Tropical forests, 483, 485
Tropomyosin, 110, 111i
Troponin, 110, 111i
Troposphere, 481
Trunk, 105i, 158i, 162i, 236, 241i, 252i, 318
Trypanosoma brucei, 346, 346i
Trypsin, 128, 128t
Tryptophan (trp), 33i, 134, 134i, 412i
TSH. *See* Thyrotropin (TSH)
Tubal ligation, 300, 300i
Tubal pregnancy. *See* Ectopic/tubal pregnancy
Tuberculosis (TB), 194, 194i, 207, 306, 345, 345i, 348t
 skin test for, 194
Tubules, 220i. *See also* Distal tubule; Nephron tubule
Tufts University, 95
Tumor cells, 176, 177t, 187i, 419, 420, 421
Tumor markers, 433
Tumor necrosis factor, 176, 177t
Tumor suppressor genes, 430, 430i, 431i, 436
Tumors, 8, 133, 278, 285, 312, 409i, 419, 421, 428–430, 433, 435–438
Turner syndrome, 400–401, 401i
Twins, 79, 303, 315, 317, 332, 383, 421
Tympanic membrane, 256
Typhus, 477
Tyrosine, 394

U

Ulcers, 74, 124, 283t, 305, 305i
Ulna, 91i, 94, 94i. *See also* Forearm bone
Ulnar nerve, 234i
Ultrasound, 323, 330, 433
Ultraviolet light, 408, 432
Ultraviolet (UV) radiation, 22, 79, 432, 439, 481
Umami, 254, 255
Umbilical arteries, 325, 325i
Umbilical cord, 320, 321, 322, 322i, 325i, 326, 326i, 330, 330i
Umbilical vein, 325, 325i
United States Department of Agriculture (USDA), 135, 135i
United States Food and Drug Administration (FDA), 7, 11, 135, 136t, 137t, 138
Universal donors, 148
Universal recipients, 148
Unsaturated fats, 30i
Uracil, 410, 410i, 413
Urea, 63, 127, 127t, 213, 216, 216t, 217, 217i, 218
Ureter, 214, 214i, 221, 223, 253i
Urethra, 68t, 214, 214i, 217, 220, 221, 290i, 291, 292i, 294, 295i, 299, 308, 309, 335, 340
Uric acid, 213, 217, 221
Urinalysis, 211, 220, 223, 283
Urinary bladder, 72i, 214, 214i, 253i, 290i, 291i, 294i, 295i, 325i
Urinary disorders, 220
Urinary excretion, 212, 213
Urinary incontinence, 335
Urinary sphincter, 335
Urinary system, 75, 77i, 159, 159i, 211–224
 aging and, 335
 cancer in, 220, 438
 digestive system and, 121, 121i
 early development and, 314t
 gas exchange and, 198, 198i
 kidney disorders and, 221
Urinary tract, 180, 220, 304, 333, 348
Urinary tract infections, 220, 221, 345. *See also* Bladder infections
Urination, 80, 217, 221, 304, 305, 309
Urine
 formation of, 216–217, 216i, 216t, 217i
 pH of, 180
 pollutants in, 143
Urine leakage, 335
U.S. Environmental Protection Agency, 37
USDA. *See* United States Department of Agriculture (USDA)
USDA nutritional guidelines, 135
Uterine cancer, 434t, 437
Uterine cavity, 317i
Uterine lining, 277t, 437

Uterine muscles, 81
Uterine tumors, 8
Uterus, 69i, 81, 236i, 274i, 274t, 275, 277t, 284, 294–296, 298, 298i, 299, 316, 316i, 317, 326, 326i, 327, 390i, 397
 assisted reproductive technologies and, 302, 303
 detecting birth defects and, 330
 ectopic pregnancy and, 338
 prenatal development and, 320, 321i, 324, 325i
 STDs and, 304
Utricle, 258, 258i
UV radiation. *See* Ultraviolet (UV) radiation

V

Vacancies, electrons, 18, 18i, 20
Vaccinations, 188, 194, 329, 339, 347, 349
Vaccines, 175, 188–189
Vagina, 68t, 294, 294i, 294t, 295i, 297, 299, 300, 301, 302, 306, 309, 316, 316i, 330, 340, 346, 390i
Vaginal canal, 326
Vaginal discharge, STDs and, 304, 305, 308, 309
Vaginal opening, 294, 323i
Vaginal sponge, 300i
Vagus nerve, 204i, 236i
Valine, 134, 134i
Variable, experimental, 7
Variant Creutzfeldt-Jakob disease (vCJD), 343, 343i
Varicella, 188i
Varicose vein, 167
Vas deferens, 290i, 291, 291t, 292i, 299, 300
Vasclip, 300
Vascular tunic, 260t
Vasectomy, 300, 300i
Vasoconstriction, 82i, 167, 173, 236i, 277t, 280, 284
Vasodilation, 82i, 83, 167, 299
Vasopressin, 275
vCJD. *See* Variant Creutzfeldt-Jakob disease (vCJD)
Veins, 125, 125i, 151, 158, 158i, 161i, 163, 166i, 167, 168, 179, 203i, 325, 325i
 aging and, 335
 coronary, 161i
 DVT and, 174
 hepatic, 127, 127i, 163, 325i
 hepatic portal, 127, 127i, 158i, 162i, 163, 325i
 iliac, 158i
 jugular, 92, 158i
 pulmonary, 158i, 160i, 162, 162i, 203i, 325i
 renal, 158i, 214i
 structure of, 166
 systemic, 203i
 umbilical, 325, 325i
 varicose, 167
Venous duct, 325, 325i
Venous pressure, 167

Applications Index

The letter i *designates illustrations;* t *designates tables.*

Global water cycle, 469, 470, 470i
Glomerulonephritis, 221
Glucose
  diabetes and, 288
  diet and, 134
  disorders and, 283, 283t
  kidney disorders and, 221
GM food. See Genetically
  modified (GM) food
Goiter, 137t, 278, 279i
Gonorrhea, 185, 304–305, 309,
  312t, 341, 344, 345
Good cholesterol, 134, 170, 195,
  283
Gout, 213
Green Revolution, 422, 482
Greenhouse effect, 473, 474–475,
  474i, 475i, 483, 484, 485
Growth
  detecting birth defects and, 330i
  development disorders and,
  329
  impaired, 136t, 137t
  minerals and, 136
  normal, 137t
  population, 480, 480i
  retarded, 136t
  stunted, 99, 137t
  vitamins and, 136, 136t
Growth hormone (GH), 88, 101,
  272t, 274t, 275, 276, 282, 288
Growth spurts, 331
Gum disease, 194, 283t. See also
  Gingivitis

H

Haemophilus influenzae booster,
  188i
Hair
  CHH and, 380, 380i
  graying/white, 332i, 333
  lack of, 398, 398i
  porphyria and, 86
Hair loss, 435
Hallucinations, 245, 404
Hallucinogen, 245
Hay fever, 185, 190
Hb. See Hemoglobin (Hb)
HBV. See Hepatitis B virus
  (HBV)
HCG. See Human chorionic
  gonadotropin (HCG)
HDL cholesterol, 163, 195, 283
HDLs. See High-density
  lipoproteins (HDLs)
Head cold, 254
Headaches, 37, 136t, 154, 244
Hearing loss, 259
Heart attacks, 11, 67, 159, 163,
  165, 170, 171, 194, 253, 283,
  283t, 285, 423
Heart contractions, 165
Heart damage, 171
Heart defects, 329
Heart disease, 30, 31, 119, 134,
  135, 138, 206i, 332, 333, 395,
  419
Heart disorders, 138i
Heart failure (HF), 137t, 139,
  171, 283, 313, 381i

Heartburn, 124, 283t
Heat
  ankle sprains and, 98–99
  calories and, 138
Heat exhaustion, 83
Heat stress, 83, 83t
Heat stroke, 83
Heimlich maneuver, 201, 201i
Hemoglobin (Hb)
  blood disorders and, 154
  sickle-cell anemia and, 380,
  381i, 386
  STDs and, 308
Hemolytic anemias, 154, 154i,
  419i
Hemolytic bacteria, 174
Hemolytic disease of the
  newborn, 150
Hemophilia, 153, 330, 396, 397i,
  420, 446
Hemophilia A, 396, 396i, 397i,
  404, 423
Hemophilia B, 396
Hemorrhage, 136t, 361
Hemorrhagic fevers, 341
Hemorrhoids, 132
Hepatitis, 127, 151, 220, 308, 342,
  348t
Hepatitis A, 188i
Hepatitis B virus (HBV), 188i,
  189, 308, 438
Hepatitis C virus (HVC), 189,
  308
Herbal remedies, 87
Herbal supplements, 11, 87
Herbicides, 37, 143, 269, 422,
  423, 432
Heredity, 374, 374i, 374t, 375,
  375i, 388, 407, 430i, 431, 435
Heroin, 210, 239, 245
Herpes, 244, 265, 308, 308i, 309,
  312t, 342
HiB booster. See *Haemophilus
  influenzae* booster
High blood pressure, 13, 135,
  137t, 138, 165, 169, 173, 223,
  283t
High-density lipoproteins
  (HDLs), 163, 170
High fructose corn syrup, 28
Hip disorders, 386
Hip replacement, 99
Histoplasmosis, 207
HIV. See Human
  immunodeficiency virus
  (HIV)
Hives, 37, 185
H5N1 virus, 339
Hodgkin's disease, 438
Home pregnancy tests, 189, 317
Hookworms, 346, 348t
Hormonal implant, 300i
Hormone replacement therapy
  (HRT), 276, 288, 333
Hormone secretion, 125i, 159,
  277, 277t, 279
Hormones
  aging and, 333
  birth control and, 301
  calories and, 138, 139
  cancer and, 429, 433, 436

growth, 88, 101, 272t, 274t,
  275, 276, 282, 288
"Hot flashes," 333
HPV. See Human
  papillomavirus (HPV)
HRT. See Hormone replacement
  therapy (HRT)
Human chorionic gonadotropin
  (HCG), 317, 320, 429, 433
Human evolution, 3i, 93, 443,
  454–455, 454i, 454t, 455i,
  456–457, 456i, 457i, 462
Human genetics, 388–404
Human growth hormone (GH),
  423, 423i
Human immunodeficiency virus
  (HIV), 151, 155, 175, 189, 191,
  244, 306–307, 306t, 307i, 309,
  309i, 321, 341, 342i, 343, 351,
  352, 431, 438, 478
Human papillomavirus (HPV),
  308, 308i, 312t, 360
Human population. See
  Population
Human sewage, 57
Hunger, 239
Huntington disease (HD), 392,
  393, 393i, 394–395, 409
Hurricane Katrina, 57
Hyperextension, 97i
Hyperopia. See Farsightedness
Hyperparathyroidism, 279
Hyperplasia, 428
Hypertension, 13, 135, 165, 165i,
  165t, 170, 170t, 171, 173, 220,
  221, 281. See also High blood
  pressure
Hyperthermia, 83
Hyperthyroidism, 278
Hyperventilation, 199, 209
Hypnosis, 253
Hypoglycemia, 280, 288
Hypotension, 165, 165t
Hypothermia, 83
Hypothyroidism, 278
Hysterectomy, 304

I

IBS. See Irritable bowel
  syndrome (IBS)
ICSI. See Intracytoplasmic
  sperm injection (ICSI)
Identical twins, 79, 317, 383, 421
Immune system. See also
  Immunity
  AD and, 334
  aging and, 335
  cancer and, 428, 430i, 431, 438
  CHH and, 380, 380i
  disorders of, 190–191
  HIV/AIDS and, 155, 175, 306,
  307
  infectious diseases and, 341,
  345, 348
  irradiation and, 361
  respiratory disorders and, 207
  rheumatoid arthritis and, 98
  smoking and, 206i
  TB and, 194
  transplants and, 187

Immunity. See also Immune
  system
  cancer and, 431
  chemical weapons of, 177t
  CHH and, 380i
  impaired, 136t
Immunization, 188–189
Immunodeficiency, 191
Immunotherapy, 189
Immunotoxins, 189
Impotence, 169
In vitro fertilization (IVF),
  302–303, 302i, 330, 332
Incompatible blood types,
  148–149
Indigestion, 171
Infant mortality, 478, 479i
Infant respiratory distress
  syndrome, 202
Infections, 77i, 99, 132, 136t, 142,
  145, 153, 155, 170, 170t, 174,
  176, 177t, 179, 182, 185,
  189–191, 220, 221, 248,
  264–265, 304, 328, 330, 393,
  395
  bacterial, 132, 180, 211, 221,
  244, 265, 283t, 294t, 329
  chlamydial, 304
  cystic fibrosis and, 387
  ear, 259, 268, 345
  fungal, 283t
  HIV (See Human
  immunodeficiency virus
  (HIV))
  innate immunity and, 180, 181
  intestinal, 132
  irradiation and, 361
  nervous system disorders and,
  244
  respiratory, 206, 207, 259, 344,
  348t, 438
  risk of, 328
  SCID and, 421i
  STD, 304–305, 308–309
  strep, 348t
  surgical-wound, 345
  transplants and, 187
  urinary tract, 220, 221, 345
  vaginal, 309
  viral, 155, 191, 244, 283, 328i,
  341, 430, 431, 438
  yeast, 306, 309, 346
Infectious diseases, 1, 191,
  339–352, 480
Infectious mononucleosis,
  154–155, 342
Infertility, 295, 302–303, 304, 312,
  317
Inflammation, 98, 124, 127, 132,
  133, 144i, 170, 170t, 174, 176,
  177t, 180–182, 184, 190, 191,
  194, 206, 207, 221, 239, 244,
  259, 264, 278, 280, 284, 305,
  308, 309, 333, 341
Inflammatory disorders, 133,
  280
Influenza, 185, 188i, 206, 348,
  348i, 348t
Inherited abnormalities, 264
Inherited diseases, 387
Inherited disorders, 221, 392

Inherited traits, 373–386, 394
Inner ear, ear disorders in, 259, 259i, 268
Insanity, 305
Insecticides, 37, 432
Insomnia, 281
Insulin, 34–35, 63, 120i, 131, 134, 191, 270, 271i, 272t, 273, 277t, 282, 282i, 283, 288, 423
Intercourse. *See* Sexual intercourse
Interferon therapy, 435
Internal bleeding, 155
Intervertebral disks, 91i, 93, 93i
Intestinal disease, 344
Intestinal disorder, 133
Intestinal disturbances, 346i
Intestinal infection, 132
Intracytoplasmic sperm injection (ICSI), 303
Intrauterine device (IUD), 300i, 301
Intrinsic clotting mechanism, 152, 152i, 153
Introns, 411
Involuntary muscle twitches, 114
Iodine-deficient diets, 278, 279i. *See also* Goiter
Iris scanning technology, 249
Iron-deficiency anemia, 154
Irritable bowel syndrome (IBS), 133
Islets of Langerhans, 126i
Itching, 252, 308, 309, 438
IUD. *See* Intrauterine device (IUD)
IVF. *See* In vitro fertilization (IVF)
IVF with embryo transfer, 303

J

Jaundice, 136t, 308
Joint pain, 37
Joints
  aging and, 332, 333
  arthritic, 98
  autoimmune disorders and, 191
  gout in, 213
  sickle-cell anemia and, 381i
"Junk" foods, 28
Juvenile-onset diabetes, 283

K

Kaposi's sarcoma, 306i
Kcal. *See* Kilocalories (kcal)
Kidney damage, 136t
Kidney dialysis, 221
Kidney disease, 155, 220, 221, 283t
Kidney disorders, 221
Kidney failure, 132, 139, 283t, 381i
Kidney stones, 137t, 138, 221, 279
Kidney transplants, 221
Kidneys
  aging and, 335, 335t
  autoimmune disorders and, 191

blood disorders and, 155
cancer in, 211, 434t, 438
creatine and, 103
diet and, 134
disease in, 155, 220, 221, 283t
disorders in, 221
glucose disorders and, 283, 283t
hypertension and, 165i
low-carb diets and, 135
nicotine and, 169
sickle-cell anemia and, 381i
transplanted, 221
Kilocalories (kcal), 138, 139, 139t
Klinefelter syndrome, 401, 401i

L

Lab-grown epidermis, 78
Labor, 81, 275, 284, 326–327. *See also* Childbirth
Lactation, 327, 327i
Lactic acid, 63, 112
Lactose intolerance, 133
Larynx, cancer of, 206i
Lasek. *See* Laser-assisted subepithelial keratectomy
Laser angioplasty, 171
Laser-assisted in situ keratomilieusis, 265
Laser-assisted subepithelial keratectomy, 265
Lasik. *See* Laser-assisted in situ keratomilieusis
LDL cholesterol, 163, 195
Lead, as health risk, 143
Learning, 37, 227, 231
Lens (eye)
  aging and, 335
  eye disorders and, 265
  glucose disorders and, 283t
Leukemias, 154–155, 399, 419i, 420, 421i, 432t, 434, 434t, 435, 439
Life span, 332
Lifestyle, 145, 328, 331, 335, 435
Lipids
  cardiovascular disorders and, 170, 170t
  development disorders and, 328
  diet and, 134
Lipitor, 163
Lipoplex therapy, 421
Liver
  cardiovascular disorders and, 170
  diet and, 134
  drugs and, 211, 245
  infectious diseases and, 347, 347i
  STDs and, 305, 308
  vitamins and, 136t, 137
Liver cancer, 431, 432t, 434t, 435, 438
Liver damage, 103, 136t, 137t
Liver transplant, 127
Logistic growth (of population), 480, 480i
Long-term memory, 242, 242i
Low blood pressure, 165, 171, 283

Low-carb diets, 135
Luft's syndrome, 41
Lumpectomy, 436
Lung cancer, 133, 195, 206, 206i, 269, 420, 429, 431, 432t, 434t, 435, 435i, 436, 437, 438
Lung failure, 313, 387
Lung infections, 395
Lungs
  aging and, 335, 335t
  allergies and, 191
  cancer of (*See* Lung cancer)
  collapsed, 201
  cystic fibrosis and, 387
  embolism in, 153
  infectious diseases and, 345
  respiratory disorders and, 206, 207
  sickle-cell anemia and, 381i
  smoking and, 195, 206i, 207i
Luteinizing hormone (LH)
  aging and, 333
  birth control and, 301
Lyme disease, 341, 345, 345i
Lymph nodes
  cancer and, 437, 438, 439
  HIV and, 307
  STDs and, 304
Lymphatic cancer, 380, 380i, 428t, 429, 438–439
Lymphocytes
  cancer and, 431
  HIV and, 307i
  immunodeficiency and, 191
  transplants and, 187
Lymphomas, 380, 434, 434t, 438

M

M. tuberculosis. See Mycobacterium tuberculosis
Macular degeneration, 265
"Mad cow disease," 343
Magnetic resonance imaging (MRI), 361, 433, 433i
Malabsorption disorders, 133
Malaria, 154, 154i, 347, 347i, 348, 348t, 380, 477
Malignant cancer, 430
Malignant melanoma, 265, 420, 434t, 439, 439i
Malignant tumors, 133, 285, 428t
Malnourishment, 142, 395
Marfan syndrome, 395, 395i
Marijuana, 211, 245
Mass extinction, 452, 453
Mastectomy, 427, 436
MDMA, in Ecstasy, 225, 231, 245
Measles, 188i, 189, 329, 341, 348t
Measles, Mumps, Rubella (MMR) booster, 188i
Medical imaging, 433
Mediterranean diet, 135
Memory, 188, 188i, 189, 194, 221, 225, 227, 231, 241, 241i, 242, 242i, 244, 254, 334, 335
Memory loss, 221, 225, 242, 244, 334
Meningitis, 239, 244, 248
Menopause, 89, 211, 295, 296, 333, 372, 436, 437

Menstrual cycle, 294–295, 295t, 296, 296i, 297, 297i, 298, 298i, 316
Menstrual periods, 255, 284, 301, 312, 318, 326, 333, 436
Menstruation, 294, 295, 295t, 298i, 308, 312, 317, 330
Mental illness, 383
Mental retardation, 329, 394, 400, 401
Metabolic disorders, 41, 137t, 330i
Metabolic poisons, 155
Metabolic wastes, 80, 115, 144, 213, 214, 324
Metabolism
  aging and, 334
  cancer and, 431
  fat, 120i
  infectious diseases and, 344
  vitamins and, 136t, 137
  wastes and, 159
Metastasis, 428i, 428t, 429, 429i
Mice, cancer experiments in, 8
Middle ear
  disorders in, 259
  infection in, 268
Migraine headaches, 244
Mineral ions, 89i, 120i, 213
Mineral salts, 280, 458
Mineral supplements, 137
Minerals, in diet, 88, 129, 134, 136–137, 137t, 142, 146i, 327, 328, 333, 469, 469i, 471, 471i. *See also specific minerals*
Miscarriages, 305, 313, 323, 329, 330, 338, 361, 400, 401
Mitochondrial disorders, 41
MMR booster. *See* Measles, Mumps, Rubella (MMR) booster
Mobility, loss of, 93
Moles, 428
Monkeypox, 351, 351i
Morning-after pills, 301
Morphine, 245, 253, 255
Mosquitoes, malaria from, 154, 477
Motion sickness, 259
MRI. *See* Magnetic resonance imaging (MRI)
MS. *See* Multiple sclerosis (MS)
Mucous membranes
  aging and, 335
  infectious diseases and, 344
  STDs and, 305, 308
Mucus
  birth control and, 312
  infectious diseases and, 340
  respiratory disorders and, 207
  smoking and, 206
Multiple births, 303, 313
Multiple myeloma, 434t
Multiple sclerosis (MS), 189, 244
Mumps, 188i
Muscle contraction
  disorders and, 114–115
  tetanus, 248
Muscle tics, 114
Muscle-wasting diseases, 114

Tumors, 8, 133, 278, 285, 312, 409i, 419, 421, 428–430, 433, 435–438
Turner syndrome, 400–401, 401i
Twins, 79, 303, 315, 317, 332, 383, 421
Typhus, 477